ADVANCED ELECTROMAGNETIC WAVE PROPAGATION METHODS

ADVANCED ELECTROMAGNETIC WAVE PROPAGATION METHODS

Guillermo Gonzalez

CRC Press
Taylor & Francis Group
Boca Raton London New York

CRC Press is an imprint of the
Taylor & Francis Group, an **informa** business

First edition published 2022
by CRC Press
6000 Broken Sound Parkway NW, Suite 300, Boca Raton, FL 33487-2742

and by CRC Press
2 Park Square, Milton Park, Abingdon, Oxon, OX14 4RN

CRC Press is an imprint of Taylor & Francis Group, LLC

Library of Congress Cataloging-in-Publication Data
Names: Gonzalez, Guillermo, 1944- author.
Title: Advanced electromagnetic wave propagation methods / Guillermo Gonzalez.
Description: First edition. | Boca Raton : CRC Press, [2022] | Includes bibliographical references and index. |
Summary: "This textbook provides a solid foundation into the approaches used in the analysis of complex electromagnetic problems and wave propagation. The techniques discussed are essential to obtain closed-form solutions or asymptotic solutions and meet an existing need for instructors and students in electromagnetic theory"-- Provided by publisher.
Identifiers: LCCN 2021025797 (print) | LCCN 2021025798 (ebook) | ISBN 9781032113708 (hbk) | ISBN 9781032114002 (pbk) | ISBN 9781003219729 (ebk)
Subjects: LCSH: Electromagnetic waves--Transmission--Mathematical models. | Electromagnetic theory Mathematics. | Radio wave propagation.
Classification: LCC QC665.T7 G67 2022 (print) | LCC QC665.T7 (ebook) | DDC 539.2--dc23
LC record available at https://lccn.loc.gov/2021025797
LC ebook record available at https://lccn.loc.gov/2021025798

ISBN: 978-1-032-11370-8 (hbk)
ISBN: 978-1-032-11400-2 (pbk)
ISBN: 978-1-003-21972-9 (ebk)

DOI: 10.1201/9781003219729

Contents

Preface

The field of electromagnetic wave propagation has its origins in the work of James Clerk Maxwell, and the common forms used for his equations are due to Oliver Heaviside. The four basic Maxwell's equations, together with the continuity equation, are the foundation of many studies in electrical engineering and physics. His equations are the underlying basis of amazing advances in the field of optics, cosmology, antennas, electromagnetic wave propagation, quantum electrodynamics, microwaves, etc.

This book is written for graduate students and for researchers in this field. It is suitable for a graduate course in electromagnetic wave propagation for students with a good background in undergraduate electromagnetics and mathematics. Graduate students, as well as researchers, usually have some difficulty following some of the published material in professional journals. This book closes such gaps by providing a comprehensive view of some of the advanced techniques used in electromagnetic wave propagation.

The book is comprehensive in nature, and the Table of Contents illustrates the topics covered. The analysis and methods used for the large variety of electromagnetic propagation problems discussed in this book show that a formal procedure can be followed to determine the resulting fields. Also, various mathematical techniques used in the evaluation of the fields are discussed in detail showing again the procedures needed. The aim is to teach the techniques that should be followed in order to obtain closed-form and asymptotic solutions for the fields in electromagnetic propagation problems that involve rectangular, cylindrical, and spherical geometries.

A short highlight of the various chapters follows. Starting in Chapter 1, the various forms that Maxwell's equations appear are introduced. This is followed with a discussion of the auxiliary vector potentials **A** and **F**, as well as the Hertz vectors, and the wave equations that they satisfy. The various forms of expressing the wave equation in terms of different forms of the propagation constants are presented. The Lorentz conditions are discussed in detail. Boundary conditions are discussed, including the Leontovitch boundary condition. Chapter 2 discusses the fundamental radiators. Electric and magnetic Hertzian dipoles are described, as well as the resulting fields in lossless and lossy media. The fields due to an infinite-length line source are presented. Duality relations are obtained. The useful applications of images are presented for both electric and magnetic radiators. Chapter 3 discusses the propagation of plane waves. The Fresnel reflection coefficients are derived and expressed in various forms which depend on the way that the propagation constant is defined. The topics of reflection and transmission from a dielectric and from a lossy region are included. Other topics covered in Chapter 3 are the plane wave reflection and transmission from interfaces, the Zenneck, lateral, and trapped waves. A discussion of plane waves in various regions where the dielectric and conductivity changes as a function of position (i.e., inhomogeneous regions) is presented for both rectangular and cylindrical regions. The chapter ends with a discussion of the WKBJ method. Chapter 4 is concerned with the solution to the wave equation in rectangular, cylindrical and spherical coordinates, with examples of wave propagation in the various coordinates systems. In rectangular coordinates: the partially-filled and empty waveguides, cavity, and dielectric waveguides. In cylindrical coordinates: the cylindrical waveguide and cavity, sectoral waveguide, radial waveguide, wedge, sectoral horn, bend waveguide and dielectric resonator. In spherical coordinates, the Debye gauge is discussed and the resulting wave equations in terms of the Debye potentials are obtained. Examples include the conical waveguide, cavity and the Earth-Ionosphere cavity. The wave transformation from rectangular to cylindrical coordinates and from rectangular to spherical coordinates is included. In Chapter 5, the Green functions are discussed. Green functions provide a convenient method to solve inhomogeneous wave equations. First, the Green functions are obtained using the eigenfunctions method, followed by the direct method. Delta functions representations are also discussed. Situations where the use of Green functions are appropriate are discussed. Green functions in rectangular geometries are derived for both electric and magnetic sources. The same is done for Green functions in cylindrical and spherical coordinates. Several forms of Green functions in spherical coordinates are derived, including the Green function in terms of the Schelkunoff-Bessel functions. Chapter 6 shows how Green functions can be obtained using integral transforms. Spatial Fourier transforms are used to develop integral representations of Green functions in rectangular, cylindrical and spherical coordinates. The use of the Hankel transform in the solution of the wave equation is shown for cylindrical coordinates, and the spherical Hankel transform for spherical coordinates. Integral representations of the delta function are also given. The required radiation conditions that the fields must satisfy are derived for problems in rectangular, cylindrical and spherical coordinates. In Chapter 7, the mathematical methods required for obtaining asymptotic solutions for the fields are presented. It contains an extensive discussion of the Watson transformations, the stationary point method and the method of steepest descent (i.e., the saddle point method). The emphasis is on the use of these methods in the evaluation of the resulting field expressions in electromagnetic wave propagation. The classical problem of a pole near a saddle point and the resulting solution in terms of the complementary error function is discussed in detail. In Chapter 8, further problems of wave propagation in rectangular coordinates are considered. Using the knowledge from the previous chapters, closed-form solutions and asymptotic expressions are obtained for the fields of many problems, such as the parallel-plate waveguide, the narrow slit, the vertical electric dipole and the magnetic dipole above a conducting and lossy surface. The radiation for an aperture in a conducting plane is analyzed using spatial Fourier transforms, as well as using the Equivalence Theorem. Chapter 9 deals with propagation problems in cylindrical geometries. The classical problems of the diffraction of a plane wave by a conducting and by a lossy cylinder are first presented. This is followed by the fields produced by an electric line source and a dipole in the presence of a cylindrical surface, where Green function solutions are used and an asymptotic evaluation of the fields is obtained. The fields from vertical electric and magnetic dipoles above a lossy surface are discussed in detail, as well as the fields due to a horizontal electric dipole. The chapter ends with a discussion of the radiation from an aperture in a circular cylinder. In Chapter 10, further problems of electromagnetic wave propagation in spherical geometries are analyzed. It begins with the classical problem of plane-wave diffraction by a conducting and by a dielectric sphere. Then, the fields due to a vertical electric dipole above a conducting sphere and the fields above a spherical surface are obtained and asymptotically evaluated.

The problem of a vertical electric dipole and of a vertical magnetic dipole in a spherical waveguide are solved in terms of zonal harmonics. This is followed by the modal solution of the problem. Appendices associated with vector relations, Bessel functions, Airy functions, Legendre functions, error functions, certain integral transformations, and orthogonality properties are included.

I had the privilege to study under Prof. George Tyras at the University of Arizona. Two other distinguished faculty members who taught there were Prof. Donald G. Dudley and Prof. James R. Wait. It was an honor to know them and to have some interaction with them. I also wish to mention Prof. Vern R. Johnson for his guidance and friendship.

My gratitude goes to Profs. Manuel A. Huerta and James C. Nearing in the Physics Department at the University of Miami, for their help over the years. Their clarity in electromagnetics and mathematics will always be appreciated. Also, thanks to Dr. Kamal Premaratne for his help with the graphs of the special functions in the Appendices.

My love goes to my wife Pat, my children Alex and Donna, my daughter- and son-in-law Samantha and Larry, and my grandkids Tyler, Analise, Mia, and Nina. Thank you for being there for me!

Guillermo Gonzalez, PhD
Professor Emeritus
University of Miami
Department of Electrical and Computer Engineering

Author

Guillermo Gonzalez, PhD is a professor emeritus in the Department of Electrical Engineering at the University of Miami. He earned an M.S. in electrical engineering from the University of Miami and a PhD from the University of Arizona. His research interests include RF and microwave electronics and electromagnetic theory.

1 Maxwell's Equations

1.1 MAXWELL'S EQUATIONS IN VACUUM

We begin with a discussion of Maxwell's equations in free space (i.e., in vacuum). That is,

$$\oint_C \mathbf{E} \cdot d\mathbf{l} = -\frac{d}{dt}\int_S \mathbf{B} \cdot \hat{\mathbf{n}}\, dS \tag{1.1.1}$$

$$\oint_C \mathbf{B} \cdot d\mathbf{l} = \mu_o \int_S \mathbf{J} \cdot \hat{\mathbf{n}} dS + \mu_o \varepsilon_o \frac{d}{dt}\int_S \mathbf{E} \cdot \hat{\mathbf{n}}\, dS \tag{1.1.2}$$

$$\oint_S \mathbf{E} \cdot \hat{\mathbf{n}} dS = \frac{1}{\varepsilon_o}\int_V \rho\, dV \tag{1.1.3}$$

$$\oint_S \mathbf{B} \cdot \hat{\mathbf{n}}\, dS = 0 \tag{1.1.4}$$

where $\mathbf{J}$ (i.e., the electric current density) is the source of the $\mathbf{E}$ and $\mathbf{B}$ fields and ρ the associated charge density. The nomenclature and units for the terms used in this book are listed in Appendix A.

The fields and sources are functions of position and time (i.e., $\mathbf{E} = \mathbf{E}(\mathbf{r}, t)$, $\mathbf{B} = \mathbf{B}(\mathbf{r}, t)$, $\rho = \rho(\mathbf{r}, t)$ and $\mathbf{J} = \mathbf{J}(\mathbf{r}, t)$). When the dependence of the quantities in terms of the arguments is clear, the arguments are omitted as in (1.1.1) to (1.1.4). A supplemental equation that relates $\mathbf{J}$ and ρ is the continuity equation:

$$\oint_S \mathbf{J} \cdot \hat{\mathbf{n}} dS = -\frac{d}{dt}\int_V \rho dV \tag{1.1.5}$$

In (1.1.1) and (1.1.2), C is the closed contour (or line integral) around the boundary of the open surface S. In (1.1.3), (1.1.4) and (1.1.5) the integral is over the closed surface S enclosing the volume V. The direction of $\hat{\mathbf{n}}$ in (1.1.1) and (1.1.2) (and in (1.1.6) below) follows the right-hand rule after the direction of $d\mathbf{l}$ is selected. The direction of $\hat{\mathbf{n}}$ in (1.1.3) and (1.1.4) (and in (1.1.7) below) points outward from the closed surface.

Using Stokes' theorem:

$$\oint_C \mathbf{T} \cdot d\mathbf{l} = \int_S \nabla \times \mathbf{T} \cdot \hat{\mathbf{n}} dS \tag{1.1.6}$$

DOI: 10.1201/9781003219729-1

and Gauss's theorem (also known as the divergence theorem):

$$\oint_S \mathbf{T} \cdot \hat{\mathbf{n}} dS = \int_V \nabla \cdot \mathbf{T} dV \tag{1.1.7}$$

(1.1.1) to (1.1.4) are written in point form as

$$\nabla \times \mathbf{E} = -\frac{\partial \mathbf{B}}{\partial t} \tag{1.1.8}$$

$$\nabla \times \mathbf{B} = \mu_o \mathbf{J} + \mu_o \varepsilon_o \frac{\partial \mathbf{E}}{\partial t} \tag{1.1.9}$$

$$\nabla \cdot \mathbf{E} = \frac{\rho}{\varepsilon_o} \tag{1.1.10}$$

$$\nabla \cdot \mathbf{B} = 0 \tag{1.1.11}$$

and the continuity equation in point form is

$$\nabla \cdot \mathbf{J} = -\frac{\partial \rho}{\partial t} \tag{1.1.12}$$

Equations (1.1.10) and (1.1.11) are not independent equations since they can be derived from (1.1.8), (1.1.9) and the continuity equation in (1.1.12). Taking the divergence of (1.1.8) makes the left-hand side zero (since $\nabla \cdot \nabla \times \mathbf{E} = 0$), and the right-hand side is simply (1.1.11). Similarly, taking the divergence of (1.1.9) the left-hand side is zero, and using (1.1.12) gives

$$0 = \nabla \cdot \mathbf{J} + \varepsilon_o \frac{\partial}{\partial t} \nabla \cdot \mathbf{E} \quad \Rightarrow \quad 0 = -\frac{\partial \rho}{\partial t} + \varepsilon_o \frac{\partial}{\partial t} \nabla \cdot \mathbf{E} \quad \Rightarrow \quad \nabla \cdot \mathbf{E} = \frac{\rho}{\varepsilon_o}$$

which is (1.1.10).

In the part of a region where there is no source (i.e., away from the source) the fields satisfy (1.1.8) to (1.1.11) with $\mathbf{J} = 0$ and $\rho = 0$.

If the electric source $\mathbf{J}$ varies sinusoidally at the frequency ω, then the fields are also sinusoidal functions of time. For example, the position and time dependence of the $\mathbf{E}$ field can be expressed in the form

$$\begin{aligned} \mathbf{E}(\mathbf{r}, t) &= \mathbf{E}(\mathbf{r}) \cos[\omega t + \varphi(\mathbf{r})] \\ &= \mathrm{Re}[\mathbf{E}(\mathbf{r}) e^{j\varphi(\mathbf{r})} e^{j\omega t}] \\ &= \mathrm{Re}[\tilde{\mathbf{E}}(\mathbf{r}) e^{j\omega t}] \end{aligned} \tag{1.1.13}$$

where

$$\tilde{\mathbf{E}}(\mathbf{r}) = \mathbf{E}(\mathbf{r})e^{j\varphi(\mathbf{r})}$$

The term $\tilde{\mathbf{E}}(\mathbf{r})$ is known as a phasor, where $\mathbf{E}(\mathbf{r})$ describes its amplitude variation as a function of position and $\varphi(\boldsymbol{r})$ describes its phase variation as a function of position. Similarly,

$$\mathbf{B}(\mathbf{r}, t) = \mathrm{Re}[\tilde{\mathbf{B}}(\mathbf{r})e^{j\omega t}] \tag{1.1.14}$$

where $\tilde{\mathbf{B}}(\mathbf{r})$ is the phasor term.

Substituting (1.1.13) and (1.1.14) into (1.1.8) to (1.1.11) and observing that

$$\frac{\partial}{\partial t}e^{j\omega t} = j\omega e^{j\omega t}$$

results in the following phasor forms of Maxwell's equations in the region away from the source:

$$\nabla \times \tilde{\mathbf{E}}(\mathbf{r}) = -j\omega\tilde{\mathbf{B}}(\mathbf{r}) \tag{1.1.15}$$

$$\nabla \times \tilde{\mathbf{B}}(\mathbf{r}) = j\omega\mu_o\varepsilon_o\tilde{\mathbf{E}}(\mathbf{r}) \tag{1.1.16}$$

$$\nabla \bullet \tilde{\mathbf{E}}(\mathbf{r}) = 0 \tag{1.1.17}$$

$$\nabla \bullet \tilde{\mathbf{B}}(\mathbf{r}) = 0 \tag{1.1.18}$$

It is common to simply write $\mathbf{E}$ and $\mathbf{B}$ instead of $\tilde{\mathbf{E}}(\mathbf{r})$ and $\tilde{\mathbf{B}}(\mathbf{r})$ in the previous equations, with the understanding that $\mathbf{E}$ and $\mathbf{B}$ represent phasor quantities. A $j\omega$ term in an equation shows that it is a phasor equation. Therefore, we write (1.1.15) to (1.1.18) in the form:

$$\nabla \times \mathbf{E} = -j\omega\mathbf{B} \tag{1.1.19}$$

$$\nabla \times \mathbf{B} = j\omega\mu_o\varepsilon_o\mathbf{E} \tag{1.1.20}$$

$$\nabla \bullet \mathbf{E} = 0 \tag{1.1.21}$$

$$\nabla \bullet \mathbf{B} = 0 \tag{1.1.22}$$

and the continuity equation reads

$$\nabla \cdot \mathbf{J} = -j\omega\rho \tag{1.1.23}$$

These equations are solved for **E** by first taking the curl of (1.1.19) and substituting (1.1.20). That is,

$$\nabla \times \nabla \times \mathbf{E} = -j\omega\nabla \times \mathbf{B} = \omega^2\mu_o\varepsilon_o\mathbf{E}$$

Using the identity

$$\nabla \times \nabla \times \mathbf{T} = \nabla(\nabla \cdot \mathbf{T}) - \nabla^2\mathbf{T} \tag{1.1.24}$$

to expand $\nabla \times \nabla \times \mathbf{E}$ and using (1.1.21), gives

$$\nabla^2\mathbf{E} + \omega^2\mu_o\varepsilon_o\mathbf{E} = 0$$

or

$$\nabla^2\mathbf{E} + \frac{\omega^2}{c^2}\mathbf{E} = 0 \tag{1.1.25}$$

where

$$c = \frac{1}{\sqrt{\mu_o\varepsilon_o}} = 299,\ 792,\ 458 \approx 3 \times 10^8\ \text{m/s}$$

is the speed of light. (1.1.25) is recognized as a wave equation for **E**. This type of wave equation is also known as Helmholtz's equation.

Defining the propagation constant associated with (1.1.25) as the free space propagation constant β_o, namely

$$\beta_o = \frac{\omega}{c} = \omega\sqrt{\mu_o\varepsilon_o} \tag{1.1.26}$$

we write

$$\nabla^2\mathbf{E} + \beta_o^2\mathbf{E} = 0 \tag{1.1.27}$$

Similarly, the equation for **B** is

$$\nabla^2\mathbf{B} + \beta_o^2\mathbf{B} = 0$$

Consider a solution to (1.1.27) that consists of an x component of **E** that varies as a function of z, namely

$$\mathbf{E} = E_x(z)\hat{\mathbf{a}}_x \tag{1.1.28}$$

Then, from (1.1.27):

$$\frac{d^2E_x}{dz^2} + \beta_o^2 E_x = 0$$

whose solutions are

$$E_x = C_1 e^{-j\beta_o z} + C_2 e^{j\beta_o z} \tag{1.1.29}$$

where C_1 and C_2 are constants in V/m. When a constant requires a unit (e.g., $C_1 = 10$ V/m) it will be represented with a capital letter, and those with no units will be represented with lower-case letters. Equation (1.1.29) is the phasor representation of the field, therefore the time dependent form follows from (1.1.13). That is,

$$\begin{aligned} E_x(z, t) &= \text{Re}[(C_1 e^{-j\beta_o z} + C_2 e^{j\beta_o z})e^{j\omega t}] \\ &= C_1 \cos(\omega t - \beta_o z) + C_2 \cos(\omega t + \beta_o z) \end{aligned} \tag{1.1.30}$$

The sinusoidal functions in (1.1.30) are known as wave functions and the exponential functions in (1.1.29) are the wave functions in phasor form.

The terms $e^{-j\beta_o z}$ or $\cos(\omega t - \beta_o z)$ represent a wave traveling at the speed of light (since β_o is given by (1.1.26)) away from an observer located at $z = -\infty$ as z increases. This wave is called an "outgoing wave."

The waves in (1.1.30) have a period T given by

$$T = \frac{2\pi}{\omega} = \frac{1}{f}$$

wavelength given by

$$\lambda = \frac{2\pi}{\beta_o} = \frac{c}{f}$$

and phase velocity given by

$$v_p = \frac{\omega}{\beta_o} = \frac{1}{\sqrt{\mu_o \varepsilon_o}} = c$$

which shows that the electromagnetic field travels at the speed of light in free space.

Similarly, $e^{j\beta_o z}$ or $\cos(\omega t + \beta_o z)$ represent a wave traveling towards an observer located at $z = -\infty$ as z decreases. This wave travels at the speed of light and it is called an "incoming wave."

Consider the outgoing wave in (1.1.29):

$$E_x = C_1 e^{-j\beta_o z} \tag{1.1.31}$$

Then, from (1.1.19) the associated magnetic field is

$$\nabla \times (E_x \hat{\mathbf{a}}_x) = \frac{\partial E_x}{\partial z}\hat{\mathbf{a}}_y = -j\omega \mathbf{B}$$

or

$$B_y = -\frac{1}{j\omega}\frac{\partial E_x}{\partial z} = C_1 \frac{\beta_o}{\omega} e^{-j\beta_o z} = \frac{E_x}{c}$$

This relation show that $\mathbf{B} = B_y(z)\hat{\mathbf{a}}_y$ is perpendicular to $\mathbf{E} = E_x(z)\hat{\mathbf{a}}_x$, and also perpendicular to the direction of propagation, which is z. These types of waves are called plane waves.

The type of wave is determined by its source. A plane wave requires a planar source of infinite planar dimensions and infinite amount of energy to generate the constant amplitude field in (1.1.31). Such infinite source is not physically realizable. Although a plane wave solution satisfies Maxwell's equations, it is not a physically realizable wave. However, the study of plane waves is very important since the fields produced by many practical sources are spherical waves, and several properties of the spherical waves can be analyzed using a plane wave approximation. This will be discussed in Chapter 2.

Equations (1.1.19) to (1.1.23) apply to any sinusoidal frequency. Of course, a periodic excitation can be represented by a Fourier series expansion, so these equations apply to each term in the Fourier series and the resulting fields are obtained using superposition.

For non-periodic excitation, the Fourier or Laplace transforms can be used in (1.1.8) to (1.1.12). For example, the Fourier transform of (1.1.8) is obtained by multiplying both sides by $e^{-j\omega t}$ and forming:

$$\mathbf{E}(\mathbf{r}, \omega) = \int_{-\infty}^{\infty} \mathbf{E}(\mathbf{r}, t) e^{-j\omega t} dt \tag{1.1.32}$$

where the inverse transform is

$$\mathbf{E}(\mathbf{r}, t) = \frac{1}{2\pi}\int_{-\infty}^{\infty} \mathbf{E}(\mathbf{r}, \omega) e^{j\omega t} d\omega \tag{1.1.33}$$

with similar expressions for **B**.

The Fourier transform of (1.1.8) also requires the Fourier transform of a derivative, which for $\partial\mathbf{B}/\partial t$ is $j\omega\, \mathbf{B}(\mathbf{r}, \omega)$, and we simply obtain an equation identical in form to (1.1.19). Hence, (1.1.19) to (1.1.23) can also be considered to represent the

Fourier transform of the fields, with **E** and **B** now representing Fourier transformed quantities. Of course, the calculation of $\mathbf{E}(\mathbf{r}, t)$ or $\mathbf{B}(\mathbf{r}, t)$ requires the use of the inverse transform in (1.1.33).

1.2 POLARIZATION, MAGNETIZATION, AND CONDUCTIVITY

The expressions for Maxwell's equations in matter are now introduced. To this end, we consider the permittivity, permeability, and conductivity of matter and how these parameters are included in Maxwell's equations.

In a dielectric, the electric field produces dipoles. For a dipole, its dipole moment **p** is defined as $\mathbf{p} = (q\Delta L)\hat{\mathbf{u}}$ where ΔL is the dipole separation distance (see Fig. 1.1). The distribution of a number of dipoles per unit volume is described by the polarization vector **P**.

In Fig. 1.1, if $q = q_o \sin \omega t$ the associated current is

$$i(t) = \frac{dq}{dt} = \omega q_o \cos \omega t = I \cos \omega t \tag{1.2.1}$$

where $I = \omega q_o$ is the amplitude. Then,

$$[i(t)\Delta L]\hat{\mathbf{u}} = (I\Delta L)\cos \omega t \hat{\mathbf{u}} \tag{1.2.2}$$

The term $(I\Delta L)$ is known as the current moment of the dipole.

Since,

$$\begin{aligned}\frac{d\mathbf{p}}{dt} = \frac{d[(q\Delta L)\hat{\mathbf{u}}]}{dt} &= \omega(q_o\Delta L)\cos \omega t \hat{\mathbf{u}} \\ &= (I\Delta L)\cos \omega t \hat{\mathbf{u}}\end{aligned} \tag{1.2.3}$$

Comparing (1.2.2) and (1.2.3), and since the charges are fixed in position, it follows that

$$[i(t)\Delta L]\hat{\mathbf{u}} = \frac{d\mathbf{p}}{dt} \tag{1.2.4}$$

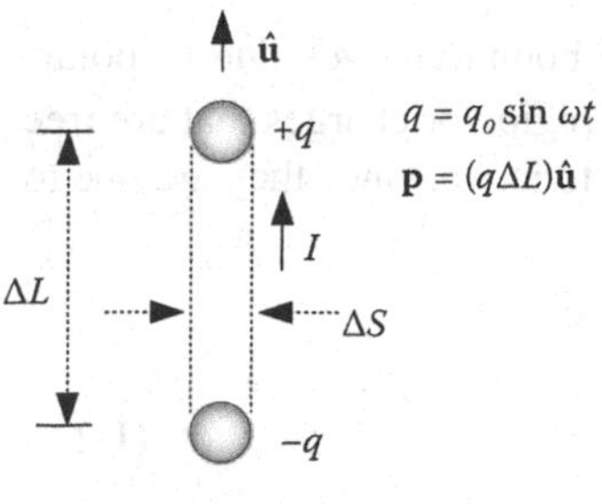

FIGURE 1.1 Model for an electric dipole.

Associated with the dipole current in Fig. 1.1 we can define a polarization current density $\mathbf{J}_p(t)$ as

$$\mathbf{J}_p(t) = J_p(t)\hat{\mathbf{u}} = \frac{i(t)}{\Delta S}\hat{\mathbf{u}}$$

Then

$$\mathbf{J}_p(t)\Delta V = [i(t)\Delta L]\hat{\mathbf{u}} = \frac{d\mathbf{p}}{dt} \tag{1.2.5}$$

A distribution of N dipoles per unit volume (N can be a function of position) is described by the polarization vector $\mathbf{P}(\mathbf{r}, t)$, where $\mathbf{P} = N\mathbf{p}$. A current density associated with the polarization vector is

$$\mathbf{J}_p(\mathbf{r}, t) = \frac{\partial(N\mathbf{p})}{\partial t} = \frac{\partial \mathbf{P}(\mathbf{r}, t)}{\partial t} \tag{1.2.6}$$

This relation will soon be included as part of $\mathbf{J}$ in Maxwell equations.

Charge is conserved, so associated with $\mathbf{J}_p(\mathbf{r}, t)$ there is a continuity equation of the form

$$\nabla\bullet \mathbf{J}_p = -\frac{\partial \rho_p}{\partial t}$$

or, using (1.2.6),

$$\frac{\partial}{\partial t}\nabla\bullet \mathbf{P} = -\frac{\partial \rho_p}{\partial t}$$

and it follows that the bound polarization charge distribution ρ_p is related to $\mathbf{P}$ by

$$\rho_p = -\nabla\bullet \mathbf{P} \tag{1.2.7}$$

The total charge density in a dielectric is written as

$$\rho = \rho_f + \rho_p \tag{1.2.8}$$

where ρ_f represents free electric charges and ρ_p the bound charges due to polarization. Free electric charges are charges added to the region or charges that are free to move in the region. The bound charges are not free to move, since they are due to the atomic effects of polarization.

Using (1.2.8), Gauss's law in (1.1.10) reads

$$\nabla\bullet \mathbf{E} = \frac{\rho_f + \rho_p}{\varepsilon_o} \tag{1.2.9}$$

Substituting (1.2.7) into (1.2.9) gives

$$\nabla \cdot (\varepsilon_o \mathbf{E} + \mathbf{P}) = \rho_f \tag{1.2.10}$$

The displacement vector **D** is defined as

$$\mathbf{D} = \varepsilon_o \mathbf{E} + \mathbf{P} \tag{1.2.11}$$

and (1.2.10) is written in the form

$$\nabla \cdot \mathbf{D} = \rho_f \tag{1.2.12}$$

which shows that the polarization effects are taken care by the **D** vector.

In many types of dielectrics, **P** is proportional to **E**. The proportionality is expressed in the form

$$\mathbf{P} = \varepsilon_o \chi_e \mathbf{E} \tag{1.2.13}$$

where χ_e is the electric susceptibility. Substituting (1.2.13) into (1.2.11) gives

$$\mathbf{D} = \varepsilon_o (1 + \chi_e) \mathbf{E} \tag{1.2.14}$$

The term $(1 + \chi_e)$ is known as the relative permittivity ε_r (also known as the dielectric constant). The permittivity of the dielectric ε is given by

$$\varepsilon = \varepsilon_o \varepsilon_r = \varepsilon_o (1 + \chi_e) \tag{1.2.15}$$

and (1.2.14) is written as

$$\mathbf{D} = \varepsilon \mathbf{E} = \varepsilon_o \varepsilon_r \mathbf{E} \tag{1.2.16}$$

From (1.2.16) the displacement current is defined as

$$\mathbf{J}_d = \frac{\partial \mathbf{D}}{\partial t} = \varepsilon \frac{\partial \mathbf{E}}{\partial t} \tag{1.2.17}$$

It is interesting to note that in electrostatics $\nabla \times \mathbf{E} = 0$, but from (1.2.11) $\nabla \times \mathbf{D} = \nabla \times \mathbf{P} \neq 0$. Hence, while the electrostatic field **E** can be expressed as the gradient of a potential, in a dielectric the same cannot be said for **D**.

Matter contains atoms and atoms contain electrons. The spin and orbital motion of electrons around the nucleus give rise to a loop current that can be described by a time-varying magnetic dipole. The current loops are associated with the motion of these electrons in atoms. If there are N atoms per unit volume there is a net magnetization **M**, which represents the total magnetic dipole moment ($N\mathbf{m}$) per unit volume, and this gives rise to a magnetic current density. A magnetic dipole **m** is represented by time-varying magnetic charges $\pm q_m(t)$ separated by a distance ΔL. In undergraduate courses it is shown that the magnetization is related to a magnetic current density by

$$\mathbf{J}_M = \nabla \times \mathbf{M} \tag{1.2.18}$$

where $\mathbf{J}_M$ is a bound current. The magnetization current is different from any impressed current. Observe that (1.2.18) shows that $\nabla \bullet \mathbf{J}_M = 0$, which is an expected result since in either spinning or orbiting electrons the same current enters and leaves any atomic region.

In Section 2.6, the model of a time-varying magnetic dipole is discussed in detail, as well as its relation to the fields produced by a small-loop radiator.

Using (1.2.17) and (1.2.18), (1.1.9) is modified to read

$$\begin{aligned} \nabla \times \frac{\mathbf{B}}{\mu_o} &= \mathbf{J} + \mathbf{J}_d + \mathbf{J}_M \\ &= \mathbf{J} + \frac{\partial \mathbf{D}}{\partial \mathbf{t}} + \nabla \times \mathbf{M} \end{aligned} \tag{1.2.19}$$

where **J** in the right-hand side of (1.2.19) now represents currents which are different from displacement and magnetization currents. Hence, (1.2.19) reads

$$\nabla \times \left(\frac{\mathbf{B}}{\mu_o} - \mathbf{M} \right) = \mathbf{J} + \frac{\partial \mathbf{D}}{\partial t} \tag{1.2.20}$$

Define the **H** vector as

$$\mathbf{H} = \frac{\mathbf{B}}{\mu_o} - \mathbf{M} \tag{1.2.21}$$

Then, (1.2.20) is expressed in the form

$$\nabla \times \mathbf{H} = \mathbf{J} + \frac{\partial \mathbf{D}}{\partial t} \tag{1.2.22}$$

which shows that the magnetization effects are taken care by the **H** vector.

Observe that (1.2.21), since $\nabla \bullet \mathbf{B} = 0$, shows that $\nabla \bullet \mathbf{H} = -\nabla \bullet \mathbf{M}$ which is zero only if $\mathbf{M} = 0$.

In many types of magnetic materials, the magnetization is proportional to the applied **H** field. Letting

$$\mathbf{M} = \chi_m \mathbf{H}$$

where χ_m is the magnetic susceptibility, it follows from (1.2.21) that

$$\mathbf{B} = \mu_o(1 + \chi_m)\mathbf{H}$$

Defining the permeability of the material as

$$\mu = \mu_o(1 + \chi_m)$$

Then,

$$\mathbf{B} = \mu\mathbf{H} = \mu_o\mu_r\mathbf{H} \tag{1.2.23}$$

where $\mu_r = 1 + \chi_m$ is the relative permeability.

In addition to dielectrics and magnetic effects we have conductive effects, described by the conductivity of the material. In a conductive material or conductive region there is a conduction current, whose current density $\mathbf{J}_c$ is related to the $\mathbf{E}$ field by

$$\mathbf{J}_c = \sigma_s\mathbf{E} \tag{1.2.24}$$

where σ_s is known as the static conductivity of the material or region. The static conductivity is associated with the free electrons. It describes how the free electrons move under the influence of the $\mathbf{E}$ field.

The conduction current in (1.2.24) is included in (1.2.22) by writing

$$\mathbf{J} \rightarrow \mathbf{J}_e + \mathbf{J}_c \tag{1.2.25}$$

where $\mathbf{J}_e$ represent the impressed electric currents (or source currents) in the region and $\mathbf{J}_c$ is the conduction currents. Impressed currents are not affected by the fields, while conduction current are affected by the fields. An example of an impressed electric current is the current in an antenna (i.e., a source) placed in the region. Substituting (1.2.25) into (1.2.22):

$$\nabla \times \mathbf{H} = \mathbf{J}_e + \sigma_s\mathbf{E} + \frac{\partial \mathbf{D}}{\partial t} \tag{1.2.26}$$

In certain dielectrics the atomic behavior of the molecules due to an applied field is such that there is a phase shift between $\mathbf{E}$ and $\mathbf{D}$ due to a loss in the dielectric. This loss can be accounted at the macroscopic level by introducing the complex permittivity for the dielectric ε_{cd}, namely

$$\mathbf{D} = \varepsilon_{cd}\mathbf{E}$$

where

$$\varepsilon_{cd} = \varepsilon_o(\varepsilon_r' - j\varepsilon_r'') \tag{1.2.27}$$

Equation (1.2.27) shows that the complex relative permittivity of the dielectric has a real part ε_r' and an imaginary part ε_r''. The imaginary part is associated with the dipole losses at it moves, accelerate, and de-accelerate to align itself with the applied field. For such regions, the displacement current can be expressed in the form

$$\frac{\partial \mathbf{D}}{\partial t} = j\omega \mathbf{D} = j\omega\varepsilon_{cd}\mathbf{E} = \omega\varepsilon_o\varepsilon_r''\mathbf{E} + j\omega\varepsilon_o\varepsilon_r'\mathbf{E} \tag{1.2.28}$$

The real part of (1.2.28) can be combined with the static conduction current, so (1.2.26) reads

$$\nabla \times \mathbf{H} = \mathbf{J}_e + (\sigma_s + \omega\varepsilon_o\varepsilon_r'')\mathbf{E} + j\omega\varepsilon_o\varepsilon_r'\mathbf{E} \tag{1.2.29}$$

Hence, an effective conductivity σ_{eff} is defined as

$$\sigma_{eff} = \sigma_s + \omega\varepsilon_o\varepsilon_r''$$

which includes the ohmic losses due to σ_s plus the dielectric losses due to $\omega\varepsilon_o\varepsilon_r''$. Then, (1.2.29) can be expressed in the form

$$\nabla \times \mathbf{H} = \mathbf{J}_e + \sigma_{eff}\mathbf{E} + j\omega\varepsilon_o\varepsilon_r'\mathbf{E} \tag{1.2.30}$$

This equation shows that in a region with static conductivity and complex permittivity, the macroscopic properties are represented by an effective conductivity σ_{eff} and permittivity $\varepsilon_o\varepsilon_r'$.

For most wave propagation problems, the specified conductivity is the value of the effective conductivity and the specified permittivity is the real part of the permittivity (i.e., $\varepsilon = \varepsilon_o\varepsilon_r'$). The reason is that when measuring the conductivity and permittivity, one usually measures the effective conductivity and the relative permittivity ε_r'. Hence, we simply use the notation of σ for σ_{eff}, and ε_r for ε_r', and (1.2.30) is written in the form

$$\nabla \times \mathbf{E} = \mathbf{J}_e + \sigma\mathbf{E} + j\omega\varepsilon_o\varepsilon_r\mathbf{E}$$

In a metal, the value of σ is mainly due to the collision of electrons as they move in the region. In metals the displacement current is negligible since $\sigma \gg \omega\varepsilon_o\varepsilon_r$. On the other hand, in a good dielectric the conduction current can be neglected since $\omega\varepsilon_o\varepsilon_r \gg \sigma$. The ratio of the conduction current to the displacement current is known as the loss tangent. That is,

$$\tan\varsigma = \frac{\sigma}{\omega\varepsilon_o\varepsilon_r}$$

The parameters μ, ε and σ are known as the constitutive parameters. They determine how the electromagnetic fields behave in a region described by μ, ε, and σ.

Typical values of the constitutive parameters for dry ground are: μ_o, $\varepsilon_r = 5$ and $\sigma \approx 10^{-5}$S/m; for wet ground: μ_o, $\varepsilon_r = 15$ and $\sigma \approx 10^{-2}$ S/m; for sea water: μ_o, $\varepsilon_r = 80$ and $\sigma \approx 4$ S/m. Metals have high values of conductivity, with typical values of: $\sigma \approx 10^7$ S/m and $\varepsilon_r = 1$.

In this book, we do not treat propagation of waves in lossy magnetic materials, such as those with high values of permeability. The values of μ_r used in this book have real values.

The characteristics of the parameters μ, ε, and σ describe the macroscopic electromagnetic response of a region to the applied fields. For example, a spatially inhomogeneous region is one where a constitutive parameter or parameters vary as a function of position (e.g., $\varepsilon = \varepsilon(r)$). A non-linear region is one where a constitutive parameter or parameters vary as a function of the applied field (e.g., $\varepsilon = \varepsilon(E_x)$. In a time-variant region, a constitutive parameter or parameters vary as a function of time (e.g., $\varepsilon = \varepsilon(t)$). In a dispersive region, a constitutive parameter or parameters vary as a function of frequency (e.g., $\varepsilon = \varepsilon(f)$. In an anisotropic region, a constitutive parameter or parameters are represented by a 3×3 matrix. Many practical problems occur in regions that are homogeneous, isotropic, linear, time invariant, non-dispersive, and non-magnetic. In these regions the parameters μ, ε, and σ are constants.

1.3 MAXWELL'S EQUATIONS IN TERMS OF THE CONSTITUTIVE PARAMETERS

In a region described by constant values of μ, ε, and σ, Maxwell's equations are expressed in the form

$$\nabla \times \mathbf{E} = -\mu \frac{\partial \mathbf{H}}{\partial t} \tag{1.3.1}$$

$$\nabla \times \mathbf{H} = \mathbf{J}_e + \sigma \mathbf{E} + \varepsilon \frac{\partial \mathbf{E}}{\partial t} \tag{1.3.2}$$

The phasor representation of (1.3.2) is

$$\nabla \times \mathbf{H} = \mathbf{J}_e + (\sigma + j\omega\varepsilon)\mathbf{E} \tag{1.3.3}$$

which can be written in the form

$$\nabla \times \mathbf{H} = \mathbf{J}_e + j\omega\left(\varepsilon - j\frac{\sigma}{\omega}\right)\mathbf{E} = \mathbf{J}_e + j\omega\varepsilon_c \mathbf{E} \tag{1.3.4}$$

where ε_c is the complex permittivity of the region. That is,

$$\sigma + j\omega\varepsilon = j\omega\varepsilon_c$$

where

$$\varepsilon_c = \varepsilon - j\frac{\sigma}{\omega} = \varepsilon\left(1 - j\frac{\sigma}{\omega\varepsilon}\right) = \varepsilon_o\left(\varepsilon_r - j\frac{\sigma}{\omega\varepsilon_o}\right) \tag{1.3.5}$$

Equation (1.3.5) shows that ε_c can be expressed as

$$\varepsilon_c = \varepsilon_o \varepsilon_{cr}$$

where ε_{cr} is the complex relativity permittivity of the region, namely

$$\varepsilon_{cr} = \varepsilon_r - j\frac{\sigma}{\omega\varepsilon_o}$$

which in the lossless case: $\varepsilon_{cr} = \varepsilon_r$.

Taking the divergence of (1.3.4) gives

$$\nabla\bullet\mathbf{J}_e + j\omega\nabla\bullet(\varepsilon_c\mathbf{E}) = 0 \tag{1.3.6}$$

The continuity equation associated with $\mathbf{J}_e$ is

$$\nabla\bullet\mathbf{J}_e = -j\omega\rho_e$$

where ρ_e is the impressed charge density associated with the impressed current density $\mathbf{J}_e$. Then, (1.3.6) can be expressed in the form

$$\nabla\bullet\mathbf{D} = \rho_e \tag{1.3.7}$$

where

$$\mathbf{D} = \varepsilon_c\mathbf{E} \tag{1.3.8}$$

Equation (1.3.8) is a more inclusive representation of the displacement vector. Sometimes in this form it is referred as the electric induction vector. If $\sigma = 0$, then $\varepsilon_c = \varepsilon$ and $\mathbf{D}$ in (1.3.8) is the standard displacement vector.

From (1.3.1), (1.3.4), and (1.3.7), Maxwell's equations in phasor form are

$$\nabla\times\mathbf{E} = -j\omega\mu\mathbf{H} \tag{1.3.9}$$

$$\nabla\times\mathbf{H} = \mathbf{J}_e + (\sigma + j\omega\varepsilon)\mathbf{E} = \mathbf{J}_e + j\omega\varepsilon_c\mathbf{E} \tag{1.3.10}$$

$$\nabla\bullet\mathbf{D} = \rho_e \Rightarrow \nabla\bullet\mathbf{E} = \frac{\rho_e}{\varepsilon_c} \tag{1.3.11}$$

$$\nabla \bullet \mathbf{B} = 0 \Rightarrow \nabla \bullet \mathbf{H} = 0 \tag{1.3.12}$$

The equation satisfied by the electric field is obtained by taking the curl of (1.3.9) and substituting (1.3.10). That is,

$$\begin{aligned} \nabla \times \nabla \times \mathbf{E} &= -j\omega\mu\nabla \times \mathbf{H} \\ &= -j\omega\mu\mathbf{J}_e - j\omega\mu(\sigma + j\omega\varepsilon)\mathbf{E} \end{aligned}$$

or

$$\nabla(\nabla \bullet \mathbf{E}) - \nabla^2\mathbf{E} = -j\omega\mu\mathbf{J}_e - j\omega\mu(\sigma + j\omega\varepsilon)\mathbf{E}$$

Then, using (1.3.11) and the continuity equation:

$$\begin{aligned} \nabla^2\mathbf{E} - \gamma^2\mathbf{E} &= j\omega\mu\mathbf{J}_e + \frac{\nabla\rho_e}{\varepsilon_c} \\ &= j\omega\mu\mathbf{J}_e - \frac{\nabla(\nabla \bullet \mathbf{J}_e)}{j\omega\varepsilon_c} \end{aligned} \tag{1.3.13}$$

where γ is the complex propagation constant, namely

$$\gamma^2 = j\omega\mu(\sigma + j\omega\varepsilon) \tag{1.3.14}$$

Similarly, taking the curl of (1.3.10) and substituting (1.3.9) gives

$$\nabla \times \nabla \times \mathbf{H} = \nabla \times \mathbf{J}_e - j\omega\mu(\sigma + j\omega\varepsilon)\mathbf{H}$$

or

$$\nabla(\nabla \bullet \mathbf{H}) - \nabla^2\mathbf{H} = \nabla \times \mathbf{J}_e - j\omega\mu(\sigma + j\omega\varepsilon)\mathbf{H}$$

Using (1.3.12) results in

$$\nabla^2\mathbf{H} - \gamma^2\mathbf{H} = -\nabla \times \mathbf{J}_e \tag{1.3.15}$$

The real and imaginary parts of γ are denoted by

$$\gamma = \alpha + j\beta \tag{1.3.16}$$

where α is the attenuation constant and β is the propagation constant. From (1.3.14) and (1.3.16) it follows that

$$\alpha = \omega\sqrt{\frac{\mu\varepsilon}{2}}\left[\sqrt{1+\left(\frac{\sigma}{\omega\varepsilon}\right)^2} - 1\right]^{1/2} \tag{1.3.17}$$

$$\beta = \omega\sqrt{\frac{\mu\varepsilon}{2}}\left[\sqrt{1+\left(\frac{\sigma}{\omega\varepsilon}\right)^2} + 1\right]^{1/2} \tag{1.3.18}$$

where α and β are positive real quantities. The ratio $\sigma/\omega\varepsilon$ represents the ratio of conduction current to displacement current.

Observe that for a lossless region:

$$\gamma^2 = -\omega^2\mu\varepsilon = -\beta^2$$

or

$$\gamma = j\beta = j\omega\sqrt{\mu\varepsilon}$$

The propagation properties of some regions can be described as follows:

a. Free space and air: $\mu = \mu_o$, $\varepsilon = \varepsilon_o$, $\sigma = 0$. Then, $\alpha = 0$ and $\beta = \beta_o = \omega\sqrt{\mu_o\varepsilon_o}$.
b. Lossless dielectric: $\mu = \mu_o\mu_r$, $\varepsilon = \varepsilon_o\varepsilon_r$, $\sigma = 0$. Then, $\alpha = 0$ and $\beta = \omega\sqrt{\mu\varepsilon}$.
c. Good dielectric: $\mu = \mu_o\mu_r$, $\omega\varepsilon >> \sigma$. Then, $\alpha \approx \frac{\sigma}{\varepsilon}\sqrt{\frac{\mu}{\varepsilon}}$ and $\beta \approx \omega\sqrt{\mu\varepsilon}$.
d. Good conductor: $\mu = \mu_o\mu_r$, $\sigma >> \omega\varepsilon$. Then, $\alpha \approx \sqrt{\frac{\omega\mu\sigma}{2}}$ and $\beta \approx \sqrt{\frac{\omega\mu\sigma}{2}}$.

The reason that the constitutive relations of air resembles those of free space is that for air: $\mu = \mu_o$, $\varepsilon_r = 1.0004$ and $\sigma \approx 0$, which are close to the free space values. The closest region that resembles free space is outer space. That is why the radiation emitted from distance stars can be seen or measured. In some of the problems in this book it will be convenient to introduce a slight loss to describe the region.

In the literature, other forms of the propagation constant are found. For example, (1.3.14) can be written as

$$\gamma^2 = -\omega^2\mu\varepsilon\left(1 - j\frac{\sigma}{\omega\varepsilon}\right) = -\omega^2\mu\varepsilon_c = -\omega^2\mu_o\varepsilon_o\mu_r\varepsilon_{cr} = -\beta_o^2\mu_r\varepsilon_{cr}$$

where ε_c and ε_{cr} are given in (1.3.5) and β_o in (1.1.26).

The complex index of refraction is defined by

$$n = \sqrt{\mu_r\varepsilon_{cr}} = \sqrt{\mu_r\left(\varepsilon_r - j\frac{\sigma}{\omega\varepsilon_o}\right)} \tag{1.3.19}$$

In terms of n, γ^2 is expressed as

$$\gamma^2 = -\beta_o^2 \mu_r \varepsilon_{cr} = -\beta_o^2 n^2 \tag{1.3.20}$$

In a lossless region the index of refraction is given by

$$n = \sqrt{\mu_r \varepsilon_r} = \sqrt{\frac{\mu\varepsilon}{\mu_o \varepsilon_o}} = \frac{c}{v}$$

where v is the velocity of propagation in a region with μ and ε values (i.e., $v = 1/\sqrt{\mu\varepsilon}$). If $\mu_r = 1$ then $n = \sqrt{\varepsilon_r}$.

The free space propagation constant also appears using the notation k_o instead of β_o. That is, (1.3.20) also appears in the form

$$\gamma^2 = -k_o^2 n^2$$

or

$$\gamma^2 = -k^2$$

where

$$k = k_o n$$

The term k, like γ, is called the complex propagation constant. In terms of k, (1.3.13) and (1.3.15) are written as

$$\nabla^2 \mathbf{E} + k^2 \mathbf{E} = j\omega\mu \mathbf{J}_e - \frac{\nabla(\nabla \cdot \mathbf{J}_e)}{j\omega\varepsilon_c} \tag{1.3.21}$$

and

$$\nabla^2 \mathbf{H} + k^2 \mathbf{H} = -\nabla \times \mathbf{J}_e \tag{1.3.22}$$

The forms for the wave equations in terms of γ (see (1.3.13) and (1.3.15)) or in terms of k (see (1.3.21) and (1.3.22)) are both used in the literature and will be used in this book.

The relation between γ and k is

$$\gamma^2 = -k^2 \quad \Rightarrow \quad \gamma = jk$$

where $k = k' - jk''$ with $k' > 0$ and $k'' > 0$. Hence,

$$\gamma = \alpha + j\beta = j(k' - jk'') = k'' + jk'$$

So $k'' = \alpha$ and $k' = \beta$, and (1.3.17) and (1.3.18) can be used to calculate k' and k''.

A simple example is to consider the solution to (1.3.13) (or (1.3.21)) in a region away from the source where there is an x component of the field that varies as a function of z. With $\gamma = \alpha + j\beta$ the electric field is given by

$$E_x = C_1 e^{-\gamma z} + C_2 e^{\gamma z}$$

Consider the outgoing wave. That is,

$$E_x = C_1 e^{-\gamma z} = C_1 e^{-\alpha z} e^{-j\beta z}$$

The time domain field is

$$E_x(z, t) = \mathrm{Re}(E_x e^{j\omega t}) = C_1 e^{-\alpha z} \cos(\omega t - \beta z)$$

which represents a field that is attenuated by $e^{-\alpha z}$ as it propagates in the positive z direction. Also, $\cos(\omega t - \beta z)$ can be written as $\cos \omega(t - z/v)$ where $v = \omega/\beta$ is the velocity of propagation.

The outgoing wave in terms of the propagation constant k is

$$E_x = C_1 e^{-\gamma z} = C_1 e^{-jkz} = C_1 e^{-k'' z} e^{-jk' z}$$

which, again, justify the form of k as $k = k' - jk''$.

From (1.3.9) the magnetic field is given by

$$H_y = -\frac{1}{j\omega\mu}\frac{\partial E_x}{\partial z} = C_1 \frac{\gamma}{j\omega\mu} e^{-\gamma z} = \frac{E_x}{\eta}$$

where η is the complex intrinsic impedance of the medium, namely

$$\eta = \frac{j\omega\mu}{\gamma} = \sqrt{\frac{j\omega\mu}{\sigma + j\omega\varepsilon}} = \sqrt{\frac{\mu}{\varepsilon - j\frac{\sigma}{\omega}}} = \sqrt{\frac{\mu}{\varepsilon_c}} \tag{1.3.23}$$

Since E_x is in V/m and H_y in A/m it is expected that their ratio produces an impedance. In free space (or air), where $\sigma = 0$, the impedance of free space is denoted by η_o where

$$\eta_o = \sqrt{\frac{\mu_o}{\varepsilon_o}} \approx 377\Omega \tag{1.3.24}$$

In general, η is of the form $\eta = |\eta| e^{j\theta}$ and the resulting H_y field is

$$H_y = \mathrm{Re}\left[\frac{C}{|\eta| e^{j\theta}} e^{-\alpha z} e^{-j\beta z} e^{j\omega t}\right] = \frac{C}{|\eta|} e^{-\alpha z} \cos(\omega t - \beta z - \theta)$$

Next, consider the propagation of a plane wave with $\mathbf{E} = E_x(x, y, z)\hat{\mathbf{a}}_x$ in a lossy region. Then, from (1.3.13):

$$\frac{\partial^2 E_x}{\partial x^2} + \frac{\partial^2 E_x}{\partial y^2} + \frac{\partial^2 E_x}{\partial z^2} - \gamma^2 = 0 \tag{1.3.25}$$

whose solution for the outgoing wave is

$$E_x = C_1 e^{-\gamma_x x} e^{-\gamma_y y} e^{-\gamma_z z} \tag{1.3.26}$$

Substituting (1.3.26) into (1.3.25) shows that (1.3.26) is the solution if

$$\gamma^2 = \gamma_x^2 + \gamma_y^2 + \gamma_z^2$$

The parameters γ_x, γ_y and γ_z are called separation constants. The solution (1.3.25) is commonly obtained using the method of separation of variables, which is discussed in Chapter 4 (see Section 4.2).

A complex propagation vector $\boldsymbol{\gamma}$ can be defined to express (1.3.26). That is, let

$$\boldsymbol{\gamma} = \gamma_x \hat{\mathbf{a}}_x + \gamma_x \hat{\mathbf{a}}_y + \gamma_x \hat{\mathbf{a}}_z$$

so (1.3.26) reads

$$E_x = C_1 e^{-\boldsymbol{\gamma} \cdot \mathbf{r}}$$

where

$$\mathbf{r} = x\hat{\mathbf{a}}_x + y\hat{\mathbf{a}}_y + z\hat{\mathbf{a}}_z$$

In terms of k:

$$E_x = Ce^{-j\mathbf{k} \cdot \mathbf{r}}$$

where

$$k^2 = k_x^2 + k_y^2 + k_z^2$$

and

$$\mathbf{k} = k_x \hat{\mathbf{a}}_x + k_y \hat{\mathbf{a}}_y + k_z \hat{\mathbf{a}}_z$$

1.4 OTHER FORMS OF MAXWELL'S EQUATIONS

In the previous section, we found Maxwell's equations in the forms shown in (1.3.1) and (1.3.2). These equations can be made symmetrical if an impressed magnetic current density $\mathbf{J}_m$ and an impressed magnetic charge density ρ_m are introduced. Of course, there are no real magnetic charges (since $\nabla \bullet \mathbf{B} = 0$). However, in some electromagnetic problems the introduction of equivalent (i.e., fictitious) magnetic currents and magnetic charges can be used to simplify the fields calculations. As we will see, radiation problems from loop antennas and apertures can be conveniently solved using such concept. Therefore, with $\mathbf{J}_m$ and ρ_m, Maxwell's equations read

$$\nabla \times \mathbf{E} = -\mathbf{J}_m - \mu \frac{\partial \mathbf{H}}{\partial t} \tag{1.4.1}$$

$$\nabla \times \mathbf{H} = \mathbf{J}_e + \sigma \mathbf{E} + \varepsilon \frac{\partial \mathbf{E}}{\partial t} \tag{1.4.2}$$

$$\nabla \bullet \mathbf{D} = \rho_e \tag{1.4.3}$$

$$\nabla \bullet \mathbf{B} = \rho_m \tag{1.4.4}$$

where $\mathbf{D} = \varepsilon_c \mathbf{E}$ and $\mathbf{B} = \mu \mathbf{H}$. The continuity equations are

$$\nabla \bullet \mathbf{J}_e = -\frac{\partial \rho_e}{\partial t}$$

and

$$\nabla \bullet \mathbf{J}_m = -\frac{\partial \rho_m}{\partial t}$$

Again, it is important to point out that $\mathbf{J}_m$ and ρ_m are only "equivalent" magnetic sources that can be used to solve certain problems. The units of $\mathbf{J}_m$ are V/m^2 and ρ_m is given in Wb/m^3. In the literature, one also finds the notation $\mathbf{M}_i$ instead of $\mathbf{J}_m$ to represent the impressed magnetic current.

The phasor forms of (1.4.1) to (1.4.4) are

$$\nabla \times \mathbf{E} = -\mathbf{J}_m - j\omega\mu \mathbf{H} \tag{1.4.5}$$

$$\nabla \times \mathbf{H} = \mathbf{J}_e + (\sigma + j\omega\varepsilon)\mathbf{E} \tag{1.4.6}$$

$$\nabla \bullet \mathbf{D} = \rho_e \quad \Rightarrow \quad \nabla \bullet \mathbf{E} = \frac{\rho_e}{\varepsilon_c} \tag{1.4.7}$$

$$\nabla \bullet \mathbf{B} = \rho_m \quad \Rightarrow \quad \nabla \bullet \mathbf{H} = \frac{\rho_m}{\mu} \tag{1.4.8}$$

and the continuity equations are

$$\nabla \bullet \mathbf{J}_e = -j\omega\rho_e \tag{1.4.9}$$

$$\nabla \bullet \mathbf{J}_m = -j\omega\rho_m \tag{1.4.10}$$

Taking the curl of (1.4.5), substituting (1.4.6) and using (1.1.24) and (1.4.7) we obtain

$$\nabla^2\mathbf{E} - \gamma^2\mathbf{E} = \nabla \times \mathbf{J}_m + j\omega\mu\mathbf{J}_e + \frac{\nabla\rho_e}{\varepsilon_c}$$

and using (1.4.9) gives

$$\nabla^2\mathbf{E} - \gamma^2\mathbf{E} = \nabla \times \mathbf{J}_m + j\omega\mu\mathbf{J}_e - \frac{\nabla(\nabla \bullet \mathbf{J}_e)}{j\omega\varepsilon_c} \tag{1.4.11}$$

Similarly, starting with (1.4.6) we obtain

$$\nabla^2\mathbf{H} - \gamma^2\mathbf{H} = -\nabla \times \mathbf{J}_e + (\sigma + j\omega\varepsilon)\mathbf{J}_m - \frac{\nabla(\nabla \bullet \mathbf{J}_m)}{j\omega\mu} \tag{1.4.12}$$

Direct solution of (1.4.11) and (1.4.12) for the fields **E** and **H** involve spatial differentiation of the sources, followed by a spatial integration. This procedure is, in general, complicated. In the next section it is shown that the calculations of the fields in terms of the sources is greatly simplified using vector potentials.

1.5 VECTOR POTENTIALS

Vector potentials are used to simplify the calculation of the field quantities. We begin with the introduction of the magnetic vector potential **A**. If there are no magnetic sources (i.e., with $\mathbf{J}_m$ and ρ_m set to zero), (1.4.8) reads

$$\nabla \bullet \mathbf{B} = 0$$

Since the divergence of a curl is zero, we can express **B** as the curl of a vector, named the magnetic vector potential **A** (or simply the vector potential **A**). That is,

$$\mathbf{B} = \mu\mathbf{H} = \nabla \times \mathbf{A} \tag{1.5.1}$$

Substituting (1.5.1) into (1.4.5) gives

$$\nabla \times (\mathbf{E} + j\omega\mathbf{A}) = 0 \tag{1.5.2}$$

Using the identity $\nabla \times (\pm\nabla\varphi) = 0$, it follows from (1.5.2) that

$$\mathbf{E} = -\nabla\varphi - j\omega\mathbf{A} \tag{1.5.3}$$

where φ is the electric scalar potential. For time independent fields, (1.5.3) reduces to the well-known static form: $\mathbf{E} = -\nabla\varphi$.

The equation satisfied by the vector potential **A** is obtained by substituting (1.5.1) and (1.5.3) into (1.4.6), namely

$$\nabla \times \nabla \times \mathbf{A} = \mu\mathbf{J}_e - \mu(\sigma + j\omega\varepsilon)(\nabla\varphi) - j\omega\mu(\sigma + j\omega\varepsilon)\mathbf{A}$$

and using (1.1.24):

$$\nabla^2\mathbf{A} - \gamma^2\mathbf{A} = -\mu\mathbf{J}_e + \nabla[(\nabla\bullet\mathbf{A}) + \mu(\sigma + j\omega\varepsilon)\varphi] \tag{1.5.4}$$

Equation (1.5.1) specifies the curl of **A**, and from Helmholtz's theorem one can specify its divergence. That is, in (1.5.4) let

$$\nabla\bullet\mathbf{A} = -\mu(\sigma + j\omega\varepsilon)\varphi = -\frac{\gamma^2}{j\omega}\varphi \tag{1.5.5}$$

This relation is known as Lorenz's condition (named after Ludvig Lorenz), and it's analyzed in the next section. Since

$$\frac{\gamma^2}{j\omega} = j\omega\mu\varepsilon_c$$

the Lorenz's condition also appears in the form

$$\nabla\bullet\mathbf{A} = -j\omega\mu\varepsilon_c\varphi$$

In the lossless case, Lorenz's condition is

$$\nabla\bullet\mathbf{A} = -j\omega\mu\varepsilon\varphi$$

Using Lorenz's condition, (1.5.4) reduces to

$$\nabla^2\mathbf{A} - \gamma^2\mathbf{A} = -\mu\mathbf{J}_e \tag{1.5.6}$$

which shows that the source of the **A** vector is the impressed electric current $\mathbf{J}_e$. In terms of the complex propagation constant k, replace $-\gamma^2$ by k^2 in (1.5.6).

Once **A** is determined from (1.5.6), the **H** field follows from (1.5.1),

$$\mathbf{H} = \frac{1}{\mu}\nabla \times \mathbf{A} \tag{1.5.7}$$

and the **E** field from (1.4.6), namely

$$\begin{aligned}\mathbf{E} &= \frac{1}{\mu(\sigma + j\omega\varepsilon)}\nabla \times \nabla \times \mathbf{A} - \frac{1}{(\sigma + j\omega\varepsilon)}\mathbf{J}_e \\ &= \frac{1}{j\omega\mu\varepsilon_c}(\nabla \times \nabla \times \mathbf{A} - \mu\mathbf{J}_e) \\ &= \frac{j\omega}{\gamma^2}\nabla \times \nabla \times \mathbf{A} - \frac{j\omega\mu}{\gamma^2}\mathbf{J}\end{aligned} \tag{1.5.8}$$

From (1.5.3) and (1.5.5), the **E** field can also be expressed in the form

$$\mathbf{E} = -j\omega\mathbf{A} + \frac{j\omega}{\gamma^2}\nabla(\nabla\bullet\mathbf{A}) \tag{1.5.9}$$

Equation (1.5.9) can also be derived by expanding $\nabla \times \nabla \times \mathbf{A}$ in (1.5.8) and using (1.5.6).

Away from the source, from (1.4.6) or (1.5.8), the **E** field is given by

$$\mathbf{E} = \frac{1}{(\sigma + j\omega\varepsilon)}\nabla \times \mathbf{H} = \frac{j\omega}{\gamma^2}\nabla \times \nabla \times \mathbf{A} \tag{1.5.10}$$

which is the region where the fields are usually evaluated.

In a lossless region: $\gamma^2 = -\omega^2\varepsilon\mu = -\beta^2$ and (1.5.6) is written as

$$\nabla^2\mathbf{A} + \beta^2\mathbf{A} = -\mu\mathbf{J}_e \tag{1.5.11}$$

This expression also appears as

$$\nabla^2\mathbf{A} + k^2\mathbf{A} = -\mu\mathbf{J}_e$$

where k is real in a lossless region (i.e., $k = k'$) and given by $k = \omega\sqrt{\mu\varepsilon}$.

The **H** field follows from (1.5.7) and the **E** field from (1.5.10) with $\sigma = 0$, namely

$$\mathbf{E} = \frac{1}{j\omega\varepsilon}\nabla \times \mathbf{H} = \frac{1}{j\omega\mu\varepsilon}\nabla \times \nabla \times \mathbf{A} \tag{1.5.12}$$

Observe that Lorenz's condition has reduced the field calculations to the determination of **A**. There is no need to use the differential equation satisfied by the scalar potential φ.

For completeness, the equation satisfied by φ is obtained by taking the divergence of (1.5.3) and using (1.4.7) and (1.5.5). That is,

$$\nabla\bullet\mathbf{E} = -\nabla^2\varphi - j\omega\nabla\bullet\mathbf{A} = \frac{\rho_e}{\varepsilon_c} \quad \Rightarrow \quad \nabla^2\varphi - \gamma^2\varphi = -\frac{\rho_e}{\varepsilon_c} \tag{1.5.13}$$

which for a lossless region reads

$$\nabla^2\varphi + \beta^2\varphi = -\frac{\rho_e}{\varepsilon}$$

Equation (1.5.13) can also be derived directly from (1.5.6) using Lorenz's condition, taking a divergence and using the continuity equation for $\nabla\bullet \mathbf{J}_e$.

Next, we derive the relations for the electric vector potential **F** (or simply the vector potential **F**). If $\mathbf{J}_e = \rho_e = 0$ (i.e., no electric sources) then from (1.4.7) we have that $\nabla\bullet \mathbf{D} = 0$. Therefore, **D** can be expressed as the curl of a vector, denoted by **F**, or

$$\mathbf{D} = \varepsilon_c \mathbf{E} = -\nabla \times \mathbf{F} \tag{1.5.14}$$

Substituting (1.5.14) into (1.4.6) gives

$$\nabla \times (\mathbf{H} + j\omega\mathbf{F}) = 0$$

or

$$\mathbf{H} = -\nabla\varphi_m - j\omega\mathbf{F} \tag{1.5.15}$$

where φ_m is the scalar magnetic potential.

The equation satisfied by the vector potential **F** is obtained by substituting (1.5.14) and (1.5.15) into (1.4.5), and using (1.1.24). The steps in the derivation are similar to those used in deriving (1.5.4). Hence, we obtain

$$\nabla^2\mathbf{F} - \gamma^2\mathbf{F} = -\varepsilon_c\mathbf{J}_m + \nabla(\nabla\bullet \mathbf{F} + j\omega\mu\varepsilon_c\varphi_m) \tag{1.5.16}$$

Lorenz's condition in this case is

$$\nabla\bullet \mathbf{F} = -j\omega\mu\varepsilon_c\varphi_m = -\frac{\gamma^2}{j\omega}\varphi_m \tag{1.5.17}$$

and (1.5.16) reduces to

$$\nabla^2\mathbf{F} - \gamma^2\mathbf{F} = -\varepsilon_c\mathbf{J}_m \tag{1.5.18}$$

In terms of the complex propagation constant k, replace $-\gamma^2$ by k^2 in (1.5.18).

The **E** field follows from (1.5.14):

$$\mathbf{E} = -\frac{1}{\varepsilon_c}\nabla \times \mathbf{F} \tag{1.5.19}$$

and **H** follows from (1.4.5):

$$\mathbf{H} = -\frac{1}{j\omega\mu}\nabla \times \mathbf{E} - \frac{\mathbf{J}_m}{j\omega\mu}$$

$$= \frac{1}{j\omega\mu\varepsilon_c}\nabla \times \nabla \times \mathbf{F} - \frac{\mathbf{J}_m}{j\omega\mu} \tag{1.5.20}$$

From (1.5.15) and (1.5.17), the **H** field can also be expressed in the form

$$\mathbf{H} = -j\omega\mathbf{F} + \frac{j\omega}{\gamma^2}\nabla(\nabla\bullet\mathbf{F}) \tag{1.5.21}$$

Away from the source, **H** is given by

$$\mathbf{H} = -\frac{1}{j\omega\mu}\nabla \times \mathbf{E} = \frac{1}{j\omega\mu\varepsilon_c}\nabla \times \nabla \times \mathbf{F} \tag{1.5.22}$$

In the lossless case, with $\varepsilon_c = \varepsilon$, (1.5.18) is written as

$$\nabla^2\mathbf{F} + \beta^2\mathbf{F} = -\varepsilon\mathbf{J}_m \tag{1.5.23}$$

Then, **E** is given by (1.5.19) with $\varepsilon_c = \varepsilon$ and **H** by (1.5.20) or away from the source by (1.5.22). This expression also appears in terms of k by replacing β^2 by k^2, where k is real in a lossless region (i.e., $k = k'$) and given by $k = \omega\sqrt{\mu\varepsilon}$.

The fields due to $\mathbf{J}_m$ are determined by the **F** vector. There is no need to use the differential equation satisfied by φ_m. For completeness the equation for φ_m is

$$\nabla^2\varphi_m - \gamma^2\varphi_m = -\frac{\rho_m}{\mu} \tag{1.5.24}$$

which for the lossless case is

$$\nabla^2\varphi_m + \beta^2\varphi_m = -\frac{\rho_m}{\mu}$$

The total **E** and **H** fields are obtained using superposition. That is, the superposition of the fields due to **A** and **F**. Denoting the fields due to **A** by $\mathbf{E}_A$ and $\mathbf{H}_A$, and those by **F** by $\mathbf{E}_F$ and $\mathbf{H}_F$, we can write

$$\mathbf{E} = \mathbf{E}_A + \mathbf{E}_F$$

and

$$\mathbf{H} = \mathbf{H}_A + \mathbf{H}_F$$

where $\mathbf{E}_A$ and $\mathbf{H}_A$ are given by (1.5.8) and (1.5.7), respectively; and $\mathbf{E}_F$ and $\mathbf{H}_F$ are given by (1.5.19) and (1.5.20), respectively.

1.6 GAUGE TRANSFORMATIONS AND LORENZ'S CONDITIONS

Consider the fields due to the **A** vector. The $\mathbf{E}_A$ and $\mathbf{H}_A$ fields are determined by the potentials **A** and φ, but the reverse is not true. That is, the potentials are not unique. To show this, consider new potentials $\mathbf{A}_n$ and φ_n (n for new) be given by

$$\mathbf{A}_n = \mathbf{A} + \nabla f \tag{1.6.1}$$

and

$$\varphi_n = \varphi - j\omega f \tag{1.6.2}$$

where f is an arbitrary scalar. Then, a new electric field $\mathbf{E}_n$, from (1.5.3) is

$$\begin{aligned}\mathbf{E}_n &= -\nabla\varphi_n - j\omega\mathbf{A}_n = -\nabla(\varphi - j\omega f) - j\omega(\mathbf{A} + \nabla f)\\ &= -\nabla\varphi - j\omega\mathbf{A}\end{aligned}$$

which is identical to the original field, or $\mathbf{E} = \mathbf{E}_n$.

Similarly, a new $\mathbf{B}_n$ field is

$$\begin{aligned}\mathbf{B}_n &= \nabla \times \mathbf{A}_n = \nabla \times (\mathbf{A} + \nabla f)\\ &= \nabla \times \mathbf{A}\end{aligned}$$

which is identical to the original field, or $\mathbf{B} = \mathbf{B}_n$.

The transformations in (1.6.1) and (1.6.2) are known as "gauge transformations," and they do not affect the original electric and magnetic fields. The gauge transformations show that the potentials cannot uniquely be defined by the observable electric and magnetic fields.

The gauge transformations introduce a freedom in the selection of the potentials. In fact, this freedom allows the selection of a set of potentials that satisfy Lorenz's conditions. Say that the selected **A** and φ does not satisfy Lorenz's condition in (1.5.5). The gauge transformations allow us to find a scalar f such that the resulting vector and scalar potentials satisfy Lorenz's condition. That is, replacing **A** by $\mathbf{A} + \nabla f$ and φ by $\varphi - j\omega f$ we find from (1.5.5) that

$$\nabla \bullet (\mathbf{A} + \nabla f) + \frac{\gamma^2}{j\omega}(\varphi - j\omega f) = 0$$

or

$$\nabla \bullet \mathbf{A} + \frac{\gamma^2}{j\omega}\varphi = -(\nabla^2 f - \gamma^2 f)$$

Since it is possible to find functions f that makes

$$\nabla^2 f - \gamma^2 f = 0$$

it follows that vector and scalar potentials can be found that satisfy Lorenz's condition. A similar derivation can be performed with Lorenz's condition in (1.5.17).

The fact that if the curl of a vector is specified one is still free to select the divergence is known as Helmholtz's theorem. The derivation of Helmholtz's theorem is a bit involved. An informal presentation of the theorem is presented. Basically, Helmholtz considered the question of whether the divergence and curl of a vector field are sufficient to determine the vector field.

Let the vector field be written as the sum of an irrotational field $\mathbf{I}(\mathbf{r})$ (i.e., $\nabla \times \mathbf{I}(\mathbf{r}) = 0$), also known as a conservative field) plus a solenoidal fields $\mathbf{S}(\mathbf{r})$ (i.e., $\nabla \bullet \mathbf{S}(r) = 0$). That is, let the vector field $\mathbf{T}(\mathbf{r})$ be

$$\mathbf{T}(\mathbf{r}) = \mathbf{I}(\mathbf{r}) + \mathbf{S}(\mathbf{r}) \tag{1.6.3}$$

where

$$\nabla \bullet \mathbf{T}(\mathbf{r}) = \nabla \bullet \mathbf{I}(\mathbf{r}) \quad (\text{since } \nabla \bullet \mathbf{S}(r) = 0) \tag{1.6.4}$$

and

$$\nabla \times \mathbf{T}(\mathbf{r}) = \nabla \times \mathbf{S}(\mathbf{r}) \quad (\text{since } \nabla \times \mathbf{I}(\mathbf{r}) = 0) \tag{1.6.5}$$

Equation (1.6.4) shows that the divergence of $\mathbf{T}$ provides information about the irrotational component, but not about its solenoidal component. Similarly, the curl of $\mathbf{T}$ in (1.6.5) provides information about the solenoidal component but not about its irrotational component. Hence, to completely specify a vector field both curl and divergence must be specified.

In terms of a scalar φ and a vector $\mathbf{Q}$, $\mathbf{T}$ can be expressed in the form

$$\mathbf{T}(\mathbf{r}) = \nabla\varphi + \nabla \times \mathbf{Q}(\mathbf{r}) \tag{1.6.6}$$

where the gradient represents the irrotational part (since $\nabla \bullet \nabla \times \mathbf{Q} = 0$) and the curl represents the solenoidal part (since $\nabla \times \nabla\varphi = 0$).

The formal proof of Helmholtz's theorem consists in showing that if a field $\mathbf{T}$ decays as $1/r$ at infinity, $\mathbf{T}$ can be expressed by a form similar to (1.6.6). Helmholtz's theorem can also be derived for a vector field in a bounded region. That is, with a surface that encloses a volume. For this case, an additional surface integral for $\mathbf{T}$ occur and the theorem holds.

There are cases where both the divergence and curl of a vector are zero. That is, $\nabla \times \mathbf{T} = 0$ and $\nabla \bullet \mathbf{T} = 0$. Such cases are not covered by Helmholtz's theorem. For such cases, since the curl is zero, $\mathbf{T}$ can be expressed as the gradient of a scalar or $\mathbf{T} = \nabla\varphi$, and it follows that

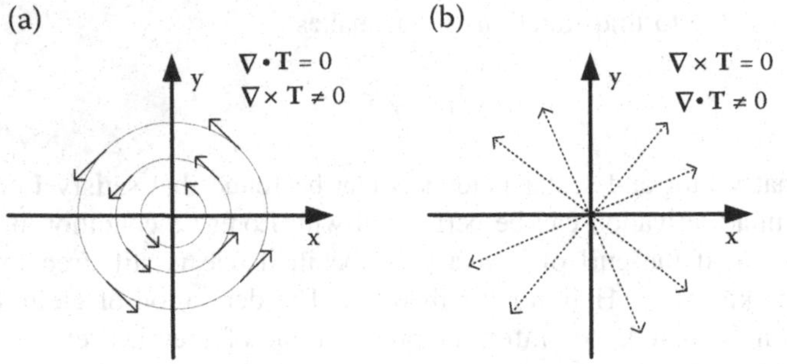

FIGURE 1.2 (a) A solenoidal field; (b) an irrotational field.

$$\nabla \cdot \mathbf{T} = \nabla \cdot \nabla \varphi = \nabla^2 \varphi = 0$$

which shows that the field can be represented by Laplace's equation (i.e., a Laplacian field). These are the field patterns that occur in electrostatic.

Figure 1.2a shows an example of a solenoidal field and Fig. 1.2b shows an irrotational field. Figure 1.2a illustrates the typical behavior of the rotational nature of the field. Figure 1.2b shows the divergent nature of the field, such as that produced by a point source.

1.7 HERTZ'S VECTORS

Hertz's vectors are like the **A** and **F** vector potentials. There are two Hertz's vectors, the electric Hertz's vector $\mathbf{\Pi}^e$ and the magnetic Hertz's vector $\mathbf{\Pi}^m$.

Lorenz's condition for the **A** vector in (1.5.5) can be used to define the electric Hertz's vector $\mathbf{\Pi}^e$. If we let

$$\varphi = -\nabla \cdot \mathbf{\Pi}^e$$

and

$$\mathbf{A} = j\omega\mu\varepsilon_c \mathbf{\Pi}^e = \frac{\gamma^2}{j\omega}\mathbf{\Pi}^e \quad \Rightarrow \quad \mathbf{\Pi}^e = \frac{j\omega}{\gamma^2}\mathbf{A} \tag{1.7.1}$$

Lorentz's condition in (1.5.5) is satisfied.

Substituting (1.7.1) into (1.5.6) produces the wave equation satisfied by $\mathbf{\Pi}^e$, namely

$$\nabla^2 \mathbf{\Pi}^e - \gamma^2 \mathbf{\Pi}^e = -\frac{j\omega\mu}{\gamma^2}\mathbf{J}_e = -\frac{\mathbf{J}_e}{\sigma + j\omega\varepsilon} \tag{1.7.2}$$

This equation shows that the source of $\mathbf{\Pi}^e$ is $\mathbf{J}_e$ which is the reason why $\mathbf{\Pi}^e$ is called the electric Hertz's vector. Looking at (1.7.1), the nomenclature seems confusing since it is the "magnetic" vector potential that is related to the "electric" Hertz's vector.

The **H** and **E** fields, from (1.5.7) and (1.5.8), namely

$$\mathbf{H} = \frac{\gamma^2}{j\omega\mu}\nabla \times \mathbf{\Pi}^e = (\sigma + j\omega\varepsilon)\nabla \times \mathbf{\Pi}^e \tag{1.7.3}$$

and

$$\mathbf{E} = \nabla \times \nabla \times \mathbf{\Pi}^e - \frac{j\omega\mu}{\gamma^2}\mathbf{J}_e \tag{1.7.4}$$

Away from the source, one sets $\mathbf{J}_e = 0$ in (1.7.4) to obtain

$$\mathbf{E} = \nabla \times \nabla \times \mathbf{\Pi}^e = \frac{j\omega\mu}{\gamma^2}\nabla \times \mathbf{H} = \frac{1}{\sigma + j\omega\varepsilon}\nabla \times \mathbf{H}$$

which is the expected result from (1.4.6).

For the lossless case (1.7.2) reads

$$\nabla^2\mathbf{\Pi}^e + \beta^2\mathbf{\Pi}^e = -\frac{\mathbf{J}_e}{j\omega\varepsilon} \tag{1.7.5}$$

and the fields, away from the source, are calculated using

$$\mathbf{E} = \nabla \times \nabla \times \mathbf{\Pi}^e = \frac{1}{j\omega\varepsilon}\nabla \times \mathbf{H} \tag{1.7.6}$$

and

$$\mathbf{H} = j\omega\varepsilon\, \nabla \times \mathbf{\Pi}^e \tag{1.7.7}$$

In the literature, one also finds (1.7.2) expressed in terms of an impressed polarization current source $\mathbf{P}_e$, where in terms of $\mathbf{J}_e$ is given by

$$\mathbf{J}_e = \frac{\partial \mathbf{P}_e}{\partial t} = j\omega\mathbf{P}_e$$

Then, (1.7.2) reads

$$\nabla^2\mathbf{\Pi}^e - \gamma^2\mathbf{\Pi}^e = -\frac{j\omega\mathbf{P}_e}{\sigma + j\omega\varepsilon}$$

If the impressed current is due to a dipole source at the origin. Then,

$$\mathbf{P}_e = \mathbf{p} = (q_o \Delta L)\sin\omega t\,\delta(x)\delta(y)\delta(z)\hat{\mathbf{a}}_z$$

and

$$\mathbf{J}_e(\mathbf{r}, t) = \omega(q_o \Delta L)\cos\omega t\,\delta(x)\delta(y)\delta(z)\hat{\mathbf{a}}_z = \mathrm{Re}[(I\Delta L)e^{j\omega t}\delta(x)\delta(y)\delta(z)]\hat{\mathbf{a}}_z$$

or in phasor form

$$\mathbf{J}_e(\mathbf{r}) = j\omega\mathbf{P}_e = (I\Delta L)\delta(x)\delta(y)\delta(z)\,\hat{\mathbf{a}}_z$$

There are no benefits in using the wave equation for the vector **A** over that of the vector $\mathbf{\Pi}^e$ to calculate the fields. In either case the source for these vectors is $\mathbf{J}_e$ and the wave equations are similar. Some authors prefer to use the **A** vector formulation, others the $\mathbf{\Pi}^e$ vector formulation. The same conclusions apply to the use of the **F** vector or $\mathbf{\Pi}^m$ vector, which is discussed next.

Lorenz's condition for the electric vector potential **F** in (1.5.17) can be used to define the magnetic Hertz vector $\mathbf{\Pi}^m$. If we let

$$\varphi_m = -\nabla\bullet\mathbf{\Pi}^m$$

and

$$\mathbf{F} = j\omega\mu\varepsilon_c\mathbf{\Pi}^m = \frac{\gamma^2}{j\omega}\mathbf{\Pi}^m \quad\Rightarrow\quad \mathbf{\Pi}^m = \frac{j\omega}{\gamma^2}\mathbf{F} \tag{1.7.8}$$

the Lorentz's condition in (1.5.17) is satisfied.

Substituting (1.7.8) into (1.5.18) produces the wave equation satisfied by $\mathbf{\Pi}^m$, namely

$$\nabla^2\mathbf{\Pi}^m - \gamma^2\mathbf{\Pi}^m = -\frac{\mathbf{J}_m}{j\omega\mu} \tag{1.7.9}$$

Away from the source, the **E** and **H** fields follow from (1.5.19) and (1.5.22), which in terms of $\mathbf{\Pi}^m$ reads

$$\mathbf{E} = -j\omega\mu\nabla\times\mathbf{\Pi}^m \tag{1.7.10}$$

and

$$\mathbf{H} = -\frac{1}{j\omega\mu}\nabla\times\mathbf{E} = \nabla\times\nabla\times\mathbf{\Pi}^m \tag{1.7.11}$$

For the lossless case, (1.7.9) reads

$$\nabla^2 \mathbf{\Pi}^{\mathrm{m}} + \beta^2 \mathbf{\Pi}^{\mathrm{m}} = -\frac{\mathbf{J}_m}{j\omega\mu} \tag{1.7.12}$$

and, away from the source, **E** and **H** are given by (1.7.10) and (1.7.11).

One also finds (1.7.9) expressed in terms of an impressed magnetization current density $\mathbf{M}_m$. Referring to (1.4.1) the appropriate relation is

$$\mathbf{J}_m = \mu \frac{\partial \mathbf{M}_m}{\partial t}$$

or in phasor form

$$\mathbf{J}_m = j\omega\mu \mathbf{M}_m$$

and (1.7.9) reads

$$\nabla^2 \mathbf{\Pi}^{\mathrm{m}} - \gamma^2 \mathbf{\Pi}^{\mathrm{m}} = -\mathbf{M}_m$$

where in the lossless case, $-\gamma^2$ is replaced by β_o^2

1.8 BOUNDARY CONDITIONS

The most common regions for the propagation of electromagnetic waves are those where there are no magnetic currents or charges, (i.e., $\mathbf{J}_m = \rho_m = 0$). In those regions, the boundary conditions satisfied by the fields at the interface of two regions with no surface electric currents ($\mathbf{J}_s$) and surface electric charges (ρ_s) are

$$\hat{\mathbf{n}} \times \mathbf{E}_1 = \hat{\mathbf{n}} \times \mathbf{E}_2 \tag{1.8.1}$$

$$\hat{\mathbf{n}} \times \mathbf{H}_1 = \hat{\mathbf{n}} \times \mathbf{H}_2 \tag{1.8.2}$$

$$\hat{\mathbf{n}} \cdot \mathbf{D}_1 = \hat{\mathbf{n}} \cdot \mathbf{D}_2 \tag{1.8.3}$$

$$\hat{\mathbf{n}} \cdot \mathbf{B}_1 = \hat{\mathbf{n}} \cdot \mathbf{B}_2 \tag{1.8.4}$$

where $\hat{\mathbf{n}}$ is a normal to the interface, as shown in Fig. 1.3. The unit vector is sometimes denoted by $\hat{\mathbf{n}}_{21}$, to indicate that it points from region 2 to 1.

Equations (1.8.1) and (1.8.2) indicate that the tangential components of **E** and **H** are continuous at the interface, and (1.8.3) and (1.8.4) show that the normal component of **D** and **B** are continuous at the interface. Equation (1.8.3) also shows that

$$\varepsilon_{c1} \hat{\mathbf{n}} \cdot \mathbf{E}_1 = \varepsilon_{c2} \hat{\mathbf{n}} \cdot \mathbf{E}_2 \quad \Rightarrow \quad (\sigma_1 + j\omega\varepsilon_1) \hat{\mathbf{n}} \cdot \mathbf{E}_1 = (\sigma_2 + j\omega\varepsilon_2) \hat{\mathbf{n}} \cdot \mathbf{E}_2 \tag{1.8.5}$$

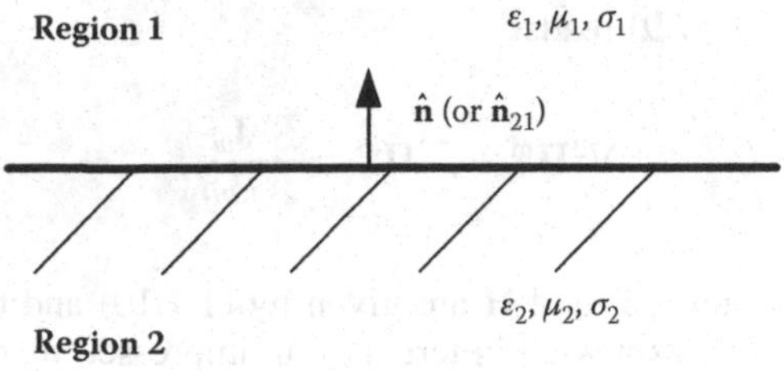

FIGURE 1.3 Boundary between two regions.

which in the lossless case is

$$\varepsilon_1 \hat{\mathbf{n}} \bullet \mathbf{E}_1 = \varepsilon_2 \hat{\mathbf{n}} \bullet \mathbf{E}_2$$

Equation (1.8.4) in terms of **H** at the interface reads

$$\mu_1 \hat{\mathbf{n}} \bullet \mathbf{H}_1 = \mu_2 \hat{\mathbf{n}} \bullet \mathbf{H}_2$$

These last two relations show that the normal components of **E** and **H** are discontinuous by an amount proportional to permittivity and permeability of the regions, respectively.

If Region 2 is a perfect electric conductor (PEC), there is no electric field inside the conductor ($\mathbf{E}_2 = 0$) and, therefore, there can be no time-varying magnetic field ($\mathbf{B}_2 = 0$). However, a perfect conductor supports a surface current density $\mathbf{J}_s$ and a surface charge density ρ_s. Therefore, the boundary conditions in the presence of a perfect conductor are

$$\hat{\mathbf{n}} \times \mathbf{E}_1 = 0 \tag{1.8.6}$$

$$\hat{\mathbf{n}} \times \mathbf{H}_1 = \mathbf{J}_s \tag{1.8.7}$$

$$\hat{\mathbf{n}} \bullet \mathbf{D}_1 = \rho_s$$

$$\hat{\mathbf{n}} \bullet \mathbf{B}_1 = 0$$

If the interface is not a perfect conductor but can support a surface current and charge density, the boundary conditions are

$$\hat{\mathbf{n}} \times (\mathbf{E}_1 - \mathbf{E}_2) = 0 \tag{1.8.8}$$

$$\hat{\mathbf{n}} \times (\mathbf{H}_1 - \mathbf{H}_2) = \mathbf{J}_s \tag{1.8.9}$$

$$\hat{\mathbf{n}} \bullet (\mathbf{D}_1 - \mathbf{D}_2) = \rho_s \tag{1.8.10}$$

$$\hat{\mathbf{n}} \bullet (\mathbf{B}_1 - \mathbf{B}_2) = 0 \tag{1.8.11}$$

The time-varying fields are related by Maxwell's equations, and it follows that the boundary conditions are interrelated. For example, from (1.4.5) it follows that

$$\frac{\partial E_y}{\partial x} - \frac{\partial E_x}{\partial y} = -j\omega B_z$$

which shows that once two components of **E** are known then a component of **B** is known. In fact, the specification of the tangential components of **E** at the interface (say E_x and E_y) determine the normal components of **B** at the interface (in this case B_z). Similarly, from (1.4.6) we obtain

$$\frac{\partial H_y}{\partial x} - \frac{\partial H_x}{\partial y} = (\sigma + j\omega\varepsilon)E_z$$

which shows that knowledge of the tangential component of **H** (say H_x and H_y) determines the normal component of **E** (in this case E_z). The previous considerations show that it is sufficient for the tangential **E** and **H** fields across an interface to satisfy the boundary conditions.

Lastly, we consider the case of an equivalent magnetic current along an interface. Then, (1.8.8) to (1.8.11) are modified to read:

$$\hat{\mathbf{n}} \times (\mathbf{E}_1 - \mathbf{E}_2) = -\mathbf{J}_{ms}$$
$$\hat{\mathbf{n}} \times (\mathbf{H}_1 - \mathbf{H}_2) = 0$$
$$\hat{\mathbf{n}} \bullet (\mathbf{D}_1 - \mathbf{D}_2) = 0$$
$$\hat{\mathbf{n}} \bullet (\mathbf{B}_1 - \mathbf{B}_2) = \rho_{ms}$$

where $\mathbf{J}_{ms}$ is the surface magnetic current and ρ_{ms} is the associated surface magnetic charge.

The boundary conditions on a perfect magnetic conductor (PMC) are the duals of the boundary conditions on a PEC. That is, in a PMC the tangential components of **H** is zero and the normal component of **E** is zero. Therefore, in a perfect magnetic conductor

$$\hat{\mathbf{n}} \times \mathbf{H}_1 = 0$$

and

$$\hat{\mathbf{n}} \times \mathbf{E}_1 = -\mathbf{J}_{ms}$$

In certain problems considerable simplification is obtained by approximating the boundary condition with a surface impedance (Z_s) that relates the tangential **E** and **H** fields at the interface. This is known as the Leontovitch boundary condition.

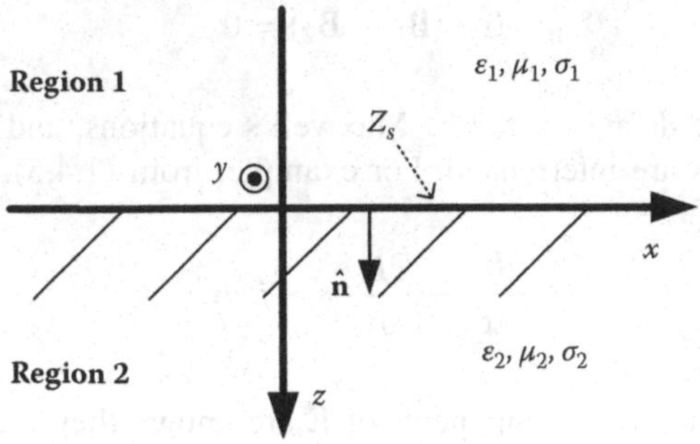

FIGURE 1.4 An example of the Leontovich boundary condition.

The common form that the Leontovich boundary condition appears is

$$\mathbf{E}_t = Z_s\, \mathbf{H}_t \times \hat{\mathbf{n}} \tag{1.8.12}$$

where $\mathbf{E}_t$ and $\mathbf{H}_t$ are the tangential electric field at the surface, and $\hat{\mathbf{n}}$ points into the surface.

For example, consider the geometry shown in Fig. 1.4, where the tangential fields in Region 1 are E_x, E_y, H_x and H_y. At the surface, $\hat{\mathbf{n}} = \hat{\mathbf{a}}_z$, and it follows from (1.8.12) that

$$\begin{aligned} E_x\hat{\mathbf{a}}_x + E_y\hat{\mathbf{a}}_y &= Z_s(H_x\hat{\mathbf{a}}_x + H_y\hat{\mathbf{a}}_y) \times \hat{\mathbf{a}}_z\Big|_{z=0} \\ &= Z_s(-H_x\hat{\mathbf{a}}_y + H_y\hat{\mathbf{a}}_x)\Big|_{z=0} \end{aligned}$$

or

$$E_x = Z_s H_y\Big|_{z=0}$$

and

$$E_y = -Z_s H_x\Big|_{z=0}$$

1.9 THE UNIQUENESS THEOREM

In this section, the Uniqueness Theorem is reviewed. This theorem guarantees that the solution to an electromagnetic boundary value problem is the only one possible (i.e., unique). Consider two solutions to Maxwell equations generated by the sources $\mathbf{J}_e$ and $\mathbf{J}_m$, namely

$$\begin{cases} \nabla \times \mathbf{E}_1 = -j\omega\mu\mathbf{H}_1 - \mathbf{J}_m \\ \nabla \times \mathbf{E}_2 = -j\omega\mu\mathbf{H}_2 - \mathbf{J}_m \end{cases} \tag{1.9.1}$$

$$\begin{cases} \nabla \times \mathbf{H}_1 = (\sigma + j\omega\varepsilon)\,\mathbf{E}_1 + \mathbf{J}_e \\ \nabla \times \mathbf{H}_2 = (\sigma + j\omega\varepsilon)\mathbf{E}_2 + \mathbf{J}_e \end{cases} \tag{1.9.2}$$

where the medium is assumed to be isotropic and homogeneous, although the following derivation also applies to spatially inhomogeneous media.

Denoting the difference of the fields by $\Delta\mathbf{E} = \mathbf{E}_1 - \mathbf{E}_2$ and $\Delta\mathbf{H} = \mathbf{H}_1 - \mathbf{H}_2$, a subtraction in (1.9.1) and in (1.9.2) shows that the difference satisfies

$$\nabla \times \Delta\mathbf{E} = -j\omega\mu\Delta\mathbf{H} \tag{1.9.3}$$

and

$$\nabla \times \Delta\mathbf{H} = (\sigma + j\omega\varepsilon)\,\Delta\mathbf{E} \tag{1.9.4}$$

Forming the dot product of (1.9.3) with $\Delta\mathbf{H}^*$ and the dot product of the conjugate of (1.9.4) with $\Delta\mathbf{E}$:

$$\Delta\mathbf{H}^* \bullet (\nabla \times \Delta\mathbf{E}) = -j\omega\mu\Delta\mathbf{H}^* \bullet \Delta\mathbf{H} = -j\omega\mu\,|\Delta\mathbf{H}|^2 \tag{1.9.5}$$

and

$$\Delta\mathbf{E} \bullet (\nabla \times \Delta\mathbf{H}^*) = (\sigma - j\omega\varepsilon)\,\Delta\mathbf{E} \bullet \Delta\mathbf{E}^* = (\sigma - j\omega\varepsilon)\,|\Delta\mathbf{E}|^2 \tag{1.9.6}$$

Subtracting (1.9.6) from (1.9.5):

$$\Delta\mathbf{H}^* \bullet (\nabla \times \Delta\mathbf{E}) - \Delta\mathbf{E} \bullet (\nabla \times \Delta\mathbf{H}^*) = -j\omega\,(\mu\,|\Delta\mathbf{H}|^2 - \varepsilon\,|\Delta\mathbf{E}|^2) - \sigma|\Delta\mathbf{E}|^2 \tag{1.9.7}$$

Using the identity:

$$\nabla \bullet (\mathbf{A} \times \mathbf{B}) = \mathbf{B} \bullet (\nabla \times \mathbf{A}) - \mathbf{A} \bullet (\nabla \times \mathbf{B})$$

(1.9.7) reads

$$\nabla \bullet (\Delta\mathbf{E} \times \Delta\mathbf{H}^*) = -j\omega\,(\mu\,|\Delta\mathbf{H}|^2 - \varepsilon\,|\Delta\mathbf{E}|^2) - \sigma\,|\Delta\mathbf{E}|^2 \tag{1.9.8}$$

Integrating (1.9.8) over a volume:

$$\int_V \nabla \bullet (\Delta\mathbf{E} \times \Delta\mathbf{H}^*)\; dV = -j\omega \int_V (\mu\,|\Delta\mathbf{H}|^2 - \varepsilon\,|\Delta\mathbf{E}|^2)\; dV - \sigma \int_V |\Delta\mathbf{E}|^2 dV \tag{1.9.9}$$

Using the divergence theorem, the left side of (1.9.9) can be expressed in terms of a surface integral

$$\oint_S \Delta\mathbf{E} \times \Delta\mathbf{H}^* \bullet \hat{\mathbf{n}} dS = -j\omega \int_V (\mu|\Delta\mathbf{H}|^2 - \varepsilon|\Delta\mathbf{E}|^2) dV - \sigma \int_V |\Delta\mathbf{E}|^2 dV \tag{1.9.10}$$

where S is the surface bounding the volume.

In (1.9.10), the cross product in the surface integral shows that only the tangential components of $\Delta\mathbf{E}$ and $\Delta\mathbf{H}^*$ contribute to the surface integral. Therefore, the surface integral is zero when the tangential components of $\mathbf{E}_1$ and $\mathbf{E}_2$ are equal (i.e., $\Delta\mathbf{E} = 0$) over the surface, or when the tangential components of $\mathbf{H}_1$ and $\mathbf{H}_2$ are equal (i.e., $\Delta\mathbf{H} = 0$) over the surface. The surface integral can also be zero if $\mathbf{E}_1$ and $\mathbf{E}_2$ are equal over part of the surface and $\mathbf{H}_1$ and $\mathbf{H}_2$ are equal over the rest of the surface.

When the surface integral in (1.9.10) is zero, the right-hand side must also be zero. Separating the real and imaginary parts shows that

$$\int_V |\Delta\mathbf{E}|^2 dV = 0 \quad \Rightarrow \quad \Delta\mathbf{E} = 0 \quad \Rightarrow \quad \mathbf{E}_1 = \mathbf{E}_2$$

and

$$\int_V (\mu|\Delta\mathbf{H}|^2 + \varepsilon|\Delta\mathbf{E}|^2)\, dV = 0 \quad \Rightarrow \quad \Delta\mathbf{H} = 0 \quad \Rightarrow \quad \mathbf{H}_1 = \mathbf{H}_2$$

which show that the fields are unique inside the volume. The lossless case can be considered as the limit of a lossy case as $\sigma \to 0$.

The Uniqueness Theorem shows that the fields are unique in a volume (i.e., a region) when the fields due to the sources $\mathbf{J}_e$ and $\mathbf{J}_m$ satisfy Maxwell equations as described in (1.9.1) and (1.9.2) and one of the following boundary conditions over the surface of the volume is satisfied:

1. The tangential component of the electric field is specified over the surface.
2. The tangential component of the magnetic field is specified over the surface.
3. The tangential component of the electric field is satisfied over part of the surface, and the tangential magnetic field is specified over the rest of the surface.

Problems

P1.1 In Appendix A, the nomenclature and units of the terms used in this book are listed. Verify that the units match in the following equations:

a. (1.1.15) and (1.1.16).
b. (1.2.11) and (1.2.19).
c. (1.3.5), (1.3.10), and (1.3.13).
d. (1.5.5), (1.5.6), and (1.5.8).
e. (1.5.18), (1.5.19), and (1.5.20).
f. (1.7.1), (1.7.2), (1.7.8), and (1.7.9).

P1.2 Derive (1.3.17) and (1.3.18).

P1.3 In (1.3.17) and (1.3.18), α and β are expressed in terms of ε, μ and σ.
a. Derive the expressions for the real and imaginary parts of the index of refraction (i.e., $n = n' - jn''$) in terms of ε, μ and σ.
b. Find an approximate expression for the complex impedance η (i.e., η' and η'') in a good dielectric.

P1.4 Derive (1.4.12).

P1.5 Derive (1.5.24).

P1.6 Write the expressions in (1.5.6), (1.5.18), (1.7.2), and (1.7.9) in terms of the complex propagation constant k where $k = k' - jk''$.

2 Radiation Fields

2.1 THE HERTZIAN ELECTRIC DIPOLE

A Hertzian electric dipole is an infinitesimal current element having a sinusoidal current with amplitude I (i.e., $I\cos(\omega t)$). It is also called a constant current element or a delta function element of strength $I\Delta L$, where ΔL denotes the small length of the Hertzian electric dipole, as shown in Fig. 2.1. The Hertzian electric dipole is not a physically realizable antenna, but is a basic radiator from which the fields from many linear antennas can be obtained using the principle of superposition.

Wave equations for the calculation of the potentials $\mathbf{A}$ and $\mathbf{\Pi}^e$ were obtained in Sections 1.5 and 1.7. These equations can be used to determine the radiation fields from the Hertzian dipole. The Hertzian electric dipole is represented by the following current density phasor:

$$\mathbf{J}_e = J_{e,z}\hat{\mathbf{a}}_z = (I\Delta L)\delta(x)\delta(y)\delta(z)\hat{\mathbf{a}}_z$$

Then, from (1.5.11), the z-component of the $\mathbf{A}$ vector in a lossless region satisfies

$$\begin{aligned}\nabla^2 A_z + \beta^2 A_z &= -\mu J_{e,z} \\ &= -\mu(I\Delta L)\delta(x)\delta(y)\delta(z)\end{aligned} \tag{2.1.1}$$

where $\beta = \omega\sqrt{\mu\varepsilon}$. Since $J_{e,z}$ is a point source, the A_z field depends only on the radial distance r, and not on ϕ or θ. Expanding the Laplacian in spherical coordinates, A_z satisfies the homogeneous equation

$$\frac{1}{r^2}\frac{\partial}{\partial r}\left(r^2\frac{\partial A_z}{\partial r}\right) + \beta^2 A_z = 0 \tag{2.1.2}$$

Letting

$$A_z = \frac{f(r)}{r}$$

it follows that $f(r)$ satisfies

DOI: 10.1201/9781003219729-2

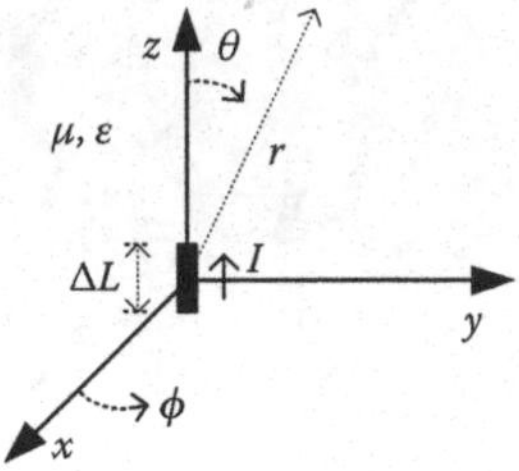

FIGURE 2.1 The Hertzian electric dipole.

$$\frac{d^2f}{dr^2} + \beta^2 f = 0$$

whose solutions are

$$f(r) = C_1 e^{-j\beta r} + C_2 e^{j\beta r}$$

Therefore, the solution to (2.1.2) is

$$A_z(r) = C_1 \frac{e^{-j\beta r}}{r} + C_2 \frac{e^{j\beta r}}{r}$$

Since the wave must be an outgoing wave, set $C_2 = 0$ and it follows that

$$A_z(r) = C_1 \frac{e^{-j\beta r}}{r} \tag{2.1.3}$$

The constant C_1 is determined by the source excitation in (2.1.1). To evaluate C_1, integrate (2.1.1) over a small volume of radius ε enclosing the Hertzian dipole and take the limit as $\varepsilon \to 0$, namely

$$\int_V \nabla^2 A_z \, dV + \beta^2 \int_V A_z \, dV = -\mu\,(I\Delta L) \int_V \delta(x)\,\delta(y)\,\delta(z) dV$$

The delta function integral is equal to one, so the right-hand side is $-\mu\,(I\Delta L)$. In the second integral on the left, since $r = \varepsilon \to 0$ then $e^{-j\beta r} \to 1$ and A_z is approximated by $A_z \approx C_1/r$. Hence,

$$\lim_{\varepsilon \to 0} \beta^2 \int_V A_z \, dV \approx \lim_{\varepsilon \to 0} \beta^2 C_1 \int_0^{\varepsilon} \frac{1}{r} \, 4\pi r^2 \, dr = \lim_{\varepsilon \to 0} \beta^2 C_1 2\pi\varepsilon^2 = 0$$

The exact evaluation would give the same result. That is,

$$\lim_{\varepsilon\to 0} \beta^2 \int_{r=0}^{\varepsilon} \int_{\theta=0}^{\pi} \int_{\phi=0}^{2\pi} C_1 \frac{e^{-j\beta r}}{r} r^2 \sin\theta dr d\theta d\phi = \lim_{\varepsilon\to 0} \beta^2 C_1 4\pi \int_0^{\varepsilon} e^{-j\beta r} r\, dr$$

$$= \lim_{\varepsilon\to 0} \beta^2 C_1 4\pi \left[\frac{e^{-j\beta r}}{\beta^2} (j\beta r + 1) \right]\Bigg|_0^{\varepsilon} = 0$$

The first integral on the left is evaluated as follows

$$\int_V \nabla^2 A_z dV = \int_V \nabla \bullet \nabla A_z dV = \oint_S \nabla A_z \bullet \hat{\mathbf{n}} dS$$

$$= \int_{\theta=0}^{\pi} \int_{\phi=0}^{2\pi} \frac{\partial}{\partial r}\left(\frac{C_1}{r}\right) r^2 \sin\theta d\theta d\phi = -4\pi C_1$$

Hence,

$$-4\pi C_1 = -\mu(I\Delta L) \quad \Rightarrow \quad C_1 = \frac{\mu(I\Delta L)}{4\pi} \tag{2.1.4}$$

The exact evaluation of ∇A_z is $-C_1(1 + j\beta r)(e^{-j\beta r}/r^2)$. Then, after performing the surface integration, in the limit as $\varepsilon \to 0$, the answer for C_1 is the same.

From (2.1.3) and (2.1.4), A_z for the Hertzian electric dipole is given by

$$A_z = \mu(I\Delta L)\frac{e^{-j\beta r}}{4\pi r} \tag{2.1.5}$$

Due to the spherical nature of the solution, the spherical components of **A** are:

$$A_r = A_z \cos\theta = \mu(I\Delta L)\frac{e^{-j\beta r}}{4\pi r} \cos\theta$$

$$A_\theta = -A_z \sin\theta = -\mu(I\Delta L)\frac{e^{-j\beta r}}{4\pi r} \sin\theta$$

$$A_\phi = 0$$

The magnetic fields components are calculated using (1.5.7). That is,

$$\mathbf{H} = \frac{1}{\mu}\nabla \times \mathbf{A}$$

which in spherical coordinates gives

$$H_r = H_\theta = 0$$

$$H_\phi = \frac{1}{\mu r}\left[\frac{\partial}{\partial r}(rA_\theta) - \frac{\partial A_r}{\partial \theta}\right]$$
$$= j\beta (I\Delta L)\left(1 + \frac{1}{j\beta r}\right)\frac{e^{-j\beta r}}{4\pi r}\sin\theta \tag{2.1.6}$$

The electric field components are calculated using (1.5.12). That is,

$$\mathbf{E} = \frac{1}{j\omega\varepsilon}\nabla \times (H_\phi \hat{\mathbf{a}}_\phi)$$

where

$$\nabla \times H_\phi \hat{\mathbf{a}}_\phi = \frac{1}{r\sin\theta}\left[\frac{\partial}{\partial\theta}(H_\phi \sin\theta)\right]\hat{\mathbf{a}}_r - \frac{1}{r}\frac{\partial}{\partial r}(rH_\phi)\hat{\mathbf{a}}_\theta$$

Therefore,

$$E_r = \eta (I\Delta L)\left(1 + \frac{1}{j\beta r}\right)\frac{e^{-j\beta r}}{2\pi r^2}\cos\theta \tag{2.1.7}$$

$$E_\theta = j\beta\eta (I\Delta L)\left[1 + \frac{1}{j\beta r} - \frac{1}{(\beta r)^2}\right]\frac{e^{-j\beta r}}{4\pi r}\sin\theta \tag{2.1.8}$$

$$E_\phi = 0$$

where $\eta = \beta/\omega\varepsilon$ was used.

Equations (2.1.6) to (2.1.8) contain terms that vary as $1/r$, $1/r^2$, and $1/r^3$. The field terms that vary as $1/r^3$ terms are known as the electrostatic field, the field that varies as $1/r^2$ is the induction field, and the field that varies as $1/r$ is the far field. Usually, the electrostatic and induction field are labeled the near field. In propagation work we are usually interested in the far field. That is, the fields very far from the source or, in general, the region where $\beta r \gg 1$. In this region, the fields vary as $1/r$. From (2.1.8) and (2.1.6), the far fields are

$$E_\theta = j\beta\eta (I\Delta L)\frac{e^{-j\beta r}}{4\pi r}\sin\theta \tag{2.1.9}$$

and

$$H_\phi = j\beta (I\Delta L)\frac{e^{-j\beta r}}{4\pi r}\sin\theta \tag{2.1.10}$$

which shows that the far fields are transverse to the direction of propagation, called a transverse electromagnetic field (i.e., a TEM field). For an observer in the far field the spherical TEM field looks like a plane wave approaching the observer. For example, think of a spherical wave as a very-large inflating balloon approaching an observer. An observer in the far field will see the large balloon surface as an approaching plane wave, where the field components are transverse to the direction of propagation. This is one reason why the properties of plane waves are important.

The ratio of (2.1.9) and (2.1.10) is

$$\frac{E_\theta}{H_\phi} = \eta$$

which is simply determined by the intrinsic impedance of the region.

We also observe that (2.1.9) follows from the relation

$$E_\theta \approx -j\omega A_\theta = j\beta\eta(I\Delta L)\frac{e^{-j\beta r}}{4\pi r}\sin\theta$$

where $j\omega\mu = j\beta\eta$ was used. In Section 2.3 it is shown that $E_\theta \approx -j\omega A_\theta$ is an appropriate expression for the far field.

The power density in the far field is calculated using the Poynting vector relation for the time-average power density **P**, where

$$\mathbf{P} = \frac{1}{2}\text{Re}(\mathbf{E} \times \mathbf{H}^*)$$

From (2.19) and (2.1.10):

$$\mathbf{P} = \frac{1}{2}\text{Re}\left[E_\theta \hat{\mathbf{a}}_\theta \times H_\phi^* \hat{\mathbf{a}}_\phi\right] = \frac{1}{2}\text{Re}\left[E_\theta H_\phi^*\right]\hat{\mathbf{a}}_r = \frac{1}{2}\frac{|E_\theta|^2}{\eta}\hat{\mathbf{a}}_r \qquad (2.1.11)$$

which shows the radial nature of the power flow.

Substituting (2.1.9) into (2.1.11) gives

$$P_r = \frac{\eta}{8}\left(\frac{I\Delta L}{\lambda}\right)^2\frac{\sin^2\theta}{r^2}$$

The radiated power W is then

$$W = \oint_S \mathbf{P} \bullet \hat{\mathbf{n}} dS$$

$$= \oint_S P_r \hat{\mathbf{a}}_r \bullet \hat{\mathbf{a}}_r dS = \int_{\theta=0}^{\pi}\int_{\phi=0}^{2\pi} P_r r^2 \sin\theta d\phi d\theta = \frac{\pi}{3}\eta\left(\frac{I\Delta L}{\lambda}\right)^2$$

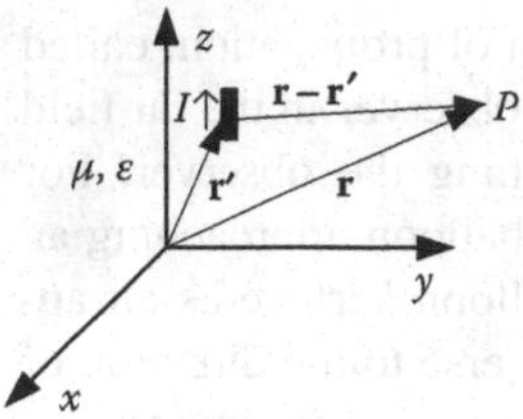

FIGURE 2.2 Hertzian dipole at position **r′**.

If the Hertzian dipole is located at position **r′**, as shown in Fig. 2.2, then A_z is given by

$$A_z = \mu(I\Delta L)\frac{e^{-j\beta|\mathbf{r}-\mathbf{r}'|}}{4\pi|\mathbf{r}-\mathbf{r}'|} \tag{2.1.12}$$

Equation (2.1.2) is recognized as a spherical Bessel differential equation, whose solutions can be written in terms of spherical Hankel functions of order zero (see Appendix C). Thus, (2.1.5) and (2.1.12) can be expressed in terms of these functions. Using,

$$h_0^{(2)}(\beta r) = j\frac{e^{-j\beta r}}{\beta r}$$

(2.1.5) reads

$$A_z = -j\frac{\beta\mu(I\Delta L)}{4\pi}h_0^{(2)}(\beta r)$$

and (2.1.12) reads

$$A_z = -j\frac{\beta\mu(I\Delta L)}{4\pi}h_0^{(2)}(\beta|\mathbf{r}-\mathbf{r}'|)$$

Next, the fields due to the electric Hertzian dipole in Fig. 2.1 are calculated for the case that the region is lossy and described by μ, ε and σ. Let us formulate the problem in terms of the $\mathbf{\Pi}^{\mathrm{e}}$ vector. From (1.7.2), the z component of the electric Hertz vector satisfies

$$[\nabla^2 - \gamma^2]\Pi_z^{\mathrm{e}}(x, y, z) = -\frac{1}{\sigma + j\omega\varepsilon}(I\Delta L)\delta(x)\delta(y)\delta(z) \tag{2.1.13}$$

The solution of (2.1.13) is like that of (2.1.1), and Π_z^{e} for the outgoing wave is given by

$$\Pi_z^{\mathrm{e}} = B_1 \frac{e^{-\gamma r}}{r}$$

The evaluation of B_1 is similar to that of C_1 in (2.1.4), and we obtain

$$B_1 = \frac{(I\Delta L)}{4\pi(\sigma + j\omega\varepsilon)}$$

Hence,

$$\Pi_z^{\mathrm{e}} = \frac{(I\Delta L)}{(\sigma + j\omega\varepsilon)} \frac{e^{-\gamma r}}{4\pi r} \tag{2.1.14}$$

The spherical components of $\mathbf{\Pi}^{\mathrm{e}}$ are

$$\Pi_r^{\mathrm{e}} = \Pi_z^{\mathrm{e}} \cos\theta$$
$$\Pi_\theta^{\mathrm{e}} = -\Pi_z^{\mathrm{e}} \sin\theta$$
$$\Pi_\phi^{\mathrm{e}} = 0$$

and the $\mathbf{H}$ and $\mathbf{E}$ vectors follow from (1.7.4) and (1.4.6), namely

$$\mathbf{H} = (\sigma + j\omega\varepsilon)\nabla \times \mathbf{\Pi}^{\mathrm{e}} \tag{2.1.15}$$

and

$$\mathbf{E} = \frac{1}{\sigma + j\omega\varepsilon}\nabla \times \mathbf{H} \tag{2.1.16}$$

The indicated operations in (2.1.15) and (2.1.16) in spherical coordinates are similar to the ones performed in (2.1.6) to (2.1.8). Hence, we obtain:

$$H_r = H_\theta = E_\phi = 0$$

$$H_\phi = \gamma(I\Delta L)\left(1 + \frac{1}{\gamma r}\right)\frac{e^{-\gamma r}}{4\pi r}\sin\theta = \frac{(I\Delta L)}{4\pi r^2}(1 + \gamma r)e^{-\gamma r}\sin\theta \tag{2.1.17}$$

$$E_r = \eta(I\Delta L)\left(1 + \frac{1}{\gamma r}\right)\frac{e^{-\gamma r}}{2\pi r^2}\cos\theta = \frac{(I\Delta L)}{2\pi(\sigma + j\omega\varepsilon)r^3}(1 + \gamma r)e^{-\gamma r}\cos\theta \tag{2.1.18}$$

$$E_\theta = \gamma\eta(I\Delta L)\left[1 + \frac{1}{\gamma r} + \frac{1}{(\gamma r)^2}\right]\frac{e^{-\gamma r}}{4\pi r}\sin\theta$$
$$= \frac{(I\Delta L)}{4\pi(\sigma + j\omega\varepsilon)r^3}(1 + \gamma r + \gamma^2 r^2)e^{-\gamma r}\sin\theta \tag{2.1.19}$$

where the relation

$$\frac{\eta}{\gamma} = \frac{1}{\sigma + j\omega\varepsilon}$$

was used. Observe that in the lossless case, since $\gamma = j\beta$, (2.1.17) to (2.1.19) reduce to the form in (2.1.6) to (2.1.8).

If the Hertzian dipole is located at position $\mathbf{r}'$, then

$$\Pi_z^e = \frac{(I\Delta L)}{(\sigma + j\omega\varepsilon)}\frac{e^{-\gamma|\mathbf{r}-\mathbf{r}'|}}{4\pi|\mathbf{r} - \mathbf{r}'|}$$

2.2 THE LINE CURRENT SOURCE OF INFINITE LENGTH

A line current source of infinite length is illustrated in Fig. 2.3. The sinusoidal current is assumed to be of constant amplitude. Using (1.5.11), A_z satisfies

$$[\nabla^2 + \beta^2]A_z(\rho) = -\mu J_{e,z} = -\mu I\frac{\delta(\rho)}{2\pi\rho} \tag{2.2.1}$$

where $\beta = \omega\sqrt{\mu\varepsilon}$. The symmetry of the problem shows that there is no z or ϕ variation of the field due to the infinite-length source. Then, writing the Laplacian in cylindrical coordinates, the homogeneous form of (2.2.1) is

$$\frac{1}{\rho}\frac{\partial}{\partial\rho}\left(\rho\frac{\partial A_z}{\partial\rho}\right) + \beta^2 A_z = 0 \tag{2.2.2}$$

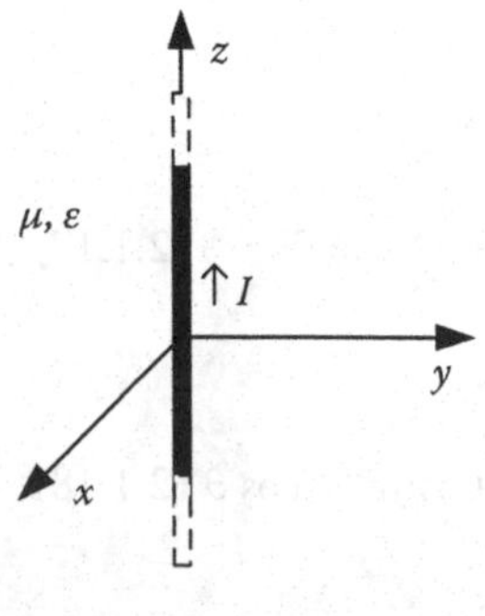

FIGURE 2.3 Line source of infinite length.

or

$$\rho^2 \frac{\partial^2 A_z}{\partial \rho^2} + \rho \frac{\partial A_z}{\partial \rho} + (\beta\rho)^2 A_z = 0$$

which is Bessel equation (see Appendix C) whose solution for an outgoing wave is

$$A_z = B_1 H_0^{(2)}(\beta\rho) \tag{2.2.3}$$

From (1.5.7), the magnetic fields are

$$H_\rho = H_z = 0$$

$$H_\phi = -\frac{1}{\mu}\frac{\partial A_z}{\partial \rho} = -B_1 \frac{\beta}{\mu} H_0^{(2)\prime}(\beta\rho) = B_1 \frac{\beta}{\mu} H_1^{(2)}(\beta\rho) \tag{2.2.4}$$

Since **H** has only a ϕ component, it follows from (1.5.12) that

$$E_\rho = E_\phi = 0$$

$$E_z = \frac{1}{j\omega\varepsilon}\frac{1}{\rho}\frac{\partial}{\partial \rho}(\rho H_\phi) = -\frac{1}{j\omega\mu\varepsilon}\frac{1}{\rho}\frac{\partial}{\partial \rho}\left(\rho \frac{\partial A_z}{\partial \rho}\right)$$

Using (2.2.2), E_z above is expressed in the form

$$E_z = \frac{\beta^2}{j\omega\mu\varepsilon} A_z = -j\omega A_z = -j\omega\, B_1 H_0^{(2)}(\beta\rho) \tag{2.2.5}$$

The constant B_1 depends on the source and is evaluated using Ampere's law. Since $\mathbf{H} = H_\phi \hat{\mathbf{a}}_\phi$ and $d\mathbf{l} = \rho d\phi \hat{\mathbf{a}}_\phi$ it follows that

$$I = \lim_{\rho\to 0} \oint_C \mathbf{H} \bullet d\,\mathbf{l} = \lim_{\rho\to 0} \int_0^{2\pi} H_\phi \rho d\phi = B_1 \frac{\beta}{\mu}(2\pi) \lim_{\rho\to 0}\left[\rho H_1^{(2)}(\beta\rho)\right] \tag{2.2.6}$$

where C is a closed contour enclosing the line source. The small argument approximation of $H_1^{(2)}(\beta\rho)$ is

$$\lim_{\rho\to 0} H_1^{(2)}(\beta\rho) = \lim_{\rho\to 0}[J_1(\beta\rho) - jY_1(\beta\rho)] \approx \lim_{\rho\to 0}\left[\frac{\beta\rho}{2} + j\frac{1}{\pi}\left(\frac{2}{\beta\rho}\right)\right] \approx j\frac{1}{\pi}\left(\frac{2}{\beta\rho}\right) \tag{2.2.7}$$

Substituting (2.2.7) into (2.2.6):

$$I = jB_1 \frac{4}{\mu} \quad \Rightarrow \quad B_1 = -j\frac{\mu I}{4}$$

Therefore, (2.2.3) to (2.2.5) read:

$$A_z = -j\frac{\mu I}{4} H_0^{(2)}(\beta\rho) \tag{2.2.8}$$

$$E_z = -\frac{\omega\mu I}{4} H_0^{(2)}(\beta\rho) = -\frac{\eta\beta I}{4} H_0^{(2)}(\beta\rho) \tag{2.2.9}$$

$$H_\phi = -j\frac{\beta I}{4} H_1^{(2)}(\beta\rho) \tag{2.2.10}$$

The far-field approximation requires the asymptotic expansions of the Hankel functions for large $\beta\rho$ values (see Appendix C). That is,

$$\lim_{\beta\rho\to\infty} H_0^{(2)}(\beta\rho) \approx \sqrt{\frac{2}{\pi\beta\rho}}\, e^{-j(\beta\rho-\pi/4)}$$

$$\lim_{\beta\rho\to\infty} H_1^{(2)}(\beta\rho) \approx j\sqrt{\frac{2}{\pi\beta\rho}}\, e^{-j(\beta\rho-\pi/4)}$$

Hence, in the far field (2.2.9) and (2.2.10) read:

$$E_z \approx -\eta I \sqrt{\frac{\beta}{8\pi}}\, \frac{e^{-j(\beta\rho-\pi/4)}}{\sqrt{\rho}}$$

$$H_\phi \approx I \sqrt{\frac{\beta}{8\pi}}\, \frac{e^{-j(\beta\rho-\pi/4)}}{\sqrt{\rho}}$$

which shows that in the far field the waves represent a TEM field, and

$$\frac{E_z}{H_\phi} = -\eta$$

The Hertzian dipole in the far field produces a spherical wave whose field terms decay as $1/r$, while the infinite-length line source produces a cylindrical wave whose field terms decay as $1/\sqrt{\rho}$.

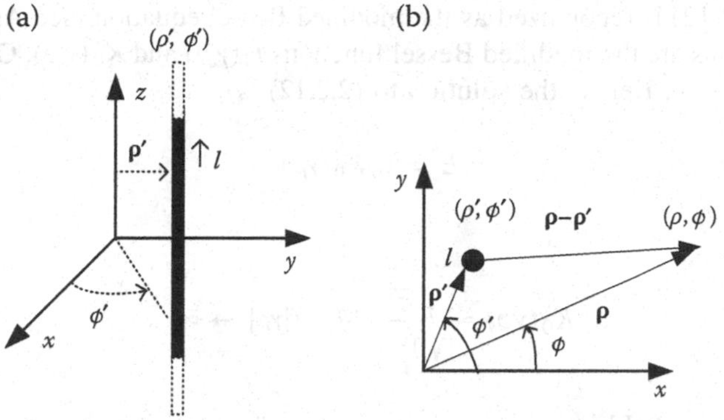

FIGURE 2.4 (a) Line source at position $\boldsymbol{\rho}'$; (b) top view from the z axis.

If the line source is located at position $\boldsymbol{\rho}'$, as shown in Fig. 2.4, then A_z is given by

$$A_z = -j\frac{\mu I}{4}H_0^{(2)}(\beta|\boldsymbol{\rho} - \boldsymbol{\rho}'|) \tag{2.2.11}$$

where

$$|\boldsymbol{\rho} - \boldsymbol{\rho}'| = \sqrt{(x - x')^2 + (y - y')^2} = \sqrt{\rho^2 + \rho'^2 - 2\rho\rho'\cos(\phi - \phi')}$$

Next, the fields from the line source in Fig. 2.3 when the region is lossy are determined. Instead of using the **A** vector formulation, as previously done, the fields will be determined using the wave equation satisfied by E_z. The fields do not vary with z or ϕ; therefore, the line source produces an H_ϕ and an E_z field. From Maxwell's equations, H_ϕ and E_z are related by

$$H_\phi = \frac{1}{j\omega\mu}\frac{\partial E_z}{\partial \rho}$$

The other field components are zero. That is, $E_\rho = E_\phi = H_\rho = H_z = 0$.

The field E_z satisfies the homogeneous equation

$$[\nabla^2 - \gamma^2]E_z(\rho) = 0$$

or

$$\frac{1}{\rho}\frac{\partial}{\partial \rho}\left(\rho\frac{\partial E_z}{\partial \rho}\right) - \gamma^2 E_z = 0 \tag{2.2.12}$$

Equation (2.2.12) is recognized as the modified Bessel equation (see Appendix C), whose solutions are the modified Bessel functions $I_0(\gamma\rho)$ and $K_0(\gamma\rho)$. Only $K_0(\gamma\rho)$ decays as $\rho \to \infty$. Hence, the solution to (2.2.12) is

$$E_z = C_1 K_0(\gamma\rho) \tag{2.2.13}$$

where

$$K_0(\gamma\rho) \approx \sqrt{\frac{\pi}{\gamma\rho}}\, e^{-\gamma\rho} \quad (|\gamma\rho| \to \infty)$$

and the magnetic field is

$$H_\phi = \frac{C_1}{j\omega\mu}\frac{\partial K_0(\gamma\rho)}{\partial\rho} = -\frac{C_1\gamma}{j\omega\mu}K_1(\gamma\rho) \tag{2.2.14}$$

To evaluate the constant C_1, let $\rho \to 0$ in (2.2.14) and equate the result to the field obtained using Ampere's law. That is,

$$H_\phi = \frac{I}{2\pi\rho} \quad \text{for} \quad \rho \to 0 \tag{2.2.15}$$

and from (2.2.14), using the small argument approximation:

$$K_1(\gamma\rho) \approx \frac{1}{\gamma\rho}$$

it follows that as $\rho \to 0$:

$$H_\phi \approx -\frac{C_1}{j\omega\mu\rho} \tag{2.2.16}$$

Equating (2.2.15) and (2.2.16) shows that

$$C_1 = -\frac{j\omega\mu I}{2\pi}$$

From (2.2.13) and (2.2.14) the fields are

$$E_z = -\frac{j\omega\mu I}{2\pi}K_0(\gamma\rho) \tag{2.2.17}$$

and

$$H_\phi = \frac{\gamma I}{2\pi}K_1(\gamma\rho) \tag{2.2.18}$$

In the lossless case, with $\gamma = j\beta$ the relations in (2.2.17) and (2.2.18) should reduce to the lossless-case fields in (2.2.9) and (2.2.10). This is shown by using the relation (see Appendix C):

$$K_\nu(jx) = \frac{\pi}{2}(-j)^{\nu+1}H_\nu^{(2)}(x)$$

Then, it follows that

$$K_0(j\beta\rho) = -j\frac{\pi}{2}H_0^{(2)}(\beta\rho)$$

and

$$K_1(j\beta\rho) = -\frac{\pi}{2}H_1^{(2)}(\beta\rho)$$

Hence, in the lossless case (2.2.17) and (2.2.18) read

$$E_z = -\frac{\omega\mu I}{4}H_0^{(2)}(\beta\rho)$$

and

$$H_\phi = -j\frac{\beta I}{4}H_1^{(2)}(\beta\rho)$$

which are identical to (2.2.9) and (2.2.10).

2.3 FAR-FIELD RELATIONS

The electric Hertzian dipole, as well as the magnetic Hertzian dipole (discussed in Section 2.6), are basic radiators from which the fields of many antennas can be derived using superposition. For electrical radiators of finite-dimensions, the general form of the **A** vector in the far field is

$$\begin{aligned}\mathbf{A} &= A_r\hat{\mathbf{a}}_r + A_\theta\hat{\mathbf{a}}_\theta + A_\phi\hat{\mathbf{a}}_\phi \\ &= [f(\theta,\phi)\hat{\mathbf{a}}_r + g(\theta,\phi)\hat{\mathbf{a}}_\theta + h(\theta,\phi)\hat{\mathbf{a}}_\phi]\frac{e^{-j\beta r}}{r}\end{aligned}$$

where $f(\theta, \phi)$, $g(\theta, \phi)$, and $h(\theta, \phi)$ denote the angular dependence of the **A** vector. The term $e^{-j\beta r}/r$ shows the spherical nature of the outgoing wave.

From (1.5.9), the far-field form of the **E** field can be calculated using

$$\mathbf{E} = -j\omega\mathbf{A} + \frac{1}{j\omega\mu\varepsilon}\nabla(\nabla \bullet \mathbf{A})$$

Performing the indicated operations in spherical coordinates and keeping only the terms that vary as $1/r$ gives

$$\mathbf{E} \approx -j\omega[g(\theta, \phi)\hat{\mathbf{a}}_\theta + h(\theta, \phi)\hat{\mathbf{a}}_\phi]\frac{e^{-j\beta r}}{r} \tag{2.3.1}$$

Similarly, from (1.5.7), the far-field form for **H** is

$$\mathbf{H} \approx -\frac{j\omega}{\eta}[h(\theta, \phi)\hat{\mathbf{a}}_\theta - g(\theta, \phi)\hat{\mathbf{a}}_\phi]\frac{e^{-j\beta r}}{r} \tag{2.3.2}$$

The above far-field expressions show that the fields are transverse to the direction of propagation. These expressions can be written as

$$\mathbf{E} \approx -j\omega(A_\theta\hat{\mathbf{a}}_\theta + A_\phi\hat{\mathbf{a}}_\phi)$$

and

$$\mathbf{H} = \frac{j\omega}{\eta}(A_\phi\hat{\mathbf{a}}_\theta - A_\theta\hat{\mathbf{a}}_\phi)$$

which show that in the far field:

$$\begin{cases} E_r = 0 & H_r = 0 \\ E_\theta = -j\omega A_\theta & H_\theta = -\dfrac{E_\phi}{\eta} \\ E_\phi = -j\omega A_\phi & H_\phi = \dfrac{E_\theta}{\eta} \end{cases} \tag{2.3.3}$$

Similarly, the general form of the **F** vector in the far field is

$$\mathbf{F} = F_r\hat{\mathbf{a}}_r + F_\theta\hat{\mathbf{a}}_\theta + F_\phi\hat{\mathbf{a}}_\phi = [f_1(\theta, \phi)\hat{\mathbf{a}}_r + g_1(\theta, \phi)\hat{\mathbf{a}}_\theta + h_1(\theta, \phi)\hat{\mathbf{a}}_\phi]\frac{e^{-j\beta r}}{r}$$

From (1.5.21) and (1.5.19), the far-field forms for the **H** and **E** field are calculated. Keeping only the terms that vary as $1/r$, it follows that

$$\mathbf{H} \approx -j\omega[g_1(\theta, \phi)\hat{\mathbf{a}}_\theta + h_1(\theta, \phi)\hat{\mathbf{a}}_\phi]\frac{e^{-j\beta r}}{r} = -j\omega(F_\theta\hat{\mathbf{a}}_\theta + F_\phi\hat{\mathbf{a}}_\phi)$$

and

$$\mathbf{E} \approx -j\omega\eta\,[h_1(\theta, \phi)\hat{\mathbf{a}}_\theta - g_1(\theta, \phi)\hat{\mathbf{a}}_\phi]\frac{e^{-j\beta r}}{r} = -j\omega\eta\,(F_\phi\hat{\mathbf{a}}_\theta - F_\theta\hat{\mathbf{a}}_\phi)$$

The far-field relations associated with the **F** vector are

$$\begin{cases} H_r = 0 & E_r = 0 \\ H_\theta = -j\omega F_\theta & E_\theta = \eta H_\phi \\ H_\phi = -j\omega F_\phi & E_\phi = -\eta H_\theta \end{cases} \tag{2.3.4}$$

An example of the **F** vector form in the far field is found in Section 2.6 where the far-field relations of the magnetic Hertzian dipole are expressed in terms of the **F** vector.

Equations (2.3.3) and (2.3.4) show the TEM wave nature of the far fields. That is, the **E** and **H** fields are perpendicular to each other and to the direction of propagation, and related by the intrinsic impedance of the region.

2.4 SOLUTION TO THE INHOMOGENEOUS WAVE EQUATION

The solution to a differential equation with a delta function excitation is known as the Green function (or Green's function) solution. For example, the Green function $G(\mathbf{r}, \mathbf{r}')$ associated with (2.1.1) is

$$[\nabla^2 + \beta^2]G(\mathbf{r}, \mathbf{r}') = -\delta(x)\delta(y)\delta(z) \tag{2.4.1}$$

where **r** denotes the observation point and **r′** the source location. The electric Hertzian dipole is located at the origin (or $\mathbf{r}' = 0$) and the observation point at $\mathbf{r} = r\,\hat{\mathbf{a}}_r$ Thus, with $G(\mathbf{r}, \mathbf{r}') = G(r, 0) = G(r)$ in (2.4.1) it is expressed as

$$[\nabla^2 + \beta^2]G(r) = -\delta(x)\delta(y)\delta(z) \tag{2.4.2}$$

The solution to (2.4.2) is similar to the solution of (2.1.1), since if we set $\mu(I\Delta L) = 1$ in (2.1.1) the equations are identical. The solution to (2.1.1) was obtained in (2.1.5). Hence, with $\mu(I\Delta L) = 1$ in (2.1.5) the result is the Green function solution to (2.4.2). That is,

$$G(r) = \frac{e^{-j\beta r}}{4\pi r} \tag{2.4.3}$$

Similarly, referring to (2.1.12) the Green function associated with a source located at **r′** is

$$G(\mathbf{r}, \mathbf{r}') = \frac{e^{-j\beta|\mathbf{r}-\mathbf{r}'|}}{4\pi|\mathbf{r} - \mathbf{r}'|} \tag{2.4.4}$$

In the case of lossy media, the Green function associated with (2.1.13) follows by setting $(I\Delta L)/(\sigma + j\omega\varepsilon) = 1$ in (2.1.14). That is,

$$G(r) = \frac{e^{-\gamma r}}{4\pi r}$$

and for a source at $\mathbf{r}'$:

$$G(\mathbf{r}, \mathbf{r}') = \frac{e^{-\gamma|\mathbf{r}-\mathbf{r}'|}}{4\pi|\mathbf{r} - \mathbf{r}'|}$$

The theory of Green functions and its applications are discussed in Chapter 5 where it is shown that a most important property of the Green function is that once the Green function is known, the solution to the associated inhomogeneous wave equation is readily obtained. Basically, the Green function is the impulse response (i.e., a unit delta function excitation) solution to an inhomogeneous linear differential equation. Since using superposition a source (i.e., $J_{e,z}$) can be represented as a sum of delta functions, the response involves a sum of Green function solutions. Hence, the solution to

$$[\nabla^2 + \beta^2]\mathbf{A} = -\mu\mathbf{J}_e$$

is

$$\mathbf{A}(\mathbf{r}) = \mu \int_{V'} \mathbf{J}_e(\mathbf{r}')\, G(\mathbf{r}, \mathbf{r}')\, dV' \tag{2.4.5}$$

In the case that the Green function is given by (2.4.4), (2.4.5) reads

$$\mathbf{A}(\mathbf{r}) = \frac{\mu}{4\pi} \int_{V'} \mathbf{J}_e(\mathbf{r}') \frac{e^{-j\beta|\mathbf{r}-\mathbf{r}'|}}{|\mathbf{r} - \mathbf{r}'|}\, dV' \tag{2.4.6}$$

For the Hertzian dipole located at the origin, as in Fig. 2.1, the current density is given by $\mathbf{J}_e(\mathbf{r}) = J_{e,z}(x, y, z)\hat{\mathbf{a}}_z = (I\Delta L)\delta(x)\delta(y)\delta(z)\hat{\mathbf{a}}_z$. Then, from (2.4.6) with $\mathbf{r} = r\hat{\mathbf{a}}_r$ and $\mathbf{r}' = 0$:

$$\begin{aligned} A_z &= \frac{\mu}{4\pi} \int_{V'} J_{e,z}(x', y', z') \frac{e^{-j\beta r}}{r} dx'dy'dz' \\ &= \mu(I\Delta L)\frac{e^{-j\beta r}}{4\pi r} \end{aligned}$$

which, as expected, is identical to (2.1.5).

In the case that the excitation is a current flowing in a wire, where $\mathbf{J}_e dV = I d\mathbf{l}$, (2.4.6) can be expressed in the form

$$\mathbf{A}(\mathbf{r}) = \frac{\mu}{4\pi} \int_{l'} I \frac{e^{-j\beta|\mathbf{r}-\mathbf{r}'|}}{|\mathbf{r} - \mathbf{r}'|} d\mathbf{l}' \tag{2.4.7}$$

In the next section, (2.4.7) is used to calculate the fields from a loop radiator.

Equation (2.4.7) can also be used to calculate the fields of the Hertzian dipole in Fig. 2.1. In this case, $d\mathbf{l}' = dl'\hat{\mathbf{a}}_z$, $\mathbf{r}' = 0$ and (2.4.7) reads

$$A_z = \frac{\mu}{4\pi} \int_{-\Delta L/2}^{\Delta L/2} I \frac{e^{-j\beta r}}{r} dl' = \mu(I\Delta L)\frac{e^{-j\beta r}}{4\pi r}$$

which, of course, is the expected result.

For the electric current source of infinite length the Green function expression follows by setting $\mu I = 1$ in (2.2.1), leaving the excitation as a delta function. Therefore, the Green function solution follows from the expression for A_z in (2.2.8) with $\mu I = 1$. That is,

$$G(\rho) = -\frac{j}{4} H_0^{(2)}(\beta\rho)$$

From (2.2.11), for an excitation at $\boldsymbol{\rho} = \boldsymbol{\rho}'$:

$$G(\boldsymbol{\rho}, \boldsymbol{\rho}') = -\frac{j}{4} H_0^{(2)}(\beta|\boldsymbol{\rho} - \boldsymbol{\rho}'|) \tag{2.4.8}$$

2.5 THE SMALL-LOOP RADIATOR

The small-loop radiator (or antenna) can be modeled as a square loop or a circular loop. The square loop is simpler to analyze since we already know from Section 2.1 how to evaluate the field of an electric Hertzian dipole. The analysis of the small loop radiator will show that it can be represented by a magnetic Hertzian dipole.

The geometry of the small square loop is shown in Fig. 2.5a where the sinusoidal loop current amplitude is I and the length of each of the four elements forming the loop are ΔL. Each element is a single electric Hertzian dipole. Dipole **1** is shown in Fig. 2.5b. The distance R_1 is

$$R_1 = \sqrt{(x - x')^2 + (y - y')^2 + (z - z')^2} \tag{2.5.1}$$

where

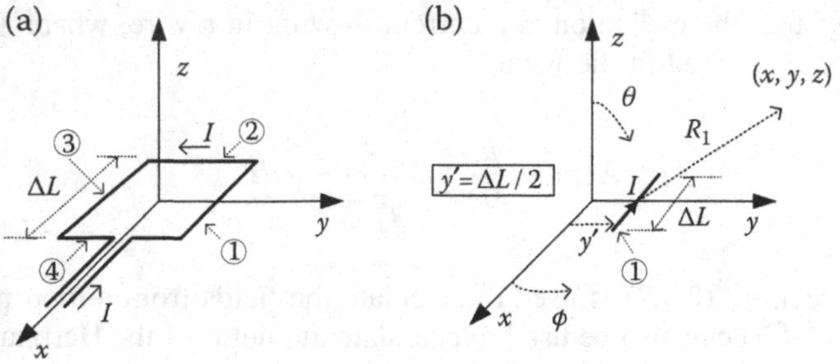

FIGURE 2.5 (a) A small square loop; (b) dipole **1** geometry. (Note: the drawings are not to scale since each element is an infinitesimal Hertzian dipole.)

$$\begin{aligned} x &= r \sin\theta \cos\phi \\ y &= r \sin\theta \sin\phi \\ z &= r \cos\theta \\ x' &= 0 \\ y' &= \frac{\Delta L}{2} \\ z' &= 0 \end{aligned}$$

From (2.5.1), neglecting the $(\Delta L)^2$ terms, gives

$$R_1 \approx \sqrt{r^2\left(1 - \frac{\Delta L}{r}\sin\theta\sin\phi\right)} \approx r - \frac{\Delta L}{2}\sin\theta\sin\phi$$

Similarly, the distances from dipoles **2**, **3**, and **4** are

$$R_2 \approx r + \frac{\Delta L}{2}\sin\theta\cos\phi$$

$$R_3 \approx r + \frac{\Delta L}{2}\sin\theta\sin\phi$$

$$R_4 \approx r - \frac{\Delta L}{2}\sin\theta\cos\phi$$

Dipole **1** produces a component of **A** in the $-\,\hat{\mathbf{a}}_x$ direction (A_{x1}), dipole **3** in the $\hat{\mathbf{a}}_x$ direction (A_{x3}), dipole **2** in the $-\,\hat{\mathbf{a}}_y$ direction (A_{y2}), and dipole **4** in the $\hat{\mathbf{a}}_y$ direction (A_{y4}). Therefore,

$$A_x = A_{x1} + A_{x3} = -\mu(I\Delta L)\frac{e^{-j\beta R_1}}{4\pi R_1} + \mu(I\Delta L)\frac{e^{-j\beta R_3}}{4\pi R_3}$$

In the far field the previous approximations for the radial distances R_1 and R_3 are used in the phase terms, and in the magnitude terms $R_1 \approx R_3 \approx r$. Then,

$$\begin{aligned} A_x &\approx \frac{\mu(I\Delta L)}{4\pi r}\left[-e^{-j\beta\left(r-\frac{\Delta L}{2}\sin\theta\sin\phi\right)} + e^{-j\beta\left(r+\frac{\Delta L}{2}\sin\theta\sin\phi\right)}\right] \\ &= -j\mu(I\Delta L)\frac{e^{-j\beta r}}{2\pi r}\sin\left(\beta\frac{\Delta L}{2}\sin\theta\sin\phi\right) \end{aligned}$$

Similarly, for the y-component of **A** due to dipoles **2** and **4:**

$$\begin{aligned} A_y &= A_{y2} + A_{y4} \\ &= j\mu(I\Delta L)\frac{e^{-j\beta r}}{2\pi r}\sin\left(\beta\frac{\Delta L}{2}\sin\theta\cos\phi\right) \end{aligned}$$

The small square loop approximation implies that $\beta\Delta L \ll 1$, so using $\sin x \approx x$ we obtain

$$A_x = -j\beta\mu(I\Delta S)\frac{e^{-j\beta r}}{4\pi r}\sin\theta\sin\phi$$

$$A_y = j\beta\mu(I\Delta S)\frac{e^{-j\beta r}}{4\pi r}\sin\theta\cos\phi$$

where $\Delta S = (\Delta L)^2$ is the area of the loop. Therefore,

$$\mathbf{A} = A_x\hat{\mathbf{a}}_x + A_y\hat{\mathbf{a}}_y = j\beta\mu(I\Delta S)\frac{e^{-j\beta r}}{4\pi r}\sin\theta(-\sin\phi\,\hat{\mathbf{a}}_x + \cos\phi\,\hat{\mathbf{a}}_y)$$

or

$$\mathbf{A} = A_\phi\hat{\mathbf{a}}_\phi = j\beta\mu(I\Delta S)\frac{e^{-j\beta r}}{4\pi r}\sin\theta\,\hat{\mathbf{a}}_\phi$$

which is the desired expression. The far fields of **E** and **H** are calculated using (2.3.3), namely

$$E_\phi = -j\omega A_\phi = \eta\beta^2(I\Delta S)\frac{e^{-j\beta r}}{4\pi r}\sin\theta \tag{2.5.2}$$

and

$$H_\theta = -\frac{E_\phi}{\eta} = -\beta^2(I\Delta S)\frac{e^{-j\beta r}}{4\pi r}\sin\theta \tag{2.5.3}$$

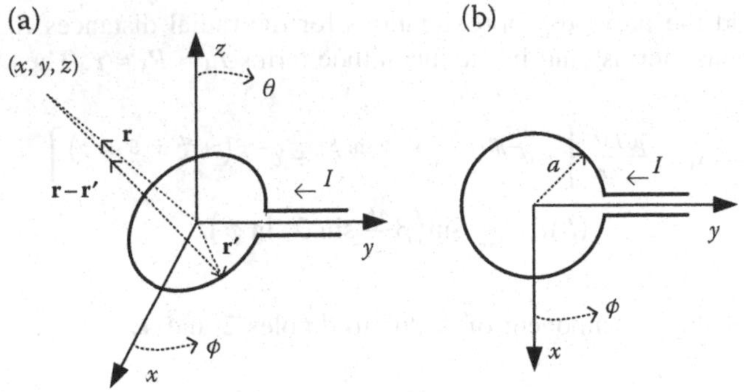

FIGURE 2.6 (a) The small loop antenna; (b) top view.

For completeness, we now evaluate the fields from a small circular loop antenna, as shown in Fig. 2.6. The field from the loop will be symmetrical in ϕ, so the observation point can be selected in the x-z plane, where $\phi = 0$. The distance R to an observation point in the x-z plane is given by (2.5.1) with

$$\begin{aligned} x &= r\sin\theta \\ y &= 0 \\ z &= r\cos\theta \\ x' &= a\cos\phi' \\ y' &= a\sin\phi' \\ z' &= 0 \end{aligned}$$

Then, it follows that

$$R = |\mathbf{r} - \mathbf{r}'| = \sqrt{r^2 + a^2 - 2ar\sin\theta\cos\phi'}$$

which for $r \gg a$ it is approximated by

$$R \approx r - a\sin\theta\cos\phi'$$

The $\mathbf{A}$ vector in the x-z plane is in the $\hat{\mathbf{a}}_\phi$ direction. From (2.4.7), with $d\mathbf{l}' = a\,d\phi'\hat{\mathbf{a}}_{\phi'}$, we obtain

$$\mathbf{A}(r, \theta, 0) = \frac{\mu I}{4\pi}\int_0^{2\pi} \frac{e^{-j\beta R}}{R} a\,d\phi'\hat{\mathbf{a}}_{\phi'}$$

Since the unit vector $\hat{\mathbf{a}}_{\phi'}$ changes direction around the loop, then

$$A_\phi(r, \theta, 0) = \hat{\mathbf{a}}_\phi \bullet \mathbf{A}(r, \theta, 0)$$

and it follows that

$$A_\phi(r, \theta, 0) = \frac{\mu I a}{4\pi} \int_0^{2\pi} (\hat{\mathbf{a}}_\phi \bullet \hat{\mathbf{a}}_{\phi'}) \frac{e^{-j\beta R}}{R} \, d\phi' \tag{2.5.4}$$

Using the far-field approximation:

$$\frac{e^{-j\beta R}}{R} \approx \frac{e^{-j\beta(r - a\sin\theta\cos\phi')}}{r}$$

and

$$\hat{\mathbf{a}}_\phi \bullet \hat{\mathbf{a}}_{\phi'} = (-\hat{\mathbf{a}}_x \sin\phi + \hat{\mathbf{a}}_y \cos\phi) \bullet (-\hat{\mathbf{a}}_x \sin\phi' + \hat{\mathbf{a}}_y \cos\phi')$$

shows that for the observation point at $\phi = 0$:

$$\hat{\mathbf{a}}_\phi \bullet \hat{\mathbf{a}}_{\phi'}|_{\phi=0} = \cos\phi'$$

Hence, (2.5.4) reads

$$A_\phi(r, \theta) = \mu I a \frac{e^{-j\beta r}}{4\pi r} \int_0^{2\pi} \cos\phi' e^{j\beta a \sin\theta \cos\phi'} d\phi'$$

which from the symmetry is valid for any ϕ.

Using the Bessel function relation:

$$J_1(x) = \frac{1}{2\pi j} \int_0^{2\pi} \cos\phi' e^{jx\cos\phi'} d\phi'$$

A_ϕ is expressed in the form

$$A_\phi = j\frac{\mu I a}{2} \frac{e^{-j\beta r}}{r} J_1(\beta a \sin\theta)$$

For a small loop $\beta a \ll 1$, so using the small argument approximation

$$J_1(\beta a \sin\theta) \approx \frac{1}{2}\beta a \sin\theta$$

the far-field relations are

$$E_\phi = -j\omega A_\phi = \eta\beta^2 (I\Delta S) \frac{e^{-j\beta r}}{4\pi r} \sin\theta \tag{2.5.5}$$

and

$$H_\theta = -\frac{E_\phi}{\eta} = -\beta^2 (I\Delta S)\frac{e^{-j\beta r}}{4\pi r}\sin\theta \tag{2.5.6}$$

where $\Delta S = \pi a^2$. As expected, (2.5.5) and (2.5.6) are the same as (2.5.2) and (2.5.3).

The time-average power density in the far field is given by

$$\mathbf{P} = \frac{1}{2}\operatorname{Re}(E_\phi \hat{\mathbf{a}}_\phi \times H_\theta^* \hat{\mathbf{a}}_\theta) = \frac{|E_\phi|^2}{2\eta}\hat{\mathbf{a}}_r$$
$$= \frac{\eta\beta^4}{32\pi^2 r^2}(I\Delta S)^2 \sin^2\theta \hat{\mathbf{a}}_r$$

and the radiated power is

$$W = \oint P_r \hat{\mathbf{a}}_r \bullet \hat{\mathbf{a}}_r dS = \frac{\eta\beta^4}{32\pi^2}(I\Delta S)^2 \int_{\phi=0}^{2\pi}\int_{\theta=0}^{\pi} \sin^3\theta d\theta d\phi$$
$$= \frac{\eta\beta^4}{12\pi}(I\Delta S)^2$$

Loop antennas find uses as receivers. For example, if the incident field is normally incident on a one-turn loop, the open-circuit voltage induced in the loop is

$$V_{oc} = j\omega\mu H_i(\pi a^2)$$

which shows its dependence on μ. To increase the induced voltage a ferrite loop with a large value of μ is used and the number of turns in the loop is increased.

2.6 THE HERTZIAN MAGNETIC DIPOLE

In this section, it is shown that a small loop with current I can be modeled by an equivalent Hertzian magnetic dipole with current moment $I_m\Delta L$, as shown in Fig. 2.7. Figure 2.7a shows that the magnetic dipole is represented by magnetic charges q_m and $-q_m$ separated by a distance ΔL and producing the current I_m. The equivalent representation is shown in Fig. 2.7b.

Two units have been used for q_m. If $q_m = q_o \sin\omega t$ is in Webers (Wb), then

$$i_m(t) = \frac{dq_m}{dt} = q_o\omega\cos\omega t = I_m\cos\omega t$$

where $I_m = q_o\omega$ is in volts. If $q_m = q_o \sin\omega t$ is in Ampere·meter (A·m), then

$$i_m(t) = \mu\frac{dq_m}{dt} = \mu q_o\omega\cos\omega t = I_m\cos\omega t$$

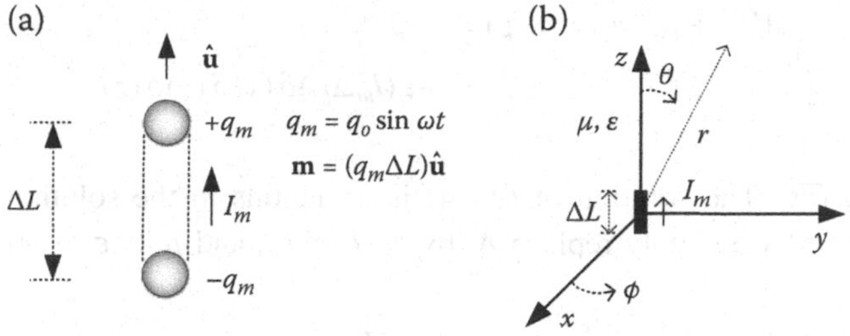

FIGURE 2.7 (a) A magnetic dipole; (b) the Hertzian magnetic dipole.

where I_m is in volts and it is given by

$$I_m = \mu q_o \omega \tag{2.6.1}$$

The conversion between the units is $q_o\,[\text{Wb}] = \mu q_o\,[\text{A·m}]$.

The magnetic dipole moment, defined in terms of an electric current in a small loop is

$$\mathbf{m} = [i(t)\Delta S]\mathbf{u}$$

where $i(t) = I\cos\omega t$. If we equate this relation to the definition of the magnetic dipole moment in terms of q_m (in A·m), namely

$$\mathbf{m} = [q_m(t)\Delta L]\mathbf{u}$$

it follows that

$$q_m(t)\Delta L = i(t)\Delta S$$

which in phasor form reads

$$-jq_o\Delta L = I\Delta S \tag{2.6.2}$$

Substituting (2.6.1) into (2.6.2) for q_o gives

$$(I_m\Delta L) = j\omega\mu(I\Delta S) \tag{2.6.3}$$

where $(I_m\Delta L)$ is the magnetic current moment in V·m. This relation will now be derived by equating the fields of a Hertzian magnetic dipole to those of the small loop.

The fields due to the magnetic Hertzian dipole in Fig. 2.7 with current moment $(I_m\Delta L)$ are determined in terms of the **F** vector. From (1.5.23):

$$\begin{aligned}[\nabla^2 + \beta^2]F_z(x, y, z) &= -\varepsilon J_{m,z} \\ &= -\varepsilon(I_m \Delta L)\delta(x)\delta(y)\delta(z)\end{aligned} \tag{2.6.4}$$

where $\beta = \omega\sqrt{\mu\varepsilon}$. The solution of (2.6.4) is analogous to the solution of (2.1.1). Hence, in (2.1.5) we simply replace A_z by F_z, I by I_m, and μ by ε to obtain

$$F_z = \varepsilon(I_m \Delta L)\frac{e^{-j\beta r}}{4\pi r} \tag{2.6.5}$$

This relation can also be expressed in terms of a spherical Hankel function as

$$F_z = -j\frac{\beta\varepsilon(I_m \Delta L)}{4\pi}h_0^{(2)}(\beta r)$$

Equation (2.6.5) can also be obtained using the Green function method. The Green function associated with (2.6.4) is given by (2.4.4), and from (2.6.4) it follows that

$$\mathbf{F}(\mathbf{r}) = \frac{\varepsilon}{4\pi}\int_{V'} \mathbf{J}_m(\mathbf{r}')\frac{e^{-j\beta|\mathbf{r}-\mathbf{r}'|}}{|\mathbf{r}-\mathbf{r}'|}dV' \tag{2.6.6}$$

Substituting $\mathbf{J}_m(\mathbf{r}) = J_{m,z}\hat{\mathbf{a}}_z = (I_m \Delta l)\delta(x)\delta(y)\delta(z)\hat{\mathbf{a}}_z$, $\mathbf{r}' = 0$ and $\mathbf{r} = r\hat{\mathbf{a}}_r$ in (2.6.6) leads to the result shown in (2.6.5).

The spherical components are

$$\begin{aligned}F_r &= F_z \cos\theta \\ F_\theta &= -F_z \sin\theta \\ F_\phi &= 0\end{aligned}$$

and the **E** and **H** fields follow from (1.5.19) and (1.5.22), namely

$$E_\phi = -j\beta(I_m \Delta L)\left(1 + \frac{1}{j\beta r}\right)\frac{e^{-j\beta r}}{4\pi r}\sin\theta \tag{2.6.7}$$

$$H_r = \frac{(I_m \Delta L)}{\eta}\left(1 + \frac{1}{j\beta r}\right)\frac{e^{-j\beta r}}{2\pi r^2}\cos\theta \tag{2.6.8}$$

$$H_\theta = j\frac{\beta(I_m \Delta L)}{\eta}\left[1 + \frac{1}{j\beta r} - \frac{1}{(\beta r)^2}\right]\frac{e^{-j\beta r}}{4\pi r}\sin\theta \tag{2.6.9}$$

These results can also be obtained using the duality between the electric Hertzian dipole and the magnetic Hertzian dipole (see Section 2.7).

The far fields follow by keeping the $1/r$ terms in (2.6.7) to (2.6.9), or they can be calculated using (2.3.4). That is,

$$H_\theta \approx -j\omega F_\theta = j\frac{\beta(I_m\Delta L)}{\eta}\frac{e^{-j\beta r}}{4\pi r}\sin\theta \tag{2.6.10}$$

and

$$E_\phi \approx -\eta H_\theta = -j\beta(I_m\Delta L)\frac{e^{-j\beta r}}{4\pi r}\sin\theta \tag{2.6.11}$$

Comparing (2.6.10) and (2.6.11) with (2.5.5) and (2.5.6) shows that the small magnetic dipole produces the same fields as those of a small electric loop if

$$(I_m\Delta L) = j\beta\eta(I\Delta S) = j\omega\mu(I\Delta S) \tag{2.6.12}$$

which is identical to the relation derived in (2.6.3).

To obtain the far fields in terms of $(I\Delta S)$, substitute (2.6.12) into (2.6.10) and (2.6.11) to obtain

$$E_\phi = \eta\beta^2(I\Delta S)\frac{e^{-j\beta r}}{4\pi r}\sin\theta$$

and

$$H_\theta = -\beta^2(I\Delta S)\frac{e^{-j\beta r}}{4\pi r}\sin\theta$$

in agreement with (2.5.5) and (2.5.6).

When the surrounding medium to the magnetic Hertzian dipole in Fig. 2.7 is described by ε, μ, and σ, simply replace $j\beta$ by γ in (2.6.7) to (2.6.9), to obtain

$$E_\phi = -\gamma(I_m\Delta L)\left(1 + \frac{1}{\gamma r}\right)\frac{e^{-\gamma r}}{4\pi r}\sin\theta = -\frac{(I_m\Delta L)}{4\pi r^2}(1+\gamma r)e^{-\gamma r}\sin\theta \tag{2.6.13}$$

$$H_r = \frac{(I_m\Delta L)}{\eta}\left(1 + \frac{1}{\gamma r}\right)\frac{e^{-\gamma r}}{2\pi r^2}\cos\theta = \frac{(I_m\Delta L)}{j\omega\mu 2\pi r^3}(1+\gamma r)e^{-\gamma r}\cos\theta \tag{2.6.14}$$

$$H_\theta = \frac{\gamma(I_m\Delta L)}{\eta}\left[1 + \frac{1}{\gamma r} + \frac{1}{(\gamma r)^2}\right]\frac{e^{-\gamma r}}{4\pi r}\sin\theta = \frac{(I_m\Delta L)}{j\omega\mu 4\pi r^3}(1+\gamma r+\gamma^2 r^2)e^{-\gamma r}\sin\theta \tag{2.6.15}$$

Another way to derive (2.6.12) is to write Maxwell's equations for an electric current source excitation as

$$\nabla \times \mathbf{E}_A = -j\omega\mu\mathbf{H}_A$$

and

$$\nabla \times \mathbf{H}_A = \mathbf{J}_e + (\sigma + j\omega\varepsilon)\mathbf{E}_A$$

Then, it follows that $\mathbf{E}_A$ satisfies

$$\nabla \times \nabla \times \mathbf{E}_A + \gamma^2\mathbf{E}_A = -j\omega\mu\mathbf{J}_e \tag{2.6.16}$$

Similarly, for a magnetic current source excitation we write

$$\nabla \times \mathbf{E}_F = -\mathbf{J}_m - j\omega\mu\mathbf{H}_F$$

and

$$\nabla \times \mathbf{H}_F = (\sigma + j\omega\varepsilon)\mathbf{E}_A$$

Then, it follows that $\mathbf{E}_F$ satisfies

$$\nabla \times \nabla \times \mathbf{E}_F + \gamma^2\mathbf{E}_F = -\nabla \times \mathbf{J}_m \tag{2.6.17}$$

Equations (2.6.16) and (2.6.17) show that the fields $\mathbf{E}_A$ and $\mathbf{E}_F$ are the same if

$$j\omega\mu\mathbf{J}_e = \nabla \times \mathbf{J}_m \tag{2.6.18}$$

and, of course, if they satisfy the same boundary conditions.

Integrating (2.6.18) over the surface S' in Fig. 2.8:

$$j\omega\mu\int_{S'} \mathbf{J}_e \bullet \hat{\mathbf{n}}'dS' = \int_{S'} \nabla \times \mathbf{J}_m \bullet \hat{\mathbf{n}}'dS' = \oint_{C'} \mathbf{J}_m \bullet d\,\mathbf{l}' \tag{2.6.19}$$

where C' is the contour associated with S', whose normal is $\hat{\mathbf{n}}'$. The magnetic current only occurs in part of the loop (between q_m and $-q_m$). The evaluation of (2.6.19) gives

$$j\omega\mu I = J_m\Delta L \tag{2.6.20}$$

Since J_m is in V/m^2 and the magnetic current I_m in volts, then $I_m = J_m\Delta S$. Multiplying (2.6.20) by the area of the current loop ΔS, gives

$$j\omega\mu(I\Delta S) = (I_m\Delta L)$$

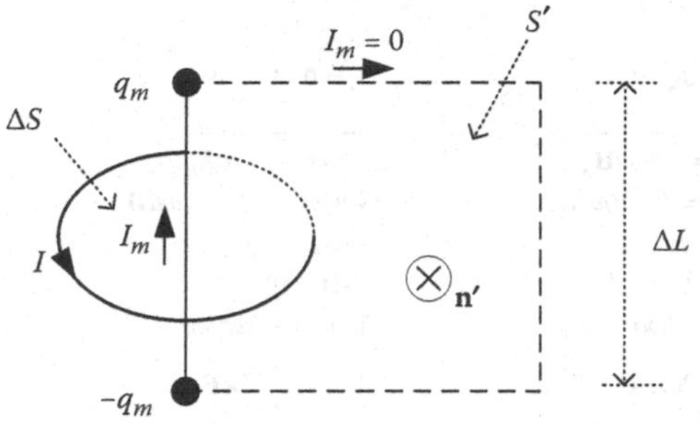

FIGURE 2.8 Magnetic dipole current moment and its equivalent electric loop current I.

which is identical to (2.6.12).

2.7 DUALITY

Duality allow us to determine the fields due to magnetic sources from known fields of electric sources and vice versa, since they satisfy dual equations. For example, duality was used to obtain the electric vector potential F_z in (2.6.5) from a knowledge of the magnetic vector potential A_z in (2.1.5).

Figure 2.9a shows the duality of the electromagnetic equations in a lossless region. The left column shows the field relations due to $\mathbf{J}_e$, and the right column the field relations due to $\mathbf{J}_m$. Figure 2.9b shows the dual quantities. In the case of a lossy region, the dual entries in Fig. 2.9c for the constitutive parameters are used.

For example, the fields from an electric Hertzian dipole are given in (2.1.6) to (2.1.8). Using the dual relations in Fig. 2.9b, the fields due to a magnetic dipole are obtained. From (2.1.6) with $H_\phi \to -E_\phi$ and $I \to I_m$, the E_ϕ component of the magnetic Hertzian dipole is

$$E_\phi = -j\beta\,(I_m \Delta L)\left(1 + \frac{1}{j\beta r}\right)\frac{e^{-j\beta r}}{4\pi r}\sin\theta$$

From (2.1.7), using the dual relations: $E_r \to H_r$, $I \to I_m$ and $\eta \to 1/\eta$, it follows that the H_r component of the magnetic Hertzian dipole is

$$H_r = \frac{(I_m \Delta L)}{\eta}\left(1 + \frac{1}{j\beta r}\right)\frac{e^{-j\beta r}}{2\pi r^2}\cos\theta$$

Similarly, from (2.1.8) it follows that H_θ for the magnetic dipole is

$$H_\theta = j\frac{\beta\,(I_m \Delta L)}{\eta}\left[1 + \frac{1}{j\beta r} - \frac{1}{(\beta r)^2}\right]\frac{e^{-j\beta r}}{4\pi r}\sin\theta$$

(a)

$\mathbf{J}_e \neq 0 \quad \mathbf{J}_m = 0$	$\mathbf{J}_e = 0 \quad \mathbf{J}_m \neq 0$
$\nabla\times\mathbf{E}_A = -j\omega\mu\mathbf{H}_A$	$\nabla\times\mathbf{H}_F = j\omega\varepsilon\mathbf{E}_F$
$\nabla\times\mathbf{H}_A = \mathbf{J}_e + j\omega\varepsilon\mathbf{E}_A$	$\nabla\times\mathbf{E}_F = -\mathbf{J}_m - j\omega\mu\mathbf{H}_F$
$\nabla\cdot\mathbf{D}_A = \rho_e$	$\nabla\cdot\mathbf{B}_F = \rho_m$
$\nabla\cdot\mathbf{B}_A = 0$	$\nabla\cdot\mathbf{D}_F = 0$
$\nabla\cdot\mathbf{J}_e = -j\omega\rho_e$	$\nabla\cdot\mathbf{J}_m = -j\omega\rho_m$
$\mathbf{H}_A = \frac{1}{\mu}\nabla\times\mathbf{A}$	$\mathbf{E}_F = -\frac{1}{\varepsilon}\nabla\times\mathbf{F}$
$\mathbf{E}_A = \frac{1}{j\omega\mu\varepsilon}\nabla\times\nabla\times\mathbf{A} - \frac{\mathbf{J}_e}{j\omega\varepsilon}$	$\mathbf{H}_F = \frac{1}{j\omega\mu\varepsilon}\nabla\times\nabla\times\mathbf{F} - \frac{\mathbf{J}_m}{j\omega\mu}$
$= -j\omega\mathbf{A} + \frac{1}{j\omega\mu\varepsilon}\nabla(\nabla\cdot\mathbf{A})$	$= -j\omega\mathbf{F} + \frac{1}{j\omega\mu\varepsilon}\nabla(\nabla\cdot\mathbf{F})$
$(\nabla^2 + \beta^2)\mathbf{A} = -\mu\mathbf{J}_e$	$(\nabla^2 + \beta^2)\mathbf{F} = -\varepsilon\mathbf{J}_m$
$\mathbf{\Pi}^e = j\omega\mu\varepsilon\mathbf{A}$	$\mathbf{\Pi}^m = j\omega\mu\varepsilon\mathbf{F}$

(b)

Electric	$\mathbf{E}_A$	$\mathbf{H}_A$	$\mathbf{J}_e$	$\mathbf{D}_A$	$\mathbf{B}_A$	$\mathbf{A}$	$\mathbf{\Pi}^e$	ρ_e	I	ε	μ	β	η
Magnetic	$\mathbf{H}_F$	$-\mathbf{E}_F$	$\mathbf{J}_m$	$\mathbf{B}_F$	$\mathbf{D}_F$	$\mathbf{F}$	$\mathbf{\Pi}^m$	ρ_m	I_m	μ	ε	β	$1/\eta$

(c)

Electric	$j\omega\mu$	$\sigma + j\omega\varepsilon$	γ	μ	ε_c
Magnetic	$\sigma + j\omega\varepsilon$	$j\omega\mu$	γ	ε_c	μ

FIGURE 2.9 (a) Dual electromagnetic equations for the lossless case; (b) the dual quantities for the lossless case; (c) dual quantities for the constitutive parameters in the lossy case.

These expressions obtained using duality are identical to those in (2.6.7) to (2.6.9).

For lossy media, the dual equations can be used to derive the fields of a magnetic Hertzian dipole in (2.6.13) to (2.6.15) from the results the electric Hertzian dipole in (2.1.17) to (2.1.19).

Another example of duality is in the calculation of the far fields due to a magnetic current in a small loop. From the results for the far fields of an electric current in a small loop in (2.5.5) and (2.5.6), using duality, the far fields for a magnetic current in a small loop are

$$H_\phi = \frac{\beta^2}{\eta}(I_m \Delta S)\frac{e^{-j\beta r}}{4\pi r}\sin\theta \tag{2.7.1}$$

and

$$E_\theta = \beta^2 (I_m \Delta S) \frac{e^{-j\beta r}}{4\pi r} \sin\theta \tag{2.7.2}$$

A final example of the use of duality is the determination of the fields due to a magnetic line source of infinitely length. From the field relations of the electric line source of infinite length in (2.2.9) and (2.2.10), using duality, the fields due to a magnetic current of infinite length are

$$H_z = -\frac{\omega\varepsilon I_m}{4} H_0^{(2)}(\beta\rho)$$

and

$$E_\phi = j\frac{\beta I_m}{4} H_1^{(2)}(\beta\rho)$$

In the lossy case, applying the dual relations to (2.2.17) and (2.2.18) the fields due a magnetic current of infinite length are

$$H_z = -\frac{(\sigma + j\omega\mu) I_m}{2\pi} K_0(\gamma\rho)$$

and

$$E_\phi = -\frac{\gamma I_m}{2\pi} K_1(\gamma\rho)$$

2.8 IMAGES

A brief review of the image method as it applies to time-varying fields is given. The image method deals with the fields from a source above a PEC or PMC. In the case of a PEC the boundary conditions are the vanishing of the tangential components of the electric fields at the surface of the PEC, and the tangential magnetic fields induce surface currents on the PEC.

The fields above the PEC are equivalent to those that are produced by the source and its image. The fields in the PEC are zero, so the fields produced by the image source are only valid for the determination of the total fields above the PEC.

In Fig. 2.10a, the images of a vertical electric dipole (VED) and a horizontal electric dipole (HED) above a PEC are shown, and in Fig. 2.10b the images of a vertical magnetic dipole (VMD) and a horizontal magnetic dipole (HMD) are shown.

In the case of a PMC, the tangential components of the magnetic fields are zero, and the tangential components of the electric fields induce magnetic surface

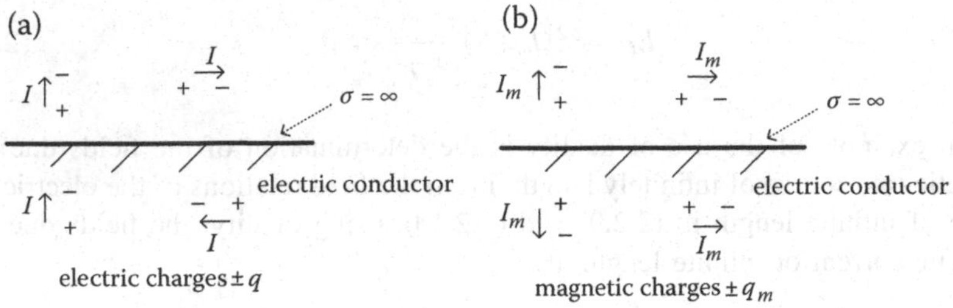

FIGURE 2.10 (a) Images of a VED and HED above a PEC; (b) images of a VMD and HMD above a PEC.

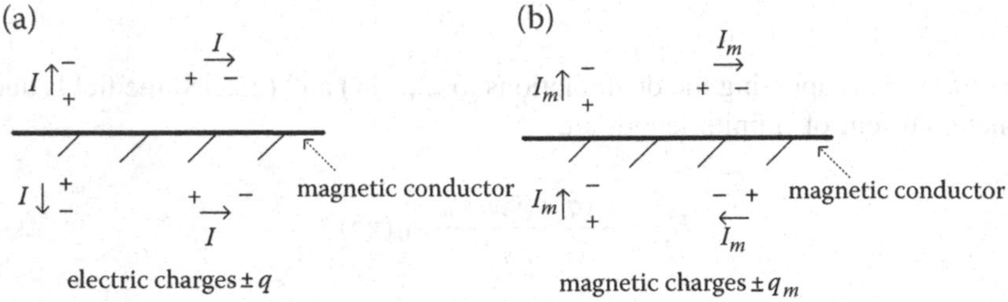

FIGURE 2.11 (a) Images of a VED and HED above a PMC; (b) images of a VMD and HMD above a PMC.

currents in the PMC. Figure 2.11a shows the images of a VED and a HED above a PMC, and in Fig. 2.11b the images of a VMD and a HMD.

The image method has many applications in antenna work. For example, when a wire antenna is end-fed above a good ground, a first-order approximation to the radiating system is to assume that the ground is a PEC and the fields above ground are calculated in terms of the source and its image.

As an example, consider the HED above the conducting plane shown in Fig. 2.10a. The fields due to the HED in Fig. 2.12a are first obtained. In order to use the results obtained for the VED, define the polar angle $\tilde{\theta}$ measured from the y-axis, as shown in Fig. 2.12a. Then, from (2.1.9) the far-field expression for $E_{\tilde{\theta}}$ is

$$E_{\tilde{\theta}} = j\beta\eta(I\Delta L)\frac{e^{-j\beta r}}{4\pi r}\sin\tilde{\theta} \tag{2.8.1}$$

where $\sin\tilde{\theta}$ can be expressed in terms of θ and ϕ. That is,

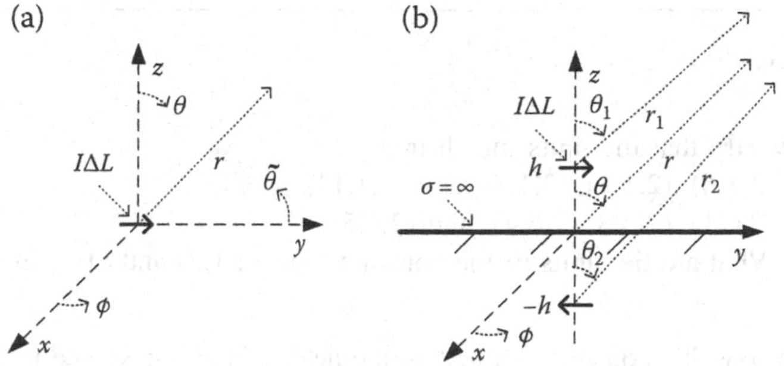

FIGURE 2.12 (a) A HED located along the *y*-axis; (b) a HED and its image.

$$\sin\tilde{\theta} = \sqrt{1 - \cos^2\tilde{\theta}}$$
$$= \sqrt{1 - (\hat{\mathbf{a}}_y \bullet \hat{\mathbf{a}}_r)^2}$$
$$= \sqrt{1 - \sin^2\theta\sin^2\phi}$$

Hence, from (2.8.1):

$$E_{\tilde{\theta}} = j\beta\eta(I\Delta L)\frac{e^{-j\beta r}}{4\pi r}\sqrt{1 - \sin^2\theta\sin^2\phi} \tag{2.8.2}$$

In Fig. 2.12b, the HED is shown above a conducting plane, as well as its image. The total field above the conducting plane is due to the HED and its image. Using (2.8.2) the total far field above the conducting plane (i.e., for $z \geq 0$) is given by

$$E_{\tilde{\theta}} = E_{\tilde{\theta},1} + E_{\tilde{\theta},2} = j\beta\eta(I\Delta L)\sqrt{1 - \sin^2\theta\sin^2\phi}\left(\frac{e^{-j\beta r_1}}{4\pi r_1} - \frac{e^{-j\beta r_2}}{4\pi r_2}\right) \tag{2.8.3}$$

In the far field, with $\theta \approx \theta_1 \approx \theta_2$, the phase terms in (2.8.3) are approximated by

$$r_1 \approx r - h\cos\theta$$
$$r_2 \approx r + h\cos\theta$$

and the magnitude term by

$$\frac{1}{r} \approx \frac{1}{r_1} \approx \frac{1}{r_2}$$

Hence, (2.8.3) reads

$$E_{\tilde{\theta}} = -\beta\eta(I\Delta L)\sqrt{1 - \sin^2\theta\sin^2\phi}\frac{e^{-j\beta r}}{2\pi r}\sin(\beta h\cos\theta)$$

Problems

P2.1 Verify that the units match in:

a. (2.1.6), (2.1.7), (2.1.8), and (2.1.14).

b. (2.6.1), (2.6.3, (2.6.4), and (2.6.5).

c. What are the units of the constant C_1 in (2.1.3) and of B_1 in (2.2.3).

P2.2 Maxwell's equations show that the fields of the line source in Fig. 2.3 can be derived in terms of the wave equation satisfied by E_z in (2.2.12). Use Maxwell's equations to verify that $E_\rho = E_\phi = H_\rho = H_z = 0$.

P2.3 Derive (2.3.1) and (2.3.2).

P2.4

a. Use duality to derive the fields of a magnetic Hertzian dipole in lossy media using the results shown in (2.1.17) to (2.1.19).

b. Verify (2.7.1) and (2.7.2).

P2.5 Verify that the image configurations for a VED and HED shown in Fig. 2.10a satisfy the boundary conditions at the surface of the PEC.

P2.6

a. A VED above a PEC is shown in Fig. P2.6. Show that in the far field at the observation point P:

$$r_1 \approx r - h\cos\theta$$

where $\theta \approx \theta_1$.

b. Show that E_θ in the far field is given by

$$E_\theta = j\eta\beta(I\Delta L)\frac{e^{-j\beta r}}{2\pi r}\sin\theta\cos(\beta h\cos\theta)$$

c. Determine H_ϕ.

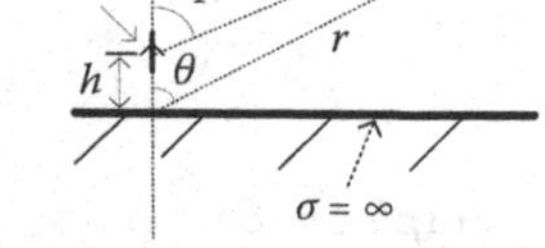

FIGURE P2.6

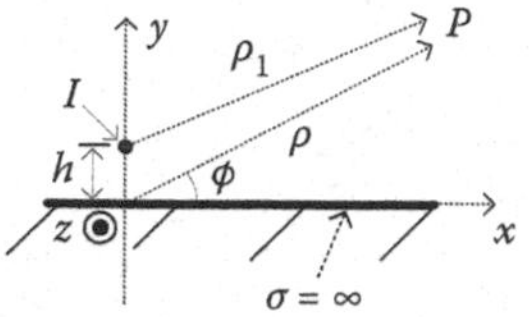

FIGURE P2.8

P2.7

a. In Problem 2.6, the VED is replaced by a small electrical loop radiator. Determine E_ϕ in the far field.
b. Determine the associated magnetic field.

P2.8

a. The line source in Fig. 2.3 is placed above a perfect conductor, as shown in the two-dimensional view in Fig. P2.8. The height of the line source is h, and $\phi' = \pi/2$. Show that in the far field:

$$\rho_1 \approx \rho - h \sin \phi$$

b. Use the method of images to show that

$$E_z = -\frac{\eta \beta I}{4}\left[H_0^{(2)}(\beta\rho_1) - H_0^{(2)}(\beta\rho_2)\right] \quad (y \geq 0)$$

where

$$\rho_2 \approx \rho + h \sin \phi$$

c. Use the asymptotic approximation of the Hankel functions to show that

$$E_z \approx -j\eta I \sqrt{\frac{\beta}{2\pi}} \frac{e^{-j(\beta p - \pi/4)}}{\sqrt{p}} \sin(\beta h \sin \phi) \quad (y \geq 0)$$

3 Plane Waves

3.1 PROPERTIES OF PLANE WAVES

Figure 3.4 shows a rotated x', y' axis with respect to the x, z axis. This is a two-dimensional rotation by an angle θ. A plane wave traveling in the positive z direction is given by

$$E_y = Ce^{-\gamma z}$$

or with $\gamma = jk$

$$E_y = Ce^{-jkz}$$

If a plane wave is traveling in the positive z' direction in Fig. 3.1, the electric field is given by

$$E_y = Ce^{-\gamma z'}$$

or

$$E_y = Ce^{-jkz'}$$

A plane wave in the z' direction can be expressed in terms of the x, z coordinates by observing that the unit vector $\hat{\mathbf{a}}_{z'}$ in the direction of propagation can be expressed as

$$\hat{\mathbf{a}}_{z'} = \sin\theta\,\hat{\mathbf{a}}_x + \cos\theta\,\hat{\mathbf{a}}_z$$

Since

$$\mathbf{r} = x\hat{\mathbf{a}}_x + z\hat{\mathbf{a}}_z$$

it follows that

$$z' = \hat{\mathbf{a}}_{z'} \bullet \mathbf{r} = x\sin\theta + z\cos\theta$$

and a plane-wave traveling in the positive z' direction is of the form

$$e^{-\gamma z'} = e^{-\gamma(\hat{\mathbf{a}}_{z'}\bullet\mathbf{r})} = e^{-\gamma(x\sin\theta + z\cos\theta)}$$

This relation suggest that we can think of the propagation constant as a vector in the direction of propagation. That is, let $\boldsymbol{\gamma} = \gamma\hat{\mathbf{a}}_{z'}$ and it follows that

DOI: 10.1201/9781003219729-3

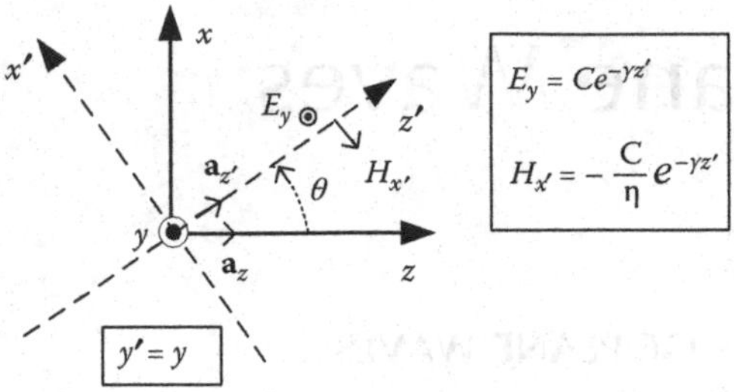

FIGURE 3.1 A plane wave traveling along the z' direction.

$$e^{-\gamma(\hat{\mathbf{a}}_{z'}\cdot\mathbf{r})} = e^{-\boldsymbol{\gamma}\cdot\mathbf{r}}$$

Hence,

$$\begin{aligned}\boldsymbol{\gamma} &= \gamma(\sin\theta\hat{\mathbf{a}}_x + \cos\theta\hat{\mathbf{a}}_z)\\ &= \gamma_y\hat{\mathbf{a}}_x + \gamma_y\hat{\mathbf{a}}_z\end{aligned}$$

where

$$\begin{aligned}\gamma_x &= \gamma\sin\theta\\ \gamma_z &= \gamma\cos\theta\end{aligned}$$

and

$$e^{-\boldsymbol{\gamma}\cdot\mathbf{r}} = e^{-\gamma_x x}e^{-\gamma_z z}$$

If written in terms of the propagation constant k, replace γ by jk. In terms of $\mathbf{k}$, the form is $e^{-j\mathbf{k}\cdot\mathbf{r}}$ where $\mathbf{k} = k_x\hat{\mathbf{a}}_x + k_z\hat{\mathbf{a}}_z$.

Maxwell's equations can be expressed in terms of z'. The gradient operation of a function f in terms of z' is expressed in the form

$$\begin{aligned}\nabla f &= \frac{\partial z'}{\partial x}\frac{\partial f}{\partial z'}\hat{\mathbf{a}}_x + \frac{\partial z'}{\partial z}\frac{\partial f}{\partial z'}\hat{\mathbf{a}}_z\\ &= \left(\frac{\partial z'}{\partial x}\hat{\mathbf{a}}_x + \frac{\partial z'}{\partial z}\hat{\mathbf{a}}_z\right)\frac{\partial f}{\partial z'}\\ &= (\sin\theta\hat{\mathbf{a}}_x + \cos\theta\hat{\mathbf{a}}_z)\frac{\partial f}{\partial z'} = \hat{\mathbf{a}}_{z'}\frac{\partial f}{\partial z'}\end{aligned}$$

and the divergence and curl operations on a vector $\mathbf{T}$ are

$$\nabla\cdot\mathbf{T} = \hat{\mathbf{a}}_{z'}\cdot\frac{\partial\mathbf{T}}{\partial z'} \tag{3.1.1}$$

and

$$\nabla \times \mathbf{T} = \hat{\mathbf{a}}_{z'} \times \frac{\partial \mathbf{T}}{\partial z'} \tag{3.1.2}$$

Using (3.1.1) and (3.1.2), Maxwell's equations for a plane wave propagating along the z' direction can be written in the form

$$\hat{\mathbf{a}}_{z'} \times \frac{\partial \mathbf{E}}{\partial z'} = -j\omega\mu\mathbf{H} \tag{3.1.3}$$

$$\hat{\mathbf{a}}_{z'} \times \frac{\partial \mathbf{H}}{\partial z'} = (\sigma + j\omega\varepsilon)\mathbf{E} \tag{3.1.4}$$

$$\hat{\mathbf{a}}_{z'} \bullet \frac{\partial \mathbf{H}}{\partial z'} = 0 \tag{3.1.5}$$

$$\hat{\mathbf{a}}_{z'} \bullet \frac{\partial \mathbf{E}}{\partial z'} = 0 \tag{3.1.6}$$

Equations (3.1.3) to (3.1.6) show that **E** and **H** are perpendicular to each other and to the direction of propagation $\hat{\mathbf{a}}_{z'}$.

Performing the operation $\hat{\mathbf{a}}_{z'}\times$ in (3.1.3) and expanding $\hat{\mathbf{a}}_{z'} \times \hat{\mathbf{a}}_{z'}\times$ gives

$$\begin{aligned} &\hat{\mathbf{a}}_{z'} \times \hat{\mathbf{a}}_{z'} \times \frac{\partial \mathbf{E}}{\partial z'} = -j\omega\mu\hat{\mathbf{a}}_{z'} \times \mathbf{H} \\ &\hat{\mathbf{a}}_{z'}\left(\hat{\mathbf{a}}_{z'} \bullet \frac{\partial \mathbf{E}}{\partial z'}\right) - (\hat{\mathbf{a}}_{z'} \bullet \hat{\mathbf{a}}_{z'})\frac{\partial \mathbf{E}}{\partial z'} = -j\omega\mu\hat{\mathbf{a}}_{z'} \times \mathbf{H} \\ &\frac{\partial \mathbf{E}}{\partial z'} = j\omega\mu\hat{\mathbf{a}}_{z'} \times \mathbf{H} \end{aligned} \tag{3.1.7}$$

where (3.1.6) was used. Taking $\partial/\partial z'$ in (3.1.7) and substituting (3.1.4) gives

$$\frac{\partial^2 \mathbf{E}}{\partial z'^2} - \gamma^2 \mathbf{E} = 0$$

Similarly, **H** satisfy

$$\frac{\partial^2 \mathbf{H}}{\partial z'^2} - \gamma^2 \mathbf{H} = 0$$

If $\mathbf{E} = E_y\hat{\mathbf{a}}_y$ (where $E_y = E_{y'}$) the solution for the outgoing wave along z' is

$$E_y = Ce^{-\gamma z'} = Ce^{-\gamma(\hat{\mathbf{a}}_{z'}\bullet\mathbf{r})} = Ce^{-\gamma(x\sin\theta + z\cos\theta)} = Ce^{-(\gamma_x x + \gamma_z z)} \tag{3.1.8}$$

In terms of the x, y coordinates, (3.1.8) satisfies

$$\frac{\partial^2 E_y}{\partial x^2} + \frac{\partial^2 E_y}{\partial z^2} - \gamma^2 E_y = 0$$

where

$$\gamma^2 = \gamma_x^2 + \gamma_z^2$$

The field in (3.1.8) can also be expressed in terms of the real and imaginary parts of γ:

$$\mathbf{E} = Ce^{-\gamma(\hat{\mathbf{a}}_{z'}\bullet\mathbf{r})}\hat{\mathbf{a}}_y = Ce^{-\alpha(\hat{\mathbf{a}}_{z'}\bullet\mathbf{r})}e^{-j\beta(\hat{\mathbf{a}}_{z'}\bullet\mathbf{r})}\hat{\mathbf{a}}_y$$

The associated magnetic field follows from (3.1.3) and (3.1.8), and is in the $-\hat{\mathbf{a}}_{x'}$ direction. That is,

$$\mathbf{H} = -\frac{1}{j\omega\mu}\hat{\mathbf{a}}_{z'} \times \frac{\partial E_y}{\partial z'}\hat{\mathbf{a}}_y = -\frac{C\gamma}{j\omega\mu}e^{-\gamma z'}\hat{\mathbf{a}}_{x'}$$

or

$$H_{x'} = -\frac{E_y}{\eta}$$

The fields E_y and $H_{x'}$ are shown in Fig. 3.1. The direction of the power flow (i.e., $E_y\hat{\mathbf{a}}_y \times H_{x'}\hat{\mathbf{a}}_{x'}$) is in the positive z' direction.

The extension of the above formulation to a plane wave traveling in a three-dimensional geometry is as follows. In three dimensions, the **E** field satisfies

$$[\nabla^2 - \gamma^2]\mathbf{E} = 0 \Rightarrow \begin{cases} \nabla^2 E_x - \gamma^2 E_x = 0 \\ \nabla^2 E_y - \gamma^2 E_y = 0 \\ \nabla^2 E_z - \gamma^2 E_z = 0 \end{cases} \tag{3.1.9}$$

whose plane wave solutions for an outgoing wave are

$$E_x = C_1 e^{-(\gamma_x x + \gamma_y y + \gamma_z z)} = C_1 e^{-\gamma\bullet\mathbf{r}}$$
$$E_y = C_2 e^{-(\gamma_x x + \gamma_y y + \gamma_z z)} = C_2 e^{-\gamma\bullet\mathbf{r}}$$
$$E_z = C_3 e^{-(\gamma_x x + \gamma_y y + \gamma_z z)} = C_3 e^{-\gamma\bullet\mathbf{r}}$$

where

$$\gamma^2 = \gamma_x^2 + \gamma_y^2 + \gamma_z^2$$

and

$$\boldsymbol{\gamma} = \gamma \hat{\mathbf{n}}$$

The unit normal $\hat{\mathbf{n}}$ points in the direction of propagation. For example, in the two-dimensional case shown in Fig. 3.1: $\hat{\mathbf{n}}$ is $\hat{\mathbf{a}}_{z'}$. In three dimensions, $\hat{\mathbf{n}}$ is usually denoted by the radial unit vector $\hat{\mathbf{a}}_r$.

The field $\mathbf{E}$ can be expressed as

$$\begin{aligned}\mathbf{E} &= (C_1 \hat{\mathbf{a}}_x + C_2 \hat{\mathbf{a}}_y + C_3 \hat{\mathbf{a}}_z) e^{-\boldsymbol{\gamma} \bullet \mathbf{r}} \\ &= \mathbf{E}_o e^{-\boldsymbol{\gamma} \bullet \mathbf{r}}\end{aligned}$$

where

$$\mathbf{E}_o = C_1 \hat{\mathbf{a}}_x + C_2 \hat{\mathbf{a}}_y + C_3 \hat{\mathbf{a}}_z$$

denotes a constant amplitude vector.

Observing that

$$\begin{aligned}\frac{\partial}{\partial x} e^{-(\gamma_x x + \gamma_y y + \gamma_z z)} &= -\gamma_x e^{-\boldsymbol{\gamma} \bullet \mathbf{r}} \\ \frac{\partial}{\partial y} e^{-\boldsymbol{\gamma} \bullet \mathbf{r}} &= -\gamma_y e^{-\boldsymbol{\gamma} \bullet \mathbf{r}} \\ \frac{\partial}{\partial z} e^{-\boldsymbol{\gamma} \bullet \mathbf{r}} &= -\gamma_z e^{-\boldsymbol{\gamma} \bullet \mathbf{r}}\end{aligned}$$

it follows that the gradient operation is

$$\nabla e^{-\boldsymbol{\gamma} \bullet \mathbf{r}} = -\boldsymbol{\gamma} e^{-\boldsymbol{\gamma} \bullet \mathbf{r}}$$

The plane wave solutions that satisfy (3.1.9) must also satisfy the relation $\nabla \bullet \mathbf{E} = 0$. Therefore,

$$\nabla \bullet (\mathbf{E}_o e^{-\boldsymbol{\gamma} \bullet \mathbf{r}}) = \mathbf{E}_o \bullet (\nabla e^{-\boldsymbol{\gamma} \bullet \mathbf{r}}) + e^{-\boldsymbol{\gamma} \bullet \mathbf{r}} \nabla \bullet \mathbf{E}_o = 0 \qquad (3.1.10)$$

and since $\nabla \bullet \mathbf{E}_o = 0$ and $\boldsymbol{\gamma} = \gamma \hat{\mathbf{n}}$, (3.1.10) reads

$$\nabla \bullet (\mathbf{E}_o e^{-\boldsymbol{\gamma} \bullet \mathbf{r}}) = \mathbf{E}_o \bullet (-\boldsymbol{\gamma} e^{-\boldsymbol{\gamma} \bullet \mathbf{r}}) = -\gamma \mathbf{E}_o e^{-\boldsymbol{\gamma} \bullet \mathbf{r}} \bullet \hat{\mathbf{n}} = 0 \Rightarrow \mathbf{E} \bullet \hat{\mathbf{n}} = 0$$

which shows that **E** is perpendicular to the direction of propagation $\hat{\mathbf{n}}$.

From Maxwell's equation:

$$\mathbf{H} = -\frac{1}{j\omega\mu}\nabla \times \mathbf{E} = -\frac{1}{j\omega\mu}\nabla \times (\mathbf{E}_o e^{-\boldsymbol{\gamma}\bullet\mathbf{r}})$$

$$= -\frac{1}{j\omega\mu}[\nabla e^{-\boldsymbol{\gamma}\bullet\mathbf{r}} \times \mathbf{E}_o + e^{-\boldsymbol{\gamma}\bullet\mathbf{r}}(\nabla \times \mathbf{E}_o)]$$

Since $\nabla \times \mathbf{E}_o = 0$ it follows that

$$\begin{aligned}\mathbf{H} &= -\frac{1}{j\omega\mu}(\nabla e^{-\boldsymbol{\gamma}\bullet\mathbf{r}} \times \mathbf{E}_o) = \frac{1}{j\omega\mu}(\boldsymbol{\gamma} e^{-\boldsymbol{\gamma}\bullet\mathbf{r}} \times \mathbf{E}_o)\\ &= \frac{\gamma}{j\omega\mu}\hat{\mathbf{n}} \times \mathbf{E}_o e^{-\boldsymbol{\gamma}\bullet\mathbf{r}} = \frac{1}{\eta}\hat{\mathbf{n}} \times \mathbf{E}\end{aligned} \tag{3.1.11}$$

which shows that **H** is perpendicular to **E** and to the direction of propagation $\hat{\mathbf{n}}$.

As an example, consider the two-dimensional case shown in Fig. 3.1 where the region is lossless with $\gamma = j\beta$ where $\beta = \omega\sqrt{\mu\varepsilon}$, and $\hat{\mathbf{n}}$ is in the direction $\hat{\mathbf{a}}_{z'}$. In this case:

$$E_y = C\,e^{-j\beta z'} = C\,e^{-j\beta_x x}e^{-j\beta_z z} = C\,e^{-j\boldsymbol{\beta}\bullet\mathbf{r}}$$

where

$$\beta_x = \beta \sin\theta$$

$$\beta_z = \beta \cos\theta$$

$$\boldsymbol{\beta} = \beta_x\hat{\mathbf{a}}_x + \beta_z\hat{\mathbf{a}}_z$$

$$\beta^2 = \beta_x^2 + \beta_z^2$$

The magnetic field follows from (3.1.11). The direction of propagation is determined by the propagation constants β_x and β_z. That is,

$$\tan\theta = \frac{\beta_x}{\beta_z} \Rightarrow \theta = \tan^{-1}\left(\frac{\beta_x}{\beta_z}\right)$$

and the planes of constant phase are given by $\boldsymbol{\beta} \bullet \mathbf{r} = c$, where c is a constant.

The power flow is in the direction of propagation (i.e., $\hat{\mathbf{n}} = \hat{\mathbf{a}}_{z'}$). Using (3.1.11) for **H**, the time average power density is

$$\mathbf{P} = \frac{1}{2}\text{Re}(\mathbf{E} \times \mathbf{H}^*) = \frac{1}{2\eta}\text{Re}(E_y \hat{\mathbf{a}}_y \times \hat{\mathbf{a}}_{z'} \times E_y^* \hat{\mathbf{a}}_y)$$

$$= \frac{|E_y|^2}{2\eta}\hat{\mathbf{a}}_{z'}$$

3.2 REFLECTION AND TRANSMISSION OF PLANE WAVES

In this section, the reflection and transmission of plane waves from plane surfaces are studied. The development of a surface impedance boundary condition is also discussed.

3.2.1 PERPENDICULAR POLARIZATION

Consider the reflection and transmission of a plane wave with the **E** field perpendicular to the plane of incidence, called perpendicular polarization, as show in Fig. 3.2. This polarization is also called transverse electric (TE) polarization (Note: other names are also used to refer to this polarization).

The incident field $\mathbf{E}_i = E_{yi}\hat{\mathbf{a}}_y$ satisfies

$$\frac{\partial^2 E_{yi}}{\partial x^2} + \frac{\partial^2 E_{yi}}{\partial z^2} - \gamma^2 E_{yi} = 0$$

The unit normal in the direction of propagation is

$$\hat{\mathbf{a}}_i = \sin\theta_i \hat{\mathbf{a}}_x + \cos\theta_i \hat{\mathbf{a}}_z$$

Hence, the solution is

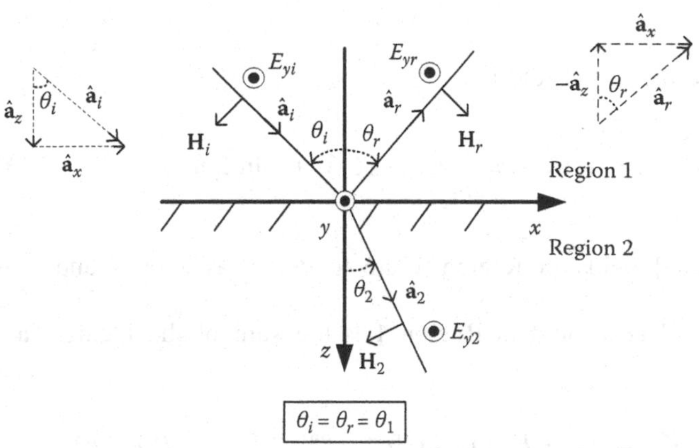

FIGURE 3.2 Perpendicular polarized plane wave.

$$\mathbf{E}_i = E_{yi}\hat{\mathbf{a}}_y = Ce^{-\gamma_1(\hat{\mathbf{a}}_i \bullet \mathbf{r})}\hat{\mathbf{a}}_y = Ce^{-\gamma_1(x \sin\theta_i + z \cos\theta_i)}\hat{\mathbf{a}}_y \tag{3.2.1}$$

where θ_i is the angle of incidence and γ_1 the complex propagation constant of Region 1.

From Maxwell's equation, the incident magnetic field is given by

$$\mathbf{H}_i = \frac{-1}{j\omega\mu}\nabla \times (E_{yi}\hat{\mathbf{a}}_y) = \frac{1}{j\omega\mu}\left(\frac{\partial E_{yi}}{\partial z}\hat{\mathbf{a}}_x - \frac{\partial E_{yi}}{\partial x}\hat{\mathbf{a}}_z\right)$$

or

$$\mathbf{H}_i = H_{xi}\hat{\mathbf{a}}_x + H_{zi}\hat{\mathbf{a}}_z = \frac{C}{\eta_1}(-\cos\theta_i\hat{\mathbf{a}}_x + \sin\theta_i\hat{\mathbf{a}}_z)e^{-\gamma_1(x \sin\theta_i + z \cos\theta_i)}$$

where $\eta_1 = j\omega\mu_1/\gamma_1$ is the intrinsic impedance of Region 1.

For the reflected field the unit normal in the direction of propagation is

$$\hat{\mathbf{a}}_r = \sin\theta_r\hat{\mathbf{a}}_x - \cos\theta_r\hat{\mathbf{a}}_z$$

Hence, the reflected field is given by

$$\mathbf{E}_r = E_{yr}\hat{\mathbf{a}}_y = CR_\perp e^{-\gamma_1(\hat{\mathbf{a}}_r \bullet \mathbf{r})} = CR_\perp e^{-\gamma_1(x \sin\theta_r - z \cos\theta_r)}\hat{\mathbf{a}}_y$$

where θ_r is the angle of reflection and $R_\perp$ is the reflection coefficient for perpendicular polarization (also denoted by R^{TE} in the literature). That is,

$$R_\perp = \left.\frac{E_{yr}}{E_{yi}}\right|_{z=0}$$

The reflected magnetic field is

$$\mathbf{H}_r = H_{xr}\hat{\mathbf{a}}_x + H_{zr}\hat{\mathbf{a}}_z = \frac{C}{\eta_1}R_\perp(\cos\theta_r\hat{\mathbf{a}}_x + \sin\theta_r\hat{\mathbf{a}}_z)e^{-\gamma_1(x \sin\theta_r - z \cos\theta_r)}$$

If the reflected fields in Region 1 are to decay as $x \to \infty$ and $z \to -\infty$ then $\text{Re}[\gamma_1] > 0$.

The total electric field in Region 1 is the sum of the incident and reflected fields, or

$$\mathbf{E}_1 = E_{y1}\hat{\mathbf{a}}_y = (E_{yi} + E_{yr})\hat{\mathbf{a}}_y = C\left[e^{-\gamma_1(x \sin\theta_i + z \cos\theta_i)} + R_\perp e^{-\gamma_1(x \sin\theta_r - z \cos\theta_r)}\right]$$

and the total magnetic field is

$$\mathbf{H}_1 = H_{x1}\hat{\mathbf{a}}_x + H_{z1}\hat{\mathbf{a}}_z = (H_{xi} + H_{xr})\hat{\mathbf{a}}_x + (H_{zi} + H_{zr})\hat{\mathbf{a}}_z$$

where

$$H_{xi} + H_{xr} = \frac{C}{\eta_1}\left[-\cos\theta_i e^{-\gamma_1(x\sin\theta_i + z\cos\theta_i)} + R_\perp \cos\theta_r e^{-\gamma_1(x\sin\theta_r - z\cos\theta_r)}\right]$$

and

$$H_{zi} + H_{zr} = \frac{C}{\eta_1}\left[\sin\theta_i e^{-\gamma_1(x\sin\theta_i + z\cos\theta_i)} + R_\perp \sin\theta_r e^{-\gamma_1(x\sin\theta_r - z\cos\theta_r)}\right]$$

The propagation constant in Region 2 is γ_2, the unit vector in the direction of propagation is $\hat{\mathbf{a}}_2$, and the angle of transmission (also called the angle of refraction) is θ_2. The transmitted fields in Region 2 is expressed in the form

$$\mathbf{E}_2 = E_{y2}\hat{\mathbf{a}}_y = CT_\perp e^{-\gamma_2(\hat{\mathbf{a}}_2 \bullet \mathbf{r})} = CT_\perp e^{-\gamma_2(x\sin\theta_2 + z\cos\theta_2)}\hat{\mathbf{a}}_y$$

and

$$\mathbf{H}_2 = H_{x2}\hat{\mathbf{a}}_x + H_{z2}\hat{\mathbf{a}}_z = \frac{C}{\eta_2}T_\perp(-\cos\theta_2\hat{\mathbf{a}}_x + \sin\theta_2\hat{\mathbf{a}}_z)e^{-\gamma_2(x\sin\theta_2 + z\cos\theta_2)}$$

where η_2 is the intrinsic impedance of Region 2, and $T_\perp$ represents the transmission coefficient (also denoted by T^{TE} in the literature). That is,

$$T_\perp = \left.\frac{E_{y2}}{E_{yi}}\right|_{z=0}$$

If the transmitted fields in Region 2 are to decay as $x \to \infty$ and $z \to \infty$ then $\mathrm{Re}[\gamma_2] > 0$.

At $z = 0$ the continuity of the tangential electric and magnetic fields are

$$E_{y1} = E_{y2}|_{z=0} \Rightarrow E_{yi} + E_{yr} = E_{y2}|_{z=0}$$

and

$$H_{x1} = H_{x2}|_{z=0} \Rightarrow H_{xi} + H_{xr} = H_{x2}|_{z=0}$$

which result in

$$e^{-\gamma_1 x \sin\theta_i} + R_\perp e^{-\gamma_1 x \sin\theta_r} = T_\perp e^{-\gamma_2 x \sin\theta_2} \tag{3.2.2}$$

and

$$\frac{1}{\eta_1}\left(-\cos\theta_i e^{-\gamma_1 x \sin\theta_i} + R_\perp \cos\theta_r e^{-\gamma_1 x \sin\theta_r}\right) = \frac{-1}{\eta_2} T_\perp \cos\theta_2 e^{-\gamma_2 x \sin\theta_2} \tag{3.2.3}$$

These relations must be satisfied for all values of x at the interface. Therefore, it follows that the x dependence must be identical, or

$$\gamma_1 \sin\theta_i = \gamma_1 \sin\theta_r = \gamma_2 \sin\theta_2 \tag{3.2.4}$$

which show that

$$\sin\theta_i = \sin\theta_r \Rightarrow \theta_i = \theta_r \tag{3.2.5}$$

The angles θ_i and θ_r, which are equal, will be labeled θ_1. Equation (3.2.4) also shows that

$$\gamma_1 \sin\theta_1 = \gamma_2 \sin\theta_2 \tag{3.2.6}$$

Equation (3.2.5) shows that the angle of incidence is equal to the angle of reflection, which describes Snell's law of reflection. The relation (3.2.6) represents a complex Snell's law of refraction, since the angle θ_2 will be complex. For the lossless case, (3.2.6) reads

$$\sqrt{\mu_1\varepsilon_1}\sin\theta_1 = \sqrt{\mu_2\varepsilon_2}\sin\theta_2 \Rightarrow \sqrt{\mu_{r1}\varepsilon_{r1}}\sin\theta_1 = \sqrt{\mu_{r2}\varepsilon_{r2}}\sin\theta_2$$

or in terms of the index of refraction,

$$n_1 \sin\theta_1 = n_2 \sin\theta_2$$

which is recognized as Snell's law of refraction. If $\mu_1 = \mu_2$, Snell's law of refraction reads

$$\sqrt{\varepsilon_1}\sin\theta_1 = \sqrt{\varepsilon_2}\sin\theta_2 \Rightarrow \sqrt{\varepsilon_{r1}}\sin\theta_1 = \sqrt{\varepsilon_{r2}}\sin\theta_2$$

Using (3.2.5) and (3.2.6) in (3.2.2) and (3.2.3), the exponentials cancel out, and it follows that

$$1 + R_\perp = T_\perp \tag{3.2.7}$$

and

$$\frac{\cos\theta_1}{\eta_1}(1 - R_\perp) = \frac{\cos\theta_2}{\eta_2}T_\perp \tag{3.2.8}$$

Solving (3.2.7) and (3.2.8) for $R_\perp$ and $T_\perp$ gives

$$R_\perp = \frac{\eta_2\cos\theta_1 - \eta_1\cos\theta_2}{\eta_2\cos\theta_1 + \eta_1\cos\theta_2} \tag{3.2.9}$$

and

$$T_\perp = \frac{2\eta_2\cos\theta_1}{\eta_2\cos\theta_1 + \eta_1\cos\theta_2} \tag{3.2.10}$$

From (3.2.6):

$$\sqrt{1 - \cos^2\theta_2} = \frac{\gamma_1}{\gamma_2}\sin\theta_1$$

or

$$\cos\theta_2 = \sqrt{1 - \gamma_{12}^2\sin^2\theta_1} \Rightarrow \cos\theta_2 = \sqrt{1 - n_{12}^2\sin^2\theta_1} \tag{3.2.11}$$

where $\gamma_{12} = \gamma_1/\gamma_2 = n_1/n_2$ and $n_{12} = n_1/n_2$. Using (3.2.11), $R_\perp$ and $T_\perp$ can be expressed only in terms of the angle of incidence as

$$R_\perp = \frac{\eta_2\cos\theta_1 - \eta_1\sqrt{1 - \gamma_{12}^2\sin^2\theta_1}}{\eta_2\cos\theta_1 + \eta_1\sqrt{1 - \gamma_{12}^2\sin^2\theta_1}} \tag{3.2.12}$$

and

$$T_\perp = \frac{2\eta_2\cos\theta_1}{\eta_2\cos\theta_1 + \eta_1\sqrt{1 - \gamma_{12}^2\sin^2\theta_1}} \tag{3.2.13}$$

Equations (3.2.9) to (3.2.12) are known as the Fresnel reflection coefficients for perpendicular polarization.

Convenient forms of expressing the fields are

$$\begin{cases} E_{y1} = C(e^{-\gamma_{z1}z} + R_\perp e^{\gamma_{z1}z})e^{-\gamma_{x1}x} & \text{or } E_{y1} = C(e^{-jk_{z1}z} + R_\perp e^{jk_{z1}z})e^{-jk_{x1}x} \\ H_{x1} = -\frac{C}{\eta_1}\cos\theta_1(e^{-\gamma_{z1}z} - R_\perp e^{\gamma_{z1}z})e^{-\gamma_{x1}x} & \text{or } H_{x1} = -\frac{C}{\eta_1}\cos\theta_1(e^{-jk_{z1}z} - R_\perp e^{jk_{z1}z})e^{-jk_{x1}x} \\ H_{z1} = \frac{C}{\eta_1}\sin\theta_1(e^{-\gamma_{z1}z} + R_\perp e^{\gamma_{z1}z})e^{-\gamma_{x1}x} & \text{or } H_{z1} = \frac{C}{\eta_1}\sin\theta_1(e^{-jk_{z1}z} + R_\perp e^{jk_{z1}z})e^{-jk_{x1}x} \\ E_{y2} = CT_\perp e^{-\gamma_{z2}z}e^{-\gamma_{x1}x} & \text{or } E_{y2} = CT_\perp e^{-jk_{z2}z}e^{-jk_{x1}x} \\ H_{x2} = -\frac{C}{\eta_2}T_\perp\cos\theta_2 e^{-\gamma_{z2}z}e^{-\gamma_{x1}x} & \text{or } H_{x2} = -\frac{C}{\eta_2}T_\perp\cos\theta_2 e^{-jk_{z2}z}e^{-jk_{x1}x} \\ H_{z2} = \frac{C}{\eta_2}T_\perp\sin\theta_2 e^{-\gamma_{z2}z}e^{-\gamma_{x1}x} & \text{or } H_{z2} = \frac{C}{\eta_2}T_\perp\sin\theta_2 e^{-jk_{z2}z}e^{-jk_{x1}x} \end{cases} \tag{3.2.14}$$

where

$$\begin{aligned} \gamma_1^2 &= \gamma_{x1}^2 + \gamma_{z1}^2 \quad \text{or} \quad k_1^2 = k_{x1}^2 + k_{z1}^2 \\ \gamma_2^2 &= \gamma_{x1}^2 + \gamma_{z2}^2 \quad \text{or} \quad k_2^2 = k_{x1}^2 + k_{z2}^2 \\ \gamma_{z1} &= \gamma_1\cos\theta_1 \quad \text{or} \quad k_{z1} = k_1\cos\theta_1 \\ \gamma_{z2} &= \gamma_2\cos\theta_2 \quad \text{or} \quad k_{z2} = k_2\cos\theta_2 \end{aligned}$$

and

$$\gamma_{x1} = \gamma_{x2} = \gamma_1\sin\theta_1 = \gamma_2\sin\theta_2 \text{ or } k_{x1} = k_{x2} = k_1\sin\theta_1 = k_2\sin\theta_2$$

If Region 1 is lossless, the magnitude of the electric field in Region 1, assuming that C is real, is

$$|E_{y1}| = C|1 + R_\perp e^{j2k_{z1}z}|$$

Letting $R_\perp = |R_\perp|e^{j\psi}$, we obtain

$$\begin{aligned} |E_{y1}| &= C|1 + |R_\perp|e^{j(2k_{z1}z+\psi)}| = C|1 + |R_\perp|\cos(2k_{z1}z + \psi) + j|R_\perp|\sin(2k_{z1}z + \psi)| \\ &= C\sqrt{1 + |R_\perp|^2 + 2|R_\perp|\cos(2k_{z1}z + \psi)} \end{aligned}$$

which shows that $|E_{y1}|$ has a maximum amplitude of

$$|E_{y1}|_{max} = C(1 + |R_\perp|)$$

when $\cos(2k_{z1}z + \psi) = 1$ (or $z = -(n + \psi/2\pi)\lambda_{z1}/2$, $n = 0, 1, 2, ...$) where

$\lambda_{z1} = 2\pi/k_1$. The minimum amplitude value occurs when $\cos(2k_{z1}z + \psi) = -1$ (or $z = -(n + 1/2 + \psi/2\pi)\lambda_{z1}/2$, $n = 0, 1, 2, \ldots$). Its value is

$$|E_{y1}|_{\min} = C(1 - |R_\perp|)$$

The time-average power density is Region 1 is given by

$$\mathbf{P}_1 = \frac{1}{2}\mathrm{Re}(\mathbf{E}_1 \times \mathbf{H}_1^*) = \frac{1}{2}\mathrm{Re}[E_{y1}\hat{\mathbf{a}}_y \times (H_{x1}^*\hat{\mathbf{a}}_x + H_{z1}^*\hat{\mathbf{a}}_z)]$$

$$= \frac{C^2}{2\eta_1}\cos\theta_1(1 - |R_\perp|^2)\hat{\mathbf{a}}_z + \frac{C^2}{2\eta_1}\sin\theta_1[1 + |R_\perp|^2 + 2|R_\perp|\cos(2k_{z1}z + \psi)]\hat{\mathbf{a}}_x$$

If C is a complex constant change C^2 to $|C|^2$. The factor $C^2/2\eta_1$ represents the magnitude of the incident power density, and the factor $\cos\theta_1$ account for its z component. The factor $|R_\perp|^2$ accounts for the reflected power in the $-z$ direction. The component in the x direction is accounted by the $\sin\theta_1$ factor and the factor in bracket, which is periodic in z, has a maximum value of $(1 + |R_\perp|)^2$ and a minimum value of $(1 - |R_\perp|)^2$.

In Region 2 the time-average power density is

$$\mathbf{P}_2 = \frac{C^2|T_\perp|^2}{2\eta_2}\cos\theta_2\hat{\mathbf{a}}_z + \frac{C^2|T_\perp|^2}{2\eta_2}\sin\theta_2\hat{\mathbf{a}}_x$$

For normal incidence (i.e., with $\theta_1 = \theta_2 = 0$) the incident and transmitted time-average power densities are

$$\mathbf{P}_1 = \frac{C^2}{2\eta_1}(1 - |R_\perp|^2)\hat{\mathbf{a}}_z$$

and

$$\mathbf{P}_2 = \frac{C^2|T_\perp|^2}{2\eta_2}\hat{\mathbf{a}}_z$$

Hence, it follows that

$$\frac{C^2}{2\eta_1}(1 - |R_\perp|^2) = \frac{C^2|T_\perp|^2}{2\eta_2} \Rightarrow \frac{C^2}{2\eta_1} = \frac{C^2|R_\perp|^2}{2\eta_1} + \frac{C^2|T_\perp|^2}{2\eta_2}$$

which show that the time-average incident power density is equal to the sum of the time-average reflected and transmitted power density. A statement that is also valid for oblique incidence.

The relations for $R_\perp$ and $T_\perp$ can be written in different forms. Since

$$\frac{\eta_1}{\eta_2} = \eta_{12} = \frac{\mu_1}{\mu_2}\frac{\gamma_2}{\gamma_1} = \mu_{12}\gamma_{21} = \mu_{12}n_{21} = \mu_{12}k_{21}$$

(3.2.9) and (3.2.10) can also be expressed as

$$R_\perp = \begin{cases} \dfrac{\cos\theta_1 - \mu_{12}\gamma_{21}\cos\theta_2}{\cos\theta_1 + \mu_{12}\gamma_{21}\cos\theta_2} \\ \dfrac{\cos\theta_1 - \mu_{12}n_{21}\cos\theta_2}{\cos\theta_1 + \mu_{12}n_{21}\cos\theta_2} \\ \dfrac{\gamma_{z1} - \mu_{12}\gamma_{z2}}{\gamma_{z1} + \mu_{12}\gamma_{z2}} = \dfrac{k_{z1} - \mu_{12}k_{z2}}{k_{z1} + \mu_{12}k_{z2}} \end{cases} \tag{3.2.15}$$

and

$$T_\perp = \begin{cases} \dfrac{2\cos\theta_1}{\cos\theta_1 + \mu_{12}\gamma_{21}\cos\theta_2} \\ \dfrac{2\cos\theta_1}{\cos\theta_1 + \mu_{12}n_{21}\cos\theta_2} \\ \dfrac{2\gamma_{z1}}{\gamma_{z1} + \mu_{12}\gamma_{z2}} = \dfrac{2k_{z1}}{k_{z1} + \mu_{12}k_{z2}} \end{cases} \tag{3.2.16}$$

where

$$k_{z1} = k_o n_1 \cos\theta_1$$
$$k_{z2} = k_o n_2 \cos\theta_2$$

Also, using (3.2.11), the relations in (3.2.15) and (3.2.16) can be expressed only in terms of θ_1.

A convenient form (easy to remember) for $R_\perp$ and $T_\perp$ is to write (3.2.9) and (3.2.10) in the forms

$$R_\perp = \frac{\eta_2^\perp - \eta_1^\perp}{\eta_2^\perp + \eta_1^\perp} \tag{3.2.17}$$

and

$$T_\perp = \frac{2\eta_2^\perp}{\eta_2^\perp + \eta_1^\perp} \tag{3.2.18}$$

where

$$\eta_1^{\perp} = \frac{\eta_1}{\cos\theta_1}$$

and

$$\eta_2^{\perp} = \frac{\eta_2}{\cos\theta_2}$$

For normal incidence $\theta_1 = \theta_2 = 0$ and (3.2.17) and (3.2.18) read

$$R_{\perp} = \frac{\eta_2 - \eta_1}{\eta_2 + \eta_1}$$

and

$$T_{\perp} = \frac{2\eta_2}{\eta_2 + \eta_1}$$

At the interface, a surface impedance boundary condition is

$$Z_s = -\frac{E_{y2}}{H_{x2}}\bigg|_{z=0} = -\frac{E_{y1}}{H_{x1}}\bigg|_{z=0} = \frac{\eta_2}{\cos\theta_2} = \eta_2^{\perp} = \frac{\eta_2}{\sqrt{1 - \gamma_{12}^2 \sin^2\theta_1}} \tag{3.2.19}$$

which is the impedance looking into Region 2. Knowledge of the surface impedance simplifies the analysis of many problem since Region 2 is represented by Z_s. Using (3.2.19), (3.2.17) reads

$$R_{\perp} = \frac{Z_s - \eta_1^{\perp}}{Z_s + \eta_1^{\perp}}$$

The surface impedance relation in (3.2.19) shows that in the case of a perpendicular polarized plane wave (i.e., TE wave) at grazing incidence (i.e., $\theta_1 \approx \pi/2$) the surface impedance is

$$Z_s \approx \frac{\eta_2}{\sqrt{1 - \gamma_{12}^2}}$$

A source for such wave is a horizontal dipole, where far from the dipole, in many situations, the field on a surface can be considered to be at grazing incidence. Furthermore, for $\sigma_1 = 0$ and $\mu_1 = \mu_2 = \mu_o$:

$$\gamma_{12} = \sqrt{\frac{j\omega\varepsilon_1}{\sigma_2 + j\omega\varepsilon_2}} = \sqrt{\frac{\varepsilon_1}{\varepsilon_2\left(1 - j\frac{\sigma_2}{\omega\varepsilon_2}\right)}}$$

Hence, under certain conditions such that $|\gamma_{12}| \to 0$, it follows from (3.2.11) that: $\cos\theta_2 \to 1$ (or $\theta_2 \approx 0$) and the surface impedance is approximately η_2. This commonly happens when $\varepsilon_2 >> \varepsilon_1$ and $\sigma_2/\omega\varepsilon_2 >> 1$.

Surface impedances have been used to represent many surfaces (such as planar, cylindrical and spherical surfaces) with great success and their use simplifies considerably the field calculations in problems involving boundaries that can be represented by its surface impedance.

3.2.2 Parallel Polarization

The parallel polarization of the **E** field is illustrated in Fig. 3.3. It is also known as transverse magnetic (TM) polarization. The analysis is similar to that of perpendicular polarization. That is, we write

$$\mathbf{H}_1 = H_{y1}\hat{\mathbf{a}}_y = (H_{yi} + H_{yr})\hat{\mathbf{a}}_y = D\left(e^{-\gamma_{z1}z} + R_{\|}^{H} e^{\gamma_{z1}z}\right) e^{-\gamma_{x1}x}\hat{\mathbf{a}}_y \tag{3.2.20}$$

where

$$\gamma_{z1} = \gamma_1 \cos\theta_1$$
$$\gamma_{x1} = \gamma_1 \sin\theta_1$$

and the angle of reflection was set equal to the angle of incidence (like in (3.2.5)). Note that $R_{\|}^{H}$ is the reflection coefficient associated with the magnetic field, or

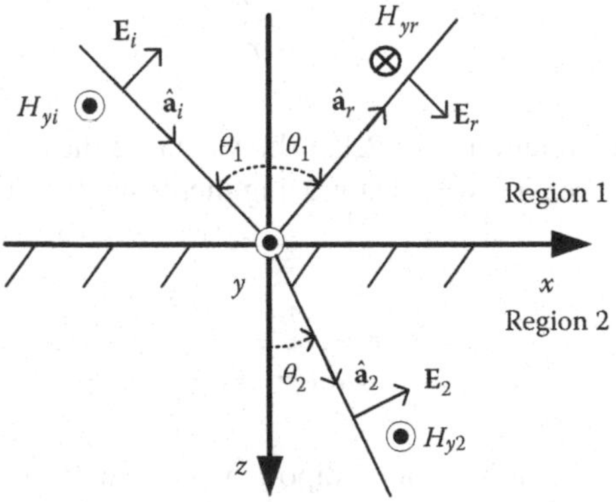

FIGURE 3.3 Parallel polarized plane wave.

$$R_{\parallel}^{H} = \left.\frac{H_{yr}}{H_{yi}}\right|_{z=0}$$

From Maxwell's equation in (1.3.10), the tangential electric field in Region 1 is

$$\begin{aligned} E_{x1} &= E_{xi} + E_{xr} = \frac{-1}{\sigma_1 + j\omega\varepsilon_1}\frac{\partial H_{y1}}{\partial z} = \frac{D\gamma_{z1}}{\sigma_1 + j\omega\varepsilon_1}\left(e^{-\gamma_{z1}z} - R_{\parallel}^{H} e^{\gamma_{z1}z}\right) e^{-\gamma_{x1}x} \\ &= D\eta_1 \cos\theta_1 \left(e^{-\gamma_{z1}z} - R_{\parallel}^{H} e^{\gamma_{z1}z}\right) e^{-\gamma_{x1}x} \end{aligned}$$

and E_{z1} is given by

$$E_{z1} = E_{zi} + E_{zr} = \frac{1}{\sigma_1 + j\omega\varepsilon_1}\frac{\partial H_{y1}}{\partial x} = -D\eta_1 \sin\theta_1 \left(e^{-\gamma_{z1}z} + R_{\parallel}^{H} e^{\gamma_{z1}z}\right) e^{-\gamma_{x1}x}$$

The electric field in Region 1 is then

$$\begin{aligned} \mathbf{E}_1 &= E_{x1}\hat{\mathbf{a}}_x + E_{z1}\hat{\mathbf{a}}_z \\ &= D\eta_1(\cos\theta_1\hat{\mathbf{a}}_x - \sin\theta_1\hat{\mathbf{a}}_z)e^{-\gamma_{z1}z}e^{-\gamma_{x1}x} - DR_{\parallel}^{H}\eta_1(\cos\theta_1\hat{\mathbf{a}}_x + \sin\theta_1\hat{\mathbf{a}}_z)e^{\gamma_{z1}z}e^{-\gamma_{x1}x} \end{aligned} \tag{3.2.21}$$

The transmitted fields are

$$\mathbf{H}_2 = H_{y2}\hat{\mathbf{a}}_y = DT_{\parallel}^{H} e^{-\gamma_{z2}z}e^{-\gamma_{x1}x}\hat{\mathbf{a}}_y$$

and

$$\mathbf{E}_2 = E_{x2}\hat{\mathbf{a}}_x + E_{z2}\hat{\mathbf{a}}_z = DT_{\parallel}^{H}\eta_2(\cos\theta_2\hat{\mathbf{a}}_x - \sin\theta_2\hat{\mathbf{a}}_z)e^{-\gamma_{z2}z}e^{-\gamma_{x1}x} \tag{3.2.22}$$

where $T_{\parallel}^{H}$ denotes the transmission coefficient associated with the magnetic field, or

$$T_{\parallel}^{H} = \left.\frac{H_{y2}}{H_{yi}}\right|_{z=0}$$

The term $DT_{\parallel}^{H}\eta_2$ represents the amplitude of the transmitted electric field. The x dependence of the transmitted fields was set to be identical to that in Region 1, or

$$\gamma_1 \sin\theta_1 = \gamma_2 \sin\theta_2$$

which is the complex form of Snell's law (see (3.2.6)).

At $z = 0$, the continuity of the tangential magnetic field gives

$$H_{y1} = H_{y2}\Big|_{z=0} \Rightarrow 1 + R_{\|}^{H} = T_{\|}^{H}$$

and the continuity of the tangential electric field, from (3.2.21) and (3.2.22), gives

$$E_{x1} = E_{x2}|_{z=0} \Rightarrow \eta_1 \cos\theta_1 (1 - R_{\|}^{H}) = T_{\|}^{H} \eta_2 \cos\theta_2$$

whose solutions are

$$R_{\|}^{H} = \frac{\eta_1 \cos\theta_1 - \eta_2 \cos\theta_2}{\eta_1 \cos\theta_1 + \eta_2 \cos\theta_2} \tag{3.2.23}$$

and

$$T_{\|}^{H} = \frac{2\eta_1 \cos\theta_1}{\eta_1 \cos\theta_1 + \eta_2 \cos\theta_2} \tag{3.2.24}$$

Equations (3.2.23) and (3.2.24) are the Fresnel reflection and transmission coefficients associated with the magnetic field for parallel polarization.

In the literature one also finds the reflection coefficient defined in terms of the amplitude ratio of the reflected to incident electric field. Denoting the amplitude ratio of the reflected to incident electric field at $z = 0$ by $R_{\|}$, the field $\mathbf{E}_1$ is written as

$$\mathbf{E}_1 = E_o(\cos\theta_1 \hat{\mathbf{a}}_x - \sin\theta_1 \hat{\mathbf{a}}_z)e^{-\gamma_{z1}z}e^{-\gamma_{x1}x} + E_o R_{\|}(\cos\theta_1 \hat{\mathbf{a}}_x + \sin\theta_1 \hat{\mathbf{a}}_z)e^{\gamma_{z1}z}e^{-\gamma_{x1}x} \tag{3.2.25}$$

and the associated magnetic field is

$$\mathbf{H}_1 = H_{y1}\hat{\mathbf{a}}_y = \left(\frac{E_o}{\eta_1}e^{-\gamma_{z1}z}e^{-\gamma_{x1}x} - \frac{E_o}{\eta_1}R_{\|}e^{\gamma_{z1}z}e^{-\gamma_{x1}x}\right)\hat{\mathbf{a}}_y = \frac{E_o}{\eta_1}\left(e^{-\gamma_{z1}z} - R_{\|}e^{\gamma_{z1}z}\right)e^{-\gamma_{x1}x}\hat{\mathbf{a}}_y \tag{3.2.26}$$

The fields in Region 2 are

$$\mathbf{E}_2 = E_o T_{\|}(\cos\theta_2 \hat{\mathbf{a}}_x - \sin\theta_2 \hat{\mathbf{a}}_z)e^{-\gamma_{z2}z}e^{-\gamma_{x1}x} \tag{3.2.27}$$

and

$$\mathbf{H}_2 = \frac{E_o}{\eta_2}T_{\|}\, e^{-\gamma_{z2}z}e^{-\gamma_{x1}x}\hat{\mathbf{a}}_y \tag{3.2.28}$$

where $T_{\|}$ is the transmission coefficient that relates the amplitude of the transmitted electric field to that of the incident electric field.

Using the boundary conditions at $z = 0$, from (3.6.25) to (3.6.28) we obtain

$$E_{x1} = E_{x2}|_{z=0} \Rightarrow \cos\theta_1 + R_{\|}\cos\theta_1 = T_{\|}\cos\theta_2$$

and

$$H_{y1} = H_{y2}\bigg|_{z=0} \Rightarrow \frac{1}{\eta_1} - \frac{1}{\eta_1}R_{\|} = \frac{1}{\eta_2}T_{\|}$$

whose solutions are

$$R_{\|} = \frac{\eta_2\cos\theta_2 - \eta_1\cos\theta_1}{\eta_2\cos\theta_2 + \eta_1\cos\theta_1} \tag{3.2.29}$$

and

$$T_{\|} = \frac{2\eta_2\cos\theta_1}{\eta_2\cos\theta_2 + \eta_1\cos\theta_1} \tag{3.2.30}$$

An alternate procedure for obtaining $R_{\|}$ and $T_{\|}$ is to compare (3.2.25) and (3.2.21) which show that we have let the amplitude of the incident electric field E_o in (3.2.25) be $E_o = D\eta_1$, and that of the reflected field as $E_oR_{\|} = -DR_{\|}^H\eta_1$. Therefore, $R_{\|}$ and $R_{\|}^H$ are related by

$$R_{\|} = \frac{-DR_{\|}^H\eta_1}{E_o} = \frac{-DR_{\|}^H\eta_1}{D\eta_1} = -R_{\|}^H$$

which results in (3.2.29).

Similarly, comparing (3.2.27) and (3.2.22) shows that the amplitude of the transmitted field is $E_oT_{\|} = DT_{\|}^H\eta_2$. Therefore, $T_{\|}$ and $T_{\|}^H$ are related by

$$T_{\|} = \frac{DT_{\|}^H\eta_2}{E_o} = \frac{DT_{\|}^H\eta_2}{D\eta_1} = T_{\|}^H\frac{\eta_2}{\eta_1}$$

which results in (3.2.30).

The reflections and transmission coefficients in (3.2.29) and (3.2.30) are the Fresnel reflection coefficients associated with the electric field for parallel polarization. Other forms of (3.2.29) and (3.2.30) are

$$R_{\|} = \begin{cases} \dfrac{\cos\theta_2 - \mu_{12}\gamma_{21}\cos\theta_1}{\cos\theta_2 + \mu_{12}\gamma_{21}\cos\theta_1} \\ \dfrac{\cos\theta_2 - \mu_{12}n_{21}\cos\theta_1}{\cos\theta_2 + \mu_{12}n_{21}\cos\theta_1} \\ \dfrac{\gamma_{z2} - \mu_{12}n_{21}^2\gamma_{z1}}{\gamma_{z2} + \mu_{12}n_{21}^2\gamma_{z1}} = \dfrac{k_{z2} - \mu_{12}n_{21}^2 k_{z1}}{k_{z2} + \mu_{12}n_{21}^2 k_{z1}} \end{cases} \tag{3.2.31}$$

and

$$T_{\|} = \begin{cases} \dfrac{2\cos\theta_1}{\cos\theta_2 + \mu_{12}\gamma_{21}\cos\theta_1} \\ \dfrac{2\cos\theta_1}{\cos\theta_2 + \mu_{12}n_{21}\cos\theta_1} \\ \dfrac{2n_{21}\gamma_{z1}}{\gamma_{z2} + \mu_{12}n_{21}^2\gamma_{z1}} = \dfrac{2n_{21}k_{z1}}{k_{z2} + \mu_{12}n_{21}^2 k_{z1}} \end{cases} \tag{3.2.32}$$

Using (3.2.11), the previous equations can be expressed only in terms of the angle of incidence.

A convenient form of expressing $R_{\|}$ in (3.2.29) is

$$R_{\|} = \frac{\eta_2^{\|} - \eta_1^{\|}}{\eta_2^{\|} + \eta_1^{\|}} \tag{3.2.33}$$

and $T_{\|}$ in (3.2.30) is

$$T_{\|} = \frac{\eta_2}{\eta_1}\frac{2\eta_1^{\|}}{\eta_2^{\|} + \eta_1^{\|}} \tag{3.2.34}$$

where

$$\eta_1^{\|} = \eta_1 \cos\theta_1$$

and

$$\eta_2^{\|} = \eta_2 \cos\theta_2$$

In the case of parallel polarization, the surface impedance is

$$Z_s = \frac{E_{x2}}{H_{y2}} = \frac{E_{x1}}{H_{y1}} = \eta_2 \cos\theta_2 = \eta_2^{\|} = \eta_2\sqrt{1-\gamma_{12}^2 \sin^2\theta_1} \tag{3.2.35}$$

and (3.2.33) can be expressed as

$$R_{\|} = \frac{Z_s - \eta_1^{\|}}{Z_s + \eta_1^{\|}}$$

Equation (3.2.35) shows that in the case of a parallel-polarized plane wave (TM wave) at grazing incidence (i.e., $\theta_1 \approx \pi/2$) the surface impedance is

$$Z_s \approx \eta_2\sqrt{1-\gamma_{12}^2} \tag{3.2.36}$$

A source for such wave is a vertical dipole, where far from the dipole, in some situations, the field on a surface can be considered to be at grazing incidence.

As previously discussed, under certain conditions such that $|\gamma_{12}| \to 0$, then $\cos\theta_2 \to 1$ (or $\theta_2 \approx 0$) and the surface impedance is approximately η_2.

3.3 REFLECTION AND TRANSMISSION IN DIELECTRIC REGIONS

Consider two dielectric regions with $\sigma_1 = \sigma_2 = 0$ and $\mu_1 = \mu_2 = \mu_o$. Figures 3.2 and 3.3 illustrate the different polarizations where in the case of dielectric regions: $\gamma_1 = j\beta_1 = j\beta_o n_1$ ($n_1 = \sqrt{\varepsilon_{r1}}$) and $\gamma_2 = j\beta_2 = j\beta_o n_2$ ($n_2 = \sqrt{\varepsilon_{r2}}$). Then,

$$\eta_{12} = \gamma_{21} = n_{21} = \frac{\sqrt{\varepsilon_{r2}}}{\sqrt{\varepsilon_{r1}}} = \frac{\sqrt{\varepsilon_2}}{\sqrt{\varepsilon_1}}$$

and the relations for the reflection and transmission coefficients, using (3.2.11), are written as

$$R_{\perp} = \frac{\cos\theta_1 - \sqrt{n_{21}^2 - \sin^2\theta_1}}{\cos\theta_1 + \sqrt{n_{21}^2 - \sin^2\theta_1}} \tag{3.3.1}$$

$$T_{\perp} = \frac{2\cos\theta_1}{\cos\theta_1 + \sqrt{n_{21}^2 - \sin^2\theta_1}} \tag{3.3.2}$$

$$R_{\|} = \frac{\sqrt{n_{21}^2 - \sin^2\theta_1} - n_{21}^2\cos\theta_1}{\sqrt{n_{21}^2 - \sin^2\theta_1} + n_{21}^2\cos\theta_1} \tag{3.3.3}$$

$$T_{\parallel} = \frac{2n_{21}\cos\theta_1}{\sqrt{n_{21}^2 - \sin^2\theta_1} + n_{21}^2\cos\theta_1} \tag{3.3.4}$$

Several interesting properties follow from (3.3.1) to (3.3.4). If $n_2 > n_1$, then the square root terms in (3.3.1) to (3.3.4) are positive and real, and if $R_{\perp} > 0$ and $R_{\parallel} > 0$ the reflected electric fields are in phase with the incident fields. If $R_{\perp} < 0$ and $R_{\parallel} < 0$ there is a 180° phase shift in the reflected electric fields.

If $n_2 < n_1$ and $n_{21} > \sin\theta_1$ the square root is positive, and if $R_{\perp} > 0$ and $R_{\parallel} > 0$ there is no phase shift in the reflected electric fields. If $R_{\perp} < 0$ and $R_{\parallel} < 0$, there is a 180° phase shift in the reflected electric fields.

If $n_2 < n_1$ and $n_{21} = \sin\theta_1$ the square root is zero, then $R_{\perp} = 1$ and $R_{\parallel} = -1$. In such cases there is total reflection. This occurs at the angle $\theta_1 = \theta_c$, known as the critical angle, namely

$$n_{21}^2 - \sin^2\theta_c = 0 \Rightarrow \sin\theta_c = n_{21} = \sqrt{\frac{\varepsilon_{r2}}{\varepsilon_{r1}}} = \sqrt{\frac{\varepsilon_2}{\varepsilon_1}} \tag{3.3.5}$$

Note that if $\mu_1 \neq \mu_2$, the critical angle is given by

$$\sin\theta_c = n_{21} = \sqrt{\frac{\mu_2\varepsilon_2}{\mu_1\varepsilon_1}}$$

Equation (3.3.5) requires that $\varepsilon_1 > \varepsilon_2$ for a real θ_c. Thus, Region 1 must be denser than Region 2, such as propagation from glass to air. At the critical angle there is no phase shift in the reflected field for a perpendicular polarized wave, and a 180° phase shift in a parallel polarized wave.

The transmission angle when $\theta_1 = \theta_c$, from Snell's law of refraction, is

$$n_1\sin\theta_c = n_2\sin\theta_2 \Rightarrow \sin\theta_2 = n_{12}n_{21} = 1 \Rightarrow \theta_2 = 90°$$

The transmitted fields for perpendicular polarization follow from (3.2.14) with $T_{\perp} = 2$ (see (3.3.2)). For example,

$$E_{y2} = 2Ce^{-j\beta_2 x}$$

and

$$H_{z2} = \frac{2C}{\eta_2}e^{-j\beta_2 x}$$

which represents a wave traveling along the interface. Since $\mu_2 = \mu_o$, η_2 is

$$\eta_2 = \sqrt{\frac{\mu_o}{\varepsilon_2}}$$

The time-average power density along the interface is

$$\mathbf{P}_2 = \frac{1}{2}\,\mathrm{Re}\left[E_{y2}\hat{\mathbf{a}}_y \times H_{z2}^{*}\hat{\mathbf{a}}_z\right] = \frac{2C^2}{\eta_2}\hat{\mathbf{a}}_x$$

which shows that the power flows along the interface, but not into Region 2. In other words, there is a lateral wave (also referred as a surface wave). If C is a complex constant change C^2 to $|C|^2$. Since no power flows into Region 2 at the critical angle, the incident and reflected power densities are identical.

If $\theta_1 > \theta_c$ then $\sin\theta_1 > n_{21}$ and the square root terms in (3.3.1) to (3.3.4) become imaginary. Consider (3.3.1) when $\theta_1 > \theta_c$. That is,

$$R_{\perp} = \frac{\cos\theta_1 + j\sqrt{\sin^2\theta_1 - n_{21}^2}}{\cos\theta_1 - j\sqrt{\sin^2\theta_1 - n_{21}^2}} \tag{3.3.6}$$

which shows that $|R_{\perp}| = 1$ and the $-j$ branch was selected for the square root, as discussed below. Hence, total reflection occurs. This effect is known as total internal reflection. When total internal reflection occurs, the angle of refraction is complex since

$$\cos\theta_2 = \sqrt{1 - n_{12}^2\sin^2\theta_1} = -j\sqrt{\frac{\varepsilon_1}{\varepsilon_2}\sin^2\theta_1 - 1} \tag{3.3.7}$$

and the field E_{y2} in Region 2 is of the form:

$$E_{y2} = CT_{\perp}e^{-j\beta_2(z\cos\theta_2 + x\sin\theta_2)} = CT_{\perp}e^{-\beta_2 z\sqrt{\frac{\varepsilon_1}{\varepsilon_2}\sin^2\theta_1 - 1}}\,e^{-j\beta_2 x\sqrt{\frac{\varepsilon_1}{\varepsilon_2}}\sin\theta_1} \tag{3.3.8}$$

The root with the minus sign was selected in (3.3.7) (and in (3.3.6)) so the wave in Region 2 decays as z increases. This wave is known as an evanescent wave (or lateral-evanescent wave). It decays rapidly in the region with low index of refraction. However, the lateral wave that exists in Region 2 is necessary to match the boundary conditions. A similar analysis for $\theta_1 > \theta_c$ applies to the horizontal polarization case.

Equation (3.3.8) can be expressed in the form

$$E_{y2} = CT_{\perp}e^{-\alpha_t z}e^{-j\beta_t x}$$

where

$$\alpha_t = \beta_2\sqrt{\frac{\varepsilon_1}{\varepsilon_2}\sin^2\theta_1 - 1}$$

$$\beta_t = \beta_2\sqrt{\frac{\varepsilon_1}{\varepsilon_2}}\sin\theta_1$$

The parameters α_t and β_t represent the attenuation and propagation constants of the transmitted field. For a wave attenuation of the form $e^{-\alpha_t z}$, the penetration depth or skin depth is defined as the distance $z = \delta$ where $\delta = 1/\alpha_t$. So, the wave is confined to a few multiples of the penetration depth.

The magnetic field in Region 2 is

$$\mathbf{H}_2 = \frac{-1}{j\omega\mu}\nabla \times E_{y2}\hat{\mathbf{a}}_y = \frac{CT_\perp}{\omega\mu}(j\alpha_t\hat{\mathbf{a}}_x + \beta_t\hat{\mathbf{a}}_z)e^{-\alpha_t z}e^{-j\beta_t x}$$

and the time-average power density is

$$\mathbf{P}_2 = \frac{1}{2}\text{Re}[\mathbf{E}_2 \times \mathbf{H}_2^*] = \frac{C^2|T_\perp|^2}{2\omega\mu}\text{Re}\left[(j\alpha_t\hat{\mathbf{a}}_z + \beta_t\hat{\mathbf{a}}_x)e^{-2\alpha_t z}\right]$$

If C is complex change C^2 to $|C|^2$. The first term in parenthesis represents a reactive power flowing in the z direction, which shows that no real power flows into Region 2. The second term shows that the real power density flows along the interface (i.e., the x direction).

In the case of parallel polarization, there is an angle where $R_\parallel = 0$. This angle of incidence is known as the Brewster angle θ_B (or polarization angle). At this angle, there is total transmission. For the case that $\mu_1 \neq \mu_2$ the Brewster angle (i.e., with $\theta_1 = \theta_B$) follows from (3.2.29) or (3.3.3). From (3.2.29), $R_\parallel = 0$ when

$$\eta_1\cos\theta_B = \eta_2\cos\theta_2 \Rightarrow \cos\theta_B = \sqrt{\frac{\mu_2\varepsilon_1}{\mu_1\varepsilon_2}}\cos\theta_2 \tag{3.3.9}$$

Squaring (3.3.9):

$$1 - \sin^2\theta_B = \mu_{21}\varepsilon_{12}(1 - \sin^2\theta_2)$$

Using Snell's law of refraction (i.e., $\sin\theta_2 = n_{12}\sin\theta_B$) to express θ_2 in terms of θ_B gives

$$1 - \sin^2\theta_B = \mu_{21}\varepsilon_{12}(1 - \mu_{12}\varepsilon_{12}\sin^2\theta_B) = \mu_{21}\varepsilon_{12} - \varepsilon_{12}^2\sin^2\theta_B$$

Solving θ_B shows that

$$\theta_B = \sin^{-1}\sqrt{\frac{\varepsilon_{21} - \mu_{21}}{\varepsilon_{21} - \varepsilon_{12}}}$$

In the case of a dielectric regions with $\mu_1 = \mu_2 = \mu_o$ this relation shows that the Brewster angle is given by

$$\theta_B = \sin^{-1}\sqrt{\frac{\varepsilon_2}{\varepsilon_1 + \varepsilon_2}} \tag{3.3.10}$$

which can be also be expressed as

$$\theta_B = \tan^{-1}\sqrt{\frac{\varepsilon_2}{\varepsilon_1}}$$

The Brewster angle in (3.3.10) occurs for either $\varepsilon_1 > \varepsilon_2$ or $\varepsilon_1 < \varepsilon_2$. Observe that in a dielectric region with $\mu_1 = \mu_2$ there is no real incident angle that makes $R_\perp = 0$, so for this case the Brewster angle only occurs in parallel polarization.

The angle of refractions when $\theta_1 = \theta_B$ is given by Snell's law. With $\mu_1 = \mu_2$:

$$\sin\theta_2 = n_{12}\sin\theta_B \Rightarrow \theta_2 = \sin^{-1}\sqrt{\frac{\varepsilon_1}{\varepsilon_1 + \varepsilon_2}} = \tan^{-1}\sqrt{\frac{\varepsilon_1}{\varepsilon_2}}$$

Using the identity

$$\tan^{-1}x + \tan^{-1}\left(\frac{1}{x}\right) = \frac{\pi}{2} \text{ for } x > 0$$

it follows that

$$\theta_B + \theta_2 = \frac{\pi}{2}$$

It is observed that if $\varepsilon_2 > \varepsilon_1$ then $\theta_B > 45°$, and if $\varepsilon_1 > \varepsilon_2$ then $\theta_B < 45°$. From (3.3.5) and (3.3.10) it follows that $\theta_c > \theta_B$. In summary, while the Brewster angle occurs for only parallel polarization and either $\varepsilon_1 > \varepsilon_2$ or $\varepsilon_2 > \varepsilon_1$, total internal reflection occurs for both polarizations if $\theta_1 > \theta_c$ but only for $\varepsilon_1 > \varepsilon_2$.

Figure 3.4 illustrates an example of the behavior of the reflection coefficients for an air-glass interface. A typical n for glass is $n = 1.5$, therefore with $n_{21} = 1.5$ the Brewster angle from (3.3.10) is $\theta_B = 56.3°$. In Fig. 3.4a, $n_{21} = 1.5$ and it is seen that since $R_\perp < 0$ its phase is 180°. Also, for $\theta_1 < \theta_B$, $R_\parallel < 0$ so its phase is 180°, while for $\theta_1 > \theta_B$ its phase is 0°. At normal incidence, from (3.3.1) and (3.3.3): $R_\perp = R_\parallel = -0.2$.

In Fig. 3.4b, $n_{21} = 1/1.5$ and the Brewster angle is $\theta_B = 33.7°$. It is seen that there is no phase shift in $R_\perp$ since $R_\perp > 0$. Also, $R_\parallel$ goes from no phase shift for $\theta_1 < \theta_B$ to a phase shift of 180° for $\theta_1 > \theta_B$. From (3.3.5), the critical angle is $\theta_c = 41.8°$. At the critical angle, $R_\perp = 1$ and $R_\parallel = -1$, with total internal reflection for $\theta_1 > \theta_c$. At normal incidence, from (3.3.1) and (3.3.3): $R_\perp = R_\parallel = 0.2$.

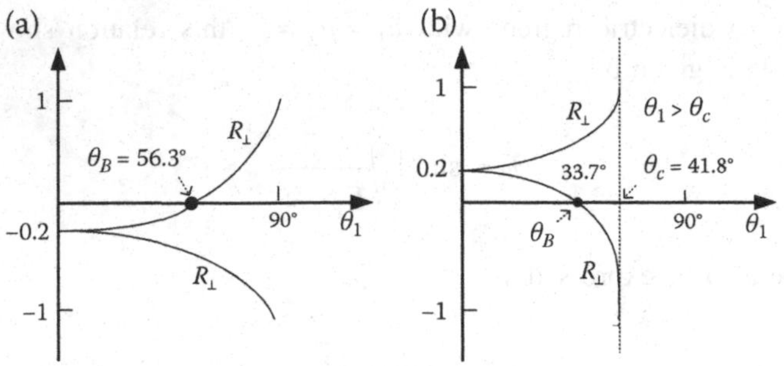

FIGURE 3.4 (a) An example of $R_\perp$ and $R_\parallel$ for $n_{21} = n_2/n_1 > 1$ with $n_{21} = 1.5$; (b) example of $R_\perp$ and $R_\parallel$ for $n_{21} = n_2/n_1 < 1$ with $n_{21} = 1/1.5$.

3.4 REFLECTION AND TRANSMISSION IN LOSSY REGIONS

3.4.1 Normal Incidence

First, we consider the case of normal incidence between two lossy regions, as shown in Fig. 3.5. At normal incidence $\theta_1 = 0$ and it follows from (3.2.17), (3.2.18), (3.2.33), and (3.2.34) that the reflection and transmission coefficients are

$$R_\perp = R_\parallel = \frac{\eta_2 - \eta_1}{\eta_2 + \eta_1}$$

and

$$T_\perp = T_\parallel = \frac{2\eta_2}{\eta_2 + \eta_1}$$

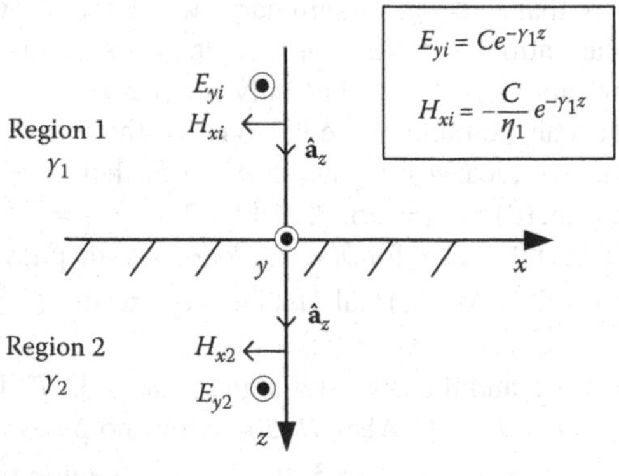

FIGURE 3.5 Plane wave normally incident between two lossy regions.

At normal incidence there is no difference between the perpendicular and parallel polarizations. In the analysis of normal incidence it is common to simply write the reflection and transmissions coefficients as R and T (instead of $R_\perp$, $R_\parallel$, $T_\perp$, and $T_\parallel$).

The fields in Regions 1 and 2 are

$$\begin{cases} \mathbf{E}_1 = C\left(e^{-\gamma_1 z} + Re^{\gamma_1 z}\right)\hat{\mathbf{a}}_y \\ \mathbf{H}_1 = -\dfrac{C}{\eta_1}\left(e^{-\gamma_1 z} - Re^{\gamma_1 z}\right)\hat{\mathbf{a}}_x \\ \mathbf{E}_2 = CTe^{-\gamma_2 z}\hat{\mathbf{a}}_y \\ \mathbf{H}_2 = -\dfrac{C}{\eta_2}Te^{-\gamma_2 z}\hat{\mathbf{a}}_x \end{cases} \tag{3.4.1}$$

where $\gamma_1 = \alpha_1 + j\beta_1$ and $\gamma_2 = \alpha_2 + j\beta_2$. The constant C can be real or complex.

The surface impedance is

$$Z_s = -\left.\frac{E_{y1}}{H_{x1}}\right|_{z=0} = -\left.\frac{E_{y2}}{H_{x2}}\right|_{z=0} = \eta_2$$

From (3.4.1), the time-average power density associated with the incident, reflected, and transmitted waves can be calculated. For example, for the incident wave:

$$E_{yi} = Ce^{-\alpha_1 z}e^{-j\beta_1 z}$$

and

$$H_{xi} = -\frac{C}{\eta_1}e^{-\alpha_1 z}e^{-j\beta_1 z}$$

The time-average power density for the incident wave is

$$\mathbf{P}_i = \frac{1}{2}\mathrm{Re}\left[E_{yi}\hat{\mathbf{a}}_y \times H_{x1}^*\hat{\mathbf{a}}_x\right] = \frac{|C|^2}{2|\eta_1|}e^{-2\alpha_1 z}\cos\theta_{\eta_1}\hat{\mathbf{a}}_z$$

where θ_{η_1} is the phase of η_1 (i.e., $\eta_1 = |\eta_1|e^{j\theta_{\eta_1}}$). In lossy regions, power is absorbed (i.e., dissipated as heat due to the conductivity). At $z = 0$ the incident power is equal to the reflected power plus the transmitted power. The transmitted power is dissipated as $z \to \infty$ due to the conductivity.

In the case that Region 1 is a dielectric and Region 2 is a perfect conductor (i.e., $\sigma_2 = \infty$ and $\eta_2 = 0$). Then, $R = -1$, $T = 0$, the fields in Region 2 are zero and in Region 1:

$$E_{y1} = C\left(e^{-j\beta_1 z} - e^{j\beta_1 z}\right) = -2jC \sin \beta_1 z$$

and

$$H_{x1} = -\frac{C}{\eta_1}\left(e^{-j\beta_1 z} + e^{j\beta_1 z}\right) = -\frac{2C}{\eta} \cos \beta_1 z$$

The tangential magnetic field at the surface of the conductor produces a surface current density, which can be calculated using (1.8.7).

Referring to Fig. 3.5, consider the case that Region 1 is a dielectric and Region 2 is a good conductor. For a good conductor: $\sigma_2 > > \omega\varepsilon_2$, and with and $\mu_2 = \mu_o$ the propagation constant in Region 2 is approximated by

$$\gamma_2 = \alpha_2 + j\beta_2 \approx \sqrt{j\omega\mu_o\sigma_2} = \sqrt{\omega\mu_o\sigma_2}\, e^{j\frac{\pi}{4}} = \sqrt{\frac{\omega\mu_o\sigma_2}{2}}\,(1 + j)$$

or

$$\alpha_2 = \beta_2 \approx \sqrt{\frac{\omega\mu_o\sigma_2}{2}}$$

The skin depth or depth of penetration δ in a good conductor is

$$\delta = \frac{1}{\alpha_2} = \sqrt{\frac{2}{\omega\mu_o\sigma_2}}$$

and the phase velocity is

$$v_p = \frac{\omega}{\beta_2} = \sqrt{\frac{2\omega}{\mu_o\sigma_2}}$$

The surface impedance of a good conductor at normal incidence is

$$Z_s = \eta_2 = \sqrt{\frac{j\omega\mu_o}{\sigma_2 + j\omega\varepsilon_o}} \approx \eta_o\sqrt{\frac{j\omega\varepsilon_o}{\sigma_2}} \approx \eta_o\sqrt{\frac{\omega\varepsilon_o}{\sigma_2}}\, e^{j\frac{\pi}{4}} \approx \eta_o\sqrt{\frac{\omega\varepsilon_o}{2\sigma_2}}\,(1 + j)$$

The fields in the good conductor are

$$E_{y2} = CTe^{-\alpha_2 z}e^{-j\beta_2 z}$$

and

$$H_{x2} = -\frac{CT}{\eta_2} e^{-\alpha_2 z} e^{-j\beta_2 z}$$

where

$$T = \frac{2\eta_2}{\eta_2 + \eta_o} \approx \frac{2\eta_2}{\eta_o} = \sqrt{\frac{2\omega\varepsilon_o}{\sigma_2}}\,(1 + j)$$

3.4.2 Oblique Incidence

Next, we analyze the case of oblique incidence from air to a lossy region for perpendicular polarization, as shown in Fig. 3.6. The electric field in Region 1 is given by

$$E_{y1} = C\left(e^{-j\beta_{z1}z} + R_{\perp} e^{j\beta_{z1}z}\right) e^{-j\beta_{x1}x}$$

where $\beta_o = \omega\sqrt{\mu_o \varepsilon_o}$, $\beta_{x1} = \beta_o \sin\theta_1$ and $\beta_{z1} = \beta_o \cos\theta_1$.

The transmitted electric field in the lossy region is of the form

$$E_{y2} = CT_{\perp} e^{-\gamma_{z2}z} e^{-\gamma_{x2}x} = CT_{\perp} e^{-(\alpha_{x2}x + \alpha_{z2}z)} e^{-j(\beta_{x2}x + \beta_{z2}z)} \quad (3.4.2)$$

where the constant amplitude planes of the transmitted field are given by

$$\alpha_{x2}x + \alpha_{z2}z = d_1$$

and the constant phase planes by

$$\beta_{x2}x + \beta_{z2}z = d_2$$

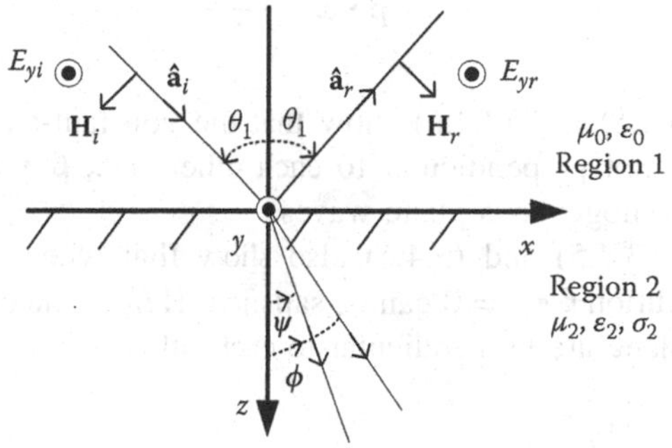

FIGURE 3.6 A perpendicular polarized plane wave at oblique incidence.

where d_1 and d_2 are constants.

The constant amplitude planes refract at an angle given by

$$\tan \psi = \frac{\alpha_{x2}}{\alpha_{z2}} \tag{3.4.3}$$

and the constant phase planes at the angle

$$\tan \phi = \frac{\beta_{x2}}{\beta_{z2}} \tag{3.4.4}$$

Figure 3.6 shows that the planes of constant amplitude and phase are not perpendicular to each other. These types of waves are also known as inhomogeneous (or complex) plane waves. The power flow is in a direction orthogonal to the constant phase planes.

In general, the waves in lossy media are of the form

$$e^{-\boldsymbol{\alpha}\bullet\mathbf{r}}e^{-j\boldsymbol{\beta}\bullet\mathbf{r}}$$

The constant amplitude plane is given by $\boldsymbol{\alpha} \bullet \mathbf{r} = d_1$ and the constant phase plane by $\boldsymbol{\beta} \bullet \mathbf{r} = d_2$, where d_1 and d_2 are constants.

Since

$$\boldsymbol{\gamma} \bullet \boldsymbol{\gamma} = \gamma^2 = (\boldsymbol{\alpha} + j\boldsymbol{\beta}) \bullet (\boldsymbol{\alpha} + j\boldsymbol{\beta}) = -\omega^2\mu\varepsilon + j\omega\mu\sigma$$

it follows that

$$\boldsymbol{\alpha} \bullet \boldsymbol{\alpha} - \boldsymbol{\beta} \bullet \boldsymbol{\beta} = -\omega^2\mu\varepsilon \Rightarrow |\boldsymbol{\beta}|^2 - |\boldsymbol{\alpha}|^2 = \omega^2\mu\varepsilon \tag{3.4.5}$$

and

$$\boldsymbol{\beta} \bullet \boldsymbol{\alpha} = \frac{\omega\mu\sigma}{2} \tag{3.4.6}$$

If $\sigma \neq 0$, then (3.4.5) and (3.4.6) show that the constant-amplitude and constant-phase planes are not perpendicular to each other (i.e., $\boldsymbol{\beta} \bullet \boldsymbol{\alpha} \neq 0$), which are the properties of inhomogeneous plane waves.

The relations (3.4.5) and (3.4.6) also show that when $\sigma = 0$ if $\boldsymbol{\beta}$ and $\boldsymbol{\alpha}$ are complex the condition $\boldsymbol{\beta} \bullet \boldsymbol{\alpha} = 0$ can be satisfied. Hence, the constant amplitude and constant phase plane are perpendicular to each other.

3.5 OTHER TYPES OF PLANE WAVES

Different combinations of propagation constants result in a variety of plane-wave types. Some wave types are the Zenneck wave, the lateral waves, and the trapped wave. Thus far, we have considered plane waves in lossless and lossy media, where a typical outgoing plane wave in terms of the propagation constant $\mathbf{k} = k_x\hat{\mathbf{a}}_x + k_z\hat{\mathbf{a}}_z$ is of the form $e^{-jk_x x}e^{-jk_z z}$. In lossless media, the propagation constants k_x and k_z have positive real values, while in lossy media $k_x = k_x' - jk_x''$ and $k_z = k_z' - jk_z''$ have imaginary parts that account for the attenuation of the waves.

3.5.1 The Zenneck Wave

Consider the geometry shown in Fig. 3.3 where Region 1 is air ($n_1 = 1$) and Region 2 is lossy with $\mu_2 = \mu_o$. From (3.2.31), with $\mu_1 = \mu_2 = \mu_o$:

$$R_\| = \frac{k_{z2} - n_2^2 k_{z1}}{k_{z2} + n_2^2 k_{z1}}$$

where

$$k_{z1} = k_o \cos\theta_1$$

and

$$k_{z2} = k_o n_2 \cos\theta_2$$

This relation shows that since n_2 is complex, $R_\|$ cannot be zero with real values of k_{z2} and k_{z1}. However, $R_\|$ can be zero if k_{z2} and k_{z1} are complex. These are the conditions for a Zenneck wave solution, also known as a Zenneck surface wave solution. The Zenneck surface wave solution is a plane wave solution that in the lossless region (i.e., air) is satisfied with a complex propagation constant.

The reflection coefficient $R_\|$ is zero when

$$k_{z2} = k_{z1}n_2^2 \Rightarrow \cos\theta_2 = n_2 \cos\theta_1 \tag{3.5.1}$$

which for the Zenneck wave solution is satisfied by a complex angle of incidence and transmission. A complex angle of incidence is equivalent to having complex values of the k_{x1} and k_{z1} in the incident wave, where $k_{x1} = k_o \sin\theta_1$. However, even with complex values of the propagation constants, since Region 1 is lossless the planes of constant phase and constant amplitudes must be orthogonal in Region 1.

Since $R_\| = 0$, the magnetic fields in Regions 1 and 2 are of the form

$$H_{y1} = Ce^{-jk_{z1}z}e^{-jk_x x} = Ce^{-(k_{z1}''z + k_x''x)}e^{-j(k_{z1}'z + k_x'x)} \qquad (z \le 0) \tag{3.5.2}$$

and

$$H_{y2} = Ce^{-jk_{z2}z}e^{-jk_x x} = Ce^{-(k_{z2}''z+k_x''x)}e^{-j\left(k_{z2}'z+k_x'x\right)} \qquad (z \geq 0) \qquad (3.5.3)$$

where the x-dependence is the same in both regions (i.e., denoted by $k_x = k_{x1} = k_{x2}$). Also, (3.5.2) and (3.5.3) satisfy $H_{y1} = H_{y2}$ at $z = 0$, and observe that z decreases in Region 1 and increases in Region 2.

The propagation constants in Regions 1 and 2 satisfy:

$$k_1^2 = k_o^2 = k_x^2 + k_{z1}^2 \qquad (3.5.4)$$

and

$$k_2^2 = k_o^2 n_2^2 = k_x^2 + k_{z2}^2 \qquad (3.5.5)$$

Expressions for k_x, k_{z1} and k_{z2} are obtained from (3.5.1), (3.5.4), and (3.5.5). Substituting (3.5.1) into (3.5.5) gives

$$k_o^2 n_2^2 = k_x^2 + n_2^4 k_{z1}^2 \qquad (3.5.6)$$

and substituting (3.5.4) into (3.5.6) gives

$$k_{z1}^2 = \frac{k_o^2}{1 + n_2^2} \Rightarrow k_{z1} = k_{z1}' - jk_{z1}'' = \frac{k_o}{\sqrt{1 + n_2^2}} \qquad (3.5.7)$$

From (3.5.1) and (3.5.7):

$$k_{z2} = k_{z2}' - jk_{z2}'' = \frac{k_o n_2^2}{\sqrt{1 + n_2^2}} \qquad (3.5.8)$$

and from (3.5.5) and (3.5.8):

$$k_x = k_x' - jk_x'' = \frac{k_o n_2}{\sqrt{1 + n_2^2}} \qquad (3.5.9)$$

The square root branch (i.e., the sign) in (3.5.7) to (3.5.9) are selected so that the fields in (3.5.2) and (3.5.3) propagate in the x direction and do not grow as $z \to \infty$ and $x \to \infty$. These conditions can produce a wave that is "tied" to the interface if the attenuation in the z direction is much larger than the attenuation in the x direction. Thus, $k_x'' > 0$, $k_{z1}'' < 0$, and $k_{z2}'' > 0$.

Inserting numerical values in (3.5.7) to (3.5.9) will show that in some cases the above conditions are satisfied. Hence, a wave that attenuates along the x and z axis, but the attenuation along the x axis is much smaller that the attenuation along the z axis is possible. This type of wave is tied to the interface, and occurs when $\sigma_2/\omega\varepsilon_2 >> 1$. For example, consider an air-water interface at $f = 100$ MHz where $\varepsilon_{r2} = 80$ and $\sigma_2 = 4$ S/m. Then, selecting the positive sign of the square root in (3.5.7) to (3.5.9):

$$n_2^2 = 80 - j\frac{4}{2\pi(100 \times 10^6)8.854 \times 10^{-12}} = 80 - j719.02$$

$$\sqrt{1 + n_2^2} = \sqrt{1 + 80 - j719.02} = 20.057 - j17.925$$

$$k_o = \frac{\omega}{c} = \frac{2\pi(100 \times 10^6)}{3 \times 10^8} = 2.0944$$

$$k_2 = k_o n_2 = 41.978 - j37.567$$

$$k_x = k_x' - jk_x'' = 2.0942 - j0.00144$$
$$k_{z1} = k_{z1}' - jk_{z1}'' = 0.0581 + j0.0519$$
$$k_{z2} = k_{z2}' - jk_{z2}'' = 41.949 - j37.593$$

which show that the attenuation in the z direction (i.e., $k_{z1}'' < 0$ and $k_{z2}'' > 0$) is much larger than in the x direction (i.e., $k_z'' > 0$). Such waves are known as Zenneck surface waves.

In the air region the constant phase planes should be perpendicular to the constant amplitude planes. This can be seen when (3.4.6) is applied to this case, which in the air region in terms of $\mathbf{k}$ is

$$\mathbf{k}' \bullet \mathbf{k}'' = 0 \Rightarrow k_x'k_x'' + k_{z1}'k_{z1}'' = 0 \qquad (3.5.10)$$

Substituting the above numerical values into (3.5.10) shows that it is satisfied. Similarly, it follows that in Region 2 these planes are not orthogonal. In Region 2, the power flow is in a direction orthogonal to the planes of constant phase.

3.5.2 The Lateral Wave

A lateral wave occurs when a plane wave is incident at the critical angle between two regions where $n_1 > n_2$. For example, Region 1 can be water and Region 2 air. Another example is when both regions are lossless such that $\varepsilon_1 > \varepsilon_2$, as discussed below.

Consider the perpendicular polarized plane wave in Fig. 3.7 incident at the critical angle $\theta_1 = \theta_c$ where Region 1 is air (i.e., $\varepsilon_1 = \varepsilon_o$) and in Region 2 represents

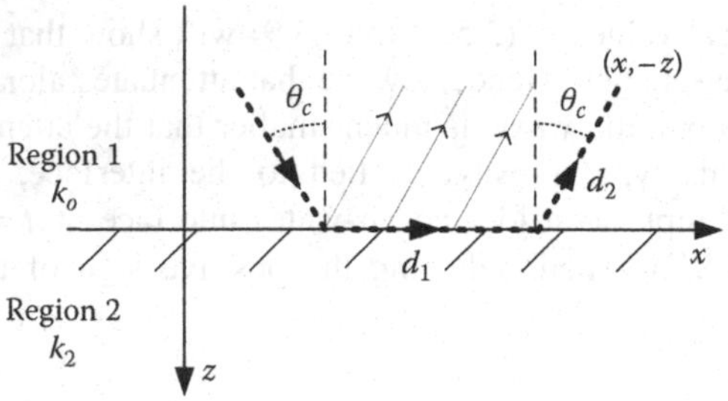

FIGURE 3.7 A lateral wave.

the Ionosphere with index of refraction is $n_2 = \sqrt{\varepsilon_{r2}} < 1$ (see Section 9.9). Since $\varepsilon_1 > \varepsilon_2$, the condition for a critical angle to occur is satisfied. At the critical angle there is total reflection (i.e., $R_\perp = 1$) and from Snell's law: $\sin\theta_c = n_2$ and $\theta_2 = \pi/2$. The reflected field is

$$E_{yr} = Ce^{jk_{z1}z}e^{-jk_{x1}x} = Ce^{jk_oz\cos\theta_c}e^{-jk_ox\sin\theta_c} \tag{3.5.11}$$

Referring to the geometry shown in Fig. 3.7, let

$$x = d_1 + d_2\sin\theta_c$$

and

$$-z = d_2\cos\theta_c$$

Then, the phase of (3.5.11) reads

$$\begin{aligned} jk_{z1}z - jk_{x1}x &= j[k_o\cos\theta_c(-d_2\cos\theta_c) - k_o\sin\theta_c(d_1 + d_2\sin\theta_c)] \\ &= -j(k_od_1\sin\theta_c + k_od_2) = -j(k_on_2d_1 + k_od_2) = -j(k_2d_1 + k_od_2) \end{aligned} \tag{3.5.12}$$

The first term in (3.5.12) shows the phase change as the wave propagates a distance d_1 in Region 2, and the second term is the phase change as the wave propagates a distance d_2 in Region 1. Hence, an incident wave at the critical angle propagates along the surface with propagation constant k_2. As the wave propagates along the surface it attenuates and radiates energy in the direction of the critical angle. This type of wave is known as a "lateral wave."

With $\theta_1 = \theta_c$, $\theta_2 = \pi/2$, $k_{z2} = k_on_2\cos\theta_2 = 0$, $k_{x1} = k_{x2} = k_o\sin\theta_c = k_on_2$ and $T_\perp = 2$. The phase of the transmitted electric fields vary as

$$-jk_{x2}x = -jk_on_2x = -jk_2x$$

which describes the lateral propagation of the fields in Region 2.

3.5.3 The Trapped Wave

The next case is that of a plane wave traveling along an interface such that the fields decay exponentially away from the surface. An example of a trapped wave is a plane-wave traveling in the x direction along an air-dielectric interface. Region 1 (i.e., $z \le 0$) represents the air region, and Region 2 (i.e., $z \ge 0$) the dielectric region. Let the magnetic fields be given by

$$H_{y1} = Ce^{jk_{z1}z}e^{-jk_x x} \qquad (z \le 0) \tag{3.5.13}$$

and

$$H_{y2} = Ce^{-jk_{z2}z}e^{-jk_x x} \qquad (z \ge 0) \tag{3.5.14}$$

where

$$k_1^2 = k_o^2 = k_x^2 + k_{z1}^2 \tag{3.5.15}$$

and

$$k_2^2 = k_o^2 n_2^2 = k_x^2 + k_{z2}^2 \tag{3.5.16}$$

The magnetic fields in (3.5.13) and (3.5.14) satisfy the boundary condition at $z = 0$. The tangential electric fields follow from

$$E_x = -\frac{1}{j\omega\varepsilon}\frac{\partial H_y}{\partial z}$$

Therefore,

$$E_{x1} = -C\frac{k_{z1}}{\omega\varepsilon_o}e^{jk_{z1}z}e^{-jk_x x} \qquad (z \le 0) \tag{3.5.17}$$

and

$$E_{x2} = C\frac{k_{z2}}{\omega\varepsilon_2}e^{-jk_{z2}z}e^{-jk_x x} \qquad (z \ge 0) \tag{3.5.18}$$

Matching the tangential electric fields at $z = 0$, it follows from (3.5.17) and (3.5.18) that

$$\frac{k_{z1}}{\varepsilon_o} = -\frac{k_{z2}}{\varepsilon_2} \Rightarrow n_2^2 k_{z1} = -k_{z2} \tag{3.5.19}$$

From (3.5.15), (3.5.16), and (3.5.19), the expressions for k_x, k_{z1} and k_{z2} are obtained. The derivation is similar to that leading to (3.5.7) to (3.5.9), and it follows that

$$k_{z1} = \frac{k_o}{\sqrt{1 + n_2^2}} = \frac{k_o}{\sqrt{1 + \varepsilon_{r2}}} \tag{3.5.20}$$

$$k_{z2} = -\frac{k_o n_2^2}{\sqrt{1 + n_2^2}} = -\frac{k_o \varepsilon_{r2}}{\sqrt{1 + \varepsilon_{r2}}} \tag{3.5.21}$$

and

$$k_x = \frac{k_o n_2}{\sqrt{1 + n_2^2}} = \frac{k_o \sqrt{\varepsilon_{r2}}}{\sqrt{1 + \varepsilon_{r2}}} \tag{3.5.22}$$

One type of a trapped wave occurs if ε_{r2} has a real-negative value such that $|\varepsilon_{r2}| > 1$. Then, the square root in the above relations is imaginary. Selecting the positive branch of the square root we obtain

$$k_{z1} = -j\frac{k_o}{\sqrt{|\varepsilon_{r2}| - 1}} = -ju_1 \qquad (u_1 > 0)$$

$$k_{z2} = -j\frac{k_o|\varepsilon_{r2}|}{\sqrt{|\varepsilon_{r2}| - 1}} = -ju_2 \qquad (u_2 > 0)$$

and

$$k_x = \frac{k_o\sqrt{|\varepsilon_{r2}|}}{\sqrt{|\varepsilon_{r2}| - 1}}$$

which show that k_{z1} and k_{z2} are imaginary numbers and k_x is real. Hence, (3.5.13) and (3.5.14) can be expressed in the form:

$$H_{y1} = Ce^{u_1 z}e^{-jk_x x} \qquad (z \leq 0)$$

and

$$H_{y2} = Ce^{-u_2 z}e^{-jk_x x} \qquad (z \geq 0)$$

These fields represent a wave that is traveling along the boundary and is attenuated above and below the interface. It is referred to as a trapped surface wave. The selection of the negative imaginary root (i.e., $-j$ branch) would have resulted in an improper solution since the wave would have increased in amplitude above and below the interface. A real-negative ε_{r2} can represent the dielectric constant of an ionized gas when the frequency is below the plasma frequency and there is negligible loss.

In Chapter 4, we will see cases when a trapped wave occurs with positive values of the dielectric constant. Such cases occur in planar and cylindrical air-dielectric interfaces.

It is of interest to observe that while (3.5.1) was derived from the condition $R_{\|} = 0$, the condition in (3.5.19) corresponds to $R_{\|} = \infty$. In terms of incident and reflected waves, an infinite reflection coefficient can be interpreted as having a reflected field in the absence of an incident field (since $H_r = R_{\|}H_i$ can be finite).

3.5.4 Other Types of Surface Waves

There are other types of surface waves, such as the "leaky wave" and the "surface plasmon." An example of a leaky wave is a wave propagating in a rectangular waveguide that has a slot in its upper face. As the wave propagate along the waveguide energy is radiated by the slot. The wave decays in the direction of propagation and radiates in a direction perpendicular to the slots.

Surface plasmons occur under certain conditions in a metal-dielectric interface when the dielectric constant of the metal is complex and its real part is a negative number. The term "surface plasmon" refers to the electromagnetic excitation and response of the surface electrons (i.e., a plasma) in the metal, which results in a surface wave. The study of surface plasmons is of interest to those working at optical frequencies. They are used to analyze the surface properties of metals. At optical frequencies, metals have relative dielectric constants with negative real parts. For example, for silver: $\varepsilon_{cr,2} = \varepsilon_{cr,2}' + j\varepsilon_{cr,z}'' \approx -17 - j0.5$ at $f \approx 500$ THz

The analysis of surface plasmons is very similar to the trapped-wave analysis in Section 3.5.3. The details are left to the problems.

3.6 REFLECTION AND TRANSMISSION FROM INTERFACES

We first consider the reflection and transmission of a plane wave normally incident on the three-region configuration shown in Fig. 3.8. As previously discussed, for normal incidence the reflection and transmission coefficients can simply be written as R and T, since they are the same for both polarizations. The incident fields in Region 1 ($z \le 0$) are given by

$$E_{y1} = C\left(e^{-\gamma_1 z} + Re^{\gamma_1 z}\right) \tag{3.6.1}$$

and

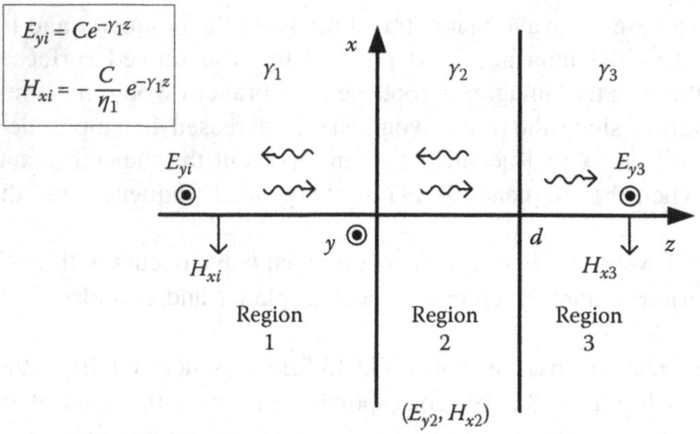

FIGURE 3.8 Plane wave normally incident on a three-region configuration.

$$H_{x1} = -\frac{C}{\eta_1}\left(e^{-\gamma_1 z} - Re^{\gamma_1 z}\right) \tag{3.6.2}$$

In Region 2 ($0 \le z \le d$) the fields are of the form

$$E_{y2} = C\left(fe^{-\gamma_2 z} + ge^{\gamma_2 z}\right) \tag{3.6.3}$$

and

$$H_{x2} = -\frac{C}{\eta_2}\left(fe^{-\gamma_2 z} - ge^{\gamma_2 z}\right) \tag{3.6.4}$$

and in Region 3 ($z \ge d$) the transmitted fields are

$$E_{y3} = CTe^{-\gamma_3 z} \tag{3.6.5}$$

and

$$H_{x3} = -\frac{CT}{\eta_3}e^{-\gamma_3 z} \tag{3.6.6}$$

Using (3.6.1) and (3.6.2), the surface impedance at z=0 is

$$Z_s = -\left.\frac{E_{y1}}{H_{x1}}\right|_{z=0} = \eta_1\frac{1+R}{1-R} \tag{3.6.7}$$

or

$$R = \frac{Z_s - \eta_1}{Z_s + \eta_1} \tag{3.6.8}$$

which shows that R is determined by Z_s. To determine Z_s, from (3.6.3) and (3.6.4) at $z = 0$:

$$Z_s = -\frac{E_{y2}}{H_{x2}}\bigg|_{z=0} = -\frac{E_{y1}}{H_{x1}}\bigg|_{z=0} = \eta_2\frac{1 + \frac{g}{f}}{1 - \frac{g}{f}} \tag{3.6.9}$$

where g/f in (3.6.9) is evaluated using the boundary conditions at $z = d$, namely

$$E_{y2} = E_{y3}\bigg|_{z=d} \Rightarrow fe^{-\gamma_2 d} + ge^{\gamma_2 d} = Te^{-\gamma_3 d} \tag{3.6.10}$$

and

$$H_{x2} = H_{x3}|_{z=d} \Rightarrow \frac{1}{\eta_2}\left(fe^{-\gamma_2 d} - ge^{\gamma_2 d}\right) = \frac{1}{\eta_3}Te^{-\gamma_3 d} \tag{3.6.11}$$

From (3.6.10) and (3.6.11) it follows that

$$\frac{g}{f} = \frac{\eta_3 - \eta_2}{\eta_3 + \eta_2}e^{-2\gamma_2 d} = R_2 e^{-2\gamma_2 d} \tag{3.6.12}$$

and

$$\frac{T}{f} = \frac{2\eta_3}{\eta_3 + \eta_2}e^{-\gamma_2 d}e^{\gamma_3 d} \tag{3.6.13}$$

where

$$R_2 = \frac{\eta_3 - \eta_2}{\eta_3 + \eta_2} \tag{3.6.14}$$

Substituting (3.6.12) into (3.6.9), and using (3.6.14), results in

$$Z_s = \eta_2\frac{1 + R_2 e^{-2\gamma_2 d}}{1 - R_2 e^{-2\gamma_2 d}} = \eta_2\frac{\eta_3 + \eta_2 \tanh \gamma_2 d}{\eta_2 + \eta_3 \tanh \gamma_2 d} \tag{3.6.15}$$

which determines the reflection coefficient in (3.6.8).

To determine the transmission coefficient, the boundary conditions at $z = 0$ are used. From (3.6.1) to (3.6.4) and (3.6.12), we obtain

$$E_{y1} = E_{y2}\Big|_{z=0} \Rightarrow 1 + R = f\left(1 + R_2 e^{-2\gamma_2 d}\right)$$

and

$$H_{x1} = H_{x2}|_{z=0} \Rightarrow \frac{1}{\eta_1}(1 - R) = \frac{f}{\eta_2}\left(1 - R_2 e^{-2\gamma_2 d}\right)$$

Solving for f gives

$$f = \left(\frac{2\eta_2}{\eta_2 + \eta_1}\right)\frac{1}{1 - R_1 R_2 e^{-2\gamma_2 d}} \tag{3.6.16}$$

where

$$R_1 = \frac{\eta_1 - \eta_2}{\eta_1 + \eta_2} \tag{3.6.17}$$

Substituting (3.6.16) into (3.6.13) gives

$$T = \frac{\left(\frac{2\eta_2}{\eta_2 + \eta_1}\right)\left(\frac{2\eta_3}{\eta_3 + \eta_2}\right)e^{-\gamma_2 d}e^{\gamma_3 d}}{1 - R_1 R_2 e^{-2\gamma_2 d}} \tag{3.6.18}$$

The transmission coefficient is interpreted as follows. The first term in parenthesis represent a transmission factor as the wave passes the boundary at $z = 0$, the second term in parenthesis is a transmission factor as the wave passes the boundary at $z = d$. The propagation from $z = 0$ to $z = d$ is represented by $e^{-\gamma_2 d}$, and the factor $e^{\gamma_3 d}$ shows that the wave propagates in Region 3 according to $e^{-\gamma_3(z-d)}$. The denominator is interpreted as the multiple reflections that occur in Region 2. If $\text{Re}(2\gamma_2 d) >> 1$ the denominator is approximate 1, which represents the case when internal reflections can be neglected. If such approximation cannot be made, then the denominator, using the binomial expansion, shows that the term $R_1 R_2$ accounts for the multiple reflections of the wave in Region 2. The reflection coefficient R_2 (see (3.6.14) represents the reflection from $z = d$ for a wave traveling in the positive z direction, and R_1 (see (3.6.17) is the reflection from $z = 0$ for a wave traveling in the negative z direction.

The previous results can also be obtained using a "transmission matrix" approach. To this end, we write (3.6.3) and (3.6.4) at $z = 0$ as

$$E_{y2}(0) = C(f + g) \tag{3.6.19}$$

$$H_{x2}(0) = -\frac{C}{\eta_2}(f - g) \tag{3.6.20}$$

and at $z = d$:

$$E_{y2}(d) = C\left(f\,e^{-\gamma_2 d} + g\,e^{\gamma_2 d}\right) \tag{3.6.21}$$

$$H_{x2}(d) = -\frac{C}{\eta_2}\left(f\,e^{-\gamma_2 d} - g\,e^{\gamma_2 d}\right) \tag{3.6.22}$$

Adding and subtracting (3.6.21) and (3.6.22) gives

$$f = \frac{e^{\gamma_2 d}}{2C}[E_{y2}(d) - \eta_2 H_{x2}(d)] \tag{3.6.23}$$

$$g = \frac{e^{-\gamma_2 d}}{2C}[E_{y2}(d) + \eta_2 H_{x2}(d)] \tag{3.6.24}$$

Substituting (3.6.23) and (3.6.24) into (3.6.19) and (3.6.20):

$$E_{y2}(0) = E_{y2}(d)\cosh\gamma_2 d - H_{x2}(d)\eta_2 \sinh\gamma_2 d \tag{3.6.25}$$

$$H_{x2}(0) = -E_{y2}(d)\frac{\sinh\gamma_2 d}{\eta_2} + H_{x2}(d)\cosh\gamma_2 d \tag{3.6.26}$$

which can be written in matrix form, known as the transmission matrix, as

$$\begin{bmatrix} E_{y2}(0) \\ H_{x2}(0) \end{bmatrix} = \begin{bmatrix} \cosh\gamma_2 d & -\eta_2 \sinh\gamma_2 d \\ -\frac{\sinh\gamma_2 d}{\eta_2} & \cosh\gamma_2 d \end{bmatrix} \begin{bmatrix} E_{y2}(d) \\ H_{x2}(d) \end{bmatrix} \tag{3.6.27}$$

The transmission matrix relates the input and output fields in a region
The surface impedance at $z = 0$ follows from (3.6.25) and (3.6.26). That is,

$$Z_s = -\frac{E_{y2}(0)}{H_{x2}(0)} = \frac{E_{y2}(d)\cosh\gamma_2 d - H_{x2}(d)\eta_2 \sinh\gamma_2 d}{E_{y2}(d)\frac{\sinh\gamma_2 d}{\eta_2} - H_{x2}(d)\cosh\gamma_2 d} \tag{3.6.28}$$

From (3.6.5) and (3.6.6), the ratio of $E_{y3}(d)$ to $H_{x3}(d)$ (which is the impedance $Z(d)$) is

$$Z(d) = -\frac{E_{y3}(d)}{H_{x3}(d)} = -\frac{E_{y2}(d)}{H_{x2}(d)} = \eta_3 \tag{3.6.29}$$

Substituting (3.6.29) into (3.6.28) gives

$$Z_s = \eta_2 \frac{\eta_3 + \eta_2 \tanh \gamma_2 d}{\eta_2 + \eta_3 \tanh \gamma_2 d}$$

which is identical to (3.6.15). The reflection coefficient R follows from (3.6.8).

The inverse matrix of (3.6.27) relates the fields at $z = d$ to those at $z = 0$. That is,

$$\begin{bmatrix} E_{y2}(d) \\ H_{x2}(d) \end{bmatrix} = \begin{bmatrix} \cosh \gamma_2 d & \eta_2 \sinh \gamma_2 d \\ \frac{\sinh \gamma_2 d}{\eta_2} & \cosh \gamma_2 d \end{bmatrix} \begin{bmatrix} E_{y2}(0) \\ H_{x2}(0) \end{bmatrix} \tag{3.6.30}$$

The transmission coefficient was derived in (3.6.18). In terms of the transmission matrix, the transmission coefficient can be calculated by letting $E_{y2}(d) = E_{y3}(d) = 1$. Then, from (3.6.29): $H_{x2}(d) = -1/\eta_3$ and from (3.6.27) the values $E_{y2}(0)$ and $H_{x2}(0)$ are found. Since

$$E_{y2}(0) = E_{y1}(0) = C(1 + R)$$
$$H_{x2}(0) = H_{x1}(0) = -\frac{C}{\eta_1}(1 - R)$$

it follows that

$$C = \frac{1}{2}[E_{y2}(0) - \eta_1 H_{x2}(0)]$$

and from (3.6.5) the transmission coefficient is given by

$$T = \frac{1}{C} e^{\gamma_3 d}$$

In the case of a lossless region, $\gamma_2 = j\beta_2$, $\cosh(j\beta_2 d) = \cos \beta_2 d$, $\sinh(j\beta_2 d) = j \sin \beta_2 d$ and (3.6.15), (3.6.27), and (3.6.30) change accordingly. For example, (3.6.15) reads

$$Z_s = \eta_2 \frac{\eta_3 + j\eta_2 \tan \gamma_2 d}{\eta_2 + j\eta_3 \tan \gamma_2 d}$$

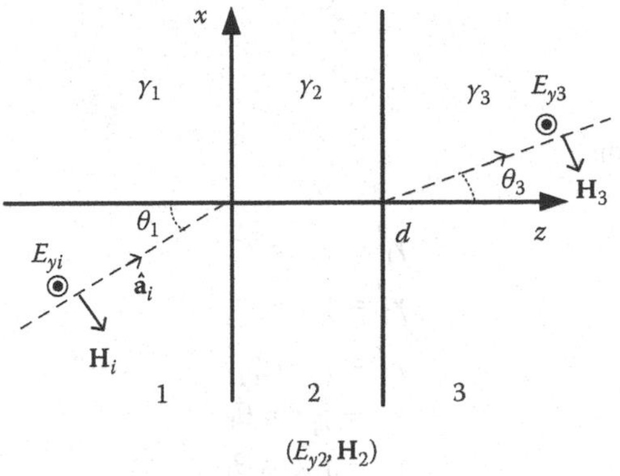

FIGURE 3.9 A perpendicular polarized wave incident on a three-layer region. The incident fields in Region 1 are E_{yi} and $\mathbf{H}_i$.

The case of a perpendicular polarized wave incident at an angle θ_1 to a three-region configuration is illustrated in Fig. 3.9. In this case the appropriate tangential fields in Region 1 ($z \le 0$) are

$$E_{y1} = C\left(e^{-\gamma_{z1}z} + R_\perp e^{\gamma_{z1}z}\right)e^{-\gamma_{x1}x}$$

and

$$H_{x1} = -\frac{C}{\eta_1^\perp}\left(e^{-\gamma_{z1}z} - R_\perp e^{\gamma_{z1}z}\right)e^{-\gamma_{x1}x}$$

In Region 2 ($0 \le z \le d$) the fields are of the form

$$E_{y2} = C\left(f e^{-\gamma_{z2}z} + g e^{\gamma_{z2}z}\right)e^{-\gamma_{x1}x}$$

and

$$H_{x2} = -\frac{C}{\eta_2^\perp}\left(f e^{-\gamma_{z2}z} - g e^{\gamma_{z2}z}\right)e^{-\gamma_{x1}x}$$

In Region 3 ($z \ge d$) the transmitted fields are

$$E_{y3} = CT_\perp e^{-\gamma_{z3}z}e^{-\gamma_{x1}x}$$

and

$$H_{x3} = -\frac{CT_\perp}{\eta_3^\perp} e^{-\gamma_{z3}z} e^{-\gamma_{x1}x}$$

In the above relations:

$$\gamma_1^2 = \gamma_{x1}^2 + \gamma_{z1}^2$$
$$\gamma_2^2 = \gamma_{x1}^2 + \gamma_{z2}^2$$
$$\gamma_3^2 = \gamma_{x1}^2 + \gamma_{z3}^2$$
$$\gamma_{x1} = \gamma_1 \sin\theta_1$$
$$\gamma_{z1} = \gamma_1 \cos\theta_1$$

$$\gamma_{z2} = \gamma_2 \cos\theta_2$$
$$\gamma_{z3} = \gamma_2 \cos\theta_3$$

$$\eta_1^\perp = \frac{j\omega\mu_1}{\gamma_{z1}} = \frac{j\omega\mu_1}{\gamma_1 \cos\theta_1} = \frac{\eta_1}{\cos\theta_1}$$

$$\eta_2^\perp = \frac{j\omega\mu_2}{\gamma_{z2}} = \frac{j\omega\mu_2}{\gamma_2 \cos\theta_2} = \frac{\eta_2}{\cos\theta_2}$$

$$\eta_3^\perp = \frac{j\omega\mu_3}{\gamma_{z3}} = \frac{j\omega\mu_3}{\gamma_3 \cos\theta_3} = \frac{\eta_3}{\cos\theta_3}$$

where from Snell law, the angles of refractions in Region 2 (i.e., θ_2) and Region 3 (i.e., θ_3) satisfy

$$\gamma_2 \sin\theta_2 = \gamma_1 \sin\theta_1$$

and

$$\gamma_3 \sin\theta_3 = \gamma_2 \sin\theta_2$$

Following the same steps that lead to (3.6.27), or simply changing η_2 to $\eta_2^\perp$ and γ_2 to γ_{z2} in (3.6.27) gives

$$\begin{bmatrix} E_{y2}(x, 0) \\ H_{x2}(x, 0) \end{bmatrix} = \begin{bmatrix} \cosh\gamma_{z2}d & -\eta_2^\perp \sinh\gamma_{z2}d \\ -\frac{\sinh\gamma_{z2}d}{\eta_2^\perp} & \cosh\gamma_{z2}d \end{bmatrix} \begin{bmatrix} E_{y2}(x, d) \\ H_{x2}(x, d) \end{bmatrix} \tag{3.6.31}$$

For parallel polarization, the appropriate tangential fields in Region 1 ($z \le 0$) are

$$H_{y1} = C\left(e^{-\gamma_{z1}z} + R_{\|}^{H} e^{\gamma_{z1}z}\right) e^{-\gamma_{x1}x}$$

and

$$E_{x1} = C\eta_1^{\|}\left(e^{-\gamma_{z1}z} - R_{\|}^{H} e^{\gamma_{z1}z}\right) e^{-\gamma_{x1}x}$$

In Region 2 ($0 \le z \le d$) the fields are of the form

$$H_{y2} = C\left(f e^{-\gamma_{z2}z} + g e^{\gamma_{z2}z}\right) e^{-\gamma_{x1}x}$$

and

$$E_{x2} = C\eta_2^{\|}\left(f e^{-\gamma_{z2}z} - g e^{\gamma_{z2}z}\right) e^{-\gamma_{x1}x}$$

In Region 3 ($z \ge d$) the transmitted fields are

$$H_{y3} = CT_{\|}^{H} e^{-\gamma_{z3}z} e^{-\gamma_{x1}x}$$

and

$$E_{x3} = C\eta_3^{\|} T_{\|}^{H} e^{-\gamma_{z3}z} e^{-\gamma_{x1}x}$$

where

$$\eta_1^{\|} = \frac{\gamma_{z1}}{\sigma_1 + j\omega\mu_1} = \frac{\gamma_1 \cos\theta_1}{\sigma_1 + j\omega\mu_1} = \eta_1 \cos\theta_1$$

$$\eta_2^{\|} = \frac{\gamma_{z2}}{\sigma_2 + j\omega\mu_2} = \eta_2 \cos\theta_2$$

$$\eta_3^{\|} = \frac{\gamma_{z3}}{\sigma_3 + j\omega\mu_3} = \eta_3 \cos\theta_3$$

The transmission matrix in this case is

$$\begin{bmatrix} H_{y2}(x,0) \\ E_{x2}(x,0) \end{bmatrix} = \begin{bmatrix} \cosh\gamma_{z2}d & \frac{1}{\eta_2^{\|}}\sinh\gamma_{z2}d \\ \eta_2^{\|}\sinh\gamma_{z2}d & \cosh\gamma_{z2}d \end{bmatrix} \begin{bmatrix} H_{y2}(x,d) \\ E_{x2}(x,d) \end{bmatrix} \tag{3.6.32}$$

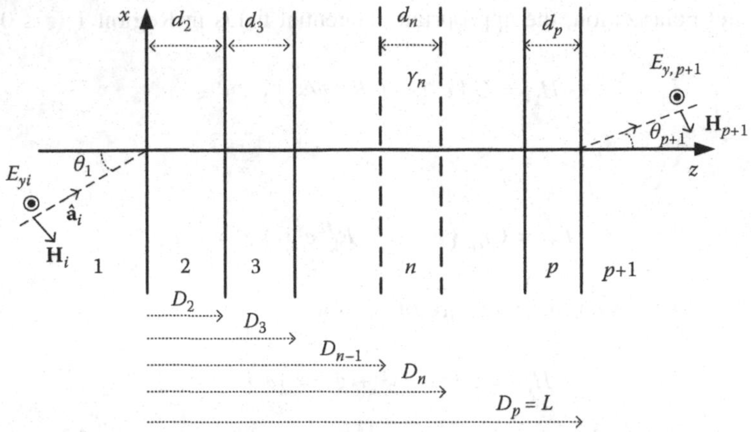

FIGURE 3.10 Model for the calculation of the transmission matrix of n-layers, where $n = 2, 3, \ldots ,p$.

The transmission matrix approach can be used to analyze a system of parallel regions. In the case of multiple interfaces, as shown in Fig. 3.10, the transmission matrix of the individual layers are denoted by $[M_n^{\perp}(d_n)]$ for perpendicular polarization and $[M_n^{\parallel}(d_n)]$ for horizontal polarization where $n = 2, 3, \ldots ,p$ and $d_n = D_n - D_{n-1}$ with $D_1 = 0$. The individual layer matrix $[M_n^{\perp}(d_n)]$ follows from (3.6.31) with the substitution of D_{n-1} for $z = 0$, D_n for $z = d$, γ_{zn} for γ_2 and $\eta_n^{\perp}$ for $\eta_2^{\perp}$. Similarly, the individual layer matrix $[M_n^{\parallel}(d_n)]$ follows from (3.6.32). Specifically, the perpendicular-polarization transmission matrix for the n-layer is

$$\begin{bmatrix} E_{yn}(x, D_{n-1}) \\ H_{xn}(x, D_{n-1}) \end{bmatrix} = \begin{bmatrix} \cosh \gamma_{zn} d_n & -\eta_n^{\perp} \sinh \gamma_{zn} d_n \\ -\frac{\sinh \gamma_{zn} d_n}{\eta_n^{\perp}} & \cosh \gamma_{zn} d_n \end{bmatrix} \begin{bmatrix} E_{yn}(x, D_n) \\ H_{xn}(x, D_n) \end{bmatrix}$$

where

$$\gamma_n^2 = \gamma_{x1}^2 + \gamma_{zn}^2$$

$$\gamma_{x1} = \gamma_1 \sin \theta_1$$

$$\gamma_1 \sin \theta_1 = \gamma_n \sin \theta_n$$

$$\eta_n^{\perp} = \frac{\eta_n}{\cos \theta_n}$$

The angle θ_n is the angle of refraction in the nth layer.

The relation between the fields at the two ends of the region is

$$\begin{bmatrix} E_{y2}(x, 0) \\ H_{x2}(x, 0) \end{bmatrix} = [M^{\perp}(L)] \begin{bmatrix} E_{y2}(x, L) \\ H_{x2}(x, L) \end{bmatrix}$$

where L is the length of the region (i.e., $L = D_p$) and

$$[M^{\perp}(L)] = [M_2^{\perp}(d_2)][M_3^{\perp}(d_3)]\cdots\cdots\cdots[M_p^{\perp}(d_p)]$$

Similarly, for the parallel polarization:

$$\begin{bmatrix} H_{y2}(x, 0) \\ E_{x2}(x, 0) \end{bmatrix} = [M^{\parallel}(L)] \begin{bmatrix} H_{y2}(x, L) \\ E_{x2}(x, L) \end{bmatrix}$$

and

$$[M^{\parallel}(L)] = \left[M_2^{\parallel}(d_2)\right]\left[M_3^{\parallel}(d_3)\right]\cdots\cdots\left[M_p^{\parallel}(d_p)\right]$$

The transmission matrix approach can also be used when the layers have inhomogeneous profiles specified by a complex propagation constant that is a function of position (i.e., $\gamma(z)$).

3.7 WAVES IN INHOMOGENEOUS REGIONS

3.7.1 Rectangular Coordinates

Inhomogeneous regions are encountered in a variety of media, such as the Ionosphere, the Earth's crust, and in optical applications. First, we consider inhomogeneous regions in rectangular coordinates where the complex propagation constant is a function of position. If the inhomogeneity in ε and σ vary as a function of position in one dimension, say z, the medium is characterized by the complex propagation constant

$$\gamma^2(z) = j\omega\mu[\sigma(z) + j\omega\varepsilon(z)] = -\omega^2\mu\varepsilon_c(z) = -\beta_o^2\mu_r\varepsilon_{cr}(z) = -\beta_o^2 n^2(z)$$

or

$$\gamma(z) = j\beta_o\sqrt{\mu_r\varepsilon_{cr}(z)} = j\beta_o n(z)$$

where

$$\varepsilon_{cr}(z) = \varepsilon_r(z) - j\frac{\sigma(z)}{\omega\varepsilon_o}$$

and

$$n(z) = \sqrt{\mu_r \varepsilon_{cr}(z)}$$

The wave equation satisfied by the electric field in such inhomogeneous region follows from (1.3.9) and (1.3.10) where the complex dielectric constant is a function of z. Taking the curl of (1.3.9) and substituting (1.3.10) gives

$$\nabla(\nabla \bullet \mathbf{E}) - \nabla^2\mathbf{E} = -j\omega\mu\mathbf{J}_e + \beta_o^2\mu_r\varepsilon_{cr}(z)\mathbf{E} \tag{3.7.1}$$

From:

$$\nabla\bullet \mathbf{D} = \rho_e \Rightarrow \nabla\bullet [\varepsilon_c(z)\mathbf{E}] = \rho_e$$

it follows that

$$\nabla \bullet \mathbf{E} = \frac{\rho_e}{\varepsilon_c(z)} - \frac{\nabla\varepsilon_{cr}(z)}{\varepsilon_{cr}(z)} \bullet \mathbf{E}$$

which substituting into (3.7.1) and using the continuity equation, gives

$$\nabla^2\mathbf{E} + \beta_o^2\mu_r\varepsilon_{cr}\mathbf{E} + \nabla\left(\frac{\nabla\varepsilon_{cr}}{\varepsilon_{cr}} \bullet \mathbf{E}\right) = j\omega\mu\mathbf{J}_e - \frac{\nabla(\nabla\bullet \mathbf{J}_e)}{j\omega\varepsilon_c}$$

Away from the source, the above relation is

$$\nabla^2\mathbf{E} + \beta_o^2\mu_r\varepsilon_{cr}\mathbf{E} + \nabla\left(\frac{\nabla\varepsilon_{cr}}{\varepsilon_{cr}} \bullet \mathbf{E}\right) = 0 \tag{3.7.2}$$

and in terms of $\gamma(z)$ and $n(z)$ is

$$\nabla^2\mathbf{E} - \gamma^2(z)\mathbf{E} + \nabla\left(\frac{\nabla n^2(z)}{n^2(z)} \bullet \mathbf{E}\right) = 0 \tag{3.7.3}$$

Similarly, the equation satisfied by the magnetic field is

$$\nabla^2\mathbf{H} + \beta_o^2\mu_r\varepsilon_{cr}\mathbf{H} + \frac{\nabla\varepsilon_{cr}}{\varepsilon_{cr}} \times \nabla \times \mathbf{H} = -\nabla \times \mathbf{J}_e + \frac{\nabla\varepsilon_{cr}}{\varepsilon_{cr}} \times \mathbf{J}_e$$

and away from the source reads

$$\nabla^2 \mathbf{H} + \beta_o^2 \mu_r \varepsilon_{cr} \mathbf{H} + \frac{\nabla \varepsilon_{cr}}{\varepsilon_{cr}} \times \nabla \times \mathbf{H} = 0 \tag{3.7.4}$$

or

$$\nabla^2 \mathbf{H} - \gamma^2(z)\mathbf{H} + \frac{\nabla n^2(z)}{n^2(z)} \times \nabla \times \mathbf{H} = 0$$

Consider the case of a perpendicular polarized plane wave incident on a lossless inhomogeneous dielectric region that varies as a function of z with $\mu_2 = \mu_o$ and $\sigma_2 = 0$, as shown in Fig. 3.11a. Then, in Region 2: $\varepsilon_{cr,2}(z) = \varepsilon_{r2}(z)$.

The incident fields in the air region, with $\gamma_1 = j\beta_o$, are

$$E_{y1} = C\left(e^{-j\beta_{z1}z} + R_\perp e^{j\beta_{z1}z}\right)e^{-j\beta_{x1}x}$$

$$H_{x1} = -\frac{C\cos\theta_1}{\eta_o}\left(e^{-j\beta_{z1}z} - R_\perp e^{j\beta_{z1}z}\right)e^{-j\beta_{x1}x}$$

$$H_{z1} = \frac{C\sin\theta_1}{\eta_o}\left(e^{-j\beta_{z1}z} + R_\perp e^{j\beta_{z1}z}\right)e^{-j\beta_{x1}x}$$

where

$$\beta_o^2 = \beta_{x1}^2 + \beta_{z1}^2$$
$$\beta_{x1} = \beta_o \sin\theta_1$$

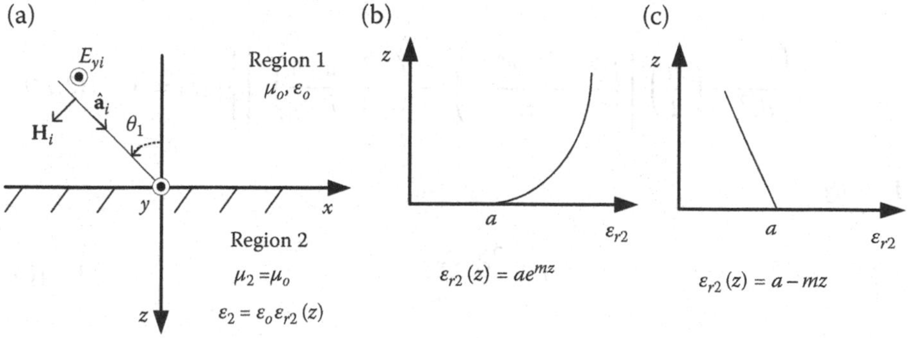

FIGURE 3.11 (a) Perpendicular polarized plane wave incident on an inhomogeneous dielectric: (b) an exponential dielectric profile: (c) a linear dielectric profile.

$$\beta_{z1} = \beta_o \cos \theta_1$$

In Region 2 the gradient term in (3.7.2) is zero since

$$\nabla\left(\frac{1}{\varepsilon_{r2}(z)} \frac{d\varepsilon_{r2}(z)}{dz}\hat{\mathbf{a}}_z \bullet E_y\hat{\mathbf{a}}_y\right) = 0$$

and, therefore, the electric field in Region 2 satisfies

$$\left[\frac{\partial^2}{\partial x^2} + \frac{\partial^2}{\partial z^2} + \beta_o^2\varepsilon_{r2}(z)\right]E_{y2} = 0 \tag{3.7.5}$$

The x-dependence of E_{y2} is the same as that of Region 1 (i.e., Snell's law). The solution of (3.7.5) is of the form

$$E_{y2} = CT_{\perp}f(z)e^{-j\beta_{x1}x} \tag{3.7.6}$$

where $T_{\perp}$ is the transmission coefficient. Substituting (3.7.6) into (3.7.5), the equation satisfied by $f(z)$ is

$$\left[\frac{d^2}{dz^2} + \beta_o^2\varepsilon_{r2}(z) - (\beta_o \sin\theta_1)^2\right]f(z) = 0 \tag{3.7.7}$$

There are several forms of $\varepsilon_{r2}(z)$ for which (3.7.7) has closed form solutions. We consider the analysis of an exponential and of a linear dielectric profile. The form of an exponential increasing profile is shown in Fig. 3.11b. It is described by

$$\varepsilon_{r2}(z) = ae^{mz} \tag{3.7.8}$$

where $a > 0$ and $m > 0$ are constants. If $a = 1$, the dielectric is continuous at $z = 0$.

Substituting (3.7.8) into (3.7.7), $f(z)$ satisfy

$$\left\{\frac{d^2}{dz^2} + \left(\frac{m}{2}\right)^2\left[\left(\frac{2\beta_o\sqrt{a}\,e^{mz/2}}{m}\right)^2 - \left(\frac{2\beta_o \sin\theta_1}{m}\right)^2\right]\right\}f(z) = 0 \tag{3.7.9}$$

Letting

$$t = \frac{2\beta_o\sqrt{a}\,e^{mz/2}}{m} \tag{3.7.10}$$

and

$$\nu = \frac{2\beta_o \sin\theta_1}{m} \tag{3.7.11}$$

in (3.7.9) we obtain

$$t^2\frac{d^2f}{dt^2} + t\frac{df}{dt} + (t^2 - \nu^2)f = 0 \tag{3.7.12}$$

Equation (3.7.12) is recognized as Bessel differential equation whose solutions can be written in terms of Hankel functions, or

$$f = C_1 H_\nu^{(1)}(t) + C_2 H_\nu^{(2)}(t)$$

Only $H_\nu^{(2)}(t)$ has the proper behavior at $z \to \infty$, since it represents an outgoing wave. That is,

$$H_\nu^{(2)}(t) \approx \sqrt{\frac{2}{\pi t}}\, e^{-j(t-\nu\pi/2-\pi/4)} \quad (t \to \infty)$$

If the dielectric has a small loss, then a has a small negative imaginary part, which can be represented by small imaginary part in the propagation constant in (3.7.10) (or $t \to t' - jt''$). Then, the wave decays as $z \to \infty$.

From (3.7.6), the transmitted electric field is

$$E_{y2} = CT_\perp H_\nu^{(2)}(t)e^{-j\beta_{x1}x} \tag{3.7.13}$$

and the tangential magnetic field is

$$H_{x2} = \frac{1}{j\omega\mu_o}\frac{\partial E_{y2}}{\partial z} = CT_\perp \frac{\beta_o\sqrt{a}}{j\omega\mu_o} e^{mz/2} H_\nu^{(2)\prime}(t)e^{-j\beta_{x1}x} \tag{3.7.14}$$

From (3.7.13) and (3.7.14), the surface impedance is

$$Z_s = -\left.\frac{E_{y2}}{H_{x2}}\right|_{z=0} = -j\frac{\omega\mu_o}{\beta_o\sqrt{a}}\frac{H_\nu^{(2)}\left(\frac{2\beta_o\sqrt{a}}{m}\right)}{H_\nu^{(2)\prime}\left(\frac{2\beta_o\sqrt{a}}{m}\right)} \tag{3.7.15}$$

which determines the reflection properties of the inhomogeneous dielectric. Observe that (3.7.11) could have been selected as $-\nu$, but since $H_{-\nu}^{(2)}(x)/H_{-\nu}^{(2)\prime}(x) = H_\nu^{(2)}(x)/H_\nu^{(2)\prime}(x)$ the result in (3.7.15) remains the same.

The form of a linearly decreasing profile is shown in Fig. 3.11c. It is given by

$$\varepsilon_{r2}(z) = a - mz$$

where $a > 0$ and $m > 0$. If `$a = 1$, the profile in Fig. 3.11c is continuous at $z = 0$, since in this case: $\varepsilon_{r1} = \varepsilon_{r2}(0) = 1$. For the liner profile, (3.7.7) reads

$$\begin{aligned}&\left[\frac{d^2}{dz^2} + \beta_o^2(a - mz) - (\beta_o \sin\theta_1)^2\right] f(z) = 0\\&\left\{\frac{d^2}{dz^2} - \beta_o^2 m\left[z - \frac{1}{m}(a - \sin^2\theta_1)\right]\right\} f(z) = 0\end{aligned} \tag{3.7.16}$$

Letting

$$t = (\beta_o^2 m)^{1/3}\left[z - \frac{1}{m}(a - \sin^2\theta_1)\right] \tag{3.7.17}$$

Then, since

$$\frac{df}{dz} = \frac{dt}{dz}\frac{df}{dt} = (\beta_o^2 m)^{1/3}\frac{df}{dt}$$

and

$$\frac{d^2f}{dz^2} = (\beta_o^2 m)^{2/3}\frac{d^2f}{dt^2}$$

(3.7.16) transforms to

$$\left(\frac{d^2}{dt^2} - t\right) f(t) = 0$$

which is recognized as Airy's differential equation (see Appendix D), whose solutions can be written in terms of Airy functions:

$$f(t) = C_1 Ai(t) + C_2 Bi(t) \tag{3.7.18}$$

The appropriate solution that decays as $z \to \infty$ (i.e., $t \to \infty$) is $Ai(t)$, since

$$Ai(t) \approx \frac{1}{2\sqrt{\pi}\, t^{1/4}} e^{-\frac{2}{3}t^{3/2}} \quad (t \to \infty)$$

The function $Bi(t)$ increases exponentially as $z \to \infty$ and, therefore, cannot be used.

The transmitted electric field, from (3.7.6) and (3.7.18), is

$$E_{y2} = CT_{\perp} Ai(t) e^{-j\beta_{x1}x} \tag{3.7.19}$$

and the tangential magnetic field is

$$H_{x2} = \frac{1}{j\omega\mu_o}\frac{\partial E_{y2}}{\partial z} = CT_{\perp}\frac{(\beta_o^2 m)^{1/3}}{j\omega\mu_o} Ai'(t) e^{-j\beta_{x1}x} \tag{3.7.20}$$

The surface impedance, using (3.7.19) and (3.7.20), is

$$Z_s = -\left.\frac{E_{y2}}{H_{x2}}\right|_{z=0} = -j\omega\mu_o(\beta_o^2 m)^{-1/3}\left.\frac{Ai(t)}{Ai'(t)}\right|_{z=0}$$

which determine the reflection properties of the inhomogeneous dielectric.

As an example, consider a linearly decreasing profile with $a = 1$. Then, (3.7.17), read

$$t = (\beta_o^2 m)^{1/3}\left(z - \frac{\cos^2\theta_1}{m}\right)$$

and the surface impedance is

$$Z_s = -j\omega\mu_o(\beta_o^2 m)^{-1/3}\frac{Ai(t_o)}{Ai'(t_o)} \tag{3.7.21}$$

where

$$t_o = -\left(\frac{\beta_o}{m}\right)^{2/3}\cos^2\theta_1$$

Observe that Z_s is imaginary and represents a capacitive susceptance.

The reflection coefficient at $z = 0$ is

$$R_{\perp} = \frac{Z_s - \eta_1^{\perp}}{Z_s + \eta_1^{\perp}} = \frac{Z_s - \dfrac{\eta_o}{\cos\theta_1}}{Z_s + \dfrac{\eta_o}{\cos\theta_1}}$$

Using (3.7.21), observing that $(-t_o)^{1/2} = (\beta_o/m)^{1/3}\cos\theta_1$ and $\mu_o\omega/\beta_o = \eta_o$, gives

$$R_{\perp} = \frac{(-t_o)^{1/2}Ai(t_o) - jAi'(t_o)}{(-t_o)^{1/2}Ai(t_o) + jAi'(t_o)}$$

Hence, $|R_\perp| = 1$ which shows that there is total reflection since for a linearly decreasing profile there is a θ_1 which result in a critical angle of incidence.

The exponential and linear dielectric profiles have been used to study the characteristics of different types of inhomogeneous regions. Other types of dielectric profiles have also been analyzed using similar techniques.

In the case that displacement effects are negligible, exponential and linear conducting profiles lead to closed form solutions. Consider an exponential increasing conductivity profile represented by

$$\sigma_2(z) = ae^{mz} \tag{3.7.22}$$

where

$$\gamma_2^2(z) \approx j\omega\mu_o\sigma_2(z)$$

For this profile, (3.7.3) shows that

$$\left[\frac{\partial^2}{\partial x^2} + \frac{\partial^2}{\partial z^2} - \gamma_2^2(z)\right]E_{y2} = 0 \tag{3.7.23}$$

The solution of (3.7.23) is of the form

$$E_{y2} = CT_\perp f(z)e^{-j\beta_{x1}x} \tag{3.7.24}$$

Substituting (3.7.22) and (3.7.24) into (3.7.23), the function $f(z)$ satisfies

$$\begin{gathered}\left[\frac{d^2}{dz^2} - j\omega\mu_o ae^{mz} - (\beta_o \sin\theta_1)^2\right]f(z) = 0\\ \left\{\frac{d^2}{dz^2} - \left(\frac{m}{2}\right)^2\left[\left(\frac{2\sqrt{\omega\mu_o a}\,e^{mz/2}e^{j\pi/4}}{m}\right)^2 + \left(\frac{2\beta_o\sin\theta_1}{m}\right)^2\right]\right\}f(z) = 0\end{gathered} \tag{3.7.25}$$

Equation (3.7.25) is solved by first letting

$$t = \frac{2\sqrt{\omega\mu_o a}\,e^{mz/2}e^{j\pi/4}}{m}$$

and

$$\nu = \frac{2\beta_o \sin\theta_1}{m}$$

Hence, (3.7.25) becomes

$$t^2\frac{d^2f}{dt^2} + t\,\frac{df}{dt} - (t^2 + \nu^2)f = 0$$

which is recognized as the modified Bessel equation whose solutions are

$$f(t) = C_1 I_\nu(t) + C_2 K_\nu(t)$$

The appropriate solution that decays as $z \to \infty$ is $K_\nu(t)$, since

$$K_\nu(t) \approx \sqrt{\frac{\pi}{2t}}\, e^{-t}\ (t \to \infty)$$

Therefore,

$$E_{y2} = CT_\perp K_\nu(t) e^{-j\beta_{x1}x} \tag{3.7.26}$$

and

$$H_{x2} = \frac{1}{j\omega\mu_o}\frac{\partial E_{y2}}{\partial z} = CT_\perp \sqrt{\frac{a}{\omega\mu_o}}\, e^{mz/2} e^{-j\pi/4} K_\nu{}'(t) e^{-j\beta_{x1}x} \tag{3.7.27}$$

The surface impedance for this profile, from (3.7.26) and (3.7.27), is

$$Z_s = -\left.\frac{E_{y2}}{H_{x2}}\right|_{z=0} = -\sqrt{\frac{\omega\mu_o}{a}}\, e^{j\pi/4}\frac{K_\nu(t_o)}{K_\nu{}'(t_o)}$$

where

$$t_o = \frac{2\sqrt{\omega\mu_o a}\, e^{j\pi/4}}{m}$$

At normal incidence: $\theta_1 = 0$ and $\nu = 0$, and the surface impedance (with $K_\nu{}'(x) = -K_1(x)$) is given by

$$Z_s = \sqrt{\frac{\omega\mu_o}{a}}\, e^{j\pi/4}\frac{K_0(t_o)}{K_1(t_o)} \tag{3.7.28}$$

which determines the reflection coefficient.

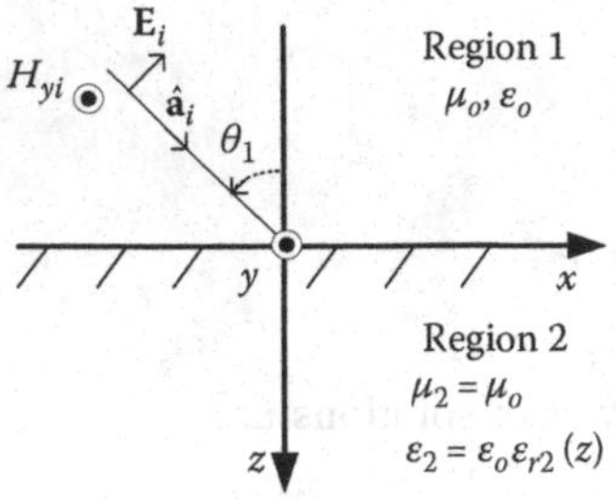

FIGURE 3.12 Parallel-polarized plane wave incident on an inhomogeneous dielectric.

Next, we consider the case of a parallel polarized plane wave incident on an inhomogeneous dielectric that varies as a function of z, as shown in Fig. 3.12. In Region 2, the magnetic field satisfy (3.7.4). Since $n_2^2(z) = \varepsilon_{r2}(z)$ and

$$\frac{\nabla \varepsilon_{r2}}{\varepsilon_{r2}} \times \nabla \times \mathbf{H} = \frac{-1}{n_2^2(z)} \frac{d\, n_2^2(z)}{dz} \frac{\partial H_{y2}}{\partial z} \hat{\mathbf{a}}_y$$

it follows that H_{y2} satisfies

$$\left[\frac{\partial^2}{\partial x^2} + \frac{\partial^2}{\partial z^2} - \frac{1}{n_2^2(z)} \frac{d\, n_2^2(z)}{dz} \frac{\partial}{\partial z} + \beta_o^2 n_2^2(z) \right] H_{y2} = 0 \tag{3.7.29}$$

Equation (3.7.29) is simplified by letting

$$H_{y2}(x, z) = n_2(z) V(x, z) \tag{3.7.30}$$

Then, (3.7.29) becomes

$$\left[\frac{\partial^2}{\partial x^2} + \frac{\partial^2}{\partial z^2} + \beta_o^2 N^2(z) \right] V = 0 \tag{3.7.31}$$

where

$$\beta_o^2 N^2 = \beta_o^2 n_2^2 + \frac{n_2''}{n_2} - \frac{2(n_2')^2}{n_2^2} \tag{3.7.32}$$

Equation (3.7.31) has the same form as (3.7.5), except that $\beta_o^2 n_2^2(z)$ is replaced by $\beta_o^2 N^2(z)$. Letting

$$V(x, z) = C\, g(z) e^{-j\beta_{x1} x}$$

and substituting into (3.7.31) results in

$$\left[\frac{d^2}{dz^2} + \beta_o^2 N^2(z) - (\beta_o \sin\theta_1)^2\right] g(z) = 0 \tag{3.7.33}$$

Referring to (3.7.30), the field H_{y2} is given by

$$H_{y2} = CT_{\|}^{H} n_2(z) g(z) e^{-j\beta_{x1}x}$$

Usually, numerical techniques are used to solve (3.7.33), since given a profile $n_2^2(z) = \varepsilon_{r2}(z)$ and calculating the associated $\beta_o^2 N^2(z)$ in (3.7.32), the resulting differential equation does not reduce (in general) to a recognized differential equation. The exponential and linear profiles lead to closed form solutions of (3.7.33). Consider the exponential dielectric profile in (3.7.8) where $n_2(z) = \sqrt{\varepsilon_{r2}} = \sqrt{a}\, e^{mz/2}$. Then, from (3.7.32):

$$\beta_o^2 N^2 = \beta_o^2 a e^{mz} - \left(\frac{m}{2}\right)^2$$

and (3.7.33) reads

$$\left\{\frac{d^2}{dz^2} + \left(\frac{m}{2}\right)^2\left[\left(\frac{2\beta_o\sqrt{a}\, e^{mz/2}}{m}\right)^2 - \left(\left[\frac{2\beta_o \sin\theta_1}{m}\right]^2 + 1\right)\right]\right\} g(z) = 0$$

which is similar in form to (3.7.9). Hence, the appropriate solution is $H_\nu^{(2)}(t)$ where t is given by

$$t = \frac{2\beta_o\sqrt{a}\, e^{mz/2}}{m} = \frac{2\beta_o n_2(z)}{m}$$

and

$$\nu = \sqrt{\left(\frac{2\beta_o \sin\theta_1}{m}\right)^2 + 1}$$

Therefore, H_{y2} is given by

$$H_{y2} = CT_{\|}^{H} n_2(z) H_\nu^{(2)}(t) e^{-j\beta_{x1}x}$$

and the tangential component of the electric field is

$$E_{x2} = -\frac{1}{j\omega\varepsilon_o\varepsilon_{r2}(z)}\frac{\partial H_{y2}}{\partial z} = -\frac{CT_{\|}^{H}}{j\omega\varepsilon_o a e^{mz}}\frac{dt}{dz}\frac{d}{dt}[n_2 H_{\nu}^{(2)}(t)] = j\frac{CT_{\|}^{H} m}{2\omega\varepsilon_o n_2}\frac{d}{dt}[tH_{\nu}^{(2)}(t)]$$
$$= j\frac{CT_{\|}^{H}\eta_o}{t}\frac{d}{dt}[tH_{\nu}^{(2)}(t)]$$

where we used $n_2 = mt/2\beta_o$.

The surface impedance is

$$Z_s = \left.\frac{E_{x2}}{H_{y2}}\right|_{z=0} = j\frac{\eta_o}{\sqrt{a}}\left.\frac{\frac{d}{dt}[tH_{\nu}^{(2)}(t)]}{t_o H_{\nu}^{(2)}(t_o)}\right|_{z=0} \tag{3.7.34}$$

where at $z = 0$ the value of t, denoted by t_o, is $t_o = 2\beta_o\sqrt{a}/m$.

Next, consider the exponential conductive profile in (3.7.22). In this case:

$$n_2^2(z) = \varepsilon_{cr,2} = -j\frac{\sigma_2(z)}{\omega\varepsilon_o} = -j\frac{a\,e^{mz}}{\omega\varepsilon_o}$$

or

$$n_2(z) = \sqrt{\frac{a}{\omega\varepsilon_o}}\;e^{mz/2}e^{-j\pi/4}$$

and from (3.7.32):

$$\beta_o^2 N^2 = -j\frac{\beta_o^2 a\,e^{mz}}{\omega\varepsilon_o} - \left(\frac{m}{2}\right)^2$$

Then, (3.7.33) reads

$$\left\{\frac{d^2}{dz^2} - \left(\frac{m}{2}\right)^2\left[\left(\frac{2\beta_o\sqrt{a}\;e^{mz/2}e^{j\pi/4}}{m\sqrt{\omega\varepsilon_o}}\right)^2 + \left(\left[\frac{2\beta_o\sin\theta_1}{m}\right]^2 + 1\right)\right]\right\}g(z) = 0$$

which is similar in form to (3.7.25). Hence, the appropriate solution is $K_{\nu}(t)$ where t is given

$$t = \frac{2\beta_o\sqrt{a}\;e^{mz/2}e^{j\pi/4}}{m\sqrt{\omega\varepsilon_o}} = \frac{2\beta_o n_2(z)e^{j\pi/2}}{m}$$

and

$$\nu = \sqrt{\left(\frac{2\beta_o \sin\theta_1}{m}\right)^2 + 1}$$

Therefore, H_{y2} is given by

$$H_{y2} = BT_{\|}^{H} n_2(z) K_\nu(t) e^{-j\beta_{x1}x}$$

The tangential component of the electric field is

$$E_{x2} = -\frac{1}{j\omega\varepsilon_o\varepsilon_{r2}(z)}\frac{\partial H_{y2}}{\partial z} = -\frac{1}{j\omega\varepsilon_o\varepsilon_{r2}(z)}\frac{dt}{dz}\frac{dH_{y2}}{dt} = j\frac{BT_{\|}^{H} m}{2\omega\varepsilon_o n_2}\frac{d[tK_\nu(t)]}{dt} = -\frac{BT_{\|}^{H}\eta_o}{t}\frac{d[tK_\nu(t)]}{dt}$$

where we used $n_2 = (mt/2\beta_o)e^{-j\pi/2}$.

The surface impedance is

$$Z_s = \left.\frac{E_{x2}}{H_{y2}}\right|_{z=0} = -\eta_o\sqrt{\frac{\omega\varepsilon_o}{a}}\, e^{j\pi/4} \left.\frac{\frac{d}{dt}[tK_\nu(t)]}{t_o K_\nu(t_o)}\right|_{z=0} \tag{3.7.35}$$

where

$$t_o = t|_{z=0} = \frac{2\beta_o\sqrt{a}\, e^{j\pi/4}}{m\sqrt{\omega\varepsilon_o}}$$

The numerator of (3.7.35) is of similar form to that in (3.7.34). These numerators can be expanded using the relations (see Appendix C):

$$H_\nu^{(2)\prime}(t) = H_{\nu-1}^{(2)}(t) - \left(\frac{\nu}{t}\right) H_\nu^{(2)}(t)$$

and

$$K_\nu{}'(t) = -K_{\nu-1}(t) - \left(\frac{\nu}{t}\right) K_\nu(t)$$

For example, (3.7.35) reads

$$Z_s = -\eta_o\sqrt{\frac{\omega\varepsilon_o}{a}}\, e^{j\pi/4}\left[\frac{K_\nu{}'(t_o)}{K_\nu(t_o)} + \frac{1}{t_o}\right] = \eta_o\sqrt{\frac{\omega\varepsilon_o}{a}}\, e^{j\pi/4}\left[\frac{K_{\nu-1}(t_o)}{K_\nu(t_o)} + \frac{1-\nu}{t_o}\right]$$

At normal incidence (i.e., with $\sin\theta_1 = 0$) it follows that $\nu = 1$ and

$$Z_s = \eta_o \sqrt{\frac{\omega \varepsilon_o}{a}} e^{j\pi/4} \frac{K_0(t_o)}{K_1(t_o)}$$

which agrees with (3.7.28). This is expected since at normal incidence there is no difference between vertical and horizontal polarization.

The transmission matrix of an inhomogeneous layer in Fig. 3.9 is calculated using the technique developed in Section 3.6. We will illustrate the process for perpendicular polarization. Say that Region 2 in Fig. 3.9 is inhomogeneous with $\varepsilon_{r2}(z)$ given by

$$\varepsilon_{r2}(z) = a e^{mz} \, (0 \leq z \leq d)$$

In such case, from (3.7.9) to (3.7.12), the fields in Region 2 are

$$E_{y2}(x, z) = [BH_\nu^{(1)}(t) + CH_\nu^{(2)}(t)] e^{-j\beta_{x1}x} \tag{3.7.36}$$

and

$$\begin{aligned} H_{x2}(x, z) &= \frac{1}{j\omega\mu} \frac{dt}{dz} [BH_\nu^{(1)\prime}(t) + CH_\nu^{(2)\prime}(t)] e^{-j\beta_{x1}x} \\ &= \frac{\beta_o \sqrt{a} e^{mz/2}}{j\omega\mu} [BH_\nu^{(1)\prime}(t) + CH_\nu^{(2)\prime}(t)] e^{-j\beta_{x1}x} \end{aligned} \tag{3.7.37}$$

where

$$t = \frac{2\beta_o \sqrt{a} \, e^{mz/2}}{m} \tag{3.7.38}$$

and

$$\nu = \frac{2\beta_o \sin \theta_1}{m} \tag{3.7.39}$$

Evaluating (3.7.36) and (3.7.37) at $z = d$ and solving for B and C gives

$$B = \frac{E_{y2}(x, d) H_\nu^{(2)\prime}(t_d) - \frac{1}{h_d} H_{x2}(x, d) H_\nu^{(2)}(t_d)}{W(t_d)} \tag{3.7.40}$$

and

$$C = \frac{\frac{1}{h_d} H_{x2}(x, d) H_\nu^{(1)}(t_d) - E_{y2}(x, d) H_\nu^{(1)\prime}(t_d)}{W(t_d)} \tag{3.7.41}$$

where

$$t_d = \frac{2\beta_o \sqrt{a}\, e^{md/2}}{m}$$

$$h_d = \frac{\beta_o \sqrt{a}\, e^{md/2}}{j\omega\mu}$$

and $W(t_d)$ is the Wronskian evaluated at $z = d$. That is,

$$W(t_d) = H_\nu^{(1)}(t_d)H_\nu^{(2)\prime}(t_d) - H_\nu^{(2)}(t_d)H_\nu^{(1)\prime}(t_d) = -\frac{j4}{\pi t_d} = -\frac{j2m}{\pi\beta_o\sqrt{a}\,e^{md/2}}$$

Evaluating (3.7.36) and (3.7.37) at $z = 0$ and substituting the values of B and C given in (3.7.40) and (3.7.41) results in the transmission matrix of the inhomogeneous region. That is,

$$\begin{bmatrix} E_{y2}(x, 0) \\ H_{x2}(x, 0) \end{bmatrix} = \begin{bmatrix} a_{11} & a_{12} \\ a_{21} & a_{22} \end{bmatrix} \begin{bmatrix} E_{y2}(x, d) \\ H_{x2}(x, d) \end{bmatrix}$$

where

$$a_{11} = \frac{H_\nu^{(1)}(t_0)H_\nu^{(2)\prime}(t_d) - H_\nu^{(2)}(t_0)H_\nu^{(1)\prime}(t_d)}{W(t_d)}$$

$$a_{12} = \frac{1}{h_d}\left[\frac{H_\nu^{(2)}(t_0)H_\nu^{(1)}(t_d) - H_\nu^{(1)}(t_0)H_\nu^{(2)}(t_d)}{W(t_d)}\right]$$

$$a_{21} = h_0\left[\frac{H_\nu^{(1)\prime}(t_0)H_\nu^{(2)\prime}(t_d) - H_\nu^{(2)\prime}(t_0)H_\nu^{(1)\prime}(t_d)}{W(t_d)}\right]$$

and

$$a_{22} = \frac{h_0}{h_d}\left[\frac{H_\nu^{(2)\prime}(t_0)H_\nu^{(1)}(t_d) - H_\nu^{(1)\prime}(t_0)H_\nu^{(2)}(t_d)}{W(t_d)}\right]$$

where

$$t_0 = \frac{2\beta_o\sqrt{a}}{m}$$

and

$$h_0 = \frac{\beta_o \sqrt{a}}{j\omega\mu}$$

3.7.2 Cylindrical Coordinates

Next, consider the propagation of waves in inhomogeneous cylindrical regions where the dielectric is a function of ρ (i.e., $\varepsilon = \varepsilon_o \varepsilon_r(\rho)$) with $\mu = \mu_o$ and $\sigma = 0$. In this case, the solution of (3.7.2) and (3.7.4) in cylindrical coordinates is analyzed.

Consider a wave propagating along the z-axis with propagation constant β_z. Therefore, the electric field in cylindrical coordinates is of the form:

$$\mathbf{E}(\rho, \phi, z) = \mathbf{E}(\rho, \phi)e^{-j\beta_z z} = [E_\rho(\rho, \phi)\hat{\mathbf{a}}_\rho + E_\phi(\rho, \phi)\hat{\mathbf{a}}_\phi + E_z(\rho, \phi)\hat{\mathbf{a}}_z]e^{-j\beta_z z} \tag{3.7.42}$$

where the notation $E_\rho(\rho, \phi)$, $E_\phi(\rho, \phi)$ and $E_z(\rho, \phi)$ denote the field components dependence on ρ and ϕ.

The substitution of (3.7.42) into (3.7.3) requires the following operations:

$$\nabla^2\mathbf{E}(\rho, \phi, z) = \nabla_t^2\,\mathbf{E}(\rho, \phi)e^{-j\beta_z z} + \frac{\partial^2 E_z(\rho, \phi, z)}{\partial z^2} = \left[\nabla_t^2\mathbf{E}(\rho, \phi) - \beta_z^2 E_z(\rho, \phi)\right]e^{-j\beta_z z} \tag{3.7.43}$$

and

$$\nabla\left[\frac{\nabla\varepsilon_r(\rho)}{\varepsilon_r(\rho)} \bullet \mathbf{E}(\rho, \phi, z)\right] = \nabla\left[\frac{1}{\varepsilon_r(\rho)}\frac{\partial\varepsilon_r(\rho)}{\partial\rho}\hat{\mathbf{a}}_\rho \bullet \mathbf{E}(\rho, \phi)e^{-j\beta_z z}\right]$$

$$= \left(\nabla_t - j\beta_z\hat{\mathbf{a}}_z\right)\left[\frac{1}{\varepsilon_r(\rho)}\frac{\partial\varepsilon_r(\rho)}{\partial\rho}E_\rho(\rho, \phi)e^{-j\beta_z z}\right] \tag{3.7.44}$$

where $\nabla_t^2\,\mathbf{E}$ denotes the transverse part of the Laplacian vector operator, and ∇_t the transverse gradient operator. The transverse vector Laplacian is given by

$$\nabla_t^2\,\mathbf{E} = \left(\nabla_t^2 E_\rho - \frac{E_\rho}{\rho^2} - \frac{2}{\rho^2}\frac{\partial E_\phi}{\partial\phi}\right)\hat{\mathbf{a}}_\rho + \left(\nabla_t^2 E_\phi - \frac{E_\phi}{\rho^2} + \frac{2}{\rho^2}\frac{\partial E_\rho}{\partial\phi}\right)\hat{\mathbf{a}}_\phi + \nabla_t^2\,E_z\hat{\mathbf{a}}_z \tag{3.7.45}$$

where the transverse Laplacian is

$$\nabla_t^2 = \frac{1}{\rho}\frac{\partial}{\partial\rho}\left(\rho\frac{\partial}{\partial\rho}\right) + \frac{1}{\rho^2}\frac{\partial^2}{\partial\phi^2}$$

Substituting (3.7.45) into (3.7.43) gives the expression for $\nabla^2\mathbf{E}$.

The transverse gradient operator is

$$\nabla_t = \hat{\mathbf{a}}_\rho \frac{\partial}{\partial \rho} + \hat{\mathbf{a}}_\phi \frac{1}{\rho}\frac{\partial}{\partial \phi}$$

Hence, (3.7.44) reads

$$\begin{aligned}&\left[\nabla_t - j\beta_z \hat{\mathbf{a}}_z\right]\left(\frac{1}{\varepsilon_r}\frac{d\varepsilon_r}{d\rho}E_\rho e^{-j\beta_z z}\right)\\ &= \left[\frac{\partial}{\partial\rho}\left(\frac{1}{\varepsilon_r}\frac{d\varepsilon_r}{d\rho}E_\rho\right)\hat{\mathbf{a}}_\rho + \frac{1}{\rho}\left(\frac{1}{\varepsilon_r}\frac{d\varepsilon_r}{d\rho}\frac{\partial E_\rho}{\partial\phi}\right)\hat{\mathbf{a}}_\phi - j\beta_z\left(\frac{1}{\varepsilon_r}\frac{d\varepsilon_r}{d\rho}E_\rho\right)\hat{\mathbf{a}}_z\right]e^{-j\beta_z z}\end{aligned} \tag{3.7.46}$$

Substituting (3.7.43), (3.7.45), and (3.7.46) into (3.7.2) results in the following differential equations for the fields:

$$\begin{cases}\nabla_t^2 E_\rho - \dfrac{E_\rho}{\rho^2} - \dfrac{2}{\rho^2}\dfrac{\partial E_\phi}{\partial\phi} + (\beta_o^2\varepsilon_r - \beta_z^2)E_\rho + \dfrac{\partial}{\partial\rho}\left(\dfrac{1}{\varepsilon_r}\dfrac{d\varepsilon_r}{d\rho}E_\rho\right) = 0\\[2ex] \nabla_t^2 E_\phi - \dfrac{E_\phi}{\rho^2} + \dfrac{2}{\rho^2}\dfrac{\partial E_\rho}{\partial\phi} + (\beta_o^2\varepsilon_r - \beta_z^2)E_\phi + \dfrac{1}{\rho}\left(\dfrac{1}{\varepsilon_r}\dfrac{d\varepsilon_r}{d\rho}\dfrac{\partial E_\rho}{\partial\phi}\right) = 0\\[2ex] \nabla_t^2 E_z + (\beta_o^2\varepsilon_r - \beta_z^2)E_z - j\beta_z\dfrac{1}{\varepsilon_r}\dfrac{d\varepsilon_r}{d\rho}E_\rho = 0\end{cases} \tag{3.7.47}$$

Similarly, from (3.7.4) observing that

$$\begin{aligned}&\frac{\nabla\varepsilon_r(\rho)}{\varepsilon_r(\rho)} \times \nabla \times \mathbf{H} = \frac{1}{\varepsilon_r}\frac{d\varepsilon_r}{d\rho}\hat{\mathbf{a}}_\rho \times \nabla \times [H_\rho(\rho,\phi)\hat{\mathbf{a}}_\rho + H_\phi(\rho,\phi)\hat{\mathbf{a}}_\phi + H_z(\rho,\phi)\hat{\mathbf{a}}_z]e^{-j\beta_z z}\\ &= \frac{1}{\varepsilon_r}\frac{d\varepsilon_r}{d\rho}\hat{\mathbf{a}}_\rho \times \left\{\left(\frac{1}{\rho}\frac{\partial H_z}{\partial\phi} + j\beta H_\phi\right)\hat{\mathbf{a}}_\rho - \left(\frac{\partial H_z}{\partial\rho} + j\beta H_\rho\right)\hat{\mathbf{a}}_\phi + \frac{1}{\rho}\left[\frac{\partial(\rho H_\phi)}{\partial\rho} - \frac{\partial H_\rho}{\partial\phi}\right]\hat{\mathbf{a}}_z\right\}e^{-j\beta_z z}\\ &= -\left\{\frac{1}{\varepsilon_r}\frac{d\varepsilon_r}{d\rho}\frac{1}{\rho}\left[\frac{\partial(\rho H_\phi)}{\partial\rho} - \frac{\partial H_\rho}{\partial\phi}\right]\hat{\mathbf{a}}_\phi + \frac{1}{\varepsilon_r}\frac{d\varepsilon_r}{d\rho}\left(\frac{\partial H_z}{\partial\rho} + j\beta H_\rho\right)\hat{\mathbf{a}}_z\right\}e^{-j\beta_z z}\end{aligned}$$

we obtain

$$\begin{cases} \nabla_t^2 H_\rho - \frac{H_\rho}{\rho^2} - \frac{2}{\rho^2}\frac{\partial H_\phi}{\partial \phi} + \left(\beta_o^2 \varepsilon_r - \beta_z^2\right) H_\rho = 0 \\ \nabla_t^2 H_\phi - \frac{H_\phi}{\rho^2} + \frac{2}{\rho^2}\frac{\partial H_\rho}{\partial \phi} + \left(\beta_o^2 \varepsilon_r - \beta_z^2\right) H_\phi + \frac{1}{\varepsilon_r}\frac{d\varepsilon_r}{d\rho}\frac{1}{\rho}\left[\frac{\partial H_\rho}{\partial \phi} - \frac{\partial(\rho H_\phi)}{\partial \rho}\right] = 0 \\ \nabla_t^2 H_z + \left(\beta_o^2 \varepsilon_r - \beta_z^2\right) H_z - \frac{1}{\varepsilon_r}\frac{d\varepsilon_r}{d\rho}\left(\frac{\partial H_z}{\partial \rho} + j\beta_z H_\rho\right) = 0 \end{cases} \quad (3.7.48)$$

Equations (3.7.47) and (3.7.48) are difficult to solve due to the coupling of the field component. However, in the case of TE^z and TM^z modes with no ϕ dependence the equations become uncoupled. From Maxwell's equations:

$$\mathbf{E} = \frac{1}{j\omega\varepsilon_o\varepsilon_r}\nabla \times \mathbf{H}$$

and

$$\mathbf{H} = -\frac{1}{j\omega\mu_o}\nabla \times \mathbf{E}$$

if $\partial/\partial\phi = 0$ we obtain:

$$E_\rho = \frac{\beta_z}{\omega\varepsilon_o\varepsilon_r} H_\phi \quad (3.7.49)$$

$$E_\phi = \frac{-1}{\omega\varepsilon_o\varepsilon_r}\left(\beta_z H_\rho - j\frac{\partial H_z}{\partial \rho}\right) \quad (3.7.50)$$

$$E_z = \frac{-j}{\omega\varepsilon_o\varepsilon_r}\frac{1}{\rho}\frac{\partial(\rho H_\phi)}{\partial \rho} \quad (3.7.51)$$

$$H_\rho = \frac{-\beta_z}{\omega\mu_o} E_\phi \quad (3.7.52)$$

$$H_\phi = \frac{1}{\omega\mu_o}\left(\beta_z E_\rho - j\frac{\partial E_z}{\partial \rho}\right) \quad (3.7.53)$$

$$H_z = \frac{j}{\omega\mu_o}\frac{1}{\rho}\frac{\partial(\rho E_\phi)}{\partial \rho} \quad (3.7.54)$$

From the previous relations the transverse field components can be expressed in terms of the longitudinal components (i.e., E_z and H_z). Substituting (3.7.53) into (3.7.49), and (3.7.50) into (3.7.52) gives

$$E_\rho = \frac{-j\beta_z}{\left(\beta_o^2 \varepsilon_r - \beta_z^2\right)} \frac{\partial E_z}{\partial \rho} \tag{3.7.55}$$

and

$$H_\rho = \frac{-j\beta_z}{\left(\beta_o^2 \varepsilon_r - \beta_z^2\right)} \frac{\partial H_z}{\partial \rho} \tag{3.7.56}$$

Similarly, it follows that

$$E_\phi = \frac{j\omega\mu_o}{\left(\beta_o^2 \varepsilon_r - \beta_z^2\right)} \frac{\partial H_z}{\partial \rho} \tag{3.7.57}$$

and

$$H_\phi = \frac{-j\omega\varepsilon_o\varepsilon_r}{\left(\beta_o^2 \varepsilon_r - \beta_z^2\right)} \frac{\partial E_z}{\partial \rho} \tag{3.7.58}$$

These relations show that TM^z modes are possible, since with $H_z = 0$ it follows that E_ϕ and H_ρ are zero and the TM^z field consist of the E_ρ, E_z and H_ϕ components. Also, TE^z modes are possible, since with $E_z = 0$ it follows that E_ρ and H_ϕ are zero and the TE^z field consist of the H_ρ, H_z and E_ϕ components.

For the TM^z modes, substituting (3.7.55) into the third equation in (3.7.47), with

$$\nabla_t^2 = \frac{1}{\rho}\frac{\partial}{\partial \rho}\left(\rho \frac{\partial}{\partial \rho}\right)$$

we obtain

$$\frac{1}{\rho}\frac{\partial}{\partial \rho}\left(\rho \frac{\partial E_z}{\partial \rho}\right) + \left(\beta_o^2 \varepsilon_r - \beta_z^2\right)E_z - \frac{\beta_z^2}{\left(\beta_o^2 \varepsilon_r - \beta_z^2\right)}\left(\frac{1}{\varepsilon_r}\frac{d\varepsilon_r}{d\rho}\right)\frac{\partial E_z}{\partial \rho} = 0 \tag{3.7.59}$$

Once (3.7.59) is solved, (3.7.55) gives E_ρ and (3.7.58) gives H_ϕ.

It also follows from (3.7.47) and (3.7.48) that with $\partial/\partial\phi = 0$, E_ρ and H_ϕ satisfy

$$\frac{1}{\rho}\frac{\partial}{\partial\rho}\left(\rho\frac{\partial E_\rho}{\partial\rho}\right) - \frac{E_\rho}{\rho^2} + \left(\beta_o^2\varepsilon_r - \beta_z^2\right)E_\rho + \frac{\partial}{\partial\rho}\left(\frac{1}{\varepsilon_r}\frac{d\varepsilon_r}{d\rho}E_\rho\right) = 0 \qquad (3.7.60)$$

$$\frac{1}{\rho}\frac{\partial}{\partial\rho}\left(\rho\frac{\partial H_\phi}{\partial\rho}\right) - \frac{H_\phi}{\rho^2} + \left(\beta_o^2\varepsilon_r - \beta_z^2\right)H_\phi - \left(\frac{1}{\varepsilon_r}\frac{d\varepsilon_r}{d\rho}\right)\frac{1}{\rho}\frac{\partial(\rho H_\phi)}{\partial\rho} = 0 \quad (3.7.61)$$

The solution of either (3.7.60) or (3.7.61), together with (3.7.49), (3.7.51), and (3.7.53) can also be used to obtain the field components.

The solution for the TE^z mode requires that the third equation in (3.7.48) be expressed only in terms of H_z. Using (3.7.56), it follows that H_z satisfies

$$\frac{1}{\rho}\frac{\partial}{\partial\rho}\left(\rho\frac{\partial H_z}{\partial\rho}\right) + \left(\beta_o^2\varepsilon_r - \beta_z^2\right)H_z - \frac{\beta_o^2\varepsilon_r}{\left(\beta_o^2\varepsilon_r - \beta_z^2\right)}\left(\frac{1}{\varepsilon_r}\frac{d\varepsilon_r}{d\rho}\right)\frac{\partial H_z}{\partial\rho} = 0 \quad (3.7.62)$$

Once (3.7.62) is solved, then (3.7.56) gives H_ρ and (3.7.57) gives E_ϕ.

It also follows from (3.7.47) and (3.7.48) that with $\partial/\partial\phi = 0$, E_ϕ, and H_ρ satisfy

$$\frac{1}{\rho}\frac{\partial}{\partial\rho}\left(\rho\frac{\partial E_\phi}{\partial\rho}\right) - \frac{E_\phi}{\rho^2} + \left(\beta_o^2\varepsilon_r - \beta_z^2\right)E_\phi = 0 \qquad (3.7.63)$$

$$\frac{1}{\rho}\frac{\partial}{\partial\rho}\left(\rho\frac{\partial H_\rho}{\partial\rho}\right) - \frac{H_\rho}{\rho^2} + \left(\beta_o^2\varepsilon_r - \beta_z^2\right)H_\rho = 0 \qquad (3.7.64)$$

The solution of either (3.7.63) or (3.7.64), together with (3.7.50), (3.7.52), and (3.7.54) can also be used to obtain the field components.

There are few inhomogeneous dielectric profiles that result in a closed form solution for the TM^z and TE^z modes. A profile that leads to a closed form solution for the TM^z and TE^z modes in the coaxial waveguide shown in Fig. 3.13 is

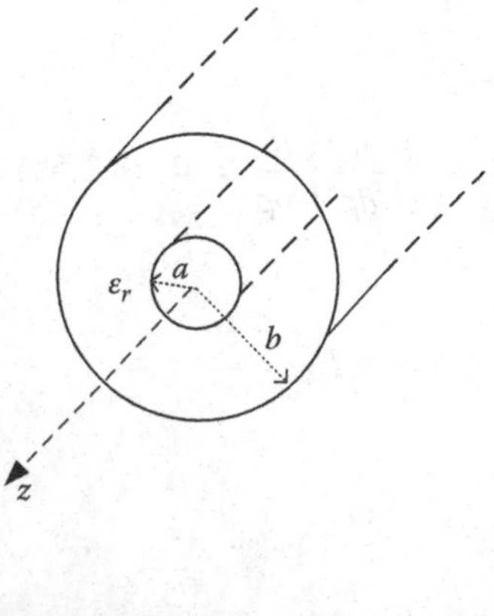

FIGURE 3.13 A coaxial waveguide.

$$\varepsilon_r(\rho) = \frac{C}{\rho^2} \qquad (a \le \rho \le b)$$

We will consider the TE^z mode. For this mode, the E_ϕ component satisfies (3.7.63) where

$$\mathbf{E}(\rho, z) = E_\phi(\rho)e^{-j\beta_z z}\mathbf{a}_\phi$$

and (3.7.63) reads

$$\frac{1}{\rho}\frac{\partial}{\partial\rho}\left(\rho\frac{\partial E_\phi}{\partial\rho}\right) - \frac{E_\phi}{\rho^2} + \left(\beta_o^2\frac{C}{\rho^2} - \beta_z^2\right)E_\phi = 0$$

which is expressed in the form

$$\rho^2\frac{\partial^2 E_\phi}{\partial\rho^2} + \rho\frac{\partial E_\phi}{\partial\rho} - \left[\left(\beta_z\rho\right)^2 + \left(1 - C\beta_o^2\right)\right]E_\phi = 0 \qquad (3.7.65)$$

The solutions to (3.7.65) are the modified Bessel functions, namely

$$E_\phi = CI_\nu(\beta_z\rho) + DK_\nu(\beta_z\rho)$$

where

$$\nu = \sqrt{1 - C\beta_o^2}$$

The boundary condition at $\rho = a$ is

$$E_\phi = 0|_{\rho=a} \Rightarrow CI_\nu(\beta_z a) + DK_\nu(\beta_z a) = 0 \Rightarrow \frac{C}{D} = -\frac{K_\nu(\beta_z a)}{I_\nu(\beta_z a)} \qquad (3.7.66)$$

and at $\rho = b$:

$$E_\phi = 0|_{\rho=b} \Rightarrow CI_\nu(\beta_z b) + DK_\nu(\beta_z b) = 0 \Rightarrow \frac{C}{D} = -\frac{K_\nu(\beta_z b)}{I_\nu(\beta_z b)} \qquad (3.7.67)$$

From (3.7.66) and (3.7.67):

$$\frac{K_\nu(\beta_z a)}{I_\nu(\beta_z a)} = \frac{K_\nu(\beta_z b)}{I_\nu(\beta_z b)} \qquad (3.7.68)$$

which determines the eigenvalue of β_z in the waveguide. Once E_ϕ is determined, the components H_ρ and H_z follow from (3.7.52) and (3.7.54), respectively.

3.8 THE WKBJ METHOD

The WKBJ method is a method for finding approximate solutions to certain differential equations. In our case, the wave equation for various types of $\gamma(z)$. The name WKBJ refer to Wentzel, Kramers, Brilloin, and Jeffreys who developed the method, although sometimes is referred as the WKB method.

3.8.1 Normal Incidence

Consider a plane wave propagating in a lossless, slowly varying inhomogeneous region where $\gamma^2(z) = -\beta_o^2\mu_r\varepsilon_r(z) = -\beta_o^2 n^2(z)$, and the electric field satisfies

$$\frac{d^2E_y}{dz^2} + \beta_o^2 n^2(z)E_y = 0 \tag{3.8.1}$$

If $n(z)$ is a positive constant (i.e., $n(z) = n$) the solution to (3.8.1) for the outgoing wave is

$$E_y = Ae^{-j\beta_o nz}$$

If $n(z)$ is positive and varies slowly as a function of z, we expect the resulting amplitude $A(z)$ and phase $\phi(z)$ to be slowly varying functions of position. Therefore, for the outgoing wave, we seek a solution of (3.8.1) in the form

$$E_y = A(z)e^{-j\phi(z)} \tag{3.8.2}$$

In order to substitute (3.8.2) into (3.8.1) we calculate

$$\frac{dE_y}{dz} = (A' - jA\phi')e^{-j\phi}$$

and

$$\frac{d^2E_y}{dz^2} = (A'' - 2jA'\phi' - jA\phi'' - \phi'^2)e^{-j\phi} \tag{3.8.3}$$

Then, substituting (3.8.3) into (3.8.1) gives

$$A'' + A\left[\beta_o^2 n^2(z) - \phi'^2\right] = 0 \tag{3.8.4}$$

and

$$2A'\phi' + A\phi'' = 0 \tag{3.8.5}$$

where (3.8.4) follows from the real part and (3.8.5) from the imaginary part. These two equations provide the exact solution to (3.8.1). However, they are coupled and, therefore, an approximate solution is sought.

We first attempt a simple approximation by assuming that only the phase changes in (3.8.2). That is, letting the amplitude be constant (i.e., $A(z) = A_o$). Then, $A'' = A' = 0$ and (3.8.4) gives

$$\phi(z) = \pm\beta_o \int_0^z n(z')\, dz' \tag{3.8.6}$$

and from (3.8.5): $\phi'' = 0$. Hence, for the outgoing wave the form of the electric field is

$$E_y = A_o e^{-j\phi(z)} = A_o e^{-j\beta_o \int_0^z n(z')\, dz'} \tag{3.8.7}$$

Equation (3.8.7) is also known as the zero order WKBJ approximation. To check under what conditions (3.8.7) satisfies (3.8.1), substitute (3.8.7) into (3.8.1). To this end, calculate

$$\frac{d^2E_y}{dz^2} = \left[-\left(\frac{d\phi}{dz}\right)^2 - j\frac{d^2\phi}{dz^2}\right] A_o e^{j\beta_o \int_0^z n(z')dz'}$$

and (3.8.1) reads

$$\left(\frac{d\phi}{dz}\right)^2 + j\frac{d^2\phi}{dz^2} = \beta_o^2 n^2 \tag{3.8.8}$$

If n is slowly varying, such that $d^2\phi/dz^2 \approx 0$, (3.8.8) satisfies (3.8.1) if

$$\left(\frac{d\phi}{dz}\right)^2 = \beta_o^2 n^2 \Rightarrow \frac{d\phi}{dz} = \pm\beta_o n \Rightarrow \phi(z) = \pm\beta_o \int_0^z n(z')dz'$$

which is (3.8.6).

Since

$$\frac{d^2\phi}{dz^2} = \pm\beta_o \frac{dn}{dz} \tag{3.8.9}$$

the approximation made in (3.8.8) is valid if

$$\frac{d^2\phi}{dz^2} << \left(\frac{d\phi}{dz}\right)^2 \Rightarrow \beta_o \frac{dn}{dz} << \beta_o^2 n^2 \Rightarrow \frac{dn}{dz} << \beta_o n^2$$

which can be expressed as

$$\frac{\lambda}{2\pi}\frac{dn}{dz} << n$$

where $\lambda = \lambda_o / n$. This condition basically states that in a distance of one wavelength the change in refractive index should be much smaller than n.

A better approximation is to observe that for a slowly varying $n(z)$, if L is the distance over which $A(z)$ change by ΔA (where $L >> \lambda$), A' and A'' vary as

$$A' \sim \frac{\Delta A}{L}$$

$$A'' \sim \frac{\Delta A}{L^2}$$

Equation (3.8.6) shows that the variation of ϕ' is proportional to $n(z)$ and $\phi'' \sim \phi'/L$. Hence, A''/A in (3.8.4) is much smaller than $\beta_o^2 n^2(z) - \phi'^2$, and we obtain

$$\phi' = \pm\beta_o n(z) \Rightarrow \phi(z) = \pm\beta_o \int_0^z n(z')dz'$$

and from (3.8.5),

$$2A'\phi' + A\phi'' = 0 \Rightarrow 2A'\beta_o n + A\beta_o n' = 0 \Rightarrow (A^2 n)' = 0$$

or

$$A = \frac{C_1}{\sqrt{n(z)}}$$

Then, the general solution for the electric field in (3.8.1) is

$$E_y = \frac{C_1}{\sqrt{n(z)}} e^{-j\beta_o \int_0^z n(z')\,dz'} + \frac{C_2}{\sqrt{n(z)}} e^{j\beta_o \int_0^z n(z')\,dz'} \tag{3.8.10}$$

where the first term represents an outgoing wave, and the second represents an incoming wave.

An alternate derivation of (3.8.10) is obtained by substituting (3.8.9) into (3.8.8) to obtain a better approximation for the phase. That is,

$$\left(\frac{d\phi}{dz}\right)^2 \pm j\beta_o \frac{dn}{dz} = \beta_o^2 n^2$$

or

$$\frac{d\phi}{dz} = \pm\sqrt{\beta_o^2 n^2 \mp j\beta_o \frac{dn}{dz}} = \pm\beta_o n\sqrt{1 \mp j\frac{1}{\beta_o n^2}\frac{dn}{dz}} \approx \pm\beta_o n\left(1 \mp j\frac{1}{2\beta_o n^2}\frac{dn}{dz}\right) = \pm\beta_o n - j\frac{1}{2n}\frac{dn}{dz}$$

$$= \pm\beta_o n - j\frac{d}{dz}\ln\sqrt{n}$$

where the binomial expansion was used. Hence, a better approximation for the phase ϕ is given by

$$\phi(z) = \pm\beta_o \int_0^z n(z')dz' - j\ln\sqrt{n(z)}$$

and E_y is

$$E_y = A_o e^{\pm j\phi(z)} = A_o e^{\pm j\beta_o \int_0^z n(z')dz'} e^{-\ln\sqrt{n(z)}} = \frac{A_o}{\sqrt{n(z)}} e^{\pm j\beta_o \int_0^z n(z')dz'}$$

which is the same as (3.8.10).

Associated with E_y there is a magnetic field. For the outgoing wave it is given by

$$H_x = \frac{1}{j\omega\mu}\frac{\partial E_y}{\partial z} = -\frac{C_1\beta_o}{\omega\mu}\sqrt{n(z)}\,e^{-j\beta_o \int_0^z n(z')dz'} + j\frac{C_1\beta_o}{2\omega\mu}\frac{n'(z)}{[n(z)]^{3/2}}e^{-j\beta_o \int_0^z n(z')dz'}$$

The second term is much smaller than the first for the slowly varying $n(z)$ and can be neglected. Therefore, H_x is given by

$$H_x = -\frac{C_1\beta_o}{\omega\mu}\sqrt{n(z)}\,e^{-j\beta_o \int_0^z n(z')\,dz'}$$

and if $\mu = \mu_o$:

$$H_x = -C_1\frac{\sqrt{n(z)}}{\eta_o}e^{-j\beta_o \int_0^z n(z')\,dz'}$$

The general solution is written as

$$H_x = -\frac{C_1\sqrt{n(z)}}{\eta_o} e^{-j\beta_o \int_0^z n(z')\,dz'} + \frac{C_2\sqrt{n(z)}}{\eta_o} e^{j\beta_o \int_0^z n(z')\,dz'} \qquad (3.8.11)$$

Equations (3.8.10) and (3.8.11) are the first-order WKBJ approximations. It can be shown that these are the first term of an asymptotic expansion. Observe that the WKBJ approximation fails in (3.8.10) and (3.8.11) at the value of z where $n(z) = 0$. It will be shown that when $n(z) = 0$ a reflected wave occurs.

The validity of the first-order WKBJ approximation can be established by substituting into (3.8.1), like what was done with (3.8.7). To this end, we calculate

$$\frac{dE_y}{dz} = C_1\left[j\beta_o n^{1/2} - \frac{1}{2}n^{-3/2}n'\right] e^{j\beta_o \int_0^z n(z')\,dz'}$$

and then calculate d^2E_y/dz^2 (a little tedious). Substituting into (3.8.1) shows that the approximation is valid if

$$\frac{1}{\beta_o^2 n^2}\left|\frac{3}{4}\left(\frac{n'}{n}\right)^2 - \frac{1}{2}\frac{n''}{n}\right| << 1 \qquad (3.8.12)$$

As the wave propagates, the amplitude of the electric field decreases by the factor $1/\sqrt{n(z)}$ and the magnetic field increases by the same factor. Hence, the Poynting vector associated with the incident wave shows that the power density in the z direction is constant. Also, the wavelength slowly changes as a function of z, since

$$\lambda(z) = \frac{2\pi}{\beta_o n(z)}$$

The WKBJ approximation fails if $n(z)$goes to zero, known as a turning point, or if $n(z)$ changes suddenly, such as an interface. Both conditions produce a reflected field. A turning point can be handled by approximating the variation in $n^2(z) = \varepsilon_r(z)$ as a linear function of z, as shown in Fig. 3.14. The turning point occurs at $z = z_o$, and the region around z_o is known as the transition region (or patching region). Outside the transition region the WKBJ approximation holds.

The WKBJ solution to the left of the transition region in Fig. 3.14 includes an incident and a reflected wave. For a wave at normal incidence, from (3.8.10), the total field for $z < z_o$ can be expressed in the form

$$\begin{aligned} E_y &= \frac{C_1}{\sqrt{n(z)}}\left(e^{-j\beta_o \int_0^z n(z')dz'} + Re^{j\beta_o \int_0^z n(z')dz'}\right) = \frac{C_1}{\sqrt{n(z)}}\left(e^{-j\beta_o \int_0^z n(z')dz'} + e^{j\phi_r} e^{j\beta_o \int_0^z n(z')dz'}\right) \\ &= \frac{2C_1}{\sqrt{n(z)}} e^{j\phi_r/2}\cos\left(\beta_o \int_0^z n(z')dz' + \frac{\phi_r}{2}\right) \end{aligned} \qquad (3.8.13)$$

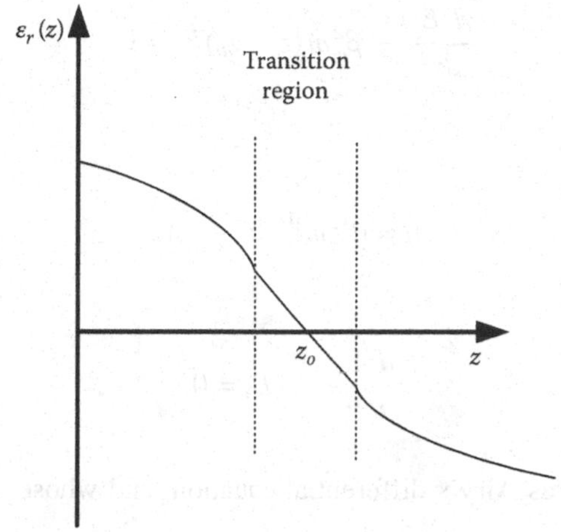

FIGURE 3.14 The transition region where $n^2(z) = \varepsilon_r(z)$ is represented by a linear variation.

where the transition region at $z = z_o$ affects the reflected wave by introducing a phase ϕ_r, denoted by the reflection coefficient $R = e^{j\phi_r}$. By energy conservation, since there are no losses or partial reflections, the magnitude of the reflection coefficient is unity

In the transition region, $n^2(z)$is represented by a linear profile, namely

$$n^2(z) = \varepsilon_r(z) = m(z_o - z) \tag{3.8.14}$$

where $m > 0$. (3.8.14) shows that $n^2 > 0$ for $z < z_o$, and $n^2 < 0$ for $z > z_o$.

Since

$$\int_0^z n(z')\,dz' = \sqrt{m}\int_0^z \sqrt{z_o - z'}\,dz' = \frac{2}{3}\sqrt{m}\,z_o^{3/2} - \frac{2}{3}\sqrt{m}\,(z_o - z)^{3/2}$$

and observing that

$$n^3(z) = m^{3/2}(z_o - z)^{3/2} \Rightarrow n^3(0) = m^{3/2}z_o^{3/2}$$

we write (3.8.13) in the form

$$E_y = \frac{2C_1}{\sqrt{n}}e^{j\phi_r/2}\cos\left\{\frac{2}{3}\frac{\beta_o}{m}[n^3 - n^3(0)] - \frac{\phi_r}{2}\right\} \tag{3.8.15}$$

The electric field in the transition region satisfies (3.8.1), which reads

$$\frac{d^2E_y}{dz^2} - \beta_o^2 m(z - z_o)E_y = 0 \tag{3.8.16}$$

Letting

$$\eta = \left(\beta_o^2 m\right)^{1/3}(z - z_o)$$

(3.8.16) is written as

$$\frac{d^2E_y}{d\eta^2} - \eta E_y = 0$$

which is recognized as Airy's differential equation, and whose solutions are

$$E_y = D_1 Ai(\eta) + D_2 Bi(\eta)$$

where D_1 and D_2 are constants.

Only the function $Ai(\eta)$ is suitable to match a sinusoidal behavior to the left of the transition region, and have an exponential decay to the right of the transition region, as seen in the plot of $Ai(\eta)$ in Appendix D. Thus, the function $Ai(\eta)$ can be made to match the WKBJ solutions on either side of the transition region. Hence,

$$E_y = D_1 Ai(\eta) = D_1 Ai\left[\left(\beta_o^2 m\right)^{1/3}(z - z_o)\right] \tag{3.8.17}$$

where for large negative η values ($\eta < -2$) the asymptotic approximation shows that

$$Ai(\eta) \approx \frac{1}{\sqrt{\pi}(-\eta)^{1/4}} \cos\left[\frac{2}{3}(-\eta)^{3/2} - \frac{\pi}{4}\right] \tag{3.8.18}$$

and for large positive η values, the asymptotic approximation is

$$Ai(\eta) \approx \frac{1}{2\sqrt{\pi}\eta^{1/4}} e^{-2/3\eta^{3/2}} \tag{3.8.19}$$

In (3.8.18), η in terms of n is

$$\eta = -\left(\frac{\beta_o}{m}\right)^{2/3} n^2$$

Then,

$$(-\eta)^{3/2} = \frac{\beta_o}{m} n^3$$

$$(-\eta)^{1/4} = \left(\frac{\beta_o}{m}\right)^{1/6} n^{1/2}$$

and it follows that the asymptotic approximation in (3.8.17) reads

$$E_y = D_1 Ai(\eta) \approx D_1 \left(\frac{m}{\beta_o}\right)^{1/6} \frac{1}{\sqrt{\pi n}} \cos\left[\frac{2}{3}\frac{\beta_o n^3}{m} - \frac{\pi}{4}\right] \tag{3.8.20}$$

Comparing (3.8.20) with (3.8.15) shows that the amplitudes can be matched by the proper selection of D_1, and the phase matching requires that

$$\frac{2}{3}\frac{\beta_o}{m}[n^3 - n^3(0)] - \frac{\phi_r}{2} = \frac{2}{3}\frac{\beta_o n^3}{m} - \frac{\pi}{4}$$

or

$$\phi_r = \frac{\pi}{2} - \frac{4}{3}\frac{\beta_o}{m} n^3(0) \tag{3.8.21}$$

Noting that

$$\int_0^{z_o} n(z')dz' = \sqrt{m} \int_0^{z_o} \sqrt{z_o - z'}\, dz' = \frac{2}{3}\sqrt{m}\, z_o^{3/2} = \frac{2}{3}\frac{n^3(0)}{m}$$

it follows that (3.8.21) can also be expressed in the form

$$\phi_r = \frac{\pi}{2} - 2\beta_o \int_0^{z_o} n(z')\, dz' \tag{3.8.22}$$

Therefore, the reflection coefficient is

$$R = e^{j\phi_r} = je^{-j2\beta_o \int_0^{z_o} n(z')\, dz'} \tag{3.8.23}$$

Once the WKBJ approximation is matched to the transition region solution, there is no need to further refer to the connection formula in terms of $Ai(\eta)$. That is, for $z < z_o$ the electric field is given by (3.8.13) with ϕ_r given by (3.8.22).

For $z > z_o$, ε_r is negative (see Fig. 3.14) and the index of refraction is written as

$$n(z) = -jn_1(z)$$

where

$$n_1(z) = \sqrt{m(z - z_o)}$$

Then, the WKB solution for the electric field is

$$E_y = \frac{C_1 T}{\sqrt{n(z)}} e^{-j\beta_o \int_{z_o}^{z} n(z')dz'} = \frac{C_1 T}{\sqrt{n(z)}} e^{-\beta_o \int_{z_o}^{z} n_1(z')dz'} \tag{3.8.24}$$

which represents an attenuated field for $z > z_o$.

The solution in the transition region is

$$E_y = D_1 Ai(\eta)$$

which for or large positive values of η, the field varies as

$$E_y = \frac{D_1}{2\sqrt{\pi}\eta^{1/4}} e^{-2/3\eta^{3/2}} \tag{3.8.25}$$

Since

$$-\frac{2}{3}\eta^{3/2} = -\beta_o \int_{z_o}^{z} n_1(z)dz$$

and

$$\eta^{1/4} = \left(\frac{\beta_o}{m}\right)^{1/6} n_1^{1/2}$$

it follows that (3.8.25) reads

$$E_y = \frac{D_1 T}{2\sqrt{\pi n(z)}} \left(\frac{m}{\beta_o}\right)^{1/6} e^{-\beta_o \int_{z_o}^{z} n_1(z')\, dz'} \tag{3.8.26}$$

A comparison of (3.8.24) and (3.8.26) shows that the amplitude constants can be matched and, therefore, the Airy solution describes correctly the field for $z > z_o$.

The transmission coefficient in (3.8.24) follows by equating the field in (3.8.13) to (3.8.24) at $z = z_o$. Hence,

$$T = \sqrt{2}\, e^{j\pi 4} e^{-j2\beta_o \int_0^{z_o} n(z')\, dz'}$$

Thus far, we have treated the case of total reflection associated with a turning point. A wave propagating in a slowly changing $n(z)$ will suffer small reflections, but these reflections in the WKBJ approximation are not there. Next, we consider the effects of partial reflections that can occur in regions where the WKBJ approximation works by letting the amplitudes be a function of z. That is, let

$$E_y = \underbrace{\frac{C_1(z)}{\sqrt{n(z)}} e^{-j\beta_o \int_0^z n(z')\,dz'}}_{E_u} + \underbrace{\frac{C_2(z)}{\sqrt{n(z)}} e^{j\beta_o \int_0^z n(z')\,dz'}}_{E_d} \tag{3.8.27}$$

and

$$\begin{aligned} H_x &= -\frac{C_1(z)}{\eta_o}\sqrt{n(z)}\, e^{-j\beta_o \int_0^z n(z')\,dz'} + \frac{C_2(z)}{\eta_o}\sqrt{n(z)}\, e^{j\beta_o \int_0^z n(z')\,dz'} \\ &= -\frac{n(z)}{\eta_o}(E_u - E_d) \end{aligned} \tag{3.8.28}$$

where E_u denotes the up wave and E_d the down wave. The down wave is due to partial reflections.

Substituting (3.8.27) and (3.8.28) into Maxwell's equations, namely

$$\frac{dE_y}{dz} = j\omega\mu_o H_x$$

and

$$\frac{dH_x}{dz} = j\omega\varepsilon_o n^2(z) E_y$$

shows that

$$\frac{d}{dz}(E_u + E_d) = -j\beta_o n(E_u - E_d) \tag{3.8.29}$$

and

$$\frac{d}{dz}(E_u - E_d) + \frac{1}{n}\frac{dn}{dz}(E_u - E_d) = -j\beta_o n(E_u + E_d) \tag{3.8.30}$$

Adding and subtracting (3.8.29) and (3.8.30) gives

$$\frac{dE_u}{dz} + j\beta_o n E_u + \frac{1}{2n}\frac{dn}{dz}E_u = \frac{1}{2n}\frac{dn}{dz}E_d \tag{3.8.31}$$

and

$$\frac{dE_d}{dz} - j\beta_o nE_d + \frac{1}{2n}\frac{dn}{dz}E_d = \frac{1}{2n}\frac{dn}{dz}E_u \tag{3.8.32}$$

which are coupled equations. Observe that if $(dn/dz)/n \to 0$, the solution to (3.8.31) is $e^{-j\beta_o nz}$, and the solution to (3.8.32) is $e^{j\beta_o nz}$, as expected. Also, if the coupling is neglected (i.e., setting $E_d = 0$ in (3.8.31) and $E_u = 0$ in (3.8.32)) the solutions are the WKBJ approximations in (3.8.27) and (3.8.28) with $C_1(z) = C_1$ and $C_2(z) = C_2$.

Substituting E_u and E_d in (3.8.27) into (3.8.31) and (3.8.32) gives

$$\frac{dC_1(z)}{dz} \approx \frac{1}{2n}\frac{dn}{dz}C_2(z)e^{j2\beta_o \int_0^z n(z')\,dz'} \tag{3.8.33}$$

and

$$\frac{dC_2(z)}{dz} \approx \frac{1}{2n}\frac{dn}{dz}C_1(z)e^{-j2\beta_o \int_0^z n(z')\,dz'} \tag{3.8.34}$$

where the terms with n'/n were neglected.

Equations (3.8.33) and (3.8.34) can be solved using iteration. That is, let $C_1(z) = C_1$ be a constant and solve (3.8.34) for $C_2(z)$, then substitute this value of $C_2(z)$ in (3.8.33) to find an approximate value of $C_1(z)$. Then, use this value of $C_1(z)$in (3.8.34) to obtain a better approximation for $C_2(z)$, and the process can be repeated. While recursive relations can be developed for this iteration method, the fact is that this method is seldom used, except to establish a limitation on the WKBJ approximation. That is, (3.8.33) and (3.8.34) produce a result of the form

$$C_1(z) = C_1 + \varepsilon C_1^1 + \ldots$$

and

$$C_2(z) = C_2 + \varepsilon C_2^1 + \ldots$$

where C_1 and C_2 are the primary wave and C_1^1 and C_2 the secondary waves. From these results one can calculate under what conditions the secondary waves can be neglected in comparison with the primary waves.

Next, the case of parallel polarization is considered. From (3.7.32), since $n(z)$ varies slowly: $\beta_o^2 N^2(z) \approx \beta_o^2 n^2(z)$. Then, (3.7.30) and (3.7.31) show that the WKBJ solution is

$$H_y = n(z)V = \sqrt{n(z)}\left[D_1 e^{-j\beta_o \int_0^z n(z')\,dz'} + D_2 e^{j\beta_o \int_0^z n(z')\,dz'}\right]$$

and the electric fields are

$$E_x = \frac{-1}{j\omega\varepsilon_o n^2(z)}\frac{\partial H_y}{\partial z} \approx \frac{\eta_o}{\sqrt{n(z)}}\left[D_1 e^{-j\beta_o \int_0^z n(z')\,dz'} - D_2 e^{j\beta_o \int_0^z n(z')\,dz'}\right]$$

and

$$E_z = -\frac{\eta_o \sin\theta_1}{n^2(z)} H_y$$

In the derivation of E_x the terms with dq/dz and dn/dz were neglected. The above WKBJ solutions, like the ones for perpendicular polarization, fail when $n(z) = 0$.

3.8.2 Oblique Incidence

In the case of a perpendicular polarized plane wave at oblique incidence, E_y satisfies

$$\frac{d^2E_y}{dx^2} + \frac{d^2E_y}{dz^2} + \beta_o^2 n^2(z) E_y = 0 \tag{3.8.35}$$

Letting

$$E_y(x, z) = Cf(z)e^{-j\beta_o x \sin\theta_1} \tag{3.8.36}$$

it follows that $f(z)$ satisfy

$$\frac{d^2f(z)}{dz^2} + \beta_o^2 q^2(z) f(z) = 0$$

where

$$q^2(z) = n^2(z) - \sin^2\theta_1 \tag{3.8.37}$$

Using Snell's law with $n(0) = 1$ (i.e., $\sin\theta_1 = n(z)\sin\theta(z)$), $q^2(z)$ can also be expressed as

$$q^2(z) = n^2(z)\cos^2\theta(z)$$

The WKBJ solution for $f(z)$ is given by (3.8.10) with $n(z)$ replaced by $q(z)$. Then, from (3.8.36), the WKBJ solution to (3.8.35) in a region with $\mu = \mu_o$ is

$$E_y = \left[\frac{C_1}{\sqrt{q(z)}} e^{-j\beta_o \int_0^z q(z')\,dz'} + \frac{C_2}{\sqrt{q(z)}} e^{j\beta_o \int_0^z q(z')\,dz'} \right] e^{-j\beta_o x \sin\theta_1} \tag{3.8.38}$$

The magnetic fields are given by

$$H_x = \left[-\frac{\sqrt{q(z)}}{\eta_o} C_1 e^{-j\beta_o \int_0^z q(z')\,dz'} + \frac{\sqrt{q(z)}}{\eta_o} C_2 e^{j\beta_o \int_0^z q(z')\,dz'} \right] e^{-j\beta_o x \sin\theta_1}$$

and

$$H_z = \frac{\sin\theta_1}{\eta_o} E_y$$

In the case of oblique incidence, the validity of the WKBJ approximation is given by (3.8.12) with $n(z)$ replaced by $q(z)$. This result shows that the approximation fails when $q = 0$ (i.e., when $n(z) = \sin\theta_1$). The condition $q = 0$ specifies the height at which reflection occurs.

A transition region is analyzed by writing (3.8.38) in the form

$$E_y = \frac{C_1}{\sqrt{q(z)}} \left(e^{-j\beta_o \int_0^z q(z')\,dz'} + R\, e^{j\beta_o \int_0^z q(z')\,dz'} \right) e^{-j\beta_o x \sin\theta_1} \tag{3.8.39}$$

where R represents the reflection coefficient.

In the transition region, $n(z)$ is given by (3.8.14). Hence, in this region $q(z)$ can be represented by a linear profile. That is,

$$q^2(z) = a(z_o - z)$$

where $a > 0$. Hence, E_y is given by (3.8.36) where $f(z)$ satisfy

$$\frac{d^2 f(z)}{dz^2} - \beta_o^2 a(z - z_o) f(z) = 0 \tag{3.8.40}$$

Letting

$$\eta = \left(\beta_o^2 a\right)^{1/3} (z - z_o)$$

the appropriate solution to (3.8.40) is $Ai(\eta)$, and E_y reads

$$E_y = C_2 Ai(\eta) e^{-j\beta_o x \sin \theta_i} \tag{3.8.41}$$

The procedure used to determine R in (3.8.39) is the same as that performed to find R in (3.8.13), and it follows that the reflection coefficient is given by (3.8.23).

The analysis of partial reflections in the case of oblique incidence results in (3.8.33) and (3.8.34) with $n(z)$ replaced by $q(z)$.

The analysis of a parallel polarized wave at oblique incidence is left to the problems.

Problems

P3.1 Verify that (3.2.26) follows from Maxwell's equation. That is, using

$$H_y = \frac{1}{j\omega\mu}\left(\frac{\partial E_z}{\partial x} - \frac{\partial E_x}{\partial z}\right)$$

P3.2 The radiation from a source is approximated by a plane wave normally incident on a surface. The source is in air (i.e., Region 1) and a submerge detector in sea water (i.e., Region 2). At the air-sea interface the field strength is 0.1 V/m. The source can operate at 100 kHz, and 10 MHz, and in sea water $\varepsilon_r = 80$ and $\sigma = 1$ S/m. At the three frequencies:

a. Determine the amplitude of the reflected fields.
b. Determine the penetration depths.
c. Determine the incident, reflected and transmitted power.
d. What is the attenuation at a distance of three skin depths.

P3.3 In (3.2.14) expressions for the fields for a perpendicular polarized plane wave are given. Write the forms of the fields for a parallel polarized plane wave in terms of $R_{\parallel}^H$ and in terms of $R_{\parallel}$.

P3.4 Show that a Brewster angle is not possible for a perpendicular polarized plane wave when $\mu_1 = \mu_2$.

P3.5 Consider the solution to the fields in a surface plasmon of the form

$$H_{y1} = Ce^{jk_{z1}z}e^{-jk_x x} = Ce^{\left(k''_{z1}z - k''_x x\right)}e^{j\left(k'_{z1}z - k'_x x\right)} \quad (z \le 0)$$

and

$$H_{y2} = Ce^{-jk_{z2}z}e^{-jk_x x} = Ce^{-\left(k''_{z2}z + k''_x x\right)}e^{-j\left(k'_{z2}z + k'_x x\right)} (z \geq 0)$$

where $f = 500$ THz, Region 1 is air (i.e., $\varepsilon_{r1} = 1$) and Region 2 is silver where

$$\varepsilon_{cr,2} = \varepsilon_{cr,2}' + j\varepsilon_{cr,2}'' = -17 - j0.5$$

a. Show that

$$k_{z1} = k'_{z1} - jk''_{z1} = \frac{k_o \varepsilon_{r1}}{\sqrt{\varepsilon_{r1} + \varepsilon_{cr,2}}} = j\frac{k_o \varepsilon_{r1}}{\sqrt{|\varepsilon_{cr2}'| + j\,|\varepsilon_{cr2}''| - \varepsilon_{r1}}}$$

$$k_{z2} = k'_{z2} - jk''_{z2} = -\frac{k_o \varepsilon_{cr2}}{\sqrt{\varepsilon_{r1} + \varepsilon_{cr2}}} = j\frac{k_o \left(|\varepsilon_{cr2}'| + j\,|\varepsilon_{cr2}''|\right)}{\sqrt{|\varepsilon_{cr2}'| + j\,|\varepsilon_{cr2}''| - \varepsilon_{r1}}}$$

$$k_x = k'_x - jk''_x = k_o\sqrt{\frac{\varepsilon_{r1}\varepsilon_{cr2}}{\varepsilon_{r1} + \varepsilon_{cr2}}} = k_o\sqrt{\frac{\varepsilon_{r1}\left(|\varepsilon_{cr2}'| + j\,|\varepsilon_{cr2}''|\right)}{|\varepsilon_{cr2}'| + j\,|\varepsilon_{cr2}''| - \varepsilon_{r1}}}$$

b. Evaluate k_{z1}, k_{z2}, and k_x to show that the fields are closely bound to the interface, and can propagates along the x direction a significant distance.

P3.6 A perpendicular polarized plane wave in incident on the exponentially increasing dielectric profile shown in Fig. P3.6. The dielectric is described by

$$\varepsilon_{r2} = a - (a - 1)e^{-2mz}$$

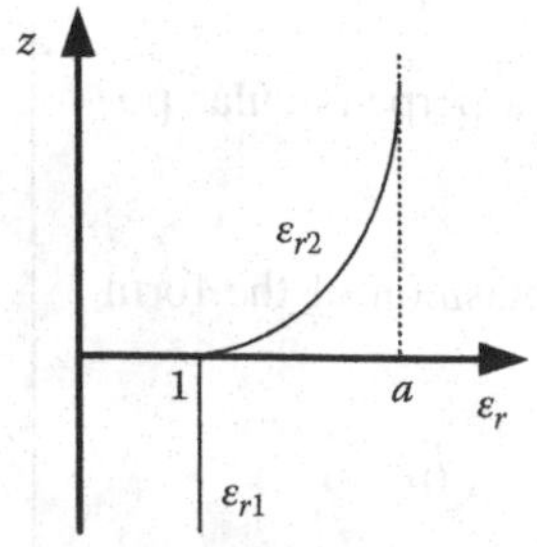

FIGURE P3.6

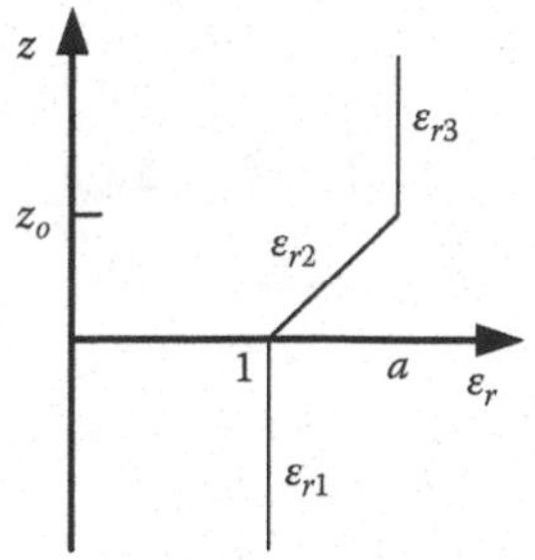

FIGURE P3.7

Determine $R_\perp$ and $T_\perp$.

P3.7 A perpendicular polarized plane wave is incident on the linearly increasing dielectric profile shown in Fig. P3.7. Determine the reflection and transmission coefficients in terms of Airy functions. In Region1: $\varepsilon_{r1} = 1$, in Region 2: $\varepsilon_{r2} = 1 + (a - 1)z/z_o$, and in Region 3: $\varepsilon_{r3} = a$.

P3.8 Determine the transmission matrix if Region 2 in Fig. 3.9 has an inhomogeneous profile given by

$$\varepsilon_{r2} = a - mz \; (0 \le z \le d)$$

P3.9 The eigenvalues relation in (3.7.68) was derived for a TE^z mode. Derive the eigenvalues relation for a TM^z mode.

P3.10 Verify (3.8.12).

P3.11

a. Show that for oblique incidence, the WKB approximation for a parallel polarized wave is given by

$$H_y = \frac{n(z)}{\sqrt{q(z)}}\left[D_1 e^{-j\beta_o \int_0^z q(z')\,dz'} + D_2 e^{j\beta_o \int_0^z q(z')\,dz'}\right] e^{-j\beta_o x \sin\theta_1}$$

b. Determine E_x and E_z.

4 Solutions to the Wave Equation

4.1 WAVE EQUATIONS

Maxwell's equations in a homogenous region and away from the source show that the **E** field satisfies a vector wave equation of the form

$$[\nabla^2 - \gamma^2]\mathbf{E} = 0 \tag{4.1.1}$$

subject to the appropriate boundary conditions. The same vector equation is satisfied by **H** and the auxiliary vectors **A**, **F**, $\mathbf{\Pi}^{\mathrm{e}}$ and $\mathbf{\Pi}^{\mathrm{m}}$. When using auxiliary vectors, the boundary conditions need to be expressed in terms of such vectors.

The vector Laplacian operator in (4.1.1) is calculated using

$$\nabla^2\mathbf{E} = \nabla(\nabla\bullet\mathbf{E}) - \nabla\times\nabla\times\mathbf{E} \tag{4.1.2}$$

In rectangular coordinates (4.1.2) gives

$$\nabla^2\mathbf{E} = \nabla^2 E_x\hat{\mathbf{a}}_x + \nabla^2 E_y\hat{\mathbf{a}}_y + \nabla^2 E_z\hat{\mathbf{a}}_z$$

which shows that each component of **E** in (4.1.1) satisfies a scalar wave equation, or

$$\begin{cases}[\nabla^2 - \gamma^2]E_x(x, y, z) = 0\\ [\nabla^2 - \gamma^2]E_y(x, y, z) = 0\\ [\nabla^2 - \gamma^2]E_z(x, y, z) = 0\end{cases} \tag{4.1.3}$$

In cylindrical coordinates, only the z-component in (4.1.2) reduces to a scalar wave equation. In spherical coordinates, none of the components of $\nabla^2\mathbf{E}$ in (4.1.2) reduce to a scalar wave equation. However, it will be shown that a scalar wave equation in terms of the radial component of an auxiliary vector is possible.

4.2 WAVE EQUATION IN RECTANGULAR COORDINATES

In rectangular coordinates, (4.1.3) shows that the scalar wave equations are of the form

$$[\nabla^2 - \gamma^2]\psi(x, y, z) = 0 \tag{4.2.1}$$

DOI: 10.1201/9781003219729-4

where ψ represents any rectangular component of a field vector or of an auxiliary field vector.

The solution of (4.2.1) can be obtained using the method of separation of variables. That is, let

$$\psi = X(x)Y(y)Z(z) \tag{4.2.2}$$

Substituting (4.2.2) into (4.2.1) and separating the variables gives

$$\frac{1}{X}\frac{d^2X}{dx^2} + \frac{1}{Y}\frac{d^2Y}{dy^2} + \frac{1}{Z}\frac{d^2Z}{dz^2} = \gamma^2$$

Each term in the left side is a function of a single variable. Therefore, this equation is satisfied if each term in the left-hand side is a constant, or

$$\begin{cases} \dfrac{1}{X}\dfrac{d^2X}{dx^2} = \gamma_x^2 & \Rightarrow \quad \dfrac{d^2X}{dx^2} - \gamma_x^2 X = 0 \\ \dfrac{1}{Y}\dfrac{d^2Y}{dy^2} = \gamma_y^2 & \Rightarrow \quad \dfrac{d^2Y}{dy^2} - \gamma_y^2 Y = 0 \\ \dfrac{1}{h}\dfrac{d^2Z}{dz^2} = \gamma_z^2 & \Rightarrow \quad \dfrac{d^2Z}{dz^2} - \gamma_z^2 Z = 0 \end{cases} \tag{4.2.3}$$

where γ_x, γ_y and γ_z are the separation constants that must satisfy

$$\gamma^2 = \gamma_x^2 + \gamma_y^2 + \gamma_z^2 \tag{4.2.4}$$

Observe that (4.2.4) shows that one of the separation constant can be written in terms of the others. For example,

$$\gamma_z = \left(\gamma^2 - \gamma_x^2 - \gamma_y^2\right)^{1/2}$$

The solutions to (4.2.3) can be expressed in different forms, such as exponential or hyperbolic trigonometric functions. Common forms are

$$\begin{cases} X = B_1 e^{-\gamma_x x} + B_2 e^{\gamma_x x} & \text{or} \quad X = B_1 \sinh\gamma_x x + B_2 \cosh\gamma_x x \\ Y = C_1 e^{-\gamma_y y} + C_2 e^{\gamma_y y} & \text{or} \quad Y = C_1 \sinh\gamma_y y + C_2 \cosh\gamma_y y \\ Z = D_1 e^{-\gamma_z z} + D_2 e^{\gamma_z z} & \text{or} \quad Z = D_1 \sinh\gamma_z z + D_2 \cosh\gamma_z z \end{cases} \tag{4.2.5}$$

In the lossless case $\gamma = j\beta$ ($\gamma^2 = -\beta^2$) and in (4.2.3) the separation constants are denoted by $-\beta_x^2$, $-\beta_y^2$ and $-\beta_z^2$, such that

$$\beta^2 = \beta_x^2 + \beta_y^2 + \beta_z^2$$

Then, the solutions are

$$\begin{cases} X = B_1 e^{-j\beta_x x} + B_2 e^{j\beta_x x} \text{ or } X = B_1 \sin\beta_x x + B_2 \cos\beta_x x \\ Y = C_1 e^{-j\beta_y y} + C_2 e^{j\beta_y y} \text{ or } Y = C_1 \sin\beta_y y + C_2 \cos\beta_y y \\ Z = D_1 e^{-j\beta_z z} + D_2 e^{j\beta_z z} \text{ or } Z = D_1 \sin\beta_z z + D_2 \cos\beta_z z \end{cases} \tag{4.2.6}$$

Boundary conditions determine the values of the separation constants. These values are known as the eigenvalues, and the functions in (4.2.5) and (4.2.6) as the eigenfunctions. As an example, if the eigenvalues γ_x and γ_y are discrete, the general solution to (4.2.1) can be written as

$$\psi = \sum_{\gamma_x} \sum_{\gamma_y} C_{\gamma_x\gamma_y} X(x) Y(y) Z(z)$$

where $C_{\gamma_x\gamma_y}$ are constants that depend on the eigenvalues γ_x and γ_y. If the eigenvalues are continuous, such as those that will be associated with unbounded region, the solution is of the form

$$\psi = \int_{\gamma_x} \int_{\gamma_y} C_{\gamma_x\gamma_y} X(x) Y(y) Z(z) d\gamma_x d\gamma_y$$

where $C_{\gamma_x\gamma_y}$ is a function of γ_x and γ_y.

Consider a field configuration with its magnetic field transverse to the z-direction, denoted by TM^z. This field can be generated by letting

$$\mathbf{A} = A_z \hat{\mathbf{a}}_z \text{ or } \mathbf{\Pi}^e = \Pi_z^e \hat{\mathbf{a}}_z$$

Then, A_z and Π_z^e satisfy (4.2.1) and the solutions are of the form in (4.2.5) or (4.2.6).

Away from the source, the fields in terms of A_z follow from (1.5.7) and (1.5.10), and the fields in terms of Π_z^e follow from (1.7.4) and (1.7.6). That is,

$$H_x = \frac{1}{\mu}\frac{\partial A_z}{\partial y} = \frac{\gamma^2}{j\omega\mu}\frac{\partial \Pi_z^e}{\partial y} \tag{4.2.7}$$

$$H_y = -\frac{1}{\mu}\frac{\partial A_z}{\partial x} = -\frac{\gamma^2}{j\omega\mu}\frac{\partial \Pi_z^e}{\partial x} \tag{4.2.8}$$

$$H_z = 0 \tag{4.2.9}$$

$$E_x = \frac{j\omega}{\gamma^2}\frac{\partial^2 A_z}{\partial x \partial z} = \frac{\partial^2 \Pi_z^e}{\partial x \partial z} \quad (4.2.10)$$

$$E_y = \frac{j\omega}{\gamma^2}\frac{\partial^2 A_z}{\partial y \partial z} = \frac{\partial^2 \Pi_z^e}{\partial y \partial z} \quad (4.2.11)$$

$$E_z = \frac{j\omega}{\gamma^2}\left(\frac{\partial^2 A_z}{\partial z^2} - \gamma^2 A_z\right) = \frac{\partial^2 \Pi_z^e}{\partial z^2} - \gamma^2 \Pi_z^e \quad (4.2.12)$$

For a field that is TM^x, let $\mathbf{A} = A_x \hat{\mathbf{a}}_x$ or $\mathbf{\Pi}^e = \Pi_x^e \hat{\mathbf{a}}_x$ and the fields are

$$H_x = 0 \quad (4.2.13)$$

$$H_y = \frac{1}{\mu}\frac{\partial A_x}{\partial z} = \frac{\gamma^2}{j\omega\mu}\frac{\partial \Pi_x^e}{\partial z} \quad (4.2.14)$$

$$H_z = -\frac{1}{\mu}\frac{\partial A_x}{\partial y} = -\frac{\gamma^2}{j\omega\mu}\frac{\partial \Pi_x^e}{\partial x} \quad (4.2.15)$$

$$E_x = \frac{j\omega}{\gamma^2}\left(\frac{\partial^2 A_x}{\partial x^2} - \gamma^2 A_x\right) = \frac{\partial^2 \Pi_x^e}{\partial x^2} - \gamma^2 \Pi_z^e \quad (4.2.16)$$

$$E_y = \frac{j\omega}{\gamma^2}\frac{\partial^2 A_x}{\partial x \partial y} = \frac{\partial^2 \Pi_x^e}{\partial x \partial y} \quad (4.2.17)$$

$$E_z = \frac{j\omega}{\gamma^2}\frac{\partial^2 A_x}{\partial x \partial z} = \frac{\partial^2 \Pi_x^e}{\partial x \partial z} \quad (4.2.18)$$

For a field that is TM^y, let $\mathbf{A} = A_y \hat{\mathbf{a}}_y$ or $\mathbf{\Pi}^e = \Pi_y^e \hat{\mathbf{a}}_y$ and the fields are

$$H_x = -\frac{1}{\mu}\frac{\partial A_y}{\partial z} = -\frac{\gamma^2}{j\omega\mu}\frac{\partial \Pi_y^e}{\partial z} \quad (4.2.19)$$

$$H_y = 0 \quad (4.2.20)$$

$$H_z = \frac{1}{\mu}\frac{\partial A_y}{\partial x} = \frac{\gamma^2}{j\omega\mu}\frac{\partial \Pi_y^e}{\partial x} \quad (4.2.21)$$

$$E_x = \frac{j\omega}{\gamma^2}\frac{\partial^2 A_y}{\partial x \partial y} = \frac{\partial^2 \Pi_y^{\mathrm{e}}}{\partial x \partial y} \tag{4.2.22}$$

$$E_y = \frac{j\omega}{\gamma^2}\left(\frac{\partial^2 A_y}{\partial y^2} - \gamma^2 A_y\right) = \frac{\partial^2 \Pi_y^{\mathrm{e}}}{\partial y^2} - \gamma^2 \Pi_y^{\mathrm{e}} \tag{4.2.23}$$

$$E_z = \frac{j\omega}{\gamma^2}\frac{\partial^2 A_y}{\partial y \partial z} = \frac{\partial^2 \Pi_y^{\mathrm{e}}}{\partial y \partial z} \tag{4.2.24}$$

The factor $\gamma^2/j\omega$ that appears in the previous relation can also be written as

$$\frac{\gamma^2}{j\omega} = j\omega\mu\varepsilon_c = j\omega\mu_o\varepsilon_o n^2$$

In the lossless case: $\varepsilon_c = \varepsilon$, $\gamma^2 = -\beta^2$ and

$$\frac{\gamma^2}{j\omega} = \frac{-\beta^2}{j\omega} = j\omega\mu\varepsilon$$

If the field configuration has its electric field transverse to the z-direction, denoted by TE^z, the field can be generated by letting

$$\mathbf{F} = F_z \hat{\mathbf{a}}_z \text{ or } \mathbf{\Pi}^{\mathrm{m}} = \Pi_z^{\mathrm{m}} \hat{\mathbf{a}}_z$$

Then, F_z and Π_z^{m} satisfy (4.2.1), and the solutions are of the form in (4.2.5) or (4.2.6).

Away from the source, the fields in terms of $\mathbf{F} = F_z \hat{\mathbf{a}}_z$ follow from (1.5.19) and (1.5.22), and in terms of $\mathbf{\Pi}^{\mathrm{m}} = \Pi_z^{\mathrm{m}} \hat{\mathbf{a}}_z$ from (1.7.10) and (1.7.11). That is,

$$E_x = -\frac{1}{\varepsilon_c}\frac{\partial F_z}{\partial y} = -j\omega\mu\frac{\partial \Pi_z^{\mathrm{m}}}{\partial y} \tag{4.2.25}$$

$$E_y = \frac{1}{\varepsilon_c}\frac{\partial F_z}{\partial x} = j\omega\mu\frac{\partial \Pi_z^{\mathrm{m}}}{\partial x} \tag{4.2.26}$$

$$E_z = 0 \tag{4.2.27}$$

$$H_x = \frac{j\omega}{\gamma^2}\frac{\partial^2 F_z}{\partial x \partial z} = \frac{\partial^2 \Pi_z^{\mathrm{m}}}{\partial x \partial z} \tag{4.2.28}$$

$$H_y = \frac{j\omega}{\gamma^2}\frac{\partial^2 F_z}{\partial y \partial z} = \frac{\partial \Pi_z^{\mathrm{m}}}{\partial y \partial z} \tag{4.2.29}$$

$$H_z = \frac{j\omega}{\gamma^2}\left(\frac{\partial^2 F_z}{\partial z^2} - \gamma^2 F_z\right) = \frac{\partial^2 \Pi_z^{\mathrm{m}}}{\partial z^2} - \gamma^2 \Pi_z^{\mathrm{m}} \tag{4.2.30}$$

For a field that is TE^x, let $\mathbf{F} = F_x \hat{\mathbf{a}}_x$ or $\mathbf{\Pi}^{\mathrm{m}} = \Pi_x^{\mathrm{m}} \hat{\mathbf{a}}_x$ and the fields are

$$E_x = 0$$

$$E_y = -\frac{1}{\varepsilon_c}\frac{\partial F_x}{\partial z} = -j\omega\mu \frac{\partial \Pi_x^{\mathrm{m}}}{\partial z}$$

$$E_z = \frac{1}{\varepsilon_c}\frac{\partial F_x}{\partial y} = j\omega\mu \frac{\partial \Pi_x^{\mathrm{m}}}{\partial y}$$

$$H_x = \frac{j\omega}{\gamma^2}\left(\frac{\partial^2 F_x}{\partial x^2} - \gamma^2 F_x\right) = \frac{\partial^2 \Pi_x^{m}}{\partial x^2} - \gamma^2 \Pi_x^{\mathrm{m}}$$

$$H_y = \frac{j\omega}{\gamma^2}\frac{\partial^2 F_x}{\partial x \partial y} = \frac{\partial^2 \Pi_x^{\mathrm{m}}}{\partial x \partial y}$$

$$H_z = \frac{j\omega}{\gamma^2}\frac{\partial^2 F_x}{\partial x \partial z} = \frac{\partial^2 \Pi_x^{\mathrm{m}}}{\partial x \partial z}$$

For a field that is TE^y, let $\mathbf{F} = F_y \hat{\mathbf{a}}_y$ or $\mathbf{\Pi}^{\mathrm{m}} = \Pi_y^{\mathrm{m}} \hat{\mathbf{a}}_y$ and the fields are

$$E_x = \frac{1}{\varepsilon_c}\frac{\partial F_y}{\partial z} = j\omega\mu \frac{\partial \Pi_y^{\mathrm{m}}}{\partial z} \tag{4.2.31}$$

$$E_y = 0 \tag{4.2.32}$$

$$E_z = -\frac{1}{\varepsilon_c}\frac{\partial F_y}{\partial x} = -j\omega\mu \frac{\partial \Pi_y^{\mathrm{m}}}{\partial x} \tag{4.2.33}$$

$$H_x = \frac{j\omega}{\gamma^2}\frac{\partial^2 F_y}{\partial x \partial y} = \frac{\partial^2 \Pi_y^{\mathrm{m}}}{\partial x \partial y} \tag{4.2.34}$$

$$H_y = \frac{j\omega}{\gamma^2}\left(\frac{\partial^2 F_y}{\partial y^2} - \gamma^2 F_y\right) = \frac{\partial^2 \Pi_y^m}{\partial y^2} - \gamma^2 \Pi_y^m \quad (4.2.35)$$

$$H_z = \frac{j\omega}{\gamma^2}\frac{\partial^2 F_y}{\partial y \partial z} = \frac{\partial^2 \Pi_y^m}{\partial y \partial z} \quad (4.2.36)$$

In the lossless case, replace ε_c by ε in the expressions (4.2.25) to (4.2.36).

4.3 WAVE PROPAGATION IN RECTANGULAR GEOMETRIES

4.3.1 Rectangular Waveguide

As an example of a TM^z field consider the rectangular waveguide shown in Fig. 4.1 where the direction of propagation is in the positive z direction. If the rectangular waveguide is air filled, then $\mu = \mu_o$ and $\varepsilon = \varepsilon_o$, and it follows that $\gamma^2 = -\beta_o^2 = -\omega^2\mu_o\varepsilon_o$. From (1.5.11), or the general form in (4.2.1), A_z satisfies

$$\left[\nabla^2 + \beta_o^2\right]A_z(x, y, z) = 0$$

and from (4.2.6) the appropriate solution in terms of A_z is

$$A_z = [B_1 \sin(\beta_x x) + B_2 \cos(\beta_x x)][C_1 \sin(\beta_y y) + C_2 \cos(\beta_y y)]e^{-j\beta_z z} \quad (4.3.1)$$

where

$$\beta_o^2 = \beta_x^2 + \beta_y^2 + \beta_z^2 \Rightarrow \beta_z = \sqrt{\beta_o^2 - \beta_x^2 - \beta_y^2}$$

The boundary conditions are the vanishing of the tangential **E** fields at $x = 0$, $x = a$, $y = 0$, and $y = b$. These boundary conditions are sufficient to satisfy the vanishing

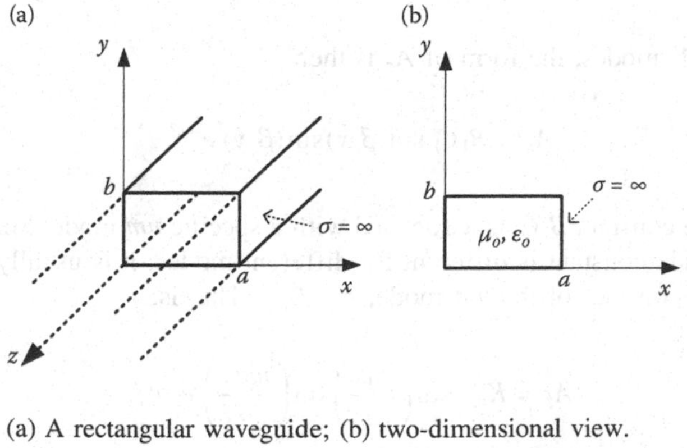

FIGURE 4.1 (a) A rectangular waveguide; (b) two-dimensional view.

of the normal components of the magnetic field at the conductor surfaces. Then, from (4.2.11):

$$E_y = \frac{-j\omega}{\beta_o^2}\frac{\partial^2 A_z}{\partial y \partial z} = \frac{-\omega\beta_y\beta_z}{\beta_o^2}[B_1 \sin(\beta_x x) + B_2 \cos(\beta_x x)][C_1 \cos(\beta_y y) - C_2 \sin(\beta_y y)]e^{-j\beta_z z} \tag{4.3.2}$$

The boundary conditions for E_y are: $E_y = 0$ at $x = 0$ and $x = a$. Therefore, in (4.3.2) the boundary condition at $x = 0$ requires that $B_2 = 0$, and the boundary condition at $x = a$ requires that

$$E_y|_{x=a} = 0 \Rightarrow \sin(\beta_x a) = 0 \Rightarrow \beta_x = \frac{n\pi}{a} \qquad (n = 1, 2, 3, \ldots)$$

which shows the discrete values that β_x can have. The reason of starting with $n = 1$ is that $n = 0$ leads to $A_z = 0$ in (4.3.1) and, therefore, no fields.

The boundary conditions for E_x are: $E_x = 0$ at $y = 0$ and $y = b$. Therefore, from (4.2.10),

$$E_x = \frac{-j\omega}{\beta_o^2}\frac{\partial^2 A_z}{\partial x \partial z} = \frac{-\omega\beta_x\beta_z}{\beta_o^2}B_1 \cos(\beta_x x)[C_1 \sin(\beta_y y) + C_2 \cos(\beta_y y)]e^{-j\beta_z z} \tag{4.3.3}$$

and in (4.3.3) the boundary condition at $y = 0$ requires that $C_2 = 0$ and at $y = b$:

$$E_x|_{y=b} = 0 \Rightarrow \sin(\beta_y b) = 0 \Rightarrow \beta_y = \frac{m\pi}{b} \qquad (m = 1, 2, 3, \ldots)$$

which shows the discrete values that β_y can have. The separation constants are required to attain the indicated discrete values in order to satisfy the boundary conditions.

For the TM^z modes, the form of A_z is then

$$A_z = B_1 C_1 \sin(\beta_x x)\sin(\beta_y y)\, e^{-j\beta_z z}$$

The amplitude constant $B_1 C_1$ is associated with a specific nm mode. Since the value of the amplitude constant is different for different modes, it is usually denoted by the amplitude constant of the nm mode, say K_{nm}. That is,

$$A_z = K_{nm} \sin\left(\frac{n\pi x}{a}\right)\sin\left(\frac{m\pi y}{b}\right) e^{-j\beta_z z}$$

This expression denotes the form of the TM^z modes, which are denoted by TM^z_{nm}. Observe that the TM^z_{0m} modes or TM^z_{n0} modes cannot exist in the rectangular waveguide.

The separation constant β_z, which represents the propagation constant of the field in the z direction, is given by

$$\beta_z = \left[\omega^2\mu_o\varepsilon_o - \left(\frac{n\pi}{a}\right)^2 - \left(\frac{m\pi}{b}\right)^2\right]^{1/2} \tag{4.3.4}$$

The fields follow from (4.2.7) to (4.2.12), which for the nm mode are

$$E_x = -K_{nm}\frac{\omega\beta_x\beta_z}{\beta_o^2}\cos\left(\frac{n\pi x}{a}\right)\sin\left(\frac{m\pi y}{b}\right)e^{-j\beta_z z}$$

$$E_y = -K_{nm}\frac{\omega\beta_y\beta_z}{\beta_o^2}\sin\left(\frac{n\pi x}{a}\right)\cos\left(\frac{m\pi y}{b}\right)e^{-j\beta_z z}$$

$$E_z = -K_{nm}\frac{j\omega\left(\beta_o^2-\beta_z^2\right)}{\beta_o^2}\sin\left(\frac{n\pi x}{a}\right)\sin\left(\frac{m\pi y}{b}\right)e^{-j\beta_z z}$$

$$H_x = K_{nm}\frac{\beta_y}{\mu_o}\sin\left(\frac{n\pi x}{a}\right)\cos\left(\frac{m\pi y}{b}\right)e^{-j\beta_z z}$$

$$H_y = -K_{nm}\frac{\beta_x}{\mu_o}\cos\left(\frac{n\pi x}{a}\right)\sin\left(\frac{m\pi y}{b}\right)e^{-j\beta_z z}$$

$$H_z = 0$$

The frequency that makes $\beta_z = 0$ is the cutoff frequency ω_c for the nm mode. From (4.3.4), in terms of f_c, the cutoff frequency for the nm mode is

$$f_c = \frac{1}{2\pi\sqrt{\mu_o\varepsilon_o}}\sqrt{\left(\frac{n\pi}{a}\right)^2 + \left(\frac{m\pi}{b}\right)^2} \tag{4.3.5}$$

The mode with the lowest f_c value is the TM^z_{11} mode. Below the cutoff frequency β_z becomes imaginary. Therefore, instead of a propagating wave (i.e., $e^{-j\beta_z z}$) the fields decay exponentially in the z direction. This occurs in any waveguide system for frequencies below its cutoff frequency.

For completeness, the fields for the TE^z_{nm} modes in the waveguide are derived in terms of F_z, and it follows that

$$F_z = Q_{nm}\cos\left(\frac{n\pi x}{a}\right)\cos\left(\frac{m\pi y}{b}\right)e^{-j\beta_z z} \tag{4.3.6}$$

where Q_{nm} is the amplitude constant associated with the nm mode, where $n = 0, 1, 2, \ldots$ and $m = 0, 1, 2, \ldots$. However, the $n = m = 0$ must be excluded, since it leads to F_z having a constant variation in x and y, and the fields will be zero. The fields are calculated using (4.2.25) to (4.2.30) and it is seen that they satisfy the boundary conditions at the waveguide walls. For example, (4.2.25) shows that E_x is proportional to $\sin(m\pi y/b)$, which is zero at $y = 0$ and $y = b$.

Equations (4.3.4) and (4.3.5) also apply to the TE^z_{nm} mode. The lowest cutoff frequency when $a > b$ is the one for the TE^z_{10} mode. For an air-filled waveguide it is given by

$$f_c\Big|_{\substack{n=1\\m=0}} = \frac{1}{2a\sqrt{\mu_o \varepsilon_o}}$$

The cutoff frequency of the TM^z_{11} mode is higher than the one for the TE^z_{10} mode. Therefore, the TE^z_{10} mode is the dominant mode when $a > b$.

Specifically, with $\gamma^2 = -\beta_o^2$ and $\gamma^2/j\omega = j\omega\mu_o\varepsilon_o$, the fields for the TE^z_{10} mode are

$$F_z\big|_{\substack{n=1\\m=0}} = Q_{10}\cos\left(\frac{\pi x}{a}\right)e^{-j\beta_z x}$$

$$E_x = E_z = H_y = 0$$

$$E_y = \frac{1}{\varepsilon_o}\frac{\partial F_z}{\partial x}\bigg|_{\substack{n=1\\m=0}} = \frac{-Q_{10}\pi}{\varepsilon_o a}\sin\left(\frac{\pi x}{a}\right)e^{-j\beta_z z}$$

$$H_x = \frac{1}{j\omega\mu_o\varepsilon_o}\frac{\partial^2 F_z}{\partial x \partial z}\bigg|_{\substack{n=1\\m=0}} = \frac{Q_{10}\beta_z\pi}{\omega\mu_o\varepsilon_o a}\sin\left(\frac{\pi x}{a}\right)e^{-j\beta_z z}$$

$$H_z = \frac{1}{j\omega\mu_o\varepsilon_o}\left(\frac{\partial^2 F_z}{\partial z^2} + \beta_o^2 F_z\right) = \frac{1}{j\omega\mu_o\varepsilon_o}\beta_x^2 F_z\bigg|_{\substack{n=1\\m=0}} = -j\frac{Q_{10}\pi^2}{\omega\mu_o\varepsilon_o a^2}\cos\left(\frac{\pi x}{a}\right)e^{-j\beta_z z}$$

Equation (4.3.5) which applies to both TE^z_{nm} and TM^z_{nm} modes show that when $m \neq 0$ and $n \neq 0$ the modes have the same cutoff frequencies. Modes that have the same cutoff frequencies but are different in their field configuration are called degenerate modes.

4.3.2 Rectangular Cavity

An air-filled rectangular cavity is shown is Fig. 4.2. It is a rectangular waveguide with conducting plates added at $z = 0$ and $z = c$. While the rectangular waveguide supports traveling waves, the waves in a rectangular cavity are standing waves. Both TM^z and TE^z modes are possible, as well as other transverse modes. Consider the TE^z mode where the form of F_z in (4.3.6) satisfy the tangential boundary

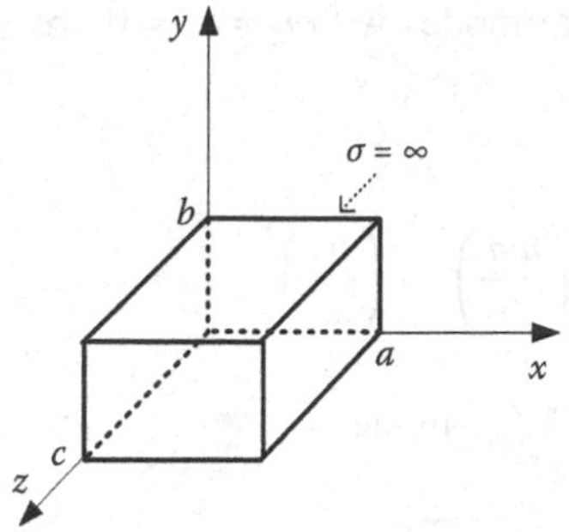

FIGURE 4.2 The rectangular cavity.

conditions at the x and y walls. Therefore, all that is needed is to satisfy the boundary conditions at $z = 0$ and $z = c$. The form of F_z in the cavity is then:

$$F_z = Q_{nm}\cos\left(\frac{n\pi x}{a}\right)\cos\left(\frac{m\pi y}{b}\right)[D_1\sin(\beta_z z) + D_2\cos(\beta_z z)] \qquad (4.3.7)$$

To satisfy the boundary condition at $z = 0$ we calculate the tangential fields, which can be either the E_x or E_y field. From (4.2.25):

$$E_x = -\frac{1}{\varepsilon_o}\frac{\partial F_z}{\partial y} = Q_{nm}\frac{m\pi}{\varepsilon_o b}\cos\left(\frac{n\pi x}{a}\right)\sin\left(\frac{m\pi y}{b}\right)[D_1\sin(\beta_z z) + D_2\cos(\beta_z z)] \qquad (4.3.8)$$

The boundary condition at $z = 0$ is

$$E_x = 0|_{z=0} \Rightarrow D_1\sin(0) + D_2\cos(0) = 0 \Rightarrow D_2 = 0$$

The boundary condition at $z = c$ is

$$E_x = 0|_{z=c} \Rightarrow D_1\sin(\beta_z c) = 0 \Rightarrow \beta_z = \frac{p\pi}{c} \qquad (p = 1, 2, 3, \ldots)$$

which represents the discrete values that β_z can have. Then, from (4.3.7) the modes function is

$$F_z = Q_{nmp}\cos(\beta_x x)\cos(\beta_y y)\sin(\beta_z z) = Q_{nmp}\cos\left(\frac{n\pi x}{a}\right)\cos\left(\frac{m\pi y}{b}\right)\sin\left(\frac{p\pi z}{c}\right) \qquad (4.3.9)$$

where the amplitude constant associated with the nmp mode is expressed as Q_{nmp}. In (4.3.9), $n = 0, 1, 2, \ldots, m = 0, 1, 2, \ldots$ and $p = 1, 2, 3, \ldots$. The fields components follow from (4.2.25) to (4.2.30) and it is observed that modes with $n = m = 0$ are not allowed since the resulting fields are zero.

For the air-filled cavity:

$$\beta_o^2 = \beta_x^2 + \beta_y^2 + \beta_z^2 \Rightarrow \omega^2\mu_o\varepsilon_o = \left(\frac{n\pi}{a}\right)^2 + \left(\frac{m\pi}{b}\right)^2 + \left(\frac{p\pi}{c}\right)^2$$

and it follows that the resonant frequency f_r of the TE^z_{nmp} mode is

$$f_r = \frac{1}{2\pi\sqrt{\mu_o\varepsilon_o}}\sqrt{\left(\frac{n\pi}{a}\right)^2 + \left(\frac{m\pi}{b}\right)^2 + \left(\frac{p\pi}{c}\right)^2} \tag{4.3.10}$$

For a cavity with ε and μ values, change μ_o to μ and ε_o to ε in (4.3.10).

If $c > a > b$ the fundamental mode is the TE^z_{101} mode. For this mode:

$$F_z\Big|_{\substack{n=1\\m=0\\p=1}} = Q_{101}\cos\left(\frac{\pi x}{a}\right)\sin\left(\frac{\pi z}{c}\right) \tag{4.3.11}$$

and from (4.3.10) the resonant frequency for the TE^z_{101} mode is

$$f_r\Big|_{\substack{n=1\\m=0\\p=1}} = \frac{1}{2\pi\sqrt{\mu_o\varepsilon_o}}\sqrt{\left(\frac{\pi}{a}\right)^2 + \left(\frac{\pi}{c}\right)^2}$$

The analysis could have also been done by starting with (4.2.1), which for the TE^z mode is

$$[\nabla^2 - \gamma^2]F_z(x, y, z) = 0$$

whose solution can be written as

$$F_z = Q\cosh(\gamma_x x)\cosh(\gamma_y y)\sinh(\gamma_z z) \tag{4.3.12}$$

where

$$\gamma^2 = \gamma_x^2 + \gamma_y^2 + \gamma_z^2$$

Substituting (4.3.12) into (4.2.25) to (4.2.30) shows that the boundary condition for the tangential electric fields at $x = 0$, $y = 0$, and $z = 0$ are satisfied. The boundary conditions at $x = a$, $y = b$, and $z = c$ require that

$$E_y = \frac{1}{\varepsilon_c}\frac{\partial F_z}{\partial x}\Big|_{x=a} = 0 \Rightarrow \sinh(\gamma_x a) = 0 \Rightarrow \gamma_x = j\frac{n\pi}{a} = j\beta_x \qquad (n = 0, 1, 2, \ldots)$$

$$E_x = -\frac{1}{\varepsilon_c}\frac{\partial F_z}{\partial y}\Big|_{y=b} = 0 \Rightarrow \sinh(\gamma_y b) = 0 \Rightarrow \gamma_y = j\frac{m\pi}{b} = j\beta_y \qquad (m = 0, 1, 2, \ldots)$$

$$E_x = -\frac{1}{\varepsilon_c}\frac{\partial F_z}{\partial y}\Big|_{z=c} = 0 \Rightarrow \sinh(\gamma_z c) = 0 \Rightarrow \gamma_z = j\frac{p\pi}{c} = j\beta_z \qquad (p = 1, 2, 3, \ldots)$$

and the mode function in (4.3.12) is written in the form

$$F_z = Q_{nmp}\cos\left(\frac{n\pi x}{a}\right)\cos\left(\frac{m\pi y}{b}\right)\sin\left(\frac{p\pi z}{c}\right)$$

which is identical to (4.3.9). Also, in the lossless case, with $\gamma^2 = -\beta_o^2$ we obtain

$$\beta_o^2 = \omega^2\mu_o\varepsilon_o = \left(\frac{n\pi}{a}\right)^2 + \left(\frac{m\pi}{b}\right)^2 + \left(\frac{p\pi}{c}\right)^2 \tag{4.3.13}$$

Equation (4.3.13) gives the same resonant frequencies as (4.3.10).

For completeness, for the TM^z_{nmp} modes the vector potential A_z is given by

$$A_z = K_{nmp}\sin\left(\frac{n\pi x}{a}\right)\sin\left(\frac{m\pi y}{b}\right)\cos\left(\frac{p\pi z}{c}\right) \qquad \begin{pmatrix} n = 1, 2, 3, \ldots \\ m = 1, 2, 3, \ldots \\ p = 0, 1, 2, \ldots \end{pmatrix}$$

where the resonant frequencies are also given by (4.3.10). For the TM^z_{nmp} modes, the TM^z_{110} mode has the lowest cutoff frequency.

4.3.3 Partially Filled Rectangular Waveguide

The cross section of a partially filled waveguide with a dielectric placed in the vertical direction is shown in Fig. 4.3a, and with a horizontal placed dielectric in Fig. 4.3b. In these types of waveguides, the TE^z and TM^z modes cannot satisfy the boundary conditions (except some TE^z modes, such as the TE^z_{n0} modes which are similar to the TE^x_{n0} modes in Fig. 4.3a). The modes that satisfy the boundary conditions in the waveguide in Fig. 4.3a are the TE^x and TM^x modes. These modes are also known as longitudinal section electric (LSE^x) and longitudinal section magnetic modes (LSM^x). For the waveguide in Fig. 4.3b, the modes that satisfy the boundary conditions are the TE^y and TM^y modes.

The fundamental mode in an empty waveguide is the TE^z_{10} mode. Hence, the analysis of the TE^x_{n0} modes (i.e., no y variation) in Fig. 4.3a will be performed. From (4.2.1) with $\psi = E_y$, the electric field in the air region (Region 1) and in the dielectric region (Region 2) are

$$E_{y1} = f_1(x)e^{-j\beta_z z}$$

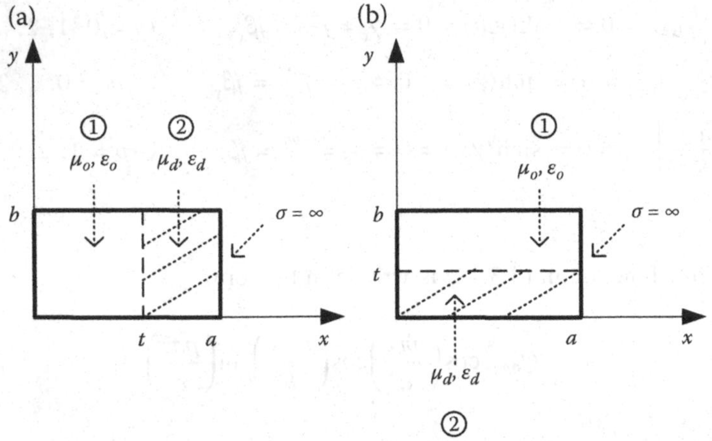

FIGURE 4.3 (a) Cross-sectional view of a rectangular waveguide partially filled with a dielectric in the vertical direction; (b) cross-sectional view of a rectangular waveguide partially filled with a dielectric in the horizontal direction.

and

$$E_{y2} = f_2(x)e^{-j\beta_z z}$$

where β_z is the same in both regions in order for the phase variation along the interface to be the same. The functions $f_1(x)$ and $f_2(x)$ satisfy the equations:

$$\left[\frac{d^2}{dx^2} + \beta_o^2 - \beta_z^2\right] f_1(x) = 0 \qquad (0 \le x \le t) \qquad (4.3.14)$$

and

$$\left[\frac{d^2}{dx^2} + \beta_d^2 - \beta_z^2\right] f_2(x) = 0 \qquad (t \le x \le a) \qquad (4.3.15)$$

where $\beta_d^2 = \beta_o^2 \mu_{rd} \varepsilon_{rd}$.

From (4.3.14) and (4.3.15), the functions $f_1(x)$ and $f_2(x)$ are given by sinusoidal functions, and the electric fields are expressed as

$$E_{y1} = [B_1 \sin(\beta_{xa} x) + B_2 \cos(\beta_{xa} x)]e^{-j\beta_z z} \qquad (4.3.16)$$

and

$$E_{y2} = \{C_1 \sin[\beta_{xd}(x - a)] + C_2 \cos[\beta_{xd}(x - a)]\}e^{-j\beta_z z} \qquad (4.3.17)$$

where in the air region,

$$\beta_o^2 = \beta_{xa}^2 + \beta_z^2 \tag{4.3.18}$$

and in the dielectric region,

$$\beta_d^2 = \beta_{xd}^2 + \beta_z^2 \tag{4.3.19}$$

In (4.3.17), the argument is written as $(x - a)$ for convenience when applying the boundary conditions.

The form of writing the constants as B_1, B_2, C_1, and C_2 in (4.3.16) and (4.3.17) is found in the literature. One should remember that the values of these constants depend on the particular mode of propagation. Thus, for the TE_{n0}^x modes being analyzed a more appropriate way of expressing the constants is to use a notation such as B_{n0}, C_{n0}, D_{n0}, and E_{n0}, which explicitly show that the constant depend on the mode n. Since the form in (4.3.16) and (4.3.17) is sometimes used, the present analysis will continue using this form; as well as in other analysis done in this book where such form for the constants is convenient.

The boundary conditions require that $E_{y1} = 0$ at $x = 0$, and $E_{y2} = 0$ at $x = a$. Then, from (4.3.16) it follows that $B_2 = 0$, and $C_2 = 0$ in (4.3.17). Therefore,

$$E_{y1} = B_1 \sin(\beta_{xa} x) e^{-j\beta_z z} \tag{4.3.20}$$

$$E_{y2} = C_1 \sin[\beta_{xd}(x - a)] e^{-j\beta_z z} \tag{4.3.21}$$

From Maxwell's equations, the magnetic fields are

$$H_x = \frac{1}{j\omega\mu}\frac{\partial E_y}{\partial z} = \begin{cases} H_{x1} = -B_1 \dfrac{\beta_z}{\omega\mu_o} \sin(\beta_{xa} x) e^{-j\beta_z z} \\ H_{x2} = -C_1 \dfrac{\beta_z}{\omega\mu_d} \sin[\beta_{xd}(x - a)] e^{-j\beta_z z} \end{cases}$$

and

$$H_z = \frac{-1}{j\omega\mu}\frac{\partial E_y}{\partial x} = \begin{cases} H_{z1} = -B_1 \dfrac{\beta_{xa}}{j\omega\mu_o} \cos(\beta_{xa} x) e^{-j\beta_z z} \\ H_{z2} = -C_1 \dfrac{\beta_{xd}}{j\omega\mu_d} \cos[\beta_{xd}(x - a)] e^{-j\beta_z z} \end{cases} \tag{4.3.22}$$

The boundary conditions at $x = t$ are the continuity of the tangential electric and magnetic fields. From (4.3.20), (4.3.21), and (4.3.22):

$$E_{y1} = E_{y2}|_{x=t} \Rightarrow B_1 \sin(\beta_{xa} t) = C_1 \sin \beta_{xd}(t - a) \tag{4.3.23}$$

and

$$H_{z1} = H_{z2}|_{x=t} \Rightarrow B_1 \frac{\beta_{xa}}{\mu_o} \cos(\beta_{xa} t) = C_1 \frac{\beta_{xd}}{\mu_d} \cos[\beta_{xd}(t-a)] \qquad (4.3.24)$$

Dividing (4.3.23) by (4.3.24), gives

$$\frac{\mu_o}{\beta_{xa}} \tan(\beta_{xa} t) = \frac{\mu_d}{\beta_{xd}} \tan[\beta_{xd}(t-a)] \qquad (4.3.25)$$

Equation (4.3.25) is known as the characteristic equation. Observe that if $t \to a$ (i.e., no dielectric), (4.3.25) gives

$$\tan(\beta_{xa} a) = 0 \Rightarrow \beta_{xa} = \frac{n\pi}{a} \qquad (n = 1, 2, 3, \ldots)$$

which, as expected, is the result for an empty waveguide.

In most dielectrics $\mu_d = \mu_o$, so in these cases (4.3.25) will not depend on μ. Equation (4.3.25) can be written in terms of β_z using (4.3.18) and (4.3.19) and solved numerically or graphically at each frequency to find the eigenvalues of β_z. Once β_z is found, β_{xa} and β_{xd} are determined from (4.3.18) and (4.3.19). The ratio B_1/C_1 for a mode follows form (4.3.23).

The cutoff frequency is determined by setting $\beta_z = 0$. Then, from (4.3.18): $\beta_{xa} = \beta_o = \omega_c\sqrt{\mu_o\varepsilon_o}$, and from (4.3.19): $\beta_{xd} = \beta_o\sqrt{\mu_{rd}\varepsilon_{rd}} = \omega_c\sqrt{\mu_d\varepsilon_d}$, and (4.3.25) can be solved for the cutoff frequency ω_c. That is,

$$\tan\left(\omega_c\sqrt{\mu_o\varepsilon_o}\, t\right) = \sqrt{\frac{\mu_{rd}}{\varepsilon_{rd}}} \tan\left[\omega_c\sqrt{\mu_d\varepsilon_d}\,(t-a)\right]$$

The solution for the TE^x_{nm} modes in Fig. 4.3a is left to the problems.

Next, consider the TE^y_{nm} modes in the configuration in Fig. 4.3b. The modes can be analyzed in terms of the auxiliary vector $\mathbf{F} = F_y\hat{\mathbf{a}}_y$. From (1.5.23), or from the general form in (4.2.1), F_y satisfies

$$\left[\frac{\partial^2}{\partial x^2} + \frac{\partial^2}{\partial y^2} + \frac{\partial^2}{\partial z^2} + \beta^2\right] F_y(x, y, z) = 0 \qquad (4.3.26)$$

The solutions of (4.3.26) in the air and dielectric regions for a traveling wave in the positive z direction are

$$F_{y1} = [B_1 \sin(\beta_{xa} x) + B_2 \cos(\beta_{xa} x)]\{C_1 \sin[\beta_{ya}(y-b)] + C_2 \cos[\beta_{ya}(y-b)]\}e^{-j\beta_z z} \qquad (4.3.27)$$

and

$$F_{y2} = [D_1 \sin(\beta_{xd}x) + D_2 \cos(\beta_{xd}x)][G_1 \sin(\beta_{yd}y) + G_2 \cos(\beta_{yd}y)]e^{-j\beta_z z} \quad (4.3.28)$$

where in the air region

$$\beta_o^2 = \beta_{xa}^2 + \beta_{ya}^2 + \beta_z^2$$

and in the dielectric region

$$\beta_d^2 = \beta_{xd}^2 + \beta_{yd}^2 + \beta_z^2$$

The boundary conditions at the conductors at $x = 0$, $x = a$, $y = 0$ and $y = b$ are the vanishing of the tangential electric field. There are two regions in Fig. 4.2.3b, so the boundary conditions must be satisfied at the conductors in each region. From (4.2.31) to (4.2.36) the boundary conditions can be expressed in different ways. Using the E_z component of the field, namely

$$E_z = -\frac{1}{\varepsilon}\frac{\partial F_y}{\partial x}$$

it follows from (4.3.27) and (4.3.28) that

$$\begin{aligned}
E_{z2}\big|_{y=0} &= 0 \Rightarrow G_1 \sin(0) + G_2 \cos(0) = 0 \Rightarrow G_2 = 0 \\
E_{z1}\big|_{y=b} &= 0 \Rightarrow C_1 \sin(0) + C_2 \cos(0) = 0 \Rightarrow C_2 = 0 \\
E_{z2}\big|_{\substack{x=0 \\ 0\le y\le t}} &= 0 \Rightarrow D_1 \cos(0) - D_2 \sin(0) = 0 \Rightarrow D_1 = 0 \\
E_{z2}\big|_{\substack{x=a \\ 0\le y\le t}} &= 0 \Rightarrow -D_2 \sin(\beta_{xd}a) = 0 \Rightarrow \beta_{xd} = \frac{n\pi}{a} \qquad (n = 0, 1, 2, \ldots) \\
E_{z1}\big|_{\substack{x=0 \\ t\le y\le b}} &= 0 \Rightarrow B_1 \cos(0) - B_2 \sin(0) = 0 \Rightarrow B_1 = 0 \\
E_{z1}\big|_{\substack{x=a \\ t\le y\le b}} &= 0 \Rightarrow -B_2 \sin(\beta_{xa}a) = 0 \Rightarrow \beta_{xa} = \beta_{xd} = \frac{n\pi}{a} \qquad (n = 0, 1, 2, \ldots)
\end{aligned}$$

where we have set $\beta_{xa} = \beta_{xd}$, a fact that follows when matching the boundary conditions at the interface (done in (4.3.33) and (4.3.34)). Hence, (4.3.27) and (4.3.28) read

$$F_{y1} = Q_{nm}^a \cos\left(\frac{n\pi x}{a}\right)\sin[\beta_{ya}(y - b)]e^{-j\beta_z z} \quad (4.3.29)$$

$$F_{y2} = Q_{nm}^d \cos\left(\frac{n\pi x}{a}\right)\sin(\beta_{yd}y)e^{-j\beta_z z} \quad (4.3.30)$$

where

$$\beta_o^2 = \omega^2 \mu_o \varepsilon_o = \left(\frac{n\pi}{a}\right)^2 + \beta_{ya}^2 + \beta_z^2 \tag{4.3.31}$$

and

$$\beta_d^2 = \omega^2 \mu_d \varepsilon_d = \left(\frac{n\pi}{a}\right)^2 + \beta_{yd}^2 + \beta_z^2 \tag{4.3.32}$$

The amplitude constants are denoted by Q_{nm}^a and Q_{nm}^d, where the superscripts denote the air and dielectric regions, respectively.

The boundary conditions at the interface are the continuity of E_z and H_z, which must be satisfied for all values of x (i.e., requiring that $\beta_{xa} = \beta_{xd}$). These boundary conditions will determine the eigenvalues of β_{ya}, β_{yd} and β_z. At $x = t$, from (4.2.29) and (4.2.30), the continuity of the tangential electric fields (see (4.2.33)) gives

$$E_{z1} = E_{z2}|_{y=t} \Rightarrow \frac{Q_{nm}^a}{\varepsilon_o}\sin[\beta_{ya}(t-b)] = \frac{Q_{nm}^d}{\varepsilon_d}\sin(\beta_{yd}t) \tag{4.3.33}$$

and from (4.2.36):

$$H_{z1} = H_{z2}|_{y=t} \Rightarrow \frac{Q_{nm}^a \beta_z \beta_{ya}}{\mu_o \varepsilon_o}\cos[\beta_{ya}(t-b)] = \frac{Q_{nm}^d \beta_z \beta_{yd}}{\mu_d \varepsilon_d}\cos(\beta_{yd}t) \tag{4.3.34}$$

Dividing (4.3.33) by (4.3.34) produces the characteristic equation:

$$\frac{\beta_{ya}}{\mu_o}\cot[\beta_{ya}(t-b)] = \frac{\beta_{yd}}{\mu_d}\cot(\beta_{yd}t) \tag{4.3.35}$$

Equations (4.3.31) and (4.3.32) show that β_{ya} and β_{yd} are functions of β_z, the value of n, and the frequency. For a given mode n and at a given frequency, (4.3.35) can be expressed in terms of β_z and its values found. Since there are many discrete values that satisfy (4.3.35) these values are denoted by m=1,2,3,…. Then, β_{ya} and β_{yd} follow from (4.3.31) and (4.3.32). The ratio Q_{nm}^a / Q_{nm}^d follow from (4.3.33).

The dominant mode is the TE_{01}^y (i.e., $n = 0$, $m = 1$). For this mode, the cutoff frequency ω_c occurs when $\beta_z = 0$. For the TE_{01}^y mode, (4.3.31) and (4.3.32) at cutoff read

$$\beta_{ya} = \omega_c \sqrt{\mu_o \varepsilon_o}$$

and

$$\beta_{yd} = \omega_c \sqrt{\mu_d \varepsilon_d}$$

Then, from (4.3.35) the cutoff frequency is determined from

$$\sqrt{\frac{\varepsilon_o}{\mu_o}} \cot[\omega_c \sqrt{\mu_o \varepsilon_o}(t - b)] = \sqrt{\frac{\varepsilon_d}{\mu_d}} \cot(\omega_c \sqrt{\mu_d \varepsilon_d} t)$$

The TM^y_{nm} modes in Fig. 4.3b can be analyzed in terms of the auxiliary vector $\mathbf{A} = A_y \hat{\mathbf{a}}_y$. A summary of the results is

$$A_{y1} = K^a_{nm} \sin(\beta_{xa} x) \cos[\beta_{ya}(y - b)] e^{-j\beta_z z}$$

$$A_{y2} = K^d_{nm} \sin(\beta_{xd} x) \cos(\beta_{yd} y) e^{-j\beta_z z}$$

where

$$\beta_{xa} = \beta_{xd} = \frac{n\pi}{a} \quad (n = 1, 2, 3, \ldots)$$
$$\beta_o^2 = \omega^2 \mu_o \varepsilon_o = \left(\frac{n\pi}{a}\right)^2 + \beta_{ya}^2 + \beta_z^2$$
$$\beta_d^2 = \omega^2 \mu_d \varepsilon_d = \left(\frac{n\pi}{a}\right)^2 + \beta_{yd}^2 + \beta_z^2$$

and the characteristic equation is

$$\frac{\beta_{ya}}{\varepsilon_o} \tan[\beta_{ya}(t - b)] = \frac{\beta_{yd}}{\varepsilon_d} \tan(\beta_{yd} t)$$

The dominant mode is the TM^y_{10} mode, where the first m-root is denoted by $m = 0$ to agree with the nomenclature of an air-filled waveguide where the dominant TM^y mode is the TM^y_{10} mode.

As a check on the previous formulation consider the TE^y_{nm} modes in a completely filled waveguide by letting $t \to b$ (or $t \to 0$) in Fig. 4.3b. If $t \to b$ and $\varepsilon_d = \varepsilon_o$ it is an air-filled waveguide. In this case (4.3.35) shows that $\beta_{yd} = m\pi/b$ where $m = 1, 2, 3, \ldots$, and can be simply denoted by β_y. Hence, the TE^y_{nm} modes for the air-filled waveguide follow from

$$F_y = Q_{nm} \cos\left(\frac{n\pi x}{a}\right) \sin\left(\frac{m\pi y}{b}\right) e^{-j\beta_z z}$$

where

$$\beta_o^2 = \beta_x^2 + \beta_y^2 + \beta_z^2 = \left(\frac{n\pi}{a}\right)^2 + \left(\frac{m\pi}{b}\right)^2 + \beta_z^2 \qquad \begin{pmatrix} n = 0, 1, 2, \ldots \\ m = 1, 2, 3, \ldots \end{pmatrix}$$

and the cutoff frequencies (i.e., when $\beta_z = 0$) follow from

$$f_c = \frac{1}{2\pi\sqrt{\mu_o \varepsilon_o}} \sqrt{\left(\frac{n\pi}{a}\right)^2 + \left(\frac{m\pi}{b}\right)^2}$$

The dominant mode with $a > b$ is the TE_{10}^y mode.

4.3.4 Rectangular Waveguide Partially Filled with an Inhomogeneous Dielectric

The partially filled waveguide with an inhomogeneous dielectric is shown in Fig. 4.4. A mode of interest in the LSE mode to x, specifically denoted by $TE_{\ell 0}^x$, where ℓ is not going to be an integer. Here, we are going to list the value of l for the dominant mode, instead of referring to the mode as TE_{m0}^x, where $m = 1$ is associated with the mode whose eigenvalue is the first eigenvalue of l. The inhomogeneous dielectric selected has an exponential variation and is given by the following relative dielectric constant:

$$\varepsilon_{rd}(x) = d e^{2m(x-t)} \qquad (t \le x \le a) \tag{4.3.36}$$

The electric fields in the two regions are

$$E_{y1} = f_1(x) e^{-j\beta_z z} \qquad (0 \le x \le t)$$

and

$$E_{y2} = f_2(x) e^{-j\beta_z z} \qquad (t \le x \le a)$$

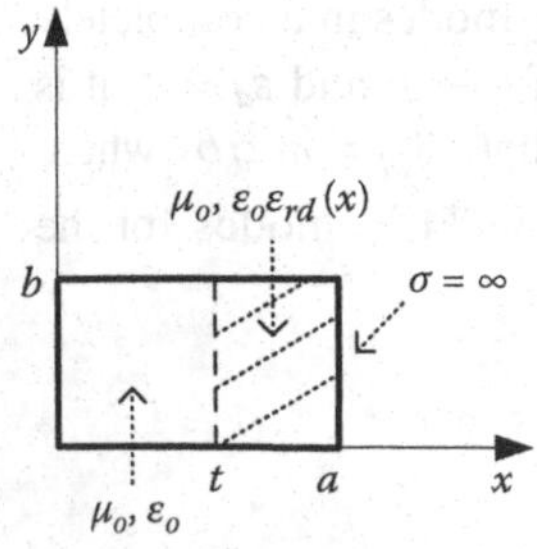

FIGURE 4.4 Rectangular waveguide partially filled with an inhomogeneous dielectric.

where from (4.3.1) with $\psi = E_y$ and $\gamma^2 = -\beta_o^2$, $f_1(x)$ satisfy

$$\left[\frac{d^2}{dx^2} + \beta_o^2 - \beta_z^2\right] f_1(x) = 0 \qquad (0 \le x \le t) \tag{4.3.37}$$

and with ε_{rd} given by (4.3.36), $f_2(x)$ satisfies

$$\left[\frac{d^2}{dx^2} + \beta_o^2 d\, e^{2m(x-t)} - \beta_z^2\right] f_2(x) = 0 \qquad (t \le x \le a) \tag{4.3.38}$$

The solution to (4.3.37) that satisfies the boundary condition $E_{y1} = 0$ at $x = 0$ is conveniently written as

$$E_{y1} = B_l \sin\left(\frac{l\pi x}{a}\right) e^{-j\beta_z z} \tag{4.3.39}$$

where

$$\beta_o^2 = \left(\frac{l\pi}{a}\right)^2 + \beta_z^2 \tag{4.3.40}$$

As it will be shown, the parameter l in (4.3.39) will not be an integer. However, as $t \to a$ (i.e., in an air-filled waveguide) we expect l to attain integer values.

The solution to an equation similar to (4.3.38) was discussed in Section 3.7. That is, write (4.3.38) as

$$\frac{d^2 f_2}{dx^2} + m^2\left[\left(\frac{\beta_o \sqrt{d}\, e^{m(x-t)}}{m}\right)^2 - \left(\frac{\beta_z}{m}\right)^2\right] f_2 = 0$$

whose solutions are Bessel functions and E_{y2} is given by

$$E_{y2} = [C_l J_\nu(\varsigma_x) + D_l Y_\nu(\varsigma_x)] e^{-j\beta_z z} \tag{4.3.41}$$

where

$$\nu = \frac{\beta_z}{m}$$

and

$$\varsigma_x = \frac{\beta_o}{m} \sqrt{d}\, e^{m(x-t)}$$

The boundary condition $E_{y2} = 0$ at $x = a$ is satisfied by setting (4.3.41) equal to zero at $x = a$. Hence,

$$\frac{C_l}{D_l} = -\frac{Y_\nu(\zeta_a)}{J_\nu(\zeta_a)} \tag{4.3.42}$$

where

$$\zeta_a = \frac{\beta_o}{m}\sqrt{d}\,e^{m(a-t)}$$

At $x = t$ the tangential fields must be continuous, or

$$E_{y1} = E_{y2}|_{x=t} \Rightarrow B_l \sin\left(\frac{l\pi t}{a}\right) = C_l J_\nu(\zeta_t) + D_l Y_\nu(\zeta_t) \tag{4.3.43}$$

and

$$H_{z1} = H_{z2}|_{x=t} \Rightarrow B_l \frac{l\pi}{a}\cos\left(\frac{l\pi t}{a}\right) = \beta_o\sqrt{d}\,[C_l J_\nu'(\zeta_t) + D_l Y_\nu'(\zeta_t)] \tag{4.3.44}$$

where

$$\zeta_t = \frac{\beta_o}{m}\sqrt{d}$$

Dividing (4.3.44) by (4.3.43) and using (4.3.42) it follows that the characteristic equation is

$$\frac{l\pi}{a\beta_o\sqrt{d}}\cot\left(\frac{l\pi t}{a}\right) = \frac{J_\nu'(\zeta_t)Y_\nu(\zeta_a) - J_\nu(\zeta_a)Y_\nu'(\zeta_t)}{J_\nu(\zeta_t)Y_\nu(\zeta_a) - J_\nu(\zeta_a)Y_\nu(\zeta_t)} \tag{4.3.45}$$

Equation (4.3.45) gives the eigenvalues of l, and from (4.3.40) the eigenvalues of β_z.

The cutoff frequency f_c for the waveguide is obtained by setting $\beta_z = 0$ in (4.3.40), or $\beta_o = l\pi/a$. Denoting the cutoff values of l by l_c it follows that the cutoff frequencies are

$$f_c = \frac{l_c}{2a\sqrt{\mu_o\varepsilon_o}} \tag{4.3.46}$$

Equation (4.3.45) at cutoff, using $\nu = \beta_z/m = 0$, $J_0'(x) = -J_1(x)$ and $Y_0'(x) = -Y_1(x)$, reads

$$\frac{1}{\sqrt{d}}\cot\left(\frac{\varsigma_a - \zeta_t}{h}t\right) = -\mathrm{ct}(\zeta_a, \zeta_t) \tag{4.3.47}$$

where

$$l_c = \frac{a}{\pi}\left(\frac{\varsigma_a - \varsigma_t}{h}\right) \tag{4.3.48}$$

$$h = \frac{\sqrt{d}}{m}\left[e^{m(a-t)} - 1\right] = \frac{\sqrt{d}}{m}\left(\frac{\varsigma_a}{\varsigma_t} - 1\right)$$

and the function $\mathrm{ct}(\zeta_a, \zeta_t)$ is the small radial cotangent function, namely

$$\mathrm{ct}(\zeta_a, \zeta_t) = \frac{J_1(\zeta_t)Y_0(\zeta_a) - J_0(\zeta_a)Y_1(\zeta_t)}{J_0(\zeta_t)Y_0(\zeta_a) - J_0(\zeta_a)Y_0(\zeta_t)}$$

Equation (4.3.47) can be solved graphically to determine the cutoff frequency. The two sides are plotted to the same scale and the intersection gives the value of value of $\varsigma_a - \varsigma_t$. Once the value of $\varsigma_a - \varsigma_t$ is obtained, the cutoff value l_c follows from (4.3.48), and f_c from (4.3.46).

As an example, consider an X-band waveguide with $a = 2.286$ cm, $t = 1$ cm and

$$\varepsilon_{r2}(x) = 3e^{0.397(x-1)}$$

Then, $\varsigma_a/\varsigma_t = 1.291$ and the resulting plot of $-\mathrm{ct}(\varsigma_a, \varsigma_t)$ is shown in Fig. 4.5. The intersection of the left side of (4.3.47) is shown as a dot. It is seen that $\varsigma_a - \varsigma_t = 2.08$, $h = 2.539$ and from (4.3.48): $l_c = 0.596$. The cutoff frequency, from (4.3.46), is $f_c = 3.91$ GHz. The lowest mode that propagates is the $\mathrm{TE}^x_{0.596,0}$.

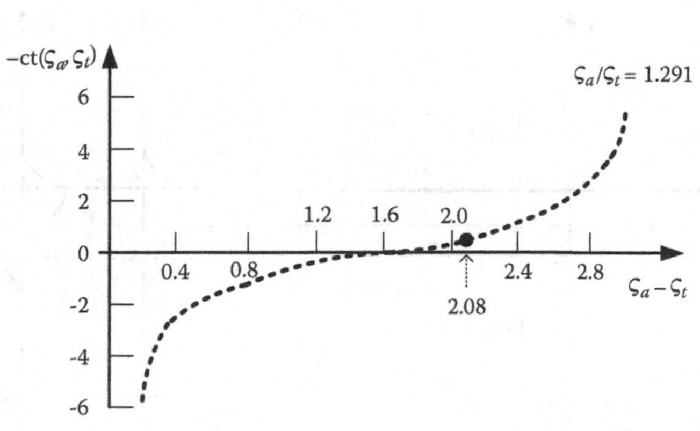

FIGURE 4.5 An example of the cutoff frequency calculation using (4.3.47).

A limiting case of the partially filled waveguide occur when $t \to 0$, which represents a completely filled waveguide with an inhomogeneous dielectric. In this case the characteristic equation in (4.3.45) with $t = 0$ becomes

$$J_\nu(\zeta_t)Y_\nu(\zeta_a) - J_\nu(\zeta_a)Y_\nu(\zeta_t) = 0$$

which at cutoff reads

$$J_0(\zeta_t)Y_0(\zeta_a) - J_0(\zeta_a)Y_0(\zeta_t) = 0$$

Another limiting case of the partially filled waveguide occur when $t \to a$, which represents an empty waveguide. For $t = a$, it follows that $\varsigma_a = \varsigma_t$ and (4.3.45) gives

$$\tan(l\pi) = 0 \Rightarrow l = n \qquad (n = 1, 2, 3, \ldots)$$

which correspond to the standard TE_{no} modes.

4.3.5 The Rectangular Dielectric Waveguide

Another waveguide system is the two-dimensional dielectric waveguide shown in Fig. 4.6a, which is also known as a slab waveguide. The dielectric extends from $-\infty < y < \infty$, so the fields do not vary with y. Solutions that propagate along the dielectric in the z-direction and attenuate in the air region are possible.

Consider a TM^z mode with no variation along y. Let $\mathbf{A} = A_z\hat{\mathbf{a}}_z$, or in the terminology of (4.2.1) let $\psi = A_z$. Then, with $\gamma = j\beta$, the wave equation for A_z is

$$\left[\frac{\partial^2}{\partial x^2} + \frac{\partial^2}{\partial z^2} + \beta^2\right]A_z(x, z) = 0$$

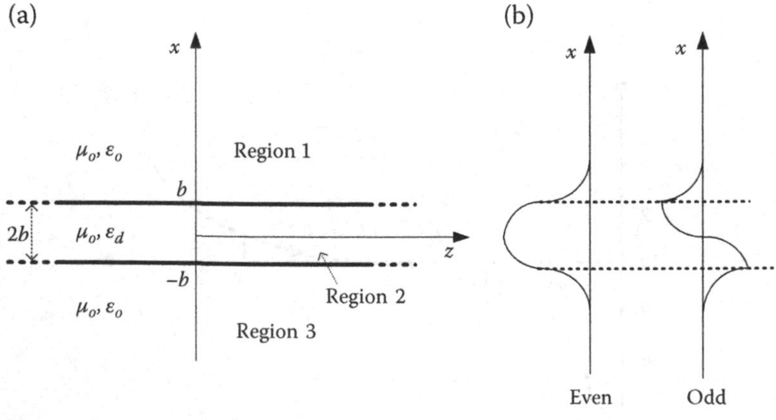

FIGURE 4.6 (a) Two-dimensional view of the dielectric waveguide; (b) an even and an odd mode.

Assume that the permeability of the dielectric is $\mu_d = \mu_o$. For $x > b$ and $x < -b$: $\beta^2 = \beta_o^2 = \omega^2\mu_o\varepsilon_o$, and for $|x| \le b$: $\beta^2 = \beta_d^2 = \omega^2\mu_o\varepsilon_d = \beta_o^2\varepsilon_{rd}$. Solutions to the wave equation that are evanescent in the air and propagate in the dielectric are given by

$$A_z = \begin{cases} B_1 e^{-\alpha_{xa}x}e^{-j\beta_z z} & x \ge b \\ \begin{bmatrix} C_1 \sin\beta_{xd}x \\ C_2 \cos\beta_{xd}x \end{bmatrix} e^{-j\beta_z z} & |x| \le b \\ B_2 e^{\alpha_{xa}x}e^{-j\beta_z z} & x \le -b \end{cases} \tag{4.3.49}$$

where

$$\beta_o^2 = \omega^2\mu_o\varepsilon_o = -\alpha_{xa}^2 + \beta_z^2 \tag{4.3.50}$$

and

$$\beta_d^2 = \beta_o^2\varepsilon_{rd} = \beta_{xd}^2 + \beta_z^2 \tag{4.3.51}$$

The fields described by (4.3.49) represent a wave propagating in the dielectric which is known as a guided wave or a trapped wave. Also, the value of the constants depends on the mode.

The solution for $|x| \le b$ was separated into odd and even modes. These modes can be analyzed separately and, if needed, superposition used. The z-dependence is the same in both regions and described by the propagation constant β_z. In the x-direction, α_{xa} is the attenuation constant associated with the air regions and β_{xd} the propagation constant associated with the dielectric region. Equation (4.3.49) shows that α_{xa} must be positive for an evanescent wave in the air, and from (4.3.50): $\beta_z > \beta_o$. In the dielectric β_{xd} must be positive, and from (4.3.51): $\beta_d > \beta_z$. Therefore, the range of β_z is

$$\beta_o < \beta_z < \beta_d \tag{4.3.52}$$

The fields are given by (4.2.7) to (4.2.12), namely

$$H_y = -\frac{1}{\mu_o}\frac{\partial A_z}{\partial x}$$

$$E_x = \frac{1}{j\omega\mu_o\varepsilon}\frac{\partial^2 A_z}{\partial z \partial x}$$

$$E_z = \frac{1}{j\omega\mu_o\varepsilon}\left(\frac{\partial^2}{\partial z^2} + \beta^2\right)A_z = \frac{1}{j\omega\mu_o\varepsilon}\left(\beta^2 - \beta_z^2\right)A_z$$

$$H_z = H_x = E_y = 0$$

The sine solutions in (4.3.49) results in odd modes, and the cosine solution in even modes. These are illustrated on Fig. 4.6b for one of the possible even and odd modes. From symmetry we expect $B_2 = B_1$ for the even modes, and $B_2 = -B_1$ for the odd modes.

We will analyze the odd modes, the analysis of the even modes is similar. For the odd modes write the solutions in (4.3.49) as

$$A_z^{\text{odd}} = \begin{cases} B_1 e^{-\alpha_{xa}x} e^{-j\beta_z z} & x \geq b \\ C_1 \sin(\beta_{xd}x) e^{-j\beta_z z} & |x| \leq b \\ B_2 e^{\alpha_{xa}x} e^{-j\beta_z z} & x \leq -b \end{cases}$$

where the fields in the three regions are

$$H_{y1} = B_1 \frac{\alpha_{xa}}{\mu_o} e^{-\alpha_{xa}x} e^{-j\beta_z z}$$

$$H_{y2} = -C_1 \frac{\beta_{xd}}{\mu_o} \cos(\beta_{xd}x) e^{-j\beta_z z}$$

$$H_{y3} = -B_2 \frac{\alpha_{xa}}{\mu_o} e^{\alpha_{xa}x} e^{-j\beta_z z}$$

$$E_{x1} = B_1 \frac{\alpha_{xa}\beta_z}{\omega\mu_o\varepsilon_o} e^{-\alpha_{xa}x} e^{-j\beta_z z}$$

$$E_{x2} = -C_1 \frac{\beta_{xd}\beta_z}{\omega\mu_o\varepsilon_d} \cos(\beta_{xd}x) e^{-j\beta_z z}$$

$$E_{x3} = -B_2 \frac{\alpha_{xa}\beta_z}{\omega\mu_o\varepsilon_o} e^{\alpha_{xa}x} e^{-j\beta_z z}$$

$$E_{z1} = B_1 \frac{\left(\beta_o^2 - \beta_z^2\right)}{j\omega\mu_o\varepsilon_o} e^{-\alpha_{xa}x} e^{-j\beta_z z} = -B_1 \frac{\alpha_{xa}^2}{j\omega\mu_o\varepsilon_o} e^{-\alpha_{xa}x} e^{-j\beta_z z}$$

$$E_{z2} = C_1 \frac{\left(\beta_d^2 - \beta_z^2\right)}{j\omega\mu_o\varepsilon_d} \sin(\beta_{xd}x) e^{-j\beta_z z} = C_1 \frac{\beta_{xd}^2}{j\omega\mu_o\varepsilon_d} \sin(\beta_{xd}x) e^{-j\beta_z z}$$

$$E_{z3} = B_2 \frac{\left(\beta_o^2 - \beta_z^2\right)}{j\omega\mu_o\varepsilon_o} e^{\alpha_{xa}x} e^{-j\beta_z z} = -B_2 \frac{\alpha_{xa}^2}{j\omega\mu_o\varepsilon_o} e^{\alpha_{xa}x} e^{-j\beta_z z}$$

The boundary conditions are the continuity of the tangential fields at $x = b$ and $x = -b$. At $x = b$:

$$H_{y1} = H_{y2}|_{x=b} \Rightarrow B_1 \alpha_{xa} e^{-\alpha_{xa} b} = -C_1 \beta_{xd} \cos(\beta_{xd} b) \tag{4.3.53}$$

$$E_{z1} = E_{z2}|_{x=b} \Rightarrow -B_1 \frac{\alpha_{xa}^2}{\varepsilon_o} e^{-\alpha_{xa} b} = C_1 \frac{\beta_{xd}^2}{\varepsilon_d} \sin(\beta_{xd} b) \tag{4.3.54}$$

and at $x = -b$:

$$H_{y3} = H_{y2}|_{x=-b} \Rightarrow B_2 \alpha_{xa} e^{-\alpha_{xa} b} = C_1 \beta_{xd} \cos(\beta_{xd} b) \tag{4.3.55}$$

$$E_{z3} = E_{z2}|_{x=-b} \Rightarrow B_2 \frac{\alpha_{xa}^2}{\varepsilon_o} e^{-\alpha_{xa} b} = C_1 \frac{\beta_{xd}^2}{\varepsilon_d} \sin(\beta_{xd} b) \tag{4.3.56}$$

From either (4.3.53) and (4.3.55), or (4.3.54) and (4.3.56) it follows that $B_2 = -B_1$. Hence, (4.3.53) is identical to (4.3.55), and (4.3.54) is identical to (4.3.55).

Dividing (4.3.54) by (4.3.53) gives the characteristic equation for the TM^z odd modes, namely

$$\beta_{xd} \tan(\beta_{xd} b) = \varepsilon_{rd} \alpha_{xa} \tag{4.3.57}$$

The characteristic equation is expressed only in terms of β_z by substituting (4.3.50) and (4.3.51) into (4.3.57).

The cutoff frequency of a mode occurs at the frequency when $\alpha_{xa} = 0$. Below that frequency, the solution in the air region become oscillatory, which represents a radiation field in the air region. Hence, below the cutoff frequency the wave is not being properly guided by the dielectric waveguide. From (4.3.50), at cutoff (i.e., with $\alpha_{xa} = 0$) it follows that $\beta_z = \beta_o = \omega_c \sqrt{\mu_o \varepsilon_o}$, and from (4.3.51):

$$\beta_{xd} = \sqrt{\omega_c^2 \mu_o \varepsilon_o \varepsilon_{rd} - \omega_c^2 \mu_o \varepsilon_o} = \omega_c \sqrt{\mu_o \varepsilon_o} \sqrt{\varepsilon_{rd} - 1} \tag{4.3.58}$$

At cutoff, (4.3.57) reads

$$\tan(\beta_{xd} b) = 0 \Rightarrow \beta_{xd} = \frac{n\pi}{b} \qquad (n = 0, 1, 2, ...) \tag{4.3.59}$$

Substituting (4.3.59) into (4.3.58) shows that the cutoff frequency is given by

$$f_c = \frac{n}{2b\sqrt{\mu_o \varepsilon_o}\sqrt{\varepsilon_{rd} - 1}} \qquad (n = 0, 1, 2, ...) \tag{4.3.60}$$

The lowest cutoff frequency is that of the TM$_0^z$ odd mode, whose cutoff frequency is zero. The field associated with the TM$_0^z$ odd mode is a TEM wave with components E_{x2} and H_{y2} traveling in the positive z direction with no attenuation.

The surface impedance is

$$Z_s = \left.\frac{E_{z1}}{H_{y1}}\right|_{x=b} = j\frac{\alpha_{xa}}{\omega\varepsilon_o} \tag{4.3.61}$$

which represents an inductive surface impedance.

For the TM^z even modes in (4.3.49) it follows that $B_1 = B_2$ and the characteristic equation is

$$-\beta_{xd}\cot(\beta_{xd}b) = \varepsilon_{rd}\alpha_{xa}$$

The cutoff frequencies for the TM^z even modes are given by

$$f_c = \frac{\left(n+\frac{1}{2}\right)}{2b\sqrt{\mu_o\varepsilon_o}\sqrt{\varepsilon_{rd}-1}} \qquad (n = 0, 1, 2, \dots) \tag{4.3.62}$$

The lowest cutoff frequency of the TM^z even modes is that of the TM$_0^z$ even mode. The surface impedance is given by (4.3.61).

The TE^z modes are analyzed using the $\mathbf{F} = F_z\hat{\mathbf{a}}_z$ vector. The resulting characteristic equations, with $\mu_d = \mu_o$, are

$$\begin{cases} -\beta_{xd}\cot(\beta_{xd}b) = \alpha_{xa} \Rightarrow \text{(for the TE}^z\text{ modes)} \\ \beta_{xd}\tan(\beta_{xd}b) = \alpha_{xa} \Rightarrow \text{(for the TE}^z\text{ odd modes)} \end{cases}$$

If the permeability of the dielectric is $\mu_d = \mu_o\mu_{rd}$, the characteristic equations read

$$\begin{cases} -\beta_{xd}\cot(\beta_{xd}b) = \mu_{rd}\alpha_{xa} \Rightarrow \text{(for the TE}^z\text{ even modes)} \\ \beta_{xd}\tan(\beta_{xd}b) = \mu_{rd}\alpha_{xa} \Rightarrow \text{(for the TE}^z\text{ odd modes)} \end{cases}$$

The cutoff frequencies for the TE^z even modes are given by (4.3.62) and for the TE^z odd modes by (4.3.60). Hence, the lowest cutoff frequency is that of the TE$_0^z$ odd mode, whose cutoff frequency is zero. The field associated with the TE$_0^z$ odd mode are those of a TEM wave.

The surface impedance for the is TE^z even and odd modes is

$$Z_s = \left.\frac{E_{z1}}{H_{y1}}\right|_{x=b} = -j\frac{\omega\mu_o}{\alpha_{xa}}$$

which represents a capacitive surface impedance.

To better understand the propagation of the modes in the dielectric consider the electric field E_y due to a TE^z mode, given by

$$E_y = C\cos(\beta_{xd}x)e^{-j\beta_z z} = \frac{C}{2}[e^{j(\beta_{xd}x-\beta_z z)} + e^{-j(\beta_{xd}x+\beta_z z)}]$$

This relation represents two plane waves in the dielectric with propagation constants given by $\mathbf{k} = \pm\beta_{xd}\hat{\mathbf{a}}_x + \beta_z\hat{\mathbf{a}}_z$. The directions of propagation make an angle with the z axis given by

$$\tan\theta = \frac{\pm\beta_{xd}}{\beta_z}$$

Hence, the guided wave is a superposition of two plane waves propagating at the angles $\pm\,\theta$ with the x, z axis. The angle associated with a given mode is a discrete angle since the characteristic equation gives discrete values of β_z. A particular mode is associated with a particular angle. From (4.3.52), a guided wave requires that

$$\beta_o < \beta_z < \beta_d \Rightarrow \frac{1}{n_2} < \frac{\beta_z}{\beta_d} < 1$$

where $n_2 = \sqrt{\varepsilon_{r2}}$. Since θ is also given by

$$\cos\theta = \frac{\beta_z}{\sqrt{\beta_{xd}^2 + \beta_z^2}} = \frac{\beta_z}{\beta_d}$$

it follows that

$$\cos\theta > \frac{1}{n_2}$$

which is the condition for total internal reflection, since the dielectric (see Fig. 4.6a) is the dense region (i.e., $n_2 > n_1$) and the angle of incidence at the upper surface is $\pi/2 - \theta$. Hence, a guided wave is a wave that experiences internal reflections at the dielectric-air interface. If $\cos\theta < 1/n_2$ then the waves refract at the interface and the fields in the air region consist of a radiating field.

In the field of optics, the previous characteristic equations are usually analyzed as follows. Consider the characteristic equation of the TM^z odd modes in (4.3.57) and write it as

$$X\tan X = \alpha_{xa}b \tag{4.3.63}$$

where

$$X = \beta_{xd}b$$

Since

$$\alpha_{xa} b = b\sqrt{\beta_z^2 - \beta_o^2}$$

and

$$\beta_z^2 = \beta_d^2 - \beta_{xd}^2$$

it follows that

$$\alpha_{xa} b = b\sqrt{\beta_d^2 - \beta_o^2 - \left(\frac{X}{b}\right)^2} = \sqrt{\beta_o^2 b^2 (\varepsilon_d - 1) - X^2} = \sqrt{V^2 - X^2}$$

where

$$V = \beta_o b\sqrt{\varepsilon_d - 1}$$

Hence, (4.3.63) can be expressed as

$$X \tan X = \sqrt{V^2 - X^2} \tag{4.3.64}$$

The term V is known as the mode volume relation.

Equation (4.3.64) can be plotted as a function of X and the intersections of the two sides of the equation gives the solutions for X and, therefore, the values of β_{xd} and β_z. The right-hand side of (4.3.64) represents a circle. That is, letting:

$$Y = \sqrt{V^2 - X^2}$$

shows that this relation represents a circle of radius V in the X-Y plane. Figure 4.7 shows a typical plot. The circle shown corresponds to $V = 2$. The intersection of Y with $X \tan X$, shown as a dot in Fig. 4.7, shows that one odd mode propagates. For larger values of V the circle can intersect the $X \tan X$ curves at several points indicating that other odd modes can propagate.

It is seen that if $0 < V < \pi/2$ there is only one TE^z odd mode. As the value of V increases more modes are possible. The total number of modes, including odd and even modes, can be expressed in terms of V.

Similar relations for the other characteristic equations can be derived to determine the solutions for the propagation constants. For example, for the TM^z even modes the characteristic equation can be expressed as

$$-X \cot X = \sqrt{V^2 - X^2}$$

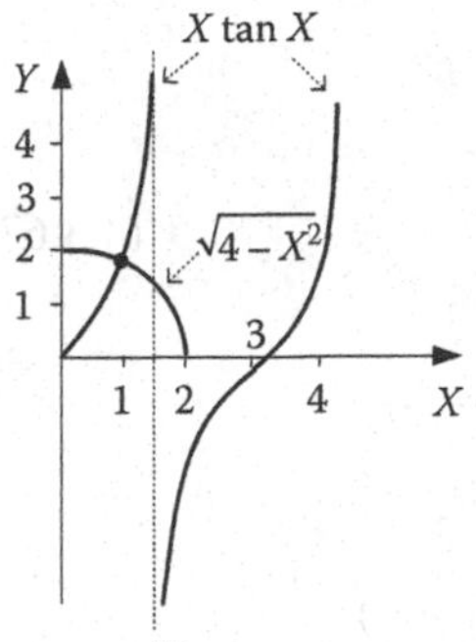

FIGURE 4.7 A typical plot of (4.3.64).

4.3.6 Dielectric Waveguide above a Conducting Plane

The geometry of a dielectric covered conducting plane is shown in Fig. 4.8. The dielectric extends from $-\infty < y < \infty$, and the fields are not functions of y. This configuration supports both TM^z and TE^z modes of propagation. The analysis of this problem is similar to the dielectric waveguide analysis except that the ground plane forces the tangential electric field at $x = 0$ to be zero.

Consider the guided TM^z modes. The TM^z even mode will not satisfy the tangential boundary condition at $x = 0$ because its cosine variation is not zero at $x = 0$. So only the TM^z odd modes are possible. For these modes, the auxiliary potential in the air and dielectric regions is given by

$$A_z^{\text{odd}} = \begin{cases} Be^{-\alpha_{xa}x}e^{-j\beta_z z} & (x \geq b) \\ C\sin(\beta_{xd}x)e^{-j\beta_z z} & (0 \leq x \leq b) \end{cases} \tag{4.3.65}$$

where

$$\beta_o^2 = \omega^2\mu_o\varepsilon_o = -\alpha_{xa}^2 + \beta_z^2 \tag{4.3.66}$$

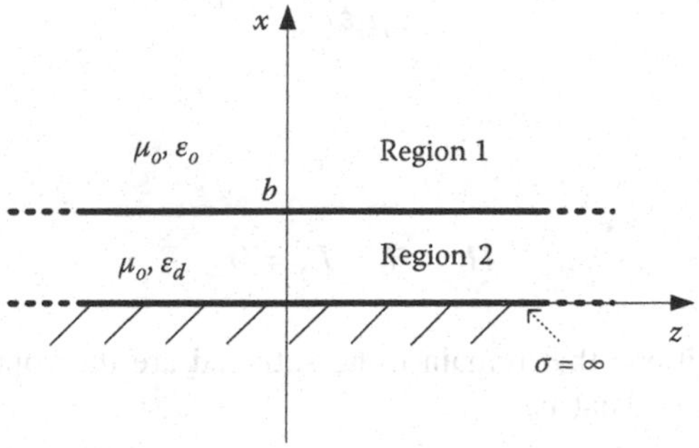

FIGURE 4.8 Dielectric waveguide above a conducting plane.

and

$$\beta_d^2 = \omega^2 \mu_o \varepsilon_d = \beta_{xd}^2 + \beta_z^2 \tag{4.3.67}$$

Also, it follows that $\beta_o < \beta_z < \beta_d$ (see (4.3.52)).

The tangential components of the fields are

$$H_y = -\frac{1}{\mu_o}\frac{\partial A_z^{\text{odd}}}{\partial x} = \begin{cases} H_{y1} = B\dfrac{\alpha_{xa}}{\mu_o}e^{-\alpha_{xa}x}e^{-j\beta_z z} & (x \geq b) \\ H_{y2} = -C\dfrac{\beta_{xd}}{\mu_o}\cos(\beta_{xd}x)e^{-j\beta_z z} & (0 \leq x \leq b) \end{cases}$$

$$E_z = \frac{1}{j\omega\mu_o\varepsilon}\left(\beta^2 - \beta_z^2\right)A_z^{\text{odd}} = \begin{cases} E_{z1} = -B\dfrac{\alpha_{xa}^2}{j\omega\mu_o\varepsilon_o}e^{-\alpha_{xa}x}e^{-j\beta_z z} & (x \geq b) \\ E_{z2} = C\dfrac{\beta_{xd}^2}{j\omega\mu_o\varepsilon_d}\sin(\beta_{xd}x)e^{-j\beta_z z} & (0 \leq x \leq b) \end{cases}$$

The component E_z can also be calculated using

$$E_z = \frac{1}{j\omega\varepsilon}\frac{\partial H_y}{\partial x}$$

The other field components are

$$E_x = \frac{1}{j\omega\mu_o\varepsilon}\frac{\partial^2 A_z^{\text{odd}}}{\partial x \partial z} = \begin{cases} E_{x1} = B\dfrac{\alpha_{xa}\beta_z}{\omega\mu_o\varepsilon_d}e^{-\alpha_{xa}x}e^{-j\beta_z z} & (x \geq b) \\ E_{x2} = -C\dfrac{\beta_{xd}\beta_z}{\omega\mu_o\varepsilon_d}\cos(\beta_{xd}x)e^{-j\beta_z z} & (0 \leq x \leq b) \end{cases}$$

and

$$H_z = H_x = E_y = 0$$

The boundary conditions that remain to be satisfied are the continuity of the tangential fields at $x = b$. That is,

$$H_{y1} = H_{y2}|_{x=b} \Rightarrow B\alpha_{xa}e^{-\alpha_{xa}b} = -C\beta_{xd}\cos(\beta_{xd}b) \tag{4.3.68}$$

and

$$E_{z1} = E_{z2}|_{x=b} \Rightarrow -B\frac{\alpha_{xa}^2}{\varepsilon_o}e^{-\alpha_{xa}b} = C\frac{\beta_{xd}^2}{\varepsilon_d}\sin(\beta_{xd}b) \tag{4.3.69}$$

Dividing (4.3.69) by (4.3.68) gives the characteristic equation, or

$$\beta_{xd}\tan\beta_{xd}b = \varepsilon_{rd}\alpha_{xa} \tag{4.3.70}$$

The ratio C/B follows from (4.3.68).

The cutoff frequency derivation is similar to that in (4.3.60), and we obtain

$$f_c = \frac{n}{2b\sqrt{\mu_o\varepsilon_o}\sqrt{\varepsilon_{rd}-1}} \qquad (n = 0, 1, 2, \ldots)$$

The dominant mode is the TM_0^z odd mode.

A similar analysis shows that only TE^z even modes are possible. The characteristic equation for the TE^z even modes is

$$-\beta_{xd}\cot(\beta_{xd}b) = \alpha_{xa} \tag{4.3.71}$$

and the cutoff frequencies are given by

$$f_c = \frac{\left(n+\frac{1}{2}\right)}{2b\sqrt{\mu_o\varepsilon_o}\sqrt{\varepsilon_{rd}-1}} \qquad (n = 0, 1, 2, \ldots) \tag{4.3.72}$$

4.4 WAVE PROPAGATION IN CYLINDRICAL GEOMETRIES

In cylindrical coordinates, the expansion of the wave equation in (4.1.1), using (4.1.2), produces the following coupled equations:

$$\nabla^2E_\rho - \left(\frac{E_\rho}{\rho^2} + \frac{2}{\rho^2}\frac{\partial E_\phi}{\partial\phi}\right) - \gamma^2E_\rho = 0$$

$$\nabla^2E_\phi - \left(\frac{E_\phi}{\rho^2} - \frac{2}{\rho^2}\frac{\partial E_\rho}{\partial\phi}\right) - \gamma^2E_\phi = 0$$

$$\nabla^2E_z - \gamma^2E_z = 0$$

Since

$$\nabla^2(E_\rho \hat{\mathbf{a}}_\rho) \neq (\nabla^2 E_\rho)\hat{\mathbf{a}}_\rho$$

$$\nabla^2(E_\phi \hat{\mathbf{a}}_\phi) \neq (\nabla^2 E_\phi)\hat{\mathbf{a}}_\phi$$

but

$$\nabla^2(E_z \hat{\mathbf{a}}_z) = (\nabla^2 E_z)\hat{\mathbf{a}}_z$$

it follows that only the z component satisfies the scalar wave equation. That is,

$$[\nabla^2 - \gamma^2]\psi(\rho, \phi, z) = 0 \tag{4.4.1}$$

where ψ represents the z component of **E**, **H**, **A**, **F**, $\mathbf{\Pi}^e$, or $\mathbf{\Pi}^m$. Equation (4.4.1) in cylindrical coordinates reads

$$\frac{\partial^2\psi}{\partial\rho^2} + \frac{1}{\rho}\frac{\partial\psi}{\partial\rho} + \frac{1}{\rho^2}\frac{\partial^2\psi}{\partial\phi^2} + \frac{\partial^2\psi}{\partial z^2} - \gamma^2\psi = 0 \tag{4.4.2}$$

The method of separation of variable is used to solve (4.4.2). Letting

$$\psi = R(\rho)\Phi(\phi)Z(z) \tag{4.4.3}$$

substituting (4.4.3) into (4.4.2), and separating the variables gives

$$\frac{1}{R}\frac{\partial^2 R}{\partial\rho^2} + \frac{1}{R}\frac{1}{\rho}\frac{\partial R}{\partial\rho} + \frac{1}{\Phi}\frac{1}{\rho^2}\frac{\partial^2\Phi}{\partial\phi^2} + \frac{1}{Z}\frac{\partial^2 Z}{\partial z^2} = \gamma^2 \tag{4.4.4}$$

The fourth term only depend on z, so is set equal to a constant. That is,

$$\frac{1}{Z}\frac{d^2Z}{dz^2} = -\beta_z^2 \Rightarrow \frac{d^2Z}{dz^2} + \beta_z^2 Z = 0 \tag{4.4.5}$$

where β_z^2 is the separation constant. Equation (4.4.4) is rearranged to read

$$\frac{\rho^2}{R}\frac{\partial^2 R}{\partial\rho^2} + \frac{\rho}{R}\frac{\partial R}{\partial\rho} + \frac{1}{\Phi}\frac{\partial^2\Phi}{\partial\phi^2} = \left(\gamma^2 + \beta_z^2\right)\rho^2 \tag{4.4.6}$$

The third term is now only a function of ϕ and is set equal to a constant, or

$$\frac{1}{\phi}\frac{d^2\Phi}{d\phi^2} = -\nu^2 \Rightarrow \frac{d^2\Phi}{d\phi^2} + \nu^2\Phi = 0 \tag{4.4.7}$$

where ν^2 is the separation constant. Then, (4.4.6) reads

$$\rho^2\frac{\partial^2 R}{\partial \rho^2} + \rho\frac{\partial R}{\partial \rho} - [(\gamma_\rho \rho)^2 + \nu^2]R = 0 \tag{4.4.8}$$

where

$$\gamma_\rho^2 = \gamma^2 + \beta_z^2 \tag{4.4.9}$$

Equation (4.4.8) is the modified Bessel's differential equation. Typical solutions to (4.4.5), (4.4.7), and (4.4.8) are

$$R = b_1 I_\nu(\gamma_\rho \rho) + b_2 K_\nu(\gamma_\rho \rho)$$

$$\Phi = \begin{cases} c_1 e^{-j\nu\phi} + c_2 e^{j\nu\phi} \\ c_1 \sin \nu\phi + c_2 \cos \nu\phi \end{cases}$$

and

$$Z = \begin{cases} d_1 e^{-j\beta_z z} + d_2 e^{j\beta_z z} \\ d_1 \sin \beta_z z + d_2 \cos \beta_z z \end{cases}$$

In a lossless region, $\gamma^2 = -\beta^2$ and (4.4.8), in this case, reads

$$\rho^2\frac{\partial^2 R}{\partial \rho^2} + \rho\frac{\partial R}{\partial \rho} + [(\beta_\rho \rho)^2 - \nu^2]R = 0$$

where

$$\beta_\rho^2 = \beta^2 - \beta_z^2 \tag{4.4.10}$$

whose solutions are usually in the forms

$$R = \begin{cases} b_1 J_\nu(\beta_\rho \rho) + b_2 Y_\nu(\beta_\rho \rho) \\ b_1 J_\nu(\beta_\rho \rho) + b_2 H_\nu^{(2)}(\beta_\rho \rho) \\ b_1 H_\nu^{(1)}(\beta_\rho \rho) + b_2 H_\nu^{(2)}(\beta_\rho \rho) \end{cases}$$

These solutions are expected since the modified Bessel functions with complex arguments (i.e., with $\gamma_\rho = j\beta_\rho$) are related to the Bessel functions. That is,

$$I_\nu(jx) = j^{\,\nu} J_\nu(x)$$

and

$$K_\nu(jx) = \frac{\pi}{2}(-j)^{\nu+1} H_\nu^{(2)}(x)$$

Away from the source, the field components in terms of A_z follow from (1.5.7) and (1.5.10), and in terms of Π_z^{e} from (1.7.6) and (1.7.7). That is,

$$H_\rho = \frac{1}{\mu}\frac{1}{\rho}\frac{\partial A_z}{\partial \phi} = \frac{\gamma^2}{j\omega\mu}\frac{1}{\rho}\frac{\partial \Pi_z^{\mathrm{e}}}{\partial \phi} \tag{4.4.11}$$

$$H_\phi = -\frac{1}{\mu}\frac{\partial A_z}{\partial \rho} = -\frac{\gamma^2}{j\omega\mu}\frac{\partial \Pi_z^{\mathrm{e}}}{\partial \rho} \tag{4.4.12}$$

$$H_z = 0 \tag{4.4.13}$$

$$E_\rho = \frac{j\omega}{\gamma^2}\frac{\partial^2 A_z}{\partial \rho\, \partial z} = \frac{\partial^2 \Pi_z^{\mathrm{e}}}{\partial \rho\, \partial z} \tag{4.4.14}$$

$$E_\phi = \frac{j\omega}{\gamma^2}\frac{1}{\rho}\frac{\partial^2 A_z}{\partial \phi\, \partial z} = \frac{1}{\rho}\frac{\partial^2 \Pi_z^{\mathrm{e}}}{\partial \phi\, \partial z} \tag{4.4.15}$$

$$E_z = \frac{j\omega}{\gamma^2}\left(\frac{\partial^2 A_z}{\partial z^2} - \gamma^2 A_z\right) = \frac{\partial^2 \Pi_z^{\mathrm{e}}}{\partial z^2} - \gamma^2 \Pi_z^{\mathrm{e}} \tag{4.4.16}$$

The field components in terms of F_z follow from (1.5.19) and (1.5.22), and in terms of Π_z^{m} from (1.7.10) and (1.7.11). That is,

$$E_\rho = -\frac{1}{\varepsilon_c}\frac{1}{\rho}\frac{\partial F_z}{\partial \phi} = -j\omega\mu\frac{1}{\rho}\frac{\partial \Pi_z^{\mathrm{m}}}{\partial \phi} \tag{4.4.17}$$

$$E_\phi = \frac{1}{\varepsilon_c}\frac{\partial F_z}{\partial \rho} = j\omega\mu\frac{\partial \Pi_z^{\mathrm{m}}}{\partial \rho} \tag{4.4.18}$$

$$E_z = 0 \tag{4.4.19}$$

$$H_\rho = \frac{j\omega}{\gamma^2}\frac{\partial^2 F_z}{\partial\rho\,\partial z} = \frac{\partial^2 \Pi_z^{\mathrm{m}}}{\partial\rho\,\partial z} \tag{4.4.20}$$

$$H_\phi = \frac{j\omega}{\gamma^2}\frac{1}{\rho}\frac{\partial^2 F_z}{\partial\phi\,\partial z} = \frac{1}{\rho}\frac{\partial^2 \Pi_z^{\mathrm{m}}}{\partial\phi\,\partial z} \tag{4.4.21}$$

$$H_z = \frac{j\omega}{\gamma^2}\left(\frac{\partial^2 F_z}{\partial z^2} - \gamma^2 F_z\right) = \frac{\partial^2 \Pi_z^{\mathrm{m}}}{\partial z^2} - \gamma^2 \Pi_z^{\mathrm{m}} \tag{4.4.22}$$

Recall that in the above equations the factor $\gamma^2/j\omega$ can also be written as

$$\frac{\gamma^2}{j\omega} = j\omega\mu\varepsilon_c = j\omega\mu_o\varepsilon_o n^2$$

4.4.1 The Cylindrical Waveguide

The geometry of a cylindrical waveguide is shown in Fig. 4.9. The constitutive parameters inside the lossless waveguide are μ_o and ε. Usually, the cylindrical waveguide is air-filled (then $\varepsilon = \varepsilon_o$). This waveguide supports TE^z and TM^z modes of propagation. Consider the TM^z mode with $\psi = \Pi_z^{\mathrm{e}}$ in (4.4.1) and $\gamma = j\beta = j\omega\sqrt{\mu_o\varepsilon}$. From (4.4.2) it follows that the modes function is

$$\Pi_z^{\mathrm{e}} = J_n(\beta_\rho\rho)[A_n\cos(n\phi) + B_n\sin(n\phi)]e^{-j\beta_z z}$$

The separation constant ν was set equal to an integer n since the fields are periodic in ϕ with period of 2π. The function $Y_n(\beta_\rho\rho)$ goes to minus infinity as $\rho \to 0$, so only $J_n(\beta_\rho\rho)$ is used. Also, from (4.4.10):

$$\beta^2 = \beta_\rho^2 + \beta_z^2$$

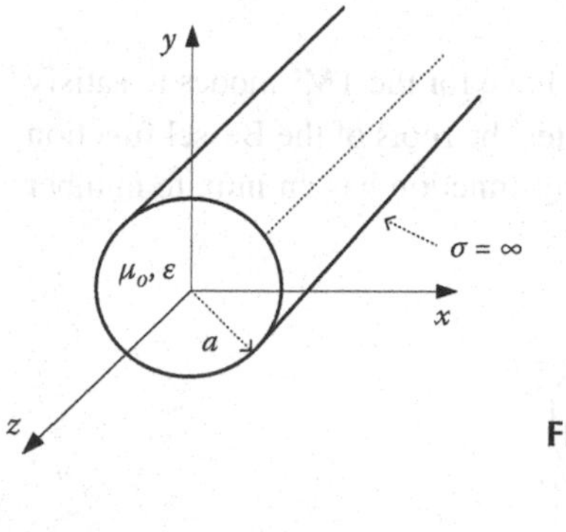

FIGURE 4.9 The cylindrical waveguide.

The linear combination of the angular solutions can be expressed in the form

$$A_n \cos(n\phi) + B_n \sin(n\phi) = C_n \cos[n(\phi - \phi_o)]$$

where

$$A_n = C_n \cos(n\phi_o)$$

and

$$B_n = C_n \sin(n\phi_o)$$

A value of C_n and ϕ_o can be found for a given A_n and B_n. Hence, Π_z^e can be expressed as

$$\Pi_z^e = C_n J_n(\beta_\rho \rho)\cos[n(\phi - \phi_o)]e^{-j\beta_z z} \tag{4.4.23}$$

The value of ϕ_o represents the angular reference from which ϕ is measured. The amplitude C_n of a mode is determined by the source. If $\phi_o = 0$, the angular field pattern varies as $\cos(n\phi)$. For an arbitrary ϕ_o the angular field pattern is the same but at a different angular reference.

The boundary condition at $\rho = a$ is the vanishing of the tangential electric field, which is satisfied by setting $E_z = 0$ at $\rho = a$. From (4.4.16), the boundary condition in terms of Π_z^e requires that $\Pi_z^e = 0$ at $\rho = a$:

$$\Pi_z^e(\rho, \phi, z)|_{\rho=a} = 0 \Rightarrow J_n(\beta_\rho a) = 0 \Rightarrow \beta_\rho a = \chi_{nl} \qquad \begin{pmatrix} n = 0, 1, 2, \ldots \\ l = 1, 2, 3, \ldots \end{pmatrix}$$

or

$$\beta_\rho = \frac{\chi_{nl}}{a}$$

This relation shows the discrete values that β_ρ can have for the TM^z modes to satisfy the boundary condition at $\rho = a$. That is, χ_{nl} denotes the roots of the Bessel function (i.e., $J_n(\chi_{nl}) = 0$), since for each order n the Bessel function has an infinite number of zeros denoted by l.

The propagation constant β_z is then given by

$$\beta_z = \sqrt{\beta^2 - \left(\frac{\chi_{nl}}{a}\right)^2}$$

and (4.4.23) is expressed as

$$\Pi_z^{\mathrm{e}} = C_{nl} J_n\left(\frac{\chi_{nl}\rho}{a}\right)\cos[n(\phi - \phi_o)]e^{-j\beta_z z}$$

where the constant is now explicitly denoted as C_{nl}.

The cutoff frequency of the nl mode is determined when $\beta_z = 0$. That is,

$$f_c = \frac{\chi_{nl}}{2\pi a\sqrt{\mu_o \varepsilon}}$$

The modes in a cylindrical waveguide are simply referred as TM_{nl} (i.e., the superscript is usually omitted) since for each n, there are l different modes. The lowest TM_{nl} mode is the TM_{01}, where $\chi_{01} = 2.405$ (see Appendix C for the roots χ_{nl}).

The TM_{nl} fields in the cylindrical waveguide, from (4.4.11) to (4.4.16), are

$$\begin{cases} H_\rho = -C_{nl}\dfrac{j\omega\varepsilon n}{\rho}J_n(\beta_\rho\rho)\sin[n(\phi - \phi_o)]e^{-j\beta_z z} \\ H_\phi = -C_{nl}j\omega\varepsilon\beta_\rho J_n'(\beta_\rho\rho)\cos[n(\phi - \phi_o)]e^{-j\beta_z z} \\ H_z = 0 \\ E_\rho = -C_{nl}j\beta_\rho\beta_z J_n'(\beta_\rho\rho)\cos[n(\phi - \phi_o)]e^{-j\beta_z z} \\ E_\phi = C_{nl}\dfrac{jn\beta_z}{\rho}J_n(\beta_\rho\rho)\sin[n(\phi - \phi_o)]e^{-j\beta_z z} \\ E_z = C_{nl}\beta_\rho^2 J_n(\beta_\rho\rho)\cos[n(\phi - \phi_o)]e^{-j\beta_z z} \end{cases}$$

For the TE^z modes, referred also as theTE_{nl} modes in a cylindrical waveguide, we let $\psi = \Pi_z^{\mathrm{m}}$ in (4.4.1). Hence, the magnetic Hertz vector for the modes is

$$\Pi_z^{\mathrm{m}} = B_n J_n(\beta_\rho\rho)\cos[n(\phi - \phi_o)]e^{-j\beta_z z}$$

The boundary condition requires that $E_\phi = 0$ at $\rho = a$, which from (4.4.18) requires that

$$\frac{\partial \Pi_z^{\mathrm{m}}}{\partial\rho} = 0|_{\rho=a} \Rightarrow J'_n(\beta_\rho a) = 0 \Rightarrow \beta_\rho a = \tilde{\chi}_{nl} \qquad \begin{pmatrix} n = 0, 1, 2, \ldots \\ l = 1, 2, 3\ldots \end{pmatrix}$$

or

$$\beta_\rho = \frac{\tilde{\chi}_{nl}}{a}$$

where $\tilde{\chi}_{nl}$ denotes the roots of the derivative of the Bessel function (i.e., $J_n'(\tilde{\chi}_{nl}) = 0$). This relation shows the discrete values that β_ρ can have for the TE_{nl} modes to satisfy the boundary condition at $\rho = a$. Then, Π_z^{m} is expressed as

$$\Pi_z^{\mathrm{m}} = B_{nl} J_n\left(\frac{\tilde{\chi}_{nl}\rho}{a}\right)\cos[n(\phi - \phi_o)]e^{-j\beta_z z} \tag{4.4.24}$$

where the constant is now explicitly denoted as B_{nl}.

The propagation constant β_z is given by

$$\beta_z = \sqrt{\beta^2 - \left(\frac{\tilde{\chi}_{nl}}{a}\right)^2}$$

and the cutoff frequency is

$$f_c = \frac{\tilde{\chi}_{nl}}{2\pi a\sqrt{\mu_o \varepsilon}}$$

The lowest TE_{nl} mode is the TE_{11}, where $\tilde{\chi}_{11} = 1.841$ (see Appendix C for the roots $\tilde{\chi}_{nl}$).

The TE_{nl} fields in the cylindrical waveguide, from (4.4.17) to (4.4.22) are

$$\begin{cases} E_\rho = B_{nl}\dfrac{j\omega\mu_o n}{\rho}J_n(\beta_\rho\rho)\sin[n(\phi - \phi_o)]e^{-j\beta_z z} \\ E_\phi = B_{nl} j\omega\mu_o\beta_\rho J_n'(\beta_\rho\rho)\cos[n(\phi - \phi_o)]e^{-j\beta_z z} \\ E_z = 0 \\ H_\rho = -B_{nl} j\beta_\rho\beta_z J_n'(\beta_\rho\rho)\cos[n(\phi - \phi_o)]e^{-j\beta_z z} \\ H_\phi = B_{nl}\dfrac{jn\beta_z}{\rho}J_n(\beta_\rho\rho)\sin[n(\phi - \phi_o)]e^{-j\beta_z z} \\ H_z = B_{nl}\beta_\rho^2 J_n(\beta_\rho\rho)\cos[n(\phi - \phi_o)]e^{-j\beta_z z} \end{cases}$$

There are degenerate modes in the cylindrical waveguide. Since $J_0'(x) = -J_1(x)$ it is seen that $\tilde{\chi}_{0n} = \chi_{1n}$. Hence, the TE_{0n} modes and the TM_{1n} modes have the same cutoff frequencies.

4.4.2 The Cylindrical Cavity

A cylindrical cavity with constitutive parameters μ and ε is shown is Fig. 4.10. It is a cylindrical waveguide with conducting plates added at $z = 0$ and $z = d$. Both TM^z and TE^z modes are possible. Consider the TE^z mode, where Π_z^{m} given by (4.4.24) satisfies the tangential boundary conditions at the $\rho = a$ wall. Therefore, all that is

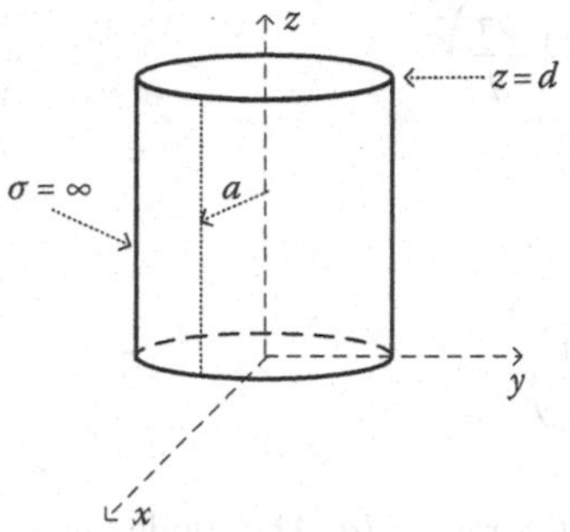

FIGURE 4.10 The cylindrical cavity.

needed is to satisfy the boundary conditions at $z = 0$ and $z = d$. The form of $\Pi_z^{\rm m}$ in the cavity is then

$$\Pi_z^{\rm m} = B_{nl} J_n(\beta_\rho \rho)\cos[n(\phi - \phi_o)][c_1 \sin(\beta_z z) + c_2 \cos(\beta_z z)] \qquad (4.4.25)$$

where

$$\beta_\rho = \frac{\tilde{\chi}_{nl}}{a}$$

To satisfy the boundary condition at $z = 0$ the tangential electric field is calculated, which from (4.4.18) is

$$E_\phi = j\omega\mu\frac{\partial \Pi_z^{\rm m}}{\partial \rho} = B_{nl} j\omega\mu\beta_\rho J_n'(\beta_\rho \rho)\cos[n(\phi - \phi_o)][c_1 \sin(\beta_z z) + c_2 \cos(\beta_z z)]$$

Hence, at $z = 0$:

$$E_\phi = 0|_{z=0} \Rightarrow c_1 \sin(0) + c_2 \cos(0) = 0 \Rightarrow c_2 = 0$$

and at $z = d$:

$$E_\phi = 0|_{z=d} \Rightarrow c_1 \sin(\beta_z d) = 0 \Rightarrow \beta_z = \frac{p\pi}{d} \qquad (p = 1, 2, 3, \ldots)$$

Therefore, the modes function for $\Pi_z^{\rm m}$ in (4.4.25) is written as

$$\Pi_z^{\rm m} = B_{nlp} J_n\left(\frac{\tilde{\chi}_{nl}\rho}{a}\right)\cos[n(\phi - \phi_o)]\sin\left(\frac{p\pi z}{d}\right)$$

where the constant is now denoted as B_{nlp}, and

$$\beta^2 = \omega^2\mu\varepsilon = \beta_\rho^2 + \beta_z^2 = \left(\frac{\tilde{\chi}_{nl}}{a}\right)^2 + \left(\frac{p\pi}{d}\right)^2$$

The resonant frequencies are given by

$$f_r = \frac{1}{2\pi\sqrt{\mu\varepsilon}}\sqrt{\left(\frac{\tilde{\chi}_{nl}}{a}\right)^2 + \left(\frac{p\pi}{d}\right)^2}$$

The resonant frequency of a given mode depends on the ratio d/a. The modes in a cylindrical cavity are denoted by TE_{nlp}. The mode with the lowest resonant frequency is the TE_{111} mode where $\tilde{\chi}_{11} = 1.841$. Then,

$$f_r\big|_{\substack{n=1\\ l=1\\ p=1}} = \frac{1}{2\pi\sqrt{\mu\varepsilon}}\sqrt{\left(\frac{1.841}{a}\right)^2 + \left(\frac{\pi}{d}\right)^2}$$

For the TM^z modes in the cavity (i.e., TM_{nlp} modes) the electric Hertz vector is

$$\Pi_z^e = Q_{nlp}J_n\left(\frac{\chi_{nl}}{a}\rho\right)\cos[n(\phi - \phi_o)]\cos\left(\frac{p\pi z}{d}\right) \qquad \begin{pmatrix} n = 0, 1, 2, \ldots \\ l = 1, 2, 3, \ldots \\ p = 0, 1, 2, \ldots \end{pmatrix}$$

and the resonant frequencies are given by

$$f_r = \frac{1}{2\pi\sqrt{\mu\varepsilon}}\sqrt{\left(\frac{\chi_{nl}}{a}\right)^2 + \left(\frac{p\pi}{d}\right)^2}$$

The lowest cutoff frequency is that of the TM_{010} mode. It is interesting to note that the TM_{010} and TE_{111} modes are degenerate modes if $d/a \approx 2.03$. If $d/a < 2.03$ the dominant mode is the TM_{010} mode, and if $d/a > 2.03$ the dominant mode is the TE_{111} mode.

As an example, the fields of the TM_{010} mode follow from (4.4.11) to (4.4.16) with

$$\Pi_z^e = Q_{010}J_0(\beta_\rho\rho)$$

where

$$\beta_\rho = \frac{\chi_{01}}{a} = \frac{2.405}{a}$$

Then,

$$E_\rho = E_\phi = H_\rho = H_z = 0$$

$$E_z = Q_{010}\beta_\rho^2 J_0(\beta_\rho \rho)$$
$$H_\phi = -Q_{010} j\omega\varepsilon\beta_\rho J_0'(\beta_\rho \rho)$$

4.4.3 The Sectoral Waveguide

The sectoral waveguide, shown in Fig. 4.11, is a sector of the cylindrical waveguide. In this configuration the ϕ dependence is not periodic. Both TM^z and TE^z modes are possible. For the TM^z mode in an air-filled sectoral waveguide (i.e., $\beta = \beta_o = \omega\sqrt{\mu_o \varepsilon_o}$) the z component of the electric Hertz vector satisfy (4.4.2) and the modes function is of the form

$$\Pi_z^e = J_\nu(\beta_\rho \rho)[B_1 \sin(\nu\phi) + B_2 \cos(\nu\phi)]e^{-j\beta_z z} \tag{4.4.26}$$

where

$$\beta_o^2 = \beta_\rho^2 + \beta_z^2$$

The boundary conditions, using (4.4.16), are

$$E_z|_{\phi=0} = 0 \Rightarrow \Pi_z^e|_{\phi=0} = 0 \tag{4.4.27}$$

$$E_z|_{\phi=\phi_o} = 0 \Rightarrow \Pi_z^e|_{\phi=\phi_o} = 0 \tag{4.4.28}$$

$$E_z|_{\rho=a} = 0 \Rightarrow \Pi_z^e|_{\rho=a} = 0 \tag{4.4.29}$$

Boundary condition (4.4.27) requires that $B_2 = 0$, and (4.4.28) requires that

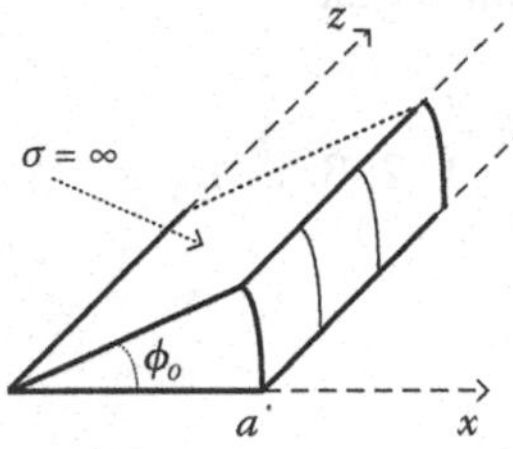

FIGURE 4.11 The sectoral waveguide.

$$\sin(\nu\phi_o) = 0 \Rightarrow \nu = \nu_n = \frac{n\pi}{\phi_o} \qquad (n = 1, 2, 3\ldots)$$

which denotes the discrete values that ν can have in order to satisfy the boundary condition at $\phi = \phi_o$.

The condition in (4.3.29) is satisfied by (4.4.26) with

$$J_{\nu_n}(\beta_\rho a) = 0 \Rightarrow \beta_\rho a = \chi_{\nu_n l} \qquad (l = 1, 2, 3\ldots)$$

or

$$\beta_\rho = \frac{\chi_{\nu_n l}}{a}$$

where $\chi_{\nu_n l}$ denotes the zeros (i.e., $J_{\nu_n}(\chi_{\nu_n l}) = 0$ where $\nu_n = n\pi/\phi_o$). Observe that for each l the order of the Bessel function is $n\pi/\phi_o$ (not an integer order).

The modes function for the electric Hertz vector in (4.4.26) is then

$$\Pi_z^{\mathrm{e}} = B_{nl} J_{\nu_n}\left(\frac{\chi_{\nu_n l}\rho}{a}\right)\sin\left(\frac{n\pi\phi}{\phi_o}\right)e^{-j\beta_z z} \tag{4.4.30}$$

The fields follow from (4.4.11) to (4.4.16), and the cutoff frequencies are given by

$$f_c = \frac{\chi_{\nu_n l}}{2\pi a\sqrt{\mu_o \varepsilon_o}}$$

For the TE^z modes in the sectoral waveguide, the magnetic Hertz vector is

$$\Pi_z^{\mathrm{m}} = Q_n J_{\nu_n}(\beta_\rho \rho)\cos\left(\frac{n\pi\phi}{\phi_o}\right)e^{-j\beta_z z} \qquad (n = 0, 1, 2, \ldots)$$

The boundary condition $E_\phi = 0$ at $\rho = a$, using (4.4.18), requires that

$$\frac{\partial \Pi_z^{\mathrm{m}}}{\partial \rho} = 0\big|_{\rho=a} \Rightarrow J_{\nu_n}'(\beta_\rho a) = 0 \Rightarrow \beta_\rho a = \tilde{\chi}_{\nu_n l} \qquad (l = 1, 2, 3, \ldots)$$

or

$$\beta_\rho = \frac{\tilde{\chi}_{\nu_n l}}{a}$$

where $\tilde{\chi}_{\nu_n l}$ denotes the zeros (i.e., $J'_{\nu_n}(\tilde{\chi}_{\nu_n l}) = 0$). Therefore, the mode function for Π_z^{m} is

$$\Pi_z^{\mathrm{m}} = Q_{nl} J_{\nu_n}\left(\frac{\tilde{\chi}_{\nu_n l}\rho}{a}\right)\cos\left(\frac{n\pi\phi}{\phi_o}\right)e^{-j\beta_z z}$$

The fields follow from (4.4.17) to (4.4.22), and the cutoff frequencies are given by

$$f_c = \frac{\tilde{\chi}_{\nu_n l}}{2\pi a\sqrt{\mu_o \varepsilon_o}}$$

4.4.4 The Radial Waveguide

A radial waveguide is illustrated in Fig. 4.12. The excitation for such waveguides is a current source along the z-axis between $z = 0$ and $z = a$. The resulting fields consist of a wave propagating in the ρ direction. From (4.4.2), with $\psi = \Pi_z^{\mathrm{e}}$ and $\gamma = j\beta_o = j\omega\sqrt{\mu_o \varepsilon_o}$ the TM^z modes for the air-filled radial waveguide in Fig. 4.12 are of the form

$$\Pi_z^{\mathrm{e}} = C_n \cos(\beta_z z)\cos[n(\phi - \phi_o)]H_n^{(2)}(\beta_\rho \rho) \tag{4.4.31}$$

where

$$\beta_o^2 = \beta_\rho^2 + \beta_z^2$$

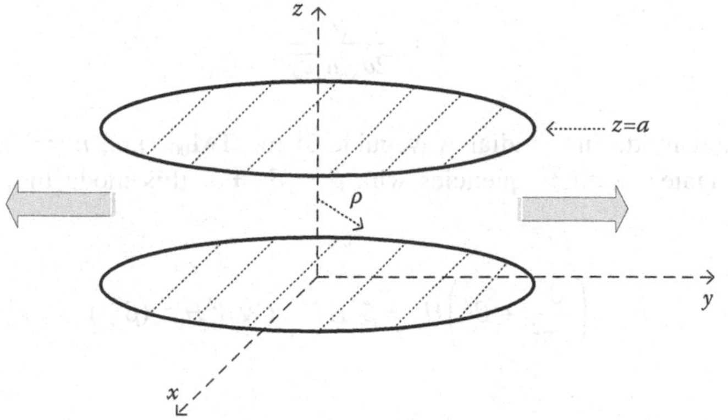

FIGURE 4.12 The radial waveguide.

The asymptotic behavior of $H_n^{(2)}(\beta_\rho \rho)$ represent an outgoing wave.

A common excitation is a constant current between $z = 0$ and $z = a$. In this case there is no ϕ variation (i.e., $n = 0$ mode) and the TM^z modes follow from

$$\Pi_z^{\mathrm{e}} = C_0 \cos(\beta_z z) H_0^{(2)}(\beta_\rho \rho) \tag{4.4.32}$$

The boundary condition $E_\rho = 0$ at $z = 0$ is satisfied by the cosine function in (4.4.32). That is,

$$E_\rho = \frac{\partial^2 \Pi_z^{\mathrm{e}}}{\partial \rho \, \partial z} = -C_0 \beta_z \beta_\rho \sin(\beta_z z) H_0^{(2)\prime}(\beta_\rho \rho)$$

which shows that $E_\rho = 0$ at $z = 0$. The boundary condition at $z = a$ requires that

$$E_\rho = 0|_{z=a} \Rightarrow \sin(\beta_z a) = 0 \Rightarrow \beta_z = \frac{p\pi}{a} \qquad (p = 0, 1, 2, \ldots)$$

From (4.4.32), the TM^z modes function is

$$\Pi_z^{\mathrm{e}} = C_{0p} \cos\left(\frac{p\pi z}{a}\right) H_0^{(2)}(\beta_\rho \rho)$$

where

$$\beta_o^2 = \beta_\rho^2 + \left(\frac{p\pi}{a}\right)^2 \tag{4.4.33}$$

and the constant is expressed as C_{0p}.

The cutoff frequency occurs when $\beta_\rho = 0$ and it is given by

$$f_c = \frac{p}{2a\sqrt{\mu_o \varepsilon_o}} \tag{4.4.34}$$

The dominant mode in a radial waveguide is the TM$_{00}$ (i.e., $n = 0$ and $p = 0$) which propagates at all frequencies with $\beta_\rho = \beta_o$. For this mode the field components are

$$E_z = \left(\frac{\partial^2}{\partial z^2} + \beta_o^2\right) \Pi_z^{\mathrm{e}} = \beta_o^2 \Pi_z^{\mathrm{e}} = C_{00} \beta_o^2 H_0^{(2)}(\beta_o \rho)$$

and

$$H_\phi = -j\omega\varepsilon_o \frac{\partial \Pi_z^e}{\partial \rho} = -jC_{00}\frac{\beta_o^2}{\eta_o}H_0^{(2)\prime}(\beta_o\rho) = jC_{00}\frac{\beta_o^2}{\eta_o}H_1^{(2)}(\beta_o\rho)$$

which is a TEM field to ρ.

For the TM_{0p} modes with $p > 0$ if the frequency of operation is below f_c, the propagation constant β_ρ in (4.4.33) becomes imaginary. That is,

$$\beta_\rho = -j\sqrt{\left(\frac{p\pi}{a}\right)^2 - \beta_o^2} = -j\alpha$$

where the negative root is used. This produces a wave that attenuates as ρ increases since

$$H_n^{(2)}(\beta_\rho\rho) \Rightarrow H_n^{(2)}(-j\alpha\rho) = \frac{2}{\pi}j^{n+1}K_n(\alpha\rho)$$

and the fields attenuate because for large values of ρ:

$$K_n(\alpha\rho) \approx \sqrt{\frac{\pi}{2}}\, e^{-\alpha\rho}$$

In the case that the excitation produces a ϕ variation the fields are calculated using (4.4.11) to (4.4.16) with Π_z^e given by (4.4.31).

For the TE^z modes with no ϕ variations, the Hertz vector is of the form

$$\Pi_z^m = B_{0p}\sin\left(\frac{p\pi z}{a}\right)H_0^{(2)}(\beta_\rho\rho)$$

The cutoff frequencies are given by (4.4.34) with $p = 1, 2, 3, \ldots$.

4.4.5 The Wedge Waveguide

The wedge waveguide configuration is shown in Fig. 4.13. It consists of two inclined conducting planes, extending from $-\infty < z < \infty$. Hence, the problem is two dimensional with the fields being a function of ρ and ϕ. From the previous analysis in Sections 4.4.3 and 4.4.4, the TM^z modes for an outgoing wave can be calculated using

$$\Pi_z^e = C_n\sin\left(\frac{n\pi\phi}{\phi_o}\right)H_{n\pi/\phi_o}^{(2)}(\beta_\rho\rho) \qquad (n = 1, 2, 3, \ldots)$$

and for the TE^z modes using

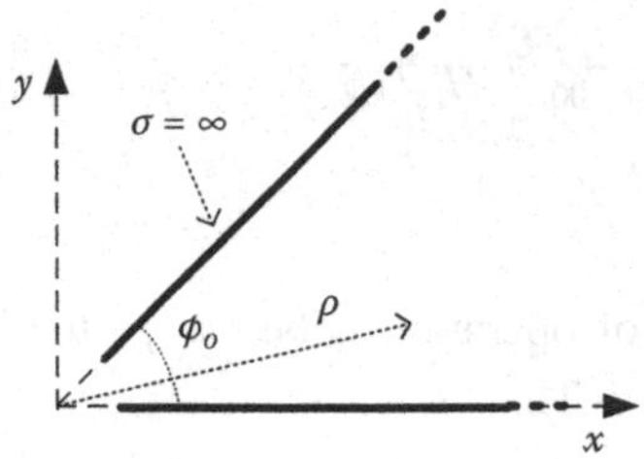

FIGURE 4.13 The wedge waveguide.

$$\Pi_z^{\mathrm{m}} = D_n \cos\left(\frac{n\pi\phi}{\phi_o}\right) H_{n\pi/\phi_o}^{(2)}(\beta_\rho \rho) \qquad (n = 0, 1, 2, \ldots)$$

where

$$\beta_o^2 = \beta_\rho^2$$

The mode excitation determines C_n and D_n.

The dominant mode is the TE_0 mode, which is a TEM mode to ρ. From (4.4.18) and (4.4.22) with $n = 0$ and $\beta_\rho = \beta_o$, we get

$$E_\phi = j\omega\mu_o \frac{\partial \Pi_z^{\mathrm{m}}}{\partial \rho} = D_0 j\omega\mu_o \beta_o H_0^{(2)\prime}(\beta_o \rho) = -D_0 j\eta_o \beta_o^2 H_1^{(2)}(\beta_o \rho)$$

and

$$H_z = \beta_o^2 \Pi_z^m = D_0 \beta_o^2 H_0^{(2)}(\beta_o \rho)$$

4.4.6 The Sectoral Horn

The configuration of an air-filled sectoral horn is shown in Fig. 4.14. The fields in the horn are usually generated by connecting the horn to a rectangular waveguide. From (4.4.2) with $\psi = A_z$ and $\gamma = j\beta_o = j\omega\sqrt{\mu_o \varepsilon_o}$ the TM^z modes for an outgoing wave are of the form

$$A_z = C_{np} \cos\left(\frac{p\pi z}{h}\right) \sin\left(\frac{n\pi\phi}{\phi_o}\right) H_{n\pi/\phi_o}^{(2)}(\beta_\rho \rho) \qquad \begin{pmatrix} n = 1, 2, 3, \ldots \\ p = 0, 1, 2, \ldots \end{pmatrix}$$

where

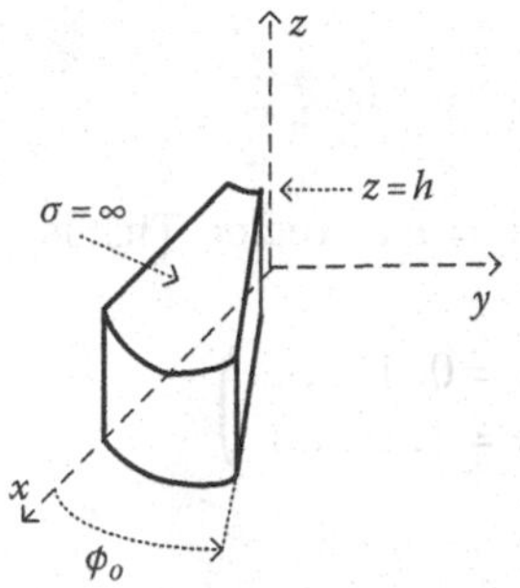

FIGURE 4.14 The sectoral-horn waveguide.

$$\beta_o^2 = \beta_\rho^2 + \left(\frac{p\pi}{h}\right)^2 \Rightarrow \beta_\rho = \sqrt{\beta_o^2 - \left(\frac{p\pi}{h}\right)^2} \tag{4.4.35}$$

The fields are given by (4.4.11) to (4.4.16), namely

$$E_\rho = \frac{1}{j\omega\mu_o\varepsilon_o}\frac{\partial^2 A_z}{\partial\rho\partial z} = -C_{pn}\frac{\beta_\rho}{j\omega\mu_o\varepsilon_o}\left(\frac{p\pi}{h}\right)\sin\left(\frac{p\pi z}{h}\right)\sin\left(\frac{n\pi\phi}{\phi_o}\right)H^{(2)\prime}_{n\pi/\phi_o}(\beta_\rho\rho)$$

$$E_\phi = \frac{1}{j\omega\mu_o\varepsilon_o}\frac{1}{\rho}\frac{\partial^2 A_z}{\partial\phi\partial z} = -C_{pn}\frac{1}{j\omega\mu_o\varepsilon_o}\frac{1}{\rho}\left(\frac{p\pi}{h}\right)\left(\frac{n\pi}{\phi_o}\right)\sin\left(\frac{p\pi z}{h}\right)\cos\left(\frac{n\pi\phi}{\phi_o}\right)H^{(2)}_{n\pi/\phi_o}(\beta_\rho\rho)$$

$$E_z = \frac{1}{j\omega\mu_o\varepsilon_o}\left(\frac{\partial^2}{\partial z^2} + \beta_o^2\right)A_z = C_{pn}\frac{\beta_\rho^2}{j\omega\mu_o\varepsilon_o}\cos\left(\frac{p\pi z}{h}\right)\sin\left(\frac{n\pi\phi}{\phi_o}\right)H^{(2)}_{n\pi/\phi_o}(\beta_\rho\rho)$$

$$H_\rho = \frac{1}{\mu_o\rho}\frac{\partial A_z}{\partial\phi} = C_{pn}\frac{1}{\mu_o\rho}\left(\frac{n\pi}{\phi_o}\right)\cos\left(\frac{p\pi z}{h}\right)\cos\left(\frac{n\pi\phi}{\phi_o}\right)H^{(2)}_{n\pi/\phi_o}(\beta_\rho\rho)$$

$$H_\phi = -\frac{1}{\mu_o}\frac{\partial A_z}{\partial\rho} = -C_{pn}\frac{\beta_\rho}{\mu_o}\cos\left(\frac{p\pi z}{h}\right)\sin\left(\frac{n\pi\phi}{\phi_o}\right)H^{(2)\prime}_{n\pi/\phi_o}(\beta_\rho\rho)$$

$$H_z = 0$$

It is simple to verify that the fields satisfy the boundary conditions in the horn waveguide. The modes are denoted by TM_{np} (i.e., z superscript omitted), although one also finds the modes denoted by TM_{pn}.

Since at a given ρ the cross-section of the waveguide looks like a rectangular waveguide, only the TM_{n0} mode propagates if $h < \lambda/2$. For this mode:

$$A_z = C_{n0}\sin\left(\frac{n\pi\phi}{\phi_o}\right)H^{(2)}_{n\pi/\phi_o}(\beta_\rho\rho) \qquad (n = 1, 2, 3\ldots)$$

The TM_{10} is considered to be the dominant mode. If $\lambda/2 < h < \lambda$ the TM_{n1} and TE_{n1} can propagate, and so on for higher-order modes to propagate. As ρ increases, there is a value of ρ where cutoff occurs. This value of ρ occurs when the cross section of the horn is about the same size of a rectangular waveguide at cutoff. From (4.4.35) cutoff occurs when $\beta_\rho = 0$, which for the TM_{10} mode is

$$\beta_o = \frac{\pi}{h} \Rightarrow f_c = \frac{1}{2h\sqrt{\mu_o \varepsilon_o}}$$

The TE^z modes, denoted by TE$_{np}$ can be obtained using the $\mathbf{F} = F_z \hat{\mathbf{a}}_z$ vector. That is,

$$F_z = D_{np} \sin\left(\frac{p\pi z}{h}\right)\cos\left(\frac{n\pi\phi}{\phi_o}\right)H^{(2)}_{n\pi/\phi_o}(\beta_\rho \rho) \qquad \begin{pmatrix} n = 0, 1, 2, \ldots \\ p = 1, 2, 3, \ldots \end{pmatrix}$$

and the fields follow from (4.4.17) to (4.4.22).

4.4.7 The Bend Waveguide

The configuration of an air-filled bend waveguide is show in Fig. 4.15. This type of waveguide is used to provide a bend path by connecting them in series with a rectangular waveguide. The waveguide modes have an azimuthal dependence, and since the propagation is along the ϕ direction the TE$^\phi$ modes function is of the form

$$\Pi_z^e = [C_{\nu p} J_\nu(\beta_\rho \rho) + B_{\nu p} Y_\nu(\beta_\rho \rho)]\cos\left(\frac{p\pi}{h}z\right)e^{-j\nu\phi} \qquad (p = 0, 1, 2, \ldots) \tag{4.4.36}$$

where

$$\beta_o^2 = \beta_\rho^2 + \left(\frac{p\pi}{h}\right)^2$$

In the bend waveguide important TE$^\phi$ modes are those with $p = 0$, which are a continuation of the TE$^z_{n0}$ modes when connected to a rectangular waveguide. For these modes, from (4.4.22):

$$E_z = \beta_o^2 \Pi_z^e = \beta_o^2 [C_{\nu 0} J_\nu(\beta_\rho a) + B_{\nu 0} Y_\nu(\beta_\rho a)]e^{-j\nu\phi}$$

The boundary condition at $\rho = a$ requires that

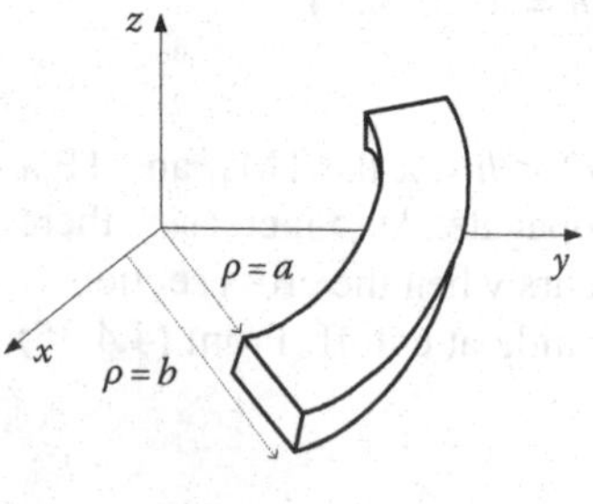

FIGURE 4.15 The bend waveguide.

$$E_z|_{\rho=a} = 0 \Rightarrow \beta_o^2 [C_{\nu 0} J_\nu(\beta_\rho a) + B_{\nu 0} Y_\nu(\beta_\rho a)] = 0$$

or

$$\frac{B_{\nu 0}}{C_{\nu 0}} = -\frac{J_\nu(\beta_\rho a)}{Y_\nu(\beta_\rho a)}$$

and (4.4.36) is written in the form

$$\Pi_z^e = \frac{C_{\nu 0}}{Y_\nu(\beta_\rho a)} [J_\nu(\beta_\rho \rho) Y_\nu(\beta_\rho a) - J_\nu(\beta_\rho a) Y_\nu(\beta_\rho \rho)] e^{-j\nu\phi}$$

The boundary condition at $\rho = b$ is satisfied if

$$E_z|_{\rho=b} = 0 \Rightarrow \frac{C_{\nu 0}}{Y_\nu(\beta_\rho a)} \beta_o^2 [J_\nu(\beta_\rho b) Y_\nu(\beta_\rho a) - J_\nu(\beta_\rho a) Y_\nu(\beta_\rho b)] = 0$$

or

$$\frac{J_\nu(\beta_\rho a)}{Y_\nu(\beta_\rho a)} = \frac{J_\nu(\beta_\rho b)}{Y_\nu(\beta_\rho b)} \tag{4.4.37}$$

which determines the eigenvalues of ν (i.e., $\nu = \nu_n$, $n = 1, 2, 3, ..$) and, therefore, the propagation constant in the ϕ direction. The modes are denoted as TE_{n0}^{ϕ}, or simply the TE_{n0} modes in a bend waveguide. The fields follow from (4.4.17) to (4.4.22) with

$$\Pi_z^e = \frac{C_{n0}}{Y_{\nu_n}(\beta_\rho a)} \left[J_{\nu_n}(\beta_\rho \rho) Y_{\nu_n}(\beta_\rho a) - J_{\nu_n}(\beta_\rho a) Y_{\nu_n}(\beta_\rho \rho)\right] e^{-j\nu_n \phi}$$

where the constant is expressed as C_{n0}. The fundamental mode is the TE_{10}^z where ν_1 is the first root of (4.4.37).

4.4.8 The Cylindrical Dielectric Waveguide

A cylindrical dielectric waveguide is shown in Fig. 4.16. The constitutive parameters of the dielectric are μ_o and $\varepsilon_d = \varepsilon_o \varepsilon_{rd}$. TE^z and TM^z modes can propagate in this waveguide when there is no azimuthal dependence (i.e., $\partial/\partial\phi = 0$), and with azimuthal dependence the modes are hybrid modes.

From (4.4.2), for the TE^z modes with no ϕ dependence the vector potential F_z in the two regions are

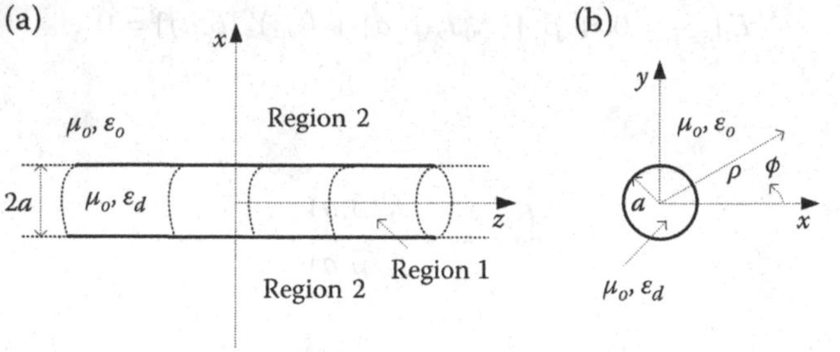

FIGURE 4.16 (a) A cylindrical dielectric waveguide; (b) cross-sectional view.

$$F_{z1} = f_1(\rho)e^{-j\beta_z z} \qquad (\rho \le a)$$

$$F_{z2} = f_2(\rho)e^{-j\beta_z z} \qquad (\rho \ge a)$$

where f_1 and f_2 satisfy

$$\left[\rho^2\frac{\partial^2}{\partial\rho^2} + \rho\frac{\partial}{\partial\rho} + \left(\beta_o^2\varepsilon_{rd} - \beta_z^2\right)\rho^2\right]f_1(\rho) = 0 \tag{4.4.38}$$

and

$$\left[\rho^2\frac{\partial^2}{\partial\rho^2} + \rho\frac{\partial}{\partial\rho} + \left(\beta_o^2 - \beta_z^2\right)\rho^2\right]f_2(\rho) = 0 \tag{4.4.39}$$

For the fields to propagate in the dielectric region and to be evanescent in the air region the radial propagation constant in the dielectric (i.e., $\beta_{\rho1}$) must be of the form

$$\beta_{\rho1}^2 = \beta_o^2\varepsilon_{rd} - \beta_z^2 > 0 \tag{4.4.40}$$

In the air region for the fields to attenuate in the ρ direction the propagation constant β_z must be greater than β_o. Hence, the attenuation constant in the air region (i.e., $\alpha_{\rho2}$) is defined as

$$\alpha_{\rho2}^2 = \beta_z^2 - \beta_o^2 > 0 \tag{4.4.41}$$

Then, (4.3.38) and (4.3.39) read

$$\left[\rho^2\frac{\partial^2}{\partial\rho^2} + \rho\frac{\partial}{\partial\rho} + (\beta_{\rho1}\rho)^2\right]f_1(\rho) = 0 \tag{4.4.42}$$

and

$$\left[\rho^2\frac{\partial^2}{\partial\rho^2}+\rho\frac{\partial}{\partial\rho}-(\alpha_{\rho2}\rho)^2\right]f_2(\rho)=0 \tag{4.4.43}$$

The solution to (4.4.42) that is finite at $\rho = 0$ is $J_0(\beta_{\rho1}\rho)$, and the solution to (4.4.43) that decays as $\rho \to \infty$ is $K_0(\alpha_{\rho2}\rho)$. These solutions are another example of a guided wave or trapped wave. That is,

$$F_{z1} = CJ_0(\beta_{\rho1}\rho)e^{-j\beta_z z} \qquad (\rho \le a)$$

and

$$F_{z2} = BK_0(\alpha_{\rho2}\rho)e^{-j\beta_z z} \qquad (\rho \ge a)$$

From (4.4.17) to (4.4.22), the fields in the dielectric regions are

$$E_{\rho1} = E_{z1} = H_{\phi1} = 0$$

$$E_{\phi1} = \frac{1}{\varepsilon_d}\frac{\partial F_{z1}}{\partial\rho} = C\frac{\beta_{\rho1}}{\varepsilon_d}J_0'(\beta_{\rho1}\rho)e^{-j\beta_z z} = -C\frac{\beta_{\rho1}}{\varepsilon_d}J_1(\beta_{\rho1}\rho)e^{-j\beta_z z}$$

$$H_{\rho1} = \frac{1}{j\omega\mu_o\varepsilon_d}\frac{\partial F_{z1}}{\partial\rho\,\partial z} = -C\frac{\beta_{\rho1}\beta_z}{\omega\mu_o\varepsilon_d}J_0'(\beta_{\rho1}\rho)e^{-j\beta_z z} = C\frac{\beta_{\rho1}\beta_z}{\omega\mu_o\varepsilon_d}J_1(\beta_{\rho1}\rho)e^{-j\beta_z z}$$

$$H_{z1} = \frac{1}{j\omega\mu_o\varepsilon_d}\left(\frac{\partial^2 F_{z1}}{\partial z^2}+\beta_o^2\varepsilon_{rd}F_{z1}\right) = \frac{\beta_{\rho1}^2}{j\omega\mu_o\varepsilon_d}F_{z1} = C\frac{\beta_{\rho1}^2}{j\omega\mu_o\varepsilon_d}J_0(\beta_{\rho1}\rho)e^{-j\beta_z z}$$

and in the air region:

$$E_{\rho2} = E_{z2} = H_{\phi2} = 0$$

$$E_{\phi2} = \frac{1}{\varepsilon_o}\frac{\partial F_{z2}}{\partial\rho} = -B\frac{\alpha_{\rho2}}{\varepsilon_o}K_1(\alpha_{\rho2}\rho)e^{-j\beta_z z}$$

$$H_{\rho2} = \frac{1}{j\omega\mu_o\varepsilon_o}\frac{\partial F_{z1}}{\partial\rho\,\partial z} = B\frac{\alpha_{\rho2}\beta_z}{\omega\mu_o\varepsilon_o}K_1(\alpha_{\rho2}\rho)e^{-j\beta_z z}$$

$$H_{z2} = \frac{1}{j\omega\mu_o\varepsilon_o}\left(\frac{\partial^2 F_{z2}}{\partial z^2} + \beta_o^2 F_{z2}\right) = -B\frac{\alpha_{\rho 2}^2}{j\omega\mu_o\varepsilon_o}K_0(\alpha_{\rho 2}\rho)e^{-j\beta_z z}$$

The boundary conditions are the continuity of the tangential fields at $\rho = a$. That is,

$$E_{\phi 1} = E_{\phi 2}|_{\rho=a} \Rightarrow C\frac{\beta_{\rho 1}}{\varepsilon_d}J_1(\beta_{\rho 1}a) = B\frac{\alpha_{\rho 2}}{\varepsilon_o}K_1(\alpha_{\rho 2}a) \tag{4.4.44}$$

and

$$H_{z1} = H_{z2}|_{\rho=a} \Rightarrow C\frac{\beta_{\rho 1}^2}{\varepsilon_d}J_0(\beta_{\rho 1}a) = -B\frac{\alpha_{\rho 2}^2}{\varepsilon_o}K_0(\alpha_{\rho 2}a) \tag{4.4.45}$$

Forming the ratio of (4.4.44) and (4.4.45):

$$-\frac{1}{\beta_{\rho 1}}\frac{J_1(\beta_{\rho 1}a)}{J_0(\beta_{\rho 1}a)} = \frac{1}{\alpha_{\rho 2}}\frac{K_1(\alpha_{\rho 2}a)}{K_0(\alpha_{\rho 2}a)} \tag{4.4.46}$$

which is the characteristic equation that determines the discrete values of the propagation constants β_z, since from (4.4.40) and (4.4.41), (4.4.46) can be expressed in terms of β_z and solved. Then, $\beta_{\rho 1}$ and $\alpha_{\rho 2}$ can be calculated. The solutions to (4.4.46) can be obtained graphically. The ratio of $J_1(x)/J_0(x)$ behaves like a cotangent function, and the ratio $K_1(x)/K_0(x)$ like an exponential function. A typical plot of (4.4.46) is shown in Fig. 4.17. The asymptotes of the left-hand side of (4.4.46) occur at the zeros of $J_0(\beta_{\rho 1}a) = 0$ or $\beta_{\rho 1}a = \chi_{0l}$ where χ_{0l} are the values of the zeros. The first zero is $\chi_{01} = 2.405$, denoted by X_1 in Fig. 4.17. The asymptote of the right-hand side (due to $K_0(\alpha_{\rho 2}a)$) occurs when $\alpha_{\rho 2} = 0$. Then, from (4.4.40) and (4.4.41):

$$\beta_{\rho 1}^2 = \beta_o^2\varepsilon_{rd} - \beta_o^2 \Rightarrow \beta_{\rho 1}a = \beta_o a\sqrt{\varepsilon_{rd} - 1} \tag{4.4.47}$$

which represents the location of the asymptote of the right-hand side of (4.4.46) along the $\beta_{\rho 1}a$ axis, denoted by X_2.

The graph in Fig. 4.17 shows one intersection as a dot (i.e., the TE_{01} mode). It is observed that if (4.4.47) is less than the first root of $J_0(\beta_{\rho 1}a) = 0$ (i.e., less than $\chi_{01} = 2.405$) there are no intersections, which means there are no propagating modes. Hence, the condition for a propagating mode is

$$\beta_o a\sqrt{\varepsilon_{rd} - 1} > 2.405 \Rightarrow f > \frac{2.405}{2\pi a\sqrt{\mu_o\varepsilon_o}\sqrt{\varepsilon_{rd} - 1}}$$

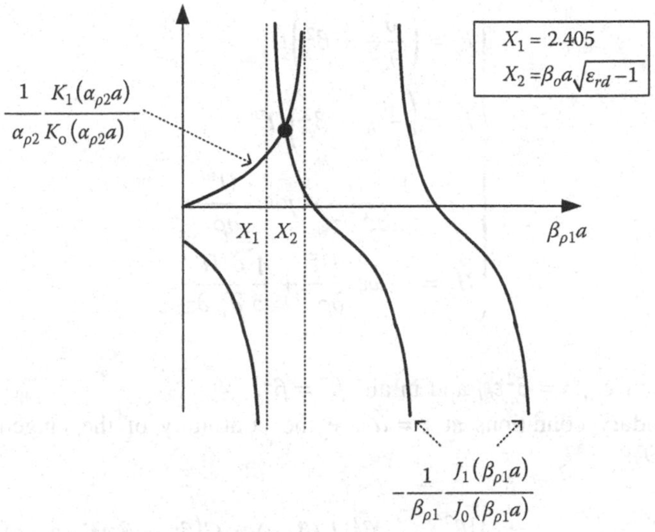

FIGURE 4.17 A typical plot of the characteristic equation in (4.4.44).

In Fig. 4.17 it is seen that as the frequency increases the point X_2 moves to the right and the curves can intersect a second time. Hence, a second mode (i.e., the TE_{02} mode) can propagate. Further increases in X_2 generates other modes.

The hybrid modes are a superposition of TE^z and TM^z modes. They are both needed to satisfy the boundary conditions when there is a ϕ dependence. The analysis can be performed using the Hertz's vectors as follows. Let

$$\begin{cases} \Pi_{z1}^{e} = AJ_n(\beta_{\rho 1}\rho)e^{-jn\phi}e^{-j\beta_z z} & (\rho \le a) \\ \Pi_{z1}^{m} = BJ_n(\beta_{\rho 1}\rho)e^{-jn\phi}e^{-j\beta_z z} & (\rho \le a) \\ \Pi_{z2}^{e} = CK_n(\alpha_{\rho 2}\rho)e^{-jn\phi}e^{-j\beta_z z} & (\rho \ge a) \\ \Pi_{z2}^{m} = DK_n(\alpha_{\rho 2}\rho)e^{-jn\phi}e^{-j\beta_z z} & (\rho \ge a) \end{cases} \tag{4.4.48}$$

where

$$\beta_{\rho 1}^2 = \beta_o^2\varepsilon_{rd} - \beta_z^2$$

and

$$\alpha_{\rho 2}^2 = \beta_z^2 - \beta_o^2$$

The fields follow from (4.4.11) to (4.4.22), where the tangential fields are given by

$$\begin{cases} E_z = \left(\dfrac{\partial^2}{\partial z^2} + \beta^2\right)\Pi_z^{\mathrm{e}} \\ H_z = \left(\dfrac{\partial^2}{\partial z^2} + \beta^2\right)\Pi_z^{\mathrm{m}} \\ E_\phi = \dfrac{1}{\rho}\dfrac{\partial^2 \Pi_z^{\mathrm{e}}}{\partial\phi\,\partial z} + j\omega\mu\dfrac{\partial \Pi_z^{\mathrm{m}}}{\partial\rho} \\ H_\phi = -j\omega\varepsilon\dfrac{\partial \Pi_z^{\mathrm{e}}}{\partial\rho} + \dfrac{1}{\rho}\dfrac{\partial^2 \Pi_z^{\mathrm{m}}}{\partial\phi\,\partial z} \end{cases} \tag{4.4.49}$$

In the dielectric: $\beta^2 = \beta_o^2 \varepsilon_{rd}$ and in air: $\beta^2 = \beta_o^2$.

The boundary conditions at $z = a$ are the continuity of the tangential fields. From (4.4.49):

$$E_{z1} = E_{z2}|_{\rho=a} \Rightarrow A\left(\beta_o^2\varepsilon_{rd} - \beta_z^2\right)J_n(\beta_{\rho1}a) = C\left(\beta_o^2 - \beta_z^2\right)K_n(\alpha_{\rho2}a)$$

which is expressed as

$$A\beta_{\rho1}^2 J_n(\beta_{\rho1}a) = -C\alpha_{\rho2}^2 K_n(\alpha_{\rho2}a) \tag{4.4.50}$$

and

$$H_{z1} = H_{z2}|_{\rho=a} \Rightarrow B\beta_{\rho1}^2 J_n(\beta_{\rho1}a) = -D\alpha_{\rho2}^2 K_n(\alpha_{\rho2}a) \tag{4.4.51}$$

Substituting (4.4.50) and (4.4.51) into (4.4.48) for C and D shows that (4.4.48) can be written in terms of A and B only. That is,

$$\begin{cases} \Pi_{z1}^{\mathrm{e}} = A\, J_n(\beta_{\rho1}\rho)e^{-jn\phi}e^{-j\beta_z z} \\ \Pi_{z1}^{\mathrm{m}} = B\, J_n(\beta_{\rho1}\rho)e^{-jn\phi}e^{-j\beta_z z} \\ \Pi_{z2}^{\mathrm{e}} = -A\dfrac{\beta_{\rho1}^2}{\alpha_{\rho2}^2}\dfrac{J_n(\beta_{\rho1}a)}{K_n(\alpha_{\rho2}a)}K_n(\alpha_{\rho2}\rho)e^{-jn\phi}e^{-j\beta_z z} \\ \Pi_{z2}^{\mathrm{m}} = -B\dfrac{\beta_{\rho1}^2}{\alpha_{\rho2}^2}\dfrac{J_n(\beta_{\rho1}a)}{K_n(\alpha_{\rho2}a)}K_n(\alpha_{\rho2}\rho)e^{-jn\phi}e^{-j\beta_z z} \end{cases} \tag{4.4.52}$$

Furthermore, if we let

$$A = A_o\frac{1}{\beta_{\rho1}^2 J_n(\beta_{\rho1}a)}$$

and

$$B = B_o \frac{1}{\beta_{\rho 1}^2 J_n(\beta_{\rho 1} a)}$$

(4.4.52) is expressed in a convenient form in terms of A_o and B_o. That is,

$$\begin{cases} \Pi_{z1}^{\mathrm{e}} = \dfrac{A_o}{\beta_{\rho 1}^2} \dfrac{J_n(\beta_{\rho 1}\rho)}{J_n(\beta_{\rho 1} a)} e^{-jn\phi} e^{-j\beta_z z} \\ \Pi_{z1}^{\mathrm{m}} = \dfrac{B_o}{\beta_{\rho 1}^2} \dfrac{J_n(\beta_{\rho 1}\rho)}{J_n(\beta_{\rho 1} a)} e^{-jn\phi} e^{-j\beta_z z} \\ \Pi_{z2}^{\mathrm{e}} = -\dfrac{A_o}{\alpha_{\rho 2}^2} \dfrac{K_n(\alpha_{\rho 2}\rho)}{K_n(\alpha_{\rho 2} a)} e^{-jn\phi} e^{-j\beta_z z} \\ \Pi_{z2}^{\mathrm{m}} = -\dfrac{B_o}{\alpha_{\rho 2}^2} \dfrac{K_n(\alpha_{\rho 2}\rho)}{K_n(\alpha_{\rho 2} a)} e^{-jn\phi} e^{-j\beta_z z} \end{cases} \tag{4.4.53}$$

Next, we apply the boundary conditions for E_ϕ and H_ϕ. From (4.4.49) and (4.4.53):

$$E_{\phi 1} = E_{\phi 2}\Big|_{\rho=a} \Rightarrow -\frac{A_o}{\beta_{\rho 1}^2}\left(\frac{n\beta_z}{a}\right) + \frac{B_o}{\beta_{\rho 1}} j\omega\mu \frac{J_n'(\beta_{\rho 1} a)}{J_n(\beta_{\rho 1} a)} = \frac{A_o}{\alpha_{\rho 2}^2}\left(\frac{n\beta_z}{a}\right) - \frac{B_o}{\alpha_{\rho 2}} j\omega\mu \frac{K_n'(\alpha_{\rho 2} a)}{K_n(\alpha_{\rho 2} a)}$$

or

$$A_o\left(\frac{n\beta_z}{a}\right)\left(\frac{1}{\beta_{\rho 1}^2} + \frac{1}{\alpha_{\rho 2}^2}\right) - B_o j\omega\mu_o\left[\frac{1}{\beta_{\rho 1}} \frac{J_n'(\beta_{\rho 1} a)}{J_n(\beta_{\rho 1} a)} + \frac{1}{\alpha_{\rho 2}} \frac{K_n'(\alpha_{\rho 2} a)}{K_n(\alpha_{\rho 2} a)}\right] = 0 \tag{4.4.54}$$

and

$$H_{\phi 1} = H_{\phi 2}|_{\rho=a} \Rightarrow A_o j\omega\varepsilon_o\left[\frac{\varepsilon_{rd}}{\beta_{\rho 1}} \frac{J_n'(\beta_{\rho 1} a)}{J_n(\beta_{\rho 1} a)} + \frac{1}{\alpha_{\rho 2}} \frac{K_n'(\alpha_{\rho 2} a)}{K_n(\alpha_{\rho 2} a)}\right] + B_o\left(\frac{n\beta_z}{a}\right)\left(\frac{1}{\beta_{\rho 1}^2} + \frac{1}{\alpha_{\rho 2}^2}\right) = 0 \tag{4.4.55}$$

Then, equating the ratio of A_o/B_o in (4.4.54) to that in (4.4.55), the characteristic equation is

$$\left(\frac{n\beta_z}{a}\right)^2\left(\frac{1}{\beta_{\rho 1}^2}+\frac{1}{\alpha_{\rho 2}^2}\right)^2=\beta_o^2\left[\frac{1}{\beta_{\rho 1}}\frac{J_n'(\beta_{\rho 1}a)}{J_n(\beta_{\rho 1}a)}+\frac{1}{\alpha_{\rho 2}}\frac{K_n'(\alpha_{\rho 2}a)}{K_n(\alpha_{\rho 2}a)}\right]$$
$$\left[\frac{\varepsilon_{rd}}{\beta_{\rho 1}}\frac{J_n'(\beta_{\rho 1}a)}{J_n(\beta_{\rho 1}a)}+\frac{1}{\alpha_{\rho 2}}\frac{K_n'(\alpha_{\rho 2}a)}{K_n(\alpha_{\rho 2}a)}\right]$$

The characteristic equation shows that when $n = 0$ (i.e., no ϕ dependence) the left-hand side is zero so the right-hand side must also be zero, which can be satisfied by either

$$\frac{1}{\beta_{\rho 1}}\frac{J_n'(\beta_{\rho 1}a)}{J_n(\beta_{\rho 1}a)}+\frac{1}{\alpha_{\rho 2}}\frac{K_n'(\alpha_{\rho 2}a)}{K_n(\alpha_{\rho 2}a)}=0 \tag{4.4.56}$$

or

$$\frac{\varepsilon_{rd}}{\beta_{\rho 1}}\frac{J_n'(\beta_{\rho 1}a)}{J_n(\beta_{\rho 1}a)}+\frac{1}{\alpha_{\rho 2}}\frac{K_n'(\alpha_{\rho 2}a)}{K_n(\alpha_{\rho 2}a)}=0 \tag{4.4.57}$$

Equation (4.4.56) corresponds to the characteristic equation for the TE^z modes, as previously obtained in (4.4.46) for the case that $n = 0$. Equation (4.4.57) corresponds to the characteristic equation for the TM^z modes.

4.4.9 The Dielectric Resonator

A dielectric resonator with constitutive parameters μ_d and ε_d is shown in Fig. 4.18. They are made with high values of relative permittivity ($\varepsilon_{rd} > 20$). A dielectric resonator acts like a cylindrical cavity having a perfect conducting magnetic wall. The magnetic-wall boundary condition at $\rho = a$ provides an approximate (but useful) analysis of the resonator. A magnetic wall resembles an open circuit, in which the current is zero. In terms of the magnetic field, the magnetic wall makes the tangential magnetic field equal to zero.

The reason that a high permittivity resembles a magnetic wall is that for a plane wave the reflection coefficient at the boundary between air and a dielectric (assuming $\mu_d = \mu_o$) is

$$R=\frac{\eta_o-\eta_d}{\eta_o+\eta_d}=\frac{\sqrt{\frac{\mu_o}{\varepsilon_o}}-\sqrt{\frac{\mu_o}{\varepsilon_d}}}{\sqrt{\frac{\mu_o}{\varepsilon_o}}+\sqrt{\frac{\mu_o}{\varepsilon_d}}}=\frac{\sqrt{\varepsilon_{rd}}-1}{\sqrt{\varepsilon_{rd}}+1}\approx 1$$

which is the reflection coefficient associated with an open circuit, or a magnetic wall. The use of the magnetic-wall approximation in the resonator in Fig. 4.18 gives a good insight on the behavior of such resonator.

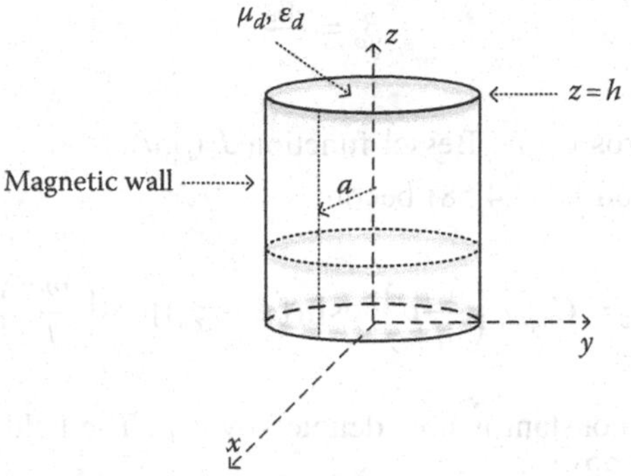

FIGURE 4.18 The dielectric resonator.

Consider the TE^z modes in the dielectric resonator in Fig. 4.18, where the modes function F_z, from (4.2.2) with $\gamma = j\beta = j\omega\sqrt{\mu_d \varepsilon_d}$, is given by

$$F_z = C_n J_n(\beta_\rho \rho)\cos[n(\phi - \phi_o)][b_1 \sin(\beta_z z) + b_2 \cos(\beta_z z)] \tag{4.4.58}$$

where

$$\beta^2 = \omega^2 \mu_d \varepsilon_d = \beta_\rho^2 + \beta_z^2 \tag{4.4.59}$$

The magnetic-wall requires that H_ϕ be zero at $z = 0$, $z = h$ and $\rho = a$. From (4.4.21):

$$H_\phi = \frac{1}{j\omega\mu_o\varepsilon_d}\frac{1}{\rho}\frac{\partial^2 F_z}{\partial\phi\,\partial z} = -C_n\frac{\beta_z n}{j\omega\mu_o\varepsilon_d\rho}J_n(\beta_\rho\rho)\sin[n(\phi - \phi_o)][b_1\cos(\beta_z z) - b_2 \sin(\beta_z z)]$$

The boundary conditions at $z = 0$ is zero if $b_1 = 0$, and at $z = h$ if

$$\sin(\beta_z h) = 0 \Rightarrow \beta_z = \frac{p\pi}{h} \qquad (p = 0, 1, 2, \ldots)$$

The boundary condition at $\rho = a$ is the vanishing of H_ϕ, which is satisfied if

$$J_n(\beta_\rho a) = 0 \Rightarrow \beta_\rho a = \chi_{nl} \qquad \begin{pmatrix} n = 0, 1, 2, \ldots \\ l = 1, 2, 3, \ldots \end{pmatrix}$$

or

$$\beta_\rho = \frac{\chi_{nl}}{a}$$

where χ_{nl} are the zeros of the Bessel function $J_n(\beta_\rho a)$.

The modes function in (4.4.58) becomes

$$F_z = C_{nlp} J_n\left(\frac{\chi_{nl}}{a}\rho\right)\cos[n(\phi - \phi_o)]\cos\left(\frac{p\pi z}{h}\right)$$

where the amplitude constant is now denoted by C_{nlp}. The field components follow from (4.4.17) to (4.4.22).

From (4.4.59), the resonant frequencies are given by

$$f_r = \frac{1}{2\pi\sqrt{\mu_d \varepsilon_d}}\sqrt{\left(\frac{\chi_{nl}}{a}\right)^2 + \left(\frac{p\pi}{h}\right)^2}$$

The lowest resonant frequency is that of the TE_{010} mode, namely

$$f_r\Big|_{\substack{n=0\\ l=1\\ p=0}} = \frac{\chi_{01}}{2\pi a\sqrt{\mu_d \varepsilon_d}} = \frac{2.405}{2\pi a\sqrt{\mu_d \varepsilon_d}}$$

The TM^z modes are obtained using

$$A_z = B_{nlp} J_n\left(\frac{\tilde{\chi}_{nl}\rho}{a}\right)\cos[n(\phi - \phi_o)]\sin\left(\frac{p\pi z}{h}\right) \tag{4.4.60}$$

where $\tilde{\chi}_{nl}$ are the zeros of the Bessel function $J_n'(\beta_\rho a)$. The resonant frequencies for the TM^z modes, from (4.4.59), are given by

$$f_r = \frac{1}{2\pi\sqrt{\mu_d \varepsilon_d}}\sqrt{\left(\frac{\tilde{\chi}_{nl}}{a}\right)^2 + \left(\frac{p\pi}{h}\right)^2} \tag{4.4.61}$$

The lowest resonant frequency of the TM^z modes is that of the TM_{111} mode, where $\tilde{\chi}_{11} = 1.841$.

4.5 WAVE TRANSFORMATION FROM RECTANGULAR TO CYLINDRICAL COORDINATES

In this section, the relations that allow us to represent plane waves in terms of cylindrical functions are derived. Such relations are convenient to use in the analysis of diffraction of a plane wave by cylindrical structures.

Consider a plane wave traveling in the x direction, given by $e^{-j\beta x}$. This wave is represented in cylindrical coordinates by

$$e^{-j\beta x} = e^{-j\beta\rho \cos\phi}$$

and can be represented in terms of cylindrical functions as

$$e^{-j\beta\rho \cos\phi} = \sum_{n=-\infty}^{\infty} a_n J_n(\beta\rho) e^{jn\phi} \tag{4.5.1}$$

The factor a_n is found by multiplying both sides of (4.5.1) by $e^{-jm\phi}$ and integrating between 0 and 2π, or

$$\int_0^{2\pi} e^{-j\beta\rho\cos\phi} e^{-jm\phi} d\phi = \sum_{n=-\infty}^{\infty} a_n J_n(\beta\rho) \int_0^{2\pi} e^{j(n-m)\phi} d\phi \tag{4.5.2}$$

The right-hand side integral is equal to $2\pi\delta_{nm}$. The left-hand side of (4.5.2) can be expressed in terms of Bessel functions using

$$J_n(\beta\rho) = \frac{j^n}{2\pi} \int_0^{2\pi} e^{-j(\beta\rho\cos\phi + m\phi)} d\phi$$

and it follows from (4.5.2) that

$$2\pi j^{-n} J_n(\beta\rho) = 2\pi a_n J_n(\beta\rho) \Rightarrow a_n = j^{-n}$$

The expansion of the plane wave in (4.5.1) is

$$e^{-j\beta x} = e^{-j\beta\rho\cos\phi} = \sum_{n=-\infty}^{\infty} j^{-n} J_n(\beta\rho) e^{jn\phi} \tag{4.5.3}$$

4.6 WAVE EQUATION IN SPHERICAL COORDINATES

In spherical coordinates we will select modes that are TE and TM to the radial distance r, called TE^r modes. For these modes, either F_r or Π_r^{m} are not solution to the scalar wave equation. That is, for $F_r(r, \theta, \phi)$:

$$\nabla^2(F_r \hat{\mathbf{a}}_r) \neq \hat{\mathbf{a}}_r \nabla^2 F_r$$

So, we have to work with (1.5.16), namely

$$\nabla \times \nabla \times (F_r \hat{\mathbf{a}}_r) + \gamma^2 (F_r \hat{\mathbf{a}}_r) = \varepsilon_c J_{m,r} \hat{\mathbf{a}}_r - j\omega\mu\varepsilon_c \nabla\varphi_m \tag{4.6.1}$$

where $J_{m,r}$ is the radial component of the magnetic current. Performing the operations indicated in (4.6.1), three equations are obtained. For the $\mathbf{a}_r$ component:

$$-\frac{1}{r^2 \sin\theta}\frac{\partial}{\partial\theta}\left(\sin\theta\frac{\partial F_r}{\partial\theta}\right) - \frac{1}{r^2\sin^2\theta}\frac{\partial^2 F_r}{\partial\phi^2} + \gamma^2 F_r = \varepsilon_c J_{m,r} - j\omega\mu\varepsilon_c\frac{\partial\varphi_m}{\partial r} \tag{4.6.2}$$

For the $\mathbf{a}_\theta$ component:

$$\frac{\partial}{\partial\theta}\left(\frac{\partial F_r}{\partial r}\right) = -j\omega\mu\varepsilon_c\frac{\partial\varphi_m}{\partial\theta} \tag{4.6.3}$$

and for the $\mathbf{a}_\phi$ component:

$$\frac{\partial}{\partial\phi}\left(\frac{\partial F_r}{\partial r}\right) = -j\omega\mu\varepsilon_c\frac{\partial\varphi_m}{\partial\phi} \tag{4.6.4}$$

Equations (4.6.3) and (4.6.4) are satisfied if we let

$$\frac{\partial F_r}{\partial r} = -j\omega\mu\varepsilon_c\varphi_m = -\frac{\gamma^2}{j\omega}\varphi_m \tag{4.6.5}$$

which is known as the Debye gauge. This is the gauge that will reduce (4.6.2) to a wave equation for F_r. Substituting (4.6.5) into (4.6.2) (i.e., for the term $\partial\varphi_m/\partial r$) gives

$$\frac{\partial^2 F_r}{\partial r^2} + \frac{1}{r^2\sin\theta}\frac{\partial}{\partial\theta}\left(\sin\theta\frac{\partial F_r}{\partial\theta}\right) + \frac{1}{r^2\sin^2\theta}\frac{\partial^2 F_r}{\partial\phi^2} - \gamma^2 F_r = -\varepsilon_c J_{m,r} \tag{4.6.6}$$

The left side is not exactly the Laplacian of F_r in spherical coordinates. However, letting

$$F_r = r\psi$$

Then,

$$\frac{\partial^2 F_r}{\partial r^2} = r\frac{\partial^2\psi}{\partial r^2} + 2\frac{\partial\psi}{\partial r} = r\left(\frac{\partial^2\psi}{\partial r^2} + \frac{2}{r}\frac{\partial\psi}{\partial r}\right) = r\left[\frac{1}{r^2}\frac{\partial}{\partial r}\left(r^2\frac{\partial\psi}{\partial r}\right)\right]$$

and (4.6.6) reads

$$r\left[\frac{1}{r^2}\frac{\partial}{\partial r}\left(r^2\frac{\partial\psi}{\partial r}\right)\right] + r\left[\frac{1}{r^2\sin\theta}\frac{\partial}{\partial\theta}\left(\sin\theta\frac{\partial\psi}{\partial\theta}\right) + \frac{1}{r^2\sin^2\theta}\frac{\partial^2\psi}{\partial\phi^2}\right] - \gamma^2 r\psi = -\varepsilon_c J_{m,r}$$

Dividing by r:

$$\frac{1}{r^2}\frac{\partial}{\partial r}\left(r^2\frac{\partial\psi}{\partial r}\right) + \frac{1}{r^2\sin\theta}\frac{\partial}{\partial\theta}\left(\sin\theta\frac{\partial\psi}{\partial\theta}\right) + \frac{1}{r^2\sin^2\theta}\frac{\partial^2\psi}{\partial\phi^2} - \gamma^2\psi = -\frac{\varepsilon_c J_{m,r}}{r}$$

which is recognized as the wave equation for ψ, where $\psi = F_r/r$. Therefore, the resulting wave equation is

$$[\nabla^2 - \gamma^2]\frac{F_r}{r} = -\frac{\varepsilon_c J_{m,r}}{r} \tag{4.6.7}$$

where ∇^2 is the Laplacian in spherical coordinates.

Once (4.6.7) is solved for F_r, the fields away from the source follow from

$$\mathbf{E} = -\frac{1}{\varepsilon_c}\nabla \times (F_r\hat{\mathbf{a}}_r)$$

and

$$\mathbf{H} = -\frac{1}{j\omega\mu}\nabla \times \mathbf{E} = \frac{1}{j\omega\mu\varepsilon_c}\nabla \times \nabla \times (F_r\hat{\mathbf{a}}_r)$$

Similarly, for the TM^r mode, A_r satisfy

$$\nabla \times \nabla \times (A_r\hat{\mathbf{a}}_r) + \gamma^2(A_r\hat{\mathbf{a}}_r) = \mu J_{e,r}\hat{\mathbf{a}}_r - j\omega\mu\varepsilon_c\nabla\varphi_e \tag{4.6.8}$$

where $J_{e,r}$ is the radial component of the electric current. The expansion of (4.6.8) is similar to that in (4.6.2) to (4.6.4), and it follows that in this case the Debye gauge is

$$\frac{\partial A_r}{\partial r} = -j\omega\mu\varepsilon_c\varphi_e = -\frac{\gamma^2}{j\omega}\varphi_e$$

Then, the wave equation satisfied by A_r/r is

$$[\nabla^2 - \gamma^2]\frac{A_r}{r} = -\frac{\mu J_{e,r}}{r} \tag{4.6.9}$$

Once (4.6.9) is solved for A_r, the fields away from the source follow from

$$\mathbf{H} = \frac{1}{\mu}\nabla \times (A_r \hat{\mathbf{a}}_r)$$

and

$$\mathbf{E} = \frac{1}{(\sigma + j\omega\varepsilon)}\nabla \times \mathbf{H} = \frac{j\omega}{\gamma^2}\nabla \times \nabla \times (A_r \hat{\mathbf{a}}_r)$$

Since the auxiliary potentials F_r and A_r in (4.6.7) and (4.6.9) require the use of the Debye gauges, these auxiliary potentials are known as Debye potentials. These are not the only way that the Debye potentials appear in the literature. In some instances, Debye potentials U and V are defined as

$$V = \frac{1}{j\omega\mu\varepsilon_c}\frac{F_r}{r} = \frac{j\omega}{\gamma^2}\frac{F_r}{r} \tag{4.6.10}$$

and

$$U = \frac{1}{j\omega\mu\varepsilon_c}\frac{A_r}{r} = \frac{j\omega}{\gamma^2}\frac{A_r}{r} \tag{4.6.11}$$

In terms of the Debye potentials U and V in (4.6.10) and (4.6.11), (4.6.7) and (4.6.9) read

$$[\nabla^2 - \gamma^2]V = -\frac{J_{m,r}}{j\omega\mu r} \tag{4.6.12}$$

and

$$[\nabla^2 - \gamma^2]U = -\frac{J_{e,r}}{(\sigma + j\omega\varepsilon)r} \tag{4.6.13}$$

One also finds the Debye potentials U and V referred as the radial Hertz vectors, using the notation Π_r^{e} for U and Π_r^{m} for V. Other notations have been used for the Debye potentials, but it should be simple to relate them to the forms in (4.6.7) and (4.6.9), or (4.6.12) and (4.6.13), which are the forms used in this book.

Away from the source, simply set $J_{e,r}$ and $J_{m,r}$ equal to zero in (4.6.7), (4.6.9), (4.6.12) and (4.6.13), and in the lossless case set $\gamma = j\beta$.

The fields in terms of the Debye potentials A_r, F_r, U and V, away from the sources, are given by

$$E_r = \begin{cases} \dfrac{j\omega}{\gamma^2}\left(\dfrac{\partial^2}{\partial r^2} - \gamma^2\right)A_r \\ \left(\dfrac{\partial^2}{\partial r^2} - \gamma^2\right)(rU) \end{cases} \tag{4.6.14}$$

$$E_\theta = \begin{cases} \dfrac{j\omega}{\gamma^2}\dfrac{1}{r}\dfrac{\partial^2 A_r}{\partial r \partial \theta} - \dfrac{1}{\varepsilon_c}\dfrac{1}{r\sin\theta}\dfrac{\partial F_r}{\partial \phi} \\ \dfrac{1}{r}\dfrac{\partial^2 (rU)}{\partial r \partial \theta} - j\omega\mu\dfrac{1}{r\sin\theta}\dfrac{\partial (rV)}{\partial \phi} \end{cases} \tag{4.6.15}$$

$$E_\phi = \begin{cases} \dfrac{j\omega}{\gamma^2}\dfrac{1}{r\sin\theta}\dfrac{\partial^2 A_r}{\partial r \partial \phi} + \dfrac{1}{\varepsilon_c}\dfrac{1}{r}\dfrac{\partial F_r}{\partial \theta} \\ \dfrac{1}{r\sin\theta}\dfrac{\partial^2 (rU)}{\partial r \partial \phi} + \dfrac{j\omega\mu}{r}\dfrac{\partial (rV)}{\partial \theta} \end{cases} \tag{4.6.16}$$

$$H_r = \begin{cases} \dfrac{j\omega}{\gamma^2}\left(\dfrac{\partial^2}{\partial r^2} - \gamma^2\right)F_r \\ \left(\dfrac{\partial^2}{\partial r^2} - \gamma^2\right)(rV) \end{cases} \tag{4.6.17}$$

$$H_\theta = \begin{cases} \dfrac{1}{\mu}\dfrac{1}{r\sin\theta}\dfrac{\partial A_r}{\partial \phi} + \dfrac{j\omega}{\gamma^2}\dfrac{1}{r}\dfrac{\partial^2 F_r}{\partial r \partial \theta} \\ \dfrac{\gamma^2}{j\omega\mu}\dfrac{1}{r\sin\theta}\dfrac{\partial (rU)}{\partial \phi} + \dfrac{1}{r}\dfrac{\partial^2 (rV)}{\partial r \partial \theta} \end{cases} \tag{4.6.18}$$

$$H_\phi = \begin{cases} -\dfrac{1}{\mu}\dfrac{1}{r}\dfrac{\partial A_r}{\partial \theta} + \dfrac{j\omega}{\gamma^2}\dfrac{1}{r\sin\theta}\dfrac{\partial^2 F_r}{\partial r \partial \phi} \\ -\dfrac{\gamma^2}{j\omega\mu}\dfrac{1}{r}\dfrac{\partial (rU)}{\partial \theta} + \dfrac{1}{r\sin\theta}\dfrac{\partial^2 (rV)}{\partial r \partial \phi} \end{cases} \tag{4.6.19}$$

Recall that in the above relations the factor $\gamma^2/j\omega$ can also be written as

$$\frac{\gamma^2}{j\omega} = j\omega\mu\varepsilon_c = j\omega\mu_o\varepsilon_o n^2$$

The homogeneous form of the wave equations in (4.6.7), (4.6.9), (4.6.12), and (4.6.13) are all in the form

$$[\nabla^2 - \gamma^2]\psi(r, \theta, \phi) = 0 \tag{4.6.20}$$

where ψ represents either A_r/r, F_r/r, U or V. Expanding the Laplacian in (4.6.20) gives

$$\frac{1}{r^2}\frac{\partial}{\partial r}\left(r^2\frac{\partial \psi}{\partial r}\right) + \frac{1}{r^2 \sin\theta}\frac{\partial}{\partial \theta}\left(\sin\theta\frac{\partial \psi}{\partial \theta}\right) + \frac{1}{r^2 \sin^2\theta}\frac{\partial^2 \psi}{\partial \phi^2} - \gamma^2\psi = 0 \tag{4.6.21}$$

Equation (4.6.21) is solved using separation of variables. Letting

$$\psi = R(r)\Theta(\theta)\Phi(\phi)$$

substituting into (4.6.21) and separating the variables, gives

$$\frac{\sin^2\theta}{R}\frac{\partial}{\partial r}\left(r^2\frac{\partial R}{\partial r}\right) + \frac{\sin\theta}{\Theta}\frac{\partial}{\partial \theta}\left(\sin\theta\frac{\partial \Theta}{\partial \theta}\right) + \frac{1}{\Phi}\frac{\partial^2 \Phi}{\partial \phi^2} = (\gamma r \sin\theta)^2 \tag{4.6.22}$$

The third term is a function of only ϕ, therefore we set

$$\frac{1}{\Phi}\frac{\partial^2 \Phi}{\partial \phi^2} = -m^2 \Rightarrow \frac{\partial^2 \Phi}{\partial \phi^2} + m^2\Phi = 0 \tag{4.6.23}$$

where m is a separation constant. In this book only integer values of m are considered.

Substituting (4.6.23) into (4.6.22), and dividing by $\sin^2\theta$, gives

$$\frac{1}{R}\frac{\partial}{\partial r}\left(r^2\frac{\partial R}{\partial r}\right) - (\gamma r)^2 + \frac{1}{\Theta \sin\theta}\frac{\partial}{\partial \theta}\left(\sin\theta\frac{\partial \Theta}{\partial \theta}\right) - \left(\frac{m}{\sin\theta}\right)^2 = 0 \tag{4.6.24}$$

The last two terms in (4.6.24) are functions of θ, so they can be set equal to a constant, namely

$$\begin{aligned} &\frac{1}{\Theta \sin\theta}\frac{\partial}{\partial \theta}\left(\sin\theta\frac{\partial \Theta}{\partial \theta}\right) - \left(\frac{m}{\sin\theta}\right)^2 = -\nu(\nu+1) \\ &\frac{1}{\sin\theta}\frac{\partial}{\partial \theta}\left(\sin\theta\frac{\partial \Theta}{\partial \theta}\right) + \left[\nu(\nu+1) - \left(\frac{m}{\sin\theta}\right)^2\right]\Theta = 0 \end{aligned} \tag{4.6.25}$$

where $\nu(\nu + 1)$ is a separation constant, and (4.6.25) is recognized as the associated Legendre's differential equation (see Appendix E).

Substituting (4.6.25) into (4.6.24) gives

$$\frac{\partial}{\partial r}\left(r^2\frac{\partial R}{\partial r}\right) - [(\gamma r)^2 + \nu(\nu+1)]R = 0 \tag{4.6.26}$$

which is the modified spherical Bessel's differential equation. In the lossless case, $\gamma = j\beta$ and (4.2.26) reads

$$\frac{\partial}{\partial r}\left(r^2\frac{\partial R}{\partial r}\right) + [(\beta r)^2 - \nu(\nu + 1)]R = 0 \tag{4.6.27}$$

which is the spherical Bessel's differential equation. Also, in the lossy case when the complex propagation constant k is used (i.e., $\gamma = jk$ where $k = k' - jk''$), (4.6.26) reads

$$\frac{\partial}{\partial r}\left(r^2\frac{\partial R}{\partial r}\right) + [(kr)^2 - \nu(\nu + 1)]R = 0 \tag{4.6.28}$$

which is the spherical Bessel's differential equation. The argument of the spherical Bessel solutions in (4.6.28) is complex if k is complex, and real when $k'' = 0$ (i.e., $k = k'$).

Solutions to (4.6.23), (4.6.25), and (4.6.26) can be expressed in different ways depending on γ, ν, and m. Some typical forms are shown:

(a) If $\gamma = j\beta$, $m = 0$ (no ϕ dependence) and $\nu = n$ (n an integer), typical solutions are:

$$R = \begin{cases} b_1 j_n(\beta r) + b_2 y_n(\beta r) \\ b_1 j_n(\beta r) + b_2 h_n^{(2)}(\beta r) \\ b_1 h_n^{(1)}(\beta r) + b_2 h_n^{(2)}(\beta r) \end{cases}$$

$$\Theta = c_1 P_n(\cos\theta) + c_2 Q_n(\cos\theta)$$

(b) If $\gamma = j\beta$, $\nu = n$ (n an integer) and m an integer, typical solutions are

$$R = \begin{cases} b_1 j_n(\beta r) + b_2 y_n(\beta r) \\ b_1 j_n(\beta r) + b_2 h_n^{(2)}(\beta r) \\ b_1 h_n^{(1)}(\beta r) + b_2 h_n^{(2)}(\beta r) \end{cases}$$

$$\Theta = c_1 P_n^m(\cos\theta) + c_2 Q_n^m(\cos\theta)$$

$$\Phi = d_1 \cos(m\phi) + d_2 \sin(m\phi)$$

(c) If $\gamma = j\beta$, $m = 0$ and $\nu \neq n$ (not an integer), typical solutions are

$$R = \begin{cases} b_1 j_\nu(\beta r) + b_2 y_\nu(\beta r) \\ b_1 j_\nu(\beta r) + b_2 h_\nu^{(2)}(\beta r) \\ b_1 h_\nu^{(1)}(\beta r) + b_2 h_\nu^{(2)}(\beta r) \end{cases}$$

$$\Theta = \begin{cases} c_1 P_\nu(\cos\theta) + c_2 Q_\nu(\cos\theta) \\ c_1 P_\nu(\cos\theta) + c_2 P_\nu(-\cos\theta) \\ c_1 P_\nu(-\cos\theta) + c_2 Q_\nu(-\cos\theta) \end{cases}$$

In this case, $P_\nu(\cos\theta)$ becomes infinite when its argument is -1, which occurs at $\theta = \pi$, and $P_\nu(-\cos\theta)$ becomes infinite when its argument is -1, which occurs at $\theta = 0$. $Q_\nu(\cos\theta)$ becomes infinite at ± 1, or at $\theta = 0$ and $\theta = \pi$.

(d) If $\gamma = j\beta$, $\nu \neq n$ (not an integer) and m an integer, typical solutions are

$$R = \begin{cases} b_1 j_\nu(\beta r) + b_2 y_\nu(\beta r) \\ b_1 j_\nu(\beta r) + b_2 h_\nu^{(2)}(\beta r) \\ b_1 h_\nu^{(1)}(\beta r) + b_2 h_\nu^{(2)}(\beta r) \end{cases}$$

$$\Theta = \begin{cases} c_1 P_\nu^m(\cos\theta) + c_2 Q_\nu^m(\cos\theta) \\ c_1 P_\nu^m(-\cos\theta) + c_2 Q_\nu^m(-\cos\theta) \end{cases}$$

$$\Phi = d_1 \cos(m\phi) + d_2 \sin(m\phi)$$

In this case, $P_\nu^m(\cos\theta)$ becomes infinite at $\theta = \pi$ and $P_\nu^m(-\cos\theta)$ becomes infinite at $\theta = 0$. $Q_\nu^m(\cos\theta)$ becomes infinite when the argument is ± 1, or at $\theta = 0$ and $\theta = \pi$. Also, $P_\nu^m(\pm\cos\theta)$ is not zero for $m > \nu$, since ν is not an integer.

(e) If $\gamma = \alpha + j\beta$ (i.e., a lossy case), the radial solutions to (4.6.26) are the modified spherical Bessel functions. Hence, with a complex γ the radial solutions in (a) and (b) are: $i_n(\gamma r)$ and $k_n(\gamma r)$, and in (c) and (d): $i_\nu(\gamma r)$ and $k_\nu(\gamma r)$.

(f) If $\gamma = jk$ where $k = k' - jk''$, (4.6.28) shows that the radial solutions are of the same form as those shown in (a) to (d) except that βr is replaced by kr in the arguments of the Bessel functions.

If ψ in (4.6.20) represents A_r/r or F_r/r, the solution must be multiplied by r to get the solutions for the Debye potentials A_r and F_r. If ψ represents the Debye potential U and V the calculation of the fields require that the potentials be multiplied by r (i.e., rU and rV). Therefore, the common form that the spherical Bessel functions appear in the solutions with $\gamma = j\beta$ are: $\beta r j_\nu(\beta r)$, $\beta r y_\nu(\beta r)$, $\beta r h_\nu^{(1)}(\beta r)$ and $\beta r h_\nu^{(2)}(\beta r)$, where in many cases ν is an integer (i.e., $\nu = n$).

Schelkunoff is credited with using the following forms of the spherical Bessel functions in electromagnetic problems:

$$\hat{j}_\nu(\beta r) = \beta r j_\nu(\beta r) = \beta r \sqrt{\frac{\pi}{2\beta r}} J_{\nu+1/2}(\beta r) = \sqrt{\frac{\beta r \pi}{2}} J_{\nu+1/2}(\beta r)$$

$$\hat{y}_\nu(\beta r) = \beta r y_\nu(\beta r) = \sqrt{\frac{\beta r \pi}{2}} Y_{\nu+1/2}(\beta r)$$

$$\hat{h}_\nu^{(1)}(\beta r) = \beta r h_\nu^{(1)}(\beta r) = \sqrt{\frac{\beta r \pi}{2}} H_{\nu+1/2}^{(1)}(\beta r)$$

$$\hat{h}_\nu^{(2)}(\beta r) = \beta r h_\nu^{(2)}(\beta r) = \sqrt{\frac{\beta r \pi}{2}} H_{\nu+1/2}^{(2)}(\beta r)$$

These functions are referred to as the Schelkunoff-Bessel functions. The above forms are also known as the Ricatti-Bessel functions. They satisfy the differential equation (see Appendix C):

$$\left[\frac{d^2}{dr^2} + \beta^2 - \frac{\nu(\nu+1)}{r^2}\right]\hat{b}_\nu(\beta r) = 0 \qquad (4.6.29)$$

where $\hat{b}_\nu(\beta r)$ is either $\hat{j}_\nu(\beta r)$, $\hat{y}_\nu(\beta r)$, $\hat{h}_\nu^{(1)}(\beta r)$ or $\hat{h}_\nu^{(2)}(\beta r)$.

For example, if $\gamma = j\beta$, $\nu = n$ and m is an integer, the modes function solution for ψ can be of the form

$$\psi = C_{nm}[b_1 j_n(\beta r) + b_2 y_n(\beta r)][c_1 P_n^m(\cos\theta) + c_2 Q_n^m(\cos\theta)][d_1 \sin m\phi + d_2 \cos m\phi]$$

Then, in terms of the Schelkunoff-Bessel functions the solution for A_r, where $\psi = A_r/r$, is

$$A_r = \frac{C_{nm}}{\beta}[b_1 \hat{j}_n(\beta r) + b_2 \hat{y}_n(\beta r)][c_1 P_n^m(\cos\theta) + c_2 Q_n^m(\cos\theta)][d_1 \sin m\phi + d_2 \cos m\phi] \qquad (4.6.30)$$

and a similar expression for F_r. For $\nu \neq n$, replace n by ν.

A solution for U or V is of the same form as that of ψ. However, when calculating the fields rU and rV are required, so we write

$$rU = \frac{C_{nm}}{\beta}[b_1\hat{j}_n(\beta r) + b_2\hat{y}_n(\beta r)][c_1 P_n^m(\cos\theta) + c_2 Q_n^m(\cos\theta)][d_1 \sin m\phi + d_2 \cos m\phi] \tag{4.6.31}$$

and a similar form for rV. For $\nu \neq n$, replace n by ν.

The term $1/\beta$ in (4.6.30) and (4.6.31) is usually incorporated into the constants and, therefore, does not appear. Also, the constant has been expressed as C_{nm} but it also appears as C_{mn}, and the sinusoidal solutions in ϕ can be written as $\cos[m(\phi - \phi_o)]$.

In the lossy case, replace β by k ($k = k' - jk''$) in the previous Schelkunoff-Bessel functions. Of course, the functions in terms of k (i.e., $\hat{j}_\nu(kr)$, $\hat{y}_\nu(kr)$, $\hat{h}_\nu^{(1)}(kr)$, $\hat{h}_\nu^{(2)}(kr)$) have complex arguments. Another choice for the lossy case is to use γ, where $\gamma = \alpha + j\beta$, and use the modified Schelkunoff-Bessel functions $\hat{i}_\nu(\gamma r)$ and $\hat{k}_\nu(\gamma r)$. That is,

$$\hat{i}_\nu(\gamma r) = \sqrt{\frac{\gamma r \pi}{2}} I_{\nu+1/2}(\gamma r)$$

and

$$\hat{k}_\nu(\gamma r) = \sqrt{\frac{\gamma r \pi}{2}} K_{\nu+1/2}(\gamma r)$$

which satisfy

$$\left[\frac{d^2}{dr^2} - \gamma^2 - \frac{\nu(\nu+1)}{r^2}\right]\left\{\begin{matrix}\hat{i}_n(\gamma r)\\ \hat{k}_n(\gamma r)\end{matrix}\right\} = 0$$

4.6.1 The Conical Waveguide

A conical waveguide is shown in Fig. 4.19. The region in the waveguide is assumed to be lossless, or $\gamma = j\beta = j\omega\sqrt{\mu\varepsilon}$. TE^r and TM^r modes can propagate in the radial direction in this configuration. For the TM^r mode, using the Debye potential A_r, the modal solutions to for the conical waveguide can be expressed as

$$A_r = C_{m\nu}\,\hat{h}_\nu^{(2)}(\beta r) P_\nu^m(\cos\theta)\cos[m(\phi - \phi_o)] \tag{4.6.32}$$

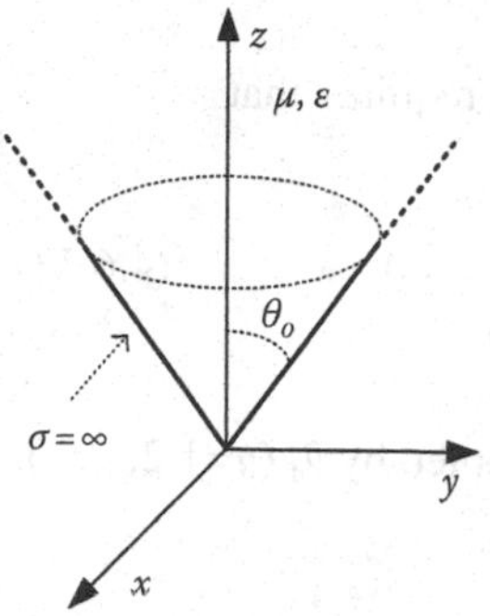

FIGURE 4.19 The conical waveguide.

where the eigenvalues of ν are determined by the boundary conditions. The radial function $\hat{h}_\nu^{(2)}(\beta r)$ is used since it represents an outgoing wave as a function of r. The function $Q_\nu^m(\cos\theta)$ is singular at $\theta = 0$ and π, so it cannot be used. The ϕ dependence is periodic in 2π and represented by $\cos[m(\phi - \phi_o)]$. The value of ϕ_o represents the angular reference for measuring ϕ.

The boundary conditions are

$$E_r = 0|_{\theta=\theta_o}$$

and

$$E_\phi = 0|_{\theta=\theta_o}$$

Either boundary condition can be used since (4.6.14) and (4.6.16) show that they produce the same results. Using (4.6.16) and observing that the boundary conditions must be satisfied for all values of r, we obtain

$$E_\phi = \frac{1}{j\omega\mu\varepsilon}\frac{1}{r\sin\theta}\frac{\partial^2 A_r}{\partial r\partial\phi}\bigg|_{\theta=\theta_o} = 0 \Rightarrow P_\nu^m(\cos\theta)|_{\theta=\theta_o} = 0 \Rightarrow P_{\nu_n}^m(\cos\theta_o) = 0$$

which is an equation in ν (difficult to solve) whose n zeros are denoted by ν_n, (n = 1,2,3,...). Therefore, the TM^r modes function is given by

$$A_r = C_{mn}\,\hat{h}_{\nu_n}^{(2)}(\beta r)P_{\nu_n}^m(\cos\theta)\cos[m(\phi - \phi_o)]$$

where C_{mn} is the amplitude constant of the mn mode.

For the TE^r modes, the modes function is

$$F_r = B_{m\nu}\,\hat{h}_\nu^{(2)}(\beta r)P_\nu^m(\cos\theta)\cos[m(\phi - \phi_o)]$$

The boundary condition for the TE^r modes, from (4.6.16), requires that

$$E_\phi = \frac{1}{\varepsilon r}\frac{\partial F_r}{\partial \theta} = 0\Bigg|_{\theta=\theta_o} \Rightarrow \frac{dP_\nu^m(\cos\theta)}{d\theta}\Bigg|_{\theta=\theta_o} = 0 \tag{4.6.33}$$

Again, in this case, there are n zeros that satisfy (4.6.33) denoted by $\tilde{\nu}_n$ (n=1,2,3,…). Therefore, the modes function is given by

$$F_r = B_{mn}\,\hat{h}_{\tilde{\nu}_n}^{(2)}(\beta r)P_{\tilde{\nu}_n}^m(\cos\theta)\cos[m(\phi - \phi_o)]$$

where B_{mn} is the amplitude constant of the mn mode.

4.6.2 The Spherical Cavity

A spherical cavity is shown in Fig. 4.20. The region in the cavity is assumed to be lossless, or $\gamma = j\beta = j\omega\sqrt{\mu\varepsilon}$. For the TM^r mode, using the Debye potential A_r we can write the modes function as

$$A_r = C_{mn}\,\hat{J}_n(\beta r)P_n^m(\cos\theta)\cos[m(\phi - \phi_o)] \tag{4.6.34}$$

where $\hat{y}_n(\beta r)$ cannot be used since the field must be finite at the origin. Also, $Q_n^m(\cos\theta)$ is singular at $\theta = 0$ and π and cannot be used.

The boundary conditions are

$$E_\theta = 0|_{r=a} \tag{4.6.35}$$

and

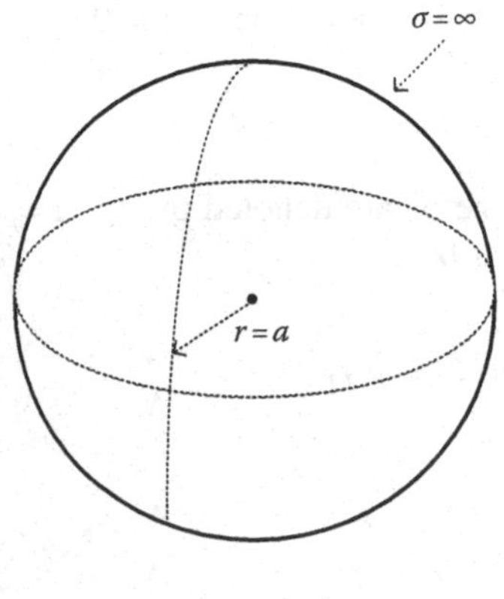

FIGURE 4.20 A spherical cavity.

$$E_\phi = 0|_{r=a} \tag{4.6.36}$$

From (4.6.15), the electric field E_θ is

$$E_\theta = \frac{1}{j\omega\mu\varepsilon}\frac{1}{r}\frac{\partial^2 A_r}{\partial r\partial\theta} = \frac{C_{mn}\beta}{j\omega\mu\varepsilon r}\hat{j}_n'(\beta r)\frac{dP_n^m(\cos\theta)}{d\theta}\cos[m(\phi - \phi_o)]$$

Then, (4.6.35) requires that

$$\hat{j}_n'(\beta a) = 0 \Rightarrow \beta a = \tilde{\kappa}_{nl} \qquad \begin{pmatrix} n = 1, 2, 3, \ldots \\ l = 1, 2, 3, \ldots \end{pmatrix}$$

or

$$\beta = \frac{\tilde{\kappa}_{nl}}{a} \tag{4.6.37}$$

where $\tilde{\kappa}_{nl}$ denotes the zeros of $j_n'(\beta a)$. The order n begins with $n = 1$, since for $n = 0$: $dP_0^m(\cos\theta)/d\theta = 0$. The boundary condition (4.6.36) gives the same information.

The modes form of A_r in (4.6.34) is then

$$A_r = C_{mnl}\hat{j}_n\left(\frac{\tilde{\kappa}_{nl}}{a}r\right)P_n^m(\cos\theta)\cos[m(\phi - \phi_o)] \tag{4.6.38}$$

where the constant is specifically expressed as C_{mnl}. The TM^r modes are denoted as the TM$_{mnl}$ modes of a spherical cavity (i.e., r superscript omitted). In the literature one also finds the modes denoted as TM$_{nlm}$.

From (4.6.37), the resonant frequencies are given by

$$f_r = \frac{\tilde{\kappa}_{nl}}{2\pi a\sqrt{\mu\varepsilon}} \tag{4.6.39}$$

The lowest mode occurs when $n = 1$ and $l = 1$ where $\tilde{\kappa}_{11} = 2.744$.

The resonant frequencies in (4.6.39) are independent of m. Therefore, for a given n and l there are several m values that produce the same resonant frequency. These are degenerate modes. Since $P_n^m(\cos\theta) = 0$ for $m > n$, the largest m is $m = n$. The lowest-order modes with the same resonance frequency are the TM$_{011}$ and TM$_{111}$ (i.e., $m = 0$ and $m = 1$). For these modes: $\tilde{\kappa}_{11} = 2.744$.

For the TM$_{011}$ mode, since $P_1^0(\cos\theta) = \cos\theta$, it follows that

$$\beta = \frac{\tilde{\kappa}_{11}}{a} = \frac{2.744}{a}$$

$$A_r|_{011} = C_{011}\,\hat{j}_1\left(\frac{2.744\,r}{a}\right)\cos\theta$$

and the fields follow from (4.6.14) to (4.6.19), namely

$$H_\phi = -\frac{1}{\mu r}\frac{\partial A_r}{\partial \theta} = \frac{C_{011}}{\mu r}\hat{j}_1\left(\frac{2.744\,r}{a}\right)\sin\theta$$

$$E_r = \frac{1}{j\omega\mu\varepsilon}\left(\frac{\partial^2}{\partial r^2} + \beta^2\right)A_r = \frac{2C_{011}}{j\omega\mu\varepsilon r^2}\hat{j}_1\left(\frac{2.744\,r}{a}\right)\cos\theta$$

$$E_\theta = \frac{1}{j\omega\mu\varepsilon}\frac{1}{r}\frac{\partial^2 A_r}{\partial r\partial\theta} = -\frac{C_{011}}{j\omega\mu\varepsilon r}\frac{2.744}{a}\hat{j}_1'(\beta r)\sin\theta$$

where (4.6.29) was used in the evaluation of E_r.

For the TM_{111} mode:

$$A_r|_{111} = C_{111}\,\hat{j}_1\left(\frac{2.744\,r}{a}\right)P_1^1(\cos\theta)\cos(\phi-\phi_o) = -C_{111}\,\hat{j}_1\left(\frac{2.744\,r}{a}\right)\sin\theta\cos(\phi-\phi_o)$$

and the fields follow from (4.6.14) to (4.6.19).

For the TE^r modes, denoted by TE_{mnl} in the spherical cavity:

$$F_r = B_{mn}\,\hat{j}_n(\beta r)P_n^m(\cos\theta)\cos[m(\phi-\phi_o)]$$

Then, from (4.6.15), the tangential field E_θ is

$$E_\theta = -\frac{1}{\varepsilon}\frac{1}{r\sin\theta}\frac{\partial F_r}{\partial\phi} = -\frac{B_{mn}\,m}{\varepsilon r\sin\theta}\hat{j}_n(\beta r)P_n^m(\cos\theta)\sin[m(\phi-\phi_o)]$$

The boundary condition in (4.6.35) (or (4.6.36)) is satisfied with

$$\hat{j}_n(\beta a) = 0 \Rightarrow \beta = \frac{\kappa_{nl}}{a} \qquad \begin{pmatrix} n = 1, 2, 3.... \\ l = 1, 2, 3.... \end{pmatrix}$$

where κ_{nl} denotes the roots of $\hat{j}_n(\beta a)$. The order n begins with $n = 1$ since for $n = 0$ the fields are zero (i.e., $mP_0^m(\cos\theta) = 0$ for any m). Therefore, the form of F_r is

$$F_r = B_{mnl}\,\hat{j}_n\left(\frac{\kappa_{nl}}{a}r\right)P_n^m(\cos\theta)\cos[m(\phi-\phi_o)]$$

where the constant is expressed as B_{mnl}.

The resonant frequencies of the TE^r modes are given by

$$f_r = \frac{\kappa_{nl}}{2\pi a\sqrt{\mu\varepsilon}} \tag{4.6.40}$$

Since (4.6.40) is independent of m, there are degenerate modes.

For example, for $n = 1$ and $l = 1$ it follows that for the TE_{m11} modes: $\kappa_{11} = 4.493$. Hence, for $m = 0$ and $m = 1$ the TE_{011} and TE_{111} are degenerate modes. For these modes:

$$F_r|_{011} = B_{011}\hat{J}_1\left(\frac{4.493r}{a}\right)P_1^0(\cos\theta) = B_{011}\hat{J}_1\left(\frac{4.493r}{a}\right)\cos\theta$$

and

$$F_r|_{111} = B_{011}\hat{J}_1\left(\frac{4.493r}{a}\right)P_1^1(\cos\theta)\cos\phi = -B_{011}\hat{J}_1\left(\frac{4.493r}{a}\right)\sin\theta\cos(\phi-\phi_o)$$

4.6.3 The Ideal Earth-Ionosphere Cavity

The ideal Earth-Ionosphere cavity is the spherical cavity between the Earth and Ionosphere when the Earth and Ionosphere are considered perfect conductors, as shown in Fig. 4.21. Such approximations can be made to analyze certain characteristics of the resulting modes for frequencies below 30 Hz. The cavity exhibits resonances for both TE^r and TM^r modes, known as Schumann resonances. One

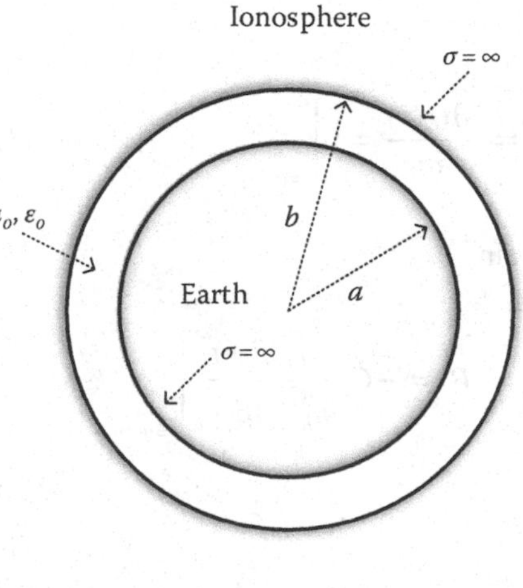

FIGURE 4.21 The ideal Earth-Ionosphere cavity.

common source of Schumann resonances are lightning discharges in the Earth-Ionosphere cavity. These resonances are observed in the 7 Hz to 30 Hz range.

Consider the TM^r modes with no ϕ variations in the cavity in terms of the Debye potential U. With $\gamma = j\beta_o$, the Debye potential is of the form

$$rU = \left[C_n \hat{j}_n(\beta_o r) + B_n \hat{h}_n^{(2)}(\beta_o r)\right] P_n(\cos\theta) \tag{4.6.41}$$

From (4.6.14) to (4.6.19), the fields components are E_r, E_θ, and H_r:

$$E_r = \left(\frac{\partial^2}{\partial r^2} + \beta_o^2\right)(rU) = \frac{n(n+1)}{r}\left[C_n \hat{j}_n(\beta_o r) + B_n \hat{h}_n^{(2)}(\beta_o r)\right] P_n(\cos\theta)$$

$$E_\theta = \frac{1}{r}\frac{\partial^2(rU)}{\partial r \partial\theta} = \frac{\beta_o}{r}\left[C_n \hat{j}_n'(\beta_o r) + B_n \hat{h}_n^{(2)\prime}(\beta_o r)\right] \frac{dP_n(\cos\theta)}{d\theta}$$

$$H_\phi = -\frac{j\omega\varepsilon_o}{r}\frac{\partial(rU)}{\partial\theta} = -\frac{j\omega\varepsilon_o}{r}\left[C_n \hat{j}_n(\beta_o r) + B_n \hat{h}_n^{(2)}(\beta_o r)\right] \frac{dP_n(\cos\theta)}{d\theta}$$

In the derivation of E_r, (4.6.29) was used. One can also write the angular derivative as (see Appendix E):

$$\frac{dP_n(\cos\theta)}{d\theta} = P_n^1(\cos\theta)$$

The boundary conditions are: $E_\theta = 0$ and $E_\phi = 0$ at $r = a$ and $r = b$. Either boundary condition can be used. Then, since the boundary conditions must be satisfied for all values of θ:

$$E_\theta = \frac{1}{r}\frac{\partial^2(rU)}{\partial r \partial\theta}\bigg|_{r=a} = 0 \Rightarrow \frac{\partial(rU)}{\partial r} = 0\bigg|_{r=a} \tag{4.6.42}$$

and

$$E_\theta = \frac{1}{r}\frac{\partial^2(rU)}{\partial r \partial\theta}\bigg|_{r=b} = 0 \Rightarrow \frac{\partial(rU)}{\partial r} = 0\bigg|_{r=b} \tag{4.6.43}$$

From (4.6.41), (4.6.42), and (4.6.43) we obtain

$$C_n \hat{j}_n'(\beta_o r) + B_n \hat{h}_n^{(2)\prime}(\beta_o r)\bigg|_{r=a} = 0 \Rightarrow B_n = -C_n \frac{\hat{j}_n'(\beta_o r)}{\hat{h}_n^{(2)\prime}(\beta_o r)}\bigg|_{r=a}$$

and

$$C_n\hat{j}_n'(\beta_o r) + B_n\hat{h}_n^{(2)\prime}(\beta_o r)\Big|_{r=b} = 0 \Rightarrow B_n = -C_n\frac{\hat{j}_n'(\beta_o r)}{\hat{h}_n^{(2)\prime}(\beta_o r)}\Bigg|_{r=b}$$

where the derivative is with respect to the argument. These relations show that

$$\frac{\hat{j}_n'(\beta_o r)}{\hat{h}_n^{(2)\prime}(\beta_o r)}\Bigg|_{r=a} = \frac{\hat{j}_n'(\beta_o r)}{\hat{h}_n^{(2)\prime}(\beta_o r)}\Bigg|_{r=b} \tag{4.6.44}$$

Equation (4.6.44) is a transcendental equation for the eigenvalues n as a function of frequency and, therefore, the resonant frequencies of the cavity. These are known as the Schumann resonances.

An approximate solution to the resonant frequencies can be obtained if we observe that the radial functions satisfy:

$$\left[\frac{d^2}{dr^2} + \beta_o^2 - \frac{n(n+1)}{r^2}\right]\left\{\begin{matrix}\hat{j}_n(\beta_o r)\\ \hat{h}_n^{(2)}(\beta_o r)\end{matrix}\right\} = 0 \tag{4.6.45}$$

Then, since in the cavity $r \approx a$, using the approximation:

$$\frac{n(n+1)}{r^2} \approx \frac{n(n+1)}{a^2} \tag{4.6.46}$$

(4.6.45) is expressed as

$$\left[\frac{d^2}{dr^2} + \beta_o^2 - \frac{n(n+1)}{a^2}\right]R(r) = 0 \tag{4.6.47}$$

where $R(r)$ represents the radial solutions when the approximation (4.6.46) is made. The radial solutions to (4.6.47) are

$$R(r) = B_1 \sin(\beta_r r) + B_2 \cos(\beta_r r)$$

where

$$\beta_r^2 = \beta_o^2 - \frac{n(n+1)}{a^2} \tag{4.6.48}$$

The boundary conditions (i.e., $E_\theta = 0$ at $r = a$ and $r = b$) in terms of $R(r)$ are

$$\frac{dR}{dr} = 0|_{r=a} \Rightarrow B_1\cos(\beta_r a) - B_2\sin(\beta_r a) = 0$$

and

$$\frac{dR}{dr} = 0|_{r=b} \Rightarrow B_1 \cos(\beta_r b) - B_2 \sin(\beta_r b) = 0$$

These relations show that the characteristic relation is

$$\frac{\sin(\beta_r a)}{\cos(\beta_r a)} = \frac{\sin(\beta_r b)}{\cos(\beta_r b)} \Rightarrow \sin[\beta_r (b - a)] = 0 \Rightarrow \beta_r = \frac{m\pi}{h} \qquad (m = 0, 1, 2, ...)$$

where $h = b - a$ is the height of the ionosphere. From (4.6.48) it follows that the resonant frequencies are

$$\beta_o^2 - \frac{n(n+1)}{a^2} = \left(\frac{m\pi}{h}\right)^2 \Rightarrow f_r = \frac{1}{2\pi\sqrt{\mu_o \varepsilon_o}} \sqrt{\frac{n(n+1)}{a^2} + \left(\frac{m\pi}{h}\right)^2}$$

The lowest resonant frequency occurs when $m = 0$. That is,

$$f_r = \frac{1}{2\pi a\sqrt{\mu_o \varepsilon_o}} \sqrt{n(n+1)}$$

This equation predicts the resonant frequencies close to the ones observed in the Earth-Ionosphere cavity due to lighting effects.

4.7 WAVE TRANSFORMATION FROM RECTANGULAR TO SPHERICAL COORDINATES

In this section, a relation that allows the representation of a plane wave in terms of spherical functions is derived. Such relation is used in the analysis of diffraction of plane waves by spherical structures.

A plane wave propagating along a given axis, selected as the z axis for easy conversion to spherical coordinates, is given by $e^{-j\beta z}$. It can be expressed in spherical coordinates as

$$e^{-j\beta z} = e^{-j\beta r \cos\theta} \qquad (4.7.1)$$

Equation (4.7.1) can be expanded in terms of spherical functions as

$$e^{-j\beta r \cos\theta} = \sum_{n=0}^{\infty} a_n j_n(\beta r) P_n(\cos\theta) \qquad (4.7.2)$$

The factor a_n is found by multiplying both sides of (4.7.2) by $P_m(\cos\theta)\sin\theta$ and integrating between 0 and π, or

$$\int_0^\pi e^{-j\beta r\cos\theta}P_m(\cos\theta)\sin\theta\,d\theta = \sum_{n=0}^{\infty} a_n j_n(\beta r)\int_0^\pi P_n(\cos\theta)P_m(\cos\theta)\sin\theta d\theta \tag{4.7.3}$$

The integral on the right-hand side is equal to $2/(2n+1)\delta_{nm}$ (see Appendix E). The integral on the left-hand side has been evaluated in terms of $j_n(\beta r)$. That is,

$$\int_0^\pi e^{-j\beta r\cos\theta}P_n(\cos\theta)\sin\theta\,d\theta = 2j^{-n}j_n(\beta r)$$

Therefore, from (4.7.3):

$$2j^{-n}j_n(\beta r) = a_n j_n(\beta r)\frac{2}{2n+1} \Rightarrow a_n = j^{-n}(2n+1)$$

and it follows that (4.7.2) reads

$$e^{-j\beta r\cos\theta} = \sum_{n=0}^{\infty} j^{-n}(2n+1)j_n(\beta r)P_n(\cos\theta) \tag{4.7.4}$$

Equation (4.7.4) is used to express the incident field associated with a plane wave in terms of spherical functions.

Problems

P4.1 Derive (4.3.6) and obtain the field expressions for the TE^z_{nm} modes.

P4.2 For the cavity shown in Fig. 4.2, derive the form of the auxiliary potential A_z associated with the TM^r_{nmp} modes.

P4.3

a. Show that TE^z_{nm} modes are not possible in the partially filled waveguide in Fig. 4.3a.
b. Determine the solution for the TE^x_{nm} modes in the partially filled waveguide in Fig. 4.3a.

P4.4

a. Derive the form of the auxiliary vector $\mathbf{A} = A_y\hat{\mathbf{a}}_y$ for the TM^y modes in the partially filled waveguide in Fig. 4.3b.

b. Determine the characteristic equation.

P4.5 In the partially filled waveguide in Fig. 4.3b let Region 1 be a dielectric and Region 2 be air. Determine the form of the auxiliary vector **F** for the TE^y_{nm} modes.

P4.6 For the dielectric waveguide in Fig. 4.6:

a. Derive the characteristic equation for the TM^z even modes.
b. Derive (4.3.62).

P4.7 Derive (4.3.71) and (4.3.72).

P4.8 Find the form of the modes functions A_z, F_z, Π^e_z, and Π^m_z in the sectoral waveguide whose cross section is shown in Fig. P4.8.

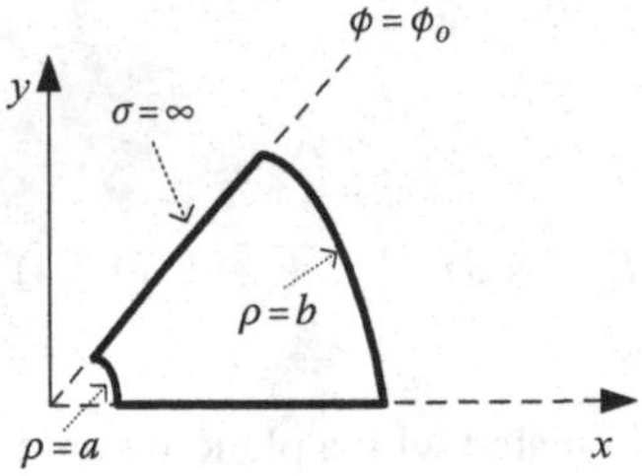

FIGURE P4.8

P4.9 For the radial waveguide in Fig. 4.12, show that the TE^z modes with no ϕ variation are calculated using

$$\Pi^{\mathrm{m}}_z = B_{0p} \sin\left(\frac{p\pi z}{a}\right) H_0^{(2)}\left(\beta_\rho \rho\right)$$

P4.10 Derive the mode function for the TE^z modes and the field expressions for the sectoral horn in Fig. 4.14.

P4.11 Derive the characteristic equation for the TM^z modes in (4.4.57) starting with the expressions for the vector potentials A_{z1} and A_{z2} associated with the two regions in Fig. 4.16.

P4.12 Derive (4.4.60) and (4.4.61).

P4.13 Derive (4.6.8) and (4.6.9).

P4.14

a. If the region inside the spherical cavity in Fig. 4.20 is lossy, write the expressions for the potentials A_r and F_r in terms of the complex propagation k.
b. Write the expressions for A_r and F_r in terms of the modified Schelkunoff-Bessel functions.

5 Sturm-Liouville Equation and Green Functions

5.1 THE STURM-LIOUVILLE EQUATION

In electromagnetism, the solution of the resulting wave equation usually involves the solution of a second-order differential equation that can be expressed in the form

$$\frac{d}{dx}\left[p(x)\frac{du}{dx}\right] + [q(x) + \lambda r(x)]u = 0 \tag{5.1.1}$$

where $x_1 \leq x \leq x_2$. Equation (5.1.1) is known as the Sturm-Liouville equation. The study of the Sturm-Liouville equation subject to boundary conditions provides a uniform method to the study of several differential equations that appear when separation of variables is used to solve the wave equation.

The solutions of (5.1.1) for certain values of λ are known as the Sturm-Liouville problems. The values of λ are called the eigenvalues and the corresponding solutions $u(x)$ are called the eigenfunctions. The cases considered are those where:

a. $p(x)$, $p'(x)$, $q(x)$, and $r(x)$ are continuous in the range $x_1 \leq x \leq x_2$.
b. $p(x)$ and $r(x)$ are positive in the range of $x_1 \leq x \leq x_2$, except possibly at the boundaries where $p(x)$ can be zero.
c. $u(x)$ is subject to specific boundary conditions (i.e., homogenous boundary conditions).

For example, for the differential equation

$$\frac{d^2\psi}{dx^2} + \beta^2\psi = 0$$

in terms of the Sturm-Liouville equation:

$$\begin{cases} u(x) = \psi(x) \\ p(x) = 1 \\ q(x) = 0 \\ r(x) = 1 \\ \lambda = \beta^2 \end{cases}$$

For the Bessel's differential equation of integer order, which can be written as

DOI: 10.1201/9781003219729-5

$$\frac{d}{dx}\left(x\frac{dy}{dx}\right)+\left(k^2x-\frac{n^2}{x}\right)y=0$$

in terms of the Sturm-Liouville equation:

$$\begin{cases} u(x)=y(x) \\ p(x)=x \\ q(x)=-\dfrac{n^2}{x} \\ r(x)=x \\ \lambda=k^2 \end{cases}$$

For the Legendre's differential equation of integer order, which can be written as

$$\frac{d}{dx}\left[(1-x^2)\frac{dy}{dx}\right]+n(n+1)y=0$$

in terms of the Sturm-Liouville equation:

$$\begin{cases} u(x)=y(x) \\ p(x)=1-x^2 \\ q(x)=0 \\ r(x)=1 \\ \lambda=n(n+1) \end{cases}$$

The parameter λ in (5.1.1) represent a separation constant, a propagation constant, or a parameter associated with an integral transform. Solutions to (5.1.1) are possible for certain values of the parameter λ (i.e., $\lambda=\lambda_n$, $n=1, 2, 3, \ldots$), called the eigenvalues of the equation. The associated solutions are the eigenfunctions, denoted by $u_n(x)$.

Several important properties of the Sturm-Liouville equation are discussed next. If u_i and u_j are two eigenfunctions of (5.1.1), we can write

$$\frac{d}{dx}\left[p(x)\frac{du_i}{dx}\right]+[q(x)+\lambda_i r(x)]u_i=0 \tag{5.1.2}$$

and

$$\frac{d}{dx}\left[p(x)\frac{du_j}{dx}\right]+[q(x)+\lambda_j r(x)]u_j=0 \tag{5.1.3}$$

where λ_i and λ_j are the associated eigenvalues. Multiplying (5.1.2) by u_j, (5.1.3) by u_i, and subtracting gives

$$u_j\frac{d}{dx}\left[p(x)\frac{du_i}{dx}\right] - u_i\frac{d}{dx}\left[p(x)\frac{du_j}{dx}\right] = (\lambda_j - \lambda_i)r(x)u_iu_j$$

or

$$\frac{d}{dx}\left[p(x)\left(u_j\frac{du_i}{dx} - u_i\frac{du_j}{dx}\right)\right] = (\lambda_j - \lambda_i)r(x)u_iu_j$$

Integrating over the range of x, say $x_1 \le x \le x_2$, gives

$$p(x)\left(u_j\frac{du_i}{dx} - u_i\frac{du_j}{dx}\right)\Bigg|_{x_1}^{x_2} = (\lambda_j - \lambda_i)\int_{x_1}^{x_2} r(x)u_iu_jdx \tag{5.1.4}$$

The left-hand side vanishes if $p(x_1) = p(x_2) = 0$, or if the boundary conditions are homogeneous, namely:

$u_i(x) = u_j(x) = 0$ at $x = x_1$ and $x = x_2$ (Dirichlet boundary condition)

or

$u_i'(x) = u_j'(x) = 0$ at $x = x_1$ and $x = x_2$ (Neumann boundary condition)

or

$$\left.\begin{aligned} a_1u_i(x) + b_1u_i'(x) = a_1u_j(x) + b_1u_j'(x) = 0 \text{ at } x = x_1 \\ a_2u_i(x) + b_2u_i'(x) = a_2u_j(x) + b_2u_j'(x) = 0 \text{ at } x = x_2 \end{aligned}\right\} \text{(Mixed boundary condition)}$$

The Dirichlet and Neumann boundary conditions are simply specific cases of the mixed boundary conditions by setting $a_1 = a_2 = 0$ for the Neumann boundary conditions, or $b_1 = b_2 = 0$ for the Dirichlet boundary conditions.

If $\lambda_i \neq \lambda_j$ and under any of the above conditions, (5.1.4) shows that

$$\int_{x_1}^{x_2} r(x)u_i(x)u_j(x)dx = 0 \tag{5.1.5}$$

Two functions that satisfy (5.1.5) are said to be orthogonal with weight $r(x)$. If $\lambda_i = \lambda_j$, then

$$\int_{x_1}^{x_2} r(x)u_i(x)u_j(x)dx \neq 0$$

If the eigenfunctions are normalized, we write

$$\int_{x_1}^{x_2} r(x)u_i(x)u_j(x)dx = 1 \tag{5.1.6}$$

In terms of the normalized functions (called orthonormal functions), (5.1.5) and (5.1.6) are written as

$$\int_{x_1}^{x_2} r(x)u_i(x)u_j(x)dx = \delta_{ij} \tag{5.1.7}$$

where δ_{ij} is the Kronecker delta.

In general, the homogenous boundary conditions satisfied by the solutions $u_n(x)$ of (5.1.1) in the range $x_1 \leq x \leq x_2$ are

$$a_1 u_n(x_1) + b_1 u_n{}'(x_1) = 0$$

and

$$a_2 u_n(x_2) + b_2 u_n{}'(x_2) = 0$$

An important property is that the eigenfunctions of (5.1.1) form a complete set. This means that a continuous function $f(x)$ (or a piecewise continuous function) can be expanded in a series of eigenfunctions in the range $x_1 \leq x \leq x_2$, namely

$$f(x) = \sum_{n=0}^{\infty} a_n u_n(x) \tag{5.1.8}$$

This equation shows the completeness property of the eigenfunction. Observe, the similarity of (5.1.8) with a Fourier series expansion.

Multiplying (5.1.8) by $r(x)u_m(x)$ and integrating between x_1 and x_2 gives

$$a_n = \int_{x_1}^{x_2} f(x')r(x')u_n(x')\,dx' \tag{5.1.9}$$

where the variable of integration has been denoted by x'. Substituting (5.1.9) into (5.1.8) gives

$$f(x) = \int_{x_1}^{x_2} f(x')\left[r(x') \sum_{n=0}^{\infty} u_n(x)u_n(x')\right]dx' \tag{5.1.10}$$

Equation (5.1.10) shows that the term in brackets is a delta function. That is

$$\delta(x - x') = r(x') \sum_{n=0}^{\infty} u_n(x) u_n(x') \tag{5.1.11}$$

This relation shows that a delta function can be expanded in a series of eigenfunctions.

Thus far, we have assumed the eigenfunctions to be real. If the eigenfunctions are complex, while $p(x)$, $q(x)$, and $r(x)$ are real in the interval $x_1 \leq x \leq x_2$, the orthonormal relation is

$$\int_{x_1}^{x_2} r(x) u_n(x) u_m^*(x) dx = \delta_{nm} \tag{5.1.12}$$

where $u_m^*(x)$ represents the complex conjugate.

In this case, multiplying (5.1.8) by $r(x) u_m^*(x)$ and integrating between x_1 and x_2 gives

$$a_n = \int_{x_1}^{x_2} f(x') r(x') u_n^*(x)\, dx' \tag{5.1.13}$$

Substituting (5.1.13) into (5.1.8) gives

$$f(x) = \int_{x_1}^{x_2} f(x') \left[r(x') \sum_{n=0}^{\infty} u_n(x) u_n^*(x') \right] dx' \tag{5.1.14}$$

which shows that the delta function is given by

$$\delta(x - x') = r(x') \sum_{n=0}^{\infty} u_n(x) u_n^*(x') \tag{5.1.15}$$

Equation (5.1.8) shows the completeness property of the eigenfunctions. That is, a piecewise continuous function can be expanded in terms of the eigenfunctions $u_n(x)$, which form a complete set. In (5.1.10) and (5.1.14), the delta function appears in the integral producing the desired $f(x)$, and shows the so-called closure property of the orthonormal eigenfunctions.

A summary of the properties in a Sturm-Liouville problem is:

a. For each eigenvalue λ_n, there is an eigenfunction $u_n(x)$.
b. The eigenfunctions satisfy homogeneous boundary conditions.
c. The eigenfunctions are orthogonal.
d. A continuous function $f(x)$ can be expanded in terms of the eigenfunctions $u_n(x)$ in the appropriate range of x.
e. The eigenvalues are real numbers. If the lowest eigenvalue is λ_1, the other eigenvalues are such that: $\lambda_1 < \lambda_2 < \lambda_3 < \ldots\ldots$

5.2 THE GREEN FUNCTIONS

When the excitation in (5.1.1) is a delta function, the solution is known as the Green function $G(x, x')$. Hence,

$$\frac{d}{dx}\left[p(x)\frac{dG}{dx}\right] + [q(x) + \lambda r(x)]G = -\delta(x - x') \tag{5.2.1}$$

where $x_1 \le x \le x_2$. One also finds (5.2.1) written with a plus sign in the right-hand side. The Green function solution to (5.2.1) is constructed by assuming that it satisfies the same homogeneous boundary conditions as those of the eigenfunction $u_n(x)$, where $u_n(x)$ satisfy,

$$\frac{d}{dx}\left[p(x)\frac{du_n}{dx}\right] + [q(x) + \lambda_n r(x)]u_n = 0 \tag{5.2.2}$$

Expanding the Green function in terms of the orthonormal functions $u_n(x)$ as

$$G(x, x') = \sum_{n=0}^{\infty} a_n u_n(x) \tag{5.2.3}$$

and substituting (5.2.3) into (5.2.1), gives

$$\sum_{n=0}^{\infty} a_n(\lambda_n - \lambda)\, r(x)u_n(x) = \delta(x - x') \tag{5.2.4}$$

where (5.2.2) was used.

Multiplying (5.2.4) by $u_m(x)$, integrating from $x_1 \le x \le x_2$ and using (5.1.7) gives

$$a_n = \frac{u_n(x')}{\lambda_n - \lambda}$$

and (5.2.3) has the form

$$G(x, x') = \sum_{n=0}^{\infty} \frac{u_n(x)u_n(x')}{\lambda_n - \lambda} \tag{5.2.5}$$

Observe that the Green function is symmetrical (i.e., $G(x, x') = G(x', x)$).

For complex eigenfuntions, (5.2.4) is multiplied by $u_m^*(x)$ and integrated from $x_1 \le x \le x_2$ to give

$$a_n = \frac{u_n^*(x')}{\lambda_n - \lambda}$$

and then, (5.2.3) has the form

$$G(x, x') = \sum_{n=0}^{\infty} \frac{u_n(x)u_n^*(x')}{\lambda_n - \lambda} \tag{5.2.6}$$

Observe that if $\lambda_n = \lambda$ the Green function is not defined.

The procedure just described to obtain the Green functions in (5.2.5) or (5.2.6) is known as the eigenfunctions method. Two other methods for obtaining the Green function are the direct method (see Section 5.6) and the Fourier transform method (see Chapter 6). These methods produce closed-form expressions for the Green functions.

Next, consider the solution to the Sturm-Liouville equation with a source $f(x)$:

$$\frac{d}{dx}\left[p(x)\frac{dy}{dx}\right] + [q(x) + \lambda r(x)]y = -f(x) \tag{5.2.7}$$

where $y(x)$ satisfies homogeneous boundary conditions. The associated homogeneous equation satisfied by the eigenfunctions $u_n(x)$ is

$$\frac{d}{dx}\left[p(x)\frac{du_n}{dx}\right] + [q(x) + \lambda_n r(x)]u_n = 0 \tag{5.2.8}$$

We seek a solution of (5.2.7) in terms of the orthonormal functions $u_n(x)$, namely

$$y(x) = \sum_{n=0}^{\infty} b_n u_n(x) \tag{5.2.9}$$

The process of finding b_n is similar to the one of finding a_n in (5.2.3). That is, substituting (5.2.9) into (5.2.7), using (5.2.8), multiplying by $u_m^*(x)$, and integrating in the range $x_1 \le x \le x_2$, gives

$$b_n = \frac{\int_{x_1}^{x_2} f(x')\, u_n^*(x')dx'}{\lambda_n - \lambda}$$

and from (5.2.9),

$$y(x) = \int_{x_1}^{x_2} f(x')\left[\sum_{n=0}^{\infty} \frac{u_n(x)u_n^*(x')}{\lambda_n - \lambda}\right]dx'$$

The quantity in the brackets is recognized as the Green function associated with (5.2.7) (see (5.2.6)). Hence, $y(x)$ can be expressed in terms of the Green function as

$$y(x) = \int_{x_1}^{x_2} f(x')G(x, x')dx' \tag{5.2.10}$$

which shows that once the Green function is found, the solution to an arbitrary excitation $f(x)$ can be calculated using (5.2.10). In the case that the eigenfunctions are real, replace $u_n^*(x)$ by $u_n(x)$.

Thus far, we have considered the one-dimensional case. In two dimensions, say x and y, the eigenfunctions representation of the Green function is

$$G(x, y, x', y') = \sum_m \sum_n \frac{u_{nm}(x, y)u_{nm}^*(x', y')}{\lambda_{nm} - \lambda} = \sum_m \sum_n \frac{u_n(x)u_m(y)u_n^*(x')u_m^*(y')}{\lambda_{nm} - \lambda} \tag{5.2.11}$$

where the summations are over all the eigenvalues. In (5.2.11) the notation

$$G(\mathbf{r}, \mathbf{r}') = G(x, x', y, y')$$

is also used, where $\mathbf{r}$ represents (x, y) and $\mathbf{r}'$ represents (x', y'). Of course, in cylindrical or spherical coordinates $\mathbf{r}$ and $\mathbf{r}'$ represent cylindrical or spherical variables.

Once the Green function is obtained, (5.2.10) shows that the solution to an excitation can be found. For example, consider the wave equation associated with a rectangular component of either $\mathbf{A}$, $\mathbf{F}$, $\mathbf{\Pi}^e$, or $\mathbf{\Pi}^m$ written in the form

$$[\nabla^2 + \beta^2]u(x) = -f(x) \tag{5.2.12}$$

where $u(x)$ represents the field component and $f(x)$ the excitation.

The associated Green function (i.e., $G(x, x')$) satisfies

$$[\nabla^2 + \beta^2]G(x, x') = -\delta(x - x') \tag{5.2.13}$$

where the excitation is a delta function located at x'.

Multiplying (5.2.13) by $u(x)$, (5.2.12) by $G(x, x')$ and subtracting, gives

$$u\nabla^2 G - G\nabla^2 u = -u\delta(x - x') + f\,G$$

Next, integrating over a volume:

$$\int_V (u\nabla^2 G - G\nabla^2 u)dV = -u(x') + \int_V f(x)G(x, x')dV \tag{5.2.14}$$

The left-hand side is evaluated using Green's theorem, which allows us to change the volume integral to one over its surface. That is,

$$\int_V (u\nabla^2 G - G\nabla^2 u)dV = \oint_S \left(u\frac{\partial G}{\partial n} - G\frac{\partial u}{\partial n}\right)dS \tag{5.2.15}$$

where $\partial/\partial n$ denotes the derivative in a direction normal to the surface. The Green function satisfies the same boundary condition as u over the surface. Hence, for homogeneous boundary conditions the surface integral in (5.2.15) vanishes. With the volume integral equal to zero, (5.2.14) shows that

$$u(x') = \int_V f(x)G(x, x')dV$$

and since $G(x, x') = G(x', x)$:

$$u(x) = \int_{V'} f(x')G(x, x')\,dV'$$

This relation shows that once the Green function is known, the response to an excitation can be calculated.

A detail approach to the development of the Green function in two and three dimensions is as follows. Consider the wave equation:

$$[\nabla^2 + \lambda]\psi(\mathbf{r}) = -f(\mathbf{r}) \tag{5.2.16}$$

and the associated Green function:

$$[\nabla^2 + \lambda]G(\mathbf{r}, \mathbf{r}') = -\delta(\mathbf{r} - \mathbf{r}') \tag{5.2.17}$$

where λ in terms of the propagation constant is $\lambda = \beta^2$ or $\lambda = k^2$. In three dimensions $\mathbf{r}$ represents (x, y, z), (ρ, ϕ, z) or (r, θ, ϕ), and $\mathbf{r}'$ the primed values.

To express the Green function in terms of eigenfunctions, the eigenfunctions associated with (5.2.16) satisfy

$$[\nabla^2 + \lambda_n]\psi_n(\mathbf{r}) = 0 \tag{5.2.18}$$

The eigenfunctions in (5.2.18) are orthogonal. This is seen by considering the eigenfunctions $\psi_m^*(x)$ which satisfy

$$[\nabla^2 + \lambda_m]\psi_m^*(\mathbf{r}) = 0 \tag{5.2.19}$$

Multiplying (5.2.18) by $\psi_m^*(x)$, (5.2.19) by $\psi_n(\mathbf{r})$, subtracting the results, and integrating over a volume V, gives

$$\int_V \left(\psi_m^* \nabla^2 \psi_n - \psi_n \nabla^2 \psi_m^*\right) dV = (\lambda_m - \lambda_n) \int_V \psi_n \psi_m^*\, dV \tag{5.2.20}$$

Using Green's theorem:

$$\int_V \left(\psi_m^* \nabla^2 \psi_n - \psi_n \nabla^2 \psi_m^*\right) dV = \oint_S \left(\psi_m^* \frac{\partial \psi_n}{\partial u} - \psi_n \frac{\partial \psi_m^*}{\partial u}\right) dS \qquad (5.2.21)$$

the volume integral is changed to a surface integral. The closed surface S encloses the volume and $\partial/\partial u$ represents the normal derivative in the direction of the unit vector $\hat{\mathbf{u}}$ on the surface.

Substituting (5.2.21) into (5.2.20) and if the eigenfunctions satisfy homogeneous boundary conditions on the surface, it follows that the left-hand side of (5.2.20) is zero. Hence,

$$(\lambda_m - \lambda_n) \int_V \psi_n \psi_m^* \, dV = 0$$

which shows that if $\lambda_m \neq \lambda_n$, the eigenfunctions are orthogonal.

The eigenfunctions can be normalized and the Green function expressed in terms of the orthonormal eigenfunctions. That is,

$$G(\mathbf{r}, \mathbf{r}') = \sum_{n=0}^{\infty} a_n \psi_n(\mathbf{r}) \qquad (5.2.22)$$

The process of finding a_n is similar to the one used to find the constants in (5.2.3) and (5.2.9). Hence,

$$a_n = \frac{\psi_n^*}{\lambda_n - \lambda}$$

and (5.2.22) reads

$$G(\mathbf{r}, \mathbf{r}') = \sum_{n=0}^{\infty} \frac{\psi_n(\mathbf{r}) \psi_n^*(\mathbf{r}')}{\lambda_n - \lambda} \qquad (5.2.23)$$

The delta function representation in terms of the eigenfunctions is

$$\delta(\mathbf{r} - \mathbf{r}') = \sum_{n=0}^{\infty} \psi_n(\mathbf{r}) \psi_n^*(\mathbf{r}')$$

Once the Green function is found in (5.2.23), the solution to (5.2.16) is

$$\psi(\mathbf{r}) = \int_{V'} f(\mathbf{r}') \, G(\mathbf{r}, \mathbf{r}') dV' \qquad (5.2.24)$$

For example, in rectangular coordinates, say x and y, (5.2.24) reads

$$\psi(x, y) = \int_{y_1}^{y_2} \int_{x_1}^{x_2} f(x', y')G(x, y, x', y')dx'dy'$$

where the Green function is given by (5.2.23) with $\psi_n(\mathbf{r}) = \psi_n(x, y)$, which can be expressed as $\sum_{n=0}^{\infty} \psi_n(x, y) = \sum_{i=0}^{\infty} \sum_{j=0}^{\infty} \psi_i(x)\psi_j(y)$. Hence, in this case, the summation in (5.2.23) is changed to a double summation (i.e., over i and j) leading to the form in (5.2.11).

The following example illustrates the eigenfunctions representation of the Green function in the solution of a differential equation. Consider the solution of the differential equation

$$\frac{d^2y}{dx^2} + \beta^2 y = -\sin\frac{\pi x}{a} \tag{5.2.25}$$

with boundary conditions given by

$$y(0) = 0$$

and

$$y(a) = 0$$

A common way of solving (5.2.25) is to write the solution as a sum of a particular (y_p) plus a complementary (y_c) solution. For the particular solution, let

$$y_p = A\sin\left(\frac{\pi x}{a}\right) + B\cos\left(\frac{\pi x}{a}\right)$$

Then,

$$y_p'' = -A\left(\frac{\pi}{a}\right)^2 \sin\left(\frac{\pi x}{a}\right) - B\left(\frac{\pi}{a}\right)^2 \cos\left(\frac{\pi x}{a}\right)$$

Substituting y_p and y_p'' into (5.2.25) gives

$$-A\left(\frac{\pi}{a}\right)^2 \sin\left(\frac{\pi x}{a}\right) - B\left(\frac{\pi}{a}\right)^2 \cos\left(\frac{\pi x}{a}\right) + \beta^2 A\sin\left(\frac{\pi x}{a}\right) + \beta^2 B\cos\left(\frac{\pi x}{a}\right)$$
$$= -\sin\left(\frac{\pi x}{a}\right)$$

which shows that $B = 0$, $\beta \neq \pi/a$ and

$$A = \frac{1}{\left(\frac{\pi}{a}\right)^2 - \beta^2}$$

Hence, the particular solution is

$$y_p = \frac{1}{\left(\frac{\pi}{a}\right)^2 - \beta^2} \sin\left(\frac{\pi x}{a}\right)$$

The complementary solution, which is the solution to the homogenous form of (5.2.25), is

$$y_c = C_1 \sin(\beta x) + C_2 \cos(\beta x)$$

and it follows that the total solution is

$$y = A \sin\left(\frac{\pi x}{a}\right) + C_1 \sin(\beta x) + C_2 \cos(\beta x) \tag{5.2.26}$$

Applying the boundary conditions to (5.2.26):

$$y(0) = 0 \Rightarrow A \sin(0) + C_1 \sin(0) + C_2 \cos(0) = 0 \Rightarrow C_2 = 0$$

and

$$y(a) = 0 \Rightarrow A \sin(\pi) + C_1 \sin(\beta a) = 0 \Rightarrow C_1 = 0$$

Hence, the total solution of (5.2.25) is

$$y = A \sin\left(\frac{\pi x}{a}\right) = \frac{1}{\left(\frac{\pi}{a}\right)^2 - \beta^2} \sin\left(\frac{\pi x}{a}\right) \tag{5.2.27}$$

The solution in (5.2.27) satisfies (5.2.25) and the boundary conditions. In (5.2.27), it is seen that β can have any value, except $\beta = \pi/a$ for which there is no solution.

Next, the Green function associated with (5.2.25) is derived for a specific value of β, say $\beta = 2$. In this case, (5.2.25) reads

$$\frac{d^2y}{dx^2} + 4y = -\sin\frac{\pi x}{a} \tag{5.2.28}$$

and the Green function satisfies

$$\frac{d^2G}{dx^2} + 4G = -\delta(x - x') \tag{5.2.29}$$

with the same boundary conditions as those of $y(x)$. That is

$$G(0) = 0$$

and

$$G(a) = 0$$

The eigenfunctions associated with (5.2.29) follow from the homogeneous equation

$$\frac{d^2u}{dx^2} + \lambda u = 0 \tag{5.2.30}$$

where λ represents β^2. This is the equation that determines the eigenfunctions $u_n(x)$ and eigenvalues λ_n. The solutions to (5.2.30) are

$$u(x) = A \sin(\sqrt{\lambda}x) + B \cos(\sqrt{\lambda}x) \tag{5.2.31}$$

Applying the boundary conditions:

$$u(0) = 0 \Rightarrow B = 0$$

and

$$u(a) = 0 \Rightarrow A \sin(\sqrt{\lambda}a) = 0 \Rightarrow \lambda = \lambda_n = \left(\frac{n\pi}{a}\right)^2 \qquad (n = 1, 2, 3, ...)$$

Therefore, (5.2.31) is expressed in the form

$$u_n(x) = A_n \sin\left(\frac{n\pi}{a}x\right) \qquad (n = 1, 2, 3, ...)$$

where the constant is now expressed as A_n. Only positive values of n are needed, since negative values produce the same eigenfunctions.

The solutions are normalized as follows:

$$A_n^2 \int_0^a \sin^2\left(\frac{n\pi}{a}x\right) = 1 \Rightarrow A_n^2\left(\frac{a}{2}\right) = 1 \Rightarrow A_n = \sqrt{\frac{2}{a}}$$

and the normalized solutions are

$$u_n(x) = \sqrt{\frac{2}{a}} \sin\left(\frac{n\pi}{a}x\right) \qquad (n = 1, 2, 3, \ldots)$$

From (5.2.5), the Green function expressed in terms of $u_n(x)$ is

$$G(x, x') = \frac{2}{a} \sum_{n=1}^{\infty} \frac{\sin\left(\frac{n\pi}{a}x\right)\sin\left(\frac{n\pi}{a}x'\right)}{\left(\frac{n\pi}{a}\right)^2 - 4}$$

and from (5.2.10), the solution to (5.2.28) is

$$y(x) = \int_0^a f(x')G(x, x')dx' = \frac{2}{a} \frac{\sin\left(\frac{n\pi}{a}x\right)}{\left(\frac{n\pi}{a}\right)^2 - 4} \int_0^a \sin\left(\frac{\pi}{a}x'\right)\sin\left(\frac{n\pi}{a}x'\right)dx \qquad (5.2.32)$$

The integral has a value of $a/2$ only for $n = 1$. Therefore, the solution is

$$y(x) = \frac{1}{\left(\frac{\pi}{a}\right)^2 - 4} \sin\left(\frac{\pi}{a}x\right) \qquad (5.2.33)$$

which is the same as the solution found in (5.2.27). (5.2.32) gives the solution for other forms of $f(x)$.

5.3 ELECTROMAGNETIC FIELD SOURCES

A Green function involves the solution of a differential equation due to a delta function excitation, which is then used to obtain the solution to the differential equation with a given source. In time-varying electromagnetics the differential equation is a wave equation, and the sources of the electromagnetic fields are the electric and magnetic currents. These sources, as well as the associated Green functions, require the use of delta functions to specify their location. Various delta function expressions are now reviewed, as well as several common sources and their delta function representation.

The delta function has the following properties:

$$\int_{V'} \delta(\mathbf{r} - \mathbf{r}')dV' = \int_V \delta(\mathbf{r} - \mathbf{r}')dV = 1 \qquad (5.3.1)$$

and

$$f(\mathbf{r}) = \int_{V'} f(\mathbf{r}')\delta(\mathbf{r} - \mathbf{r}')dV'$$

In rectangular coordinates, the delta function representation in three dimensions is

$$\delta(\mathbf{r}-\mathbf{r}') = \delta(x-x')\delta(y-y')\delta(z-z')$$

where

$$\mathbf{r}-\mathbf{r}' = (x-x')\hat{\mathbf{a}}_x + (y-y')\hat{\mathbf{a}}_y + (z-z')\hat{\mathbf{a}}_z$$

$$|\mathbf{r}-\mathbf{r}'| = \sqrt{(x-x')^2 + (y-y')^2 + (z-z')^2}$$

and the element of volume is

$$dV = dx\,dy\,dz$$

In two dimensions, say x and y, the delta function is

$$\delta(\mathbf{r}-\mathbf{r}') = \delta(x-x')\delta(y-y')$$

and in one dimension, say x, it is

$$\delta(\mathbf{r}-\mathbf{r}') = \delta(x-x')$$

In cylindrical coordinates, the delta function representation in three dimensions is

$$\delta(\boldsymbol{\rho}-\boldsymbol{\rho}') = \frac{\delta(\rho-\rho')\delta(\phi-\phi')\delta(z-z')}{\rho}$$

where

$$|\boldsymbol{\rho}-\boldsymbol{\rho}'| = \sqrt{\rho^2 + \rho'^2 - 2\rho\rho'\cos(\phi-\phi') + (z-z')^2}$$

The element of volume is

$$dV = \rho\,d\rho\,d\phi\,dz$$

In cylindrical coordinates, for two dimensions:

$$\delta(\boldsymbol{\rho}-\boldsymbol{\rho}') = \frac{\delta(\rho-\rho')\delta(\phi-\phi')}{\rho} \qquad \text{(no } z \text{ dependence)}$$

and

$$\delta(\boldsymbol{\rho} - \boldsymbol{\rho}') = \frac{\delta(\rho - \rho')\delta(z - z')}{2\pi\rho} \qquad \text{(no } \phi \text{ dependence)}$$

In one dimension:

$$\delta(\boldsymbol{\rho} - \boldsymbol{\rho}') = \frac{\delta(\rho - \rho')}{2\pi\rho} \qquad \text{(no } \phi \text{ and } z \text{ dependence)}$$

In spherical coordinates the delta function representation in three dimensions is

$$\delta(\mathbf{r} - \mathbf{r}') = \frac{\delta(r - r')\delta(\theta - \theta')\delta(\phi - \phi')}{r^2 \sin\theta}$$

where

$$|\mathbf{r} - \mathbf{r}'| = \sqrt{r^2 + r'^2 - 2rr'[\sin\theta\sin\theta'\cos(\phi - \phi') + \cos\theta\cos\theta']}$$

and for $\phi = \phi'$:

$$|\mathbf{r} - \mathbf{r}'| = \sqrt{r^2 + r'^2 - 2rr'\cos(\theta - \theta')}$$

The element of volume is

$$dV = r^2 \sin\theta dr d\theta d\phi$$

In spherical coordinates, for two dimensions:

$$\delta(\mathbf{r} - \mathbf{r}') = \frac{\delta(r - r')\delta(\phi - \phi')}{2r^2} \qquad \text{(no } \theta \text{ dependence)}$$
$$\delta(\mathbf{r} - \mathbf{r}') = \frac{\delta(r - r')\delta(\theta - \theta')}{2\pi r^2 \sin\theta} \qquad \text{(no } \phi \text{ dependence)}$$

and in one dimension:

$$\delta(\mathbf{r} - \mathbf{r}') = \frac{\delta(r - r')}{4\pi r^2} \qquad \text{(no } \theta \text{ and } \phi \text{ dependence)}$$

The factors in the denominator of the various delta function are selected so that (5.3.1) is satisfied. For example, for the delta function in one dimension in a spherical coordinate, we have:

$$\int_V \frac{\delta(r - r')}{4\pi r^2} dV = \int_{r=0}^{\infty} \int_{\theta=0}^{\pi} \int_{\phi=0}^{2\pi} \frac{\delta(r - r')}{4\pi r^2} r^2 \sin\theta dr d\theta d\phi = \int_0^{\infty} \delta(r - r') dr = 1$$

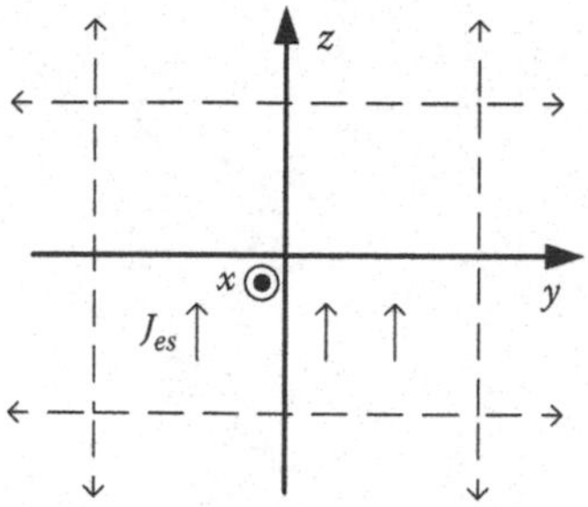

FIGURE 5.1 An infinite current sheet in the in the y, z plane.

Next, we discuss some typical sources and their delta function representation. In rectangular coordinates, an infinite current sheet in the in the yz plane is shown in Fig. 5.1. This is a theoretical current distribution which can approximate the current in a finite-width current sheet. The current density in the plane is $\mathbf{J}_e(x) = J_{es}\delta(x)\hat{\mathbf{a}}_z$, where J_{es} is the surface current density in A/m. For this configuration, the fields will have no y or z dependence.

Another configuration in rectangular coordinates is shown in Fig. 5.2. The current source is assumed to be of infinite length in the z direction and located at x', y'. The current density is given by

$$\mathbf{J}_e(x, y) = I\delta(x - x_o)\delta(y - y_o)\hat{\mathbf{a}}_z$$

The resulting fields for this configuration are two dimensional and will vary as functions of x and y.

A cylindrical surface-current distribution along the z axis is shown in Fig. 5.3. This current distribution can approximate a finite-length cylindrical current distribution. The surface current density in the cylinder along the z direction is J_{es} and is given by

$$\mathbf{J}_e(\rho) = J_{es}\delta(\rho - a)\hat{\mathbf{a}}_z = I\frac{\delta(\rho - a)}{2\pi a}\hat{\mathbf{a}}_z$$

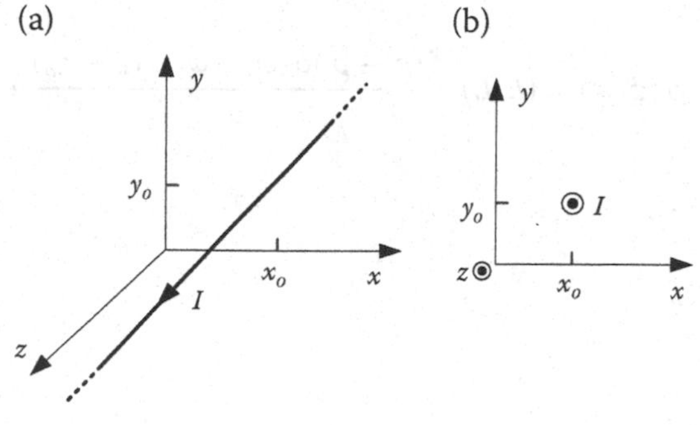

FIGURE 5.2 (a) An infinite-length current source located at x_o, y_o; (b) two-dimensional view.

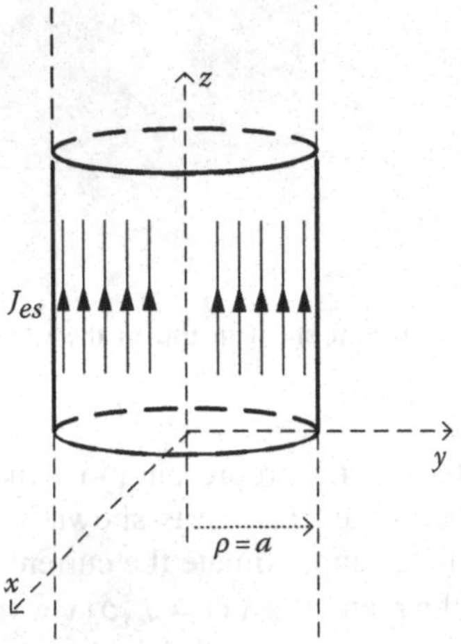

FIGURE 5.3 A cylindrical current distribution of infinite length.

such that

$$I = \int_0^{2\pi} J_{es} a d\phi = 2\pi a J_{es}$$

For this configuration, the resulting fields have no ϕ or z dependence.

The location of the infinite-length electric line current in Fig. 5.2 can also be expressed in cylindrical coordinates, as shown in Fig. 5.4. The current density is given

$$\mathbf{J}_e(\rho, \phi) = I \frac{\delta(\rho - \rho_o)\delta(\phi - \phi_o)}{\rho} \hat{\mathbf{a}}_z$$

For this configuration, the resulting fields have no z dependence.

For a Hertzian dipole located at (ρ_o, ϕ_o, z_o), pointing in the z direction, the current density in cylindrical coordinates is given by

$$\mathbf{J}_e(\rho, \phi, z) = (I\Delta L) \frac{\delta(\rho - \rho_o)\delta(\phi - \phi_o)\delta(z - z_o)}{\rho} \hat{\mathbf{a}}_z$$

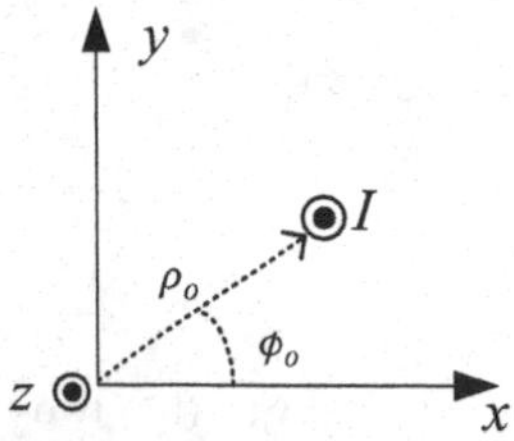

FIGURE 5.4 An infinite-length current source current at ρ_o, ϕ_o.

In spherical coordinates, for a Hertzian dipole located at (r_o, θ_o, ϕ_o), the current density is given by

$$\mathbf{J}_e(r, \theta, \phi) = (I\Delta L)\,\frac{\delta(r - r_o)\delta(\theta - \theta_o)\delta(\phi - \phi_o)}{r^2 \sin\theta}\hat{\mathbf{a}}_r \tag{5.3.2}$$

5.4 GREEN FUNCTION USING EIGENFUNCTIONS SOLUTIONS

5.4.1 One-Dimension Green Functions in Rectangular Coordinates

The Green function solution using the summation of eigenfunctions is now considered. The main drawback to the eigenfunctions solution is that involves an infinite summation, which might present computational problems.

The Green function solution to a one-dimensional wave equation is as follows. Consider the region between $x = 0$ and $x = a$, where the Green function satisfies

$$\left[\frac{d^2}{dx^2} + \beta^2\right]G(x, x') = -\delta(x - x') \tag{5.4.1}$$

The propagation constant is β and the boundary conditions are

$$G(0, x') = G(a, x') = 0$$

The eigenfunctions associated with (5.4.1) follow from the homogeneous equation

$$\left[\frac{d^2}{dx^2} + \lambda\right]u(x) = 0 \tag{5.4.2}$$

subject to the same boundary conditions as those of $G(x, x')$. The solution to (5.4.2) is

$$u(x) = C_1 \sin(\sqrt{\lambda}x) + C_2 \cos(\sqrt{\lambda}x) \tag{5.4.3}$$

Applying the boundary conditions:

$$u(x)|_{x=0} \Rightarrow C_1 \sin(0) + C_2 \cos(0) = 0 \Rightarrow C_2 = 0$$

and

$$u(x)|_{x=a} \Rightarrow C_2 \sin(\sqrt{\lambda}a) = 0 \Rightarrow \sqrt{\lambda}a = n\pi \Rightarrow \lambda = \lambda_n = \left(\frac{n\pi}{a}\right)^2 \quad (n = 1, 2, 3, \ldots) \tag{5.4.4}$$

Hence, the eigenfunctions $u_n(x)$ are written in the form

$$u_n(x) = C_n \sin\left(\frac{n\pi}{a}x\right) \tag{5.4.5}$$

where the constants are denoted by C_n since it is associated with the eigenfunctions $u_n(x)$ ($n = 1, 2, 3, \ldots$). It is seen from (5.4.4) that in terms of the Sturm-Liouville notation, λ_n are the eigenvalues.

The Green function solution in terms of eigenfunctions is given in (5.2.5), which requires normalized eigenfunctions. To normalize the eigenfunctions in (5.4.5), use the integral relation:

$$\int_0^a \sin\left(\frac{n\pi x}{a}\right)\sin\left(\frac{m\pi x}{a}\right)dx = \frac{a}{2}\delta_{nm} \tag{5.4.6}$$

Then, it follows from (5.4.5) that

$$C_n^2 \int_0^a \sin^2\left(\frac{n\pi x}{a}\right)dx = 1 \Rightarrow C_n^2 \frac{a}{2} = 1 \Rightarrow C_n = \sqrt{\frac{2}{a}}$$

and the orthonormal solutions are

$$u_n(x) = \sqrt{\frac{2}{a}}\sin\left(\frac{n\pi x}{a}\right) \tag{5.4.7}$$

From (5.2.5), the Green function is

$$G(x, x') = \frac{2}{a}\sum_{n=1}^{\infty} \frac{\sin\left(\frac{n\pi x}{a}\right)\sin\left(\frac{n\pi x'}{a}\right)}{\left(\frac{n\pi}{a}\right)^2 - \beta^2} \tag{5.4.8}$$

Also, from (5.1.11), the following delta function representation is obtained:

$$\delta(x - x') = \frac{2}{a}\sum_{n=1}^{\infty} \sin\left(\frac{n\pi x}{a}\right)\sin\left(\frac{n\pi x'}{a}\right)$$

Observe that the Green function in (5.4.8) is symmetrical (i.e., $G(x, x') = G(x', x)$), and has a singularity at

$$\beta = \frac{n\pi}{a} \tag{5.4.9}$$

The singularity occurs at the eigenvalues of the homogeneous equation, and no

Green function solution exists at the frequencies associated with (5.4.9). These frequencies are called resonant frequencies. From (5.4.9) the resonant frequencies are

$$\omega_r \sqrt{\mu\varepsilon} = \frac{n\pi}{a} \Rightarrow f_r = \frac{n}{2a\sqrt{\mu\varepsilon}}$$

At $f = f_r$, the frequency of the source matches the resonant frequency of the configuration. At resonance, the solution will increase to infinity.

Once the Green function is known, the solution to a wave equation with an excitation $f(x)$ is readily obtained. Let the wave equation be

$$\left[\frac{d^2}{dx^2} + \beta^2\right] y(x) = -f(x)$$

where $y(x)$ satisfies the same boundary conditions as those of the Green function. The function $f(x)$ represents a source and $y(x)$ is given by

$$y(x) = \int_0^a f(x')G(x, x')dx'$$

Next, we consider the solution to (5.4.1) when the boundary conditions are

$$\left.\frac{\partial G}{\partial x}\right|_{x=0} = 0$$

and

$$\left.\frac{\partial G}{\partial x}\right|_{x=a} = 0$$

In this case, applying the same boundary conditions to $u(x)$ in (5.4.3) gives

$$\left.\frac{\partial u}{\partial x}\right|_{x=0} = 0 \Rightarrow C_1 \cos(0) - C_2 \sin(0) = 0 \Rightarrow C_1 = 0$$

and

$$\left.\frac{\partial u}{\partial x}\right|_{x=a} = 0 \Rightarrow C_2 \sin(\sqrt{\lambda}\, a) = 0 \Rightarrow \lambda = \lambda_n = \left(\frac{n\pi}{a}\right)^2 \quad (n = 0, 1, 2, ...)$$

Hence, the eigenfunctions $u_n(x)$ are written in the form

$$u_n(x) = C_n \cos\left(\frac{n\pi x}{a}\right) \tag{5.4.10}$$

To obtain the Green function, the eigenfunctions in (5.4.10) are normalized. To normalize (5.4.10) we use the integral

$$\int_0^a \cos\left(\frac{n\pi x}{a}\right)\cos\left(\frac{m\pi x}{a}\right)dx = \frac{a}{\varepsilon_n}\delta_{nm} \tag{5.4.11}$$

where ε_n is Neumann's number. That is,

$$\varepsilon_n = \begin{cases} 1 \text{ for } n = 0 \\ 2 \text{ for } n \neq 0 \end{cases}$$

Then, from (5.4.10):

$$C_n^2 \int_0^a \cos^2\left(\frac{n\pi x}{a}\right)dx = 1 \Rightarrow C_n^2 \frac{a}{\varepsilon_n} = 1 \Rightarrow C_n = \sqrt{\frac{\varepsilon_n}{a}}$$

and the orthonormal solutions are

$$u_n(x) = \sqrt{\frac{\varepsilon_n}{a}}\cos\left(\frac{n\pi x}{a}\right) \tag{5.4.12}$$

From (5.2.5), the Green function is

$$G(x, x') = \frac{1}{a}\sum_{n=0}^{\infty} \varepsilon_n \frac{\cos\left(\frac{n\pi x}{a}\right)\cos\left(\frac{n\pi x'}{a}\right)}{\left(\frac{n\pi}{a}\right)^2 - \beta^2} \tag{5.4.13}$$

and from (5.1.11) the delta function representation is

$$\delta(x - x') = \frac{1}{a}\sum_{n=0}^{\infty} \varepsilon_n \cos\left(\frac{n\pi x}{a}\right)\cos\left(\frac{n\pi x'}{a}\right)$$

5.4.2 Two-Dimensional Green Functions in Rectangular Coordinates

Consider the solution to the wave equation satisfied by the electric field E_z in the rectangular region shown in Fig. 5.5. The infinite-length current excitation, given by $\mathbf{J}_e = J_{e,z}(x, y)\hat{\mathbf{a}}_z$, generates a lateral electric field that can be analyzed using either (1.3.13) or (1.4.11) with the term $\nabla[\nabla \cdot J_{e,z}(x, y)\hat{\mathbf{a}}_z] = 0$, namely

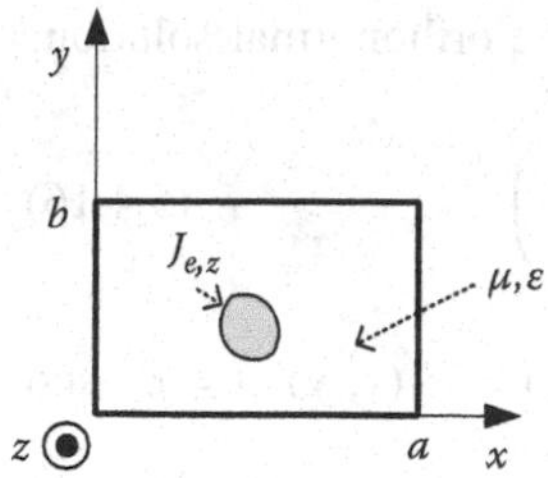

FIGURE 5.5 A rectangular region.

$$\left[\frac{\partial^2}{\partial x^2} + \frac{\partial^2}{\partial y^2} + \beta^2\right] E_z(x, y) = j\omega\mu J_{e,z}(x, y) \tag{5.4.14}$$

where $\beta = \omega\sqrt{\mu\varepsilon}$.

The solution to the homogeneous form of the equation in (5.4.14) is

$$E_z = [B_1 \sin(\beta_x x) + B_2 \cos(\beta_x x)]\,[C_1 \sin(\beta_y y) + C_2 \cos(\beta_y y)] \tag{5.4.15}$$

where

$$\beta^2 = \beta_x^2 + \beta_y^2$$

Applying the boundary conditions:

$$E_z(0, y) = E_z(a, y) = 0$$

and

$$E_z(x, 0) = E_z(x, b) = 0$$

it follows from (5.4.15) that

$$E_z(x, y) = D_{nm} \sin(\beta_x x) \sin(\beta_y y) = D_{nm} \sin\left(\frac{n\pi x}{a}\right) \sin\left(\frac{m\pi y}{b}\right) \qquad \begin{pmatrix} n = 1, 2, 3, \ldots \\ m = 1, 2, 3, \ldots \end{pmatrix}$$

where, denoting the eigenvalues of β by β_{nm}:

$$\beta_{nm}^2 = \beta_x^2 + \beta_y^2 = \left(\frac{n\pi}{a}\right)^2 + \left(\frac{m\pi}{b}\right)^2$$

Using (5.4.6) to normalize the above solution, produces the orthonormal solution:

$$E_z(x, y) = \frac{2}{\sqrt{ab}} \sin\left(\frac{n\pi x}{a}\right)\sin\left(\frac{m\pi y}{b}\right) \tag{5.4.16}$$

where in terms of the Sturm-Liouville notation: $u_n(x)u_m(y) = E_z(x, y)$, $\lambda = \beta^2$ and $\lambda_{nm} = \beta_{nm}^2$.

Using (5.4.16), the Green function (see (5.2.11) is

$$G(x, y, x', y') = \frac{4}{ab} \sum_{m=1}^{\infty} \sum_{n=1}^{\infty} \frac{\sin\left(\frac{n\pi x}{a}\right)\sin\left(\frac{n\pi x'}{a}\right)\sin\left(\frac{m\pi y}{b}\right)\sin\left(\frac{m\pi y'}{b}\right)}{\beta_{nm}^2 - \beta^2} \tag{5.4.17}$$

and from (5.1.11) the delta functions relations are

$$\delta(x - x') = \frac{2}{a} \sum_{n=1}^{\infty} \sin\left(\frac{n\pi x}{a}\right)\sin\left(\frac{n\pi x'}{a}\right) \tag{5.4.18}$$

and

$$\delta(y - y') = \frac{2}{b} \sum_{m=1}^{\infty} \sin\left(\frac{m\pi y}{b}\right)\sin\left(\frac{m\pi y'}{b}\right)$$

The field produced by the current source in (5.4.14) follows from (5.2.10). That is,

$$E_z(x, y) = -j\omega\mu \int_0^a \int_0^b J_{e,z}(x', y')\, G(x, y, x', y')dx'dy'$$

where the Green function is given by (5.4.17).

If the current source is given by

$$\mathbf{J} = J_{e,z}(x, y)\hat{\mathbf{a}}_z = I\delta(x - x_o)\delta(y - y_o)\hat{\mathbf{a}}_z$$

Then,

$$E_z(x, y) = -j\omega\mu I\, G(x, y, x_o, y_o)$$

The case of a magnetic current source in the rectangular region is shown in Fig. 5.6. The infinite-length current excitation, given by $\mathbf{J}_m = J_{m,z}(x, y)\hat{\mathbf{a}}_z$, generates a lateral magnetic field that can be analyzed using (1.4.12), namely

$$\left[\frac{\partial^2}{\partial x^2} + \frac{\partial^2}{\partial y^2} + \beta^2\right] H_z(x, y) = j\omega\varepsilon J_{m,z}(x, y) \tag{5.4.19}$$

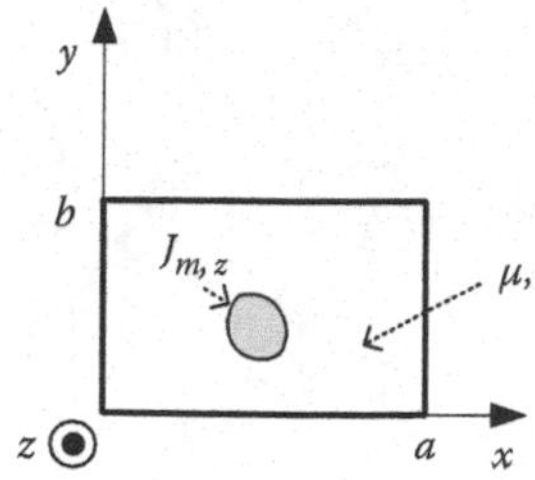

FIGURE 5.6 A rectangular region with a magnetic current source.

where $\beta = \omega\sqrt{\mu\varepsilon}$.

The boundary conditions are the vanishing tangential electric field at the boundaries, which from Maxwell's equations:

$$E_x = \frac{1}{j\omega\varepsilon}\frac{\partial H_z}{\partial y}\bigg|_{y=0,b} = 0 \Rightarrow \frac{\partial H_z}{\partial y} = 0\bigg|_{y=0,b}$$

and

$$E_y = -\frac{1}{j\omega\varepsilon}\frac{\partial H_z}{\partial x}\bigg|_{x=0,a} = 0 \Rightarrow \frac{\partial H_z}{\partial x} = 0\bigg|_{x=0,a}$$

The solution to the homogeneous equation in (5.4.19) that satisfies the boundary conditions is

$$H_z(x, y) = B_{nm}\cos\left(\frac{n\pi x}{a}\right)\cos\left(\frac{m\pi y}{b}\right) \quad \begin{pmatrix} n = 0, 1, 2, \ldots \\ m = 0, 1, 2, \ldots \end{pmatrix}$$

To normalize this expression, (5.4.11) is used to obtain

$$H_z(x, y) = \sqrt{\frac{\varepsilon_n\varepsilon_m}{ab}}\cos\left(\frac{n\pi x}{a}\right)\cos\left(\frac{m\pi y}{b}\right) \quad \begin{pmatrix} n = 0, 1, 2, \ldots \\ m = 0, 1, 2, \ldots \end{pmatrix}$$

which shows the eigenfunctions and eigenvalues required to obtain the Green function.

The Green function associated with (5.4.19), denoted by G^m, is expressed in terms of the above orthonormal functions. That is,

$$G^m(x, y, x', y') = \frac{1}{ab}\sum_{n=0}^{\infty}\sum_{m=0}^{\infty}\frac{\varepsilon_n\varepsilon_m\cos\left(\frac{n\pi x}{a}\right)\cos\left(\frac{n\pi x'}{a}\right)\cos\left(\frac{m\pi y}{b}\right)\cos\left(\frac{m\pi y'}{b}\right)}{\beta_{nm}^2 - \beta^2} \tag{5.4.20}$$

where

$$\beta_{nm}^2 = \left(\frac{n\pi}{a}\right)^2 + \left(\frac{m\pi}{b}\right)^2$$

From (5.1.11), the delta functions for the magnetic case are

$$\delta(x - x') = \frac{1}{a}\sum_{n=0}^{\infty} \varepsilon_n \cos\left(\frac{n\pi x}{a}\right)\cos\left(\frac{n\pi x'}{a}\right)$$

and

$$\delta(y - y') = \frac{1}{b}\sum_{n=0}^{\infty} \varepsilon_m \cos\left(\frac{m\pi y}{b}\right)\cos\left(\frac{m\pi y'}{b}\right)$$

For the case of the magnetic source in (5.4.19), H_z is given by

$$H_z(x, y) = -j\omega\varepsilon \int_0^a \int_0^b J_{m,z}(x', y')\, G^m(x, y, x', y')dx'dy'$$

5.4.3 Two-Dimensional Green Functions in Cylindrical Coordinates

5.4.3.1 An Angular Sector

A two-dimensional angular configuration is shown in Fig. 5.7. Consider the solution to the wave equation satisfied by the electric field E_z where the source is an infinite-length current excitation, given by $\mathbf{J}_e = J_{e,z}(x, y)\hat{\mathbf{a}}_z$. The lateral electric field can be analyzed using (1.3.13) or (1.4.11), namely

$$\left[\frac{1}{\rho}\frac{\partial}{\partial\rho}\left(\rho\frac{\partial}{\partial\rho}\right) + \frac{1}{\rho^2}\frac{\partial^2}{\partial\phi^2} + \beta^2\right]E_z(\rho, \phi) = j\omega\mu J_{e,z}(\rho, \phi) \qquad (5.4.21)$$

where $\beta = \omega\sqrt{\mu\varepsilon}$. The boundary conditions are

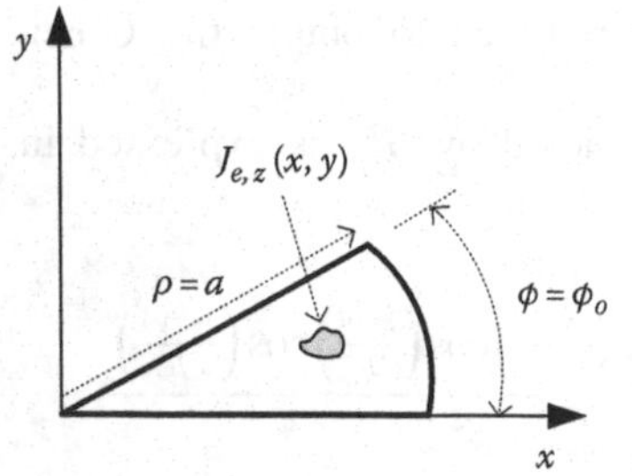

FIGURE 5.7 An angular configuration.

$$E_z(\rho, 0) = E_z(\rho, \phi_o) = 0$$

and

$$E_z(a, \phi) = 0$$

The solution to the homogeneous equation in (5.4.21) is obtained using separation of variables with $E_z(\rho, \phi) = R(\rho)\Phi(\phi)$, to obtain

$$\frac{1}{\rho}\frac{\partial}{\partial\rho}\left(\rho\frac{\partial R}{\partial\rho}\right) + \left(\beta^2 - \frac{\nu^2}{\rho^2}\right)R = 0 \tag{5.4.22}$$

and

$$\frac{d^2\Phi}{d\phi^2} + \nu^2\Phi = 0 \tag{5.4.23}$$

where ν^2 is the separation constant.

The solution of (5.4.23) that satisfies the boundary conditions at $\phi = 0$ and $\phi = \phi_o$ is

$$\Phi = a_n \sin \nu\phi = a_n \sin\left(\frac{n\pi\phi}{\phi_o}\right) \qquad (n = 1, 2, 3, ...) \tag{5.4.24}$$

which shows that the eigenvalues ν, denoted by ν_n, are

$$\nu = \nu_n = \frac{n\pi}{\phi_o} \qquad (n = 1, 2, 3, ...) \tag{5.4.25}$$

The solution to (5.4.22) that is finite at $\rho = 0$ is

$$R = b_\nu J_\nu(\beta\rho)$$

Therefore, the solution to the homogenous form of (5.4.21) is

$$E_z(\rho, \phi) = C_n \sin\left(\frac{n\pi\phi}{\phi_o}\right) J_{\nu_n}(\beta\rho) \tag{5.4.26}$$

The boundary condition at $\rho = a$ requires that

$$E_z(a, z) = 0 \Rightarrow J_{\nu_n}(\beta a) = 0 \Rightarrow \beta a = \chi_{\nu_n l} \qquad \begin{pmatrix} n = 1, 2, 3, ... \\ l = 1, 2, 3, ... \end{pmatrix}$$

or

$$\beta = \beta_{\nu_n l} = \frac{\chi_{\nu_n l}}{a}$$

where $\chi_{\nu_n l}$ is the l zeros of the Bessel function $J_{\nu_n}(\beta a)$ for $n = 1, 2, 3, \ldots$. Note that the order of the Bessel function is not an integer, so the roots $\chi_{\nu_n l}$ are associated with non-integer orders. The first n is $n = 1$ since the angular part in (5.4.26) is zero for $n = 0$. Therefore, (5.4.26) is written as

$$E_z(\rho, \phi) = C_{nl} \sin(\nu_n \phi) J_{\nu_n}\left(\frac{\chi_{\nu_n l}\rho}{a}\right)\left(\nu_n = \frac{n\pi}{\phi_o}\right) \tag{5.4.27}$$

where C_{nl} is the amplitude constant associated with the nl mode.

To normalize (5.4.27) we first normalize (5.4.24), namely

$$a_n^2 \int_0^{\phi_o} \sin^2\left(\frac{n\pi\phi}{\phi_o}\right) d\phi = 1 \Rightarrow a_n^2 \frac{\phi_o}{2} = 1 \Rightarrow a_n = \sqrt{\frac{2}{\phi_o}}$$

Hence, the normalized form of (5.4.24) is

$$\Phi = \sqrt{\frac{2}{\phi_o}} \sin\left(\frac{n\pi\phi}{\phi_o}\right) \qquad (n = 1, 2, 3, \ldots)$$

To normalize $R(\rho)$, use the integral (see Appendix C):

$$\int_0^a J_\nu(\beta_p\rho) J_\nu(\beta_q\rho)\rho d\rho = \frac{a^2}{2}[J_\nu'(\beta_p a)]^2 \delta_{pq} = \frac{a^2}{2} J_{\nu+1}^2(\beta_p a)\delta_{pq}$$

which is valid when $J_\nu(\beta_p a) = 0$. Therefore,

$$b_\nu^2 \frac{a^2}{2} J_{\nu+1}^2(\chi_{nl}) = 1 \Rightarrow b_\nu = \frac{\sqrt{2}}{a J_{\nu+1}(\chi_{nl})}$$

and the normalized form of $R(\rho)$ is

$$R(\rho) = \frac{\sqrt{2}}{a\, J_{\nu+1}(\chi_{nl})} J_\nu(\beta\rho) \tag{5.4.28}$$

With $\nu = \nu_n = n\pi/\phi_o$ and $\beta = \beta_{\nu_n l} = \chi_{\nu_n l}/a$, the orthonormal form of $E_z(\rho, \phi)$ is

$$E_z(\rho,\phi)=\frac{2}{a\sqrt{\phi_o}\,J_{\nu_n+1}(\chi_{nl})}\sin(\nu_n\phi)J_{\nu_n}\left(\frac{\chi_{\nu_n l}\rho}{a}\right)\qquad\left(\nu_n=\frac{n\pi}{\phi_o}\right)\tag{5.4.29}$$

The Green function is constructed using (5.2.11) expressed in terms of ρ and ϕ, or using the two-dimensional form of (5.2.19). (5.4.29) in terms of the Sturm-Liouville notation shows that $E_z(\rho,\phi)=u_n(\rho)u_m(\phi)$, $\lambda_{nl}=(\chi_{\nu_n l}/a)^2$ and $\lambda=\beta^2$.

The Green function associated with (5.4.21) satisfies

$$\left[\frac{1}{\rho}\frac{\partial}{\partial\rho}\left(\rho\frac{\partial}{\partial\rho}\right)+\frac{1}{\rho^2}\frac{\partial^2}{\partial\phi^2}+\beta^2\right]G=-\frac{\delta(\rho-\rho')\delta(\phi-\phi')}{\rho}\tag{5.4.30}$$

subject to the same boundary conditions. Hence, from the previous considerations the Green function is given by

$$G(\rho,\phi,\rho',\phi')=\frac{4}{a^2\phi_o}\sum_{n=1}^{\infty}\sum_{l=1}^{\infty}\frac{\sin(\nu_n\phi)\sin(\nu_n\phi')J_{\nu_n}\left(\frac{\chi_{\nu_n l}\rho}{a}\right)J_{\nu_n}\left(\frac{\chi_{\nu_n l}\rho'}{a}\right)}{\left[\left(\frac{\chi_{\nu_n l}}{a}\right)^2-\beta^2\right]J_{\nu_n+1}^2\left(\chi_{\nu_n l}\right)}\qquad\left(\nu_n=\frac{n\pi}{\phi_o}\right)\tag{5.4.31}$$

From (5.1.11), the following delta function representations are obtained:

$$\delta(\phi-\phi')=\frac{2}{\phi_o}\sum_{n=1}^{\infty}\sin\left(\frac{n\pi\phi}{\phi_o}\right)\sin\left(\frac{n\pi\phi'}{\phi_o}\right)\tag{5.4.32}$$

and

$$\frac{\delta(\rho-\rho')}{\rho}=\frac{2}{a^2}\sum_{l=1}^{\infty}\frac{J_\nu\left(\frac{\chi_{\nu l}\rho}{a}\right)J_\nu\left(\frac{\chi_{\nu l}\rho'}{a}\right)}{J_{\nu+1}^2(\chi_{\nu l})}\tag{5.4.33}$$

where (5.4.33) applies to any ν.

In the case of a magnetic current source, the lateral field H_z satisfies

$$\left[\frac{1}{\rho}\frac{\partial}{\partial\rho}\left(\rho\frac{\partial}{\partial\rho}\right)+\frac{1}{\rho^2}\frac{\partial^2}{\partial\phi^2}+\beta^2\right]H_z(\rho,\phi)=j\omega\varepsilon J_{m,z}(\rho,\phi)\tag{5.4.34}$$

where the solution of the homogenous equation associated with (5.4.34) is

$$H_z(\rho,\phi)=[B_1\sin(\nu\phi)+B_2\cos(\nu\phi)]J_\nu(\beta\rho)$$

The boundary conditions are

$$E_\rho = \frac{1}{j\omega\varepsilon}\frac{1}{\rho}\frac{\partial H_z}{\partial \phi}\bigg|_{\substack{\phi=0\\ \phi=\phi_o}} = 0$$

and

$$E_\phi = -\frac{1}{j\omega\varepsilon}\frac{\partial H_z}{\partial \rho}\bigg|_{\rho=a} = 0$$

The angular boundary conditions for E_ρ are satisfied with $B_1 = 0$ and

$$B_2 \sin(\nu\phi_o) = 0 \Rightarrow \nu = \nu_n = \frac{n\pi}{\phi_o} \qquad (n = 0, 1, 2, ...)$$

and the radial boundary conditions at $\rho = a$ are satisfied with

$$J'_{\nu_n}(\beta a) = 0 \Rightarrow \beta a = \tilde{\chi}_{\nu_n l} \qquad \begin{pmatrix} n = 0, 1, 2, ... \\ l = 1, 2, 3... \end{pmatrix}$$

or

$$\beta = \beta_{\nu_n l} = \frac{\tilde{\chi}_{\nu_n l}}{a}$$

where $\tilde{\chi}_{\nu_n l}$ (i.e., the eigenvalues) are the zeros of the Bessel function $J'_{\nu_n}(\beta a)$ for $n = 0, 1, 2, ...$ Then,

$$H_z(\rho, \phi) = B_{nl}\cos(\nu_n\phi)J_{\nu_n}\left(\frac{\tilde{\chi}_{\nu_n l}\rho}{a}\right) \qquad \left(\nu_n = \frac{n\pi}{\phi_o}\right)$$

The normalized Φ solution is

$$\Phi = \sqrt{\frac{\varepsilon_n}{\phi_o}}\cos\left(\frac{n\pi\phi}{\phi_o}\right) \qquad (n = 0, 1, 2, ...) \tag{5.4.35}$$

where ε_n is Neumann's number.

The radial solution is

$$R = b_\nu J_\nu(\beta\rho)$$

To normalize $R(\rho)$, use the integral

$$\int_0^a J_\nu(\beta_p\rho)J_\nu(\beta_q\rho)\rho\, d\rho = \frac{a^2}{2}\left(1 - \frac{\nu^2}{\beta_p^2 a^2}\right)J_\nu^2(\beta_p a)\delta_{pq}$$

which is valid when $J_\nu'(\beta_p a) = 0$.

In our case, $\beta_p a = \beta_{\nu_n} a = \tilde{\chi}_{\nu_n l}$ and $\nu = \nu_n = n\pi/\phi_o$. Hence,

$$b_\nu^2 \frac{a^2}{2}\left(1 - \frac{\nu_n^2}{\tilde{\chi}_{\nu_n l}^2}\right)J_{\nu_n}^2(\tilde{\chi}_{\nu_n l}) = 1 \Rightarrow b_\nu = \frac{\sqrt{2}}{a}\frac{\tilde{\chi}_{\nu_n l}}{\sqrt{\tilde{\chi}_{\nu_n l}^2 - \nu_n^2}}\frac{1}{J_{\nu_n}(\tilde{\chi}_{\nu_n l})}$$

and the normalized form of $R(\rho)$ is

$$R = \frac{\sqrt{2}}{a}\frac{\tilde{\chi}_{\nu_n l}}{\sqrt{\tilde{\chi}_{\nu_n l}^2 - \nu_n^2}}\frac{1}{J_{\nu_n}(\tilde{\chi}_{\nu_n l})}J_{\nu_n}\left(\frac{\tilde{\chi}_{\nu_n l}\rho}{a}\right) \tag{5.4.36}$$

Using (5.4.35) and (5.4.36), it follows from (5.2.11) or (5.2.19) that the associated Green function, denoted by G^m, is

$$G^m(\rho, \phi, \rho', \phi') = \frac{2}{a^2\phi_o}\sum_{n=0}^{\infty}\sum_{l=1}^{\infty}\frac{\varepsilon_n \cos(\nu_n\phi)\cos(\nu_n\phi')\,(\tilde{\chi}_{\nu_n l})^2 J_{\nu_n}\left(\frac{\tilde{\chi}_{\nu_n l}\rho}{a}\right)J_{\nu_n}\left(\frac{\tilde{\chi}_{\nu_n l}\rho'}{a}\right)}{\left[\left(\frac{\tilde{\chi}_{\nu_n l}}{a}\right)^2 - \beta^2\right]\left(\tilde{\chi}_{\nu_n l}^2 - \nu_n^2\right)J_{\nu_n}^2(\tilde{\chi}_{\nu_n l})}\left(\nu_n = \frac{n\pi}{\phi_o}\right) \tag{5.4.37}$$

5.4.3.2 A Cylindrical Region

A cylindrical region with an axial current $\mathbf{J}_e = J_{e,z}\hat{\mathbf{a}}_z$ is shown in Fig. 5.8. The E_z field satisfies (5.4.21) and the Green function follows from (5.4.30). The boundary condition is

$$E_z(\rho, \phi) = 0|_{\rho=a}$$

and for the Green function:

$$G(\rho, \phi, \rho', \phi') = 0|_{\rho=a}$$

For the angular solution to (5.4.30), use the eigenfunctions:

$$\Phi(\phi) = b_1 e^{jn\phi} + b_2 e^{-jn\phi} \tag{5.4.38}$$

that must be periodic in 2π. Consider the first term in (5.4.38). For a periodic solution, n must be an integer, since

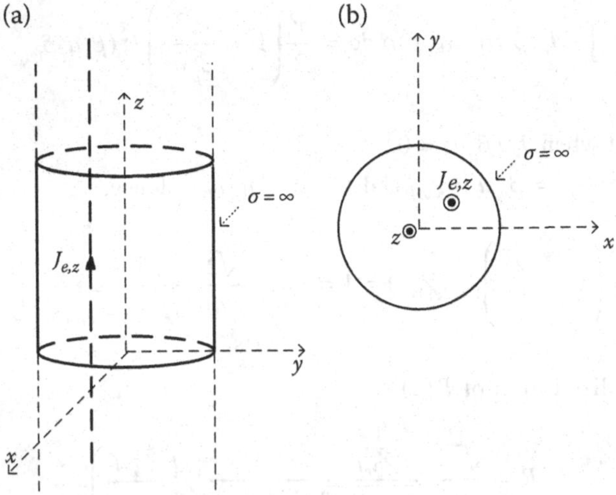

FIGURE 5.8 (a) A circular conducting region; (b) two-dimensional view.

$$\Phi(0) = \Phi(2\pi n) \Rightarrow b_1 e^{j0} = b_1 e^{jn2\pi} = b_1 \qquad (n = 0, 1, 2, ...)$$

Similar considerations apply to the second term in (5.4.38). A convenient way of writing (5.4.38) that includes the positive and negative values of n is

$$\Phi(\phi) = \sum_{n=-\infty}^{\infty} a_n e^{jn\phi}$$

Since,

$$\int_0^{2\pi} e^{jn\phi} e^{-jm\phi}\, d\phi = 2\pi\delta_{nm}$$

the normalized angular eigenfunctions are

$$\Phi(\phi) = \frac{1}{\sqrt{2\pi}} \textstyle\sum_{n=-\infty}^{\infty} e^{jn\phi} = \varepsilon_n \sqrt{\dfrac{1}{2\pi}} \displaystyle\sum_{n=0}^{\infty} \cos n\phi \qquad (5.4.39)$$

where ε_n is the Neumann's number.

The associated delta function representation has the following forms:

$$\delta(\phi - \phi') = \frac{1}{2\pi} \sum_{n=-\infty}^{\infty} e^{jn(\phi-\phi')} = \frac{\varepsilon_n}{2\pi} \sum_{n=0}^{\infty} \cos n(\phi - \phi') \qquad (5.4.40)$$

The radial part of the Green function that satisfies the boundary condition at $\rho = a$ was obtained in the (5.4.28) and it applies to the circular configuration in Fig. 5.8 with $\nu = n$. Therefore, the Green function for the TM^z modes in the circular configuration in Fig. 5.8, from (5.4.28) and (5.4.40), is

$$G(\rho, \phi, \rho', \phi') = \frac{1}{\pi a^2} \sum_{n=-\infty}^{\infty} \sum_{l=1}^{\infty} \frac{J_n\left(\frac{\chi_{nl}\rho}{a}\right) J_n\left(\frac{\chi_{nl}\rho'}{a}\right)}{\left[\left(\frac{\chi_{nl}}{a}\right)^2 - \beta^2\right] J_{n+1}^2(\chi_{nl})} e^{jn(\phi-\phi')}$$

In this case, the roots χ_{nl} are associated with integer values of the order in the Bessel functions. Also, a radial delta function representation follows from (5.4.33) with $\nu = n$.

For the TE^z mode in the circular configuration, H_z satisfies (5.4.34). The details of the derivation are similar to those leading to (5.4.37), except that $\nu = n$ and (5.4.39) is used for the angular eigenfunctions. Therefore, the Green function denoted by $G^m(\rho, \phi)$ is

$$G^m(\rho, \phi, \rho', \phi') = \frac{1}{\pi a^2} \sum_{n=-\infty}^{\infty} \sum_{l=1}^{\infty} \frac{\tilde{\chi}_{nl}^2 J_n\left(\frac{\tilde{\chi}_{nl}\rho}{a}\right) J_n\left(\frac{\tilde{\chi}_{nl}\rho'}{a}\right)}{\left[\left(\frac{\tilde{\chi}_{nl}}{a}\right)^2 - \beta^2\right] \left(\tilde{\chi}_{nl}^2 - n^2\right) J_n^2(\tilde{\chi}_{nl})} e^{jn(\phi-\phi')}$$

and the radial delta function is

$$\frac{\delta(\rho - \rho')}{\rho} = \frac{2}{a^2} \sum_{l=1}^{\infty} \frac{\tilde{\chi}_{nl}^2 J_n\left(\frac{\tilde{\chi}_{nl}\rho}{a}\right) J_n\left(\frac{\tilde{\chi}_{nl}\rho'}{a}\right)}{\left(\tilde{\chi}_{nl}^2 - n^2\right) J_n^2(\tilde{\chi}_{nl})}$$

5.4.4 Green Functions in Spherical Coordinates

Consider the Green function for the interior of the spherical surface of radius a, shown in Fig. 4.20, subject to the same boundary condition as that of E_θ. That is,

$$G(\mathbf{r}, \mathbf{r}')|_{r=a} = 0$$

The Green function satisfies the equation

$$[\nabla^2 + \beta^2] G(\mathbf{r}, \mathbf{r}') = -\delta(\mathbf{r} - \mathbf{r}')$$

which in spherical coordinates reads

$$\left[\frac{1}{r^2}\frac{\partial}{\partial r}\left(r^2\frac{\partial}{\partial r}\right) + \frac{1}{r^2 \sin\theta}\frac{\partial}{\partial\theta}\left(\sin\theta\frac{\partial}{\partial\theta}\right) + \frac{1}{r^2\sin^2\theta}\frac{\partial^2}{\partial\phi^2} + \beta^2\right] G(\mathbf{r}, \mathbf{r}')$$
$$= -\frac{\delta(r - r')\delta(\theta - \theta')\delta(\phi - \phi')}{r^2 \sin\theta} \tag{5.4.41}$$

where $\beta = \omega\sqrt{\mu\varepsilon}$.

The eigenfunctions associated with the angular variables are

$$\Theta(\theta) = a_{nm}P_n^m(\cos\theta) \tag{5.4.42}$$

and

$$\Phi(\phi) = b_m e^{jm\phi}$$

and for the radial variable:

$$R(r) = c_n j_n(\beta r) \tag{5.4.43}$$

To normalize (5.4.42), use the integral:

$$\int_0^\pi P_n^m(\cos\theta)P_q^m(\cos\theta)\sin\theta d\theta = \frac{2}{2n+1}\frac{(n+m)!}{(n-m)!}\delta_{nq}$$

and it follows from that

$$a_{nm}^2\int_0^\pi P_n^m(\cos\theta)P_q^m(\cos\theta)\sin\theta d\theta = 1 \Rightarrow a_{nm} = \sqrt{\frac{2n+1}{2}\frac{(n-m)!}{(n+m)!}}$$

Therefore, the orthonormal $\Theta(\theta)$ solutions are

$$\Theta = \sqrt{\frac{(2n+1)}{2}\frac{(n-m)!}{(n+m)!}}P_n^m(\cos\theta) \tag{5.4.44}$$

The orthonormal form of $\Phi(\phi)$ was derived in (5.4.39), namely

$$\Phi = \frac{1}{\sqrt{2\pi}}\sum_{m=-\infty}^{\infty} e^{jm\phi} \tag{5.4.45}$$

The solutions to the homogeneous equation in (5.4.41) (denoted by the eigenfunctions u_{nm}) which are finite at the origin are of the form

$$u_{nm} = B_{nm} j_n(\beta r) P_n^m(\cos\theta) e^{jm\phi} \tag{5.4.46}$$

or in terms of spherical harmonics as

$$u_{nm} = C_{nm} j_n(\beta r) Y_{nm}(\theta, \phi)$$

Both forms are commonly used.

Applying the boundary condition at $r = a$ to (5.4.46) gives

$$u_{nm}|_{r=a} = 0 \Rightarrow j_n(\beta a) = 0 \Rightarrow \beta a = \kappa_{nl} \Rightarrow \beta = \beta_{nl} = \frac{\kappa_{nl}}{a} \qquad \begin{pmatrix} n = 0, 1, 2, \ldots \\ l = 1, 2, 3, \ldots \end{pmatrix}$$

where κ_{nl} are the zeros of the spherical Bessel functions. Hence, in terms of the Sturm-Liouville notation, the eigenvalues are

$$\lambda_{nl} = \left(\frac{\kappa_{nl}}{a}\right)^2$$

and (5.4.46) is expressed in the form

$$u_{nlm} = B_{nlm} j_n\left(\frac{\kappa_{nl}}{a} r\right) P_n^m(\cos\theta) e^{jm\phi} \tag{5.4.47}$$

where the constant is now expressed as B_{nlm}.

The normalization requires that

$$\int_{r=0}^{a} \int_{\phi=0}^{2\pi} \int_{\theta=0}^{\pi} |u_{nlm}|^2 r^2 \sin\theta dr d\theta d\phi = 1$$

The normalizations for the angular variables of u_{nlm} are shown in (5.4.44) and (5.4.45). For the normalization of the radial part in (5.4.47), let

$$R(r) = c_n j_n(\beta r)$$

To normalize R, use the integral relation (see Appendix C):

$$\int_0^a j_\nu(\beta_p r) j_\nu(\beta_q r) r^2 dr = \frac{a^3}{2} [j_{\nu+1}(\beta_p a)]^2 \delta_{pq}$$

which is valid when $j_\nu(\beta_p a) = 0$. In our case, $\beta_p a = \beta_{nl} a = \kappa_{nl}$ and $\nu = n$. Hence, it follows that

$$c_n^2 \int_0^a j_n^2(\beta r)\, r^2 dr = 1$$

or

$$c_n^2 \left[\frac{a^3}{2} j_{n+1}^2(\kappa_{nl}) \right] = 1 \Rightarrow c_n = \sqrt{\frac{2}{a^3}} \frac{1}{j_{n+1}(\kappa_{nl})}$$

Hence, the normalized radial functions are

$$R(r) = \sqrt{\frac{2}{a^3}} \frac{1}{j_{n+1}(\kappa_{nl})} j_n\left(\frac{\kappa_{nl}}{a} r\right)$$

The normalized form of (5.4.47) is

$$u_{nlm} = \frac{1}{\sqrt{2\pi}} \sqrt{\frac{(2n+1)}{2} \frac{(n-m)!}{(n+m)!}} \sqrt{\frac{2}{a^3}} \frac{1}{j_{n+1}(\kappa_{nl})} j_n\left(\frac{\kappa_{nl}}{a} r\right) P_n^m(\cos\theta) e^{jm\phi}$$

and it follows that the eigenfunctions expression for the Green function is

$$G(\mathbf{r}, \mathbf{r}') = \frac{2}{a^3} \sum_{n=0}^{\infty} \sum_{l=1}^{\infty} \sum_{m=-n}^{n} \frac{(2n+1)}{4\pi} \frac{(n-m)!}{(n+m)!} \frac{j_n\left(\frac{\kappa_{nl}}{a} r\right) j_n\left(\frac{\kappa_{nl}}{a} r'\right)}{\left[\left(\frac{\kappa_{nl}}{a}\right)^2 - \beta^2\right] j_{n+1}^2(\kappa_{nl})} P_n^m(\cos\theta) P_n^m(\cos\theta') e^{jm(\phi-\phi')} \tag{5.4.48}$$

The summation in m is from $-n$ to n since $P_n^m(\cos\theta) = 0$ for $|m| > n$.

Equation (5.4.48) can also be written in terms of spherical harmonics. Since

$$Y_{nm}(\theta, \phi) = \sqrt{\frac{(2n+1)}{4\pi} \frac{(n-m)!}{(n+m)!}} P_n^m(\cos\theta) e^{jm\phi}$$

it follows that

$$G(\mathbf{r}, \mathbf{r}') = \frac{2}{a^3} \sum_{n=0}^{\infty} \sum_{l=1}^{\infty} \sum_{m=-n}^{n} \frac{j_n\left(\frac{\kappa_{nl}}{a} r\right) j_n\left(\frac{\kappa_{nl}}{a} r'\right)}{\left[\left(\frac{\kappa_{nl}}{a}\right)^2 - \beta^2\right] j_{n+1}^2(\kappa_{nl})} Y_{nm}(\theta, \phi) Y_{nm}^*(\theta', \phi') \tag{5.4.49}$$

Using the addition relation for the spherical harmonics, namely

$$P_n(\cos\gamma) = \frac{4\pi}{2n+1} \sum_{m=-n}^{n} Y_{nm}(\theta, \phi)\, Y_{nm}^*(\theta', \phi')$$

where

$$\cos\gamma = \cos\theta\cos\theta' + \sin\theta\sin\theta'\cos(\phi - \phi')$$

Then, (5.4.49) can also be expressed in the form

$$G(\mathbf{r}, \mathbf{r}') = \frac{2}{a^3}\sum_{n=0}^{\infty}\sum_{l=1}^{\infty}\frac{2n+1}{4\pi}\frac{j_n\left(\frac{\kappa_{nl}}{a}r\right)j_n\left(\frac{\kappa_{nl}}{a}r'\right)}{\left[\left(\frac{\kappa_{nl}}{a}\right)^2 - \beta^2\right]j_{n+1}^2(\kappa_{nl})}P_n(\cos\gamma)$$

5.5 GREEN FUNCTIONS WITH CONTINUOUS EIGENVALUES

In the previous section, we obtained Green functions using eigenfunctions expansions. That technique is useful when dealing with closed regions. If the region extents to infinity the eigenvalues become continuous and they can be expressed in integral forms.

For example, let's analyze the consequences of letting $a \to \infty$ in Fig. 5.5. The resulting configuration is shown in Fig. 5.9. The Green function and delta function expressions for the configuration in Fig. 5.9 can be obtained by taking the limit as $a \to \infty$ in (5.4.17) and (5.4.18), which only affects the x variable. Let the eigenvalues β_x be denoted by $\alpha_n = n\pi/a$. Therefore, the difference between two eigenvalues is

$$\Delta\alpha_n = \alpha_{n+1} - \alpha_n = \frac{\pi}{a}$$

Hence, (5.4.17) is written as

$$G(x, x', y, y') = \frac{2}{b}\sum_{m=1}^{\infty}\sin\left(\frac{m\pi y}{b}\right)\sin\left(\frac{m\pi y'}{b}\right)\left[\frac{2}{\pi}\sum_{n=1}^{\infty}\frac{\sin(\alpha_n x)\sin(\alpha_n x')}{\alpha_n^2 + \left(\frac{m\pi}{b}\right)^2 - \beta^2}\Delta\alpha_n\right] \tag{5.5.1}$$

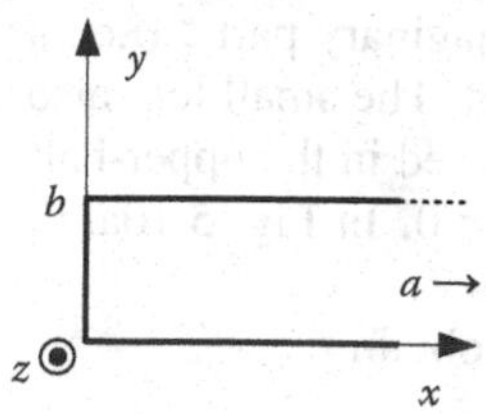

FIGURE 5.9 A rectangular region with $a \to \infty$.

where a was replaced by $\pi/\Delta\alpha_n$. Taking the limit as $a \to \infty$ in (5.5.1), the eigenvalues become continuous (i.e., $\Delta\alpha_n \to d\alpha$ and $\alpha_n \to \alpha$) and we obtain

$$G(x, y, x', y') = \frac{2}{b}\sum_{m=1}^{\infty} \sin\left(\frac{m\pi y}{b}\right)\sin\left(\frac{m\pi y'}{b}\right)\left[\frac{2}{\pi}\int_0^{\infty} \frac{\sin(\alpha x)\sin(\alpha x')}{\alpha^2 + \left(\frac{m\pi}{b}\right)^2 - \beta^2}d\alpha\right] \tag{5.5.2}$$

Similarly, the delta function representation in (5.4.18) becomes

$$\delta(x - x') = \frac{2}{\pi}\int_0^{\infty} \sin(\alpha x)\sin(\alpha x')d\alpha$$

The delta function representation for the y variable remains the same. That is,

$$\delta(y - y') = \frac{2}{b}\sum_{m=1}^{\infty} \sin\left(\frac{m\pi y}{b}\right)\sin\left(\frac{m\pi y'}{b}\right)$$

The integral in (5.5.2) can be evaluated as follows. Write (5.5.2) in the form

$$\begin{aligned} I &= \frac{2}{\pi}\int_0^{\infty} \frac{\sin(\alpha x)\sin(\alpha x')}{\alpha^2 - t^2}d\alpha \\ &= \frac{1}{4\pi}\int_{-\infty}^{\infty} \frac{e^{j\alpha(x-x')} + e^{j\alpha(-x+x')} - e^{j\alpha(x+x')} - e^{j\alpha(-x-x')}}{\alpha^2 - t^2}d\alpha \end{aligned} \tag{5.5.3}$$

where

$$t^2 = \beta^2 - \left(\frac{m\pi}{b}\right)^2$$

and consider the evaluation of the first term in (5.5.3), namely

$$\int_{-\infty}^{\infty} \frac{e^{j\alpha(x-x')}}{\alpha^2 - t^2}d\alpha$$

The function has poles at $\alpha = -t + j\varepsilon$ and $\alpha = t - j\varepsilon$, as shown in Fig. 5.10a. For convenience, we have assumed that the poles have a small imaginary part ε (i.e., a small loss) so the poles are not on top of the path of integration. The small loss also provides the proper convergence. The path of integration is closed in the upper-half plane for $x - x' > 0$, and in the lower-half plane for $x - x' < 0$. In Fig. 5.10a, C_1 denotes the semicircular part of the path.

For the path in Fig. 5.10a, using the residue theorem, we obtain

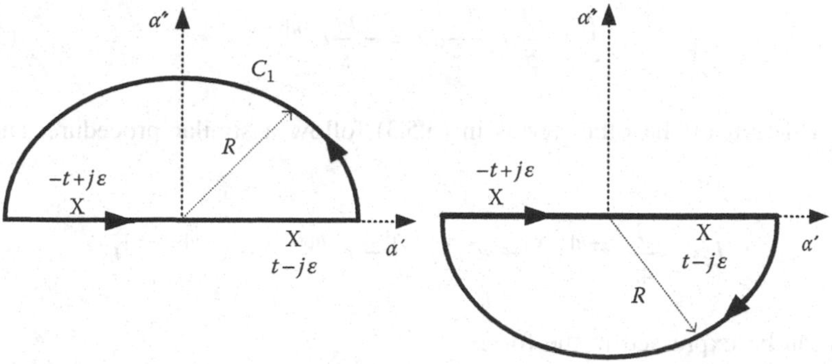

FIGURE 5.10 (a) Path of integration for $x - x' > 0$; (b) path of integration for $x - x' < 0$.

$$\int_{-\infty}^{\infty} \frac{e^{j\alpha(x-x')}}{\alpha^2 - t^2}\, d\alpha + \int_{C_1} \frac{e^{j\alpha(x-x')}}{\alpha^2 - t^2}\, d\alpha = 2\pi j \left[\text{Residue } \frac{e^{j\alpha(x-x')}}{\alpha^2 - t^2}\bigg|_{\alpha=-t+j\varepsilon} \right] \tag{5.5.4}$$

Along C_1, let $\alpha = R\, e^{j\theta}$ and take the limit as $R \to \infty$. Observing that

$$e^{j\alpha(x-x')} = e^{jRe^{j\theta}(x-x')} = e^{jR(\cos\theta + j\sin\theta)(x-x')} = e^{jR\cos\theta(x-x')} e^{-R\sin\theta(x-x')}$$

it follows that as $R \to \infty$ the integral over C_1 vanishes since $x - x' > 0$ and $\sin\theta > 0$ for $0 < \theta < \pi$, (i.e., $e^{-R\sin\theta(x-x')} \to 0$). Therefore, the evaluation of (5.5.4) is

$$\int_{-\infty}^{\infty} \frac{e^{j\alpha(x-x')}}{\alpha^2 - t^2} d\alpha = 2\pi j \frac{e^{j\alpha(x-x')}(\alpha + t)}{(\alpha - t)(\alpha + t)}\bigg|_{\alpha=-t+j\varepsilon} = -\frac{j\pi}{t} e^{-jt(x-x')} \quad \text{for } x - x' > 0 \tag{5.5.5}$$

where ε was set to zero in the final result. If included, the exponential in (5.5.5) becomes $e^{-jt(x-x')} e^{-\varepsilon(x-x')}$ which decays as $x \to \infty$.

Similarly, for $x - x' < 0$, evaluating the integral along the path in Fig. 5.10b gives

$$\int_{-\infty}^{\infty} \frac{e^{j\alpha(x-x')}}{\alpha^2 - t^2} d\alpha = -2\pi j \frac{e^{j\alpha(x-x')}(\alpha - t)}{(\alpha - t)(\alpha + t)}\bigg|_{\alpha=t-j\varepsilon} = -\frac{j\pi}{t} e^{jt(x-x')} \quad \text{for } x - x' < 0 \tag{5.5.6}$$

where the contribution from the lower semicircle vanishes as $R \to \infty$.

Equations (5.5.5) and (5.5.6) show that the first term in (5.5.3) can be expressed as

$$\int_{-\infty}^{\infty} \frac{e^{j\alpha(x-x')}}{\alpha^2 - t^2} d\alpha = -\frac{j\pi}{t} e^{-jt|x-x'|} \tag{5.5.7}$$

The evaluation of the other terms in (5.5.3) follow a similar procedure, and we obtain

$$I = -\frac{j}{4t}(e^{-jt|x-x'|} + e^{-jt|-x+x'|} - e^{-jt|x+x'|} - e^{-jt|-x-x'|})$$

which can be expressed in the form

$$I = \begin{cases} \dfrac{e^{-jtx}}{t} \sin(tx') & (x \geq x') \\ \dfrac{e^{-jtx\prime}}{t} \sin(tx) & (0 \leq x \leq x') \end{cases} \tag{5.5.8}$$

Hence, the Green function in (5.5.2) reads

$$G(x, y, x', y') = \frac{2}{b} \sum_{m=1}^{\infty} \sin\left(\frac{m\pi y}{b}\right) \sin\left(\frac{m\pi y'}{b}\right) \frac{e^{-jtx}}{t} \sin(tx') \qquad (x \geq x')$$

and

$$G(x, y, x', y') = \frac{2}{b} \sum_{m=1}^{\infty} \sin\left(\frac{m\pi y}{b}\right) \sin\left(\frac{m\pi y'}{b}\right) \frac{e^{-jtx\prime}}{t} \sin(tx) \qquad (0 \leq x \leq x')$$

5.5.1 Green Function for an Infinite-Length Angular Configuration

Let $a \to \infty$ in the angular configuration in Fig. 5.7. The resulting infinite-length angular configuration is shown in Fig. 5.11. For this configuration, the Green and delta functions follow from (5.4.31) and (5.4.33) by taking the limit as $a \to \infty$. To this end, the large argument expansion of $J_\nu(x)$ is used to determine the difference between eigenvalues. That is,

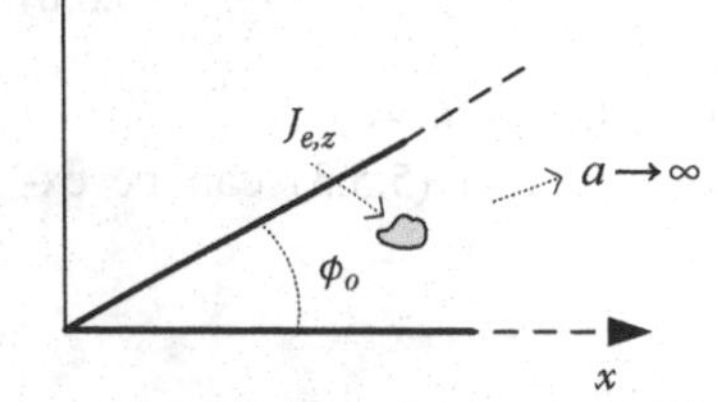

FIGURE 5.11 Infinite angular configuration.

$$J_\nu(x) \approx \sqrt{\frac{2}{\pi x}} \cos\left(x - \nu\frac{\pi}{2} - \frac{\pi}{4}\right)$$

where the zeros are given by

$$x - \nu\frac{\pi}{2} - \frac{\pi}{4} = \left(l + \frac{1}{2}\right)\pi \text{ for } l = 0, \pm 1, \pm 2... \tag{5.5.9}$$

or in (5.4.31) and (5.4.33) in terms of $\chi_{\nu l}$:

$$\chi_{\nu l} = \left(\frac{\nu}{2} + \frac{3}{4} + l\right)\pi \tag{5.5.10}$$

Hence, the difference between two eigenvalues is

$$\Delta\chi_{\nu l} = \chi_{\nu, l+1} - \chi_{\nu, l} = \pi \tag{5.5.11}$$

Also, in (5.4.31) and (5.4.33) using the large argument approximation for $J_{\nu+1}(\chi_{\nu l})$:

$$J^2_{\nu+1}(\chi_{\nu l}) \approx \frac{2}{\pi\chi_{\nu l}} \tag{5.5.12}$$

Therefore, using (5.5.11) and (5.5.12), (5.4.33) reads

$$\frac{\delta(\rho - \rho')}{\rho} = \sum_{l=1}^{\infty} \frac{\chi_{\nu l}}{a} J_\nu\left(\frac{\chi_{\nu l}\rho}{a}\right) J_\nu\left(\frac{\chi_{\nu l}\rho'}{a}\right) \frac{\Delta\chi_{\nu l}}{a} \tag{5.5.13}$$

which shows that as $a \to \infty$: $\Delta\chi_{\nu l}/a \to d\chi$ and $\chi_{\nu l}/a \to \chi$. Taking the limit as $a \to \infty$ in (5.5.13) gives

$$\frac{\delta(\rho - \rho')}{\rho} = \int_0^\infty \chi J_\nu(\chi\rho) J_\nu(\chi\rho')\, d\chi \tag{5.5.14}$$

The variable χ in (5.5.14) is usually written as λ.

Using the same procedure for the Green function in (5.4.31), the Green function for the configuration in Fig. 5.11 is

$$G(\rho, \phi, \rho', \phi') = \frac{2}{\phi_0} \sum_{n=0}^{\infty} \sin(\nu_n\phi)\sin(\nu_n\phi') \int_0^\infty \frac{\lambda J_{\nu_n}(\lambda\rho) J_{\nu_n}(\lambda\rho')}{\lambda^2 - \beta^2} d\lambda \qquad \left(\nu_n = \frac{n\pi}{\phi_o}\right) \tag{5.5.15}$$

5.6 DIRECT METHOD FOR GREEN FUNCTION SOLUTIONS

In this section, we examine the properties of Green functions and develop the direct method (also known as the continuity method or the closed form method) for deriving the Green function solution.

In the previous section, we have seen some properties of the Green functions:

a. The Green function satisfies (5.2.1).
b. The Green function satisfies a homogenous equation for $x \neq x'$.
c. The Green function satisfies homogeneous boundary conditions.
d. The Green function is continuous in $x_1 \leq x \leq x_2$, which includes $x = x'$.
e. The Green function is symmetrical, or $G(x, x') = G(x', x)$.
f. The new property that the Green function must satisfy in (5.2.1) is that its derivative has a discontinuity at $x = x'$ given by

$$\left.\frac{dG(x, x')}{dx}\right|_{x=x'+\varepsilon} - \left.\frac{dG(x, x')}{dx}\right|_{x=x'-\varepsilon} = -\frac{1}{p(x')}$$

This property is usually expressed in the following form:

$$\left.\frac{dG_>(x, x')}{dx}\right|_{x=x'} - \left.\frac{dG_<(x, x')}{dx}\right|_{x=x'} = -\frac{1}{p(x')} \tag{5.6.1}$$

where the notation $G_>$ represent the Green function solution in the region $x > x'$, and $G_<$ in the region $x < x'$.

This last property is derived by integrating (5.2.1) between $x = x' - \varepsilon$ and $x = x' + \varepsilon$. That is,

$$p(x)\left.\frac{dG}{dx}\right|_{x'-\varepsilon}^{x'+\varepsilon} + \int_{x'-\varepsilon}^{x'+\varepsilon} [q(x) + \lambda r(x)]\, G\, dx = -1 \tag{5.6.2}$$

The second term in (5.6.2) is zero, since $q(x)$, $r(x)$, and $G(x, x')$ are continuous in $x_1 \leq x \leq x_2$. Therefore, (5.6.1) follow from (5.6.2).

It is also observed that in term of $G_<$ and $G_>$, property (d) at $x = x'$ reads

$$G_<(x, x')|_{x=x'} = G_>(x, x')|_{x=x'} \tag{5.6.3}$$

In the direct method, the Green function is divided into $G_<$ and $G_>$. The solutions for $G_<$ and $G_>$ must be linearly independent. If they are linearly dependent (5.6.3), makes them equal and, therefore, the discontinuity in (5.6.2) is not satisfied.

A simple way of checking that $G_<$ and $G_>$ are linearly independent is as follows. The solution for $G_<$ satisfies the boundary condition at $x = x_1$, and the solution for $G_>$ satisfies the boundary condition at $x = x_2$. If $G_>$ does not satisfy the boundary condition at $x = x_1$, then solutions $G_<$ and $G_>$ are linearly independent.

The direct method can be summarized as follows. The solution to the Green function in (5.2.1) for $x \neq x'$ satisfies the homogeneous equation:

$$\frac{d}{dx}\left[p(x)\frac{dG}{dx}\right] + [q(x) + \lambda r(x)]G = 0$$

This equation is solved for $x < x'$ leading to the solution $G_<$ subject to the boundary condition at $x = x_1$, and for $x > x'$ leading to the solution $G_>$ subject to the boundary condition at $x = x_2$. Then, the boundary conditions at $x = x'$ (i.e., (5.6.1) and (5.6.3)) are applied, the constants associated with the solutions of $G_<$ and $G_>$ are evaluated, and closed form solutions found. Since $G_<$ and $G_>$ are continuous at $x = x'$, the closed form solution for $G_<$ is valid for $x_1 \leq x \leq x'$ and that of $G_>$ for $x' \leq x \leq x_2$.

A more precise statement on the continuity of $G_<$ and $G_>$ at $x = x'$ is that the limit value of $G_<$ as x approaches x' from the left (i.e., $x \to x' - \varepsilon$) is the same as the limit value of $G_>$ as x approaches x' from the right (i.e., $x \to x' + \varepsilon$).

Figure 5.12 shows an example of the behavior of the Green functions $G_<$ and $G_>$ in the range $0 \leq x \leq b$. Observe the continuity at $x = x'$ and the discontinuity in the derivative (i.e., (5.6.1). In this example, $G_<(x, x')$ is valid for $0 \leq x \leq x'$ and $G_>$ for $x' \leq x \leq b$.

For a source $f(x)$, the solution to the inhomogeneous equation (5.2.7) in the range $0 \leq x \leq b$ is given by

$$y(x) = \int_0^b f(x')G(x, x')dx' = \int_0^x f(x')G_>(x, x')dx' + \int_x^b f(x')G_<(x, x')dx'$$

Observe that $G_>$ is used in the integral from 0 to x since in this range $x \geq x'$, and $G_<$ is used in the integral from x to a, since in this range $x' \geq x$.

In general, if the solutions to (5.2.1) that satisfy the boundary conditions are $u_1(x)$ for $a \leq x \leq x'$ and $u_2(x)$ for $x' \leq x \leq b$; then, $G_<$ and $G_>$ are given by

$$G_<(x, x') = u_1(x)u_2(x') \qquad (a \leq x \leq x')$$

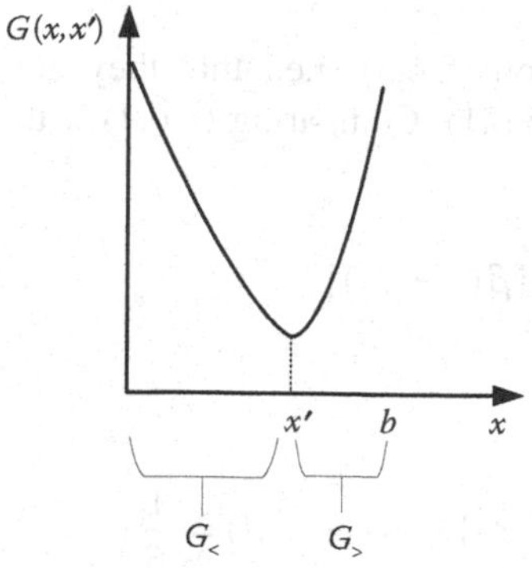

FIGURE 5.12 An example of the behavior of $G_<$ and $G_>$.

and

$$G_>(x, x') = u_1(x')u_2(x) \qquad (x' \le x \le b)$$

Then,

$$y(x) = u_2(x) \int_a^x f(x')\, u_1(x')\, dx' + u_1(x) \int_x^b f(x')\, u_2(x')\, dx'$$

The first integral uses $G_>$ since $x > x'$, and the second integral uses $G_<$ since $x' > x$.

5.6.1 One Dimension

The Green function for a one-dimensional wave equation is now derived using the direct method. Consider the region between $x = 0$ and $x = a$, where the Green function satisfies

$$\left[\frac{d^2}{dx^2} + \beta^2\right] G(x, x') = -\delta(x - x') \tag{5.6.4}$$

The propagation constant is β, and the boundary conditions are

$$G(0, x') = G(a, x') = 0$$

Using the direct method, the solutions to the homogenous form of (5.6.4) (i.e., for $x \neq x'$) that satisfy the boundary conditions are

$$G_<(x, x') = B \sin(\beta x) \qquad (0 \le x < x')$$

and

$$G_>(x, x') = C \sin[\beta(a - x)] \qquad (x' < x \le a)$$

The functions $G_<$ and $G_>$ must satisfy the condition (5.6.3) (i.e., that they are continuous at $x = x'$), and their derivatives satisfy (5.6.1). Comparing (5.6.4) with (5.2.1) shows that $p(x') = 1$. Therefore,

$$G_< = G_>|_{x=x'} \Rightarrow B \sin(\beta x') = C \sin[\beta(a - x')]$$

and from (5.6.1):

$$\left.\frac{dG_>}{dx}\right|_{x=x'} - \left.\frac{dG_<}{dx}\right|_{x=x'} = -1 \Rightarrow C \cos[\beta(a - x')] + B \cos(\beta x') = \frac{1}{\beta}$$

Solving for B and C gives

$$B = \frac{\sin[\beta(a - x')]}{\beta \sin(\beta a)}$$

and

$$C = \frac{\sin(\beta x')}{\beta \sin(\beta a)}$$

Therefore,

$$G_<(x, x') = \frac{\sin(\beta x)\sin[\beta(a - x')]}{\beta \sin(\beta a)} \qquad (0 \le x \le x') \qquad (5.6.5)$$

and

$$G_>(x, x') = \frac{\sin(\beta x')\sin[\beta(a - x)]}{\beta \sin(\beta a)} \qquad (x' \le x \le a) \qquad (5.6.6)$$

Observe that the symmetry property of the Green function shows that $G_>$ follows from $G_<$ by interchanging x and x'. It is seen from (5.6.5) and (5.6.6) that the Green functions are continuous at x', but the derivative is discontinuous by -1 (i.e., $-1/p(x') = -1$). Also, the Green functions are invalid if $\beta a = n\pi$ ($n = 0, 1, 2, \ldots$).

The value of $p(x')$ can also be calculated directly from (5.6.4). Integrating (5.6.4) between $x = x' - \varepsilon$ and $x = x' + \varepsilon$ gives

$$\int_{x'-\varepsilon}^{x'+\varepsilon} \frac{d^2G}{dx^2}\, dx + \int_{x'-\varepsilon}^{x'+\varepsilon} \beta^2 dx = -\int_{x'-\varepsilon}^{x'+\varepsilon} \delta(x - x')\, dx$$

The second term is zero, and the delta function integral is one. Therefore, we obtain

$$\left.\frac{dG}{dx}\right|_{x'+\varepsilon} - \left.\frac{dG}{dx}\right|_{x'-\varepsilon} = -1 \Rightarrow \frac{dG_>}{dx} - \frac{dG_<}{dx} = -1$$

which, as expected, shows that the discontinuity is equal to -1.

The solutions in (5.6.5) and (5.6.6) look quite different from that in (5.4.8). The summation in (5.4.8) has been replaced by explicit expressions. In general, it is not easy to see that the two forms are equivalent. Of course, one can show numerically that both forms produce the same values.

The equivalency of (5.4.8) with (5.6.5) and (5.6.6) is shown as follows. Using the trigonometric identity for the product of two sine functions, write (5.4.8) in the form

$$G(x, x') = \frac{a}{\pi^2} \sum_{n=1}^{\infty} \frac{\cos\left[\frac{n\pi}{a}(x - x')\right] - \cos\left[\frac{n\pi}{a}(x + x')\right]}{n^2 - \left(\frac{\beta a}{\pi}\right)^2} \tag{5.6.7}$$

Then, using the following identity

$$\sum_{n=-\infty}^{\infty} \frac{e^{jn\theta}}{n^2 - \alpha^2} = -\frac{n}{\alpha} \frac{\cos[\alpha(\pi - \theta)]}{\sin(\alpha\pi)} \qquad (0 \le \theta \le 2\pi)$$

which can be written as

$$\sum_{n=1}^{\infty} \frac{\cos(n\theta)}{n^2 - \alpha^2} = \frac{1}{2\alpha^2} - \frac{\pi}{2} \frac{\cos[\alpha(\theta - \pi)]}{\alpha \sin(\alpha\pi)} \tag{5.6.8}$$

it follows from (5.6.7) and (5.6.8) that $G(x, x')$ can be expressed in the form

$$G(x, x') = \frac{1}{2}\left[\frac{\cos[\beta(x - a) + \beta x'] - \cos[\beta(x - a) - \beta x']}{\beta \sin \beta a}\right]$$

Using trigonometric identities, the above expression reads

$$G(x, x') = \frac{\sin(\beta x')\sin[\beta(a - x)]}{\beta \sin \beta a}$$

which is recognized as $G_>$ in (5.6.6). Interchanging x and x' results in (5.6.5). As you can see, the equivalency of the two forms is not obvious (unless one knows the identity above).

An easier way to show the equivalency is to consider the analytical structure of (5.6.5) and (5.6.6) as a function of β. The Green function in (5.2.6) shows that the function has simple poles at $\lambda = \lambda_n$, with residues given by $-\, u_n(x)u_n^*(x')$. The sum of the residues produces a summation of the eigenfunctions. Hence, by viewing the Green function as a function of β, the eigenfunctions and eigenvalues can be determined and the infinite series form of the Green function obtained. To illustrate this process, the poles in (5.6.5) occur at

$$\sin(\beta a) = 0 \Rightarrow \beta = \beta_n = \frac{n\pi}{a} \qquad (n = \pm1, \pm2, \pm3, \ldots) \tag{5.6.9}$$

Observe that there is no pole at $\beta = 0$ (i.e., the limit as $\beta \to 0$ of $\sin(\beta x)/\beta$ is one). The residue at the nth pole in (5.6.5) is

$$\text{Residue} = \left.\frac{\sin(\beta x)\sin[\beta(a-x')]}{\beta\frac{d\sin(\beta a)}{d\beta}}\right|_{\beta=\frac{n\pi}{a}} = \frac{\sin\left(\frac{n\pi}{a}x\right)\sin\left(n\pi - \frac{n\pi}{a}x'\right)}{n\pi\cos(n\pi)}$$

$$= -\frac{1}{n\pi}\sin\left(\frac{n\pi}{a}x\right)\sin\left(\frac{n\pi}{a}x'\right)$$

Then, forming (5.2.6) by summing all the residues, one obtains

$$G = -\sum_{\substack{n=-\infty \\ n\neq 0}}^{\infty} \frac{\sin\left(\frac{n\pi x}{a}\right)\sin\left(\frac{n\pi x'}{a}\right)}{n\pi\left(\beta - \frac{n\pi}{a}\right)} \tag{5.6.10}$$

Since the numerator is an even function, (5.6.10) can be expressed in the form:

$$G = \sum_{n=1}^{\infty} \sin\left(\frac{n\pi x}{a}\right)\sin\left(\frac{n\pi x'}{a}\right)\frac{1}{n\pi}\left(\frac{1}{\beta+\frac{n\pi}{a}} - \frac{1}{\beta-\frac{n\pi}{a}}\right) = \frac{2}{a}\sum_{n=1}^{\infty}\frac{\sin\left(\frac{n\pi x}{a}\right)\sin\left(\frac{n\pi x'}{a}\right)}{\left(\frac{n\pi}{a}\right)^2 - \beta^2}$$

which is identical to (5.4.8). The same result is obtained is we started with $G_>$ in (5.6.6).

5.6.2 Rectangular Region

Consider the rectangular region shown in Fig. 5.5. The Green function satisfies

$$\left[\frac{\partial^2}{\partial x^2} + \frac{\partial^2}{\partial y^2} + \beta^2\right]G(x, x', y, y') = -\delta(x - x')\delta(y - y') \tag{5.6.11}$$

This equation must be reduced to one dimension in order to use the direct method. To this end, use the fact that $\delta(x - x')$ given by (5.4.18) suggest a solution to (5.6.11) in the form

$$G(x, y, x', y') = \frac{2}{a}\sum_{n=1}^{\infty} G(y, y')\sin\left(\frac{n\pi x}{a}\right)\sin\left(\frac{n\pi x'}{a}\right) \tag{5.6.12}$$

where $G(y, y')$ is the Green function associated with the y dependence. Substituting (5.4.18) and (5.6.12) into (5.6.11) results in the following expression for the Green function $G(y, y')$:

$$\left[\frac{d^2}{dy^2} + \beta_y^2\right]G(y, y') = -\delta(y - y') \tag{5.6.13}$$

where

$$\beta_y^2 = \beta^2 - \left(\frac{n\pi}{a}\right)^2$$

The direct method can now be applied to (5.6.13). The solutions to (5.6.13) that satisfy the boundary conditions (i.e., $G_<(0, y') = 0$ and $G_>(b, y') = 0$) are

$$G_<(y, y') = B\sin(\beta_y y) \qquad (0 \le y < y')$$

and

$$G_>(y, y') = C\sin[\beta_y(b - y)] \qquad (y' < y \le b)$$

The functions $G_<$ and $G_>$ are continuous at $y = y'$, and the derivative satisfies (5.6.1). Therefore,

$$G_< = G_>|_{y=y'} \Rightarrow B\sin(\beta_y y') = C\sin[\beta_y(b - y')] \tag{5.6.14}$$

and

$$\left.\frac{dG_>}{dy}\right|_{y=y'} - \left.\frac{dG_<}{dy}\right|_{y=y'} = -1 \Rightarrow C\cos[\beta_y(b - y')] + B\cos(\beta_y y') = \frac{1}{\beta_y} \tag{5.6.15}$$

Solving (5.6.14) and (5.6.15) for B and C gives

$$B = \frac{\sin[\beta_y(b - y')]}{\beta_y \sin(\beta_y b)}$$

and

$$C = \frac{\sin(\beta_y y')}{\beta_y \sin(\beta_y b)}$$

Therefore, the Green functions for the rectangular region in Fig. 5.5 are expressed in the forms

$$G_<(x, y, x', y') = \frac{2}{a}\sum_{n=1}^{\infty} \frac{\sin\left(\frac{n\pi x}{a}\right)\sin\left(\frac{n\pi x'}{a}\right)\sin(\beta_y y)\sin[\beta_y(b - y')]}{\beta_y \sin(\beta_y b)} \qquad (0 \le y \le y') \tag{5.6.16}$$

and

$$G_>(x, y, x', y') = \frac{2}{a}\sum_{n=1}^{\infty} \frac{\sin\left(\frac{n\pi x}{a}\right)\sin\left(\frac{n\pi x'}{a}\right)\sin(\beta_y y')\sin[\beta_y(b-y)]}{\beta_y \sin(\beta_y b)} \qquad (y' \le y \le a)$$

(5.6.17)

Compare the Green function in (5.4.17) with the forms in (5.6.16) and (5.6.17). They can be shown to be equivalent by expressing in (5.4.17) the summation in y in the form

$$\frac{2}{b}\sum_{m=1}^{\infty} \frac{\sin\left(\frac{m\pi y}{b}\right)\sin\left(\frac{m\pi y'}{b}\right)}{\beta_{nm}^2 - \beta^2} = \frac{2}{b}\sum_{m=1}^{\infty} \frac{\sin\left(\frac{m\pi y}{b}\right)\sin\left(\frac{m\pi y'}{b}\right)}{\left(\frac{m\pi}{b}\right)^2 - \beta_y^2}$$

Then, the procedure used with (5.6.7) shows that the summation is equal to

$$\frac{2}{b}\sum_{m=1}^{\infty} \frac{\sin\left(\frac{m\pi y}{b}\right)\sin\left(\frac{m\pi y'}{b}\right)}{\left(\frac{m\pi}{b}\right)^2 - \beta_y^2} = \frac{\sin(\beta_y y')\sin[\beta_y(b-y)]}{\beta_y \sin(\beta_y b)} \qquad (y' \le y \le b)$$

and for $0 \le y \le y'$ interchange y and y' in the above relation. Hence, (5.4.17) is equivalent to (5.6.16) and (5.6.17). The analysis of the analytic structure of (5.6.16) and (5.6.17) would have given the same result.

For the magnetic current source in Fig. 5.6, the resulting Green function (denoted by G^m) is

$$G_<^m(x, x', y, y') = -\frac{1}{a}\sum_{n=0}^{\infty} \frac{\varepsilon_n \cos\left(\frac{n\pi x}{a}\right)\cos\left(\frac{n\pi x'}{a}\right)\cos(\beta_y y)\cos[\beta_y(b-y')]}{\beta_y \sin(\beta_y b)} \qquad (0 \le y \le y')$$

for

$y' \le y \le a$ interchange y and y' in the above relation.

5.6.3 Angular Sector

The geometry for the angular sector is shown in Fig. 5.7. The Green function satisfies

$$\left[\frac{1}{\rho}\frac{\partial}{\partial \rho}\left(\rho\frac{\partial}{\partial \rho}\right) + \frac{1}{\rho^2}\frac{\partial^2}{\partial \phi^2} + \beta^2\right]G = -\frac{\delta(\rho-\rho')\delta(\phi-\phi')}{\rho} \qquad (5.6.18)$$

The angular delta function representation was found in (5.4.32). Therefore, the Green function solution of (5.6.18) is of the form

$$G(\rho, \phi, \rho', \phi') = \frac{2}{\phi_o}\sum_{n=1}^{\infty} G(\rho, \rho')\sin(\nu_n \phi)\sin(\nu_n \phi') \qquad \left(\nu_n = \frac{n\pi}{\phi_o}\right) \qquad (5.6.19)$$

Substituting (5.4.32) and (5.6.19) into (5.6.18) results in the following expression for the Green function $G(\rho, \rho')$:

$$\left[\frac{1}{\rho}\frac{\partial}{\partial\rho}\left(\rho\frac{\partial}{\partial\rho}\right)+\beta^2-\frac{\nu_n^2}{\rho^2}\right]G(\rho, \rho') = -\frac{\delta(\rho-\rho')}{\rho} \tag{5.6.20}$$

The solutions to (5.6.20) are

$$G_<(\rho, \rho') = B_n J_{\nu_n}(\beta\rho) \qquad (0 \le \rho < \rho') \tag{5.6.21}$$

and

$$G_>(\rho, \rho') = C_n J_{\nu_n}(\beta\rho) + D_n H_{\nu_n}^{(2)}(\beta\rho) \qquad (\rho' < \rho \le a) \tag{5.6.22}$$

If the boundary condition at $\rho = a$ requires that $G_>(a, \rho') = 0$. Then, from (5.6.22):

$$G_>(a, \rho') = 0 \Rightarrow C_n = -D_n\frac{H_{\nu_n}^{(2)}(\beta a)}{J_{\nu_n}(\beta a)} \tag{5.6.23}$$

The Green functions $G_<(\rho, \rho')$ and $G_>(\rho, \rho')$ must be continuous at $\rho = \rho'$ and their derivatives satisfy (5.6.1). Hence, from (5.6.21) and (5.6.22):

$$G_< = G_>|_{\rho=\rho'} \Rightarrow B_n J_{\nu_n}(\beta\rho') = C_n J_{\nu_n}(\beta\rho') + D_n H_{\nu_n}^{(2)}(\beta\rho') \tag{5.6.24}$$

and

$$\left.\frac{dG_>}{d\rho}\right|_{\rho=\rho'} - \left.\frac{dG_<}{d\rho}\right|_{\rho=\rho'} = -\frac{1}{\rho'} \Rightarrow C_n J'_{\nu_n}(\beta\rho') + D_n H_{\nu_n}^{(2)\prime}(\beta\rho') - B_n J'_{\nu_n}(\beta\rho') = -\frac{1}{\beta\rho'} \tag{5.6.25}$$

where $p(\rho') = \rho'$ (i.e., compare (5.6.20) with (5.2.1)). The boundary condition in (5.6.25) also follows directly from (5.6.20). Multiplying (5.6.20) by ρ and integrating between $\rho = \rho' - \varepsilon$ and $\rho = \rho' + \varepsilon$ gives

$$\int_{\rho'-\varepsilon}^{\rho'+\varepsilon}\frac{\partial}{\partial\rho}\left(\rho\frac{\partial G}{\partial\rho}\right)d\rho + \int_{\rho'-\varepsilon}^{\rho'+\varepsilon}\left(\beta^2-\frac{\nu_n^2}{\rho^2}\right)G\rho d\rho = -\int_{\rho'-\varepsilon}^{\rho'+\varepsilon}\delta(\rho-\rho')d\rho$$

Since $G(\rho, \rho')$ is continuous when the second integral is zero, and the integral on the right is unity. Therefore, we obtain

$$\rho' \frac{dG_>}{d\rho}\bigg|_{\rho=\rho'} - \rho' \frac{dG_<}{d\rho}\bigg|_{\rho=\rho'} = -1 \Rightarrow \frac{dG_>}{d\rho}\bigg|_{\rho=\rho'} - \frac{dG_<}{d\rho}\bigg|_{\rho=\rho'} = -\frac{1}{\rho'}$$

which, of course, is identical to (5.6.25).

From (5.6.24) and (5.6.2), solving for $B_n - C_n$ and D_n:

$$B_n - C_n = \frac{\begin{vmatrix} 0 & -H_{\nu_n}^{(2)}(\beta\rho') \\ \dfrac{1}{\beta\rho'} & -H_{\nu_n}^{(2)\prime}(\beta\rho') \end{vmatrix}}{-W\left[J_{\nu_n}(\beta\rho'), H_{\nu_n}^{(2)}(\beta\rho')\right]} = -\frac{j\pi}{2} H_{\nu_n}^{(2)}(\beta\rho') \tag{5.6.26}$$

and

$$D_n = \frac{\begin{vmatrix} J_{\nu_n}(\beta\rho') & 0 \\ J'_{\nu_n}(\beta\rho') & \frac{1}{\beta\rho'} \end{vmatrix}}{-W\left[J_{\nu_n}(\beta\rho'), H_{\nu_n}^{(2)}(\beta\rho')\right]} = -\frac{j\pi}{2} J_{\nu_n}(\beta\rho') \tag{5.6.27}$$

where the Wronskian is

$$W\left[J_{\nu_n}(\beta\rho'), H_{\nu_n}^{(2)}(\beta\rho')\right] = -\frac{2j}{\pi\beta\rho'}$$

Substituting (5.6.27) into (5.6.23) gives C_n, and then B_n follows from (5.6.26). The result is

$$B_n = -\frac{j\pi}{2}\left[H_{\nu_n}^{(2)}(\beta\rho') - \frac{H_{\nu_n}^{(2)}(\beta a)}{J_{\nu_n}(\beta a)} J_{\nu_n}(\beta\rho')\right]$$

and from (5.6.21) the solution valid for $0 \le \rho \le \rho'$ is

$$G_<(\rho, \rho') = -\frac{j\pi}{2} J_{\nu_n}(\beta\rho)\left[H_{\nu_n}^{(2)}(\beta\rho') - \frac{H_{\nu_n}^{(2)}(\beta a)}{J_{\nu_n}(\beta a)} J_{\nu_n}(\beta\rho')\right]$$

Hence, from (5.6.19) the Green function for the infinite angular region in Fig. 5.7, valid for $0 \le \rho \le \rho'$, is

$$G_<(\rho, \rho', \phi, \phi') = -\frac{j\pi}{\phi_o}\sum_{n=1}^{\infty} J_{\nu_n}(\beta\rho)\left[H_{\nu_n}^{(2)}(\beta\rho') - \frac{H_{\nu_n}^{(2)}(\beta a)}{J_{\nu_n}(\beta a)}J_{\nu_n}(\beta\rho')\right]$$
$$\times \sin(\nu_n\phi)\sin(\nu_n\phi') \qquad \left(\nu_n = \frac{n\pi}{\phi_o}\right) \tag{5.6.28}$$

The solution for $G_>$, valid for $\rho' \le \rho \le a$, is obtained by interchanging ρ and ρ' in (5.6.28).

In the case of magnetic current excitation, the boundary conditions are

$$\left.\frac{\partial G_>}{\partial \rho}\right|_{\rho=a} = 0$$

and

$$\left.\frac{\partial G}{\partial \phi}\right|_{\substack{\phi=0\\ \phi=\phi_o}} = 0$$

In this case, the relation between C_n and D_n is

$$C_n = -D_n\frac{H_{\nu_n}^{(2)\prime}(\beta a)}{J'_{\nu_n}(\beta a)}$$

and the Green function for $0 \le \rho \le \rho'$, denoted by $G_<^m$, is

$$G_<^m(\rho, \rho', \phi, \phi') = -\frac{j\pi}{2\phi_o}\sum_{n=1}^{\infty} \varepsilon_n J_{\nu_n}(\beta\rho)\left[H_{\nu_n}^{(2)}(\beta\rho') - \frac{H_{\nu_n}^{(2)\prime}(\beta a)}{J'_{\nu_n}(\beta a)}J_{\nu_n}(\beta\rho')\right]$$
$$\times \cos(\nu_n\phi)\cos(\nu_n\phi')$$

For $G_>^m$, valid for $\rho \le \rho \le a$, interchange ρ and ρ' in the expression above.

5.6.4 Infinite-Length Angular Configuration

The infinite-length angular configuration is shown in Fig. 5.11. The Green function satisfies (5.6.18) whose solution is of the form given in (5.6.19), where $G(\rho, \rho')$ satisfies (5.6.20). The solution of (5.6.20) for the infinite-length angular configuration is

$$G_<(\rho, \rho') = B_n J_{\nu_n}(\beta\rho) \qquad (0 \le \rho < \rho') \tag{5.6.29}$$

and

$$G_>(\rho, \rho') = C_n H_{\nu_n}^{(2)}(\beta\rho) \qquad (\rho > \rho') \tag{5.6.30}$$

Equation (5.6.29) is finite at the origin, and (5.6.30) satisfies the boundary condition at $\rho \to \infty$.

Applying the Green function conditions at $\rho = \rho'$ to (5.6.29) and (5.6.30) gives

$$G_< = G_>|_{\rho=\rho'} \Rightarrow B_n J_{\nu_n}(\beta\rho') = C_n H_{\nu_n}^{(2)}(\beta\rho') \tag{5.6.31}$$

and

$$\left.\frac{dG_>}{d\rho}\right|_{\rho=\rho'} - \left.\frac{dG_<}{d\rho}\right|_{\rho=\rho'} = -\frac{1}{\rho'} \Rightarrow C_n H_{\nu_n}^{(2)\prime}(\beta\rho') - B_n J_{\nu_n}'(\beta\rho') = -\frac{1}{\beta\rho'} \tag{5.6.32}$$

Solving (5.6.31) and (5.6.32) gives

$$B_n = -\frac{j\pi}{2} H_{\nu_n}^{(2)}(\beta\rho')$$

Therefore, from (5.6.29), for $0 \le \rho \le \rho'$, it follows that

$$G_<(\rho, \rho') = -\frac{j\pi}{2} H_{\nu_n}^{(2)}(\beta\rho') J_{\nu_n}(\beta\rho)$$

and from (5.6.19) for the configuration in Fig. 5.11:

$$G_<(\rho, \rho', \phi, \phi') = -\frac{j\pi}{\phi_o} \sum_{n=1}^{\infty} J_{\nu_n}(\beta\rho) H_{\nu_n}^{(2)}(\beta\rho') \sin(\nu_n\phi) \sin(\nu_n\phi') \qquad \left(\nu_n = \frac{n\pi}{\phi_o}\right)$$

The solution for $G_>$ (valid for $\rho' \le \rho \le \infty$) is obtained by interchanging ρ and ρ' in the expression above.

In the case of magnetic current excitation, the boundary conditions require that

$$\left.\frac{\partial G}{\partial \phi}\right|_{\substack{\phi=0 \\ \phi=\phi_o}} = 0$$

Hence, the Green function for $0 \le \rho \le \rho'$, denoted by G^m, is

$$G_<^m(\rho, \phi, \rho', \phi') = -\frac{j\pi}{2\phi_o}\sum_{n=0}^{\infty} \varepsilon_n J_{\nu_n}(\beta\rho)H_{\nu_n}^{(2)}(\beta\rho')\cos(\nu_n\phi)\cos(\nu_n\phi') \qquad \left(\nu_n = \frac{n\pi}{\phi_o}\right)$$

For $G_>^m$, valid for $\rho' \leq \rho \leq \infty$, interchange ρ and ρ' in the expression above.

5.6.5 Cylindrical Region

5.6.5.1 Source Inside the Cylinder

A cylindrical region with the source inside the cylinder is shown in Fig. 5.13a. The Green function satisfies (5.6.18) and the angular delta function representation is given in (5.4.40). Therefore, we seek a solution of the Green function in the form

$$G(\rho, \phi, \rho', \phi') = \frac{1}{2\pi}\sum_{n=-\infty}^{\infty} G(\rho, \rho')e^{jn(\phi-\phi')} \tag{5.6.33}$$

Substituting (5.6.33) and (5.4.40) into (5.6.18) shows that $G(\rho, \rho')$ satisfies

$$\left[\frac{1}{\rho}\frac{\partial}{\partial\rho}\left(\rho\frac{\partial}{\partial\rho}\right) + \beta^2 - \frac{n^2}{\rho^2}\right]G(\rho, \rho') = -\frac{\delta(\rho - \rho')}{\rho} \tag{5.6.34}$$

The solution to (5.6.34) is

$$G_<(\rho, \rho') = B_n J_n(\beta\rho) \qquad (0 \leq \rho < \rho') \tag{5.6.35}$$

and

$$G_>(\rho, \rho') = C_n J_n(\beta\rho) + D_n H_n^{(2)}(\beta\rho) \qquad (\rho' < \rho \leq a) \tag{5.6.36}$$

If the boundary condition at $\rho = a$ requires that

$$G_>(a, \rho') = 0 \Rightarrow C_n = -D_n\frac{H_n^{(2)}(\rho a)}{J_n(\rho a)} \tag{5.6.37}$$

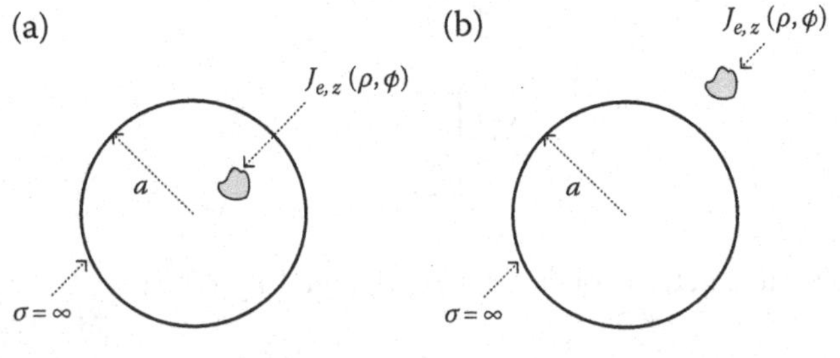

FIGURE 5.13 (a) Source inside the cylinder; (b) source outside the cylinder.

The Green function conditions at $\rho = \rho'$ are

$$G_< = G_>|_{\rho=\rho'} \Rightarrow B_n J_n(\beta\rho') = C_n J_n(\beta\rho') + D_n H_n^{(2)}(\beta\rho') \tag{5.6.38}$$

and

$$\left.\frac{dG_>}{d\rho}\right|_{\rho=\rho'} - \left.\frac{dG_<}{d\rho}\right|_{\rho=\rho'} = -\frac{1}{\rho'} \Rightarrow C_n J_n'(\beta\rho') + D_n H_n^{(2)\prime}(\beta\rho') - B_n J_n'(\beta\rho') = -\frac{1}{\beta\rho'} \tag{5.6.39}$$

The solution of (5.6.37) to (5.6.39) is similar to the solution of (5.6.23) to (5.6.25). Hence, we obtain

$$G_<(\rho, \rho') = -\frac{j\pi}{2} J_n(\beta\rho)\left[H_n^{(2)}(\beta\rho') - \frac{H_n^{(2)}(\beta a)}{J_n(\beta a)} J_n(\beta\rho')\right] \tag{5.6.40}$$

and from (5.6.33) the Green function for the source inside the cylinder in Fig. 5.13a, valid for $0 \le \rho \le \rho'$, is then

$$G_<(\rho, \phi, \rho', \phi') = -\frac{j}{4} \sum_{n=-\infty}^{\infty} J_n(\beta\rho)\left[H_n^{(2)}(\beta\rho') - \frac{H_n^{(2)}(\beta a)}{J_n(\beta a)} J_n(\beta\rho')\right] e^{jn(\phi-\phi')} \tag{5.6.41}$$

The solution for $G_>$, valid for $\rho' \le \rho \le a$, is obtained by interchanging ρ and ρ' in (5.6.41).

Had the solution for $\rho' < \rho \le a$ been written as

$$G_>(\rho, \rho') = C_n J_n(\beta\rho) + D_n Y_n(\beta\rho) \qquad (\rho' < \rho \le a)$$

the resulting Green function would have been

$$G_<(\rho, \phi, \rho', \phi') = -\frac{1}{4} \sum_{n=-\infty}^{\infty} J_n(\beta\rho)\left[Y_n(\beta\rho') - \frac{Y_n(\beta a)}{J_n(\beta a)} J_n(\beta\rho')\right] e^{jn(\phi-\phi')}$$

This expression produces the same results as (5.6.41), which can be verified by letting $H_n^{(2)}(x) = J_n(x) - jY_n(x)$ in (5.6.41).

If the boundary condition at $\rho = a$ requires that $\partial G_>/\partial\rho = 0$ (i.e., for a magnetic current excitation) then, from (5.6.36):

$$\frac{\partial G_>(a, \rho')}{\partial \rho} = 0 \Rightarrow C_n = -D_n \frac{H_n^{(2)\prime}(\rho a)}{J_n'(\rho a)}$$

and the Green function, denoted by G^m, is

$$G_<^m(\rho, \phi, \rho', \phi') = -\frac{j}{4} \sum_{n=-\infty}^{\infty} J_n(\beta\rho) \left[H_n^{(2)}(\beta\rho') - \frac{H_n^{(2)\prime}(\beta a)}{J_n'(\beta a)} J_n(\beta\rho') \right] e^{jn(\phi-\phi')} \qquad (0 \le \rho \le \rho')$$

The solution for $G_>^m$, valid for $\rho' \le \rho \le a$, is obtained by interchanging ρ and ρ' in the expression above.

In the case that $a \to \infty$, the geometry in Fig. 5.13a becomes an infinite region. When $\beta a \to \infty$; then $H_n^{(2)}(\beta a) \to 0$ and (5.6.41) reduces to

$$G_<(\rho, \phi, \rho', \phi') = -\frac{j}{4} \sum_{n=-\infty}^{\infty} J_n(\beta\rho) H_n^{(2)}(\beta\rho')\, e^{jn(\phi-\phi')} \qquad (5.6.42)$$

The solution for $G_>$ is obtained by interchanging ρ and ρ' in (5.6.42).

Recall that the two-dimensional Green function is also given by (2.4.8). That is,

$$G = -\frac{j}{4} H_0^{(2)}(\beta|\boldsymbol{\rho} - \boldsymbol{\rho}'|) \qquad (5.6.43)$$

it follows from (5.6.42) and (5.6.43) that

$$H_0^{(2)}(\beta|\boldsymbol{\rho} - \boldsymbol{\rho}'|) = \begin{cases} \sum_{n=-\infty}^{\infty} J_n(\beta\rho) H_n^{(2)}(\beta\rho') e^{jn(\phi-\phi_o)} & (\rho \le \rho') \\ \sum_{n=-\infty}^{\infty} J_n(\beta\rho') H_n^{(2)}(\beta\rho) e^{jn(\phi-\phi_o)} & (\rho \ge \rho') \end{cases} \qquad (5.6.44)$$

which is known as the addition theorem of Hankel functions. The addition theorem allows us to express the fields with the origin (i.e., $\rho = 0$) as reference, while in (5.6.43) the radial distances are measured from $|\boldsymbol{\rho} - \boldsymbol{\rho}'| = 0$.

As an example, for the infinite-length electric current source shown in Fig. 2.4, the fields can be calculated in terms of the $\mathbf{A} = A_z\hat{\mathbf{a}}_z$vector, which satisfy

$$\left[\frac{1}{\rho}\frac{d}{d\rho}\left(\rho\frac{d}{d\rho}\right) + \frac{1}{\rho^2}\frac{\partial^2}{\partial\phi^2} + \beta^2\right] A_z(\rho, \phi) = -\mu I \frac{\delta(\rho - \rho_o)\delta(\phi - \phi_o)}{\rho}$$

where the line source location is at (ρ_o, ϕ_o). Then, using (5.6.42), $A_z(\rho, \phi)$ for $\rho \le \rho_o$ is given by

$$A_z = -\mu I \int_{\phi=0}^{2\pi} \int_{\rho'=0}^{\rho_o} \frac{\delta(\rho' - \rho_o)\delta(\phi' - \phi_o)}{\rho'} G_<(\rho, \phi, \rho', \phi')\rho' d\rho' d\phi'$$
$$= -j\frac{\mu I}{4} \sum_{n=-\infty}^{\infty} \int_{\phi=0}^{2\pi} \int_{\rho'=0}^{\rho_o} \delta(\rho' - \rho_o)\delta(\phi' - \phi_o) J_n(\beta\rho) H_n^{(2)}(\beta\rho') e^{jn(\phi-\phi')} d\rho' d\phi'$$
$$= -j\frac{\mu I}{4} \sum_{n=-\infty}^{\infty} J_n(\beta\rho) H_n^{(2)}(\beta\rho_o) e^{jn(\phi-\phi_o)} \qquad (\rho \le \rho_o) \tag{5.6.45}$$

For $\rho \ge \rho_o$ interchange ρ and ρ_o in (5.6.45). Using (5.6.44), (5.6.45) can also be written as

$$A_z = -j\frac{\mu I}{4} H_0^{(2)}(\beta|\mathbf{\rho} - \mathbf{\rho}_o|)$$

which is identical to (2.2.11).

5.6.5.2 Source Outside the Cylinder

A source located outside the cylindrical surface is shown in Fig. 5.13b. In this case, the appropriate solutions to (5.6.34) are

$$G_<(\rho, \rho') = C_n J_n(\beta\rho) + B_n H_n^{(2)}(\beta\rho) \qquad (a \le \rho < \rho') \tag{5.6.46}$$

and

$$G_>(\rho, \rho') = D_n H_n^{(2)}(\beta\rho) \qquad (\rho > \rho') \tag{5.6.47}$$

For an electric current excitation, the boundary condition at $\rho = a$ requires that $G_<(a, \rho') = 0$, or

$$G_<(a, \rho') = 0 \Rightarrow C_n = -B_n \frac{H_n^{(2)}(\beta a)}{J_n(\beta a)} \tag{5.6.48}$$

The Green function conditions at $\rho = \rho'$ are

$$G_< = G_>|_{\rho=\rho'} \Rightarrow C_n J_n(\beta\rho') + B_n H_n^{(2)}(\beta\rho') = D_n H_n^{(2)}(\beta\rho') \tag{5.6.49}$$

and

$$\left.\frac{dG_>}{d\rho}\right|_{\rho=\rho'} - \left.\frac{dG_<}{d\rho}\right|_{\rho=\rho'} = -\frac{1}{\rho'} \Rightarrow D_n H_n^{(2)\prime}(\beta\rho') - C_n J_n'(\beta\rho') - B_n H_n^{(2)\prime}(\beta\rho')$$
$$= -\frac{1}{\beta\rho'} \tag{5.6.50}$$

Solving for the constants in (5.6.48) to (5.6.50) gives

$$G_>(\rho, \rho') = \frac{-j\pi}{2} H_n^{(2)}(\beta\rho)\left[J_n(\beta\rho') - \frac{J_n(\beta a)}{H_n^{(2)}(\beta a)} H_n^{(2)}(\beta\rho')\right]$$

and from (5.6.33) the Green function solution for $\rho \geq \rho'$ is

$$G_>(\rho, \rho', \phi, \phi') = \frac{-j}{4} \sum_{n=-\infty}^{\infty} H_n^{(2)}(\beta\rho)\left[J_n(\beta\rho') - \frac{J_n(\beta a)}{H_n^{(2)}(\beta a)} H_n^{(2)}(\beta\rho')\right] e^{jn(\phi-\phi')} \tag{5.6.51}$$

The solution for $G_<$, valid for $a \leq \rho \leq \rho'$, is obtained by interchanging ρ and ρ' in (5.6.51).

If the boundary condition at $\rho = a$ requires that $\partial G_</\partial\rho = 0$. Then from (5.6.46):

$$\left.\frac{\partial G_<(\rho, \rho')}{\partial\rho} = 0\right|_{\rho=a} \Rightarrow B_1 = -B_2 \frac{H_\nu^{(2)\prime}(\beta a)}{J_\nu'(\beta a)}$$

and the Green function, denoted by $G_>^m$ for $\rho \geq \rho'$, is

$$G_>^m(\rho, \phi, \rho', \phi') = -\frac{j}{4} \sum_{n=-\infty}^{\infty} H_n^{(2)}(\beta\rho)\left[J_n(\beta\rho') - \frac{J_n'(\beta a)}{H_n^{(2)\prime}(\beta a)} H_n^{(2)}(\beta\rho')\right] e^{jn(\phi-\phi')} \tag{5.6.52}$$

The solution for $G_<^m$, valid for $a \leq \rho \leq \rho'$, is obtained by interchanging ρ and ρ' in (5.6.52).

5.6.6 Spherical Region

In spherical coordinates, the Green function satisfies

$$\left[\frac{1}{r^2}\frac{\partial}{\partial r}\left(r^2\frac{\partial}{\partial r}\right) + \frac{1}{r^2 \sin\theta}\frac{\partial}{\partial\theta}\left(\sin\theta\frac{\partial}{\partial\theta}\right) + \frac{1}{r^2\sin^2\theta}\frac{\partial^2}{\partial\phi^2} + \beta^2\right] G(\mathbf{r}, \mathbf{r}') = -\frac{\delta(r-r')\delta(\theta-\theta')\delta(\phi-\phi')}{r^2\sin\theta} \tag{5.6.53}$$

The normalized eigenfunctions associated with the angular variables (see (5.4.44) and (5.4.45)) are

$$\Theta_{nm} = \sqrt{\frac{(2n+1)}{2}\frac{(n-m)!}{(n+m)!}} P_n^m(\cos\theta) \tag{5.6.54}$$

and

$$\Phi(\phi) = \frac{1}{\sqrt{2\pi}} \sum_{m=-\infty}^{\infty} e^{jm\phi} \tag{5.6.55}$$

Using (5.6.54) and (5.6.55), the angular delta function representation is

$$\frac{\delta(\theta-\theta')\delta(\phi-\phi')}{\sin\theta} = \sum_{n=0}^{\infty}\sum_{m=-n}^{n} \frac{(2n+1)}{4\pi}\frac{(n-m)!}{(n+m)!} P_n^m(\cos\theta) P_n^m(\cos\theta') e^{jm(\phi-\phi')} \tag{5.6.56}$$

or in terms of spherical harmonics as

$$\frac{\delta(\theta-\theta')\delta(\phi-\phi')}{\sin\theta} = \sum_{n=0}^{\infty}\sum_{m=-n}^{n} Y_{nm}(\theta,\phi) Y_{nm}^*(\theta',\phi')$$

The summation in m is from $-n$ to n since $P_n^m(\cos\theta) = 0$ for $|m| > n$.

The Green function in (5.6.53) is, therefore, of the form

$$\begin{aligned} G(\mathbf{r},\mathbf{r}') &= \sum_{n=0}^{\infty}\sum_{m=-n}^{n} G(r,r') Y_{nm}(\theta,\phi) Y_{nm}^*(\theta,\phi) \\ &= \sum_{n=0}^{\infty}\sum_{m=-n}^{n} G(r,r') \frac{(2n+1)}{4\pi}\frac{(n-m)!}{(n+m)!} P_n^m(\cos\theta) P_n^m(\cos\theta') e^{jm(\phi-\phi')} \end{aligned} \tag{5.6.57}$$

which is recognized as a Fourier-Legendre representation of $G(\mathbf{r},\mathbf{r}')$.

Substituting (5.6.56) and (5.6.57) into (5.6.53) gives the differential equation satisfied by the radial part of the Green function (i.e., $G(r,r')$). That is,

$$\left[\frac{d^2}{dr^2} + \frac{2}{r}\frac{d}{dr} + \left(\beta^2 - \frac{n(n+1)}{r^2}\right)\right] G(r,r') = -\frac{\delta(r-r')}{r^2}$$

or

$$\frac{d}{dr}\left[r^2 \frac{dG(r,r')}{dr}\right] + [(\beta r)^2 - n(n+1)] G(r,r') = -\delta(r-r') \tag{5.6.58}$$

which is recognized as the spherical Bessel differential equation.

Consider the solution to (5.6.58) in free space, using the conditions that the Green function must be finite at $r = 0$ and represent an outgoing wave as $r \to \infty$. The solutions to (5.6.58) that satisfy these conditions are

$$G_<(r,r') = B_n j_n(\beta r) \qquad (r < r') \tag{5.6.59}$$

and

$$G_>(r, r') = C_n h_n^{(2)}(\beta r) \qquad (r > r') \tag{5.6.60}$$

The boundary conditions for $G(r, r')$ at $r = r'$ are

$$G_< = G_>|_{r=r'} \Rightarrow B_n j_n(\beta r') = C_n h_n^{(2)}(\beta r') \tag{5.6.61}$$

and

$$\left.\frac{\partial G_>}{\partial r}\right|_{r=r'} - \left.\frac{\partial G_<}{\partial r}\right|_{r=r'} = -\frac{1}{r'^2} \Rightarrow C_n h_n^{(2)\prime}(\beta r') - B_n j_n'(\beta r') = -\frac{1}{\beta r'^2} \tag{5.6.62}$$

This last relation follows by comparing (5.6.58) with the Sturm-Liouville equation, which shows that $p(r') = -1/r'^2$. It can also be derived directly from (5.6.68) (left to the problems).

From (5.6.61) and (5.6.62), we obtain

$$B_n = -\frac{h_n^{(2)}(\beta r')/\beta r'^2}{W\left[j_n(\beta r'),\, h_n^{(2)}(\beta r')\right]} = -j\beta h_n^{(2)}(\beta r')$$

and

$$C_n = \frac{j_n(\beta r')/\beta r'^2}{W\left[j_n(\beta r'),\, h_n^{(2)}(\beta r')\right]} = -j\beta j_n(\beta r')$$

where the Wronskian is $-j/(\beta r')^2$.

Substituting B and C into (5.6.59) and (5.6.60), and then into (5.6.57) gives

$$G(\mathbf{r}, \mathbf{r}') = -j\beta \sum_{n=0}^{\infty} \sum_{m=-n}^{n} \frac{(2n+1)}{4\pi} \frac{(n-m)!}{(n+m)!} P_n^m(\cos\theta) P_n^m(\cos\theta') e^{jm(\phi-\phi')} \times \begin{cases} j_n(\beta r) h_n^{(2)}(\beta r') & (r \le r') \\ j_n(\beta r') h_n^{(2)}(\beta r) & (r \ge r') \end{cases} \tag{5.6.63}$$

which is the representation of

$$G(\mathbf{r}, \mathbf{r}') = \frac{e^{-j\beta|\mathbf{r}-\mathbf{r}'|}}{4\pi|r - \mathbf{r}'|}$$

in spherical coordinates.

Equation (5.6.63) can also be expressed as

$$G(\mathbf{r}, \mathbf{r}') = -j\beta \sum_{n=0}^{\infty} \sum_{m=-n}^{n} Y_{nm}(\theta, \phi) Y_{nm}^*(\theta', \phi') \begin{cases} j_n(\beta r) h_n^{(2)}(\beta r') & (r \le r') \\ j_n(\beta r') h_n^{(2)}(\beta r) & (r \ge r') \end{cases}$$

$$= -j\beta \sum_{n=0}^{\infty} \sum_{m=-n}^{n} \frac{2n+1}{4\pi} P_n^m(\cos\gamma) \begin{cases} j_n(\beta r) h_n^{(2)}(\beta r') & (r \le r') \\ j_n(\beta r') h_n^{(2)}(\beta r) & (r \ge r') \end{cases} \qquad (5.6.64)$$

Using the spherical Schelkunoff-Bessel functions, (5.6.63) and (5.6.64) read

$$G(\mathbf{r}, \mathbf{r}') = \frac{-j}{\beta r r'} \sum_{n=0}^{\infty} \sum_{m=-n}^{n} \frac{(2n+1)}{4\pi} \frac{(n-m)!}{(n+m)!} P_n^m(\cos\theta) P_n^m(\cos\theta') e^{jm(\phi-\phi')}$$

$$\times \begin{cases} \hat{j}_n(\beta r) \hat{h}_n^{(2)}(\beta r') & (r \le r') \\ \hat{j}_n(\beta r') \hat{h}_n^{(2)}(\beta r) & (r \ge r') \end{cases} \qquad (5.6.65)$$

and

$$G(\mathbf{r}, \mathbf{r}') = \frac{-j}{\beta r r'} \sum_{n=0}^{\infty} \sum_{m=-n}^{n} Y_{nm}(\theta, \phi) Y_{nm}^*(\theta', \phi') \begin{cases} \hat{j}_n(\beta r) \hat{h}_n^{(2)}(\beta r') & (r \le r') \\ \hat{j}_n(\beta r') \hat{h}_n^{(2)}(\beta r) & (r \ge r') \end{cases} \qquad (5.6.66)$$

Useful forms of (5.6.63) and (5.6.65) occur when there is no ϕ dependence, and the source is at $\theta' = 0$. In this case, $m = 0$, and (5.6.63) and (5.6.65) reduce to

$$G(\mathbf{r}, \mathbf{r}') = -j\beta \sum_{n=0}^{\infty} \frac{(2n+1)}{4\pi} P_n(\cos\theta) \begin{cases} j_n(\beta r) h_n^{(2)}(\beta r') & (r \le r') \\ j_n(\beta r') h_n^{(2)}(\beta r) & (r \ge r') \end{cases} \qquad (5.6.67)$$

and

$$G(\mathbf{r}, \mathbf{r}') = \frac{-j}{\beta r r'} \sum_{n=0}^{\infty} \frac{(2n+1)}{4\pi} P_n(\cos\theta) \begin{cases} \hat{j}_n(\beta r) \hat{h}_n^{(2)}(\beta r') & (r \le r') \\ \hat{j}_n(\beta r') \hat{h}_n^{(2)}(\beta r) & (r \ge r') \end{cases} \qquad (5.6.68)$$

These equations are used in the study of wave propagation in a spherical waveguide. For example, in the Earth-Ionosphere waveguide.

As an example, for an electric dipole source in free space located at $r' = r_o$, $\theta' = 0$ and $\phi' = 0$, the fields can be calculated in terms of $\mathbf{A} = A_z \hat{\mathbf{a}}_z$, where A_z satisfies

$$\left[\frac{1}{r^2} \frac{\partial}{\partial r}\left(r^2 \frac{\partial}{\partial r} \right) + \frac{1}{r^2 \sin\theta} \frac{\partial}{\partial \theta}\left(\sin\theta \frac{\partial}{\partial \theta} \right) + \beta^2 \right] A_z(r, \theta) = -\mu(I\Delta L) \frac{\delta(r - r_o)\delta(\theta)}{2\pi r^2 \sin\theta}$$

The fields are independent of ϕ and the Green function is given by (5.6.67). Therefore,

$$\begin{aligned}A_z &= -\mu(I\Delta L)\int_{\theta'=0}^{\pi}\int_{\theta'=0}^{2\pi}\int_{r'=0}^{r_o}\frac{\delta(r'-r_o)\delta(\theta')}{2\pi r'^2\sin\theta'}G_<(r,\theta,r',\theta')r'^2\sin\theta' dr'd\theta'd\phi' \\ &= -j\frac{\beta\mu(I\Delta L)}{4\pi}\sum_{n=0}^{\infty}\int_{\theta'=0}^{\pi}\int_{r'=0}^{r_o}(2n+1)P_n(\cos\theta)j_n(\beta r)h_n^{(2)}(\beta r')\delta(r'-r_o)\delta(\theta')dr'd\theta' \\ &= -j\frac{\beta\mu(I\Delta L)}{4\pi}\sum_{n=0}^{\infty}(2n+1)P_n(\cos\theta)j_n(\beta r)h_n^{(2)}(\beta r_o) \qquad (r\le r_o)\end{aligned}$$

(5.6.69)

For $r' \ge r_o$ interchange r and r_o in (5.6.69).

From Section 2.1, A_z in this example can also be expressed as

$$A_z = -j\frac{\beta\mu(I\Delta L)}{4\pi}h_0^{(2)}(\beta|\mathbf{r}-\mathbf{r}'|)$$

which upon comparing with (5.6.69) shows that

$$h_0^{(2)}(\beta|\mathbf{r}-\mathbf{r}'|) = \begin{cases}\sum_{n=0}^{\infty}(2n+1)P_n(\cos\theta)j_n(\beta r)h_n^{(2)}(\beta r_o) & (r\le r_o)\\ \sum_{n=0}^{\infty}(2n+1)P_n(\cos\theta)j_n(\beta r_o)h_n^{(2)}(\beta r) & (r\ge r_o)\end{cases}$$

which is the form of the addition theorem for the spherical Hankel function when $r' = r_o$, $\theta' = 0$ and there is no ϕ dependence.

5.6.7 Spherical Conductor

A spherical conductor is shown in Fig. 5.14 where the delta function excitation is inside the conductor. The Green function satisfies (5.6.53), subject to the boundary condition:

$$G(\mathbf{r},\mathbf{r}') = 0|_{r=a}$$

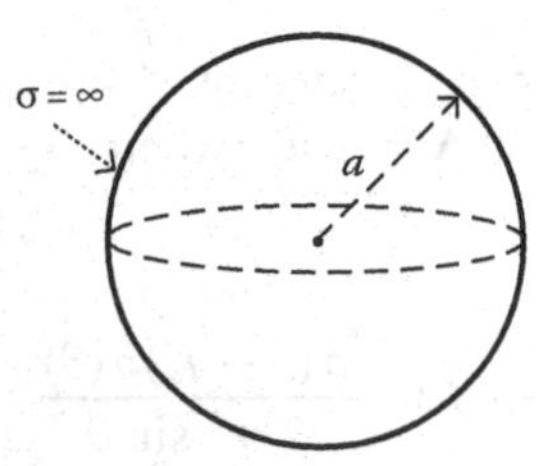

FIGURE 5.14 Spherical conductor of radius a.

The solution to (5.6.53) is of the form shown in (5.6.57) where $G(r, r')$ satisfies (5.6.58). Appropriate solutions to (5.6.58) for this case are

$$G_<(r, r') = B_n j_n(\beta r) \qquad (0 \le r < r') \tag{5.6.70}$$

and

$$G_>(r, r') = C_n j_n(\beta r) + D_n y_n(\beta r) \qquad (r' < r \le a) \tag{5.6.71}$$

The boundary condition at $r = a$ is satisfied by

$$G_> = 0|_{r=a} \Rightarrow D_n = -C_n \frac{j_n(\beta a)}{y_n(\beta a)} \tag{5.6.72}$$

Then, $G_>$ in (5.6.72) reads

$$G_> = C_n \left[j_n(\beta r) - \frac{j_n(\beta a)}{y_n(\beta a)} y_n(\beta r) \right]$$

At $r = r'$ the boundary conditions for the Green functions require that

$$G_< = G_>|_{r=r'} \Rightarrow B_n j_n(\beta r') = C_n \left[j_n(\beta r') - \frac{j_n(\beta a)}{y_n(\beta a)} y_n(\beta r') \right] \tag{5.6.73}$$

and

$$\begin{aligned} \left.\frac{\partial G_>}{\partial r}\right|_{r=r'} - \left.\frac{\partial G_<}{\partial r}\right|_{r=r'} &= -\frac{1}{r'^2} \Rightarrow C_n \left[j_n{}'(\beta r') - \frac{j_n(\beta a)}{y_n(\beta a)} y_n{}'(\beta r') \right] - B_n j_n{}'(\beta r') \\ &= -\frac{1}{\beta r'^2} \end{aligned} \tag{5.6.74}$$

Solving (5.6.73) and (5.6.74) gives

$$B_n = -\beta \left[y_n(\beta r') - \frac{y_n(\beta a)}{j_n(\beta a)} j_n(\beta r') \right] \tag{5.6.75}$$

where the Wronskian

$$W[j_n(\beta r'), y_n(\beta r')] = \frac{1}{(\beta r')^2}$$

was used.

Substituting (5.6.75) into (5.6.70) gives $0 \le r \le r'$:

$$G_{<}(r, r') = -\beta j_n(\beta r)\left[y_n(\beta r') - \frac{y_n(\beta a)}{j_n(\beta a)} j_n(\beta r')\right] \tag{5.6.76}$$

and from (5.6.57):

$$G_{<}(\mathbf{r}, \mathbf{r}') = -\beta \sum_{n=0}^{\infty} \sum_{m=-n}^{n} \frac{(2n+1)}{4\pi} \frac{(n-m)!}{(n+m)!} j_n(\beta r)\left[y_n(\beta r') - \frac{y_n(\beta a)}{j_n(\beta a)} j_n(\beta r')\right] \times P_n^m(\cos\theta) P_n^m(\cos\theta') e^{jm(\phi-\phi')} \tag{5.6.77}$$

which can also be expressed in terms of spherical harmonics as

$$G_{<}(\mathbf{r}, \mathbf{r}') = -\beta \sum_{n=0}^{\infty} \sum_{m=-n}^{n} j_n(\beta r)\left[y_n(\beta r') - \frac{y_n(\beta a)}{j_n(\beta a)} j_n(\beta r')\right] Y_{nm}(\theta, \phi) Y_{nm}^{*}(\theta', \phi') \tag{5.6.78}$$

The solution for $G_{>}$, valid for $r' \le r \le a$, is obtained by interchanging r and r' in (5.6.77) and (5.6.78).

The equivalency between (5.6.78) and (5.4.49) can be established by determining the analytic properties of (5.6.78) as a function of β. The poles of (5.6.78) are those associated with the radial part of the solution in (5.6.76). The poles occur at

$$j_n(\beta a) = 0 \Rightarrow \beta = \beta_{nl} = \frac{\kappa_{nl}}{a} \quad \begin{pmatrix} n = 0, \pm 1, \pm 2, \ldots \\ l = 1, 2, 3, \ldots \end{pmatrix}$$

where n has positive and negative values since $j_n(\beta a)$ has poles when n is positive and negative. The poles for positive values of n are located at positive values of β_{nl}, and those for negative values of n at negative values of occur β_{nl}.

The residue at the nl pole in (5.6.76) is

$$\text{Residue} = \beta \frac{y_n(\beta a)}{\frac{dj_n(\beta a)}{d\beta}} j_n(\beta r) j_n(\beta r')\Bigg|_{\beta = \frac{\kappa_{nl}}{a}} = \frac{\kappa_{nl}}{a^2} \frac{y_n(\kappa_{nl})}{j_n'(\kappa_{nl})} j_n\left(\frac{\kappa_{nl}}{a} r\right) j_n\left(\frac{\kappa_{nl}}{a} r'\right) \tag{5.6.79}$$

The Wronskian shows that with $j_n(\kappa_{nl}) = 0$:

$$j_n(\kappa_{nl}) y_n'(\kappa_{nl}) - j_n'(\kappa_{nl}) y_n(\kappa_{nl}) = \frac{1}{\kappa_{nl}^2} \Rightarrow y_n(\kappa_{nl}) = -\frac{1}{k_{nl}^2 j_n'(\kappa_{nl})}$$

Using $j_n'(\kappa_{nl}) = -j_{n+1}(\kappa_{nl})$ (see Appendix C), it follows that

$$y_n(\kappa_{nl}) = -\frac{1}{k_{nl}^2 j_{n+1}(\kappa_{nl})}$$

Then, (5.6.79) reads

$$\text{Residue} = \frac{1}{a^2\kappa_{nl}} \frac{1}{j_{n+1}^2(\kappa_{nl})} j_n\left(\frac{\kappa_{nl}}{a} r\right) j_n\left(\frac{\kappa_{nl}}{a} r'\right)$$

Adding the contribution from all the poles (i.e., at $\pm\beta_{nl}$) shows that

$$G(r, r') = \sum_{n=-\infty}^{\infty} \sum_{l=1}^{\infty} \frac{1}{a^2\kappa_{nl}} \frac{1}{j_{n+1}^2(\kappa_{nl})} \frac{j_n\left(\frac{\kappa_{nl}}{a} r\right) j_n\left(\frac{\kappa_{nl}}{a} r'\right)}{\beta - \frac{\kappa_{nl}}{a}}$$

$$= \sum_{n=0}^{\infty} \sum_{l=1}^{\infty} \frac{1}{a^2\kappa_{nl}} \frac{1}{j_{n+1}^2(\kappa_{nl})} j_n\left(\frac{\kappa_{nl}}{a} r\right) j_n\left(\frac{\kappa_{nl}}{a} r'\right)\left(\frac{1}{\beta - \frac{\kappa_{nl}}{a}} - \frac{1}{\beta + \frac{\kappa_{nl}}{a}}\right)$$

and it follows that

$$G(\mathbf{r}, \mathbf{r}') = \frac{2}{a^3} \sum_{n=0}^{\infty} \sum_{l=1}^{\infty} \sum_{m=-n}^{n} \frac{j_n\left(\frac{\kappa_{nl}}{a} r\right) j_n\left(\frac{\kappa_{nl}}{a} r'\right)}{\left[\left(\frac{\kappa_{nl}}{a}\right)^2 - \beta^2\right] j_{n+1}^2(\kappa_{nl})} Y_{nm}(\theta, \phi) Y_{nm}^*(\theta', \phi')$$

which is identical to (5.4.48).

Problems

P5.1 Consider the differential equation:

$$\frac{d^2y}{dx^2} + \frac{1}{4} y = \sin 2x$$

where $y(0) = 0$ and $y(\pi) = 0$.

a. Determine the complementary, particular, and general solution of $y(x)$.
b. Determine the Green function using the eigenvalue method.
c. Evaluate $y(x)$ using (5.2.10).

P5.2

a. Use the direct method to determine the Green functions (i.e., $G_>$ and $G_<$) associated with the differential equation:

$$\frac{d^2y}{dx^2} + y = x$$

where $y(0) = 0$ and $y(\pi/2) = 0$.

b. Determine $y(x)$.

P5.3 Verify that the units match in the Green function expression in (5.4.31) and in (5.4.37).

P5.4 Show that the analytic structure analysis of (5.6.16) and (5.6.17) produces the result in (5.4.17).

P5.5 Determine the Green function for the magnetic current source (i.e., $G^m(x, x', y, y')$) in Fig. 5.6.

P5.6 Use the direct method to determine the Green function for a magnetic current source (i.e., $G^m(\rho, \phi, \rho', \phi')$) in Fig. 5.11.

P5.7 Derive (5.6.52).

P5.8

a. Determine the Green functions $G_<$ and $G_>$ for the angular configuration shown in Fig. P5.8, where $-\infty < z < \infty$.

b. Determine the Green functions for the case of a magnetic source excitation (i.e., $G_<^m$ and $G_>^m$).

P5.9 Starting with (5.6.58) verify the Green function condition in (5.6.62), namely

$$\left.\frac{\partial G_>}{\partial r}\right|_{r=r'} - \left.\frac{\partial G_<}{\partial r}\right|_{r=r'} = -\frac{1}{r'^2}$$

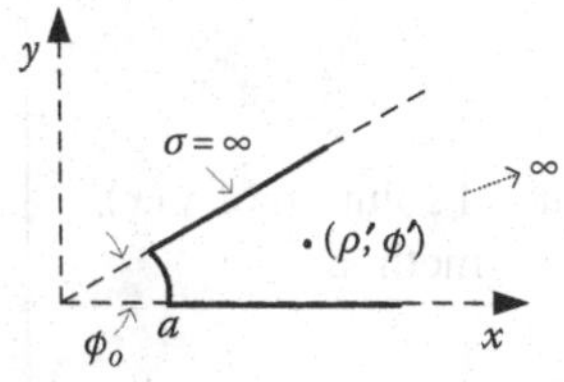

FIGURE P5.8

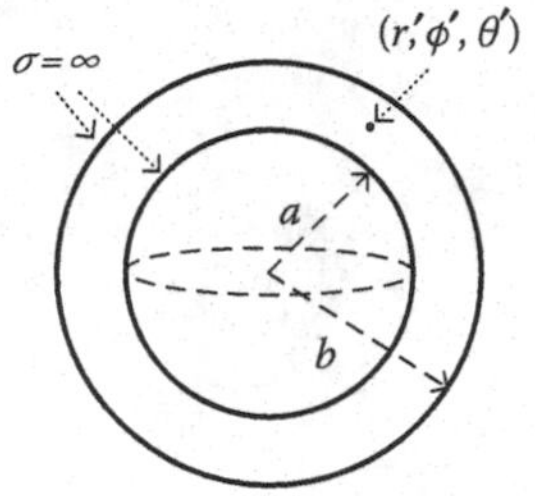

FIGURE P5.10

P5.10

a. Determine the Green functions $G_<$ and $G_>$ for the spherical configuration shown in Fig. P5.10 when the excitation is located at $(r', 0, 0)$.
b. Determine the Green functions when the excitation is located at (r', ϕ', θ').

6 Integral Transforms for Green Functions

6.1 INTEGRAL TRANSFORM FOR THE WAVE EQUATION IN RECTANGULAR COORDINATES

The utility of Fourier transforms and Laplace transforms in the solution of differential equations is well known. For example, if $f(t)$ denotes the current or voltage in an electric circuit, the Fourier transform of $f(t)$ is

$$f(\omega) = \int_{-\infty}^{\infty} f(t)\, e^{-j\omega t} dt$$

The Fourier transform changes the time domain function $f(t)$ to the frequency domain function $f(\omega)$. This process changes the differential equation satisfied by $f(t)$ to an algebraic equation for $f(\omega)$. Once $f(\omega)$ is found, the inverse Fourier transform

$$f(t) = \frac{1}{2\pi} \int_{-\infty}^{\infty} f(\omega)\, e^{j\omega t} dt$$

gives the solution to $f(t)$.

In electromagnetic wave propagation we deal with a three-dimensional space, where the fields or auxiliary potentials in phasor form, denoted by $\psi(\mathbf{r})$, satisfy the wave equation:

$$[\nabla^2 + k^2]\psi(\mathbf{r}) = -f(\mathbf{r})$$

where $k = k' - jk''$. In terms of γ the equation reads

$$[\nabla^2 - \gamma^2]\psi(\mathbf{r}) = -f(\mathbf{r})$$

where $\gamma = \alpha + j\beta$. In a lossless region: $\gamma = j\beta = j\omega\sqrt{\mu\varepsilon}$ and $k = k' = \beta = \omega\sqrt{\mu\varepsilon}$.

The three-dimensional spacial Fourier transformation of these equation is now discussed. In rectangular coordinates the three-dimensional Fourier transform is defined by

$$f(\alpha_1, \alpha_2, \alpha_3) = \int_{-\infty}^{\infty} \int_{-\infty}^{\infty} \int_{-\infty}^{\infty} f(x, y, z)\, e^{-j(\alpha_1 x + \alpha_2 y + \alpha_3 z)} dxdydz \qquad (6.1.1)$$

DOI: 10.1201/9781003219729-6

and the inverse Fourier transform is

$$f(x, y, z) = \frac{1}{(2\pi)^3}\int_{-\infty}^{\infty}\int_{-\infty}^{\infty}\int_{-\infty}^{\infty} f(\alpha_1, \alpha_2\alpha_3)\, e^{j(\alpha_1 x+\alpha_2 y+\alpha_3 z)} d\alpha_1 d\alpha_2 d\alpha_3 \quad (6.1.2)$$

Consider the wave equation in rectangular coordinates:

$$\left[\frac{\partial^2}{\partial x^2} + \frac{\partial^2}{\partial y^2} + \frac{\partial^2}{\partial z^2} + k^2\right]\psi(x, y, z) = -f(x, y, z) \quad (6.1.3)$$

where the functions $\psi(x, y, z)$ and $f(x, y, z)$ are Fourier transformable. First, the z-variable in (6.1.3) is transformed by multiplying both sides by $e^{-j\alpha_3 z}$ and integrating with respect to z from $-\infty$ to ∞. That is,

$$\left[\frac{\partial^2}{\partial x^2} + \frac{\partial^2}{\partial y^2} + k^2\right]\int_{-\infty}^{\infty}\psi(x, y, z)e^{-j\alpha_3 z}dz + \int_{-\infty}^{\infty}\frac{\partial^2\psi(x, y, z)}{\partial z^2}e^{-j\alpha_3 z}dz$$
$$= -\int_{-\infty}^{\infty} f(x, y, z)e^{-j\alpha_3 z}dz$$

or

$$\left[\frac{\partial^2}{\partial x^2} + \frac{\partial^2}{\partial y^2} + k^2\right]\psi(x, y, \alpha_3) + \int_{-\infty}^{\infty}\frac{\partial^2\psi(x, y, z)}{\partial z^2}e^{-j\alpha_3 z}dz = -f(x, y, \alpha_3) \quad (6.1.4)$$

The second term in (6.1.4) requires some manipulations. Integrating by parts with $u = e^{-j\alpha_3 z}$ and $dv = (\partial^2\psi/\partial z^2)dz$ gives

$$\int_{-\infty}^{\infty}\frac{\partial^2\psi(x, y, z)}{\partial z^2}e^{-j\alpha_3 z}dz = e^{-j\alpha_3 z}\frac{\partial\psi}{\partial z}\Bigg|_{-\infty}^{\infty} + j\alpha_3\int_{-\infty}^{\infty}\frac{\partial\psi(x, y, z)}{\partial z}e^{-j\alpha_3 z}dz$$

A second integration by parts with $u = e^{-j\alpha_3 z}$ and $dv = (\partial\psi/\partial z)dz$ gives

$$\int_{-\infty}^{\infty}\frac{\partial^2\psi(x, y, z)}{\partial z^2}e^{-j\alpha_3 z}dz = e^{-j\alpha_3 z}\left(\frac{\partial\psi}{\partial z} + j\alpha_3\psi\right)\Bigg|_{-\infty}^{\infty} - \alpha_3^2\int_{-\infty}^{\infty}\psi(x, y, z)e^{-j\alpha_3 z}dz$$
$$= e^{-j\alpha_3 z}\left(\frac{\partial\psi}{\partial z} + j\alpha_3\psi\right)\Bigg|_{-\infty}^{\infty} - \alpha_3^2\,\psi(x, y, \alpha_3) \quad (6.1.5)$$

or

$$\int_{-\infty}^{\infty}\frac{\partial^2\psi(x, y, z)}{\partial z^2}e^{-j\alpha_3 z}dz = -\alpha_3^2\,\psi(x, y, \alpha_3) \quad (6.1.6)$$

provided that the integrated part in (6.1.5) vanishes. The vanishing of the integrated term is a radiation condition, which as we will see it must also be satisfied by the x and y variables. This integrated part vanishes if ψ and $\partial\psi/\partial z$ vanish at $z = \infty$ and $-\infty$. The reason is that physical sources do not produce a field at infinity.

The use of the Fourier transform with respect to z has changed the original equation to an equation with one less independent variable. That is, substituting (6.1.6) into (6.1.4) gives

$$\left[\frac{\partial^2}{\partial x^2} + \frac{\partial^2}{\partial y^2} + k^2 - \alpha_3^2\right]\psi(x, y, \alpha_3) = -f(x, y, \alpha_3) \tag{6.1.7}$$

Continuing with (6.1.7), the y and x variables are Fourier transformed by first multiplying (6.1.7) by $e^{-j\alpha_2 y}$ and integrating from $-\infty$ to ∞ (as was done to the z variable in (6.1.4)) and then repeating the procedure with $e^{-j\alpha_1 x}$. Thus, we obtain

$$\int_{-\infty}^{\infty} \frac{\partial^2\psi(x, y, \alpha_3)}{\partial y^2} e^{-j\alpha_2 y} dy = -\alpha_2^2\, \psi(x, \alpha_2, \alpha_3)$$

and

$$\int_{-\infty}^{\infty} \frac{\partial^2\psi(x, \alpha_2, \alpha_3)}{\partial x^2} e^{-j\alpha_1 x} dx = -\alpha_1^2\, \psi(\alpha_1, \alpha_2, \alpha_3)$$

where the resulting integrated terms (as in (6.1.5)) must vanish. Hence, the radiation condition can be expressed in the general form

$$e^{-j\alpha_n X_n}\left(\frac{\partial\psi}{\partial X_n} + j\alpha_n\psi\right)\Bigg|_{-\infty}^{\infty} = 0 \quad \text{for} \quad n = 1, 2, 3 \tag{6.1.8}$$

where the notation X_n represents $X_1 = x$, $X_2 = y$ and $X_3 = z$.

Equation (6.1.7) is then written as

$$(k^2 - \alpha_1^2 - \alpha_2^2 - \alpha_3^2)\psi(\alpha_1, \alpha_2, \alpha_3) = -f(\alpha_1, \alpha_2, \alpha_3) \Rightarrow$$

$$\psi(\alpha_1, \alpha_2, \alpha_3) = \frac{f(\alpha_1, \alpha_2, \alpha_3)}{\alpha^2 - k^2} \tag{6.1.9}$$

where

$$\alpha^2 = \alpha_1^2 + \alpha_2^2 + \alpha_3^2$$

Finally, inverting (6.1.9) gives

$$\psi(x, y, z) = \frac{1}{(2\pi)^3} \int_{-\infty}^{\infty} \int_{-\infty}^{\infty} \int_{-\infty}^{\infty} \frac{f(\alpha_1, \alpha_2, \alpha_3)}{\alpha^2 - k^2} e^{j(\alpha_1 x + \alpha_2 y + \alpha_3 z)} d\alpha_1 d\alpha_2 d\alpha_3 \quad (6.1.10)$$

For a given problem, the success of finding $\psi(x, y, z)$ rests in the ability of solving (6.1.10).

If the excitation is a delta function then (6.1.3) represents the Green function, or

$$[\nabla^2 + k^2] G(\mathbf{r}, \mathbf{r}') = -\delta(x - x')\delta(x - y')\delta(x - z')$$

Since the Fourier transform of the delta function excitation is

$$\begin{aligned} f(\alpha_1, \alpha_2 \alpha_3) &= \int_{-\infty}^{\infty} \int_{-\infty}^{\infty} \int_{-\infty}^{\infty} \delta(x - x')\delta(x - y')\delta(x - z') e^{-j(\alpha_1 x + \alpha_2 y + \alpha_3 z)} dxdydz \\ &= e^{-j(\alpha_1 x' + \alpha_2 y' + \alpha_3 z')} \end{aligned}$$

Then, from (6.1.10):

$$G(\mathbf{r}, \mathbf{r}') = \frac{1}{(2\pi)^3} \int_{-\infty}^{\infty} \int_{-\infty}^{\infty} \int_{-\infty}^{\infty} \frac{e^{j[\alpha_1(x - x') + \alpha_2(y - y') + \alpha_3(z - z')]}}{\alpha^2 - k^2} d\alpha_1 d\alpha_2 d\alpha_3 \quad (6.1.11)$$

6.1.1 One-Dimensional Green Function

In a one-dimension unbounded region, say x, the Green function satisfies

$$\left[\frac{d^2}{dx^2} + k^2\right] G(x, x') = -\delta(x - x') \quad (6.1.12)$$

Using the Fourier transform to transform the x variable gives

$$(k^2 - \alpha_1^2) G(\alpha_1, x') = -e^{-j\alpha_1 x'} \quad \Rightarrow \quad G(\alpha_1, x') = \frac{e^{-j\alpha_1 x'}}{\alpha_1^2 - k^2} \quad (6.1.13)$$

provided that the radiation condition in (6.1.8) for the x variable is satisfied.

The inverse transform of (6.1.13) is

$$G(x, x') = \frac{1}{2\pi} \int_{-\infty}^{\infty} \frac{e^{j\alpha_1(x - x')}}{\alpha_1^2 - k^2} d\alpha_1 \quad (6.1.14)$$

The evaluation of the integral in (6.1.14) was done in (5.5.7). Thus, we obtain

$$G(x, x') = \frac{-j}{2k} e^{-jk|x - x'|} \quad (6.1.15)$$

which satisfies the radiation condition (see (6.1.31)).

Substituting (6.1.14) into (6.1.12) shows that the delta function is represented by

$$\delta(x - x') = \frac{1}{2\pi}\int_{-\infty}^{\infty} e^{j\alpha_1(x-x')}d\alpha_1 \tag{6.1.16}$$

an expression that can also be derived using other methods.

If the delta function representation in (6.1.16) is known a priori, one can let the Green function in (6.1.12) be in a form similar that of the delta function. That is, in (6.1.12), let

$$G(x, x') = \frac{1}{2\pi}\int_{-\infty}^{\infty} f(\alpha_1)e^{j\alpha_1(x-x')}d\alpha_1 \tag{6.1.17}$$

where $f(\alpha_1)$ is to be determined. Substituting (6.1.16) and (6.1.17) into (6.1.12) gives

$$\frac{1}{2\pi}\int_{-\infty}^{\infty}(-\alpha_1^2 + k^2)f(\alpha_1)e^{j\alpha_1(x-x')}d\alpha_1 = -\frac{1}{2\pi}\int_{-\infty}^{\infty} e^{j\alpha_1(x-x')}d\alpha_1$$

to obtain

$$f(\alpha_1) = \frac{1}{\alpha_1^2 - k^2}$$

Then, (6.1.17) reads

$$G(x, x') = \frac{1}{2\pi}\int_{-\infty}^{\infty}\frac{e^{j\alpha_1(x-x')}}{\alpha_1^2 - k^2}d\alpha_1$$

which is identical to (6.1.14).

The solution to (6.1.12) can also be obtained using the direct method. That is,

$$G_< = Be^{jkx} \quad (x < x')$$

and

$$G_> = Ce^{-jkx} \quad (x > x')$$

The boundary conditions at $x = x'$ are

$$G_< = G_>|_{x=x'} \quad \Rightarrow \quad Be^{jkx'} = Ce^{-jkx'}$$

and

$$\left.\frac{\partial G_>}{\partial x} - \frac{\partial G_<}{\partial x}\right|_{x=x'} = -1 \Rightarrow Ce^{-jkx'} + Be^{jkx'} = \frac{1}{jk}$$

Solving for the constants B and C gives

$$B = \frac{-j}{2k}e^{-jkx'}$$

and

$$C = \frac{-j}{2k}e^{jkx'}$$

Therefore,

$$G_< = \frac{-j}{2k}e^{jk(x-x')} \quad (x \leq x')$$

and for $G_>$ interchange x and x' in the above relation. Therefore, the Green function can be expressed in the form shown in (6.1.15). That is,

$$G(x, x') = \frac{-j}{2k}e^{-jk|x-x'|}$$

The expression for the Green function can also appear in terms of $\gamma = \alpha + j\beta = jk$. In terms of γ, the solution is

$$G(x, x') = \frac{1}{2\gamma}e^{-\gamma|x-x'|} \qquad (6.1.18)$$

6.1.2 Two-Dimensional Green Function

A two-dimensional unbounded region is now considered. In two dimensions, say x and y, the Green function satisfies

$$\left[\frac{\partial^2}{\partial x^2} + \frac{\partial^2}{\partial y^2} + k^2\right]G(x, y, x', y') = -\delta(x - x')\delta(x - y') \qquad (6.1.19)$$

A representation of the Green function in two-dimensional rectangular coordinates is obtained by first Fourier transforming one of the variables. Fourier transforming the x variable gives

$$\left[\frac{d^2}{dy^2} + k^2 - \alpha_1^2\right] G(\alpha_1, y, x', y') = -e^{-j\alpha_1 x'}\delta(y - y')$$

or

$$\left[\frac{d^2}{dy^2} + \tau^2\right] \frac{G(\alpha_1, y, x', y')}{e^{-j\alpha_1 x'}} = -\delta(y - y') \qquad (6.1.20)$$

where τ is defined as

$$\tau^2 = k^2 - \alpha_1^2$$

The solution of (6.1.20) is similar to the one obtained in (5.5.7) or (6.1.15). Therefore, it follows that

$$G(\alpha_1, y, x', y') = \frac{-j}{2\tau} e^{-j\alpha_1 x'} e^{-j\tau|y-y'|}$$

where $\mathrm{Im}(\tau) < 0$ for the radiation condition to be satisfied as $y \to \pm\infty$. The inverse Fourier transform is

$$G(x, y, x', y') = \frac{-j}{4\pi} \int_{-\infty}^{\infty} \frac{e^{j\alpha_1(x-x')} e^{-j\tau|y-y'|}}{\tau} d\alpha_1$$

which from (2.4.8) or (5.6.43) this integral form of the Green function is a representation of the Hankel function in terms of a spectrum of plane waves. That is,

$$G(\boldsymbol{\rho}, \boldsymbol{\rho}') = G(x, y, x', y') = -\frac{j}{4} H_0^{(2)}(k|\boldsymbol{\rho} - \boldsymbol{\rho}'|) = \frac{-j}{4\pi} \int_{-\infty}^{\infty} \frac{e^{j\alpha_1(x-x')} e^{-j\tau|y-y'|}}{\tau} d\alpha_1$$

The solution to (6.1.19) also follows from (6.1.10), where in two dimensions reads

$$G(x, y, x', y') = \frac{1}{(2\pi)^2} \int_{-\infty}^{\infty} \int_{-\infty}^{\infty} \frac{e^{j[\alpha_1(x-x') + \alpha_2(y-y')]}}{\alpha^2 - k^2} d\alpha_1 d\alpha_2 \qquad (6.1.21)$$

where

$$\alpha^2 = \alpha_1^2 + \alpha_2^2$$

Substituting (6.1.21) into (6.1.19) shows that the delta function representation in two dimensions is

$$\delta(x - x')\delta(y - y') = \frac{1}{(2\pi)^2} \int_{-\infty}^{\infty} \int_{-\infty}^{\infty} e^{j\alpha_1(x-x')} e^{j\alpha_2(y-y')} d\alpha_1 d\alpha_2$$

Equation (6.1.21) can also be expressed in terms of a Hankel function. To this end (6.1.21), is evaluated by first changing the variables to polar coordinates (see Fig. 6.1). The orientation of the transform space is arbitrary, so it is selected to coincide with the origin at $|\boldsymbol{\rho} - \boldsymbol{\rho}'| = 0$. That is, let

$$x - x' = |\boldsymbol{\rho} - \boldsymbol{\rho}'| \cos \varphi$$
$$y - y' = |\boldsymbol{\rho} - \boldsymbol{\rho}'| \sin \varphi$$

and

$$\alpha_1 = \lambda \cos \gamma$$
$$\alpha_2 = \lambda \sin \gamma$$

Then,

$$\alpha_1(x - x') + \alpha_2(y - y') = \lambda|\boldsymbol{\rho} - \boldsymbol{\rho}'| \cos(\varphi - \gamma)$$
$$\lambda^2 = \alpha^2 = \alpha_1^2 + \alpha_2^2$$
$$d\alpha_1 d\alpha_2 = \lambda d\lambda d\gamma$$

and (6.1.21), writing $G(\boldsymbol{\rho}, \boldsymbol{\rho}')$ for $G(x, y, x', y')$, reads

$$G(\boldsymbol{\rho}, \boldsymbol{\rho}') = \frac{1}{(2\pi)^2} \int_{\lambda=0}^{\infty} \int_{\gamma=0}^{2\pi} \frac{e^{j\lambda|\boldsymbol{\rho}-\boldsymbol{\rho}'| \cos(\varphi-\gamma)}}{\lambda^2 - k^2} \lambda d\lambda d\gamma$$

Using the identity (see (4.5.3)):

$$e^{j\lambda|\boldsymbol{\rho}-\boldsymbol{\rho}'| \cos(\varphi-\gamma)} = \sum_{n=-\infty}^{\infty} j^n J_n(\lambda|\boldsymbol{\rho} - \boldsymbol{\rho}'|) e^{-jn(\varphi-\gamma)}$$

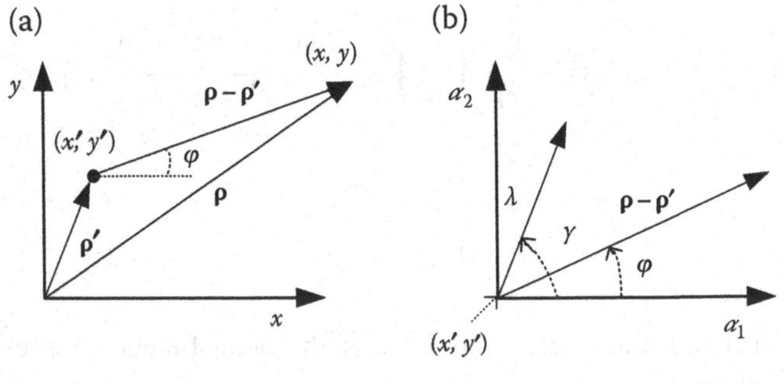

FIGURE 6.1 Variables in the α plane.

the Green function reads

$$G(\boldsymbol{\rho},\boldsymbol{\rho}')=\frac{1}{(2\pi)^2}\sum_{n=-\infty}^{\infty}\int_{\lambda=0}^{\infty}\left[\int_{\gamma=0}^{2\pi}e^{-jn(\varphi-\gamma)}d\gamma\right]\frac{j^nJ_n(\lambda|\boldsymbol{\rho}-\boldsymbol{\rho}'|)}{\lambda^2-k^2}\lambda d\lambda$$

The angular integral is $2\pi\delta_{n0}$. Therefore,

$$G(\boldsymbol{\rho},\boldsymbol{\rho}')=\frac{1}{2\pi}\int_0^{\infty}\frac{J_0(\lambda|\boldsymbol{\rho}-\boldsymbol{\rho}'|)}{\lambda^2-k^2}\lambda d\lambda \tag{6.1.22}$$

Equation (6.1.22) is evaluated as follows. Write the integral in terms of Hankel functions as

$$\frac{1}{2\pi}\int_0^{\infty}\frac{J_0(\lambda|\boldsymbol{\rho}-\boldsymbol{\rho}'|)}{\lambda^2-k^2}\lambda d\lambda=\frac{1}{4\pi}\int_0^{\infty}\frac{\left[H_0^{(1)}(\lambda|\boldsymbol{\rho}-\boldsymbol{\rho}'|)+H_0^{(2)}(\lambda|\boldsymbol{\rho}-\boldsymbol{\rho}'|)\right]}{\lambda^2-k^2}\lambda d\lambda$$

To use residue theory it is convenient if the limits of integration are from $-\infty$ to ∞, letting $\lambda=-\lambda$ in the integral with $H_0^{(1)}(\lambda|\boldsymbol{\rho}-\boldsymbol{\rho}'|)$ and using the relation

$$H_0^{(1)}(-x)=-H_0^{(2)}(x)$$

it follows that this integral is the same as the integral involving $H_0^{(2)}(\lambda|\boldsymbol{\rho}-\boldsymbol{\rho}'|)$ but with limits from $-\infty$ to 0. Therefore, combining the two integrals shows that

$$G(\boldsymbol{\rho},\boldsymbol{\rho}')=\frac{1}{2\pi}\int_0^{\infty}\frac{J_0(\lambda|\boldsymbol{\rho}-\boldsymbol{\rho}'|)}{\lambda^2-k^2}\lambda d\lambda=\frac{1}{4\pi}\int_{-\infty}^{\infty}\frac{H_0^{(2)}(\lambda|\boldsymbol{\rho}-\boldsymbol{\rho}'|)}{\lambda^2-k^2}\lambda d\lambda$$

The above integral is evaluated using residue theory. The path of integration is shown in Fig. 6.2. The poles are located at $\lambda=-k=-k'+j\varepsilon$ and $\lambda=k=k'-j\varepsilon$. A small loss in k'', denoted by $k''=\varepsilon$, was introduced to move the poles slightly away from the path of integration. A branch cut is also needed due to the Hankel function singularity.

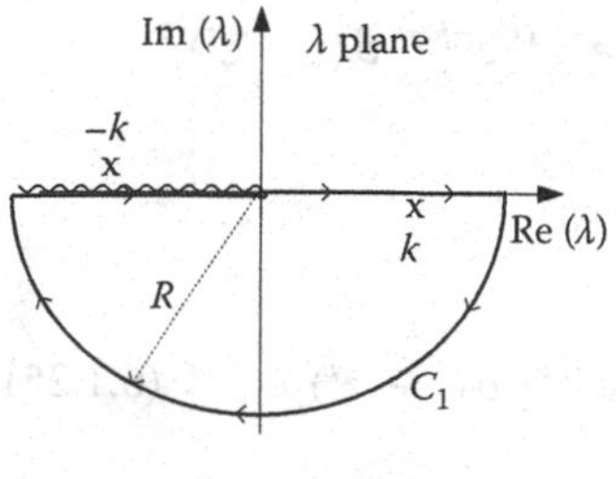

FIGURE 6.2 Path of integration in the λ plane with poles at $\lambda=-k=-k'+j\varepsilon$ and $\lambda=k=k'-j\varepsilon$.

The asymptotic expansion of the Hankel function shows that the integral along the semicircular path C_1 vanishes as $R \to \infty$. The contribution from the small semicircle around the origin is also zero. Therefore,

$$\frac{1}{4\pi}\int_{-\infty}^{\infty}\frac{H_0^{(2)}(\lambda|\boldsymbol{\rho}-\boldsymbol{\rho}'|)}{\lambda^2-k^2}\lambda d\lambda = -2\pi j\,[\text{Residue at } \lambda = k]$$

$$= -2\pi j\frac{1}{4\pi}\frac{H_0^{(2)}(\lambda|\boldsymbol{\rho}-\boldsymbol{\rho}'|)}{(\lambda+k)(\lambda-k)}\lambda(\lambda-k)|_{\lambda=k} = \frac{-j}{4}H_0^{(2)}(k|\boldsymbol{\rho}-\boldsymbol{\rho}'|)$$

The evaluation of (6.1.21) has shown that

$$G(\boldsymbol{\rho},\boldsymbol{\rho}') = \frac{-j}{4}H_0^{(2)}(k|\boldsymbol{\rho}-\boldsymbol{\rho}'|) = \frac{1}{(2\pi)^2}\int_{-\infty}^{\infty}\int_{-\infty}^{\infty}\frac{e^{j[\alpha_1(x-x')+\alpha_2(y-y')]}}{\alpha^2-k^2}d\alpha_1 d\alpha_2 \quad (6.1.23)$$

which is the expected result, and provides another representation of cylindrical waves in terms of a spectrum of plane waves.

The asymptotic expansion of the Hankel function in (6.1.23) as $\rho \to \infty$ shows that it represents an outgoing wave that satisfies the radiation condition. If k has a small loss (i.e., $k = k' - j\varepsilon$), the wave decreases exponentially, since

$$H_0^{(2)}(k|\boldsymbol{\rho}-\boldsymbol{\rho}'|) \approx \sqrt{\frac{2}{\pi k|\boldsymbol{\rho}-\boldsymbol{\rho}'|}}\,e^{-j(k|\boldsymbol{\rho}-\boldsymbol{\rho}'|-\pi/4)} \to 0 \quad (\rho\to\infty)$$

6.1.3 Three-Dimensional Green Function

The representation of the Green function in three dimensions in rectangular coordinates is obtained if we first Fourier transform two of the rectangular coordinates as follows. Starting with

$$\left[\frac{\partial^2}{\partial x^2}+\frac{\partial^2}{\partial y^2}+\frac{\partial^2}{\partial z^2}+k^2\right]G(\mathbf{r},\mathbf{r}') = -\delta(x-x')\delta(y-y')\delta(z-z') \quad (6.1.24)$$

and applying the Fourier transform to the x and y variables, gives

$$\left[\frac{d^2}{dz^2}+\beta^2-\alpha_1^2-\alpha_2^2\right]G(\alpha_1,\alpha_2,z,\mathbf{r}') = -e^{-j\alpha_1 x'}e^{-j\alpha_2 y'}\delta(z-z')$$

or

$$\left[\frac{d^2}{dz^2}+\tau^2\right]G(\alpha_1,\alpha_2,z,\mathbf{r}') = -e^{-j\alpha_1 x'}e^{-j\alpha_2 y'}\delta(z-z') \quad (6.1.25)$$

where

$$\tau^2 = k^2 - \alpha_1^2 - \alpha_2^2$$

The Fourier transform was used on the x and y variables but depending on the problem any two variables can be chosen.

The solution of (6.1.25) uses the same method employed to obtain the solution in (6.1.15). Hence, for (6.1.25) the solution is

$$G(\alpha_1, \alpha_2, z, \mathbf{r}') = \frac{-j}{2\tau} e^{-j\alpha_1 x'} e^{-j\alpha_2 y'} e^{-j\tau|z-z'|}$$

and $\text{Im}(\tau) < 0$. The inverse Fourier transform is a three-dimensional representation of the Green function, namely

$$G(\mathbf{r}, \mathbf{r}') = \frac{e^{-jk|\mathbf{r}-\mathbf{r}'|}}{4\pi|\mathbf{r} - \mathbf{r}'|} = \frac{-j}{8\pi^2} \int_{-\infty}^{\infty} \int_{-\infty}^{\infty} \frac{e^{j\alpha_1(x-x')} e^{j\alpha_2(y-y')} e^{-j\tau|z-z'|}}{\tau} d\alpha_1 d\alpha_2 \quad (6.1.26)$$

which represents the free space Green function as a spectrum of plane waves.

The solution to (6.1.24) also follows from (6.1.11), where in three dimensions can be expressed in the form

$$G(\mathbf{r}, \mathbf{r}') = \frac{1}{(2\pi)^3} \int_{-\infty}^{\infty} \int_{-\infty}^{\infty} \int_{-\infty}^{\infty} \frac{e^{j\boldsymbol{\alpha}\cdot(\mathbf{r},\mathbf{r}')}}{\alpha^2 - k^2} d\alpha_1 d\alpha_2 d\alpha_3 \quad (6.1.27)$$

where

$$\boldsymbol{\alpha} = \alpha_1 \hat{\mathbf{a}}_x + \alpha_2 \hat{\mathbf{a}}_y + \alpha_3 \hat{\mathbf{a}}_z$$

$$\mathbf{R} = \mathbf{r} - \mathbf{r}' = (x - x')\hat{\mathbf{a}}_x + (y - y')\hat{\mathbf{a}}_y + (z - z')\hat{\mathbf{a}}_z$$

The vector $\mathbf{R}$ is shown in Fig. 6.3a. The integral in (6.1.27) is in the transform-plane coordinates α_1, α_2, α_3. Since the orientation of the transform plane coordinates is arbitrary, we can select the α_3 axis to coincide with the vector $\mathbf{R}$, as shown in Fig. 6.3b.

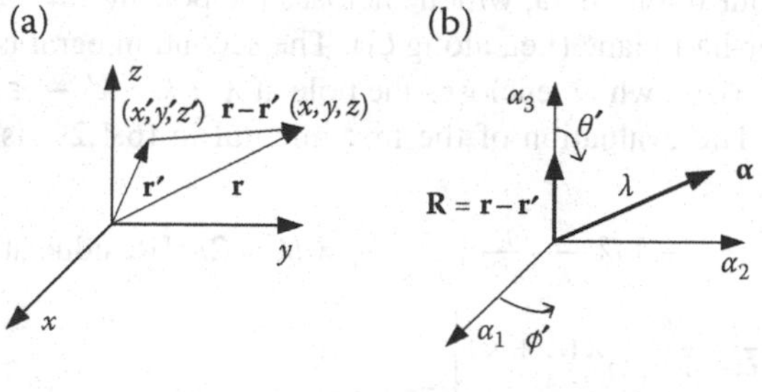

FIGURE 6.3 (a) The vector $\mathbf{R} = \mathbf{r} - \mathbf{r}'$; (b) the vector $\mathbf{R}$ in the α plane.

The integral is conveniently evaluated in spherical coordinates. The spherical variables in the α_1, α_2, α_3 plane are denoted by λ, θ', and ϕ'. Therefore, with

$$\begin{aligned}
\alpha_1 &= \lambda \sin\theta' \cos\phi' \\
\alpha_2 &= \lambda \sin\theta' \sin\phi' \\
\alpha_3 &= \lambda \cos\theta' \\
\lambda^2 &= \alpha^2 = \alpha_1^2 + \alpha_2^2 + \alpha_3^2 \\
d\alpha_1 d\alpha_2 d\alpha_3 &= \lambda^2 \sin\theta' d\lambda d\theta' d\phi'
\end{aligned}$$

it follows that

$$\boldsymbol{\alpha} \bullet \mathbf{R} = \lambda R \cos\theta'$$

where $R = |\mathbf{r} - \mathbf{r}'|$. Then, the integral in (6.1.27) becomes

$$G(\mathbf{r}, \mathbf{r}') = \frac{1}{(2\pi)^3} \int_{\lambda=0}^{\infty} \int_{\theta'=0}^{\pi} \int_{\phi'=0}^{2\pi} \frac{e^{j\lambda R \cos\theta'}}{\lambda^2 - k^2} \lambda^2 \sin\theta' d\lambda\, d\theta' d\phi' \tag{6.1.28}$$

The integration in ϕ' is 2π and

$$\int_0^{\pi} e^{j\lambda R \cos\theta'} \sin\theta'\, d\theta' = \frac{2}{\lambda R} \sin\lambda R$$

Hence, (6.1.28) reads

$$\begin{aligned}
G(\mathbf{r}, \mathbf{r}') &= \frac{1}{2\pi^2 R} \int_0^{\infty} \frac{\sin(\lambda R)}{\lambda^2 - k^2} \lambda d\lambda = \frac{1}{4\pi^2 R} \int_{-\infty}^{\infty} \frac{\sin(\lambda R)}{\lambda^2 - k^2} \lambda d\lambda \\
&= -\frac{j}{8\pi^2 R} \int_{-\infty}^{\infty} \frac{e^{j\lambda R}}{\lambda^2 - k^2} \lambda d\lambda - \frac{j}{8\pi^2 R} \int_{-\infty}^{\infty} \frac{e^{-j\lambda R}}{\lambda^2 - k^2} \lambda d\lambda
\end{aligned} \tag{6.1.29}$$

The first integral in (6.1.29) has poles in the λ plane at $\lambda = -k = -k' + j\varepsilon$ and $\lambda = k = k' - j\varepsilon$. A small imaginary part, denoted by $k'' = \varepsilon$, was introduced to move the poles slightly away from the path of integration. This integral is evaluated by using the contour in Fig. 6.4a, which encloses the pole at $\lambda = -k = -k' + j\varepsilon$ and closes in the upper-half plane (i.e., along C_1). The second integral is evaluated using the contour in Fig. 6.4b, which encloses the pole at $\lambda = k = k' - j\varepsilon$ and closes in the lower-half plane. The evaluation of the first integral in (6.1.29) is

$$\begin{aligned}
&-\frac{j}{8\pi^2 R} \int_{-\infty}^{\infty} \frac{e^{j\lambda R}}{\lambda^2 - k^2} \lambda d\lambda - \frac{j}{8\pi^2 R} \int_{C_1} \frac{e^{j\lambda R}}{\lambda^2 - k^2} \lambda d\lambda = 2\pi j\,[\text{Residue at } \lambda = -k] \\
&= 2\pi j \frac{-j}{8\pi^2 R} \frac{e^{j\lambda R}}{(\lambda + k)(\lambda - k)} \lambda(\lambda + k) \bigg|_{\lambda=-k} = \frac{e^{-jkR}}{8\pi R}
\end{aligned}$$

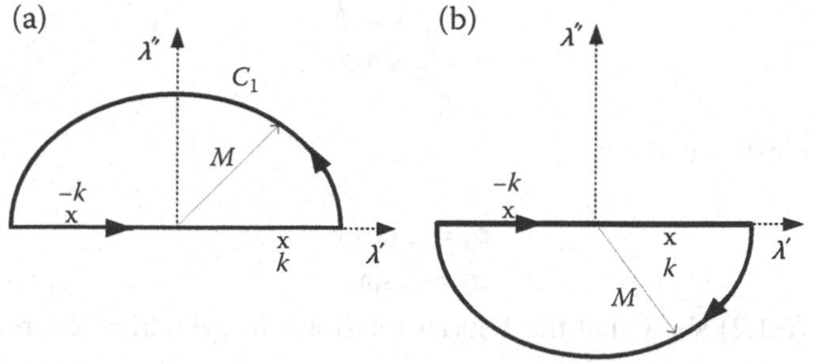

FIGURE 6.4 (a) Contour of integration that encloses the pole at $\lambda = -k = -k' + j\varepsilon$; (b) contour of integration that encloses the pole at $\lambda = k = k' - j\varepsilon$.

The integral along C_1 vanishes since with $\lambda = Me^{j\theta}$ it follows that

$$-\frac{j}{8\pi^2 R}\int_{C_1}\frac{e^{j\lambda R}}{\lambda^2 - k^2}\lambda d\lambda \approx \lim_{M\to\infty}\int_0^{\pi}\frac{e^{jMe^{j\theta}R}}{(Me^{j\theta})^2}M^2 e^{j2\theta}d\theta$$

The magnitude of the exponent varies as $e^{-MR\sin\theta}$ where $\sin\theta$ is positive between 0 and π. Hence, as $M \to \infty$, the integral along C_1 vanishes.

The evaluation of the second integral in (6.1.29) using the contour in Fig. 6.4b is similar and gives $e^{-jkR}/8\pi R$. Therefore,

$$G(\mathbf{r}, \mathbf{r}') = \frac{e^{-jkR}}{8\pi R} + \frac{e^{-jkR}}{8\pi R} = \frac{e^{-jk|\mathbf{r}-\mathbf{r}'|}}{4\pi|\mathbf{r}-\mathbf{r}'|} \tag{6.1.30}$$

which is the expected result.

The Green function in (6.1.30) represents an outgoing wave that must satisfy the radiation condition. Consider first the radiation condition as $z \to \infty$. Since α_3 is real and $\partial R/\partial z = z/R$ the radiation condition (see (6.1.8)) reads

$$e^{-j\alpha_3 z}\left(\frac{\partial G}{\partial z} + j\alpha_3 G\right)\Bigg|_{z=-\infty}^{z=\infty} = -e^{-j\alpha_3 z}\frac{e^{-jkR}}{4\pi R}\left[\frac{z}{R}\left(jk + \frac{1}{R}\right) - j\alpha_3\right]\Bigg|_{z=-\infty}^{z=\infty} = 0 \tag{6.1.31}$$

which goes to zero as $z \to \pm\infty$ (i.e., $R \to \infty$). Also, if k has a small imaginary part (i.e., $k \to k' - j\varepsilon$) the exponential decays as $R \to \infty$. The same considerations apply to the y and x components, showing that (6.1.30) satisfies the radiation condition.

6.2 INTEGRAL TRANSFORM FOR THE WAVE EQUATION IN CYLINDRICAL COORDINATES

When the problem has cylindrical symmetry, it is convenient to write (6.1.1) and (6.1.2) in terms of cylindrical variables. Letting,

$$x = \rho \cos \phi$$
$$y = \rho \sin \phi$$

and in the transform space:

$$\alpha_1 = \lambda \cos \gamma$$
$$\alpha_2 = \lambda \sin \gamma$$

(6.1.1) and (6.1.2) show that the Fourier transform in cylindrical coordinates is

$$f(\lambda, \gamma, \alpha_3) = \int_{z=-\infty}^{\infty} \int_{\phi=0}^{2\pi} \int_{\rho=0}^{\infty} f(\rho, \phi, z)\, e^{-j\lambda\rho \cos(\phi-\gamma)} e^{-j\alpha_3 z} \rho \, d\rho \, d\phi \, dz \quad (6.2.1)$$

and the inverse Fourier transform is

$$f(\rho, \phi, z) = \frac{1}{(2\pi)^3} \int_{\alpha_3=-\infty}^{\infty} \int_{\gamma=0}^{2\pi} \int_{\lambda=0}^{\infty} f(\lambda, \gamma, \alpha_3)\, e^{j\lambda\rho \cos(\phi-\gamma)} e^{j\alpha_3 z} \lambda d\lambda d\gamma d\alpha_3 \quad (6.2.2)$$

6.2.1 The Hankel Transform

In polar coordinates, (6.2.1) reads

$$f(\lambda, \gamma) = \int_{\phi=0}^{2\pi} \int_{\rho=0}^{\infty} f(\rho, \phi)\, e^{-j\lambda\rho \cos(\phi-\gamma)} \rho d\rho d\phi \quad (6.2.3)$$

and the inverse transform in (6.2.2) reads

$$f(\rho, \phi) = \frac{1}{(2\pi)^2} \int_{\gamma=0}^{2\pi} \int_{\lambda=0}^{\infty} f(\lambda, \gamma) e^{j\lambda\rho \cos(\phi-\gamma)} \lambda d\lambda d\gamma \quad (6.2.4)$$

Next, assume that $f(\rho, \phi)$ in (6.2.3) can be expressed in the form

$$f(\rho, \phi) = \sum_{n=-\infty}^{\infty} f_n(\rho)\, e^{jn\phi} \quad (6.2.5)$$

Then, (6.2.3) reads

$$f(\lambda, \gamma) = \sum_{n=-\infty}^{\infty} \int_{\rho=0}^{\infty} f_n(\rho) \left[\int_{\phi=0}^{2\pi} e^{-j\lambda\rho \cos(\phi-\gamma)} e^{jn\phi} d\phi \right] \rho d\rho$$

Using the Bessel function relation (see Appendix C):

$$J_n(\lambda\rho) = \frac{e^{-jn(\gamma-\pi/2)}}{2\pi} \int_0^{2\pi} e^{-j[\lambda\rho\cos(\phi-\gamma)]} e^{jn\phi} d\phi$$

results in

$$f(\lambda, \gamma) = 2\pi \sum_{n=-\infty}^{\infty} e^{jn(\gamma-\pi/2)} \int_0^{\infty} f_n(\rho)\, J_n(\lambda\rho)\rho d\rho \tag{6.2.6}$$

From (6.2.6), the n-order Hankel transform (or Fourier-Bessel transform) of $f_n(\rho)$ is defined as

$$f_n(\lambda) = \int_0^{\infty} f_n(\rho) J_n(\lambda\rho)\rho d\rho \tag{6.2.7}$$

and (6.2.6) can be expressed as

$$f(\lambda, \gamma) = 2\pi \sum_{n=-\infty}^{\infty} f_n(\lambda) e^{jn(\gamma-\pi/2)} \tag{6.2.8}$$

Substituting (6.2.5) and (6.2.8) into the inversion formula in (6.2.4) gives

$$\begin{aligned}\sum_{n=-\infty}^{\infty} f_n(\rho) e^{jn\phi} &= \frac{1}{2\pi} \sum_{n=-\infty}^{\infty} \int_0^{\infty} f_n(\lambda) \left[e^{-jn\pi/2} \int_0^{2\pi} e^{j\lambda\rho\cos(\phi-\gamma)} e^{jn\gamma} d\gamma \right] \lambda d\lambda \\ &= \sum_{n=-\infty}^{\infty} e^{jn\phi} \int_0^{\infty} f_n(\lambda) J_n(\lambda\rho)\lambda d\lambda\end{aligned}$$

The integral term is the inverse Hankel transform (or inverse Fourier-Bessel transform). That is

$$f_n(\rho) = \int_0^{\infty} f_n(\lambda) J_n(\lambda\rho)\lambda d\lambda \tag{6.2.9}$$

Although the transform was derived for integer values of n, it can be extended to include non-integer values of n.

Substituting (6.2.7) into (6.2.9) gives

$$\begin{aligned}f_n(\rho) &= \int_0^{\infty} \left[\int_0^{\infty} f_n(\rho') J_n(\lambda\rho')\rho' d\rho' \right] J_n(\lambda\rho)\lambda d\lambda \\ &= \int_0^{\infty} f_n(\rho') \left[\int_0^{\infty} \rho' J_n(\lambda\rho') J_n(\lambda\rho)\lambda d\lambda \right] d\rho'\end{aligned}$$

The term in brackets is recognized as a delta function. That is

$$\frac{\delta(\rho-\rho')}{\rho'}=\int_0^\infty J_n(\lambda\rho)J_n(\lambda\rho')\lambda d\lambda$$

which agrees with (5.5.14).

The Hankel transform is now used to solve the wave equation in cylindrical coordinates, namely

$$\left[\frac{1}{\rho}\frac{\partial}{\partial\rho}\left(\rho\frac{\partial}{\partial\rho}\right)+\frac{1}{\rho^2}\frac{\partial^2}{\partial\phi^2}+k^2\right]\psi(\rho,\phi)=-f(\rho,\phi) \tag{6.2.10}$$

Assume that $\psi(\rho,\phi)$ and $f(\rho,\phi)$ can be expanded in the forms

$$f(\rho,\phi)=\sum_{n=-\infty}^{\infty} f_n(\rho)e^{jn\phi}$$

and

$$\psi(\rho,\phi)=\sum_{n=-\infty}^{\infty} \psi_n(\rho)e^{jn\phi}$$

Substituting into (6.2.10) gives

$$\left[\frac{1}{\rho}\frac{d}{d\rho}\left(\rho\frac{d}{d\rho}\right)-\frac{n^2}{\rho^2}+k^2\right]\psi_n(\rho)=-f_n(\rho) \tag{6.2.11}$$

The Hankel transform can now be applied to (6.2.11). Multiplying (6.2.11) by $\rho J_n(\lambda\rho)$ and integrating from $\rho=0$ to ∞ shows that

$$\int_0^\infty\left[\frac{1}{\rho}\frac{d}{d\rho}\left(\rho\frac{d\psi_n(\rho)}{\partial\rho}\right)-\frac{n^2}{\rho^2}\psi_n(\rho)\right]\rho J_n(\lambda\rho)d\rho+k^2\psi_n(\lambda)=-f_n(\lambda) \tag{6.2.12}$$

where $\psi_n(\lambda)$ and $f_n(\lambda)$ are the Hankel transform of $\psi_n(\rho)$ and $f_n(\rho)$, respectively. Integrating the first term in (6.2.12) by parts twice shows that

$$\int_0^\infty \frac{d}{d\rho}\left(\rho\frac{d\psi_n(\rho)}{d\rho}\right)J_n(\lambda\rho)d\rho=\rho\left[J_n(\lambda\rho)\frac{d\psi_n(\rho)}{d\rho}-\frac{dJ_n(\lambda\rho)}{d\rho}\psi_n(\rho)\right]\Bigg|_0^\infty$$
$$+\int_0^\infty\psi_n(\rho)\frac{d}{d\rho}\left(\rho\frac{dJ_n(\lambda\rho)}{d\rho}\right)d\rho \tag{6.2.13}$$

If the integrated term vanishes, which represents the radiation condition for cylindrical waves, then (6.2.12) reads

$$\int_0^\infty \left[\frac{1}{\rho}\frac{d}{d\rho}\left(\rho\frac{dJ_n(\lambda\rho)}{d\rho}\right) - \frac{n^2}{\rho^2}J_n(\lambda\rho)\right]\rho\psi_n(\rho)d\rho + k^2\psi_n(\lambda) = -f_n(\lambda) \qquad (6.2.14)$$

The term in brackets, from Bessel's differential equation, is equal to $-\lambda^2 J_n(\lambda\rho)$. Substituting into (6.2.14):

$$-\lambda^2\int_0^\infty \psi_n(\rho)\rho J_n(\lambda\rho)d\rho + k^2\psi_n(\lambda) = -f_n(\lambda)$$

or

$$-\lambda^2\psi_n(\lambda) + k^2\psi_n(\lambda) = -f_n(\lambda) \quad \Rightarrow \quad \psi_n(\lambda) = \frac{f_n(\lambda)}{\lambda^2 - k^2} \qquad (6.2.15)$$

Taking the inverse Hankel transform of (6.2.15):

$$\psi_n(\rho) = \int_0^\infty \frac{f_n(\lambda)}{\lambda^2 - k^2}J_n(\lambda\rho)\lambda d\lambda$$

Therefore, the solution to (6.2.10) is

$$\psi(\rho,\phi) = \sum_{n=-\infty}^{\infty}\left[\int_0^\infty \frac{f_n(\lambda)}{\lambda^2 - k^2}J_n(\lambda\rho)\lambda d\lambda\right]e^{jn\phi}$$

and using (6.2.7):

$$\psi(\rho,\phi) = \sum_{n=-\infty}^{\infty}\left\{\int_0^\infty f_n(\rho')\left[\int_0^\infty \frac{J_n(\lambda\rho)J_n(\lambda\rho')}{\lambda^2 - k^2}\lambda d\lambda\right]\rho' d\rho'\right\}e^{jn\phi} \qquad (6.2.16)$$

The term in brackets is the Green function associated with (6.2.11). This Green function is specifically derived and evaluated in the next section.

We now return to the integrated term in (6.2.13). The Bessel functions of the first kind and its derivative is finite at $\rho = 0$, so the integrated term is zero at the lower limit. At the upper limit, if $\psi_n(\rho)$ and $d\psi_n(\rho)/d\rho$ are zero, the integrated term is zero. A more specific condition follows if we use the asymptotic relations:

$$\lim_{\lambda\rho\to\infty} J_n(\lambda\rho) \approx \sqrt{\frac{2}{\pi\lambda\rho}}\cos\left(\lambda\rho - \frac{n\pi}{2} - \frac{\pi}{4}\right)$$

and

$$\lim_{\lambda\rho\to\infty} \frac{dJ_n(\lambda\rho)}{d\rho} \approx -\sqrt{\frac{2\lambda}{\pi\rho}} \sin\left(\lambda\rho - \frac{n\pi}{2} - \frac{\pi}{4}\right)$$

Expressing the sinusoidal terms in terms of exponentials, the integrated term in (6.2.13) reads

$$\sqrt{\frac{\rho}{2\pi\lambda}}\left[\frac{d\psi_n(\rho)}{d\rho} - j\lambda\psi_n(\rho)\right] e^{j(\lambda\rho - n\pi/2 - \pi/4)}$$
$$+ \sqrt{\frac{\rho}{2\pi\lambda}}\left[\frac{d\psi_n(\rho)}{d\rho} + j\lambda\psi_n(\rho)\right] e^{-j(\lambda\rho - n\pi/2 - \pi/4)}$$

Hence, it follows that the integrated term at the upper limit vanishes if

$$\lim_{\rho\to\infty} \sqrt{\frac{\rho}{2\pi\lambda}}\left[\frac{d\psi_n(\rho)}{d\rho} \pm j\lambda\psi_n(\rho)\right] = 0$$

The radiation condition is simply stated as

$$\lim_{\rho\to\infty} \sqrt{\rho}\left[\frac{d\psi_n(\rho)}{d\rho} + j\lambda\psi_n(\rho)\right] = 0 \tag{6.2.17}$$

6.2.2 Two-Dimensional Green Function

Next, the Green function associated with the wave equation in two-dimensions cylindrical coordinates is determined using the Hankel transform. The Green function satisfies

$$\left[\frac{1}{\rho}\frac{\partial}{\partial\rho}\left(\rho\frac{\partial}{\partial\rho}\right) + \frac{1}{\rho^2}\frac{\partial^2}{\partial\phi^2} + k^2\right] G(\rho, \phi, \rho', \phi') = -\frac{\delta(\rho-\rho')\delta(\phi-\phi')}{\rho} \tag{6.2.18}$$

Since,

$$\delta(\phi - \phi') = \frac{1}{2\pi}\sum_{n=-\infty}^{\infty} e^{jn(\phi-\phi')}$$

we let the Green function be of the form

$$G(\rho, \phi, \rho', \phi') = \frac{1}{2\pi}\sum_{n=-\infty}^{\infty} G_n(\rho, \rho')\, e^{jn(\phi-\phi')} \tag{6.2.19}$$

and from (6.2.18) the Green function $G_n(\rho, \rho')$ satisfies

$$\left[\frac{1}{\rho}\frac{\partial}{\partial \rho}\left(\rho\frac{\partial}{\partial \rho}\right) - \frac{n^2}{\rho^2} + k^2\right]G_n(\rho, \rho') = -\frac{\delta(\rho - \rho')}{\rho} \tag{6.2.20}$$

The Hankel transform can now be applied to (6.2.20). Multiplying (6.2.20) by $\rho J_n(\lambda\rho)$, using the Bessel differential equation, and integrating from $\rho = 0$ to ∞, gives

$$(-\lambda^2 + k^2)G_n(\lambda, \rho') = -J_n(\lambda\rho') \Rightarrow G_n(\lambda, \rho') = \frac{J_n(\lambda\rho')}{\lambda^2 - k^2}$$

Taking the inverse transform:

$$G_n(\rho, \rho') = \int_0^\infty \frac{J_n(\lambda\rho)J_n(\lambda\rho')}{\lambda^2 - k^2}\lambda d\lambda$$

which is recognized as the term in brackets in (6.2.16). Substituting into (6.2.19) gives

$$G(\rho, \phi, \rho', \phi') = \frac{1}{2\pi}\sum_{n=-\infty}^{\infty} e^{jn(\phi-\phi')}\int_0^\infty \frac{J_n(\lambda\rho)J_n(\lambda\rho')}{\lambda^2 - k^2}\lambda d\lambda \tag{6.2.21}$$

The previous derivation could be done if the delta function representation in (6.2.18) is known a priori. That is, with

$$\frac{\delta(\rho - \rho')\delta(\phi - \phi')}{\rho} = \frac{1}{2\pi}\sum_{n=-\infty}^{\infty} e^{jn(\phi-\phi')}\int_0^\infty J_n(\lambda\rho)J_n(\lambda\rho')\lambda d\lambda \tag{6.2.22}$$

Then, letting the Green function be of the form

$$G(\rho, \phi, \rho', \phi') = \frac{1}{2\pi}\sum_{n=-\infty}^{\infty} e^{jn(\phi-\phi')}\int_0^\infty G_n(\rho, \rho')J_n(\lambda\rho)\lambda d\lambda \tag{6.2.23}$$

it follows that substituting (6.2.22) and (6.2.23) into (6.2.18) gives

$$G_n(\rho, \rho') = \frac{J_n(\lambda\rho')}{\lambda^2 - k^2} \tag{6.2.24}$$

Substituting (6.2.24) into (6.2.23) results in the same expression as that in (6.2.21).

The integral in (6.2.21) is evaluated as follows. There are two cases, $\rho > \rho'$ and $0 \le \rho < \rho'$. Consider the case where $\rho > \rho'$. Write the integral in terms of Hankel functions as

$$\int_0^\infty \frac{J_n(\lambda\rho)J_n(\lambda\rho')}{\lambda^2 - k^2}\lambda d\lambda = \frac{1}{2}\int_0^\infty \frac{[H_n^{(1)}(\lambda\rho) + H_n^{(2)}(\lambda\rho)]J_n(\lambda\rho')}{\lambda^2 - k^2}\lambda d\lambda$$
$$= \frac{1}{2}\int_0^\infty \frac{J_n(\lambda\rho')H_n^{(1)}(\lambda\rho)}{\lambda^2 - k^2}\lambda d\lambda + \frac{1}{2}\int_0^\infty \frac{J_n(\lambda\rho')H_n^{(2)}(\lambda\rho)}{\lambda^2 - k^2}\lambda d\lambda$$

To use residue theory, the limits of integration should be from $-\infty$ to ∞. Letting $\lambda = -\lambda$ in the integral with $H_n^{(1)}(\lambda\rho)$ and using the relations

$$J_n(-x) = (-1)^n J_n(x)$$
$$H_n^{(1)}(-x) = (-1)^{n+1} H_n^{(2)}(x)$$

it follows that this integral is the same as the one with $H_n^{(2)}(\lambda\rho)$ but with limits from $-\infty$ to 0. Therefore, the two integrals are combined to read

$$\int_0^\infty \frac{J_n(\lambda\rho)J_n(\lambda\rho')}{\lambda^2 - k^2}\lambda d\lambda = \frac{1}{2}\int_{-\infty}^\infty \frac{J_n(\lambda\rho')H_n^{(2)}(\lambda\rho)}{\lambda^2 - k^2}\lambda d\lambda \tag{6.2.25}$$

Equation (6.2.25) has poles at $\lambda = -k = -k' + j\varepsilon$ and $\lambda = k = k' - j\varepsilon$ (i.e., introducing a small loss: $k'' = \varepsilon$), and a branch point singularity due to the Hankel function at $z = 0$. The path of integration is shown in Fig. 6.2 and is closed in the lower-half plane. The branch cut associated with the Hankel function is selected along the negative λ axis. The contribution from the semicircular path C_1 vanishes as $R \to \infty$, as well as around the small semicircle. Hence, using residue theory:

$$\frac{1}{2}\int_{-\infty}^\infty \frac{J_n(\lambda\rho')H_n^{(2)}(\lambda\rho)}{\lambda^2 - k^2}\lambda d\lambda = -2\pi j \frac{1}{2}\frac{J_n(\lambda\rho')H_n^{(2)}(\lambda\rho)}{(\lambda + k)(\lambda - k)}\lambda(\lambda - k)\Bigg|_{\lambda = k} = -\frac{j\pi}{2}J_n(k\rho')H_n^{(2)}(k\rho) \tag{6.2.26}$$

The contribution from the large semicircle along C_1 vanishes, since the large argument approximations show that

$$J_n(\lambda\rho') \approx \sqrt{\frac{2}{\pi\lambda\rho'}}\cos\left(\lambda\rho' - \frac{\pi}{4} - \frac{n\pi}{2}\right)$$
$$= \sqrt{\frac{1}{2\pi\lambda\rho'}}\,[e^{j(\lambda\rho' - \pi/4 - n\pi/2)} + e^{-j(\lambda\rho' - \pi/4 - n\pi/2)}]$$

and

$$H_n^{(2)}(\lambda\rho) \approx \sqrt{\frac{2}{\pi\lambda\rho}}\, e^{-j(\lambda\rho - \pi/4 - n\pi/2)}$$

So, with $\lambda = Re^{j\theta}$, the exponential variation of $J_n(\lambda\rho')H_n^{(2)}(\lambda\rho)$ along C_1 goes as

$$e^{j\lambda(\rho'-\rho)} + e^{-j\lambda(\rho'+\rho)} \sim e^{-R\sin\theta(\rho'-\rho)} + e^{R\sin\theta(\rho'+\rho)} \tag{6.2.27}$$

Noting that $\rho > \rho'$, $\rho' + \rho > 0$ and $\rho' - \rho < 0$, it follows that since $\sin\theta < 0$ along C_1, (6.2.27) goes to zero as $R \to \infty$. Therefore, the integrand in (6.2.26) vanishes along C_1. The integral around the small semicircle around the origin is evaluated by letting $\lambda = \varepsilon e^{j\theta}$ where $\varepsilon \to 0$, and it follows that this integral also vanishes.

From (6.2.26) and (6.2.21), the Green function for $\rho \geq \rho'$ is

$$G(\rho, \phi, \rho', \phi') = -\frac{j}{4} \sum_{n=-\infty}^{\infty} J_n(k\rho') H_n^{(2)}(k\rho) e^{jn(\phi-\phi')} \quad (\rho \geq \rho') \tag{6.2.28}$$

The derivation for $0 \leq \rho < \rho'$ is similar. For this case, in (6.2.21) express $J_n(\lambda\rho')$ in terms of Hankel functions and follow the steps leading to (6.2.26). The resulting Green function is given by (6.2.28) with ρ and ρ' interchanged.

The addition theorem for Hankel functions in (5.6.44) shows that (6.2.28) is recognized as

$$G(\boldsymbol{\rho}, \boldsymbol{\rho}') = -\frac{j}{4} H_0^{(2)}(k|\boldsymbol{\rho} - \boldsymbol{\rho}'|)$$

6.2.3 Three-Dimensional Green Function

The transformations to cylindrical coordinates applied to (6.1.21) can also be applied to the Green function in three dimensions in (6.1.26), to obtain

$$G(\mathbf{r}, \mathbf{r}') = \frac{-j}{8\pi^2} \int_0^{\infty} \frac{e^{-j\tau|z-z'|}}{\tau} \left[\int_0^{2\pi} e^{-j\lambda\rho\cos(\phi-\gamma)} d\gamma \right] \lambda d\lambda \tag{6.2.29}$$

where

$$\tau = \sqrt{k^2 - \lambda^2}$$

and $\mathrm{Im}(\tau) < 0$. The angular integral is recognized as $2\pi J_0(\lambda\rho)$. Therefore,

$$G(\mathbf{r}, \mathbf{r}') = \frac{-j}{4\pi} \int_0^{\infty} \frac{J_0(\lambda\rho)}{\tau} e^{-j\tau|z-z'|} \lambda d\lambda \tag{6.2.30}$$

Writing $J_0(\lambda\rho)$ in terms of Hankel functions, the expression can be written only in terms of $H_0^{(2)}(\lambda\rho)$ with limits of integration between $-\infty$ to ∞. That is,

$$G(\mathbf{r}, \mathbf{r}') = \frac{-j}{8\pi} \int_{-\infty}^{\infty} \frac{H_0^{(2)}(\lambda\rho)}{\tau} e^{-j\tau|z-z'|} \lambda d\lambda \tag{6.2.31}$$

This is an important relation that appears in the calculations of the fields produced by a dipole above the Earth. Equations (6.2.30) and (6.2.31) are known as the Sommerfeld (or Sommerfeld-Ott) relations.

Equations (6.2.30) and (6.2.31) also appear in terms of the propagation constant γ (i.e., $\gamma = jk$). Equation (6.2.31) in terms of γ is

$$G(\mathbf{r}, \mathbf{r}') = \frac{1}{8\pi} \int_{-\infty}^{\infty} \frac{H_0^{(2)}(\lambda\rho)}{\tau} e^{-\tau|z-z'|} \lambda \, d\lambda$$

where

$$\tau = \sqrt{\gamma^2 + \lambda^2}$$

Another three-dimensional representation of the Green function is as follows. The Green function satisfies

$$\left[\frac{1}{\rho}\frac{\partial}{\partial\rho}\left(\rho\frac{\partial}{\partial\rho}\right) + \frac{1}{\rho^2}\frac{\partial^2}{\partial\phi^2} + \frac{\partial^2}{\partial z^2} + k^2\right] G(\mathbf{r}, \mathbf{r}') = -\frac{\delta(\rho - \rho')\delta(\phi - \phi')\delta(z - z')}{\rho} \tag{6.2.32}$$

Applying the Fourier transform to the z variable gives

$$\left[\frac{1}{\rho}\frac{\partial}{\partial\rho}\left(\rho\frac{\partial}{\partial\rho}\right) + \frac{1}{\rho^2}\frac{\partial^2}{\partial\phi^2} + \tau^2\right] G(\rho, \phi, \alpha_3, \mathbf{r}') = -e^{-j\alpha_3 z'}\frac{\delta(\rho - \rho')\delta(\phi - \phi')}{\rho} \tag{6.2.33}$$

where

$$\tau^2 = k^2 - \alpha_3^2$$

Equation (6.2.33) is similar to that in (6.2.18). Hence, its solution follows similar steps and an inverse Fourier transform gives $G(\mathbf{r}, \mathbf{r}')$ in (6.2.32) (left to the problems).

6.3 INTEGRAL TRANSFORM FOR THE WAVE EQUATION IN SPHERICAL COORDINATES

In the three-dimensional Fourier transform in (6.1.1) let

$$x = r \sin\theta \cos\phi$$
$$y = r \sin\theta \sin\phi$$
$$z = r \cos\theta$$

and

$$\alpha_1 = \lambda \sin\beta \cos\gamma$$
$$\alpha_2 = \lambda \sin\beta \sin\gamma$$
$$\alpha_3 = \lambda \cos\beta$$

$$d\alpha_1 d\alpha_2 d\alpha_3 = \lambda^2 \sin\beta \, d\lambda \, d\beta \, d\gamma$$

Then,

$$e^{-j(\alpha_1 x + \alpha_2 y + \alpha_3 z)} = e^{-jr\lambda[\cos\beta\cos\theta + \sin\beta\sin\theta\cos(\phi-\gamma)]}$$

or

$$e^{-j\boldsymbol{\alpha}\cdot\mathbf{r}} = e^{-jr\lambda\cos\varsigma}$$

where the angle ς between $\boldsymbol{\alpha}$ and $\mathbf{r}$ follows from

$$\cos\varsigma = \cos\beta\cos\theta + \sin\beta\sin\theta\cos(\phi - \gamma)$$

Substituting the above relations into (6.1.1), the Fourier transform in spherical coordinates is

$$f(\lambda, \beta, \gamma) = \int_{r=0}^{\infty}\int_{\theta=0}^{\pi}\int_{\phi=0}^{2\pi} f(r, \theta, \phi)\, e^{-jr\lambda\cos\varsigma} r^2 \sin\theta dr\, d\theta\, d\phi \qquad (6.3.1)$$

and the inverse transform in spherical coordinates follows from (6.1.2). That is

$$f(r, \theta, \phi) = \frac{1}{(2\pi)^3}\int_{\lambda=0}^{\infty}\int_{\beta=0}^{\pi}\int_{\gamma=0}^{2\pi} f(\lambda, \beta, \gamma) e^{jr\lambda\cos\varsigma}\lambda^2 \sin\beta\, d\lambda\, d\beta\, d\gamma \qquad (6.3.2)$$

6.3.1 The Spherical Hankel Transform

In (6.3.1), the completeness property of the spherical harmonics permits the expansion of a well-behaved function $f(r, \theta, \phi)$ in terms of spherical harmonics as

$$f(r, \theta, \phi) = \sum_{n=0}^{\infty}\sum_{m=-n}^{n} g_{nm}(r) Y_{nm}(\theta, \phi) \qquad (6.3.3)$$

Since,

$$\int_{\theta=0}^{\pi}\int_{\phi=0}^{2\pi} Y_{nm}(\theta, \phi)\, Y^*_{n'm'}(\theta, \phi) \sin\theta d\theta d\phi = \delta_{nn'}\delta_{mm'} \qquad (6.3.4)$$

it follows that multiplying (6.3.3) by $Y^*_{nm}(\theta, \phi)$, integrating and using (6.3.4) the coefficients $g_{nm}(r)$ in the expansions are given by

$$g_{nm}(r) = \int_{\phi=0}^{2\pi} \int_{\theta=0}^{\pi} f(r, \theta, \phi) Y^*_{nm}(\theta, \phi) \sin\theta d\theta d\phi \tag{6.3.5}$$

The angular part in (6.3.3) could have been expressed in terms of $P_n^m(\cos\theta)e^{jm\phi}$ instead of $Y_{nm}(\theta, \phi)$.

Using the identity (see (4.7.4)):

$$e^{-j\lambda r \cos\varsigma} = \sum_{n=0}^{\infty} j^{-n}(2n+1) j_n(\lambda r) P_n(\cos\varsigma)$$

where

$$\begin{aligned} P_n(\cos\varsigma) &= \sum_{m=-n}^{n} \frac{(n-m)!}{(n+m)!} P_n^m(\cos\theta) P_n^m(\cos\beta) e^{jm(\phi-\gamma)} \\ &= \frac{4\pi}{2n+1} \sum_{m=-n}^{n} Y_{nm}(\beta, \gamma) Y^*_{nm}(\theta, \phi) \end{aligned}$$

Then,

$$e^{-j\alpha \cdot \mathbf{r}} = e^{-j\lambda r \cos\varsigma} = 4\pi \sum_{n=0}^{\infty} \sum_{m=-n}^{n} j^{-n} j_n(\lambda r) Y_{nm}(\beta, \gamma) Y^*_{nm}(\theta, \phi) \tag{6.3.6}$$

The complex conjugate in the spherical harmonics can be taken in either the (θ, ϕ) or the (β, γ) variables.

Substituting (6.3.3) and (6.3.6) into (6.3.1), and using (6.3.4) to evaluate the angular integral shows that

$$f(\lambda, \beta, \gamma) = 4\pi \sum_{n=0}^{\infty} \sum_{m=-n}^{n} j^{-n} \left[\int_0^{\infty} g_{nm}(r) j_n(\lambda r) r^2 dr \right] Y_{nm}(\beta, \gamma) \tag{6.3.7}$$

The integral term in (6.3.7) is defined as the mth-order spherical Hankel transform $g_{nm}(\lambda)$. That is,

$$g_{nm}(\lambda) = \int_0^{\infty} g_{nm}(r) j_n(\lambda r)\, r^2 dr \tag{6.3.8}$$

Then, (6.3.7) reads

$$f(\lambda, \beta, \gamma) = 4\pi \sum_{n=0}^{\infty} \sum_{m=-n}^{n} j^{-n} g_{nm}(\lambda) Y_{nm}(\beta, \gamma) \tag{6.3.9}$$

Equation (6.3.9) indicates that the spherical Fourier transform $f(\lambda, \beta, \gamma)$ is expressed in terms of the spherical Hankel transform.

In applications where there is no ϕ dependence (i.e., $m = 0$), the spherical Hankel transform reads $g_{n0}(\lambda)$, or is simply written as $g_n(\lambda)$. That is,

$$g_n(\lambda) = \int_0^{\infty} g_n(r) j_n(\lambda r)\, r^2 dr$$

Substituting (6.3.9) and (6.3.6) into the inverse transform in (6.3.2) gives

$$\begin{aligned} f(r, \theta, \phi) &= \sum_{n=0}^{\infty} \sum_{m=-n}^{n} \left[\frac{2}{\pi} \int_{\alpha=0}^{\infty} g_{nm}(\lambda) j_n(\lambda r) \lambda^2 d\lambda \right] Y_{nm}(\theta, \phi) \\ &= \sum_{n=0}^{\infty} \sum_{m=-n}^{n} g_{nm}(r) Y_{nm}(\theta, \phi) \end{aligned} \tag{6.3.10}$$

where we used (6.3.4) in the form

$$\int_{\gamma=0}^{2\pi} \int_{\beta=0}^{\pi} Y_{nm}(\beta, \gamma) Y^*_{n'm'}(\beta, \gamma) \sin\beta \, d\beta \, d\gamma = \delta_{nn'} \delta_{mm'}$$

and $g_{nm}(r)$ is the m-order inverse spherical Hankel transform. That is,

$$g_{nm}(r) = \frac{2}{\pi} \int_0^{\infty} g_{nm}(\lambda) j_n(\lambda r)\, \lambda^2 d\lambda \tag{6.3.11}$$

The m-order spherical Hankel transform pair is given by (6.3.8) and (6.3.11).

For the case that $m = 0$, the inverse spherical Hankel transform is simply written as

$$g_n(r) = \frac{2}{\pi} \int_0^{\infty} g_n(\lambda) j_n(\lambda r)\, \lambda^2 d\lambda$$

As an example, the spherical Hankel transform of a delta-function source is determined by letting

$$f(r, \theta, \phi) = \frac{1}{r^2 \sin\theta} \delta(r - r_o) \delta(\phi - \phi_o) \delta(\theta - \theta_o) \tag{6.3.12}$$

Expanding (6.3.12) in terms of spherical harmonics (see (6.3.3)), the coefficients of the expansion are given by (6.3.5), namely

$$g_{nm}(r) = \int_{\phi=0}^{2\pi} \int_{\theta=0}^{\pi} \frac{1}{r^2}\delta(r - r_o)\delta(\theta - \theta_o)\delta(\phi - \phi_o)Y_{nm}^*(\theta, \phi)d\theta d\phi$$
$$= \frac{1}{r^2}\delta(r - r_o)Y_{nm}^*(\theta_o, \phi_o)$$

and from (6.3.8) the spherical Hankel transform is

$$g_{nm}(\lambda) = \int_0^\infty \delta(r - r_o)Y_{nm}^*(\theta_o, \phi_o)j_n(\lambda r)dr = j_n(\lambda r_o)Y_{nm}^*(\theta_o, \phi_o)$$

The spherical Fourier transform of $f(r, \theta, \phi)$, from (6.3.9), is

$$f(\lambda, \beta, \gamma) = 4\pi \sum_{n=0}^{\infty} \sum_{m=-n}^{n} j^{-n} j_n(\lambda r_o)Y_{nm}^*(\theta_o, \phi_o)Y_{nm}(\beta, \gamma) \qquad (6.3.13)$$

Observe that (6.3.6) shows that (6.3.13) can be expressed as

$$f(\lambda, \beta, \gamma) = e^{-j\boldsymbol{\alpha}\cdot\mathbf{r}_o} = e^{-j(\alpha_1 x_o + \alpha_2 y_o + \alpha_3 z_o)}$$

which is the expected result for the Fourier transform of the spacial delta function.

Another example is the case where the function $f(r, \theta, \phi) = f(r, \theta)$ (i.e., independent of ϕ). Then, the coefficients $g_{n0}(r)$ in (6.3.5), simply written as $g_n(r)$, are given by

$$g_n(r) = \int_0^{2\pi} \int_0^{\pi} f(r, \theta)Y_{n0}^*(\theta, \phi)\sin\theta d\theta d\phi$$
$$= \sqrt{\frac{2n+1}{4\pi}}\left[\int_0^{2\pi} d\phi\right]\int_0^{\pi} f(r, \theta)P_n(\cos\theta)\sin\theta d\theta$$
$$= \sqrt{\pi(2n+1)}\int_0^{\pi} f(r, \theta)P_n(\cos\theta)\sin\theta d\theta$$

For a given $f(r, \theta)$, the spherical Hankel transform of $g_n(r)$ follows from (6.3.8) and the spherical Fourier transform of $f(r, \theta)$ follows from (6.3.9).

The spherical Hankel transform is now applied to the wave equation in spherical coordinates:

$$\left[\frac{1}{r^2}\frac{\partial}{\partial r}\left(r^2\frac{\partial}{\partial r}\right) + \frac{1}{r^2\sin\theta}\frac{\partial}{\partial\theta}\left(\sin\theta\frac{\partial}{\partial\theta}\right) + \frac{1}{r^2\sin^2\theta}\frac{\partial^2}{\partial\phi^2} + k^2\right]\psi(r, \theta, \phi)$$
$$= -f(r, \theta, \phi) \qquad (6.3.14)$$

Letting

$$f(r, \theta, \phi) = \sum_{n=0}^{\infty} \sum_{m=-n}^{n} f_n(r) Y_{nm}(\theta, \phi) \tag{6.3.15}$$

$$\psi(r, \theta, \phi) = \sum_{n=0}^{\infty} \sum_{m=-n}^{n} \psi_n(r) Y_{nm}(\theta, \phi) \tag{6.3.16}$$

and substituting into (6.3.14) shows that $\psi_n(r)$ satisfies

$$\left[\frac{1}{r^2}\frac{d}{dr}\left(r^2\frac{d}{dr}\right) - \frac{n(n+1)}{r^2} + k^2\right]\psi_n(r) = -f_n(r) \tag{6.3.17}$$

Multiplying (6.3.17) by $r^2 j_n(\alpha r)$ and integrating over r from $-\infty$ to ∞, gives

$$\int_0^\infty \frac{1}{r^2}\frac{d}{dr}\left(r^2\frac{d\psi_n(r)}{dr}\right) r^2 j_n(\lambda r)dr - \int_0^\infty \frac{n(n+1)}{r^2}\psi_n(r) r^2 j_n(\lambda r)dr + k^2\psi_n(\lambda)$$
$$= -f_n(\lambda) \tag{6.3.18}$$

where $\psi_n(\lambda)$ and $f_n(\lambda)$ are the spherical Hankel transforms of $\psi_n(r)$ and $f_n(r)$, respectively. The first integral in (6.3.18) is evaluated using integration by parts twice, to obtain

$$\int_0^\infty \frac{d}{dr}\left(r^2\frac{d\psi_n(r)}{dr}\right) j_n(\lambda r)dr = r^2\frac{d\psi_n(r)}{dr} j_n(\lambda r)\Big|_0^\infty - \int_0^\infty r^2\frac{d\psi_n(r)}{dr}\frac{dj_n(\lambda r)}{dr}dr$$
$$= r^2\left[\frac{d\psi_n(r)}{dr} j_n(\lambda r) - \psi_n(r)\frac{dj_n(\lambda r)}{dr}\right]\Bigg|_0^\infty + \int_0^\infty \psi_n(r)\frac{d}{dr}\left[r^2\frac{dj_n(\lambda r)}{dr}\right]dr \tag{6.3.19}$$

If the integrated term in (6.3.19) vanishes, it follows that (6.3.18) reads

$$\int_0^\infty \left[\frac{1}{r^2}\frac{d}{dr}\left(r^2\frac{dj_n(r)}{dr}\right) - \frac{n(n+1)}{r^2} j_n(\lambda r)\right] r^2\psi_n(r)dr + k^2\psi_n(\lambda) = -f_n(\lambda)$$

The term in brackets, from the spherical Bessel's differential equation, is equal to $-\lambda^2 j_n(\lambda r)$. Therefore,

$$-\lambda^2\int_0^\infty \psi_n(r) j_n(\lambda r) r^2 dr + k^2\psi_n(\lambda) = -f_n(\lambda)$$

or

$$(\lambda^2 - k^2)\psi_n(\lambda) = f_n(\lambda) \Rightarrow \psi_n(\lambda) = \frac{f_n(\lambda)}{\lambda^2 - k^2} \tag{6.3.20}$$

The inverse spherical Hankel transform of (6.3.20) is

$$\psi_n(r) = \frac{2}{\pi}\int_0^\infty \frac{f_n(\lambda)}{\lambda^2 - k^2} j_n(\lambda r)\lambda^2 d\lambda$$

and $\psi(r, \theta, \phi)$ is given by (6.3.16), which completes the formal solution.

We now return to the integrated term in (6.3.19). The integrated term vanishes at the lower limit (i.e., at $r = 0$). At the upper limit ($r = \infty$) the spherical Bessel functions are replaced by their asymptotic values:

$$j_n(\lambda r) \approx \frac{1}{\lambda r}\cos\left(\lambda r - \frac{(n+1)\pi}{2}\right)$$

and

$$\frac{dj_n(\lambda r)}{dr} \approx -\frac{1}{r}\sin\left(\lambda r - \frac{(n+1)\pi}{2}\right)$$

Expressing the sinusoidal terms in terms of exponentials (like in the derivation of (6.2.17)), the integrated term vanishes if the following radiation condition is satisfied:

$$\lim_{r\to\infty} r\left[\frac{d\psi_n(r)}{dr} + j\lambda\psi_n(r)\right] = 0 \tag{6.3.21}$$

6.3.2 Three-Dimensional Green Function

The Green function for the wave equation in spherical coordinates is now derived using the Hankel transform. In spherical coordinates, the Green function satisfies

$$\left[\frac{1}{r^2}\frac{\partial}{\partial r}\left(r^2\frac{\partial}{\partial r}\right) + \frac{1}{r^2\sin\theta}\frac{\partial}{\partial\theta}\left(\sin\theta\frac{\partial}{\partial\theta}\right) + \frac{1}{r^2\sin^2\theta}\frac{\partial^2}{\partial\phi^2} + k^2\right]G(\mathbf{r}, \mathbf{r}') = -\frac{\delta(r - r')\delta(\theta - \theta')\delta(\phi - \phi')}{r^2\sin\theta} \tag{6.3.22}$$

Since the angular representation of the delta function is

$$\frac{\delta(\theta-\theta')\delta(\phi-\phi')}{\sin\theta} = \sum_{n=0}^{\infty}\sum_{m=-n}^{n} \frac{(2n+1)}{4\pi}\frac{(n-m)!}{(n+m)!} P_n^m(\cos\theta)P_n^m(\cos\theta')e^{jm(\phi-\phi')}$$
$$= \sum_{n=0}^{\infty}\sum_{m=-n}^{n} Y_{nm}(\theta,\phi)Y_{nm}^{*}(\theta',\phi') \tag{6.3.23}$$

we let the Green function solution to (6.3.22) be of the form

$$G(\mathbf{r},\mathbf{r}') = \sum_{n=0}^{\infty}\sum_{m=-n}^{n} G(r,r')\frac{(2n+1)}{4\pi}\frac{(n-m)!}{(n+m)!} P_n^m(\cos\theta)P_n^m(\cos\theta')e^{jm(\phi-\phi')}$$
$$= \sum_{n=0}^{\infty}\sum_{m=-n}^{n} G(r,r')Y_{nm}(\theta,\phi)Y_{nm}^{*}(\theta',\phi') \tag{6.3.24}$$

Substituting (6.3.23) and (6.3.24) into (6.3.22) and using the associated Legendre's differential equation satisfied by $Y_{nm}(\theta, \phi)$ (see Appendix E) shows that $G(r, r')$ satisfies

$$\left[\frac{1}{r^2}\frac{d}{dr}\left(r^2\frac{d}{dr}\right) + \left(k^2 - \frac{n(n+1)}{r^2}\right)\right]G(r,r') = -\frac{\delta(r-r')}{r^2} \tag{6.3.25}$$

Equation (6.3.25) can be solved using the spherical Hankel transform. Multiply (6.3.25) by $r^2 j_n(\lambda r)$ and integrate from 0 to ∞, to obtain (see the procedure used in (6.3.17) to (6.3.20)):

$$G(\lambda, r') = \frac{j_n(\lambda r')}{\lambda^2 - k^2}$$

where $G(\lambda, r')$ is the spherical Hankel transform of $G(r, r')$. Taking the inverse transform:

$$G(r,r') = \frac{2}{\pi}\int_0^{\infty} \frac{j_n(\lambda r)j_n(\lambda r')}{\lambda^2 - k^2}\lambda^2 d\lambda \tag{6.3.26}$$

and the Green function in (6.3.24) reads

$$G(\mathbf{r},\mathbf{r}') = \frac{2}{\pi}\sum_{n=0}^{\infty}\sum_{m=-n}^{n}\left[\int_0^{\infty} \frac{j_n(\lambda r)j_n(\lambda r')}{\lambda^2 - k^2}\lambda^2 d\lambda\right] Y_{nm}(\theta,\phi)Y_{nm}^{*}(\theta,\phi) \tag{6.3.27}$$

The evaluation of the integral in (6.3.27) is quite similar to that in (6.2.25), which led to the result in (6.2.26). Thus, for $r \geq r'$ we obtain

$$\int_0^{\infty} \frac{j_n(\lambda r) j_n(\lambda r')}{\lambda^2 - k^2} \lambda^2 d\lambda = -\frac{jk\pi}{2} j_n(kr') h_n^{(2)}(kr) \tag{6.3.28}$$

and (6.3.27) reads

$$\begin{aligned} G(\mathbf{r}, \mathbf{r}') = \frac{e^{-jk|\mathbf{r}-\mathbf{r}'|}}{4\pi|\mathbf{r} - \mathbf{r}'|} &= -jk \sum_{n=0}^{\infty} \sum_{m=-n}^{n} Y_{nm}(\theta, \phi) Y_{nm}^{*}(\theta', \phi') j_n(kr') h_n^{(2)}(kr) \\ &= -jk \sum_{n=0}^{\infty} \sum_{m=-n}^{n} \frac{(2n+1)}{4\pi} \frac{(n-m)!}{(n+m)!} P_n^m(\cos\theta) P_n^m(\cos\theta') e^{jm(\phi-\phi')} \\ &\quad j_n(kr') h_n^{(2)}(kr) \quad (r \geq r') \end{aligned} \tag{6.3.29}$$

For $0 \leq r \leq r'$ interchange r' and r in the above relation. (6.3.29) is identical to the Green function in (5.6.63), which was obtained using the direct method.

Problems

P6.1. Show that the contribution to the path integral from the small semicircle around the origin in Fig. 6.2 is zero.

P6.2. Verify (6.2.13).

P6.3. What is the relation between the Green function in (6.2.28) and the Green function in the form

$$G(\boldsymbol{\rho}, \boldsymbol{\rho}') = \frac{-j}{4} H_0^{(2)}(k|\boldsymbol{\rho} - \boldsymbol{\rho}'|)$$

when there is no ϕ dependence?

P6.4. Determine $G(\mathbf{r}, \mathbf{r}')$ in (6.2.32).

P6.5. Verify (6.3.10).

P6.6. Verify (6.3.25).

P6.7. Show all the steps leading to (6.3.28).

P6.8. Express (6.3.29) in terms of Schelkunoff-Bessel functions.

7 Some Mathematical Methods

7.1 THE WATSON TRANSFORMATION

In electromagnetic wave propagation, the Watson transformation is used to transform a slowly convergent series involving Bessel functions or Bessel and Legendre functions to a new series that converges fast. These slowly convergent series appear on many diffraction problems involving cylindrical coordinates, as well as in many propagation problems in spherical coordinates. Detailed applications of the Watson transformation are discussed in Chapters 9 and 10.

Before discussing the Watson transformation, consider the summation of series using the theory of residues. Consider the case where $f(z)$ and $g(z)$ are analytic functions except at simple poles in the complex plane where the poles of $g(z)$ are along the real axis at integer values of z (e.g., such as those that arise with $g(z) = 1/\sin(\pi z)$ where $z = n$, $n = 0, \pm 1, \pm 2, \ldots$). The poles of $f(z)$ are assumed to occur at complex values of z, denoted by z_i, $i = 1, 2, 3, .., k, .., m$ where m can be infinite. The poles of $f(z)$ can also occur at real non-integer values of z, but such case does not appear in the problems discussed in this book. Then, if C_N is the closed path shown in Fig. 7.1 that encloses the $n = 0$ to $n = \pm N$ poles of $g(z)$ along the real axis and the $z_i = 1, 2, 3, \ldots, k$ poles of $f(z)$ within the contour C_N, it follows from residue theory that

$$\oint_{C_N} g(z)f(z)dz = 2\pi j \sum_{n=-N}^{N} \text{residues of } g(z)f(z) \text{ at poles of } g(z)$$

$$+ 2\pi j \sum_{i=1}^{k} \text{residues of } g(z)f(z) \text{at poles of} f(z) \qquad (7.1.1)$$

While C_N is shown as a square closed path, any closed path enclosing the singularities can be used.

Next, let $N \to \infty$ so all the poles are enclosed. If

$$\lim_{N\to\infty} \oint_{C_N} g(z)f(z)dz = 0 \qquad (7.1.2)$$

the left-hand side of (7.1.1) is zero and we obtain

$$\sum_{n=-\infty}^{\infty} \text{residues of } g(z)f(z) \text{ at poles of } g(z) = -\sum_{i=1}^{m} \text{residues of } g(z)f(z) \text{ at poles of} f(z) \qquad (7.1.3)$$

DOI: 10.1201/9781003219729-7

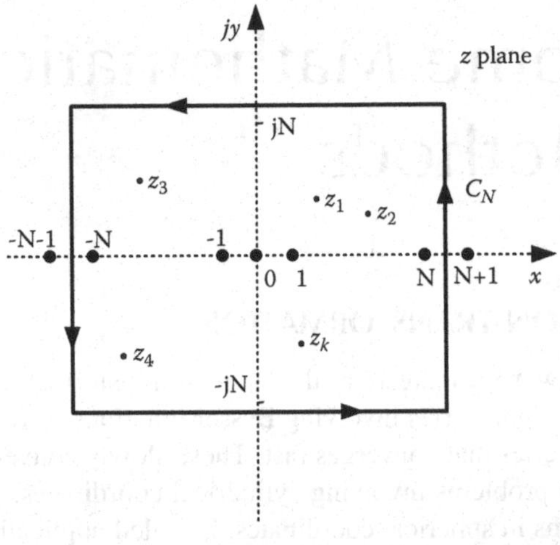

FIGURE 7.1 Contour of integration C_N showing the real axis poles up to $\pm N$, and the k poles of $f(z)$ within C_N.

where the residue of $g(z)f(z)$ at a given $z = n$ is

$$\text{residue of } g(z)f(z)|_{z=n} = [g(z)(z-n)f(z)]|_{z=n}$$

If the term $g(z)(z-n)$ at $z = n$ is made, for convenience, to be unity. Then,

$$\text{residue of } g(z)f(z)|_{z=n} = f(n)$$

and from (7.1.3):

$$\sum_{n=-\infty}^{\infty} f(n) = -\sum_{i=1}^{m} \text{residues of } g(z)f(z) \text{ at poles of } f(z) \tag{7.1.4}$$

which shows that the infinite sum of $f(n)$ can be evaluated in terms of the residues of $f(z)$.

A suitable function for $g(z)$ is $\pi\cot(\pi z)$. The multiplication by π makes the residue part associated with $g(z)$ be unity. The poles of $g(z)$ occur at $\sin(\pi z) = 0$. That is, at $z = n$ where $n = 0, \ \pm 1, \ \pm 2 \ldots.$ The residues at the poles are calculated using

$$\sum_{n=-\infty}^{\infty} \left\{ \frac{\pi\cos(\pi z)}{\frac{d[\sin(\pi z)]}{dz}} \right\} f(z) \Bigg|_{z=n} = \sum_{n=-\infty}^{\infty} f(n) \tag{7.1.5}$$

since the term in brackets is unity. Of course, to evaluate the summation we still need to evaluate the residues of $f(z)$ as shown in (7.1.4). That is,

$$\sum_{n=-\infty}^{\infty} f(n) = -\sum_{i=1}^{m} \text{residues of } [\pi \cot(\pi z) f(z)] \text{ at poles of } f(z) \tag{7.1.6}$$

The previous summation formula requires that (7.1.2) is satisfied. It can be shown that as $N \to \infty$ the value of $|g(z)f(z)| \to 0$ if $f(z)$ is of order $O(1/|z|^2)$ (e.g., such as $f(z) \approx 1/|z|^2$).

Other relations that are used to evaluate sums are

$$\begin{cases} \sum_{n=-\infty}^{\infty} (-1)^n f(n) = -\sum_{i=1}^{m} \text{residues of } [\pi \csc(\pi z) f(z)] \text{ at poles of } f(z) \\ \sum_{n=-\infty}^{\infty} f\left(\frac{2n+1}{2}\right) = \sum_{i=1}^{m} \text{residues of } [\pi \tan(\pi z) f(z)] \text{ at poles of } f(z) \\ \sum_{n=-\infty}^{\infty} (-1)^n f\left(\frac{2n+1}{2}\right) = \sum_{i=1}^{m} \text{residues of } [\pi \sec(\pi z) f(z)] \text{ at poles of } f(z) \end{cases}$$

A typical example of the previous theory is the evaluation of the infinite sum:

$$\sum_{n=-\infty}^{\infty} \frac{1}{n^2 + a^2}$$

where a is a real number.

To evaluate the sum, let $g(z) = \pi \cot(\pi z)$ and

$$f(z) = \frac{1}{z^2 + a^2}$$

whose poles are at $z = \pm ja$, and integrate (7.1.1) along the contour C in Fig. 7.2 which encloses all the poles. With $z = R\, e^{j\theta}$ it follows that $|f(z)| \approx 1/R^2$, and as $R \to \infty$ the contribution from the large radius contour vanishes. Therefore, from (7.1.5) and (7.1.6):

$$\begin{aligned} \sum_{n=-\infty}^{\infty} \frac{1}{n^2 + a^2} &= -\sum_{i=1}^{2} \text{residues of } \pi \cot(\pi z) f(z) \text{at poles of } f(z) \\ &= -\left.\frac{\pi \cot(\pi z)(z - ja)}{(z + ja)(z - ja)}\right|_{z=ja} - \left.\frac{\pi \cot(\pi z)(z + ja)}{(z + ja)(z - ja)}\right|_{z=-ja} \\ &= -\frac{\pi \cot(j\pi a)}{j2a} + \frac{\pi \cot(-j\pi a)}{j2a} \\ &= \frac{\pi \coth(\pi a)}{2a} + \frac{\pi \coth(\pi a)}{2a} = \frac{\pi}{a} \coth(\pi a) \end{aligned} \tag{7.1.7}$$

Next, the Watson transformation is analyzed. To describe the Watson transformation, consider the integral

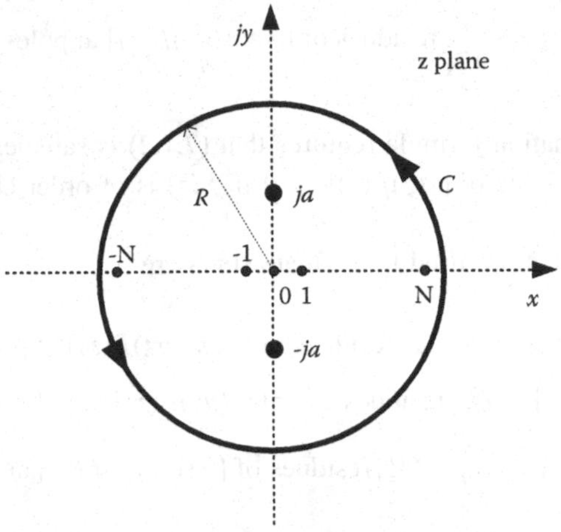

FIGURE 7.2 Contour C for the evaluation of the sum in (7.1.7).

$$\oint_C g(\nu)B(\nu)d\nu \tag{7.1.8}$$

where $C = C_1 + C_2$ is a closed contour that encloses the poles of $g(\nu)$, which are on the real axis. One such closed contour is shown in Fig. 7.3a. The term $B(\nu)$ will be an expression involving Bessel functions, spherical Bessel functions, or a combination of those functions with Legendre functions. The singularities of $g(\nu)$ on the real axis, denoted by $\nu = n$ ($n = 0, \pm1, \pm2$ ....) are different from the singularities of $B(\nu)$, which are denoted by $\pm\,\nu_p$ ($p = 1, 2, 3, \ldots$) and assumed to be complex. In Fig. 7.3a the singularities of $B(\nu)$ are shown in the fourth and second quadrant (i.e., at $\pm\,\nu_p$). These are the typical location of the singularities that occur in several of the problems considered in Chapters 9 and 10.

For example, with $g(\nu) = \pi\cot(\pi\nu)$, (7.1.8) reads

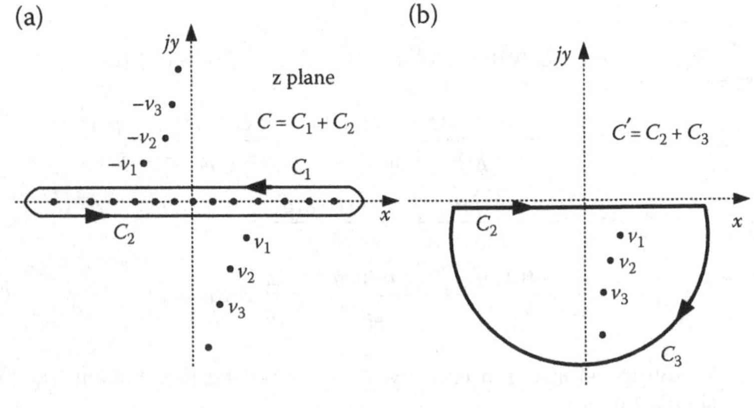

FIGURE 7.3 (a) Path of integration $C = C_1 + C_2$; (b) path of integration $C' = C_2 + C_3$.

$$\oint_C \pi \cot(\pi z) B(\nu)\, d\nu = 2\pi j \sum_{n=-\infty}^{\infty} B(n) \tag{7.1.9}$$

Next, we attempt to change the path of integration C to a new path of integration that encloses only the singularities of $B(\nu)$ in the lower-half plane. To this end, write (7.1.9) as

$$\int_{\infty+j\varepsilon}^{-\infty+j\varepsilon} g(\nu)B(\nu)d\nu + \int_{-\infty-j\varepsilon}^{\infty-j\varepsilon} g(\nu)B(\nu)d\nu = 2\pi j \sum_{n=-\infty}^{\infty} B(n) \tag{7.1.10}$$

where the contribution from the small semicircles as $\nu \to \pm\infty$ must vanish. Letting $\nu = -\nu$ in the first integral in (7.1.10) and observing that $g(\nu)$ is an odd function (i.e., $g(-\nu) = -g(\nu)$), it follows that if $B(\nu)$ is an even function (i.e., $B(-\nu) = B(\nu)$), then $g(\nu)B(\nu)$ is odd and the first integral reads

$$\int_{\infty+j\varepsilon}^{-\infty+j\varepsilon} g(\nu)B(\nu)d\nu = -\int_{-\infty-j\varepsilon}^{\infty-j\varepsilon} g(-\nu)B(-\nu)d\nu = \int_{-\infty-j\varepsilon}^{\infty-j\varepsilon} g(\nu)B(\nu)d\nu$$

which is identical to the second integral in (7.1.10). Hence, (7.1.10) reads

$$2\int_{-\infty-j\varepsilon}^{\infty-j\varepsilon} g(\nu)B(\nu)d\nu = 2\int_{-\infty-j\varepsilon}^{\infty-j\varepsilon} \pi \cot(\pi\nu)B(\nu)d\nu = 2\pi j \sum_{n=-\infty}^{\infty} B(n)$$

and since the path from $-\infty - j\varepsilon$ to $\infty - j\varepsilon$ is the path C_2 in Fig. 7.3b, we obtain

$$\int_{C_2} \pi \cot(\pi\nu)B(\nu)d\nu = \pi j \sum_{n=-\infty}^{\infty} B(n) \tag{7.1.11}$$

The path of integration C_2 is slightly below the real axis.

The integral in (7.1.11) is evaluated by closing the path of integration with an infinite semicircle in the lower-half plane, denoted by C_3 as shown in Fig. 7.3b. Of course, the contribution along the infinite semicircle must vanish, and this puts certain conditions on the function $g(\nu)B(\nu)$, since it must vanish along C_3. The integral in (7.1.11) is then evaluated along $C' = C_2 + C_3$ which is $-2\pi j$ times the residues at the poles of $B(\nu)$. Then,

$$\int_{C_2} g(\nu)B(\nu)d\nu = \oint_{C'=C_2+C_3} g(\nu)B(\nu)d\nu$$

$$\pi j \sum_{n=-\infty}^{\infty} B(n) = -2\pi j \sum_{p=1}^{\infty} \text{residues of } g(\nu)B(\nu)|_{\nu=\nu_p}$$

or

$$\sum_{n=-\infty}^{\infty} B(n) = -2 \sum_{p=1}^{\infty} \text{residues of } \pi \cot(\pi\nu)B(\nu)\Big|_{\nu=\nu_p}$$

where ν_p ($p = 1, 2, 3, \ldots$) denotes the complex poles of $B(\nu)$ in the lower-half plane.

The Watson transformation is a series to integral transformation which is evaluated in terms of the poles of $B(\nu)$. In the case illustrated in Fig. 7.3 the poles of $B(\nu)$ are in the lower-half plane and these poles determine the value of the infinite sum. In many cases, the form of $B(\nu)$ makes the sum to converge very fast. The previous manipulations could have been done by moving the path of integration slightly above the real axis. In such case, the path of integration is closed with an infinite semicircle in the upper-half plane enclosing the poles of $B(\nu)$ in the upper half plane. Of course, for this case, the function $g(\nu)B(\nu)$ must vanish along the infinite semicircle in the upper-half plane.

Other forms of the Watson transformation where $B(\nu)$ is even and has poles in the lower-half plane are

$$\begin{cases} \oint_C \frac{B(\nu)}{\sin(\pi\nu)} d\nu = 2j \sum_{n=-\infty}^{\infty} (-1)^n B(n) \\ \oint_{C'} \frac{B(\nu)}{\sin(\pi\nu)} d\nu = j \sum_{n=-\infty}^{\infty} (-1)^n B(n) = -2\pi j \sum_{p=1}^{\infty} \text{residues of } \frac{1}{\sin(\pi\nu)} B(\nu)\Big|_{\nu=\nu_p} \end{cases}$$

$$\begin{cases} \oint_C \frac{e^{j\nu(\phi-\pi)}}{\sin(\pi\nu)} B(\nu)\, d\nu = 2j \sum_{n=-\infty}^{\infty} B(n) e^{jn\phi} \\ \oint_{C'} \frac{\cos[\nu(\phi-\pi)]}{\sin(\pi\nu)} B(\nu) d\nu = j \sum_{n=-\infty}^{\infty} B(n) e^{jn\phi} \\ \qquad = -2\pi j \sum_{p=1}^{\infty} \text{residues of } \frac{\cos[\nu(\phi-\pi)]}{\sin(\pi\nu)} B(\nu)\Big|_{\nu=\nu_p} \end{cases} \tag{7.1.12}$$

The Watson transformation in (7.1.12) is discussed in Section 9.1

Some series require the sum of the terms to be from 0 to ∞. For example, the series

$$\sum_{n=0}^{\infty} (-1)^n (2n+1) B(n)$$

This type of sum appears in problems involving wave propagation in spherical geometries where the sum is slowly convergent. The sum can be evaluated using the following integral:

$$I = j\oint_C \frac{\nu}{\cos(\pi\nu)} B\left(\nu - \frac{1}{2}\right) d\nu$$

where the closed contour $C = C_1 + C_2$ encloses the positive real axis poles of $\cos(\pi\nu)$, as shown in Fig. 7.4a. The poles are located at $\nu = (n + 1/2)$ where $n = 0, 1, 2, \ldots$. Therefore, using residue theory:

$$I = j\oint_C \frac{\nu}{\cos(\pi\nu)} B\left(\nu - \frac{1}{2}\right) d\nu = 2\pi j \sum_{n=0}^{\infty} \left[j \frac{\nu B\left(\nu - \frac{1}{2}\right)}{\frac{d}{d\nu}\cos(\pi\nu)} \right]\Bigg|_{\nu=n+1/2}$$

$$= 2\pi \left[\frac{\nu B\left(\nu - \frac{1}{2}\right)}{\pi \sin(\pi\nu)} \right]\Bigg|_{\nu=n+1/2} = \sum_{n=0}^{\infty} (-1)^n (2n+1) B(n) \qquad (7.1.13)$$

The integrand is in (7.1.13) is odd if $B(\nu - 1/2)$ is even. When the integrand is odd the path of integration can be changed to $C' = C_3 + C_2$, as shown in Fig. 7.4b.

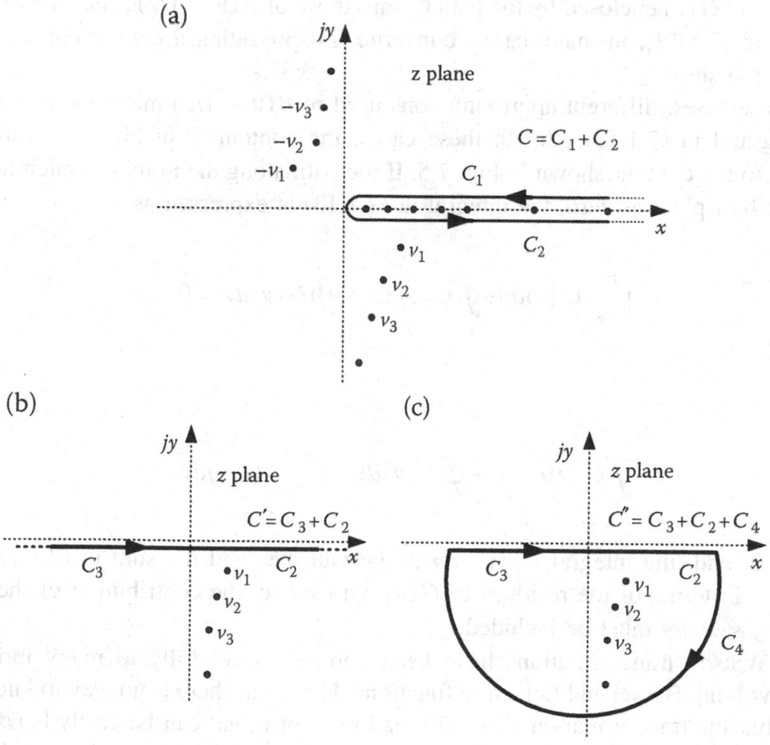

FIGURE 7.4 (a) Contour C enclosing the poles of $\cos(\pi\nu)$ along the positive real axis; (b) the path C'; (c) the closed path C'' enclosing the poles of $B(\nu - 1/2)$.

Closing the path with an infinite semicircle in the lower-half plane the integral is evaluated in terms of the poles of $B(\nu - 1/2)$ in the lower-half plane. The closed path is denoted by C'' in Fig. 7.4b. Of course, the integrand must vanish along the semicircular path C_4 as its radius goes to infinity.

In (7.1.13), letting

$$G(\nu) = j\frac{\nu}{\cos(\pi\nu)}B\left(\nu - \frac{1}{2}\right)$$

the evaluation of (7.1.13) along the closed path C'' is

$$I = \oint_C G(\nu)d\nu = \int_{C'} G(\nu)d\nu = \oint_{C''} G(\nu)d\nu = -2\pi j \sum_{p=1}^{\infty} \text{residues of } G(\nu)|_{\nu=\nu_p}$$

or

$$\sum_{n=0}^{\infty} (-1)^n(2n+1)B(n) = -2\pi j \sum_{p=1}^{\infty} \text{residues of } G(\nu)|_{\nu=\nu_p} \tag{7.1.14}$$

The poles of $G(\nu)$ enclosed by the path C'' are those of $B(\nu - 1/2)$. The sum of the residues in (7.1.14), in many cases, converge fast providing a convenient way to evaluate the sum.

In some cases, different approximations used in $B(\nu - 1/2)$ might or not make the integrand in (7.1.13) odd. In these cases, the contour C in Fig. 7.4a can be changed to the contour shown in Fig. 7.5. If the path along the infinite semicircle in the right-half-plane is zero, the integral in (7.1.13) is expressed as

$$\int_{-j\infty}^{j\infty} G(\nu)d\nu + \oint_C G(\nu)d\nu + \oint_{C'} G(\nu)d\nu = 0$$

or

$$\oint_C G(\nu)d\nu = -\oint_{C'} G(\nu)d\nu - \int_{-j\infty}^{j\infty} G(\nu)d\nu$$

If $G(\nu)$ is odd, the integral from $-j\infty$ to $j\infty$ vanishes and the sum in (7.1.13) is evaluated in terms of the residues of $G(\nu)$. Otherwise, the contribution of the integral $-j\infty$ to $j\infty$ must be included.

The Watson transformations have been applied successfully to many infinite sums involving Bessel and Legendre functions. However, there is no way to know a priory that the transformation used will lead to a form that can be easily handled.

Some problems where the Watson transformation has been used to evaluate slowly convergent infinite sums are discussed in Chapters 9 and 10.

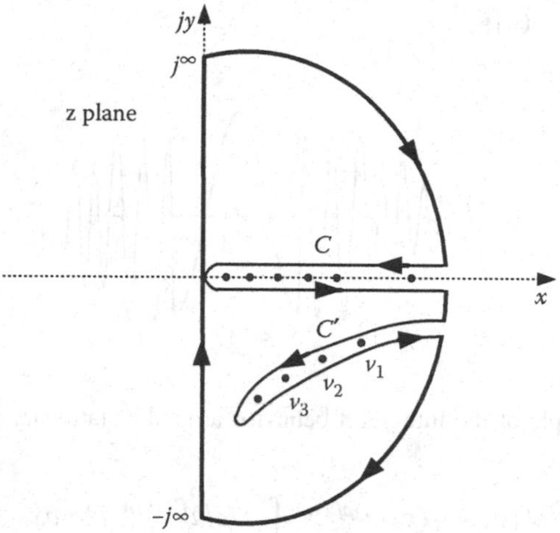

FIGURE 7.5 Contour of integration.

7.2 THE METHOD OF STATIONARY PHASE

The method of stationary phase is used to evaluate integrals of the type

$$I(\rho) = \int_a^b g(x)\, e^{j\rho f(x)} dx \tag{7.2.1}$$

where $g(x)$ and $f(x)$ are real functions. The parameter ρ is real and assumed to have a large value. The function $g(x)$ is assumed to be a slowly varying function of x in the interval $a \le x \le b$. The function

$$e^{j\rho f(x)} = \cos[\rho f(x)] + j \sin[\rho f(x)]$$

causes the integrand to fluctuate very rapidly, except at the values of x for which $f(x)$ is constant (i.e., stationary). Thus, the integral vanishes at all points except at the stationary point or points. An example of the integrand behavior around a single stationary point at $x = x_0$ is shown in Fig. 7.6.

The stationary point x_0 can be obtained from the equation $f'(x_0) = 0$. At the stationary point (or points) the integrand is approximately a constant $g(x_0)$ multiplied by the expansion of $e^{j\rho f(x)}$ around $x = x_0$. Expanding $f(x)$ around x_0 in a Taylor series gives

$$f(x) = f(x_0) + f'(x_0)(x - x_0) + \frac{f''(x_0)}{2}(x - x_0)^2 + \ldots \approx f(x_0) + \frac{f''(x_0)}{2}(x - x_0)^2 \tag{7.2.2}$$

The approximation is valid if $(x - x_0)f'''(x_0) \ll f''(x_0)$.

Substituting (7.2.2) into (7.2.1):

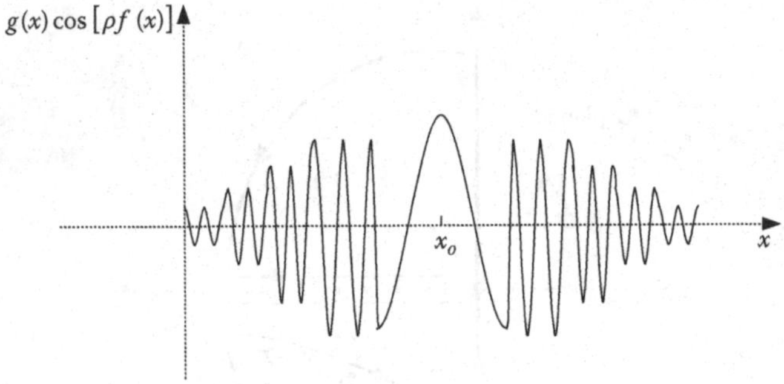

FIGURE 7.6 Example of the integrand behavior around a stationary point at $x = x_0$.

$$I(\rho) \approx g(x_0)e^{j\rho f(x_0)} \int_{-\infty}^{\infty} e^{j\frac{\rho}{2}f''(x_0)(x-x_0)^2} dx$$

where the limits were extended to $-\infty$ to ∞ because the sinusoidal contribution to the integral vanish on each side (away) from $x = x_0$. Using the integral relation:

$$\int_{-\infty}^{\infty} e^{\pm j\alpha x^2/2} dx = \sqrt{\frac{2\pi}{\alpha}}\, e^{\pm j\pi/4} \quad (\alpha > 0)$$

we obtain

$$I(\rho) \approx g(x_0)\sqrt{\frac{2\pi}{\rho|f''(x_0)|}}\, e^{j\rho f(x_0) \pm j\pi/4} \tag{7.2.3}$$

where the + or − signs in the exponent depend on whether $f''(x_0)$ is positive or negative. In some cases, x_0 coincides with one of the limits of the integral, say a. In this case the method gives

$$I(\rho) \approx \frac{1}{2}g(a)\sqrt{\frac{2\pi}{\rho|f''(a)|}}\, e^{j\rho f(a) \pm j\pi/4}$$

Other results associated with the stationary phase method are

a. If $f(x)$ has n stationary points at $x_0, x_1, \ldots, x_n$, it follows that

$$I(\rho) \approx \sum_{i=0}^{n} g(x_i)\sqrt{\frac{2\pi}{\rho|f''(x_i)|}}\, e^{j\rho f(x_i) \pm j\pi/4} \tag{7.2.4}$$

b. If $f(x)$ has a double zero at $x = x_0$ (i.e., when $f'(u_0) = f''(u_0) = 0$), then

$$I(\rho) \approx \frac{\Gamma\left(\frac{1}{3}\right)}{\sqrt{3}} g(x_0) \left[\frac{6}{\rho|f'''(x_0)|}\right]^{1/3} e^{j\rho f(x_0)} \tag{7.2.5}$$

A typical example in the use of the method of stationary phase is in the evaluation of the following Bessel function integral:

$$J_0(x) = \frac{1}{2\pi}\int_{-\pi}^{\pi} e^{jx\sin\theta} d\theta \tag{7.2.6}$$

for large values of x. In (7.2.6) in the notation of (7.2.1): $g(x) = 1$, $\rho = x$, $f(x) = f(\theta) = \sin\theta$ and $dx = d\theta$. The stationary points are located at $f'(\theta) = \cos\theta = 0$ or at $x_0 = \pi/2$ and $x_1 = -\pi/2$. Then, $f''(\pi/2) = -1$ and $f''(-\pi/2) = 1$, and from (7.2.4):

$$J_0(x) \approx \frac{1}{2\pi}\left[\sqrt{\frac{2\pi}{x}} e^{j(x-\pi/4)} + \sqrt{\frac{2\pi}{x}} e^{-j(x-\pi/4)}\right] = \sqrt{\frac{2}{\pi x}} \cos(x - \pi/4)$$

7.3 THE METHOD OF LAPLACE

The method of Laplace is used to evaluate integrals of the type

$$I(\rho) = \int_{-\infty}^{\infty} g(x) e^{\rho f(x)} dx \tag{7.3.1}$$

where ρ is real and has a large value, and $g(x)$ is a slowly varying function. If $f(x)$ has a single maximum at $x = a$, then $f'(a) = 0$ and $f''(a) < 0$. In the interval close to $x = a$ use the approximations $g(x) \approx g(a)$ and

$$f(x) \approx f(a) + \frac{f''(a)}{2}(x - a)^2$$

Therefore, (7.3.1) is approximated by

$$I(\rho) \approx g(a) e^{\rho f(a)} \int_{-\infty}^{\infty} e^{-\frac{\rho}{2}|f''(a)|(x-a)^2} dx \tag{7.3.2}$$

Since,

$$\int_{-\infty}^{\infty} e^{-\alpha x^2} dx = \sqrt{\frac{\pi}{\alpha}}$$

(7.3.2) gives

$$I(\rho) \approx g(a)e^{\rho f(a)}\sqrt{\frac{2\pi}{\rho|f''(a)|}} \tag{7.3.3}$$

The difference between the methods of Laplace and stationary phase is that in (7.2.1) the integration away from the stationary point is negligible because of the rapid oscillation of the integrand produces a cancellation, while in (7.3.1) the integration away from $x = a$ tends to zero because the integrand tends to zero.

In the case that

$$I(\rho) = \int_{x_1}^{x_2} g(x)e^{\rho f(x)}dx$$

and $f(x)$ has a single maximum at $x = a$, where a is an interior point between x_1 and x_2, the limits of integration can be extended to $\pm\infty$, and $I(\rho)$ is given by (7.3.3).

The method of Laplace also applies to integrals of the type

$$I(\rho) = \int_{x_1}^{x_2} g(x)e^{-\rho f(x)}dx \tag{7.3.4}$$

If $f(x)$ has a single minimum at $x = a$, where a is an interior point between x_1 and x_2, then $f'(a) = 0$ and $f''(a) > 0$. In the interval close to a we can use the approximations $g(x) \approx g(a)$ and

$$f(x) \approx f(a) + \frac{f''(a)}{2}(x - a)^2$$

Therefore, since ρ is a real large-value parameter the contribution to the integral in (7.3.4) outside $x \approx a$ vanishes and the limits of integration can be extended to $\pm\infty$. Hence, (7.3.4) can be approximated by

$$I(\rho) \approx g(a)e^{-\rho f(a)} \int_{-\infty}^{\infty} e^{-\frac{\rho}{2}f''(a)(x-a)^2}dz$$

or

$$I(\rho) \approx g(a)e^{-\rho f(a)}\sqrt{\frac{2\pi}{\rho f''(a)}} \tag{7.3.5}$$

As an example, consider the evaluation of $n!$ for large values of n. The factorial and gamma functions are related by

$$n! = \Gamma(n + 1)$$

where

$$\Gamma(n+1) = \int_0^\infty t^n e^{-t} dt \tag{7.3.6}$$

Letting $t = xn$ in (7.3.6) gives

$$\Gamma(n+1) = n^{n+1}\int_0^\infty x^n e^{-nx} dx = n^{n+1}\int_0^\infty e^{n\ln x} e^{-nx} dx = n^{n+1}\int_0^\infty e^{-n(x-\ln x)} dx \tag{7.3.7}$$

Comparing (7.3.7) with (7.3.4), let $f(x) = x - \ln x$, then

$$f'(x) = 1 - \frac{1}{x} = 0 \Rightarrow x = a = 1$$

and

$$f''(x) = \Rightarrow f''(a) = \frac{1}{1} = 1$$

Therefore, from (7.3.5) with $f(a) = f(1) = 1$:

$$\Gamma(n+1) \approx n^{n+1} e^{-n} \sqrt{\frac{2\pi}{n}}$$

and it follows that

$$n! = \Gamma(n+1) \approx n^n e^{-n} \sqrt{2\pi n}$$

which is known as Stirling formula.

7.4 THE METHOD OF STEEPEST DESCENT

The method of steepest descent, also known as the saddle-point method, is used to evaluate integrals of the type

$$I(\rho) = \int_C g(z) e^{\rho f(z)} dz \tag{7.4.1}$$

where ρ is a real and large-value parameter. The functions $f(z)$ and $g(z)$ are analytic functions of z in a region that contains the path of integration C.

Suppose that $\mathrm{Re}[f(z)]$ has a maximum value at $z = z_s$, and that the path on integration can be changed from the path C to a path C_1 such that $\mathrm{Re}[f(z)]$ exhibits a rapid decrease away from its maximum value at z_s. Since ρ has a large value, then it follows that $|e^{\rho f(z)}|$ has a maximum at z_s and decreases very rapidly away from z_s along the path C_1. Then, (7.4.1) can be evaluated by changing the path C to a path C_1.

Of course, provided that there are no singularities of the integrand between the two paths. If there are singularities between the two paths, they must be accounted using residue theory and/or branch cut integration.

Next, some concepts from complex variables are reviewed. The real and imaginary parts of an analytic function $f(z) = u(x, y) + jv(x, y)$ satisfy the Cauchy-Riemann equations, or

$$\frac{\partial u(x, y)}{\partial x} = \frac{\partial v(x, y)}{\partial y}$$

and

$$\frac{\partial u(x, y)}{\partial y} = -\frac{\partial v(x, y)}{\partial x}$$

From these relations, it follows that the real and imaginary parts satisfy Laplace's equation:

$$\frac{\partial^2 u(x, y)}{\partial x^2} + \frac{\partial^2 u(x, y)}{\partial y^2} = 0$$

and

$$\frac{\partial^2 v(x, y)}{\partial x^2} + \frac{\partial^2 v(x, y)}{\partial y^2} = 0$$

which show that if the second derivative with respect to x is positive, the second derivative with respect to y is negative, and vice versa. The point z_s, where $\mathrm{Re}[f(z_s)]$ is a maximum, is really an "extremum." This point is called an extremum because in the complex plane the real and imaginary part of an analytic function do not possess an absolute maximum.

In the complex plane the real and imaginary parts of an analytic function have the property that the path of most rapid decrease (or increase) of one part are paths of constant values of the other. Since the path of integration must pass along the lines of most rapid decrease of $\mathrm{Re}[f(z)]$, it must also coincide with $\mathrm{Im}[f(z)] = $ constant. A path satisfying these conditions is called a steepest descent path.

The above statement is proved by observing that

$$\nabla u(x, y) = \frac{\partial u}{\partial x}\hat{\mathbf{a}}_x + \frac{\partial u}{\partial y}\hat{\mathbf{a}}_y$$

is orthogonal to a level of curve of $u(x, y)$. That is, the gradient of $u(x, y)$ points in the normal direction to its level curves, where a level curves of $u(x, y)$ is where $u(x, y) = c$ (i.e., has a constant value c). Similarly,

$$\nabla v(x, y) = \frac{\partial v}{\partial x}\hat{\mathbf{a}}_x + \frac{\partial v}{\partial y}\hat{\mathbf{a}}_y$$

is orthogonal to a level of curve of $v(x, y)$.

The vector ∇u and ∇v are perpendicular to each other if $\nabla u \bullet \nabla v = 0$. Using the Cauchy-Riemann relations it follows that

$$\nabla u \bullet \nabla v = \frac{\partial u}{\partial x}\frac{\partial v}{\partial x} + \frac{\partial u}{\partial y}\frac{\partial v}{\partial y} = \frac{\partial u}{\partial x}\left(-\frac{\partial u}{\partial y}\right) + \frac{\partial u}{\partial y}\left(\frac{\partial u}{\partial x}\right) = 0$$

which shows that the level curves of $u(x, y)$ and $v(x, y)$ are orthogonal to each other. This shows that ∇v is tangent to the level curves of $u(x, y)$, and ∇u is tangent to the level curves of $v(x, y)$. Since the directions of the tangents are orthogonal, the level curves of the real and imaginary parts of a complex function are orthogonal to each other. Thus, a level curve where the real part has its most rapid decrease corresponds to a level curve (i.e., a constant value) of its imaginary part.

There is one exception to the orthogonality of the level curves that occur when $\nabla u = 0$ (known as a critical point), then the vectors ∇u and ∇v are not orthogonal (see the example in Fig. 7.8).

The point z_s where $\text{Re}[f(z)]$ has a maximum is called the "saddle point," and since $\text{Im}[f(z)]$ = constant along the steepest descent path, it follows that the saddle point occurs at

$$\frac{df(z)}{dz} = 0 \qquad (7.4.2)$$

Figure 7.7 shows a picture of a saddle. It shows two paths through a saddle point. One follows the center of the ridges, and the other the center of the valleys. The

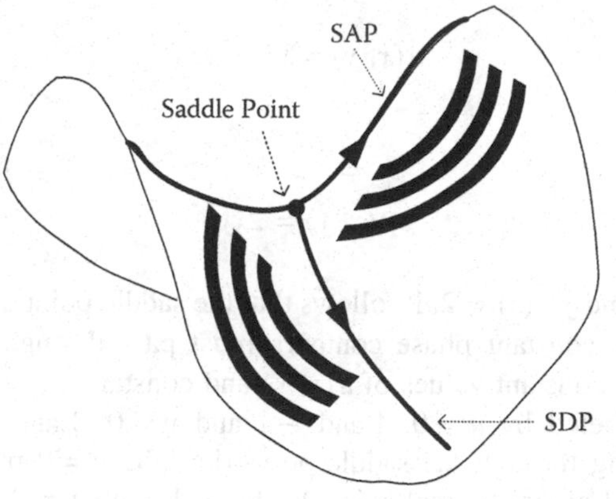

FIGURE 7.7 Illustration of a saddle point showing the steepest descent path (SDP) and the steepest ascent path (SAP).

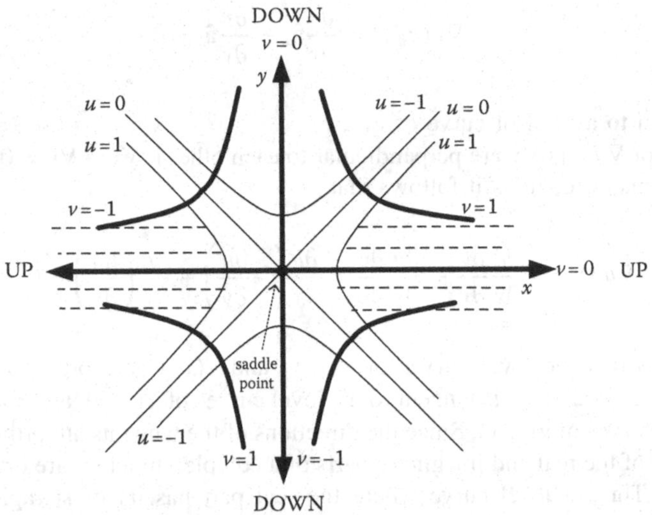

FIGURE 7.8 Constant phase contours for $f(z) = z^2$.

path along the ridge is the path of steepest ascent, and the path along the valley is the path of steepest descent. Along the two paths $\mathrm{Im}[f(z)] =$ constant. Therefore, the saddle-point region must be inspected carefully to select the steepest descent path (SDP), and not the steepest ascent path (SAP). Along the SDP, since the imaginary part is constant, there is no oscillatory behavior due to the imaginary part.

A useful example that describes the path through a saddle point is when $f(z) = z^2$ in (7.4.1). Then,

$$f(z) = z^2 = (x + jy)^2 = (x^2 - y^2) + j2xy$$

where the real and imaginary parts are

$$u(x, y) = x^2 - y^2 \tag{7.4.3}$$

and

$$v(x, y) = 2xy$$

Since $f'(z) = 2z$ and $f''(z) = 2$, it follows that the saddle point occurs at $z = 0$ (i.e., at $f'(z) = 0$). The constant phase contours must pass through $z = 0$. Figure 7.8 shows contours of constant values of $u(x, y)$ and constant values of $v(x, y)$. These are hyperbolas labeled by $u = 0$, 1 and -1 and $v = 0$, 1 and -1. The constant phase paths passing through the saddle point (i.e., the $v = 0$ paths) are along the imaginary axis and along the real axis. On the path with $x = 0$ and y positive and negative, (7.4.3) shows that $u = -y^2$, which shows that $u(x, y)$ is decreasing as one moves away from the saddle point along the y axis. Hence, this is the path of steepest descent, denoted by DOWN in Fig. 7.8. On the path with $y = 0$ and x positive and negative, (7.4.3) shows that $u = x^2$, which shows that $u(x, y)$ is increasing as one moves away from the saddle point along the x axis. Hence, this is the path of steepest ascent and denoted by UP in Fig. 7.8. The dashed lines represent the region where $u(x, y)$ increases as $|x|$ increases. Fig. 7.8 in three dimensions is in the shape of a saddle as in Fig. 7.7. The curves in Fig. 7.8 are orthogonal to each other except for the curves that intersect at the origin (i.e., the $u = 0$ and $v = 0$ curves), which is a critical point.

The solution of (7.4.2) gives the location of the saddle point (or points). In the neighborhood of the saddle point (i.e., where $f'(z_s) = 0$) we approximate $f(z)$ by

$$f(z) \approx f(z_s) + \frac{1}{2} f''(z_s)(z - z_s)^2 \tag{7.4.4}$$

Along the steepest descent path: $\text{Im}[f(z)] = \text{Im}[f(z_s)] = \text{constant}$, and $\text{Re}[f(z)] < \text{Re}[f(z_s)]$, then from (7.4.4) the term $f''(z_s)(z - z_s)^2/2$ is real and negative. Substituting (7.4.4) into (7.4.1):

$$I(\rho) \approx g(z_s) e^{\rho f(z_s)} \int_{SDP} e^{\rho f''(z_s)(z - z_s)^2/2} dz \tag{7.4.5}$$

where SDP is the steepest descent path.

To determine the SDP though z_s, let

$$z - z_s = re^{j\phi} \qquad (-\pi < \phi \leq \pi)$$

where ϕ determines the angle of inclination of the path at the saddle point. Then, with

$$f''(z_s) = |f''(z_s)| e^{j\varphi} \qquad (0 \leq \varphi < 2\pi)$$

it follows that in (7.4.5):

$$\frac{\rho}{2} |f''(z_s)| r^2 e^{j(\varphi + 2\phi)} < 0$$

which must be real and negative. This requires that

$$e^{j(\varphi + 2\phi)} = -1 \Rightarrow \varphi + 2\phi = \pm\pi \Rightarrow \phi = \pm\frac{\pi}{2} - \frac{\varphi}{2} \tag{7.4.6}$$

The + or – signs correspond to the two inclinations of the path at the saddle point. For example, if $\varphi = \pi/2$, then $\phi = \pi/4$ or $-3\pi/4$. Another way of determining the angle of inclination is to select ϕ such that $f''(z_s)e^{j2\phi}$ is real and negative.

We can then write (7.4.5) as

$$I(\rho) \approx g(z_s)e^{\rho f(z_s)}\int_{-\infty}^{\infty} e^{-\rho|f''(z_s)|r^2/2}\frac{dz}{dr}dr \approx g(z_s)e^{\rho f(z_s)}e^{j\phi}\int_{-\infty}^{\infty} e^{-\rho|f''(z_s)|r^2/2}dr$$

$$\approx g(z_s)e^{\rho f(z_s)}e^{j\phi}\sqrt{\frac{2\pi}{\rho|f''(z_s)|}} \tag{7.4.7}$$

The limits of integration were changed to ± ∞ because the integrand is essentially zero for values of r away from zero. Equation (7.4.7) is the first term of an asymptotic series and is the term usually needed to describe the solution.

Before obtaining higher order terms of the asymptotic series, (7.4.7) is derived using a more general approach. Equation (7.4.4) suggest the transformation

$$f(z) = f(z_s) - s^2 \tag{7.4.8}$$

where

$$s^2 = \frac{1}{2}f''(z_s)(z - z_s)^2$$

is a real positive variable, such that $-\infty \le s \le \infty$. The transformation in (7.4.8) also appears with $-s^2/2$ instead of $-s^2$. The saddle point corresponds to $s = 0$ in the s plane. Thus, using (7.4.8), (7.4.1) becomes

$$I(\rho) \approx e^{\rho f(z_s)}\int_{-\infty}^{\infty} G(s)e^{-\rho s^2}ds \tag{7.4.9}$$

where

$$G(s) = g(z)\frac{dz}{ds}$$

Expanding $G(s)$ about $s = 0$ gives

$$G(s) = G(0) + G'(0)s + \frac{G''(0)}{2!}s^2 + \ldots = \sum_{n=0}^{\infty}\frac{G^{[n]}(0)}{n!}s^n \tag{7.4.10}$$

The notation $[n]$ means the nth derivative. The convergence of (7.4.10) is a circle of radius r centered at $s = 0$, where r is the distance to the nearest singularity of $G(s)$. Keeping only the first term in (7.4.10), (7.4.9) reads

$$I(\rho) \approx G(0)e^{\rho f(z_s)}\int_{-\infty}^{\infty} e^{-\rho s^2}ds \approx G(0)e^{\rho f(z_s)}\sqrt{\frac{\pi}{\rho}} \tag{7.4.11}$$

where

$$G(0) = g(z_s)\frac{dz}{ds}\bigg|_{s=0}$$

From (7.4.8): $f'(z) = -2s\,(ds/dz)$ and $f''(z) = -2(ds/dz)^2$ at $s = 0$. Therefore,

$$\frac{dz}{ds}\bigg|_{s=0} = \sqrt{\frac{-2}{f''(z_s)}}$$

and (7.4.11) reads

$$I(\rho) \approx g(z_s)e^{\rho f(z_s)}\sqrt{\frac{-2\pi}{\rho f''(z_s)}} \tag{7.4.12}$$

Letting $f''(z_s) = |f''(z_s)|e^{j\varphi}$, (7.4.12) is written in the form

$$I(\rho) \approx g(z_s)e^{\rho f(z_s)}e^{j\phi}\sqrt{\frac{2\pi}{\rho|f''(z_s)|}} \tag{7.4.13}$$

where

$$\phi = \pm\frac{\pi}{2} - \frac{\varphi}{2}$$

The argument of the square root is specified by the term $e^{j\phi}$. Since ds is positive along the path of integration the argument of the square root is the argument of dz, which is ϕ. The results in (7.4.7) and (7.4.13) are, of course, identical.

The complete asymptotic expansion is obtained by substituting (7.4.10) into (7.4.9), and using

$$\int_{-\infty}^{\infty} s^n e^{-\rho s^2} ds = \begin{cases} \dfrac{\Gamma\left(\frac{n+1}{2}\right)}{\rho^{n+1/2}} & \text{(even } n\text{)} \\ 0 & \text{(odd } n\text{)} \end{cases}$$

Then,

$$I(\rho) \approx e^{\rho f(z_s)} \sum_{n=0,2,4,\ldots}^{\infty} \frac{G^{[n]}(0)}{n!}\frac{\Gamma\left(\frac{n+1}{2}\right)}{\rho^{n+1/2}} \tag{7.4.14}$$

where $G^{[n]}(0)$ denotes the nth derivative of $G(s)$ evaluated at $s = 0$.

As (7.4.14) shows, higher-order terms of the asymptotic series require the evaluation of derivatives of $G(s)$. This is a tedious process, so we will only outline the process. For example,

$$G(0) = g(z_s)\frac{dz}{ds}\bigg|_{s=0}$$

and

$$G'(0) = g'(z_s)\left(\frac{dz}{ds}\bigg|_{s=0}\right)^2 + g(z_s)\frac{d^2z}{ds^2}\bigg|_{s=0}$$

which requires derivatives of z with respect to s evaluated at $s = 0$. Such derivatives are obtained from (7.4.8) where the complete Taylor expansion of $f(z)$ shows that

$$-s^2 = \frac{1}{2!}f''(z_s)(z - z_s)^2 + \frac{1}{3!}f'''(z_s)(z - z_s)^3 + \ldots$$

This series must be inverted to obtain $(z - z_s)$ as a function of s. For example,

$$(z - z_s) = \left(\frac{dz}{ds}\bigg|_{s=0}\right)s + \left(\frac{d^2z}{ds^2}\bigg|_{s=0}\right)\frac{s^2}{2!} + \left(\frac{d^3z}{ds^3}\bigg|_{s=0}\right)\frac{s^2}{3!} + ..$$

Then, forming the terms $(z - z_s)^2$, $(z - z_s)^3$,..., substituting in the expression for $-s^2$ and equating equal powers of s gives the desired derivatives of dz/ds at $s = 0$, and (7.4.14) can be evaluated.

The first two term of the asymptotic series are

$$I(\rho) \approx g(z_s)e^{\rho f(z_s)}e^{j\phi}\sqrt{\frac{2\pi}{\rho|f''(z_s)|}} \times \left\{1 + \frac{1}{2\rho f''(z_s)}\left[\frac{f'''(z_s)}{f''(z_s)}\frac{g'(z_s)}{g(z_s)} + \frac{1}{4}\frac{f''''(z_s)}{f''(z_s)} - \frac{5}{12}\left[\frac{f'''(z_s)}{f''(z_s)}\right]^2 - \frac{g''(z_s)}{g(z_s)}\right]\right\} \tag{7.4.15}$$

The proof that the series obtained by the method of steepest descent is an asymptotic series known as Watson's Lemma.

Several examples using the method of steepest descent are considered next. The first is the evaluation of the asymptotic forms of the Hankel functions. The Sommerfeld integrals for the Hankel functions are derived in Appendix C, namely

$$H_\nu^{(1)}(x) = \frac{e^{-j\nu\pi/2}}{\pi}\int_{C_1} e^{j(x\cos\beta+\nu\beta)}d\beta \tag{7.4.16}$$

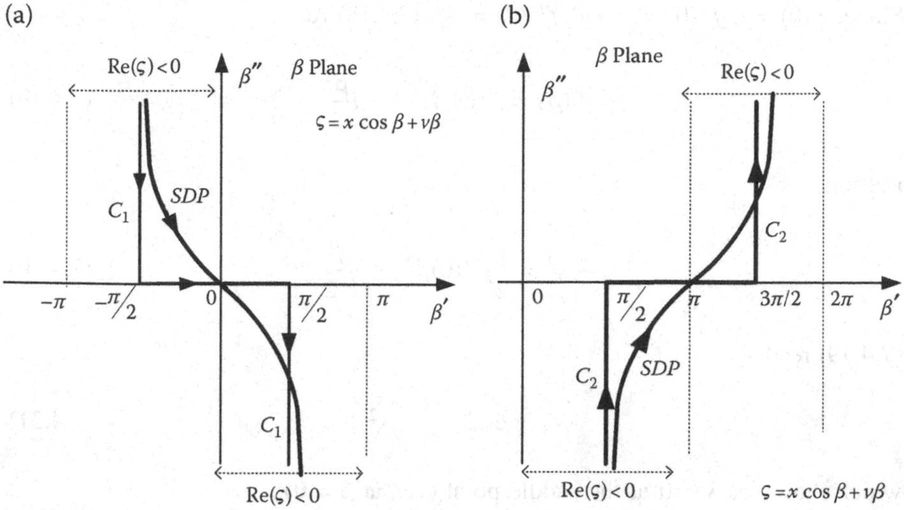

FIGURE 7.9 (a) Path C_1 and the steepest descent path for the evaluation of $H_\nu^{(1)}(x)$; (b) path C_2 and the steepest descent path for the evaluation of $H_\nu^{(2)}(x)$.

and

$$H_\nu^{(2)}(x) = \frac{e^{-j\nu\pi/2}}{\pi}\int_{C_2} e^{j(x\cos\beta+\nu\beta)}\,d\beta \tag{7.4.17}$$

where the paths C_1 and C_2 are shown in Fig. 7.9. Different asymptotic approximations to (7.4.16) and (7.4.17) are derived, depending on the ranges on x and ν.

For large values of x with $x >> \nu$ the exponential in (7.4.16) and (7.4.17) is approximated by

$$e^{j[x\cos\beta+\nu\beta]} \approx e^{xf(\beta)}$$

where

$$f(\beta) = j\cos\beta.$$

The saddle points follow from

$$f'(\beta) = -j\sin\beta = 0 \Rightarrow \beta = n\pi \qquad (n = 0, \pm1, \pm2, \ldots)$$

For $H_\nu^{(1)}(x)$, where the path C_1 is between $-\pi/2 \le \beta \le \pi/2$, the saddle point occurs at $\beta = 0$ (i.e., with $n = 0$). Expanding $f(\beta)$ around $\beta = 0$ gives

$$f(\beta) \approx f(0) + f'(0)\beta + \frac{1}{2}f''(0)\beta^2 \tag{7.4.18}$$

Since, $f(0) = j$, $f'(0) = 0$ and $f''(0) = -j$, (7.4.18) reads

$$f(\beta) = j\cos\beta \approx j - j\frac{\beta^2}{2} \tag{7.4.19}$$

Letting

$$-s^2 = \frac{1}{2}f''(0)\beta^2 = -j\frac{\beta^2}{2} \tag{7.4.20}$$

(7.4.19) reads

$$j\cos\beta = j - s^2 \tag{7.4.21}$$

which show that $s = 0$ at the saddle point (i.e., at $\beta = 0$).

The path of steepest descent follows from the imaginary part of the exponent. Since

$$jx\cos\beta = jx\cos(\beta' + j\beta'') = x\sin\beta'\sinh\beta'' + jx\cos\beta'\cosh\beta''$$

and at the saddle point: $jx\cos 0 = jx$, it follows that the constant value of the imaginary part, which is the path of steepest descent, is

$$\cos\beta'\cosh\beta'' = 1 \tag{7.4.22}$$

The path C_1 and the SDP are shown in Fig. 7.9a. The inclination of the path of integration at the saddle point follows from (7.4.6). Since $f''(0) = -j$ its phase is $\varphi = -\pi/2$, then select $\phi = -\pi/4$ (see Fig. 7.9a). Equation (7.4.22) also shows that as $\beta'' \to \pm\infty$ the real part of β goes approaches $\beta' \to \mp\pi/2$, making the ends of the SDP approach the C_1 path.

From (7.4.21) and (7.4.22) the SDP in the s plane follows from

$$s^2 = j[1 - \cos(\beta' + j\beta'')] = -\sin\beta'\sinh\beta''$$

which shows that s^2 is a positive real quantity since $\sin\beta' < 0$ when $-\pi < \beta' < 0$, $\sin\beta' > 0$ when $0 < \beta' < \pi$, $\sinh\beta'' > 0$ when $\beta'' > 0$, and $\sinh\beta'' < 0$ when $\beta'' < 0$. In terms of the real variable s, the limits of integration are from $-\infty$ to ∞ and for the case under consideration (7.4.16) reads

$$H_\nu^{(1)}(x) \approx \frac{e^{j(x-\nu\pi/2)}}{\pi}\int_{-\infty}^{\infty} e^{-xs^2}\frac{d\beta}{ds}ds \tag{7.4.23}$$

From (7.4.20):

$$\frac{d\beta}{ds} = \sqrt{2}\,e^{-j\pi/4}$$

and since

$$\int_{-\infty}^{\infty} e^{-xs^2} ds = \sqrt{\frac{\pi}{x}}$$

(7.4.23) reads

$$H_\nu^{(1)}(x) \approx \sqrt{\frac{2}{\pi x}}\, e^{j(x-\nu\pi/2-\pi/4)} \tag{7.4.24}$$

which is a well-known asymptotic approximation.

The evaluation of $H_\nu^{(2)}(x)$ for large values of x with $x >> \nu$ is similar. As shown in Fig. 7.9b, the path of integration for $H_\nu^{(2)}(x)$ is between $\pi/2 \le \beta \le 3\pi/2$. Therefore, the saddle point occurs at $\beta = \pi$ where $f'(\pi) = 0$. Expanding $f(\beta)$ around $\beta = \pi$ gives

$$f(\beta) \approx f(\pi) + \frac{1}{2} f''(\pi)\beta^2$$

Since $f(\pi) = -j$ and $f''(\pi) = j$, then

$$f(\beta) = j\cos\beta = -j + j\frac{\beta^2}{2} = -j - s^2$$

and (7.4.17) reads

$$H_\nu^{(2)}(x) \approx \frac{e^{-j(x-\nu\pi/2)}}{\pi} \int_{-\infty}^{\infty} e^{-xs^2} \frac{d\beta}{ds} ds$$

Since now $d\beta/ds = \sqrt{2}\,e^{j\pi/4}$, the above integral shows that

$$H_\nu^{(2)}(x) \approx \sqrt{\frac{2}{\pi x}}\, e^{-j(x-\nu\pi/2-\pi/4)} \tag{7.4.25}$$

Next, consider the case where both x and ν have large values. For such cases the exponential in (7.4.16) and (7.4.17) is written as

$$e^{jf(\beta)} \Rightarrow f(\beta) = x\cos\beta + \nu\beta \tag{7.4.26}$$

Thus, the saddle points follow from

$$f'(\beta) = -x\sin\beta + \nu = 0 \Rightarrow \sin\beta = \frac{\nu}{x} \tag{7.4.27}$$

which for $\nu < x$ the saddle points in the interval $-\pi/2$ to $3\pi/2$ are located at

$$\beta_1 = \sin^{-1}\left(\frac{\nu}{x}\right) \qquad (\nu < x) \tag{7.4.28}$$

and, since $\sin(\pi - \beta_1) = \sin\beta_1$:

$$\beta_2 = \pi - \beta_1 = \pi - \sin^{-1}\left(\frac{\nu}{x}\right) \qquad (\nu < x) \tag{7.4.29}$$

The saddle point β_1 is the one needed to evaluate $H_\nu^{(1)}(x)$, and the saddle point β_2 for $H_\nu^{(2)}(x)$.

For $\nu > x$, β must be complex to satisfy (7.4.27). That is,

$$\sin(\beta' + j\beta'') = \sin\beta'\cosh\beta'' + j\cos\beta'\sinh\beta'' = \frac{\nu}{x}$$

which is satisfied with $\beta' = \pi/2$ and $\cosh\beta'' = \nu/x$. Hence, one saddle points is located at

$$\beta_3 = \frac{\pi}{2} + j\cosh^{-1}\left(\frac{\nu}{x}\right) \qquad (\nu > x) \tag{7.4.30}$$

and another at

$$\beta_4 = \frac{\pi}{2} - j\cosh^{-1}\left(\frac{\nu}{x}\right) \qquad (\nu > x) \tag{7.4.31}$$

In this case, the saddle point at β_3 is the one needed to evaluate $H_\nu^{(2)}(x)$, and the saddle point β_4 for $H_\nu^{(1)}(x)$.

The evaluation of $H_\nu^{(1)}(x)$ when x and ν have large values but $\nu < x$ is as follows. The Taylor expansion of $f(\beta)$ in (7.4.26) (using two terms) around the saddle point at β_1 is

$$f(\beta) \approx f(\beta_1) + \frac{1}{2}f''(\beta_1)(\beta - \beta_1)^2 \approx (x\cos\beta_1 + \nu\beta_1) - \frac{1}{2}x\cos\beta_1(\beta - \beta_1)^2$$

or

$$jf(\beta) \approx j(x\cos\beta_1 + \nu\beta_1) - j\frac{x}{2}\cos\beta_1(\beta - \beta_1)^2$$

Observing that if one lets

$$s^2 = j\frac{x}{2}\cos\beta_1(\beta-\beta_1)^2 \tag{7.4.32}$$

Then: $jf(\beta) \approx jf(\beta_1) - s^2$, and (7.4.16) reads

$$H_\nu^{(1)}(x) \approx \frac{e^{-j\nu\pi/2}e^{jf(\beta_1)}}{\pi}\int_{-\infty}^{\infty} e^{-s^2}\frac{d\beta}{ds}ds \tag{7.4.33}$$

From (7.4.32):

$$\frac{d\beta}{ds} = \sqrt{\frac{2}{x\cos\beta_1}}\,e^{-j\pi/4}$$

and since

$$\int_{-\infty}^{\infty} e^{-s^2}ds = \sqrt{\pi}$$

(7.4.33) reads

$$H_\nu^{(1)}(x) \approx \sqrt{\frac{2}{\pi x\cos\beta_1}}\,e^{j[x\cos\beta_1+\nu(\beta_1-\pi/2)-\pi/4]} \qquad (\nu < x) \tag{7.4.34}$$

From (7.4.28):

$$\cos\beta_1 = \sqrt{1-\sin^2\beta_1} = \sqrt{1-\left(\frac{\nu}{x}\right)^2}$$

and (7.4.34) can be expressed in the form

$$H_\nu^{(1)}(x) \approx \sqrt{\frac{2}{\pi(x^2-\nu^2)^{1/2}}}\,e^{j[(x^2-\nu^2)^{1/2}-\nu\cos^{-1}(\nu/x)-\pi/4]} \qquad (\nu < x) \tag{7.4.35}$$

The derivation of $H_\nu^{(2)}(x)$ when x and ν have large values but $\nu < x$ is similar. Using the saddle point β_2 in (7.4.29) one obtains

$$H_\nu^{(2)}(x) \approx \sqrt{\frac{2}{\pi(x^2-\nu^2)^{1/2}}}\,e^{-j[(x^2-\nu^2)^{1/2}-\nu\cos^{-1}(\nu/x)-\pi/4]} \qquad (\nu < x)$$

which agrees with $H_\nu^{(2)}(x) = H_\nu^{(1)*}(x)$.

The derivation of $H_\nu^{(1)}(x)$ when x and ν have large values but $\nu > x$ uses the saddle point β_4, and the evaluation of $H_\nu^{(2)}(x)$ uses β_3. For these cases, one obtains

$$H_\nu^{(1)}(x) \approx -j\sqrt{\frac{2}{\pi(\nu^2 - x^2)^{1/2}}}\, e^{[\nu \coth^{-1}(\nu/x) - (\nu^2 - x^2)^{1/2}]} \qquad (\nu > x) \tag{7.4.36}$$

and

$$H_\nu^{(2)}(x) \approx j\sqrt{\frac{2}{\pi(\nu^2 - x^2)^{1/2}}}\, e^{[\nu \coth^{-1}(\nu/x) - (\nu^2 - x^2)^{1/2}]} \qquad (\nu > x) \tag{7.4.37}$$

The relations (7.4.34) to (7.4.37) are known as the Debye approximations (or second-order approximations) to the Hankel functions.

Observe that when ν is very close to x, the above approximations fail. What happens is that as $\nu \to x$ in the evaluation of $H_\nu^{(1)}(x)$ the saddle points β_1 and β_4 approach each other (around $\pi/2$) and, therefore, the contribution from both saddle points must added. The best way to analyze such situations is to obtain a better approximation to $f(\beta)$ around $\pi/2$. That is, let

$$f(\beta) \approx f(\pi/2) + f'(\pi/2)(\beta - \pi/2) + \frac{1}{2}f''(\pi/2)(\beta - \pi/2)^2 + \frac{1}{6}f'''(\pi/2)(\beta - \pi/2)^3 \tag{7.4.38}$$

where

$$f(\beta) = x\cos\beta + \nu\beta \Rightarrow \begin{cases} f(\pi/2) = \nu\pi/2 \\ f'(\pi/2) = \nu - x \\ f''(\pi/2) = 0 \\ f'''(\pi/2) = x \end{cases}$$

Then, (7.4.38) reads

$$f(\beta) \approx \nu\pi/2 + (\nu - x)(\beta - \pi/2) + \frac{x}{6}(\beta - \pi/2)^3 \tag{7.4.39}$$

Substituting (7.4.39) into (7.4.16) gives

$$H_\nu^{(1)}(x) \approx \frac{1}{\pi}\int_{C_1} e^{j(\nu - x)(\beta - \pi/2)} e^{j\frac{x}{6}(\beta - \pi/2)^3}\, d\beta$$

The convergence of the integral, when β is far from $\pi/2$, depends on the term $(\beta - \pi/2)^3$. The integral converges in the regions where $\mathrm{Im}(\beta - \pi/2)^3 > 0$. Letting $\beta - \pi/2 = re^{j\theta}$, then

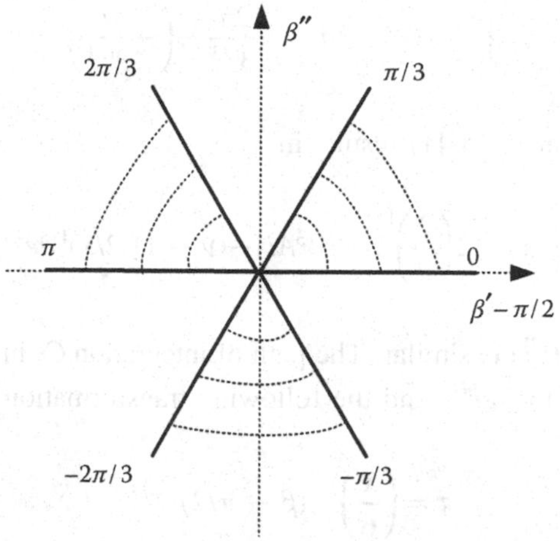

FIGURE 7.10 Regions where $\mathrm{Im}(\beta - \pi/2)^3 > 0$.

$$\mathrm{Im}[r^3 e^{j3\theta}] > 0 \Rightarrow \sin(3\theta) > 0$$

which is satisfied in the shaded dashed regions in Fig. 7.10. That is, when $0 < \theta < \pi/3$, $2\pi/3 < \theta < \pi$ and $-2\pi/3 < \theta < -\pi/3$.

The path of integration C_1 can be changed to a new path that starts in the region $2\pi/3 < \theta < \pi$ and ends in the region $-2\pi/3 < \theta < -\pi/3$. Specifically, if the path is selected from $\infty e^{j2\pi/3}$ to $\infty e^{-j\pi/3}$, then (7.4.16) reads

$$H_\nu^{(1)}(x) \approx \frac{1}{\pi} \int_{\infty e^{j2\pi/3}}^{\infty e^{-j\pi/3}} e^{j(\nu - x)(\beta - \pi/2)} e^{j\frac{x}{6}(\beta - \pi/2)^3} d\beta \tag{7.4.40}$$

Letting

$$\tau = \left(\frac{x}{6}\right)^{1/3} (\beta - \pi/2) e^{j\pi/3} \Rightarrow \beta - \pi/2 = \tau \left(\frac{6}{x}\right)^{1/3} e^{-j\pi/3}$$

(7.4.40) in terms of τ reads

$$H_\nu^{(1)}(x) \approx \frac{1}{\pi} \left(\frac{6}{x}\right)^{1/3} e^{-j\pi/3} \int_{-\infty}^{\infty} e^{j(q\tau - \tau^3)} d\tau \tag{7.4.41}$$

where

$$q = (\nu - x) \left(\frac{6}{x}\right)^{1/3} e^{-j\pi/3}$$

The integral in (7.4.41) is recognized as an integral form of an Airy function. That is,

$$\int_{-\infty}^{\infty} e^{j(q\tau-\tau^3)}d\tau = \frac{2\pi}{3^{1/3}}Ai\left(-\frac{q}{3^{1/3}}\right)$$

which substituting into (7.4.41) results in

$$H_\nu^{(1)}(x) \approx 2\left(\frac{2}{x}\right)^{1/3} e^{-j\pi/3}Ai[-(\nu - x)(2/x)^{1/3}e^{-j\pi/3}] \qquad (7.4.42)$$

The derivation for $H_\nu^{(2)}(x)$ similar. The path of integration C_2 in (7.4.17) is changed to go from $\infty e^{-j2\pi/3}$ to $\infty e^{j\pi/3}$ and the following transformation is used:

$$\tau = \left(\frac{x}{6}\right)^{1/3}(\beta - \pi/2)e^{-j\pi/3}$$

Then, the asymptotic expression for $H_\nu^{(2)}(x)$ is

$$H_\nu^{(2)}(x) \approx 2\left(\frac{2}{x}\right)^{1/3} e^{j\pi/3}Ai[-(\nu - x)(2/x)^{1/3}e^{j\pi/3}] \qquad (7.4.43)$$

It also follows, using differentiation, that

$$H_\nu^{(1)\prime}(x) \approx -2\left(\frac{2}{x}\right)^{2/3} e^{j\pi/3}Ai'[-(\nu - x)(2/x)^{1/3}e^{-j\pi/3}] \qquad (7.4.44)$$

and

$$H_\nu^{(2)\prime}(x) \approx -2\left(\frac{2}{x}\right)^{2/3} e^{-j\pi/3}Ai'[-(\nu - x)(2/x)^{1/3}e^{j\pi/3}] \qquad (7.4.45)$$

The relations in (7.4.42) to (7.4.45) are known as the third-order approximations of the Hankel functions. They are used to determine the zeros of the Hankel functions, since the zeros of the Airy functions are well known (see Appendix D). They are also used in conjunction with the Watson transformation to evaluate infinite sums involving Hankel functions (see Chapters 9 and 10).

The zeros of the Airy functions are denoted by a_p and a_p' (i.e., $Ai(a_p) = 0$ and $Ai'(a_p') = 0$) and listed in Appendix D.

From (7.4.43), the zeros of the Hankel function $H_\nu^{(2)}(x)$ are given by

$$\nu_p = x - a_p\left(\frac{x}{2}\right)^{1/3} e^{-j\pi/3} \qquad (p = 1, 2, 3, \ldots) \qquad (7.4.46)$$

and those of $H_\nu^{(2)\prime}(x)$ by

$$\nu_p' = x - a_p'\left(\frac{x}{2}\right)^{1/3} e^{-j\pi/3} \qquad (p = 1, 2, 3, \ldots) \tag{7.4.47}$$

For example, $a_1 = -2.338$ and $a_1' = -1.019$. Therefore,

$$\nu_1 = x + 2.338\left(\frac{x}{2}\right)^{1/3} e^{-j\pi/3}$$

and

$$\nu_1' = x + 1.019\left(\frac{x}{2}\right)^{1/3} e^{-j\pi/3}$$

A second example of the method of steepest descent is in the evaluation of integrals of the form

$$I = \int_{-\infty}^{\infty} F(\tau) H_n^{(2)}(\lambda\rho) e^{-j\tau z} \lambda d\lambda \tag{7.4.48}$$

where

$$\tau = \sqrt{k^2 - \lambda^2}$$

This type of integral occurs in some problems in cylindrical coordinates (see Chapter 9).

For large values of ρ, such that $\lambda\rho >> n$, the asymptotic representation of the Hankel function is

$$H_n^{(2)}(\lambda\rho) \approx \sqrt{\frac{2}{\pi\lambda\rho}}\, e^{-j(\lambda\rho - n\pi/2 - \pi/4)} \tag{7.4.49}$$

and (7.4.48) reads

$$I \approx \sqrt{\frac{2}{\pi\rho}}\, e^{j(n\pi/2 + \pi/4)} \int_{-\infty}^{\infty} F(\tau)\, e^{-j\lambda\rho} e^{-j\tau z} \sqrt{\lambda}\, d\lambda \tag{7.4.50}$$

Letting

$$\lambda = k \sin\beta \quad z = r\cos\theta \quad \rho = r\sin\theta$$

Then,

$$\tau = \sqrt{k^2 - k^2\sin^2\beta} = k\cos\beta$$

and

$$e^{-j\lambda\rho}e^{-j\tau z} = e^{-jkr\sin\beta\sin\theta}e^{-jkr\cos\beta\cos\theta} = e^{-jkr\cos(\beta-\theta)} \tag{7.4.51}$$

Hence, (7.4.50) can be expressed in the form

$$I \approx \int_{\Gamma} G(\beta)e^{krf(\beta)}d\beta \tag{7.4.52}$$

where

$$G(\beta) = j^n e^{j\pi/4} k\, F(k\cos\beta)\sqrt{\frac{2k}{\pi r\sin\theta}}\sqrt{\sin\beta}\cos\beta \tag{7.4.53}$$

and

$$f(\beta) = -j\cos(\beta-\theta) \tag{7.4.54}$$

The transformation $\lambda = k\sin\beta$ is analyzed in Appendix F. The resulting path of integration Γ in the β plane is shown in Fig. 7.11.

From (7.4.54), the saddle point occurs at

$$f'(\beta) = \frac{d}{d\beta}[-j\cos(\beta-\theta)] = 0 \Rightarrow \sin(\beta-\theta) = 0 \Rightarrow \beta = \beta_s = \theta$$

At the saddle point, $\beta = \beta_s = \theta$, $f(\beta_s) = -j$, and $f''(\beta_s) = j = e^{j\pi/2}$. The function $f''(\beta_s)$ shows that its argument is $\varphi = \pi/2$. Therefore, the inclination of the steepest descent path (SDP) as it crosses the saddle point is

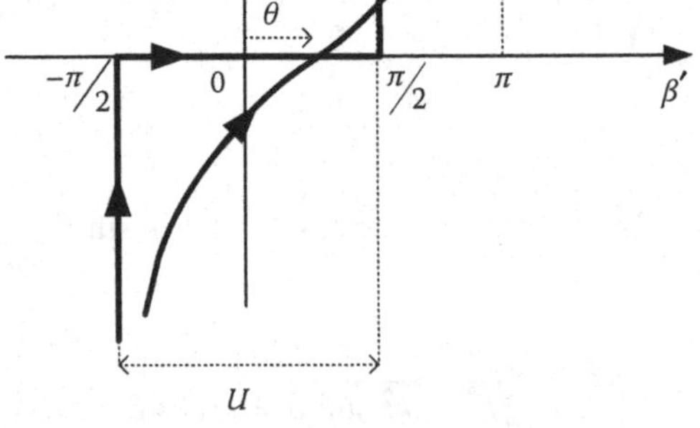

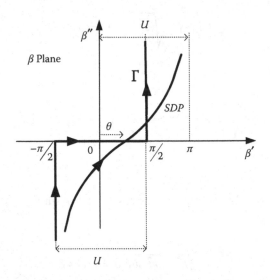

FIGURE 7.11 The path Γ according to the transformation $\lambda = k\sin\beta$ and the SDP.

$$\phi = \pm\frac{\pi}{2} - \frac{\varphi}{2} = \pm\frac{\pi}{2} - \frac{\pi}{4} = \frac{\pi}{4} \text{ or } -\frac{3\pi}{4}$$

To determine the SDP, express (7.4.54) in the form

$$f(\beta) = -j\cos(\beta' + j\beta'' - \theta) = \sin(\beta' - \theta)\sinh\beta'' - j\cos(\beta' - \theta)\cosh\beta''$$

Along the steepest descent path: $\mathrm{Im}[f(\beta)] = \mathrm{Im}[f(\beta_s)] = -1$. Hence, the path follows

$$\mathrm{Im}[f(\beta)] = -1 \Rightarrow \cos(\beta' - \theta)\cosh\beta'' = 1$$

and the real part:

$$e^{kr\sin(\beta'-\theta)\sinh\beta''}$$

has its most rapid decrease.

The SDP is shown in Fig. 7.11. Observe that the inclination of the path at the saddle point is $\phi = \pi/4$ which agrees with the change of the Γ path to the SDP.

The steepest descent path in the s plane follows from the transformation

$$f(\beta) = f(\beta_s) - s^2 \Rightarrow -j\cos(\beta - \theta) = -j - s^2$$

where

$$s^2 = -\frac{1}{2}f''(\theta)(\beta - \theta)^2 = -j\frac{1}{2}(\beta - \theta)^2$$

and

$$\frac{d\beta}{ds} = \sqrt{2}\,e^{j\pi/4}$$

Then, (7.4.52) reads

$$I \approx e^{-jkr}\int_{-\infty}^{\infty} G(\beta)|_{\beta=\theta}\frac{d\beta}{ds}e^{-krs^2}ds \approx e^{-jkr}G(\theta)\sqrt{2}\,e^{j\pi 4}\int_{-\infty}^{\infty} e^{-krs^2}ds$$

$$= G(\theta)\sqrt{\frac{2\pi}{kr}}\,e^{-j(kr-\pi/4)}$$

where from (7.4.53):

$$G(\theta) = j^n e^{j\pi/4} k\,F(k\cos\theta)\cos\theta\sqrt{\frac{2k}{\pi r}}$$

and I reads

$$I \approx 2j^{n+1} k\, F(k \cos\theta) \cos\theta \frac{e^{-jkr}}{r}$$

In terms of (7.4.48), the steepest descent evaluation of the integral is

$$\int_{-\infty}^{\infty} F(\tau) H_n^{(2)}(\lambda\rho) e^{-j\tau z} \lambda d\lambda \approx 2j^{n+1} k\, F(k \cos\theta) \cos\theta \frac{e^{-jkr}}{r} \tag{7.4.55}$$

We assumed that $F(k \cos\beta)$ had no singularities between the path Γ and the SDP. Otherwise, the singularities must be taken into account.

The integral in (7.4.48) can also be evaluated using the transformation

$$\lambda = k \cos\beta$$

In order for the exponent to be of the form in (7.4.51), define the complement angle: $\theta_c = \pi/2 - \theta$. Then,

$$z = r \cos\theta = r \sin\theta_c$$

and

$$\rho = r \sin\theta = r \cos\theta_c$$

Using the asymptotic approximation for the Hankel function in (7.4.49), and with

$$\lambda = k \cos\beta$$

it follows that

$$\tau = \sqrt{k^2 - k^2 \cos^2\beta} = k \sin\beta$$

and

$$e^{-j\lambda\rho} e^{-j\tau z} = e^{-jkr \cos\beta \cos\theta_c} e^{-jkr \sin\beta \sin\theta_c} = e^{-jkr \cos(\beta - \theta_c)}$$

Hence, (7.4.48) reads

$$I \approx \int_{\Gamma} G(\beta) e^{krf(\beta)} d\beta \tag{7.4.56}$$

where

$$G(\beta) = j^n e^{-j3\pi/4} k\, F(k \sin\beta) \sqrt{\frac{2k}{\pi r \cos\theta_c}} \sqrt{\cos\beta}\, \sin\beta \tag{7.4.57}$$

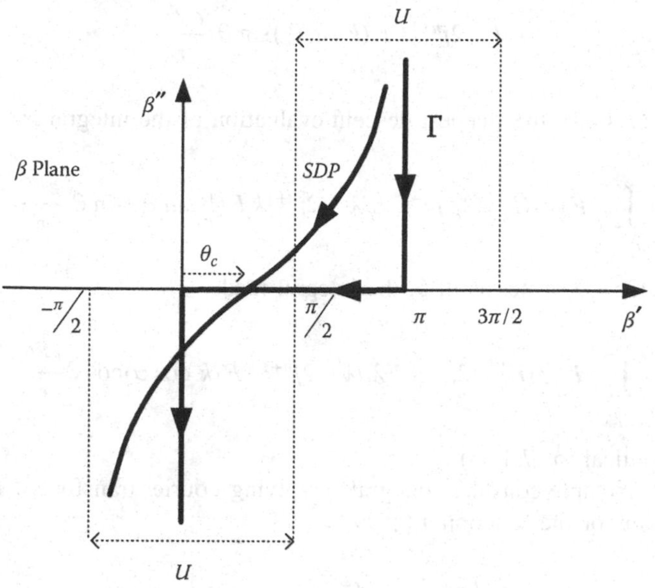

FIGURE 7.12 The path Γ according to the transformation $\lambda = k\cos\beta$ and the SDP.

and

$$f(\beta) = -j\cos(\beta - \theta_c) \tag{7.4.58}$$

The transformation $\lambda = k\cos\beta$ is analyzed in Appendix F. The resulting path of integration Γ in the β plane and the SDP are shown in Fig. 7.12. Observe that the inclination of the path at the saddle point is $\phi = -3\pi/4$ which agree with the change of the Γ path to the SDP.

The evaluation of (7.4.56) using the method of steepest descent is

$$I \approx \int_{SDP} G(\beta) e^{-jkr\cos(\beta-\theta_c)} d\beta \approx e^{-jkr} G(\theta_c)\sqrt{2}\, e^{-j3\pi/4} \int_{-\infty}^{\infty} e^{-krs^2} ds$$

$$\approx G(\theta_c)\sqrt{\frac{2\pi}{kr}}\, e^{-j(kr+3\pi/4)}$$

where from (7.4.57):

$$G(\theta_c) = j^n e^{-j3\pi/4} k\, F(k\sin\theta_c)\sin\theta_c \sqrt{\frac{2k}{\pi r}}$$

and I reads

$$I \approx 2j^{n+1} k\, F(k \sin \theta_c) \sin \theta_c \frac{e^{-jkr}}{r}$$

In terms of (7.4.48), the steepest descent evaluation of the integral is

$$\int_{-\infty}^{\infty} F(\tau) H_n^{(2)}(\lambda\rho) e^{-j\tau z} \lambda d\lambda \approx 2j^{n+1} k\, F(k \sin \theta_c) \sin \theta_c \frac{e^{jkr}}{r}$$

With $\theta_c = \pi/2 - \theta$, in terms of θ, the integral reads

$$\int_{-\infty}^{\infty} F(\tau) H_n^{(2)}(\lambda\rho) e^{-j\tau z} \lambda d\lambda \approx 2j^{n+1} k\, F(k \cos \theta) \cos \theta \frac{e^{jkr}}{r}$$

which is identical to (7.4.55).

The next example considers integrals involving Fourier transforms. The Fourier transform pair for the function $f(z)$ is

$$\begin{cases} F_1(j\alpha) = \int_{-\infty}^{\infty} f(z) e^{-j\alpha z} dz \\ f(z) = \frac{1}{2\pi} \int_{-\infty}^{\infty} F_1(j\alpha) e^{j\alpha z} d\alpha \end{cases}$$

which also appear in the form

$$\begin{cases} F_2(j\alpha) = \int_{-\infty}^{\infty} f(z) e^{j\alpha z} dz \\ f(z) = \frac{1}{2\pi} \int_{-\infty}^{\infty} F_2(j\alpha) e^{-j\alpha z} d\alpha \end{cases}$$

Either form ends up producing the same $f(z)$. In the second form $F_2(j\alpha)$ is the complex conjugate of $F_1(j\alpha)$. Since the inverse of $F_2(j\alpha)$ is formed with $e^{-j\alpha z}$ it produces the same $f(z)$ as the first form. The complex functions $F_1(j\alpha)$ and $F_2(j\alpha)$ are usually simply written as $F_1(\alpha)$ and $F_2(\alpha)$. Either form transforms a function $f(z)$ to the α domain and the inverse recovers $f(z)$. Both forms are found in the literature.

Consider the following integral:

$$I = \int_{-\infty}^{\infty} F(\alpha) H_n^{(2)}(\tau\rho) e^{-j\alpha z} d\alpha \tag{7.4.59}$$

where $z > 0$,

$$\tau = \sqrt{k^2 - \alpha^2}$$

and $\text{Im}(\tau) < 0$. Equation (7.4.59) represents the inverse transform of $F(\alpha) H_n^{(2)}(\tau\rho)$. The factor $1/2\pi$ that appears in the inverse transform has been included in $F(\alpha)$.

This type of inverse transform occurs in some diffraction problems in cylindrical coordinates.

Making the change of variable from α to $-\alpha$ in (7.4.59) we obtain

$$I = \int_{-\infty}^{\infty} F(-\alpha) H_n^{(2)}(\tau\rho) e^{j\alpha z} d\alpha \tag{7.4.60}$$

In many cases, $F(\alpha)$ is an even function, making (7.4.59) and (7.4.60) equal.

Consider the evaluation of (7.4.59). The form of (7.4.59) with $e^{-j\alpha z}$ in the integrand is convenient since it leads to the exponential form in (7.4.62).

For large values of ρ, such that $|\tau\rho| >> n$, the asymptotic representation of the Hankel function in (7.4.49) is used, and (7.4.59) reads

$$I = \sqrt{\frac{2}{\pi\rho}}\, e^{j(n\pi/2+\pi/4)} \int_{-\infty}^{\infty} \frac{F(\alpha)}{\sqrt{\tau}} e^{-j\tau\rho} e^{-j\alpha z} d\alpha \tag{7.4.61}$$

Letting

$$z = r\cos\theta$$
$$\rho = r\sin\theta$$

and using the transformation

$$\alpha = k\cos\beta$$

Then,

$$\tau = \sqrt{k^2 - k^2\cos^2\beta} = k\sin\beta$$

and

$$e^{-j\tau\rho} e^{-j\alpha z} = e^{-jkr\sin\beta\sin\theta} e^{-jkr\cos\beta\cos\theta} = e^{-jkr\cos(\beta-\theta)} \tag{7.4.62}$$

Hence, (7.4.61) reads

$$I \approx \int_{\Gamma} G(\beta) e^{krf(\beta)} d\beta \tag{7.4.63}$$

where

$$G(\beta) = j^n e^{-j3\pi/4} \sqrt{\frac{2k}{\pi r\sin\theta}}\, F(k\cos\beta)\sqrt{\sin\beta}$$

and

$$f(\beta) = -j\cos(\beta - \theta) \tag{7.4.64}$$

The resulting path of integration Γ in the β plane is shown in Fig. 7.12. The saddle point, from (7.4.64), occurs at $\beta = \beta_s = \theta$. Hence, in Fig. 7.12 replace θ_c by θ. The inclination of the steepest descent path (SDP) as it crosses the saddle point is $\phi = -3\pi/4$. Also,

$$\frac{d\beta}{ds} = \sqrt{2}\, e^{-j3\pi/4}$$

The evaluation of (7.4.63) using the method of steepest descent is

$$I = \int_{SDP} G(\beta)e^{-jkr\cos(\beta-\theta)}d\beta \approx G(\theta)e^{-jkr}e^{-j3\pi/4}\sqrt{\frac{2\pi}{kr}}$$

where

$$G(\theta) = j^n e^{-j3\pi/4}\sqrt{\frac{2k}{\pi r}}\,F(k\cos\theta)$$

Therefore, the integral I in (7.4.59) is given by

$$\int_{-\infty}^{\infty} F(\alpha)H_n^{(2)}(\tau\rho)e^{-j\alpha z}d\alpha \approx 2j^{n+1}F(k\cos\theta)\frac{e^{-jkr}}{r} \tag{7.4.65}$$

We assumed that $F(k\cos\beta)$ had no singularities between the path Γ and the SDP. Otherwise, the singularities must be taken into account. The results in (7.4.65) also apply if n is replaced by a non-integer order ν.

Equation (7.4.65) also shows that

$$\int_{-\infty}^{\infty} F(\tau)H_n^{(2)}(\tau\rho)e^{\pm j\alpha z}d\alpha \approx 2j^{n+1}F(k\sin\theta)\frac{e^{-jkr}}{r} \tag{7.4.66}$$

In the case that the integral in (7.4.65) involves the derivative of the Hankel function, whose asymptotic expansion is

$$H_n^{(2)\prime}(\tau\rho) \approx \sqrt{\frac{2}{\pi\tau\rho}}\,e^{-j(\tau\rho - n\pi/2+\pi/4)}$$

Then, it follows that

$$\int_{-\infty}^{\infty} F(\alpha)\,H_n^{(2)\prime}(\tau\rho)e^{-j\alpha z}d\alpha \approx 2j^n F(k\cos\theta)\frac{e^{-jkr}}{r}$$

The integral in (7.4.59) can also be evaluated using the transformation

$$\alpha = k \sin \beta$$

In order for the exponent to be of the form in (7.4.62), define the complement angle: $\theta_c = \pi/2 - \theta$. Then,

$$z = r \cos \theta = r \sin \theta_c \qquad \rho = r \sin \theta = r \cos \theta_c$$

and with

$$\alpha = k \sin \beta$$

it follows that

$$\tau = \sqrt{k^2 - k^2 \sin^2 \beta} = k \cos \beta$$

Using the asymptotic form of the Hankel function the resulting exponential in (7.4.59) is

$$e^{-j\tau\rho} e^{-j\alpha z} = e^{-jkr \cos\beta \cos\theta_c} e^{-jkr \sin\beta \sin\theta_c} = e^{-jkr \cos(\beta - \theta_c)}$$

and it follows that (7.4.59) reads

$$I \approx \int_{\Gamma} G(\beta) e^{krf(\beta)} d\beta \tag{7.4.67}$$

where

$$G(\beta) = j^n e^{j\pi/4} \sqrt{\frac{2k}{\pi r \cos \theta_c}} F(k \sin \beta) \sqrt{\cos \beta}$$

and

$$f(\beta) = -j \cos(\beta - \theta_c)$$

The saddle point occurs at $\beta = \theta_c$ and the inclination of the SDP at the saddle point is $\pi/4$. The resulting path of integration Γ and SDP in the β plane are shown in Fig. 7.11, where θ should be replaced by θ_c.

The evaluation of (7.4.67) using the method of steepest descent is

$$I \approx \int_{SDP} G(\beta) e^{-jkr \cos(\beta - \theta_c)} d\beta \approx G(\theta_c) e^{-jkr} e^{j\pi/4} \sqrt{\frac{2\pi}{kr}}$$

where

$$G(\theta_c) = j^n e^{j\pi/4} \sqrt{\frac{2k}{\pi r}} F(k \sin \theta_c)$$

Therefore, the integral I in terms of θ_c is

$$I \approx \int_{SDP} G(\beta) e^{-jkr \cos(\beta - \theta_c)} d\beta \approx 2j^{n+1} F(k \sin \theta_c) \frac{e^{-jkr}}{r}$$

and (7.4.59) reads

$$\int_{-\infty}^{\infty} F(\alpha) H_n^{(2)}(\lambda\rho) e^{-j\alpha z} d\alpha \approx 2j^{n+1} F(k \sin \theta_c) \frac{e^{-jkr}}{r}$$

which in terms of θ is

$$\int_{-\infty}^{\infty} F(\alpha) H_n^{(2)}(\lambda\rho) e^{-j\alpha z} d\alpha \approx 2j^{n+1} F(k \cos \theta) \frac{e^{-jkr}}{r} \tag{7.4.68}$$

As expected, this result is identical to (7.4.65).

In some of the previous examples the resulting integrals are of the form

$$I = \int_{SDP} G(\beta) e^{krf(\beta)} d\beta \tag{7.4.69}$$

where

$$f(\beta) = -j \cos(\beta - \theta)$$

and the SDP are of the form shown in Figs. 7.11 and 7.12. The saddle point method was used to evaluate (7.4.69) and the first term of the asymptotic expansion was found. The evaluation of the next term in the asymptotic expansion is now presented. To this end, (7.4.15) is used.

Since:

$$\begin{cases} f(\beta) = -j\cos(\beta - \theta) \Rightarrow f(\theta) = -j \\ f'(\beta) = j\sin(\beta - \theta) \Rightarrow f'(\theta) = 0 \\ f''(\beta) = j\cos(\beta - \theta) \Rightarrow f''(\theta) = j \\ f'''(\beta) = -j\sin(\beta - \theta) \Rightarrow f'''(\theta) = 0 \\ f''''(\beta) = -j\cos(\beta - \theta) \Rightarrow f''''(\theta) = -j \end{cases}$$

it follows from (7.4.15) that for the SDP in Fig. 7.11:

$$I \approx G(\theta)\sqrt{\frac{2\pi}{kr}}\, e^{-jkr} e^{j\pi/4}\left\{1 + \frac{j}{2kr}\left[\frac{1}{4} + \frac{G''(\theta)}{G(\theta)}\right] + \ldots\right\}$$

and for the SDP in Fig. 7.12 replace $e^{j\pi/4}$ by $e^{-j3\pi/4}$ in the above expression.
In the next example, consider the evaluation of the integral:

$$I = \int_{\Gamma} \frac{e^{-jkr\cos(\beta-\theta)}}{\beta - \beta_p} d\beta \tag{7.4.70}$$

which has a pole at $\beta = \beta_p$. The path of integration Γ is shown in Fig. 7.13a. In (7.4.70):

$$f(\beta) = -j\cos(\beta - \theta)$$

which shows that the saddle point is at $\beta = \theta$, and the SDP is given by $\mathrm{Im}(\beta) = -1$. That is,

$$\mathrm{Im}[-j\cos(\beta - \theta)] = \mathrm{Im}[-j\cos(\beta' - \theta)\cosh\beta'' - \sin(\beta' - \theta)\sinh\beta''] = -1$$

and the SDP satisfies

$$\cos(\beta' - \theta)\cosh\beta'' = 1 \Rightarrow \beta' = \theta + \cos^{-1}(\mathrm{sech}\beta'')$$

This equation describes the path of steepest descent. It passes thru the saddle point when $\beta'' = 0$, and the asymptotes of the SDP path (i.e., when $\beta'' \to \pm\infty$) are seen to be at $\beta' \to \theta \pm \pi/2$. The SDP is shown in Fig. 7.13a.

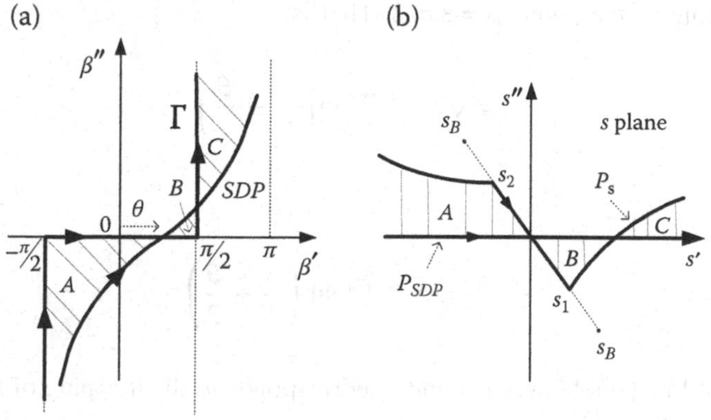

FIGURE 7.13 (a) Path of integration Γ and the SDP in the β plane; (b) path of integration P_s and the SDP (i.e., P_{SDP}) in the s plane.

If the pole β_p is located in the shaded regions, labeled A, B, and C, the pole will be enclosed when changing the path Γ into the SDP. Hence, its residue must be accounted for.

In the s plane, the transformation:

$$f(\beta) = f(\theta) - s^2 \Rightarrow \cos(\beta - \theta) = 1 - js^2 \tag{7.4.71}$$

shows that

$$s = \pm e^{-j\pi/4}\sqrt{1 - \cos(\beta - \theta)} = \pm\sqrt{2}\, e^{-j\pi/4} \sin\left(\frac{\beta - \theta}{2}\right) \tag{7.4.72}$$

Hence, it follows from (7.4.72) that:

$$\frac{ds}{d\beta} = \pm\frac{1}{\sqrt{2}} e^{-j\pi/4} \cos\left(\frac{\beta - \theta}{2}\right) \Rightarrow \left.\frac{d\beta}{ds}\right|_{s=0} = \pm\sqrt{2}\, e^{j\pi 4} \tag{7.4.73}$$

The positive sign in (4.7.71) and (4.7.72) must be used since the phase of $\beta - \theta$ is $\pi/4$ as the SDP crosses the saddle point.

Equation (7.4.72) describe how the path Γ transforms into the path P_s in the s plane, as shown in Fig. 7.13b. The locations of the regions A, B, and C in the s plane are also shown. The SDP in the s plane, labeled P_{SDP}, is along the real axis and extends from $-\infty$ to ∞. The pole at β_p transforms into the s plane according to

$$s_p = \sqrt{2}\, e^{-j\pi/4} \sin\left(\frac{\beta_p - \theta}{2}\right)$$

If s_p is located in the shaded regions in Fig. 7.13b (i.e., in A, B, and C) the pole will be enclosed when changing the path P_s into the P_{SDP}. The points s_1 and s_2 correspond to the mapping of the points $\beta = \pm\pi/2$. That is,

$$s_1 = \sqrt{2}\, e^{-j\pi/4} \sin\left(\frac{\pi}{4} - \frac{\theta}{2}\right)$$

and

$$s_2 = -\sqrt{2}\, e^{-j\pi/4} \sin\left(\frac{\pi}{4} + \frac{\theta}{2}\right)$$

The straight line path between s_1 and s_2 corresponds to the mapping of the straight Γ path between $-\pi/2$ and $\pi/2$.

The evaluation of (7.4.70) is

$$I = I_{Pole} + e^{-jkr} \int_{-\infty}^{\infty} \frac{1}{\beta - \beta_p} \frac{d\beta}{ds} e^{-krs^2} ds \tag{7.4.74}$$

where

$$I_{Pole} = \begin{cases} 2\pi j e^{-jkr\cos(\beta_p - \theta)} \text{ for pole in regions } A \text{ and } C \\ -2\pi j e^{-jkr\cos(\beta_p - \theta)} \text{ for pole in region } B \\ 0 \text{ for pole outside regions } A, B \text{ and } C \end{cases}$$

In (7.4.74), the term $d\beta/ds$ needs to be expressed in terms of s. From (7.4.73):

$$\frac{d\beta}{ds} = \frac{\sqrt{2}\, e^{j\pi/4}}{\cos\left(\frac{\beta-\theta}{2}\right)} = 2\sqrt{2}\, e^{j\pi/4} \frac{\sin\left(\frac{\beta-\theta}{2}\right)}{\sin(\beta - \theta)} \tag{7.4.75}$$

Using (7.4.71) and (7.4.72):

$$\sin\left(\frac{\beta - \theta}{2}\right) = \frac{e^{j\pi/4}}{\sqrt{2}} s$$

$$\sin(\beta - \theta) = \sqrt{1 - \cos^2(\beta - \theta)} = \sqrt{1 - (1 - js^2)^2} = s\sqrt{s^2 + j2}$$

Hence, (7.4.75) reads

$$\frac{d\beta}{ds} = \frac{2j}{\sqrt{s^2 + j2}} \approx \sqrt{2}\, e^{j\pi/4}\left(1 - j\frac{s^2}{2}\right)^{-1/2} \approx \sqrt{2}\, e^{j\pi/4}\left(1 + j\frac{s^2}{4} - \frac{3}{32}s^4 + \ldots\right) \tag{7.4.76}$$

The square root in (7.4.76) introduces branch points at $s_{BP} = \pm\sqrt{2}\, e^{-j\pi/4}$ (see Fig. 7.13b). The convergence of (7.4.74) is limited by the locations of the pole and the branch points. If $|s_p| > \sqrt{2}$ the circle of convergence is limited by $|s_{BP}| = \sqrt{2}$. If $|s_p| < \sqrt{2}$ the circle of convergence is limited by $|s_p|$.

The evaluation of (7.4.74), using the first term in (7.4.76) is

$$\begin{aligned} I &\approx I_{Pole} + e^{-jkr}\sqrt{2}\, e^{j\pi/4} \frac{1}{\theta - \beta_p} \int_{-\infty}^{\infty} e^{-krs^2} ds \\ &\approx I_{Pole} + \sqrt{\frac{2\pi}{kr}} \frac{e^{-j(kr - \pi/4)}}{\theta - \beta_p} \end{aligned} \tag{7.4.77}$$

Using (7.4.15), the first two terms of the asymptotic expansion are

$$I \approx I_{Pole} + \sqrt{\frac{2\pi}{kr}} \frac{e^{-j(kr-\pi/4)}}{\theta - \beta_p} \left\{ 1 + \frac{j}{2kr} \left[\frac{1}{4} + \frac{2}{\left(\theta - \beta_p\right)^2} \right] + \ldots \right\} \tag{7.4.78}$$

Observe in (7.4.77) and (7.4.78) that when β_p approaches θ (i.e., pole close to the saddle point), the series is of no use. What occurs is that the radius of convergence of the series approaches zero. In this case, the modified saddle point method in the next section must be used.

The effect of the pole depends on its amplitude. With $\beta_p = \beta_p' + j\beta_p''$, it follows that its amplitude depends on:

$$e^{-kr \sin(\beta_p' - \theta) \sinh \beta_p''}$$

which can be very small or, in some case (e.g., if $\beta_p'' \to 0$) its magnitude remains fairly constant and becomes the dominant term.

7.5 POLE NEAR THE SADDLE POINT

In this section, the modified saddle point method is discussed. In some cases, a pole of the integrand in

$$I(\rho) = \int_{\Gamma} G(\beta) e^{\rho f(\beta)} d\beta$$

is close to the saddle point. With

$$f(\beta) = f(\beta_s) - s^2$$

the integral reads:

$$I(\rho) = e^{\rho f(\beta_s)} \int_{P_s} G(s) \frac{d\beta}{ds} e^{-\rho s^2} ds \tag{7.5.1}$$

where P_s is the path in the s plane, such as the one shown in Fig. 7.13b.

If a pole at $s = s_p$ is close to the saddle point at $s = 0$, the Taylor expansion of the term $G(s)(d\beta/ds)$ in (7.5.1) is valid within a small radius of convergence determined by the distance from the pole at s_p to the origin at $s = 0$ (i.e., the saddle point). To handle such a situation, this term is written as a sum of a term with a pole at $s = s_p$, and a term that is analytic everywhere. That is,

$$G(s) \frac{d\beta}{ds} = \frac{a_{-1}}{s - s_p} + G_1(s) \tag{7.5.2}$$

where a_{-1} is the residue at the pole s_p, and $G_1(s)$ is the analytic part.

Expressing $G(s)$ in terms of its numerator $N(s)$ and denominator $D(s)$:

$$G(s) = \frac{N(s)}{D(s)}$$

it follows that the residue is given by

$$a_{-1} = \left. \frac{N(s)}{dD(s)/ds} \frac{d\beta}{ds} \right|_{s=s_p}$$

Substituting (7.5.2) into (7.5.1) and changing the path P_s to the SDP path (i.e., P_{SDP}) shows that

$$I(\rho) \approx a_{-1} e^{\rho f(\beta_s)} \int_{-\infty}^{\infty} \frac{1}{s - s_p} e^{-\rho s^2} ds + e^{\rho f(\beta_s)} \int_{-\infty}^{\infty} G_1(s) e^{-\rho s^2} ds \qquad (7.5.3)$$

The second integral in (7.5.3) is analytic and is evaluated using (7.4.11), namely

$$e^{\rho f(\beta_s)} \int_{-\infty}^{\infty} G_1(s) e^{-\rho s^2} ds \approx e^{\rho f(\beta_s)} G_1(0) \sqrt{\frac{\pi}{\rho}}$$

The first integral in (7.5.3) is denoted by $f(\rho)$. That is,

$$f(\rho) = \int_{-\infty}^{\infty} \frac{1}{s - s_p} e^{-\rho s^2} ds$$

Its evaluation depends on whether the pole is above or below the real axis in the s plane. The integral is not defined for $\operatorname{Im} s_p = 0$ (i.e., for real values of s_p).

The analysis of the integral as function of s_p when $\operatorname{Im} s_p \to 0$ shows that the integral is discontinuous across the real s axis. When the pole approaches the real axis from above (i.e., with $s_p = x + j\varepsilon$, $\varepsilon \to 0$) the path of integration is indented as shown Fig. 7.14a, and when approaching from below (i.e., with $s_p = x - j\varepsilon$, $\varepsilon \to 0$) the path is indented as shown in Fig. 7.14b. Next, the following difference is formed:

$$f(\rho, x + j\varepsilon) - f(\rho, x - j\varepsilon) = \int_{C_1} \frac{1}{s - s_p} e^{-\rho s^2} ds - \int_{C_2} \frac{1}{s - s_p} e^{-\rho s^2} ds$$

The contributions from the straight portions of the path cancel, leaving the circular contour around the pole at $s_p = x$. Hence, the residue at the pole is

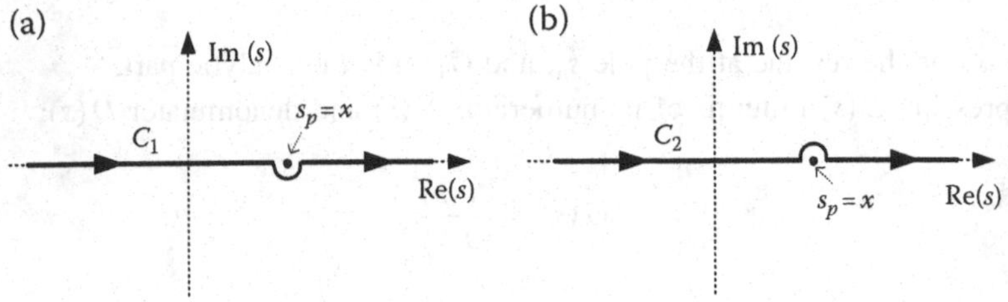

FIGURE 7.14 (a) Path of integration C_1 and pole $s_p = x + j\varepsilon$ ($\varepsilon \to 0$); (b) path of integration C_2 and pole $s_p = x - j\varepsilon$ ($\varepsilon \to 0$).

$$f(\rho, x + j\varepsilon) - f(\rho, x - j\varepsilon) = 2\pi j\, e^{-\rho x^2}$$

which shows the discontinuity of $f(\rho)$ across the real axis.

We first consider the evaluation of $f(\rho)$ for the case that $\mathrm{Im}(s_p) > 0$ and the pole at s_p is not captured when changing the path P_s to P_{SDP}. The integration contour and pole location are shown in Fig. 7.15. Multiplying and dividing the integrand of $f(\rho)$ by $s + s_p$ gives

$$f(\rho) = \int_{-\infty}^{\infty} \frac{1}{s - s_p} e^{-\rho s^2} ds = \int_{-\infty}^{\infty} \frac{s}{s^2 - s_p^2} e^{-\rho s^2} ds + \int_{-\infty}^{\infty} \frac{s_p}{s^2 - s_p^2} e^{-\rho s^2} ds$$

The first integral on the right-hand side is odd, and therefore zero. Hence,

$$f(\rho) = s_p \int_{-\infty}^{\infty} \frac{1}{s^2 - s_p^2} e^{-\rho s^2} ds \tag{7.5.4}$$

Multiplying (7.5.4) by $e^{\rho s_p^2}$:

$$f(\rho) e^{\rho s_p^2} = s_p \int_{-\infty}^{\infty} \frac{e^{-\rho (s^2 - s_p^2)}}{s^2 - s_p^2} ds \tag{7.5.5}$$

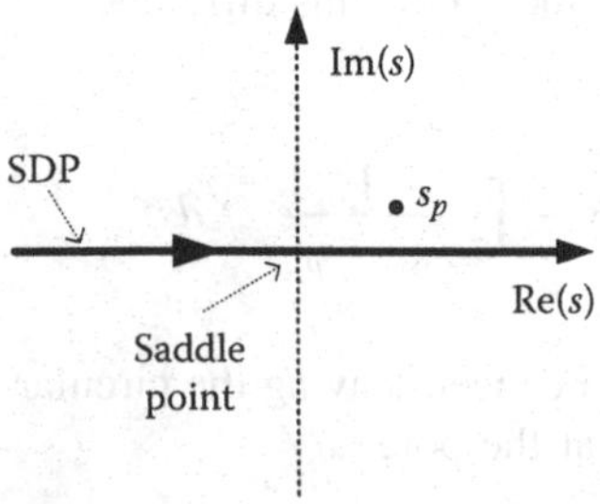

FIGURE 7.15 Path on integration and pole s_p with $\mathrm{Im}(s_p) > 0$.

Next, taking the derivative of (7.5.5) with respect to ρ shows that

$$\frac{d}{d\rho}\left[f(\rho)e^{\rho s_p^2}\right] = -s_p \int_{-\infty}^{\infty} e^{-\rho(s^2-s_p^2)}ds = -s_p e^{\rho s_p^2}\sqrt{\frac{\pi}{\rho}} \tag{7.5.6}$$

Integrating (7.5.6) over ρ from 0 to ρ, gives

$$f(\rho)e^{\rho s_p^2} - f(0) = -\sqrt{\pi}\, s_p \int_0^{\rho} \frac{e^{\rho s_p^2}}{\sqrt{\rho}} d\rho \tag{7.5.7}$$

where the integral can be expressed in terms of the error function (see Appendix G). That is, letting $\rho s_p^2 = -z^2$ in the integral. Then,

$$z = j\sqrt{\rho}\, s_p$$

and

$$d\rho = -\frac{2z}{s_p^2}dz$$

Hence, (7.5.7) reads

$$f(\rho)e^{\rho s_p^2} - f(0) = 2j\sqrt{\pi}\int_0^{j\sqrt{\rho}\, s_p} e^{-z^2}dz = j\pi\, \mathrm{erf}(j\sqrt{\rho}\, s_p)$$

or

$$f(\rho) = f(0)e^{-\rho s_p^2} + j\pi e^{-\rho s_p^2}\mathrm{erf}(j\sqrt{\rho}\, s_p) \tag{7.5.8}$$

The integral for $f(0)$ is given by

$$f(0) = s_p \int_{-\infty}^{\infty} \frac{1}{s^2 - s_p^2} ds \tag{7.5.9}$$

and is evaluated using residue theory. The pole s_p can have an imaginary part that can be either greater or less than zero. For $\mathrm{Im}\, s_p > 0$, the pole is located at $s = s_p$ and the path of integration is closed with a semicircular path in the upper-half plane, since the contribution from the semicircular path is zero as the radius of the path goes to infinity. Then, the evaluation of (7.5.9) is

$$f(0) = 2\pi j s_p\left[\frac{(s - s_p)}{(s + s_p)(s - s_p)}\right]_{s=s_p} = j\pi \qquad (\mathrm{Im}\, s_p > 0) \tag{7.5.10}$$

If Im $s_p < 0$, the pole at $s = s_p$ requires that the path of integration be closed in the lower-half plane. In this case:

$$f(0) = -j\pi \qquad (\mathrm{Im} s_p < 0) \tag{7.5.11}$$

Substituting (7.5.10) into (7.5.8) gives

$$\begin{aligned} f(\rho) &= j\pi e^{-\rho s_p^2} + j\pi e^{-\rho s_p^2}\,\mathrm{erf}(j\sqrt{\rho}\,s_p) \\ &= j\pi e^{-\rho s_p^2}[1 + \mathrm{erf}(j\sqrt{\rho}\,s_p)] \\ &= j\pi e^{-\rho s_p^2}[2 - \mathrm{erfc}(j\sqrt{\rho}\,s_p)] \\ &= j\pi e^{-\rho s_p^2}\,\mathrm{erfc}(-j\sqrt{\rho}\,s_p) \qquad (\mathrm{Im} s_p > 0) \end{aligned} \tag{7.5.12}$$

where erfc(z) is the complementary error function (see Appendix G). That is,

$$\mathrm{erfc}(z) = \frac{2}{\sqrt{\pi}}\int_z^{\infty} e^{-t^2}dt \qquad [\arg(z) \le \pi/4]$$

where the path of integration in the t plane is directed from z towards the right to ∞. The lower limit with Im $s_p > 0$ is such that $\mathrm{Re}(-j\sqrt{\rho}\,s_p) \to \infty$ as $\rho \to \infty$, making $\mathrm{erfc}(-j\sqrt{\rho}\,s_p) \to 0$ as $\rho \to \infty$.

For Im $s_p < 0$, substituting (7.5.11) into (7.5.8) gives

$$\begin{aligned} f(\rho) &= -j\pi e^{-\rho s_p^2}[1 - \mathrm{erf}(j\sqrt{\rho}\,s_p)] \\ &= -j\pi e^{-\rho s_p^2}[1 + \mathrm{erf}(-j\sqrt{\rho}\,s_p)] \\ &= -j\pi e^{-\rho s_p^2}[2 - \mathrm{erfc}(-j\sqrt{\rho}\,s_p)] \\ &= -j\pi e^{-\rho s_p^2}\,\mathrm{erfc}(j\sqrt{\rho}\,s_p) \qquad (\mathrm{Im} s_p < 0) \end{aligned} \tag{7.5.13}$$

Since Im $s_p < 0$, it follows that $\mathrm{Re}(j\sqrt{\rho}\,s_p) \to \infty$ and $\mathrm{erfc}(j\sqrt{\rho}\,s_p) \to 0$ as $\rho \to \infty$.

The functions $f(\rho)$ in (7.5.12) and (7.5.13) which represent (7.5.4) for $\mathrm{Im}(s_p) > 0$ and $\mathrm{Im}(s_p) < 0$ are discontinuous as s_p crosses the real axis. An electromagnetic propagation problem requires that its solution varies continuously as the physical parameters vary. If the constitutive parameters make the pole s_p cross the real axis, the function $f(\rho)$ must be continued analytically. For example, consider the case that $\mathrm{Im}(s_p) < 0$ shown in Fig. 7.16, where $f(\rho)$ is given by (7.5.13). If the imaginary part varies, moving the pole into the upper half plane, the path of integration must be continued analytically, as shown in Fig. 7.16. This results in a new value of $f(\rho)$ that includes the residue at the pole. The analytic continuation is

$$f(\rho) = -j\pi e^{-\rho s_p^2}\,\mathrm{erfc}(j\sqrt{\rho}\,s_p) \Rightarrow -2\pi j e^{-\rho s_p^2} + j\pi e^{-\rho s_p^2}\,\mathrm{erfc}(-j\sqrt{\rho}\,s_p)$$

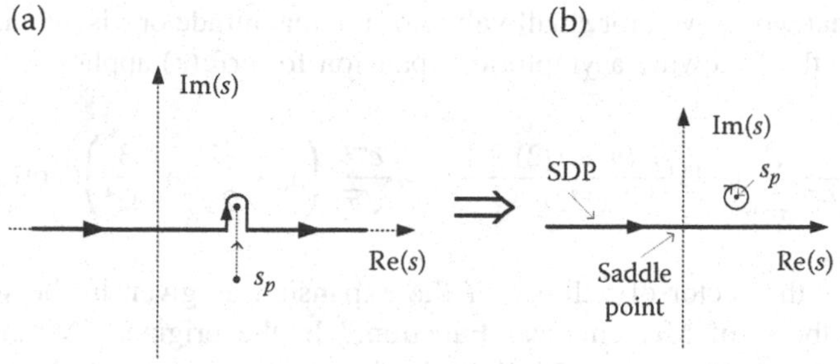

FIGURE 7.16 Analytic continuation of the path as the pole s_p varies from $\mathrm{Im}(s_p) < 0$ to $\mathrm{Im}(s_p) > 0$.

The first term represents the residue at the pole and the second term the value of the integral for $\mathrm{Im}(s_p) > 0$.

The evaluation of (7.5.3) for the case that $\mathrm{Im}(s_p) > 0$, using (7.5.12) is

$$I \approx j\pi a_{-1} e^{\rho f(\beta_s)} e^{-\rho s_p^2} \operatorname{erfc}(-j\sqrt{\rho}\, s_p) + e^{\rho f(\beta_s)} G_1(0)\sqrt{\frac{\pi}{\rho}} \qquad (\mathrm{Im} s_p > 0) \quad (7.5.14)$$

and for $\mathrm{Im}(s_p) < 0$, using (7.5.13):

$$I \approx -j\pi a_{-1} e^{\rho f(\beta_s)} e^{-\rho s_p^2} \operatorname{erfc}(j\sqrt{\rho}\, s_p) + e^{\rho f(\beta_s)} G_1(0)\sqrt{\frac{\pi}{\rho}} \qquad (\mathrm{Im} s_p < 0) \quad (7.5.15)$$

where in (7.5.14) and (7.5.15) the complementary error functions must be continued analytically in the case that Im s_p varies and crosses the real axis. In many cases the contribution from the analytic part of $G(s)$ (i.e., $G_1(s)$) is negligible. In these cases, the last terms in (7.5.14) and (7.5.15) are neglected.

Next, we consider the case where the pole at s_p (with $\mathrm{Im}(s_p) > 0$) is captured when changing the path of integration from P_s to P_{SDP}. In this case, the contribution from the pole:

$$-2\pi j\, a_{-1} e^{\rho f(\beta_s)} e^{-\rho s_p^2}$$

must be added to (7.5.14). Similarly, when $\mathrm{Im}(s_p) < 0$ and the pole is captured, its contribution is

$$2\pi j\, a_{-1} e^{\rho f(\beta_s)} e^{-\rho s_p^2}$$

and its contribution must be added to (7.5.15).

The complementary error functions that appear in the previous relations can be computed using numerical integration. However, closed form expressions are desirable since they describe the behavior of fields. In Appendix G, a uniform expansion that works well for small values of the magnitude or z is given. For larger values of $|z|$, the following asymptotic expansion for $\operatorname{erfc}(z)$ applies:

$$\operatorname{erfc}(z) = \frac{e^{-z^2}}{\sqrt{\pi}} \sum_{n=0}^{\infty} (-1)^n \frac{\Gamma(n+1/2)}{\Gamma(1/2)} \frac{1}{z^{2n+1}} = \frac{e^{-z^2}}{\sqrt{\pi}\, z}\left(1 - \frac{1}{2z^2} + \frac{3}{4z^4}\right)(|\arg(z)| \leq 3\pi/4)$$

which shows the sector of validity of the expansion as given in the well-known "NIST Handbook of Mathematical Function." In the original "Abramowitz and Stegun" handbook, the sector of validity is given as $|\arg(z)| \leq \pi/4$. As explained in the NIST handbook, for a complex z with $|\arg(z)| \leq \pi/4$ the remainder terms are bounded in magnitude by the first term neglected. Other bounded conditions are listed for the other sectors between $\pm\, \pi/4$ and $\pm\, 3\pi/4$.

The asymptotic expansion of a complex function with an essential singularity at $z = \infty$ is usually valid only in a given sector, say $|\arg(z)| \leq \alpha$, with a different asymptotic expansions in another sector. The change in form of the asymptotic expansion at the boundary of a sector is known as the Stokes phenomenon.

As an example of the Stokes phenomenon, consider the behavior of the function

$$\cosh z^2 = \frac{e^{z^2} + e^{-z^2}}{2} \tag{7.5.16}$$

as $|z| \to \infty$.

Letting $z = re^{j\theta}$, then

$$z^2 = r^2 e^{j2\theta} = r^2(\cos 2\theta + j \sin 2\theta)$$

Since $\cos 2\theta > 0$ when $-\pi/4 < \theta < \pi/4$ and $3\pi/4 < \theta < 5\pi/4$, it follows that $\operatorname{Re}(z^2) > 0$ in these ranges. Also, $\cos 2\theta < 0$ when $\pi/4 < \theta < 3\pi/4$ and $-3\pi/4 < \theta < -\pi/4$, and it follows that $\operatorname{Re}(z^2) < 0$ in these ranges.

The behavior of (7.5.16) as $|z| \to \infty$ is determined by either $e^{\operatorname{Re}(z^2)}$ or $e^{-\operatorname{Re}(z^2)}$. Since:

$$\begin{cases} e^{\operatorname{Re}(z^2)} > e^{-\operatorname{Re}(z^2)} \text{ in } \pi/4 < \theta < \pi/4 \text{ and } 3\pi/4 < \theta < 5\pi/4 \\ e^{-\operatorname{Re}(z^2)} > e^{\operatorname{Re}(z^2)} \text{ in } \pi/4 < \theta < 3\pi/4 \text{ and } -3\pi/4 < \theta < -\pi/4 \end{cases}$$

Then, as $|z| \to \infty$:

$$\cosh z^2 \approx \frac{e^{z^2}}{2} \text{ in } \pi/4 < \theta < \pi/4 \text{ and } 3\pi/4 < \theta < 5\pi/4$$

and

$$\cosh z^2 \approx \frac{e^{-z^2}}{2} \text{ in } \pi/4 < \theta < 3\pi/4 \text{ and } -3\pi/4 < \theta < -\pi/4$$

The behavior of the function changes when $\arg(z) = \pm\pi/4$ and $\arg(z) = \pm 3\pi/4$. The lines that describe these sectors are called anti-Stokes lines. Along these lines the functions e^{z^2} and e^{-z^2} change from being dominant in one sector to subdominant in the other sector. The lines where the subdominant terms are the smallest are called the Stokes lines. For this example, the lines $\arg(z) = 0,\ \pm\pi/2,$ and π are the Stokes lines.

Problems

P7.1 Use the method of residues to show that

$$\sum_{n=0}^{\infty} \frac{(-1)^n}{n^2 + a^2} = \frac{1}{2a^2} + \frac{\pi}{2a} \frac{1}{\sinh(\pi a)}$$

P7.2

a. Show that

$$\oint_C \frac{e^{j\nu\phi}}{1 - e^{j2\pi\nu}} B(\nu)d\nu = -\sum_{n=-\infty}^{\infty} B(n)e^{jn\phi}$$

where C is a closed contour that encloses the zeros of $1 - e^{j2\pi\nu}$, as in Fig 7.3a.

b. Explain how the Watson transformation is used if the poles of $B(\nu)$ are on the second and fourth quadrant.

P7.3

a. Use the method of Laplace to show that

$$\int_{-\infty}^{\infty} \frac{\sin x}{x} e^{-\rho \cosh x} dx \approx e^{-\rho} \sqrt{\frac{2\pi}{\rho}}$$

b. What is the result if the limits of integration are changed to be from –2 to 2?

P7.4

a. Use the method of stationary phase to show that

$$\int_0^{\infty} x \cos\left[\rho\left(\frac{x^3}{3} - x\right)\right] dx = \text{Re}\left[\int_0^{\infty} x\, e^{j\rho\left(\frac{x^3}{3}-x\right)} dx\right] \approx \sqrt{\frac{\pi}{\rho}} \cos\left(\frac{\pi}{4} - \frac{2\rho}{3}\right)$$

b. What is the result if the limits of integration are changed to be from 0 to 5?

P7.5 Are the following statements true or false? Explain your answers

a. The method of Laplace corresponds to having a saddle point on the real axis at $x = x_o$ with a maximum such that $\partial^2 u/\partial x^2 < 0$ and a minimum along the imaginary axis where $\partial^2 u/\partial y^2 > 0$.

b. The method of stationary phase corresponds to having a saddle point where the path of integration follows the path $\psi + 2\phi = \pi/2$.

P7.6

a. Referring to Fig. 7.9 and since $2J_\nu(x) = H_\nu^{(1)}(x) + H_\nu^{(2)}(x)$, determine the path of integration for the evaluation of $J_\nu(x)$, and the Sommerfeld integral representation for $J_\nu(x)$.

b. Perform the third-order approximation to show that

$$J_\nu(x) \approx \left(\frac{2}{x}\right)^{1/3} Ai\left[(\nu - x)\left(\frac{2}{x}\right)^{1/3}\right]$$

HINT: For $J_\nu(x)$ in Fig. 7.10, the required transformation is

$$\tau = -\left(\frac{x}{6}\right)^{1/3}\left(\beta - \frac{\pi}{2}\right)$$

P7.7

a. Show that if (7.4.59) is evaluated with $\alpha = k \cos\beta$, $z = r\cos\theta$, and $\rho = r\sin\theta$ the resulting integral is of the form

$$\int_\Gamma G(\beta) e^{jkr\cos(\beta+\theta)} d\beta$$

b. Determine $G(\beta)$.

c. Show the saddle point and draw the SDP.

d. Evaluate the integral.

P7.8 Use the method of steepest descent to evaluate the integral.

$$\int_{-\infty}^{\infty} F(\alpha)H_n^{(2)}(\tau\rho)e^{j\alpha z}d\alpha$$

P7.9 Show that

$$\int_{-\infty}^{\infty} \frac{e^{-\rho s^2}}{s - jb}ds = j\pi e^{\rho b^2}\,\mathrm{erfc}(\sqrt{\rho}\,b)$$

where $b > 0$.

8 Further Studies of Electromagnetic Waves in Rectangular Geometries

8.1 THE PARALLEL-PLATE WAVEGUIDE

8.1.1 TM Modes

The parallel-plate waveguide model is shown in Fig. 8.1, where the plates extend from $-\infty < y < \infty$. This simple waveguide configuration provides useful information to several wave propagation problems. For example, as a first approximation the Earth-Ionosphere waveguide can be analyzed at ELF and LF by such model, where the parallel plates are assumed to be perfect conductors. The TM modes that propagate in Fig. 8.1 can be derived using the **A** vector or the $\mathbf{\Pi}^e$ vector.

The z component of $\mathbf{\Pi}^e$ satisfies (1.7.5) with $\beta = \beta_o = \omega\sqrt{\mu_o \varepsilon_o}$ (or (4.2.1) with $\Psi = \Pi_z^e$ and $\gamma = j\beta_o$). That is,

$$\left[\frac{\partial^2}{\partial x^2} + \frac{\partial^2}{\partial z^2} + \beta_o^2\right]\Pi_z^e(x, z) = 0 \tag{8.1.1}$$

where the fields do not depend on y. The fields are given by (4.2.7) to (4.2.12), namely

$$E_x = \frac{\partial^2 \Pi_z^e}{\partial x \partial z} \tag{8.1.2}$$

$$E_z = \frac{\partial^2 \Pi_z^e}{\partial z^2} + \beta_o^2 \Pi_z^e = -\frac{\partial^2 \Pi_z^e}{\partial x^2} \tag{8.1.3}$$

and

$$H_y = \frac{\beta_o^2}{j\omega\mu_o}\frac{\partial \Pi_z^e}{\partial x} = -j\omega\varepsilon_o \frac{\partial \Pi_z^e}{\partial x} \tag{8.1.4}$$

The mode function solution to (8.1.1) that satisfy the boundary conditions at $z = 0$ and $z = h$ (i.e., $E_x = 0$ at $z = 0$ and $z = h$) and propagates in the x direction is

DOI: 10.1201/9781003219729-8

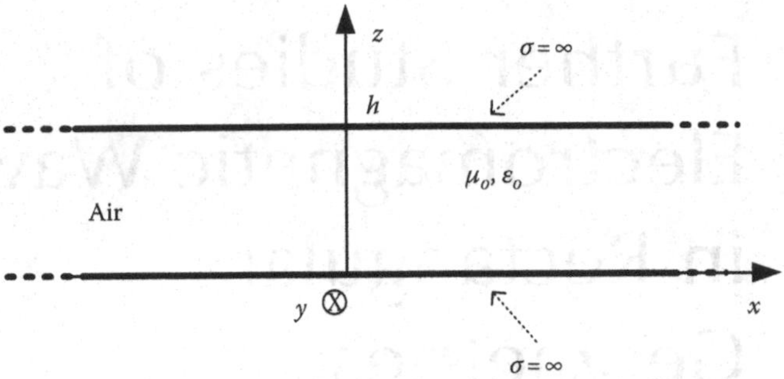

FIGURE 8.1 Geometry for a parallel-plate waveguide.

$$\Pi_z^e = C_n \cos(\beta_z z)\, e^{-j\beta_x x}$$

where

$$\beta_z = \frac{n\pi}{h} \qquad (n = 0, 1, 2, \ldots)$$

and

$$\beta_o^2 = \beta_x^2 + \beta_z^2 \Rightarrow \beta_x = \sqrt{\beta_o^2 - \left(\frac{n\pi}{h}\right)^2} \tag{8.1.5}$$

The resulting fields, from (8.1.2) to (8.1.4), are:

$$\begin{cases} E_x = jC_n\beta_x\beta_z \sin(\beta_z z)\, e^{-j\beta_x x} \\ E_z = C_n\beta_x^2 \cos(\beta_z z)\, e^{-j\beta_x x} \\ H_y = -C_n\omega\varepsilon_o\beta_x \cos(\beta_z z)\, e^{-j\beta_x x} \end{cases} \tag{8.1.6}$$

The excitation factors C_n of the modes are determined by the source that generates the fields in the waveguide.

The modes are defined in terms of the value of n, where $n = 0, 1, 2, \ldots$. The $n = 0$ mode, denoted by TM_0, from (8.1.5) occurs when $\beta_z = 0$, $\beta_x = \beta_o = \omega\sqrt{\mu_o\varepsilon_o}$ and from (8.1.6) the fields are

$$E_z = C_n\beta_o^2\, e^{-j\beta_o x}$$

and

$$H_y = -C_n\omega\varepsilon_o\beta_o e^{-j\beta_o x}$$

This mode is a TEM mode that propagates at all frequencies.

The cutoff frequency (f_c) of the n mode for $n \geq 1$ occurs when $\beta_x = 0$, and from (8.1.5) is given by

$$\beta_o = \frac{n\pi}{h} \Rightarrow \omega_c = \frac{1}{\sqrt{\mu_o \varepsilon_o}}\left(\frac{n\pi}{h}\right) \Rightarrow f_c = \frac{n}{2h\sqrt{\mu_o \varepsilon_o}}$$

The mode with $n = 1$ is the TM_1 mode and so on for the higher TM modes. For example, in the case of the Earths-Ionosphere waveguide for a height of 75 km, the cutoff for the TM_1 mode is 2 kHz, and for the TM_2 mode is 4 kHz. At a frequency below the cutoff frequency of a mode, β_x in (8.1.5) becomes imaginary and the wave attenuates as a function of x.

It is also observed that the TM modes in Fig. 8.1 can be analyzed in terms of the wave equation satisfied by H_y. That is,

$$\left[\frac{\partial^2}{\partial x^2} + \frac{\partial^2}{\partial z^2} + \beta_o^2\right] H_y(x, z) = 0$$

and E_x and E_z follow from Maxwell's equations, namely

$$E_x = -\frac{1}{j\omega\varepsilon_o}\frac{\partial H_y}{\partial z}$$

and

$$E_z = \frac{1}{j\omega\varepsilon_o}\frac{\partial H_y}{\partial x}$$

Hence, the fields that satisfy the boundary conditions are written in the form:

$$\begin{cases} H_y = B_n \cos(\beta_z z)\, e^{-j\beta_x x} \\ E_x = -jB_n \frac{1}{\omega\varepsilon_o}\beta_z \sin(\beta_z z)\, e^{-j\beta_x x} \\ E_z = -B_n \frac{1}{\omega\varepsilon_o}\beta_x \cos(\beta_z z)\, e^{-j\beta_x x} \end{cases} \tag{8.1.7}$$

The forms of the fields in (8.1.6) and (8.1.7) look different. However, if the constants are related by $B_n = -C_n \omega\varepsilon_o\beta_x$ these equations, as expected, are identical.

The time average Poynting vector of a TM mode is

$$\begin{aligned} \mathbf{P} &= \tfrac{1}{2}\mathrm{Re}(\mathbf{E} \times \mathbf{H}^*) = \tfrac{1}{2}\mathrm{Re}\left[(E_x\hat{\mathbf{a}}_x + E_z\hat{\mathbf{a}}_z) \times H_y^*\hat{\mathbf{a}}_y\right] \\ &= \tfrac{1}{2}\mathrm{Re}\left[-jC_n^2\omega\varepsilon_o\beta_x^2\beta_z \sin(\beta_z z)\cos(\beta_z z)\hat{\mathbf{a}}_z + C_n^2\omega\varepsilon_o\beta_x^3\cos^2(\beta_z z)\hat{\mathbf{a}}_x\right] \\ &= \tfrac{1}{2}C_n^2\omega\varepsilon_o\beta_x^3\cos^2(\beta_z z)\hat{\mathbf{a}}_x \end{aligned}$$

If C_n is complex, replace C_n^2 by $|C_n|^2$.

8.1.2 TE Modes

The TE modes are derived using either the **F** vector or the $\mathbf{\Pi}^{\mathrm{m}}$ vector. The z component of $\mathbf{\Pi}^{\mathrm{m}}$ satisfy (1.7.12) (or (4.2.1)), namely

$$\left[\frac{\partial^2}{\partial x^2} + \frac{\partial^2}{\partial z^2} + \beta_o^2\right] \Pi_z^{\mathrm{m}}(x, z) = 0 \tag{8.1.8}$$

where the fields are given by (4.2.19) to (4.2.24):

$$E_y = j\omega\mu_o \frac{\partial \Pi_z^{\mathrm{m}}}{\partial x}$$

$$H_x = \frac{\partial^2 \Pi_z^{\mathrm{m}}}{\partial x \partial z}$$

$$H_z = \frac{\partial^2 \Pi_z^{\mathrm{m}}}{\partial z^2} + \beta_o^2 \Pi_z^{\mathrm{m}} = -\frac{\partial^2 \Pi_z^{\mathrm{m}}}{\partial x^2}$$

The solution to (8.1.8) that satisfy the boundary conditions at $z = 0$ and $z = h$ (i.e., $E_y = 0$ at $z = 0$ and $z = h$) and propagates in the x direction is

$$\Pi_z^{\mathrm{m}} = D_n \sin(\beta_z z) e^{-j\beta_x x}$$

where

$$\beta_z = \frac{n\pi}{h} \qquad (n = 1, 2, 3, \ldots)$$

and

$$\beta_o^2 = \beta_x^2 + \beta_z^2 \Rightarrow \beta_x = \sqrt{\beta_o^2 - \left(\frac{n\pi}{h}\right)^2}$$

A magnetic source determines the excitation factor D_n of a mode.
The resulting fields are

$$E_y = D_n \omega\mu_o \beta_x \sin(\beta_z z) e^{-j\beta_x x}$$
$$H_x = -jD_n \beta_x \beta_z \cos(\beta_z z) e^{-j\beta_x x}$$
$$H_z = D_n \beta_x^2 \sin(\beta_z z) e^{-j\beta_x x}$$

In this case a TEM mode is not possible since n cannot be zero.

The cut-off frequency of the n mode occurs when $\beta_x = 0$, and is given by

$$\beta_o = \frac{n\pi}{h} \Rightarrow \omega_c = \frac{1}{\sqrt{\mu_o \varepsilon_o}}\left(\frac{n\pi}{h}\right) \Rightarrow f_c = \frac{n}{2h\sqrt{\mu_o \varepsilon_o}}$$

The cutoff frequencies equation for the TE modes is identical to that of the TM modes, except that the lowest TE mode is the TE_1, which occurs when $n = 1$.

The time average Poynting vector of a TE mode is

$$\begin{aligned}\mathbf{P} &= \frac{1}{2}\mathrm{Re}(\mathbf{E} \times \mathbf{H}^*) = \frac{1}{2}\mathrm{Re}\left[E_y \hat{\mathbf{a}}_y \times \left(H_x^* \hat{\mathbf{a}}_x + H_z^* \hat{\mathbf{a}}_z\right)\right] \\ &= \frac{1}{2}\mathrm{Re}\left[-jB_n^2 \omega\mu_o \beta_x^2 \beta_z \sin(\beta_z z)\cos(\beta_z z)\hat{\mathbf{a}}_z + C_n^2 \omega\mu_o \beta_x^3 \sin^2(\beta_z z)\hat{\mathbf{a}}_x\right] \\ &= \frac{1}{2}B_n^2 \omega\mu_o \beta_x^3 \sin^2(\beta_z z)\hat{\mathbf{a}}_x\end{aligned}$$

If B_n is complex, replace B_n^2 by $|B_n|^2$.

8.2 PARALLEL-PLATE WAVEGUIDE WITH A STEP DISCONTINUITY

A parallel-plate waveguide with a step discontinuity is shown in Fig. 8.2, where $-\infty < y < \infty$. The following analysis provides a rough approximation to the effects of a discontinuity in a waveguide system. An application of the model is in the analysis of the fields in the Earth-Ionosphere waveguide as the fields propagate between a day-night interface. For example, at LF and VLF the reflection height of the ionosphere at day time is around 70 km and at night time is around 85 km. Of course, a better approximation for the change in height is to consider a smooth linear transition. Another application is in the modeling of the transition between two waveguides of different heights. The purpose of the model used in Fig. 8.2 is to simply illustrate the matching method when multimodes are generated by the discontinuity.

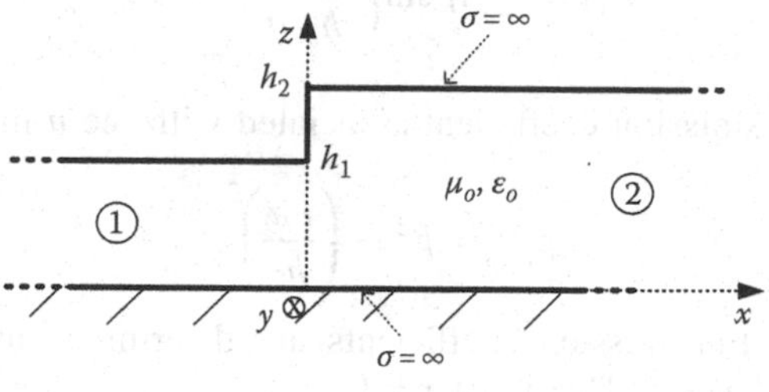

FIGURE 8.2 A parallel-plate waveguide with a step discontinuity.

Modes that can propagate in a waveguide depend on the source excitation. Propagating and non-propagating modes are excited. However, the non-propagating modes vanish at a short distance from the source. Consider a source that excites a TE_1 incident field in Region 1 given by

$$E_y^i = C_1 \sin\left(\frac{\pi z}{h_1}\right) e^{-j\beta_{x1}x}$$

where C_1 is a constant associated with the excitation of the mode and can be normalized to a unity value. The propagation constant β_{x1} is

$$\beta_{x1}^2 = \beta_o^2 - \left(\frac{\pi}{h_1}\right)^2$$

The junction generates a multimode reflected and transmitted field. The reflected field is given by

$$E_y^r = \sum_{n=1}^{\infty} R_n \sin\left(\frac{n\pi z}{h_1}\right) e^{j\beta_{x1,n}x}$$

where R_n is the reflection coefficient associated with the n-mode, and

$$\beta_{x1,n}^2 = \beta_o^2 - \left(\frac{n\pi}{h_1}\right)^2 \tag{8.2.1}$$

Hence, the total field in Region 1 is

$$E_{y1} = E_y^i + E_y^r = \sin\left(\frac{\pi z}{h_1}\right) e^{-j\beta_{x1}x} + \sum_{n=1}^{\infty} R_n \sin\left(\frac{n\pi z}{h_1}\right) e^{j\beta_{x1,n}x}$$

The transmitted field in Region 2 is

$$E_{y2} = \sum_{n=1}^{\infty} T_n \sin\left(\frac{n\pi z}{h_2}\right) e^{-j\beta_{x2,n}x}$$

where T_n is the transmission coefficient associated with the n-mode, and

$$\beta_{x2,n}^2 = \beta_o^2 - \left(\frac{n\pi}{h_2}\right)^2 \tag{8.2.2}$$

The reflection and transmission coefficients are determined using the boundary conditions at the interface. That is, at $x = 0$:

$$E_{y1} = E_{y2}|_{\substack{x=0 \\ 0 \le z \le h_1}}$$

or

$$\sin\left(\frac{\pi z}{h_1}\right) + \sum_{n=1}^{\infty} R_n \sin\left(\frac{n\pi z}{h_1}\right) = \sum_{n=1}^{\infty} T_n \sin\left(\frac{n\pi z}{h_2}\right) \tag{8.2.3}$$

Multiplying (8.2.3) by $\sin(p\pi z/h_2)$ and integrating from $0 \le z \le h_2$ results in

$$I_{1p} + \sum_{n=1}^{\infty} R_n I_{np} = T_p \frac{h_2}{2} \tag{8.2.4}$$

The right-hand side of (8.2.3) has a value only for $n = p$, which results in $T_p h_2/2$. The left-hand side used the integral:

$$\int_0^h \sin(az)\sin(bz)dz = \left. \frac{\sin(a-b)z}{2(a-b)} - \frac{\sin(a+b)z}{2(a+b)} \right|_0^h$$

to obtain

$$I_{np} = \int_0^{h_2} \sin\left(\frac{n\pi z}{h_1}\right)\sin\left(\frac{p\pi z}{h_2}\right)dz = (-1)^p \frac{p h_1^2 h_2}{\pi(n^2h_2^2 - p^2h_1^2)} \sin\left(\frac{n\pi h_2}{h_1}\right) \tag{8.2.5}$$

Next, the boundary condition satisfied by the tangential magnetic field is used. The tangential magnetic field is

$$H_z = -\frac{1}{j\omega\mu_o}\frac{\partial E_y}{\partial x}$$

Therefore,

$$H_{z1} = \frac{\beta_{x1}}{\omega\mu_o} \sin\left(\frac{\pi z}{h_1}\right) e^{-j\beta_{x1}x} - \frac{1}{\omega\mu_o} \sum_{n=1}^{\infty} \beta_{x1,n} R_n \sin\left(\frac{n\pi z}{h_1}\right) e^{j\beta_{x1,n}x}$$

which represent the incident and reflected tangential magnetic fields. The transmitted tangential magnetic field is

$$H_{z2} = \frac{1}{\omega\mu_o} \sum_{n=1}^{\infty} \beta_{x2,n} T_n \sin\left(\frac{n\pi z}{h_2}\right) e^{-j\beta_{x2,n}x}$$

The boundary condition is

$$H_{z1} = H_{z2}|_{\substack{x=0 \\ 0 \le z \le h_1}}$$

Therefore,

$$\beta_{x1} \sin\left(\frac{\pi z}{h_1}\right) - \sum_{n=1}^{\infty} \beta_{x1,n} R_n \sin\left(\frac{n\pi z}{h_1}\right) = \sum_{n=1}^{\infty} \beta_{x2,n} T_n \sin\left(\frac{n\pi z}{h_2}\right) \tag{8.2.6}$$

Multiplying (8.2.6) by $\sin(p\pi z/h_1)$, integrating from $0 \le z \le h_1$ shows that

$$\frac{h_1}{2}\left(\beta_{x1}\delta_{1p} - \beta_{x1,p}R_p\right) = \sum_{n=1}^{\infty} \beta_{x2,n} T_n \widehat{I_{np}} \tag{8.2.7}$$

where $\widehat{I_{np}}$ is given by (8.2.5) with h_1 and h_2 interchanged. The number n in the sum (8.2.4), from (8.2.1), can be restricted to a maximum value given by $n_{1,\max} = \beta_o h_1/\pi = 2h_1/\lambda$, which denotes the maximum number of propagating modes (i.e., when $\beta_{x1,n} = 0$). Similarly, from (8.2.2), n in (8.2.7) is restricted to $n_{2,\max} = \beta_o h_2/\pi = 2h_2/\lambda$ (i.e., when $\beta_{x2,n} = 0$). Equations (8.2.4) and (8.2.7) provide a set of simultaneous linear equations that can be solved for the reflections and transmission coefficients.

The time average reflected power density is

$$\mathbf{P}_r = \frac{1}{2}\mathrm{Re}[E_y^r H_z^{r*}](-\hat{\mathbf{a}}_x) = \frac{1}{2\omega\mu_o}\mathrm{Re}\left[\sum_{n=1}^{n_{1,\max}} |R_n|^2 \beta_{x1,n} \sin^2\left(\frac{n\pi z}{h_1}\right)\right](-\hat{\mathbf{a}}_x) \tag{8.2.8}$$

Integrating (8.2.8) from 0 to h_1 gives the average reflected power. That is,

$$W_r = \frac{h_1}{4\omega\mu_o} \sum_{n=1}^{n_{1,\max}} \beta_{x1,n} |R_n|^2$$

Similarly, the average transmitted power is

$$W_t = \frac{h_2}{4\omega\mu_o} \sum_{n=1}^{n_{2,\max}} \beta_{x2,n} |T_n|^2.$$

8.3 ELECTRIC LINE SOURCE ABOVE A PERFECT CONDUCTOR

The geometry of the problem is shown in Fig. 8.3. The region $z > 0$ is described by μ, ε and σ. The electric source generates TM^y fields which can be calculated in terms of $\mathbf{A} = A_y\hat{\mathbf{a}}_y$. From (1.5.6), A_y satisfies

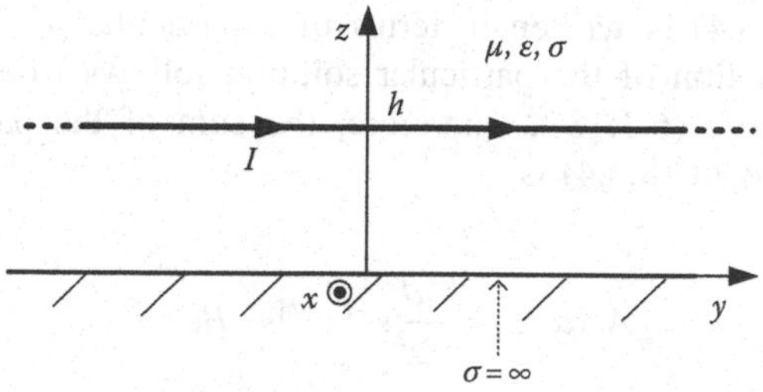

FIGURE 8.3 Electric line source above a conductor.

$$\left[\frac{\partial^2}{\partial x^2} + \frac{\partial^2}{\partial z^2} - \gamma^2\right] A_y(x, y) = -\mu J_{e,y} = -\mu I \delta(x)\delta(z - h) \tag{8.3.1}$$

The fields do not vary with y, since the source is assumed to be infinite in length. Also, it is observed that the field variation with x extents from $-\infty < x < \infty$. The fields are given by (4.2.13) to (4.2.18):

$$E_y = -j\omega A_y \tag{8.3.2}$$

$$H_x = -\frac{1}{\mu}\frac{\partial A_y}{\partial z}$$

$$H_z = \frac{1}{\mu}\frac{\partial A_y}{\partial x}$$

Equation (8.3.1) is solved by first Fourier transforming the x variable, namely

$$A_y(\alpha, z) = \int_{-\infty}^{\infty} A_y(x, z) e^{-j\alpha x} dx$$

and the inverse transform is

$$A_y(x, z) = \frac{1}{2\pi}\int_{-\infty}^{\infty} A_y(\alpha, z) e^{j\alpha x} d\alpha \tag{8.3.3}$$

Multiplying (8.3.1) by $e^{-j\alpha x}$ and integrating from $-\infty$ to ∞ gives

$$\left[\frac{d^2}{dz^2} - \tau^2\right] A_y(\alpha, z) = -\mu I \delta(z - h) \tag{8.3.4}$$

where

$$\tau^2 = \gamma^2 + \alpha^2.$$

The solution to (8.3.4) is written in terms of a particular and a complementary solution. The evaluation of the particular solution follows from (6.1.15) with jk replaced by τ (or use (6.1.18)). Therefore, the sum of the particular and complementary solutions of (8.3.4) is

$$A_y(\alpha, z) = \frac{\mu I}{2\tau} e^{-\tau|z-h|} + Be^{-\tau z}$$

where $\mathrm{Re}(\tau) > 0$ for the solution to go to zero as $z \to \infty$.

The boundary condition at the conductor is

$$E_y = 0|_{z=0} \Rightarrow A_y = 0|_{z=0} \Rightarrow \frac{\mu I}{2\tau} e^{-\tau h} + B = 0$$

or

$$B = -\frac{\mu I}{2\tau} e^{-\tau h}$$

Therefore,

$$A_y(\alpha, z) = \frac{\mu I}{2\tau} e^{-\tau|z-h|} - \frac{\mu I}{2\tau} e^{-\tau(z+h)}$$

Taking the inverse transform of $A_y(\alpha, z)$ (see (8.3.3)) and substituting into (8.3.2) gives

$$E_y(x, z) = -\frac{j\omega\mu I}{4\pi} \int_{-\infty}^{\infty} \left(\frac{e^{-\tau|z-h|}}{\tau} - \frac{e^{-\tau(z+h)}}{\tau} \right) e^{j\alpha x} d\alpha \tag{8.3.5}$$

Using the Fourier transforms pair relations:

$$\int_0^{\infty} K_0\left(\gamma\sqrt{x^2 + z^2}\right) e^{-j\alpha x} dx = \frac{\pi e^{-\sqrt{\gamma^2+\alpha^2} z}}{\sqrt{\gamma^2 + \alpha^2}}$$

and

$$K_0\left(\gamma\sqrt{x^2 + z^2}\right) = \frac{1}{2} \int_{-\infty}^{\infty} \frac{e^{-\sqrt{\gamma^2+\alpha^2} z}}{\sqrt{\gamma^2 + \alpha^2}} e^{j\alpha x} d\alpha$$

(8.3.5) is expressed in terms of the modified Bessel function K_0, namely

$$E_y = -\frac{j\omega\mu I}{2\pi} [K_0(\gamma R_1) - K_0(\gamma R_2)] \tag{8.3.6}$$

where

$$R_1 = \sqrt{x^2 + (z - h)^2}$$

and

$$R_2 = \sqrt{x^2 + (z + h)^2}$$

In the case of a lossless region, with $\gamma = j\beta$ and

$$K_0(j\beta R) = -j\frac{\pi}{2}H_0^{(2)}(\beta R)$$

(8.3.6) reads

$$E_y = \frac{-\omega\mu I}{4}\left[H_0^{(2)}(\beta R_1) - H_0^{(2)}(\beta R_2)\right] \tag{8.3.7}$$

The first term in (8.3.7) is the field due to the source (see (2.2.9)) and the second term is the field from the image source (i.e., a current $-I$ at $z = -h$). The expressions in (8.3.6) and (8.3.7) are recognized as the ones obtained using the method of images.

In the far field:

$$H_0^{(2)}(\beta R) \approx \sqrt{\frac{2}{\pi\beta R}}\, e^{-j\left(\beta R - \frac{\pi}{4}\right)}$$

and (8.3.7) reads

$$E_y \approx \frac{-\omega\mu I e^{j\pi/4}}{\sqrt{8\pi\beta}}\left[\frac{e^{-j\beta R_1}}{\sqrt{R_1}} - \frac{e^{-j\beta R_2}}{\sqrt{R_2}}\right] \tag{8.3.8}$$

Also, in the far field $R_1 >> 2h$ and $R_2 >> 2h$, so in the denominator of (8.3.8): $R_1 \approx R_2$, and in the phase: $R_2 - R_1 \approx 2h\cos\theta$. Hence, (8.3.8) can be simplified to read

$$E_y \approx \frac{-\omega\mu I e^{j\pi/4}}{(8\pi\beta R_1)^{1/2}} e^{-j\beta R_1}\left[1 - e^{-j2\beta h\cos\theta}\right]$$

The magnitude of E_y is

$$|E_y| \approx \frac{\omega\mu I}{(2\pi\beta R_1)^{1/2}}|\sin(\beta h\cos\theta)|$$

which shows how $|E_y|$ varies as a function of θ (i.e., $\sin(\beta h \cos\theta)$), which represents the radiation pattern.

8.3.1 Electric Line Source above a Lossy Surface

The geometry of the problem is shown in Fig. 8.4. The electric line source generates TM^y fields which can be calculated using (1.5.6). In Region 1, the wave equation is

$$[\nabla^2 + k_1^2]A_{y1}(x, z) = -\mu_o J_{e,y} = -\mu_o I\delta(x)\delta(z - h) \tag{8.3.9}$$

where $k_1 = k_o = \omega\sqrt{\mu_o\varepsilon_o}$. In Region 2:

$$[\nabla^2 + k_2^2]A_{y2}(x, z) = 0 \tag{8.3.10}$$

where

$$k_2 = k_o n_2 = k_o\sqrt{\varepsilon_{r2} - j\frac{\sigma_2}{\omega\varepsilon_o}}$$

The fields do not vary with y since the source is assumed to be infinite in length. Also, it is observed that the field variation with x extents from $-\infty < x < \infty$. Therefore, Fourier transforming the x variable in (8.3.9) and (8.3.10), gives

$$\left[\frac{d^2}{dz^2} + \tau_1^2\right]A_{y1}(\alpha, z) = -\mu_o I\delta(z - h) \tag{8.3.11}$$

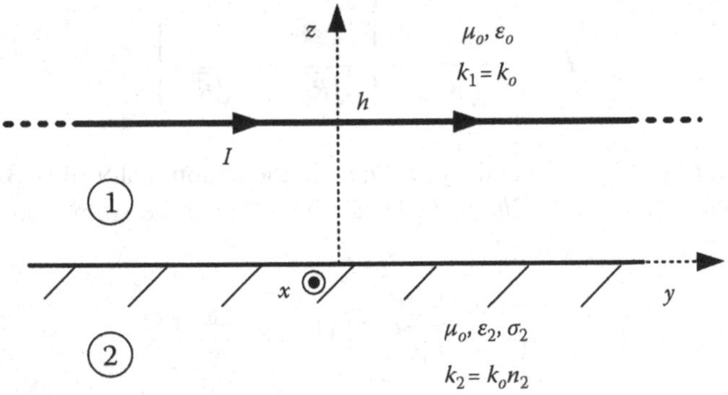

FIGURE 8.4 Line source above a lossy surface.

and

$$\left[\frac{d^2}{dz^2} + \tau_2^2\right]A_{y2}(\alpha, z) = 0 \tag{8.3.12}$$

where

$$\tau_1^2 = k_1^2 - \alpha^2$$
$$\tau_2^2 = k_2^2 - \alpha^2$$

The Fourier transform pair is

$$A(\alpha, z) = \int_{-\infty}^{\infty} A(x, y)e^{-j\alpha x}dx$$

and

$$A(x, z) = \frac{1}{2\pi}\int_{-\infty}^{\infty} A(\alpha, y)e^{j\alpha x}d\alpha$$

The solution to (8.3.11) is written in terms of a particular (see (6.1.15)) and a complementary solution as

$$A_{y1}(\alpha, z) = -\frac{j\mu_o I}{2\tau_1}e^{-j\tau_1|z-h|} + Be^{-j\tau_1 z} \tag{8.3.13}$$

and for (8.3.12):

$$A_{y2}(\alpha, z) = Ce^{j\tau_2 z} \tag{8.3.14}$$

where $\mathrm{Im}(\tau_1) < 0$ and $\mathrm{Im}(\tau_2) < 0$ to satisfy the radiation conditions as $z \to \pm\infty$.

The boundary conditions are the continuity of E_y and H_x at $z = 0$. Therefore,

$$E_{y1} = E_{y2}\Big|_{z=0} \Rightarrow A_{y1} = A_{y2}\Big|_{z=0} \Rightarrow -\frac{j\mu_o I}{2\tau_1}e^{-j\tau_1 h} + B = C \tag{8.3.15}$$

and

$$H_{x1} = H_{x2}|_{z=0} \Rightarrow \frac{\partial A_{y1}}{\partial z} = \frac{\partial A_{y2}}{\partial z}\Big|_{z=0} \Rightarrow \frac{\mu_o I}{2}e^{-j\tau_1 h} - j\tau_1 B = j\tau_2 C \tag{8.3.16}$$

Solving (8.3.15) and (8.3.16) gives

$$B = -\frac{j\mu_o I}{2\tau_1}\frac{\tau_1 - \tau_2}{\tau_1 + \tau_2}e^{-j\tau_1 h}$$

and

$$C = -j\mu_o I\frac{1}{\tau_1 + \tau_2}e^{-j\tau_1 h}$$

Then, substituting B into (8.3.13) and taking the inverse transform gives

$$A_{y1}(x, z) = \frac{-j\mu_o I}{4\pi}\int_{-\infty}^{\infty}\left[\frac{e^{-j\tau_1|z-h|}}{\tau_1} + \frac{1}{\tau_1}\left(\frac{\tau_1 - \tau_2}{\tau_1 + \tau_2}\right)e^{-j\tau_1(z+h)}\right]e^{j\alpha x}d\alpha \quad (8.3.17)$$

Writing

$$\frac{\tau_1 - \tau_2}{\tau_1 + \tau_2} = -1 + \frac{2\tau_1}{\tau_1 + \tau_2}$$

(8.3.17) reads

$$A_{y1}(x, z) = \frac{-j\mu_o I}{4\pi}\int_{-\infty}^{\infty}\left[\frac{e^{-j\tau_1|z-h|}}{\tau_1} - \frac{e^{-j\tau_1(z+h)}}{\tau_1} + \left(\frac{2}{\tau_1 + \tau_2}\right)e^{-j\tau_1(z+h)}\right]e^{j\alpha x}d\alpha \quad (8.3.18)$$

Using the Fourier transform pair relations:

$$\int_0^{\infty} H_0^{(2)}\left(k\sqrt{x^2 + z^2}\right)e^{-j\alpha x}dx = \frac{2e^{-j\sqrt{k^2-\alpha^2}\,z}}{\sqrt{k^2 - \alpha^2}}$$

and

$$H_0^{(2)}\left(k\sqrt{x^2 + z^2}\right) = \frac{1}{\pi}\int_{-\infty}^{\infty}\frac{e^{-j\sqrt{k^2-\alpha^2}\,z}}{\sqrt{k^2 - \alpha^2}}e^{j\alpha x}d\alpha \quad (8.3.19)$$

(8.3.18) can be written in terms of a Hankel function. That is,

$$A_{y1}(x, z) = -\frac{j\mu_o I}{4}H_0^{(2)}(k_1 R_1) + \frac{j\mu_o I}{4}H_0^{(2)}(k_1 R_2) - \frac{j\mu_o I}{2\pi}\int_{-\infty}^{\infty}\left(\frac{1}{\tau_1 + \tau_2}\right)e^{-j[\tau_1(z+h)-\alpha x]}d\alpha \quad (8.3.20)$$

where

$$R_1 = \sqrt{x^2 + (z - h)^2}$$

and

$$R_2 = \sqrt{x^2 + (z + h)^2}$$

The fields follow from (4.2.13) to (4.2.18). The first term in (8.3.20) represents the field from the source and the second term the image field. The third term accounts for the surface effect due to its finite conductivity. In the case of infinite conductivity, the third term in (8.3.20) vanishes and

$$E_{y1} = -j\omega A_{y1} = -\frac{\omega\mu_o I}{4}\left[H_0^{(2)}(k_1 R_1) - H_0^{(2)}(k_1 R_2)\right]$$

which is identical to (8.3.7).

The transmitted field follows by substituting C into (8.3.14) and taking the inverse transform. That is,

$$A_{y2}(x, z) = \frac{-j\mu_o I}{2\pi}\int_{-\infty}^{\infty}\left(\frac{1}{\tau_1 + \tau_2}\right)e^{-j\tau_1 h}e^{j\tau_2 z}e^{j\alpha x}d\alpha \tag{8.3.21}$$

Next, consider the evaluation of the integral in (8.3.20) when the line source and observation point are close to the surface (i.e., $h \to 0$ and $z \to 0$). In this case, $R_1 \approx R_2$ and the electric field is

$$E_{y1} = -j\omega A_{y1} \approx -\frac{\omega\mu_o I}{2\pi}\int_{-\infty}^{\infty}\frac{1}{\tau_1+\tau_2}e^{j\alpha x}d\alpha \approx -\frac{\omega\mu_o I}{\pi}\int_0^{\infty}\frac{1}{\tau_1 + \tau_2}\cos(\alpha x)d\alpha \tag{8.3.22}$$

where the integral involving the odd part of $e^{j\alpha x}$ (i.e., $\sin(\alpha x)$) is zero.

Since

$$\frac{1}{\tau_1 + \tau_2} = \frac{\tau_1 - \tau_2}{\tau_1^2 - \tau_2^2} = \frac{\tau_1 - \tau_2}{k_1^2 - k_2^2}$$

(8.3.22) is written as

$$E_{y1} \approx -\frac{\omega\mu_o I}{\pi(k_1^2 - k_2^2)}\left[\int_0^{\infty}\tau_1\cos(\alpha x)d\alpha - \int_0^{\infty}\tau_2\cos(\alpha x)d\alpha\right] \tag{8.3.23}$$

These integrals are evaluated by observing from (8.3.19) that as $z \to 0$:

$$H_0^{(2)}(k_o x) = \frac{1}{\pi}\int_{-\infty}^{\infty} \frac{1}{\sqrt{k_1^2 - \alpha^2}} e^{j\alpha x} d\alpha = \frac{2}{\pi}\int_0^{\infty} \frac{\cos(\alpha x)}{\tau_1} d\alpha$$

Then,

$$\begin{aligned}\frac{d^2 H_0^{(2)}(k_o x)}{dx^2} &= -\frac{2}{\pi}\int_0^{\infty} \frac{\alpha^2 \cos(\alpha x)}{\tau_1} d\alpha = -\frac{2}{\pi} k_1^2 \int_0^{\infty} \frac{\cos(\alpha x)}{\tau_1} d\alpha \\ &+\frac{2}{\pi}\int_0^{\infty} \tau_1 \cos(\alpha x) d\alpha = -\, k_1^2 H_0^{(2)}(k_o x) \\ &+\frac{2}{\pi}\int_0^{\infty} \tau_1 \cos(\alpha x) d\alpha\end{aligned} \tag{8.3.24}$$

Another expression for $d^2 H_0^{(2)}(k_1 x)/dx^2$ follows from (see Appendix C)

$$\begin{aligned}\frac{dH_0^{(2)}(k_1 x)}{dx} &= -k_1 H_1^{(2)}(k_1 x) \\ \frac{d^2 H_0^{(2)}(k_1 x)}{dx^2} &= -k_1^2 \frac{d H_1^{(2)}(k_1 x)}{dx} = -k_1^2 \left[H_0^{(2)}(k_1 x) - \frac{1}{k_1 x} H_1^{(2)}(k_1 x) \right]\end{aligned} \tag{8.3.25}$$

Equating (8.3.24) and (8.3.25) shows that the first integral in (8.3.23) is

$$\frac{2}{\pi}\int_0^{\infty} \tau_1 \cos(\alpha x)\, d\alpha = \frac{k_1}{x} H_1^{(2)}(k_1 x) \tag{8.3.26}$$

The evaluation of the second integral in (8.3.23) is the same except that k_1 is replaced by k_2 in (8.3.26). Hence, (8.2.23) reads

$$E_{y1} \approx \frac{\omega \mu_o I}{2(k_1^2 - k_2^2)x}\left[k_2 H_1^{(2)}(k_2 x) - k_1 H_1^{(2)}(k_1 x)\right] \tag{8.3.27}$$

Since k_2 is complex, if $\mathrm{Re}(k_2 x) \gg 0$, the exponential decaying behavior of the Hankel function shows that the first term in (8.3.27) goes to zero, and E_{y1} behaves according to

$$E_{y1} \approx -\frac{\omega \mu_o I\, k_1}{2(k_1^2 - k_2^2)x} H_1^{(2)}(k_1 x) \tag{8.3.28}$$

Furthermore, if $k_o x \gg 1$, using

$$H_1^{(2)}(k_1 x) \approx \sqrt{\frac{2}{\pi k_1 x}}\, e^{-j(k_1 x - 3\pi/4)}$$

(8.3.28) reads:

$$E_{y1} \approx \frac{\omega\mu_o I k_1}{(k_1^2 - k_2^2)} \sqrt{\frac{1}{2\pi k_1 x^3}}\, e^{-j(k_1 x + \pi/4)} \tag{8.3.29}$$

The evaluation of the magnetic field component H_{z1} is left to the problems.

8.4 RADIATION FROM A NARROW SLIT

A narrow slit of infinite length in a conducting plane is shown in Fig. 8.5. The amplitude of the sinusoidal voltage across the slit is V_o. This voltage is associated with an electric field in the x direction, which is produced by a y directed surface current in the slot conductors (see Fig. 8.5b). Since the slot is of infinite length, the fields do not vary with z and it follows from (4.2.19) to (4.2.24) that the field components generated by the vector potential $A_y(x, y)$ are $E_x(x, y)$, $E_y(x, y)$, and $H_z(x, y)$. In other words, a TM field to y.

Another way of determining the field configuration is to observe that the sinusoidal voltage V_o produces a radiated field with $E_x(x, y)$ and $E_y(x, y)$ components (i.e., with no z variation), and from Maxwell's equations an $H_z(x, y)$ field.

The radiation fields produced by the slit configuration in Fig. 8.5 can be determined by either solving the wave equation for $A_y(x, y)$ or $H_z(x, y)$. We will use the wave equation satisfied by the magnetic field $H_z(x, y)$ for $y > 0$:

$$\left[\frac{\partial^2}{\partial x^2} + \frac{\partial^2}{\partial y^2} + \beta_o^2\right] H_z(x, y) = 0$$

Fourier transforming the x variable gives

$$\left[\frac{d^2}{dy^2} + \tau^2\right] H_z(\alpha, y) = 0 \tag{8.4.1}$$

where

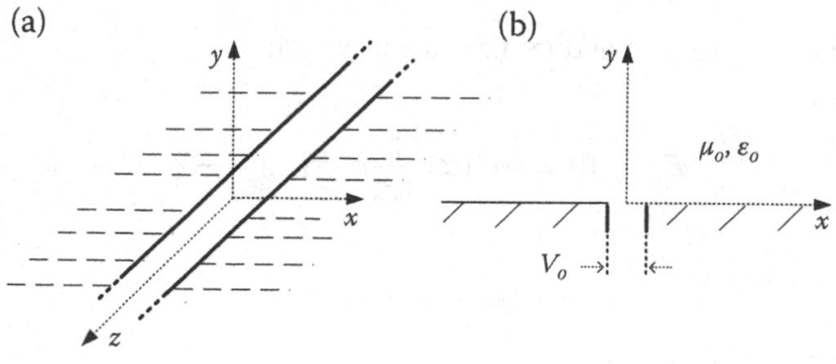

FIGURE 8.5 (a) A narrow-slit radiator; (b) two-dimensional view.

$$\tau^2 = \beta_o^2 - \alpha^2$$

The Fourier transform pair are

$$H_z(\alpha, y) = \int_{-\infty}^{\infty} H_z(x, y)e^{-j\alpha x}dx$$

and

$$H_z(x, y) = \frac{1}{2\pi}\int_{-\infty}^{\infty} H_z(\alpha, y)e^{j\alpha x}d\alpha$$

The solution to (8.4.1) is

$$H_z(\alpha, y) = C(\alpha)\, e^{-j\tau y} \tag{8.4.2}$$

which represents an outgoing wave. The radiation condition as $y \to \infty$ requires that $\mathrm{Im}(\tau) < 0$ (i.e., add a small loss to the propagation constant). The constant $C(\alpha)$ is determined by the source condition. Consider a slot excitation described by a delta function, such that

$$E_x(x, y)|_{y=0} = -V_o\delta(x) \tag{8.4.3}$$

The Fourier transform of (8.4.3) is

$$E_x(\alpha, y)|_{y=0} = -V_o\int_{-\infty}^{\infty} \delta(x)\, e^{-j\alpha x}dx = -V_o \tag{8.4.4}$$

From Maxwell's equations:

$$E_x(\alpha, y) = \frac{1}{j\omega\varepsilon_o}\frac{\partial H_z(\alpha, y)}{\partial y} = -C(\alpha)\frac{\tau}{\omega\varepsilon_o}\, e^{-j\tau y} \tag{8.4.5}$$

and it follows from (8.4.4) and (8.4.5) that at $y = 0$:

$$E_x(\alpha, 0) = -C(\alpha)\frac{\tau}{\omega\varepsilon_o}e^{-j\tau y}|_{y=0} = -V_o \tag{8.4.6}$$

or

$$C(\alpha) = \frac{V_o\omega\varepsilon_o}{\tau}$$

The solution for $C(\alpha)$ can also be obtained by observing that the inverse transform (8.4.5) is

$$E_x(x, y) = \frac{-1}{2\pi\omega\varepsilon_o}\int_{-\infty}^{\infty} C(\alpha)\,\tau e^{-j\tau y} e^{j\alpha x} d\alpha \tag{8.4.7}$$

Using the delta function representation in (6.1.16), (8.4.3) reads

$$E_x(x, y)|_{y=0} = -V_o\delta(x) = -\frac{V_o}{2\pi}\int_{-\infty}^{\infty} e^{-j\alpha x} d\alpha \tag{8.4.8}$$

Equating (8.4.7) and (8.4.8) at $y = 0$ gives

$$C = \frac{V_o\omega\varepsilon_o}{\tau}$$

as expected.

Substituting $C(\alpha)$ into (8.4.2) and taking the inverse transform gives

$$H_z(x, y) = \frac{V_o\omega\varepsilon_o}{2\pi}\int_{-\infty}^{\infty} \frac{e^{-j\tau y}}{\tau} e^{j\alpha x} d\alpha \tag{8.4.9}$$

The integral in (8.4.9) is recognized as a Hankel function (see (8.3.19)). Therefore,

$$H_z(x, y) = \frac{V_o\omega\varepsilon_o}{2} H_0^{(2)}(\beta_o\rho)$$

where $\rho = (x^2 + y^2)^{1/2}$. In the far field, using the asymptotic expression for the Hankel function:

$$H_z(x, y) \approx \frac{V_o}{\eta_o}\sqrt{\frac{\beta_o}{2\pi\rho}}\, e^{-j(\beta_o\rho - \pi/4)} \tag{8.4.10}$$

In the case that $y > 0$ represents a lossy region with propagation constant γ, replace β_o by $-j\gamma$ and the solution is

$$H_y(x, y) = \frac{jV_o\omega\varepsilon_o}{\pi} K_0(\gamma\rho)$$

where we used the identity

$$H_0^{(2)}(-j\gamma\rho) = j\frac{2}{\pi} K_0(\gamma\rho)$$

In the case of a slit with width $2d$ the electric field at $y = 0$ is

$$E_x(x, y)|_{y=0} = -\frac{V_o}{2d} \qquad (-d \leq x \leq d)$$

and (8.4.6) reads

$$E_x(\alpha, 0) = -C(\alpha)\frac{\tau}{\omega\varepsilon_o}e^{-j\tau y}|_{y=0} = -\frac{V_o}{2d}\int_{-d}^{d} e^{-j\alpha x}dx = -\frac{V_o}{2d}\frac{\sin\alpha d}{\alpha d}$$

or

$$C(\alpha) = \frac{V_o\omega\varepsilon_o}{\tau}\frac{\sin(\alpha d)}{\alpha d}$$

Then, the magnetic field is given by

$$H_z(x, y) = \frac{V_o\omega\varepsilon_o}{2\pi}\int_{-\infty}^{\infty}\frac{\sin\alpha d}{\alpha d}\frac{e^{-j\tau y}}{\tau}e^{j\alpha x}d\alpha$$

As a check, as $d \to 0$ it follows that $\sin\alpha d/\alpha d \to 1$ and the expression reduces to that in (8.4.9).

8.5 VERTICAL HERTZIAN DIPOLE ABOVE A LOSSY SURFACE

8.5.1 Electric Hertzian Dipole

This is one of the classical problems in electromagnetic wave propagation. Figure 8.6 shows an electrical Hertzian dipole above a lossy surface (such as the Earth's surface). The fields for this configuration are usually determined in terms of either the $\mathbf{A}$ vector or $\mathbf{\Pi}^e$ vector.

The expression for the $\mathbf{\Pi}^e$ vector in the air region (i.e., Region 1) is

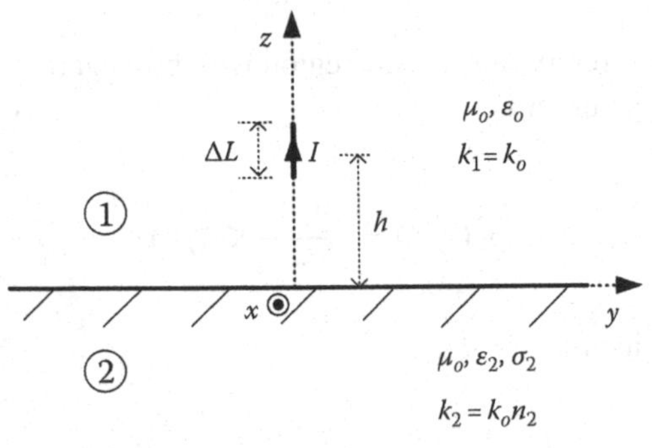

FIGURE 8.6 Vertical Hertzian dipole above a lossy surface.

$$[\nabla^2 + k_1^2]\Pi_{z1}^{e}(x, y, z) = -\frac{J_{e,z}}{j\omega\varepsilon_o} = -\frac{(I\Delta L)}{j\omega\varepsilon_o}\delta(x)\delta(y)\delta(z - h) \qquad (8.5.1)$$

where

$$k_1 = k_o = \omega\sqrt{\mu_o\varepsilon_o}$$

In the lossy region (i.e., Region 2):

$$[\nabla^2 + k_2^2]\Pi_{z2}^{e}(x, y, z) = 0 \qquad (8.5.2)$$

where, since $\mu_2 = \mu_o$, it follows that

$$k_2 = k_o n_2 = k_o\sqrt{\varepsilon_{r2} - j\frac{\sigma_2}{\omega\varepsilon_o}}$$

The fields are calculated using (4.2.7) to (4.2.12), which in terms of the propagation constant k are

$$H_x = -\frac{k^2}{j\omega\mu_o}\frac{\partial\Pi_z^{e}}{\partial y}$$

$$H_y = \frac{k^2}{j\omega\mu_o}\frac{\partial\Pi_z^{e}}{\partial x}$$

$$E_x = \frac{\partial^2\Pi_z^{e}}{\partial x\partial z}$$

$$E_y = \frac{\partial^2\Pi_z^{e}}{\partial y\partial z}$$

$$E_z = \left(\frac{\partial^2}{\partial z^2} + k^2\right)\Pi_z^{e}$$

The solution of (8.5.1) requires a particular and a complementary solution. The particular solution, which follows from the Green function associated with (8.5.1) is

$$\Pi_{z1,p}^{e} = \frac{(I\Delta L)}{j\omega\varepsilon_o}\frac{e^{-jk_1R_1}}{4\pi R_1}$$

where

$$R_1 = \sqrt{x^2 + y^2 + (z - h)^2}$$

The Green function can also be written as a spectrum of plane waves, as in (6.1.26). Hence, the particular solution can be expressed in the form

$$\Pi^e_{z1,p} = \frac{(I\Delta L)}{j\omega\varepsilon_o}\frac{e^{-jk_1R_1}}{4\pi R_1} = \frac{-(I\Delta L)}{8\pi^2\omega\varepsilon_o}\int_{-\infty}^{\infty}\int_{-\infty}^{\infty}\frac{e^{j(\alpha_1 x+\alpha_2 y-\tau_1|z-h|)}}{\tau_1}d\alpha_1 d\alpha_2 \quad (8.5.3)$$

where

$$\tau_1 = \sqrt{k_1^2 - \alpha_1^2 - \alpha_2^2}$$

and $\text{Im}(\tau_1) < 0$. The complementary solution in Region 1 is written as

$$\Pi^e_{z1,c} = \frac{-(I\Delta L)}{8\pi^2\omega\varepsilon_o}\int_{-\infty}^{\infty}\int_{-\infty}^{\infty} R(\alpha_1,\alpha_2)e^{j(\alpha_1 x+\alpha_2 y-\tau_1 z)}d\alpha_1 d\alpha_2 \quad (8.5.4)$$

where $R(\alpha_1, \alpha_2)$ represent the reflection coefficient. Therefore,

$$\Pi^e_{z1} = \Pi^e_{z1,p} + \Pi^e_{z1,c} = \frac{-(I\Delta L)}{8\pi^2\omega\varepsilon_o}\int_{-\infty}^{\infty}\int_{-\infty}^{\infty}\left[\frac{e^{-j\tau_1|z-h|}}{\tau_1} + R(\alpha_1,\alpha_2)e^{-j\tau_1 z}\right]e^{j(\alpha_1 x+\alpha_2 y)}d\alpha_1 d\alpha_2 \quad (8.5.5)$$

Similarly, the solution to (8.5.2) in Region 2 is written as

$$\Pi^e_{z2} = \frac{-(I\Delta L)}{8\pi^2\omega\varepsilon_o}\int_{-\infty}^{\infty}\int_{-\infty}^{\infty} T(\alpha_1,\alpha_2)e^{j(\alpha_1 x+\alpha_2 y+\tau_2 z)}d\alpha_1 d\alpha_2 \quad (8.5.6)$$

where

$$\tau_2 = \sqrt{k_2^2 - \alpha_1^2 - \alpha_2^2}$$

and $\text{Im}(\tau_2) < 0$. In (8.5.6), $T(\alpha_1, \alpha_2)$ represents the transmission coefficient. The factors in front of the integrals in (8.5.4) and (8.5.6) were selected (for convenience) to be the same as the factor in (8.5.3).

The boundary conditions at $z = 0$ are the continuity of the tangential fields. That is,

$$H_{x1} = H_{x2}|_{z=0} \Rightarrow k_1^2\frac{\partial \Pi^e_{z1}}{\partial y} = k_2^2\frac{\partial \Pi^e_{z2}}{\partial y}\bigg|_{z=0}$$

$$H_{y1} = H_{2}|_{z=0} \Rightarrow k_1^2\frac{\partial \Pi^e_{z1}}{\partial x} = k_2^2\frac{\partial \Pi^e_{z2}}{\partial x}\bigg|_{z=0}$$

$$E_{x1} = E_{x2}\bigg|_{z=0} \Rightarrow \frac{\partial^2 \Pi^e_{z1}}{\partial x \partial z} = \frac{\partial^2 \Pi^e_{z2}}{\partial x \partial z}\bigg|_{z=0}$$

$$E_{y1} = E_{y2}|_{z=0} \Rightarrow \frac{\partial^2 \Pi^e_{z1}}{\partial y \partial z} = \frac{\partial^2 \Pi^e_{z2}}{\partial y \partial z}\bigg|_{z=0}$$

The boundary conditions must hold for all values of x and y. Therefore, the first two boundary conditions produce the same requirement, namely

$$k_1^2 \Pi_{z1}^{e} = k_2^2 \Pi_{z2}^{e}\Big|_{z=0} \tag{8.5.7}$$

and the last two require that

$$\frac{\partial \Pi_{z1}^{e}}{\partial z} = \frac{\partial \Pi_{z2}^{e}}{\partial z}\Bigg|_{z=0} \tag{8.5.8}$$

Applying the boundary condition in (8.5.7) and (8.5.8) to (8.5.5) and (8.5.6) gives

$$\frac{e^{-j\tau_1 h}}{\tau_1} + R(\alpha_1, \alpha_2) = \frac{k_2^2}{k_1^2} T(\alpha_1, \alpha_2)$$

and

$$\frac{e^{-j\tau_1 h}}{\tau_1} - R(\alpha_1, \alpha_2) = \frac{\tau_2}{\tau_1} T(\alpha_1, \alpha_2)$$

which are solved to give

$$R(\alpha_1, \alpha_2) = \frac{k_2^2 \tau_1 - k_1^2 \tau_2}{k_2^2 \tau_1 + k_1^2 \tau_2} \frac{e^{-j\tau_1 h}}{\tau_1}$$

and

$$T(\alpha_1, \alpha_2) = \frac{2k_1^2}{k_2^2 \tau_1 + k_1^2 \tau_2} e^{-j\tau_1 h}$$

These expressions can also be written in terms of the index of refraction. That is, $k_2^2/k_1^2 = n_{21}^2 = n_2^2$ since $n_1 = 1$.

The reflection coefficient can be expressed in the forms

$$R(\alpha_1, \alpha_2) = \frac{e^{-j\tau_1 h}}{\tau_1}\left[1 - \frac{2k_1^2 \tau_2}{k_2^2 \tau_1 + k_1^2 \tau_2}\right] \tag{8.5.9}$$

and

$$R(\alpha_1, \alpha_2) = \frac{e^{-j\tau_1 h}}{\tau_1}\left[-1 + \frac{2k_2^2 \tau_1}{k_2^2 \tau_1 + k_1^2 \tau_2}\right] \tag{8.5.10}$$

The form in (8.5.9) when substituted into (8.5.5) leads to

$$\Pi_{z1}^{e} = \frac{(I\Delta L)}{j\omega\varepsilon_o}\left[\frac{e^{-jk_1R_1}}{4\pi R_1} + \frac{e^{-jk_1R_2}}{4\pi R_2} + P\right] \tag{8.5.11}$$

where

$$R_2 = \sqrt{x^2 + y^2 + (z+h)^2}$$

and

$$P = \frac{j}{4\pi^2}\int_{-\infty}^{\infty}\int_{-\infty}^{\infty}\frac{k_1^2\tau_2}{\tau_1(k_2^2\tau_1 + k_1^2\tau_2)}e^{j[\alpha_1x+\alpha_2y-\tau_1(z+h)]}d\alpha_1 d\alpha_2 \tag{8.5.12}$$

Equation (8.5.11) shows that the first term represents the field from the primary source, the second term represents the field from a secondary source at the image position, and the third term represents a factor that accounts for the lossy surface effects due to its finite conductivity. If the surface is a perfect conductor, then $k_2 \to \infty$ and $P = 0$, and (8.5.11) reads

$$\Pi_{z1}^{e} = \frac{(I\Delta L)}{j\omega\varepsilon_o}\left[\frac{e^{-jk_1R_1}}{4\pi R_1} + \frac{e^{-jk_1R_2}}{4\pi R_2}\right]$$

which is the result that one obtains using image theory for an electric dipole above a perfect conductor, as shown in Fig. 8.7.

If (8.5.10) is substituted into (8.5.5), then

$$\Pi_{z1}^{e} = \frac{(I\Delta L)}{j\omega\varepsilon_o}\left[\frac{e^{-jk_1R_1}}{4\pi R_1} - \frac{e^{-jk_1R_2}}{4\pi R_2} + Q\right] \tag{8.5.13}$$

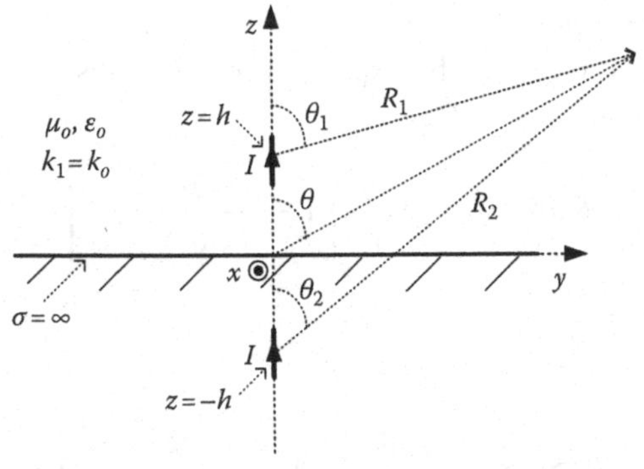

FIGURE 8.7 A vertical electric dipole above a perfect conductor.

where

$$Q = -\frac{j}{4\pi^2}\int_{-\infty}^{\infty}\int_{-\infty}^{\infty}\frac{k_2^2}{k_2^2\tau_1 + k_1^2\tau_2}e^{j[\alpha_1 x+\alpha_2 y-\tau_1(z+h)]}d\alpha_1 d\alpha_2 \qquad (8.5.14)$$

Observe that for points of observation along the lossy surface (i.e., $\theta \approx \pi/2$) the expression (8.5.13) shows that the first two terms cancel (since $R_1 \approx R_2$). Hence, the fields along the surface are described by Q. Also, if the lossy surface is replaced by a perfect conductor, then $k_2 \to \infty$ and (8.5.14) shows that

$$Q = -\frac{j}{4\pi^2}\int_{-\infty}^{\infty}\int_{-\infty}^{\infty}\frac{1}{\tau_1}e^{j[\alpha_1 x+\alpha_2 y-\tau_1(z+h)]}d\alpha_1 d\alpha_2 = 2\frac{e^{jk_1R_2}}{4\pi R_2}$$

and (8.5.13) reads

$$\Pi_{z1}^e = \frac{(I\Delta L)}{j\omega\varepsilon_o}\left[\frac{e^{-jk_1R_1}}{4\pi R_1} + \frac{e^{-jk_1R_2}}{4\pi R_2}\right]$$

as expected.

Using the transformations to cylindrical coordinates in Section 6.1.2, (8.5.12) and (8.5.14) can be expressed in the form

$$P = \frac{j}{4\pi}\int_{-\infty}^{\infty}\frac{k_1^2\tau_2}{\tau_1(k_2^2\tau_1 + k_1^2\tau_2)}H_0^{(2)}(\lambda\rho)e^{-j\tau_1(z+h)}\lambda d\lambda \qquad (8.5.15)$$

and

$$Q = -\frac{j}{4\pi}\int_{-\infty}^{\infty}\frac{k_2^2}{(k_2^2\tau_1 + k_1^2\tau_2)}H_0^{(2)}(\lambda\rho)e^{-j\tau_1(z+h)}\lambda d\lambda \qquad (8.5.16)$$

Equations (8.5.15) and (8.5.16) illustrate the cylindrical symmetry of the problem. The field is symmetrical in ϕ (i.e., $\partial/\partial\phi = 0$), and varies as a function of ρ and z. In fact, in Section 9.6 the cylindrical symmetry is used to derive (8.5.15) and (8.5.16).

The above integrals were originally evaluated in the λ plane. Consider the integral in (8.5.16). This integral can be evaluated using contour integration by closing the path of integration with a large semicircle in the lower-half plane, accounting for the singularities and using residue theory.

The integrand in (8.5.16) has branch points and poles in the λ plane. The branch points associated with $\tau_1 = \sqrt{k_1^2 - \lambda^2}$ and $\tau_2 = \sqrt{k_2^2 - \lambda^2}$ are located at $\lambda = \pm k_1$ and $\lambda = \pm k_2$. There is also a branch point at $z = 0$ due to the Hankel function. The branch points due to $\lambda = \pm k_1$ lie on the real axis (since $k_1 = k_o$). Assuming that k_1 has a small imaginary part (i.e., $k_1 = k_1' - jk_1''$) the branch points are slightly moved away from the real axis. The branch cut associated with the Hankel function can be drawn from the origin to $-\infty$ slightly above the negative portion of the real axis. The branch cuts associated with $\pm k_1$ and $\pm k_2$ must be selected so that the integral goes to zero as $|\lambda| \to \infty$ in the lower-half plane.

Consider the branch cuts shown in Fig. 8.8a associated with the branch points $\pm k_1$. The cut shown as a wiggled line is that of the Hankel function. The convergence of (8.5.16) in the lower-half plane as $|\lambda| \to \infty$ depends on the exponential behavior of $H_0^{(2)}(\lambda\rho)e^{-j\tau_1(z+h)}$. The Hankel function $H_0^{(2)}(\lambda\rho)$ goes to zero as $|\lambda| \to \infty$ in the lower-half plane. This is seen by observing that for $|\lambda| \to \infty$ and $\rho > 0$:

$$H_0^{(2)}(\lambda\rho) \approx \sqrt{\frac{2}{\pi\lambda\rho}}\, e^{-j(\lambda\rho-\pi/4)}$$

Letting $\lambda = M\, e^{j\theta}$, the magnitude of $H_0^{(2)}(\lambda\rho)$ goes as

$$\frac{e^{-j\lambda\rho}}{\sqrt{M}} \sim \frac{e^{-jM\, e^{j\theta}\rho}}{\sqrt{M}} \sim \frac{e^{M\rho\, \sin\theta}}{\sqrt{M}}$$

(a)

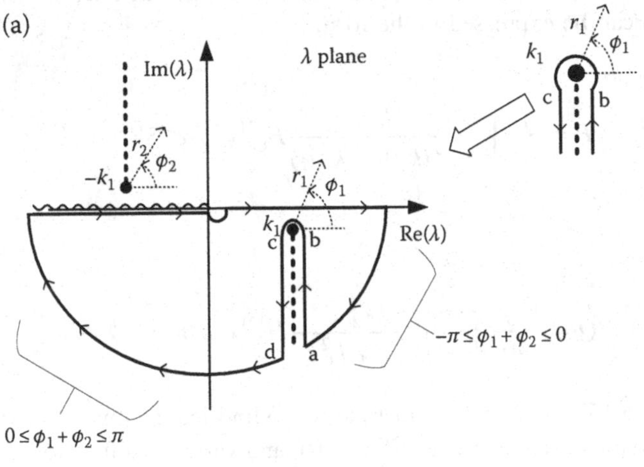

(b)

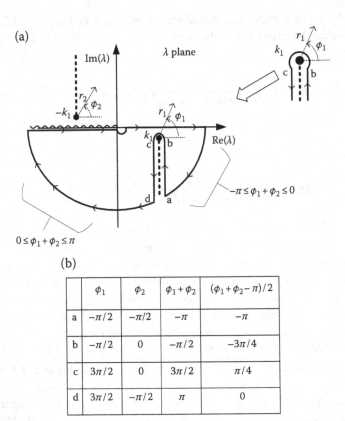

	ϕ_1	ϕ_2	$\phi_1+\phi_2$	$(\phi_1+\phi_2-\pi)/2$
a	$-\pi/2$	$-\pi/2$	$-\pi$	$-\pi$
b	$-\pi/2$	0	$-\pi/2$	$-3\pi/4$
c	$3\pi/2$	0	$3\pi/2$	$\pi/4$
d	$3\pi/2$	$-\pi/2$	π	0

FIGURE 8.8 (a) A branch cut for $\tau_1 = \sqrt{k_1^2 - \lambda^2}$; (b) the phase variation of τ_1 along the cut.

Since $\sin\theta \le 0$ in the lower-half plane, it follows that as $M \to \infty$ the above term goes to zero in the lower-half plane.

The exponential behavior of $e^{-j\tau_1(z+h)}$ in the lower-half plane is now analyzed. Referring to Fig. 8.8a, let

$$\lambda = k_1 + r_1 e^{j\phi_1}$$

and

$$\lambda = -k_1 + r_2 e^{j\phi_2}$$

Then,

$$\tau_1 = \sqrt{(k_1 - \lambda)(k_1 + \lambda)} = \sqrt{r_1 r_2}\, e^{j\left(\frac{\phi_1+\phi_2-\pi}{2}\right)}$$

and with $z + h = R_2 \cos\theta_2$ the exponent reads

$$e^{-j\tau_1(z+h)} = e^{-jR_2\cos\theta_2\sqrt{r_1r_2}\,e^{j\left(\frac{\phi_1+\phi_2-\pi}{2}\right)}} = e^{-jR_2\cos\theta_2\sqrt{r_1r_2}\left[\cos\left(\frac{\phi_1+\phi_2-\pi}{2}\right)+j\sin\left(\frac{\phi_1+\phi_2-\pi}{2}\right)\right]}$$
$$= e^{R_2\cos\theta_2\sqrt{r_1r_2}\sin\left(\frac{\phi_1+\phi_2-\pi}{2}\right)} e^{-jR_2\cos\theta_2\sqrt{r_1r_2}\cos\left(\frac{\phi_1+\phi_2-\pi}{2}\right)}$$

Since $0 \le \theta_2 \le \pi/2$ the term $\cos\theta_2 \ge 0$. Thus, the integral in (8.5.16) vanishes in the lower-half-plane along the infinite semicircle if

$$\sin\left(\frac{\phi_1 + \phi_2 - \pi}{2}\right) \le 0 \Rightarrow -\pi \le \frac{\phi_1 + \phi_2 - \pi}{2} \le 0 \Rightarrow -\pi \le \phi_1 + \phi_2 \le \pi \quad (8.5.17)$$

Figure 8.8b shows the phase of $\phi_1, \phi_2, \phi_1 + \phi_2$ and $(\phi_1 + \phi_2 - \pi)/2$ along the branch cut path at points **a** to **d**. It is observed that (8.5.17) is satisfied at points **a** and **d**. To the right of the branch cut, along the semicircle, the phase varies from 0 to $-\pi$ ($-\pi$ being the value at point **a**), and to the left of the branch point form π (π being the value at point **b**) to 0. Thus, the imaginary part of τ_1 is less than zero along the semicircle (i.e., $\mathrm{Im}(\tau_1) < 0$)) since (8.5.17) is satisfied. Observe that the phase difference between two opposite points on the branch cut is π (say at points **a** to **d** and points **b** and **c**).

The previous analysis of the behavior of $H_0^{(2)}(\lambda\rho)e^{-j\tau_1(z+h)}$ shows that it vanishes along the semicircle in the lower-half plane, as its radius goes to infinity.

Next, it is observed that the branch cut can also be selected such that $\mathrm{Im}(\tau_1) = 0$ along the cut. This is the common branch cut selected for τ_1 in the evaluation of (8.5.16). That is,

$$\mathrm{Im}(\tau_1) = \mathrm{Im}\sqrt{k_1^2 - \lambda^2} = 0 \quad (8.5.18)$$

The convergence of (8.5.16) depends on τ_1 and not on τ_2. As far as (8.5.16) is concerned, there are no requirements as to how the branch cut associated with τ_2 is drawn in the lower-half plane. However, it is convenient to select the cut in a similar way (i.e., $\text{Im}(\tau_2) = 0$). It is to be noted that the convergence of the transmitted fields (see (8.5.6)) depends on τ_2, and such a cut is appropriate for the analysis of the transmitted fields. For the branch points at $\pm k_2$, the branch cuts follow from

$$\text{Im}(\tau_2) = \text{Im}\sqrt{k_2^2 - \lambda^2} = 0 \tag{8.5.19}$$

The radiation conditions require that $\text{Im}(\tau_1) < 0$ and $\text{Im}(\tau_2) < 0$.

The branch cuts give rise to a Riemann surface with four sheets. The top sheet (also known as the upper Riemann surface) is defined by

$$\text{Im}(\tau_1) < 0$$

and

$$\text{Im}(\tau_2) < 0$$

The sign of the imaginary parts of τ_1 or τ_2 change as its branch cut is crossed.

To analyze (8.5.18) let $\tau_1 = \tau_1' + j\tau_1''$, $\lambda = \lambda' + j\lambda''$ and $k_1 = k_1' - jk_1''$. Then, $\tau_1^2 = k_1^2 - \lambda^2$ reads

$$\tau_1'^2 - \tau_1''^2 = k_1'^2 - k_1''^2 + \lambda''^2 - \lambda'^2$$

and

$$\tau_1'\tau_1'' = -\lambda'\lambda'' - k_1'k_1''$$

The condition $\text{Im}(\tau_1) = \tau_1'' = 0$, reduces the above relations to

$$\tau_1'^2 = (k_1'^2 - \lambda'^2) + (\lambda''^2 - k_1''^2) \tag{8.5.20}$$

and

$$\lambda'\lambda'' = -k_1'k_1'' \tag{8.5.21}$$

Equation (8.5.21) represents a hyperbola through the point $\lambda = k_1$ in the fourth quadrant and through the point $\lambda = -k_1$ in the second quadrant, as shown in Fig. 8.9. Along the solid portions of the hyperbolas: $k_1' > \lambda'$ and $|\lambda''| > k_1''$, which show that (8.5.20) is satisfied. In the dashed portions of the hyperbolas $\lambda' > k_1'$ and $|\lambda''| < k_1''$, which show that (8.5.20) is not satisfied. Also, note that as $k_1' \to 0$

the branch points approach the real axis. The analysis of (8.5.19) is similar, leading to branch points and branch cuts shown in Fig. 8.10.

The poles in (8.5.16), denoted by λ_p, are determined by

$$k_2^2\tau_1 + k_1^2\tau_2 = 0 \Rightarrow n_2^2\tau_1 + \tau_2 = 0 \Rightarrow n_2^2\tau_1 = -\tau_2 \tag{8.5.22}$$

where we used: $k_2/k_1 = k_2/k_o = n_2$. Equation (8.5.22) is solved by first squaring both sides. Hence, (8.5.22) reads

$$n_2^4\left(k_1^2 - \lambda_p^2\right) = k_1^2 n_2^2 - \lambda_p^2$$

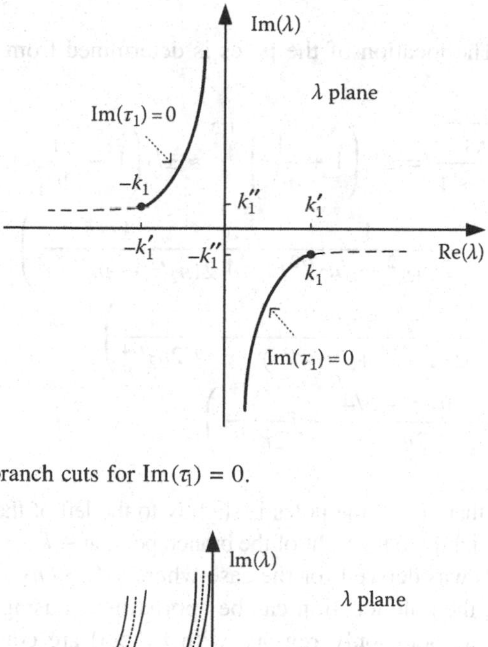

FIGURE 8.9 The branch cuts for $\mathrm{Im}(\tau_1) = 0$.

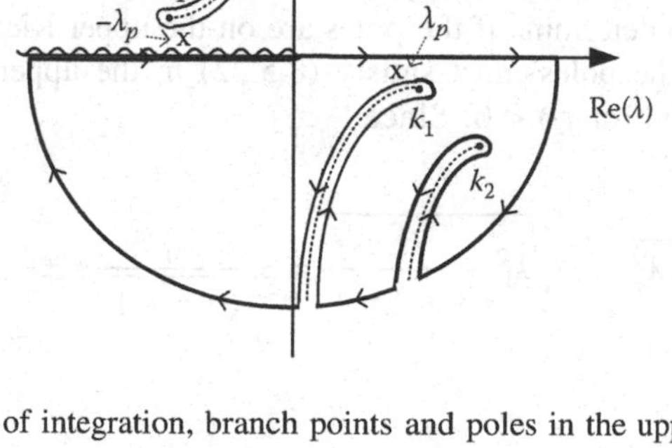

FIGURE 8.10 Path of integration, branch points and poles in the upper Riemann surface.

and the pole locations are at

$$\lambda_p = \pm \frac{k_1 n_2}{\sqrt{n_2^2 + 1}} \tag{8.5.23}$$

The squaring process introduces an addition solution corresponding to $n^2 \tau_1 = \tau_2$ that should also be considered.

In the case that the lossy surface represents a lossy ground or the sea-water surface, such that $\sigma_2/\omega\varepsilon_o >> \varepsilon_{r2}$, it follows that

$$n_2^2 = n_2'^2 - jn_2''^2 = \varepsilon_{r2} - j\frac{\sigma_2}{\omega\varepsilon_o} \Rightarrow \begin{cases} n_2'^2 = \varepsilon_{r2} \\ n_2''^2 = \dfrac{\sigma_2}{\omega\varepsilon_o} \end{cases}$$

where $n_2''^2 >> n_2'^2$. The location of the poles is determined from (8.5.23):

$$\begin{aligned} \lambda_p &= \pm k_1 \sqrt{\frac{n_2^2}{n_2^2 + 1}} = \pm k_1 \left(1 + \frac{1}{n_2^2}\right)^{-1/2} \approx \pm k_1 \left(1 - \frac{1}{2n_2^2} + \frac{3}{4}\frac{1}{2n_2^4} + ..\right) \\ &\approx \pm k_1 \left(1 - \frac{1}{2(n_2'^2 - jn_2''^2)} + \frac{3}{4}\frac{1}{2(n_2'^2 - jn_2''^2)^2}\right) \\ &\approx \pm k_1 \left(1 - \frac{n_2'^2 + jn_2''^2}{2(n_2'^2)^2 + 2(n_2''^2)^2} - \frac{3}{4}\frac{1}{2n_2''^4}\right) \\ &\approx \pm k_1 \left(1 - \frac{n_2'^2 + 3/4}{2n_2''^4} - j\frac{1}{2n_2''^2}\right) \end{aligned} \tag{8.5.24}$$

The real part shows that one of the poles is slightly to the left of the branch point at k_1, and the other pole is lightly to the right of the branch point at $-k_1$, as shown in Fig. 8.10.

Equation (8.5.24) was derived for the case where $n_2'' >> n_2'$. In other situations, such as dry ground, the pole location can be approximated using $n_2' >> n_2''$ (left to the problems). In this book only regions with $n_2' > 0$ are considered. There are some metallic regions where $n_2' < 0$ (i.e., dielectric constants having a negative real part), which require the analysis of (8.5.23) with $n_2' < 0$.

Next, we need to determine if the poles are on the upper Riemann surface. That is, the location of the poles must satisfy (8.5.22) in the upper Riemann surface, where $\mathrm{Im}(\tau_1) < 0$ and $\mathrm{Im}(\tau_2) < 0$. Since

$$\tau_1 = \pm\sqrt{k_1^2 - \lambda_p^2} = \pm\sqrt{k_1^2 - \frac{k_1^2 n_2^2}{n_2^2 + 1}} = \pm\frac{k_1}{\sqrt{n_2^2 + 1}} \approx \pm\frac{k_1}{n_2} \approx \pm\frac{k_1}{|n_2|} e^{j\pi 4}$$

where

$$n_2 = \sqrt{\varepsilon_{r2} - j\frac{\sigma_2}{\omega\varepsilon_o}} \approx \frac{\sigma_2}{\omega\varepsilon_o} e^{-j\pi/4} = |n_2| e^{-j\pi/4}$$

the root with the minus sign is selected to satisfy $\mathrm{Im}(\tau_1) < 0$. Hence,

$$\tau_1 \approx -\frac{k_1}{n_2} \approx -\frac{k_1}{|n_2|} e^{j\pi/4} \tag{8.5.25}$$

Next, we determine τ_2. That is,

$$\tau_2 = \pm\sqrt{(k_1 n_2)^2 - \lambda_p^2} = \pm\sqrt{(k_1 n_2)^2 - \frac{(k_1 n_2)^2}{n_2^2 + 1}} \approx \pm\sqrt{(k_1 n_2)^2 - k_1^2} \approx \pm k_1 n_2$$
$$= \pm k_1 |n_2| e^{-j\pi/4}$$

The root with the plus sign is selected to satisfy $\mathrm{Im}(\tau_2) < 0$. Hence,

$$\tau_2 \approx k_1 n_2 \approx k_1 |n_2| e^{-j\pi/4} \tag{8.5.26}$$

Substituting (8.5.25) and (8.5.26) into (8.5.22) show that the equation is satisfied.

Hence, the poles in (8.5.24) are on the upper Riemann surface. A similar analysis with $\mathrm{Im}(\tau_1) > 0$ and $\mathrm{Im}(\tau_2) > 0$ shows that there are poles in the Riemann sheet defined by $\mathrm{Im}(\tau_1) > 0$ and $\mathrm{Im}(\tau_2) > 0$. For the additional solution that satisfies: $n^2 \tau_1 = \tau_2$, the poles are located in the Riemann sheets defined by $\mathrm{Im}(\tau_1) > 0$ and $\mathrm{Im}(\tau_2) < 0$, and $\mathrm{Im}(\tau_1) < 0$ and $\mathrm{Im}(\tau_2) > 0$.

The path of integration and the singularities in the upper Riemann surfaces are shown in Fig. 8.10. The path of integration is closed in the lower-half plane because that is where the integrands vanish as $|\lambda| \to \infty$. In Fig. 8.11 the lower-half of the upper Riemann surface is shown. The figure also shows the values of the imaginary parts of τ_1 and τ_2 in the four sheets. with the pole in the top Riemann sheet. The pole λ_p appears in the upper Riemann surface and also in the lowest Riemann sheet. The connection between the sheets is also shown.

The poles' locations in Fig. 8.10 are for the particular case that we are considering. The poles' locations depend on n_2 and depending on the constitutive parameters of the region the poles can be located to the right or to the left of the branch point. If $k_1'' \to 0$ the branch points $\pm\, k_1$ are on the real axis and the contour of integration in Fig. 8.11 must be indented below the branch point at $-\, k_1$, and above the branch point at k_1 above the branch point.

The integration described in Fig. 8.11 in the λ-plane is involved, and it can be simplified (for example) by using the transformation $\lambda = k_1 \sin\beta$ and performing the integration in the β plane. The next chapter deals with problems in cylindrical coordinates, where this problem is revisited and the integral in (8.5.16) is evaluated in the β plane.

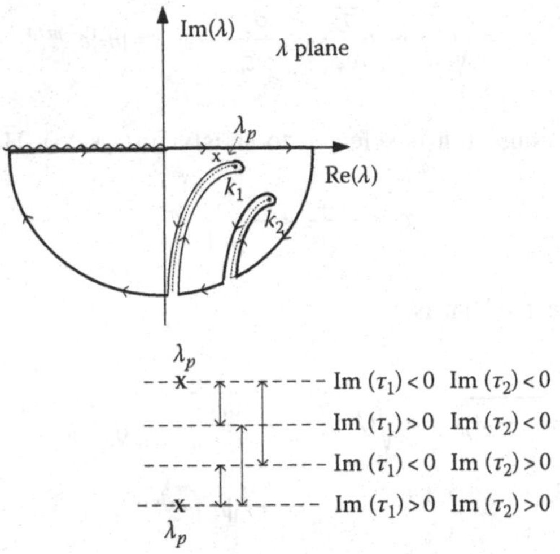

FIGURE 8.11 The lower-half of the upper Riemann surface and the values of the imaginary parts of τ_1 and τ_2 in the four sheets.

8.5.2 Magnetic Hertzian Dipole

Figure 8.12 shows a magnetic Hertzian dipole above a lossy surface. The fields for this configuration are usually determined in terms of either the **F** vector or $\mathbf{\Pi}^{\mathrm{m}}$ vector.

The expression for the $\mathbf{\Pi}^{\mathrm{m}}$ vector in the air region (Region 1) is

$$[\nabla^2 + k_1^2]\Pi_{z1}^{\mathrm{m}}(x, y, z) = -\frac{J_{m,z}}{j\omega\mu_o} = -\frac{(I_m\Delta L)}{j\omega\mu_o}\delta(x)\delta(y)\delta(z-h) \quad (8.5.27)$$

and in the lossy region (Region 2) is

$$[\nabla^2 + k_2^2]\Pi_{z2}^{\mathrm{m}}(x, y, z) = 0 \quad (8.5.28)$$

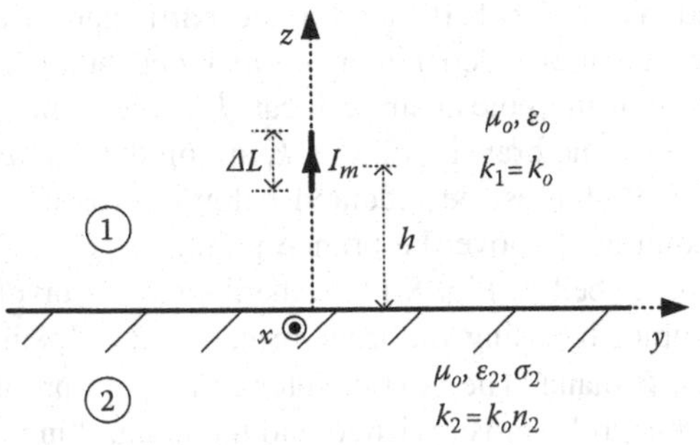

FIGURE 8.12 Magnetic Hertzian dipole above a lossy surface.

where

$$k_1 = k_o = \omega\sqrt{\mu_o \varepsilon_o}$$

and

$$k_2 = k_o n_2 = k_o\sqrt{\varepsilon_{r2} - j\frac{\sigma_2}{\omega\varepsilon_o}}$$

The fields are calculated using (4.2.19) to (4.2.24), namely

$$E_x = -j\omega\mu_o \frac{\partial \Pi_z^m}{\partial y}$$
$$E_y = j\omega\mu_o \frac{\partial \Pi_z^m}{\partial x}$$
$$H_x = \frac{\partial^2 \Pi_z^m}{\partial x \partial z}$$
$$H_y = \frac{\partial^2 \Pi_z^m}{\partial y \partial z}$$
$$H_z = \left(\frac{\partial^2}{\partial z^2} + k^2\right)\Pi_z^m$$

The solution of (8.5.27) requires a particular and a complementary solution. The solutions are

$$\Pi_{z1,p}^m = \frac{(I_m \Delta L)}{j\omega\mu_o}\frac{e^{-jkR_1}}{4\pi R_1} = \frac{-(I_m \Delta L)}{8\pi^2\omega\mu_o}\int_{-\infty}^{\infty}\int_{-\infty}^{\infty}\frac{e^{j(\alpha_1 x + \alpha_2 y - \tau_1|z-h|)}}{\tau_1}d\alpha_1 d\alpha_2$$

and

$$\Pi_{z1,c}^m = \frac{-(I_m \Delta L)}{8\pi^2\omega\mu_o}\int_{-\infty}^{\infty}\int_{-\infty}^{\infty} R^m(\alpha_1, \alpha_2)e^{j(\alpha_1 x + \alpha_2 y - \tau_1 z)}d\alpha_1 d\alpha_2$$

where

$$R_1 = \sqrt{x^2 + y^2 + (z-h)^2}$$

$$\tau_1 = \sqrt{k_1^2 - \alpha_1^2 - \alpha_2^2}$$

and $\text{Im}(\tau_1) < 0$. The term $R^m(\alpha_1, \alpha_2)$ is the reflection coefficient associated with the magnetic dipole excitation. Therefore,

$$\Pi_{z1}^{m} = \Pi_{z1,p}^{m} + \Pi_{z1,c}^{m}$$
$$= \frac{-(I_m \Delta L)}{8\pi^2 \omega \mu_o} \int_{-\infty}^{\infty} \int_{-\infty}^{\infty} \left[\frac{e^{-j\tau_1 |z-h|}}{\tau_1} + R^m(\alpha_1, \alpha_2) e^{-j\tau_1 z} \right] e^{j(\alpha_1 x + \alpha_2 y)} d\alpha_1 d\alpha_2 \quad (8.5.29)$$

The solution to (8.5.28) in Region 2 is written as

$$\Pi_{z2}^{m} = \frac{-(I_m \Delta L)}{8\pi^2 \omega \mu_o} \int_{-\infty}^{\infty} \int_{-\infty}^{\infty} T^m(\alpha_1, \alpha_2) e^{j(\alpha_1 x + \alpha_2 y + \tau_2 z)} d\alpha_1 d\alpha_2 \quad (8.5.30)$$

where

$$\tau_2 = \sqrt{k_2^2 - \alpha_1^2 - \alpha_2^2}$$

and $\text{Im}(\tau_2) < 0$. The term $T^m(\alpha_1, \alpha_2)$ represents the transmission coefficient.

The boundary conditions at $z = 0$ are the continuity of the tangential fields, or

$$E_{x1} = E_{x2}\Big|_{z=0} \Rightarrow \frac{\partial \Pi_{z1}^{m}}{\partial y} = \frac{\partial \Pi_{z2}^{m}}{\partial y}\Big|_{z=0}$$
$$E_{y1} = E_{y2}\Big|_{z=0} \Rightarrow \frac{\partial \Pi_{z1}^{m}}{\partial x} = \frac{\partial \Pi_{z2}^{m}}{\partial x}\Big|_{z=0}$$
$$H_{x1} = H_{x2}\Big|_{z=0} \Rightarrow \frac{\partial^2 \Pi_{z1}^{m}}{\partial x \partial z} = \frac{\partial^2 \Pi_{z2}^{m}}{\partial x \partial z}\Big|_{z=0}$$

$$H_{y1} = H_{y2}\Big|_{z=0} \Rightarrow \frac{\partial^2 \Pi_{z1}^{m}}{\partial y \partial z} = \frac{\partial^2 \Pi_{z2}^{m}}{\partial y \partial z}\Big|_{z=0}$$

The boundary conditions must be satisfied for all values of x and y. Hence, the above conditions on the electric fields are simply stated as

$$\Pi_{z1}^{m} = \Pi_{z2}^{m}\Big|_{z=0} \quad (8.5.31)$$

and on the magnetic fields:

$$\frac{\partial \Pi_{z1}^{m}}{\partial z} = \frac{\partial \Pi_{z2}^{m}}{\partial z}\Big|_{z=0} \quad (8.5.32)$$

Applying the boundary condition to (8.5.29) and (8.5.30):

$$\frac{e^{-j\tau_1 h}}{\tau_1} + R^m(\alpha_1, \alpha_2) = T^m(\alpha_1, \alpha_2)$$

and

$$\frac{e^{-j\tau_1 h}}{\tau_1} - R^m(\alpha_1, \alpha_2) = \frac{\tau_2}{\tau_1} T^m(\alpha_1, \alpha_2)$$

which are solved to give

$$R^m(\alpha_1, \alpha_2) = \frac{\tau_1 - \tau_2}{\tau_1 + \tau_2} \frac{e^{-j\tau_1 h}}{\tau_1}$$

and

$T^m(\alpha_1, \alpha_2) = \frac{2}{\tau_1 + \tau_2} e^{-j\tau_1 h}$

The reflection coefficient can be expressed in the forms

$$R^m(\alpha_1, \alpha_2) = \frac{e^{-j\tau_1 h}}{\tau_1}\left[1 - \frac{2\tau_2}{\tau_1 + \tau_2}\right] \tag{8.5.33}$$

and

$$R^m(\alpha_1, \alpha_2) = \frac{e^{-j\tau_1 h}}{\tau_1}\left[-1 + \frac{2\tau_1}{\tau_1 + \tau_2}\right] \tag{8.5.34}$$

The form in (8.5.33) when substituted into (8.5.29) leads to

$$\Pi_{z1}^{m} = \frac{(I_m \Delta L)}{j\omega\mu_o}\left[\frac{e^{-jkR_1}}{4\pi R_1} + \frac{e^{-jkR_2}}{4\pi R_2} + P^m\right] \tag{8.5.35}$$

where $k = k_o$ and

$$R_2 = \sqrt{x^2 + y^2 + (z + h)^2}$$

and

$$\begin{aligned} P^m &= \frac{j}{4\pi^2}\int_{-\infty}^{\infty}\int_{-\infty}^{\infty} \frac{\tau_2}{\tau_1(\tau_1 + \tau_2)} e^{j[\alpha_1 x + \alpha_2 y - \tau_1(z+h)]} d\alpha_1 d\alpha_2 \\ &= \frac{j}{2\pi}\int_{-\infty}^{\infty} \frac{\tau_2}{\tau_1(\tau_1 + \tau_2)} J_0(\lambda\rho) e^{j\tau_1(z+h)} \lambda d\lambda \end{aligned} \tag{8.5.36}$$

If the surface is a perfect conductor, then $k_2 \rightarrow \infty$, $\tau_2 \rightarrow \infty$ and (8.5.36) reads

$$P^m = \frac{j}{2\pi}\int_{-\infty}^{\infty}\frac{1}{\tau_1}J_0(\lambda\rho)e^{j\tau_1(z+h)}\lambda d\lambda = -2\frac{e^{jkR_2}}{4\pi R_2}$$

and from (8.3.35):

$$\Pi_{z1}^{m} = \frac{(I_m \Delta L)}{j\omega\mu_o}\left[\frac{e^{-jkR_1}}{4\pi R_1} - \frac{e^{-jkR_2}}{4\pi R_2}\right] \tag{8.5.37}$$

which is the result that one obtains using image theory for a magnetic dipole above a perfect conductor.

If (8.5.34) is used, substituting into (8.5.29) leads to

$$\Pi_{z1}^{m} = \frac{(I_m \Delta L)}{j\omega\mu_o}\left[\frac{e^{-jkR_1}}{4\pi R_1} - \frac{e^{-jkR_2}}{4\pi R_2} + Q^m\right] \tag{8.5.38}$$

where

$$\begin{aligned} Q^m &= -\frac{j}{4\pi^2}\int_{-\infty}^{\infty}\int_{-\infty}^{\infty}\frac{1}{\tau_1+\tau_2}e^{j[\alpha_1 x+\alpha_2 y-\tau_1(z+h)]}d\alpha_1 d\alpha_2 \\ &= -\frac{j}{2\pi}\int_0^{\infty}\frac{1}{\tau_1+\tau_2}J_0(\lambda\rho)e^{-j\tau_1(z+h)}\lambda d\lambda \end{aligned} \tag{8.5.39}$$

If the surface is a perfect conductor, then $k_2 \to \infty$, $\tau_2 \to \infty$ and $Q^m = 0$. In this case (8.5.38), reduces to (8.5.37), as expected.

Proceeding with the evaluation of (8.5.39) it is seen that the integrand has no poles in the λ plane when $\tau_1 + \tau_2 = 0$, and there is no need to write the integral in terms of $H_0^{(2)}(\lambda\rho)$ since the integral in (8.5.39) can be evaluated in closed form. Multiplying the numerator and denominator of the integrand by $\tau_1 - \tau_2$ and since $\tau_1^2 - \tau_2^2 = k_o^2(1 - n_2^2)$, (8.5.39) reads

$$Q^m = \frac{j}{2\pi k_o^2(n_2^2-1)}\left[\int_0^{\infty}\tau_1 J_o(\lambda\rho)e^{-j\tau_1(z+h)}\lambda d\lambda - \int_0^{\infty}\tau_2 J_o(\lambda\rho)e^{-j\tau_1(z+h)}\lambda d\lambda\right] \tag{8.5.40}$$

If the loop radiator is close to the lossy surface it follows that $h \approx 0$ and $R_1 \approx R_2$. Thus, from (8.5.38) and (8.5.40):

$$\Pi_{z1}^{m} = \frac{(I_m \Delta L)}{j\omega\mu_o}Q^m = \frac{(I\Delta S)}{2\pi k_o^2(n_2^2-1)}\left[j\int_0^{\infty}\tau_1 J_o(\lambda\rho)e^{-j\tau_1 z}\lambda d\lambda - j\int_0^{\infty}\tau_2 J_o(\lambda\rho)e^{-j\tau_1 z}\lambda d\lambda\right] \tag{8.5.41}$$

where (2.6.12) was used to write the expression in terms of $(I\Delta S)$.

The integrals in (8.5.41) are evaluated as follows. Using the Sommerfeld integral in (6.2.30):

$$\frac{e^{-jkR}}{R} = \int_0^\infty \frac{e^{-j\tau z}}{j\tau} J_0(\lambda\rho)\lambda\, d\lambda \tag{8.5.42}$$

where

$$R = \sqrt{x^2 + y^2 + z^2} = \sqrt{\rho^2 + z^2}$$

and

$$\tau = \sqrt{k^2 - \lambda^2}$$

it follows that taking a derivative of (8.5.42) with respect to z:

$$\frac{\partial}{\partial z}\left(\frac{e^{-jkR}}{R}\right) = -ze^{-jkR}\left(\frac{1 + jkR}{R^3}\right)$$

Hence,

$$\int_0^\infty e^{-j\tau z} J_0(\lambda\rho)\lambda\, d\lambda = ze^{-jkR}\left(\frac{1 + jkR}{R^3}\right) \tag{8.5.43}$$

Taking another derivative with respect to z in (8.5.43) and evaluating the result at $z = 0$ (which sets the observation point at the surface in (8.5.41)) gives

$$j\int_0^\infty \tau J_0(\lambda\rho)\lambda\, d\lambda = -(1 + jk\rho)\left(\frac{e^{-jk\rho}}{\rho^3}\right) \tag{8.5.44}$$

The result in (8.5.44) permits the evaluation of (8.5.41) in closed form, namely

$$\Pi_{z1}^{\mathrm{m}} = -\frac{(I\Delta S)}{2\pi k_o^2(n_2^2 - 1)\rho^3}\left[(1 + jk_o\rho)e^{-jk_o\rho} - (1 + jk_o n_2\rho)e^{-jk_o n_2\rho}\right]$$

The fields follow from (4.4.17) to (4.4.22). From (4.4.17):

$$\begin{aligned} E_{\phi 1} &= j\omega\mu_o \frac{\partial \Pi_{z1}^{\mathrm{m}}}{\partial \rho} \\ &= \frac{j\omega\mu_o(I\Delta S)}{2\pi k_o^2(n_2^2 - 1)\rho^4}\left[(3 + j3k_o\rho - k_o^2\rho^2)e^{-jk_o\rho} - (3 + j3k_o n_2\rho - k_0^2 n_2^2\rho^2)e^{-jk_o n_2\rho}\right] \end{aligned} \tag{8.5.45}$$

and from (4.4.22):

$$H_{z1} = \frac{\partial^2 \Pi_{z1}^{m}}{\partial z^2} - \gamma^2 \Pi_{z1}^{m} = -\frac{\partial^2 \Pi_{z1}^{m}}{\partial \rho^2} - \frac{1}{\rho}\frac{\partial \Pi_{z1}^{m}}{\partial \rho} = -\frac{1}{\rho}\frac{\partial}{\partial \rho}\left(\rho \frac{\partial \Pi_{z1}^{m}}{\partial \rho}\right)$$

where the wave equation in cylindrical coordinates was used to express the H_{z1} field as a function of ρ. Therefore, it follows that

$$H_{z1} = \frac{(I\Delta S)}{2\pi k_o^2 (n_2^2 - 1)\rho^5}\left[(9 + j9k_o\rho - 4k_o^2\rho^2 - jk_o^3\rho^3)e^{-jk_o\rho} - (9 + j9k_o\rho - 4k_o^2 n_2^2\rho^2 - jk_o^3 n_2^3\rho^3)e^{-jk_o n_2\rho}\right] \quad (8.5.46)$$

8.6 HORIZONTAL ELECTRIC HERTZIAN DIPOLE ABOVE A LOSSY SURFACE

The geometry of the problem is shown in Fig. 8.13. The expression for the **A** vector in the air region (i.e., Region 1) is

$$[\nabla^2 + k_1^2]A_{x1}(x, y, z) = -\mu_o J_{e,x} = -\mu_o (I\Delta L)\delta(x)\delta(y)\delta(z - h) \quad (8.6.1)$$

where

$$k_1 = k_o = \omega\sqrt{\mu_o \varepsilon_o}$$

In the lossy surface (i.e., Region 2), A_{x2} satisfies

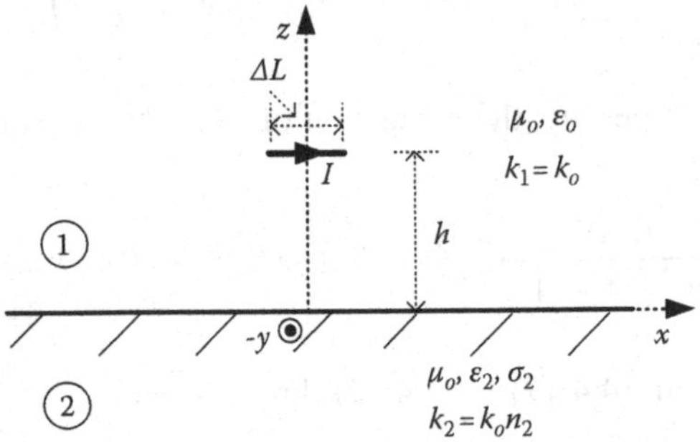

FIGURE 8.13 Horizontal electric dipole above a lossy surface.

$$[\nabla^2 + k_2^2]A_{x2}(x, y, z) = 0 \tag{8.6.2}$$

where, since $\mu_2 = \mu_o\mu_{r2}$, it follows that

$$k_2 = k_o n_2 = k_o\sqrt{\mu_{r2}\left(\varepsilon_{r2} - j\frac{\sigma_2}{\omega\varepsilon_o}\right)}$$

The vector potential $\mathbf{A} = A_x\mathbf{a}_x$ does not generate an H_x field component, and the HED generates such component. To generate such component, and to satisfy the boundary conditions, either an electric vector potential component must be included, or the use of the z component of the $\mathbf{A}$ vector. To see this point clearly, let's try to formulate the problem with only $\mathbf{A} = A_x\mathbf{a}_x$. Then, from (4.2.16) and (4.2.17), the tangential components are

$$E_x = -\frac{j\omega}{k^2}\left(\frac{\partial^2 A_x}{\partial x^2} + k^2 A_x\right) = -j\omega A_x - \frac{j\omega}{k^2}\frac{\partial^2 A_x}{\partial x^2} \tag{8.6.3}$$

and

$$E_y = -\frac{j\omega}{k^2}\frac{\partial^2 A_x}{\partial x \partial y} \tag{8.6.4}$$

The boundary conditions for E_x and E_y, which must be satisfied for all values of x and y along the surface, are

$$E_{x1} = E_{x2}|_{z=0} \Rightarrow \begin{cases} A_{x1} = A_{x2}|_{z=0} \\ \left.\dfrac{1}{k_1^2}\dfrac{\partial^2 A_{x1}}{\partial x^2} = \dfrac{1}{k_2^2}\dfrac{\partial^2 A_{x2}}{\partial x^2}\right|_{z=0} \Rightarrow \left.\dfrac{1}{k_1^2}A_{x1} = \dfrac{1}{k_2^2}A_{x2}\right|_{z=0} \end{cases} \tag{8.6.5}$$

and

$$E_{y1} = E_{y2}\Big|_{z=0} \Rightarrow \left.\frac{1}{k_1^2}\frac{\partial^2 A_{x1}}{\partial x \partial y} = \frac{1}{k_2^2}\frac{\partial^2 A_{x2}}{\partial x \partial y}\right|_{z=0} \Rightarrow \left.\frac{1}{k_1^2}A_{x1} = \frac{1}{k_2^2}A_{x2}\right|_{z=0} \tag{8.6.6}$$

Hence, (8.6.5) and (8.6.6) require that $k_1^2 = k_2^2$ or $k_1 = k_2$, which is a contradiction since $k_1 \neq k_2$.

To properly satisfy the boundary conditions we let $\mathbf{A} = A_x\mathbf{a}_x + A_z\mathbf{a}_z$ where A_z satisfies

$$[\nabla^2 + k_1^2]A_{z1}(x, y, z) = 0 \tag{8.6.7}$$

and

$$[\nabla^2 + k_2^2]A_{z2}(x, y, z) = 0 \tag{8.6.8}$$

With $\mathbf{A} = A_x\mathbf{a}_x + A_z\mathbf{a}_z$, the tangential components of the fields, from (4.2.7) to (4.2.18), are

$$E_x = -j\omega A_x + \frac{1}{j\omega\mu\varepsilon_c}\frac{\partial}{\partial x}\left(\frac{\partial A_x}{\partial x} + \frac{\partial A_z}{\partial z}\right) \tag{8.6.9}$$

$$E_y = \frac{1}{j\omega\mu\varepsilon_c}\frac{\partial}{\partial y}\left(\frac{\partial A_x}{\partial x} + \frac{\partial A_z}{\partial z}\right) \tag{8.6.10}$$

$$H_x = \frac{1}{\mu}\frac{\partial A_z}{\partial y} \tag{8.6.11}$$

$$H_y = \frac{1}{\mu}\left(\frac{\partial A_x}{\partial z} - \frac{\partial A_z}{\partial x}\right) \tag{8.6.12}$$

Applying the Fourier transform to the x and y variables in the wave equations, namely

$$A(\alpha_1, \alpha_2, z) = \int_{-\infty}^{\infty}\int_{-\infty}^{\infty} A(x, y, z)e^{-j(\alpha_1 x+\alpha_2 y)}dxdy$$

it follows that (8.6.1), (8.6.2), (8.6.7), and (8.6.8) read

$$\left[\frac{d^2}{dz^2} + \tau_1^2\right]A_{x1}(\alpha_1, \alpha_2, z) = -\mu_o(I\Delta L)\delta(z - h) \tag{8.6.13}$$

$$\left[\frac{d^2}{dz^2} + \tau_1^2\right]A_{z1}(\alpha_1, \alpha_2, z) = 0 \tag{8.6.14}$$

$$\left[\frac{d^2}{dz^2} + \tau_2^2\right]A_{x2}(\alpha_1, \alpha_2, z) = 0 \tag{8.6.15}$$

$$\left[\frac{d^2}{dz^2} + \tau_2^2\right]A_{z2}(\alpha_1, \alpha_2, z) = 0 \tag{8.6.16}$$

where

$$\tau_1 = \sqrt{k_1^2 - \alpha_1^2 - \alpha_2^2}$$

and

$$\tau_2 = \sqrt{k_2^2 - \alpha_1^2 - \alpha_2^2}$$

The solutions of (8.6.13) to (8.6.16) are

$$A_{x1}(\alpha_1, \alpha_2, z) = \frac{-j\mu_o(I\Delta L)}{2\tau_1} e^{-j\tau_1|z-h|} + B(\alpha_1, \alpha_2)e^{-j\tau_1 z}$$

$$A_{z1}(\alpha_1, \alpha_2, z) = C(\alpha_1, \alpha_2)e^{-j\tau_1 z}$$

$$A_{x2}(\alpha_1, \alpha_2, z) = D(\alpha_1, \alpha_2)e^{j\tau_2 z}$$

$$A_{z2}(\alpha_1, \alpha_2, z) = G(\alpha_1, \alpha_2)e^{j\tau_2 z}$$

where $\text{Im}(\tau_1) < 0$ and $\text{Im}(\tau_2) < 0$. The terms $B(\alpha_1, \alpha_2)$, $C(\alpha_1, \alpha_2)$, $D(\alpha_1, \alpha_2)$, and $E(\alpha_1, \alpha_2)$ will simply be denoted by B, C, D, and E.

Applying the inverse transform:

$$A_{x1}(x, y, z) = \frac{1}{(2\pi)^2}\int_{-\infty}^{\infty}\int_{-\infty}^{\infty}\left[\frac{-j\mu_o(I\Delta L)}{2\tau_1}e^{-j\tau_1|z-h|} + Be^{-j\tau_1 z}\right]e^{j(\alpha_1 x+\alpha_2 y)}d\alpha_1 d\alpha_2 \tag{8.6.17}$$

$$A_{z1}(x, y, z) = \frac{1}{(2\pi)^2}\int_{-\infty}^{\infty}\int_{-\infty}^{\infty} C\, e^{j(\alpha_1 x+\alpha_2 y-\tau_1 z)}d\alpha_1 d\alpha_2 \tag{8.6.18}$$

$$A_{x2}(x, y, z) = \frac{1}{(2\pi)^2}\int_{-\infty}^{\infty}\int_{-\infty}^{\infty} D\, e^{j(\alpha_1 x+\alpha_2 y+\tau_2 z)}d\alpha_1 d\alpha_2 \tag{8.6.19}$$

$$A_{z2}(x, y, z) = \frac{1}{(2\pi)^2}\int_{-\infty}^{\infty}\int_{-\infty}^{\infty} G\, e^{j(\alpha_1 x+\alpha_2 y+\tau_2 z)}d\alpha_1 d\alpha_2 \tag{8.6.20}$$

The boundary conditions are the continuity of the tangential electric and magnetic fields at $z = 0$ which must be satisfied for all values of x and y. From (8.6.10):

$$E_{y1} = E_{y2}\Big|_{z=0} \Rightarrow \frac{1}{\mu_o\varepsilon_o}\left(\frac{\partial A_{x1}}{\partial x} + \frac{\partial A_{z1}}{\partial z}\right)\Big|_{z=0} = \frac{1}{\mu_2\varepsilon_{c2}}\left(\frac{\partial A_{x2}}{\partial x} + \frac{\partial A_{z2}}{\partial z}\right)\Big|_{z=0}$$

and substituting (8.6.17) to (8.6.20) gives

$$\frac{1}{\mu_o \varepsilon_o}\left[-\frac{j\mu_o(I\Delta L)}{2\tau_1}\alpha_1 e^{-j\tau_1 h} + \alpha_1 B - \tau_1 C\right] = \frac{1}{\mu_2 \varepsilon_{c2}}(\alpha_1 D + \tau_2 G) \quad (8.6.21)$$

From (8.6.9), observing that the continuity of the second term is the same as the continuity of the second term in (8.6.10) (which resulted in (8.6.21)), it follows that

$$E_{x1} = E_{x2}|_{z=0} \Rightarrow A_{x1} = A_{x2}|_{z=0}$$

Then, using (8.6.17) and (8.6.19) gives

$$\frac{-j\mu_o(I\Delta L)}{2\tau_1}e^{-j\tau_1 h} + B = D \quad (8.6.22)$$

From (8.6.11):

$$H_{x1} = H_{x2}|_{z=0} \Rightarrow \frac{1}{\mu_o}A_{z1} = \frac{1}{\mu_2}A_{z2}\bigg|_{z=0} \quad (8.6.23)$$

and substituting (8.6.18) and (8.6.20) gives

$$\frac{1}{\mu_o}C = \frac{1}{\mu_2}G \quad (8.6.24)$$

From (8.6.12):

$$H_{y1} = H_{y2}\bigg|_{z=0} \Rightarrow \frac{1}{\mu_o}\left(\frac{\partial A_{x1}}{\partial z} - A_{z1}\right)\bigg|_{z=0} = \frac{1}{\mu_2}\left(\frac{\partial A_{x2}}{\partial z} - A_{z2}\right)\bigg|_{z=0} \Rightarrow \frac{1}{\mu_o}\frac{\partial A_{x1}}{\partial z}\bigg|_{z=0} = \frac{1}{\mu_2}\frac{\partial A_{x2}}{\partial z}\bigg|_{z=0}$$

where (8.6.23) was used. Then, from (8.6.17) and (8.6.19):

$$\frac{1}{\mu_o}\left[\frac{j\mu_o(I\Delta L)}{2}e^{-j\tau_1 h} + \tau_1 B\right] = -\frac{1}{\mu_2}\tau_2 D \quad (8.6.25)$$

The evaluation of the constants is as follows. From (8.6.22) and (8.6.25), we obtain

$$B = -\frac{j\mu_o(I\Delta L)}{2\tau_1}\left(\frac{\mu_{r2}\tau_1 - \tau_2}{\mu_{r2}\tau_1 + \tau_2}\right)e^{-j\tau_1 h} \quad (8.6.26)$$

and

$$D = -j\mu_o(I\Delta L)\left(\frac{\mu_{r2}}{\mu_{r2}\tau_1 + \tau_2}\right)e^{-j\tau_1 h} \tag{8.6.27}$$

Substituting (8.6.24), (8.6.26), and (8.6.27) into (8.6.21) gives

$$C = -j\mu_o(I\Delta L)\alpha_1 \frac{\mu_{r2}\varepsilon_{cr,2} - 1}{(\varepsilon_{cr,2}\tau_1 + \tau_2)(\mu_{r2}\tau_1 + \tau_2)}e^{-j\tau_1 h} \tag{8.6.28}$$

where ε_{c2} was expressed as $\varepsilon_{c2} = \varepsilon_o\varepsilon_{cr,2}$. The form of G follows from (8.6.24).
Substituting B into (8.6.17):

$$A_{x1}(x, y, z) = \frac{-j\mu_o(I\Delta L)}{(2\pi)^2}\int_{-\infty}^{\infty}\int_{-\infty}^{\infty}\left[\frac{e^{-j\tau_1|z-h|}}{2\tau_1} + \frac{1}{2\tau_1}\left(\frac{\mu_{r2}\tau_1 - \tau_2}{\mu_{r2}\tau_1 + \tau_2}\right)e^{-j\tau_1(z+h)}\right] e^{j(\alpha_1 x + \alpha_2 y)}\,d\alpha_1\,d\alpha_2 \tag{8.6.29}$$

This integral is handled using the method described in Section 6.1.2. That is, let

$$x = \rho\cos\phi$$
$$y = \rho\sin\phi$$
$$\alpha_1 = \lambda\cos\gamma$$
$$\alpha_2 = \lambda\sin\gamma$$

and (8.6.29) reads

$$A_{x1}(x, y, z) = \frac{-j\mu_o(I\Delta L)}{(2\pi)^2}\int_0^{\infty}\left\{\left[\frac{e^{-j\tau_1|z-h|}}{2\tau_1} + \frac{1}{2\tau_1}\left(\frac{\mu_{r2}\tau_1 - \tau_2}{\mu_{r2}\tau_1 + \tau_2}\right)e^{-j\tau_1(z+h)}\right] \times \int_0^{2\pi} e^{j\lambda\rho\cos(\gamma-\phi)}\,d\gamma\right\}\lambda\,d\lambda \tag{8.6.30}$$

where

$$\tau_1 = \sqrt{k_1^2 - \lambda^2}$$

and

$$\tau_2 = \sqrt{k_2^2 - \lambda^2}$$

The angular integral is recognized as $2\pi J_0(\lambda\rho)$ and (8.6.30) reads

$$A_{x1}(x, y, z) = \frac{-j\mu_o(I\Delta L)}{4\pi}\int_0^\infty\left[\frac{e^{-j\tau_1|z-h|}}{\tau_1} + \frac{1}{\tau_1}\left(\frac{\mu_{r2}\tau_1 - \tau_2}{\mu_{r2}\tau_1 + \tau_2}\right)e^{-j\tau_1(z+h)}\right]J_0(\lambda\rho)\lambda d\lambda \tag{8.6.31}$$

Similarly, substituting the expression for D into (8.6.19) gives

$$A_{x2}(x, y, z) = \frac{-j\mu_o(I\Delta L)}{2\pi}\int_0^\infty\left[\frac{\mu_{r2}}{\mu_{r2}\tau_1 + \tau_2}e^{-j\tau_1 h}e^{j\tau_2 z}\right]J_0(\lambda\rho)\lambda d\lambda \tag{8.6.32}$$

The evaluation of (8.6.18) uses the expression for C. That is,

$$A_{z1}(x, y, z) = -\frac{j\mu_o(I\Delta L)}{(2\pi)^2}\int_0^\infty\left[\frac{\mu_{r2}\varepsilon_{cr,2} - 1}{(\varepsilon_{cr,2}\tau_1 + \tau_2)(\mu_{r2}\tau_1 + \tau_2)}e^{-j\tau_1(z+h)}\right]$$
$$\times\left[\int_0^{2\pi}\cos\gamma e^{j\alpha\rho\cos(\gamma-\phi)}d\gamma\right]\lambda^2 d\lambda \tag{8.6.33}$$

The angular integral is evaluated by letting $\theta = \gamma - \phi$ and expanding $\cos(\theta + \phi)$. The odd part cancels out and using

$$J_1(x) = \frac{1}{2\pi j}\int_0^{2\pi}\cos\theta e^{jx\cos\theta}d\theta$$

or

$$\cos\phi\int_0^{2\pi}\cos\theta\, e^{jx\cos\theta}d\theta = \cos\phi[j2\pi J_1(x)]$$

we obtain:

$$A_{z1}(x, y, z) = \frac{\mu_o(I\Delta L)\cos\phi}{2\pi}\int_0^\infty\frac{\mu_{r2}\varepsilon_{cr,2} - 1}{(\varepsilon_{cr,2}\tau_1 + \tau_2)(\mu_{r2}\tau_1 + \tau_2)}e^{-j\tau_1(z+h)}J_1(\lambda\rho)\lambda^2 d\lambda \tag{8.6.34}$$

Similarly, from (8.6.20):

$$A_{z2}(x, y, z) = \frac{\mu_2(I\Delta L)\cos\phi}{2\pi}\int_0^\infty\frac{\mu_{r2}\varepsilon_{cr,2} - 1}{(\varepsilon_{cr,2}\tau_1 + \tau_2)(\mu_{r2}\tau_1 + \tau_2)}e^{-j\tau_1 h}e^{j\tau_2 z}J_1(\lambda\rho)\lambda^2 d\lambda \tag{8.6.35}$$

The asymptotic evaluation of the integrals in (8.6.31) to (8.6.35) is done using the method of steepest descent. Details are discussed in Chapter 9.

8.7 VERTICAL ELECTRIC AND MAGNETIC DIPOLES IN A LOSSY REGION

8.7.1 Electric Dipole

First, we consider the case where a vertical electric dipole is located in a lossy region such that the displacement currents can be neglected compared to the conduction current. An example of a lossy region is the sea where the conductivity is around 4 S/m and only the lower frequencies ($f < 100$ Hz) propagate to a significant distance. Another example is that of a dipole located in wet ground.

The geometry of the problem is shown in Fig. 8.14, where Region 1 is the lossy region, and Region 2 is air.

In Region 1, Maxwell equations are

$$\nabla \times \mathbf{E} = -j\omega\mu_o \mathbf{H}$$

and

$$\nabla \times \mathbf{H} \approx \sigma_1 \mathbf{E}$$

where the displacement current is neglected, since $\sigma_1 \gg \omega\varepsilon_1$. These equations can be solved in terms of the Hertz vector $\mathbf{\Pi}^e$ which in Region 1, from (1.7.2), satisfies

$$[\nabla^2 + k_1^2]\Pi_{z1}^e(x, y, z) = -\frac{J_{e,z}}{\sigma_1} = -\frac{(I\Delta L)}{\sigma_1}\delta(x)\delta(y)\delta(z-h) \qquad (8.7.1)$$

where

$$k_1^2 = -j\omega\mu_1(\sigma_1 + j\omega\varepsilon_1) \approx -j\omega\mu_1\sigma_1 \Rightarrow k_1 = \sqrt{-j\omega\mu_1\sigma_1}$$

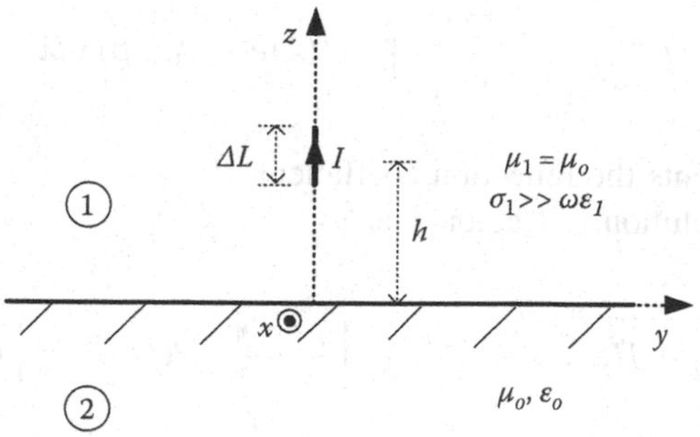

FIGURE 8.14 Geometry of the problem for a vertical dipole in a lossy region.

In the air region (Region 2):

$$[\nabla^2 + k_2^2]\Pi^e_{z2}(x, y, z) = 0 \tag{8.7.2}$$

where

$$k_2 = k_o = \omega\sqrt{\mu_o \varepsilon_o}$$

The field components in Regions 1 and 2 follow from (4.2.7) to (4.2.12). In Region 1: $\gamma_1^2/j\omega\mu_o \approx \sigma_1$, and in Region 2: $\gamma_2^2/j\omega\mu_o = j\omega\varepsilon_o$. Therefore,

$$H_{x1} = \sigma_1 \frac{\partial \Pi^e_{z1}}{\partial y} \qquad H_{x2} = j\omega\varepsilon_o \frac{\partial \Pi^e_{z2}}{\partial y}$$

$$H_{y1} = -\sigma_1 \frac{\partial \Pi^e_{z1}}{\partial x} \qquad H_{y2} = -j\omega\varepsilon_o \frac{\partial \Pi^e_{z2}}{\partial x}$$

$$H_{z1} = 0 \qquad H_{z2} = 0$$

$$E_{x1} = \frac{\partial^2 \Pi^e_{z1}}{\partial x \partial z} \qquad E_{x2} = \frac{\partial^2 \Pi^e_{z2}}{\partial x \partial z}$$

$$E_{y1} = \frac{\partial^2 \Pi^e_{z1}}{\partial y \partial z} \qquad E_{y2} = \frac{\partial^2 \Pi^e_{z2}}{\partial y \partial z}$$

$$E_{z1} = \frac{\partial^2 \Pi^e_{z1}}{\partial z^2} + k_1^2 \Pi^e_{z1} \qquad E_{z2} = \frac{\partial^2 \Pi^e_{z2}}{\partial z^2} + k_2^2 \Pi^e_{z2}$$

The particular solution of (8.7.1), using (6.2.30), is

$$\Pi^e_{z1,p} = \frac{(I\Delta L)}{\sigma_1}\frac{e^{-jk_1R_1}}{4\pi R_1} = -j\frac{(I\Delta L)}{4\pi\sigma_1}\int_0^\infty \frac{e^{-j\tau_1|z-h|}}{\tau_1} J_0(\lambda\rho)\lambda d\lambda$$

where

$$\tau_1 = \sqrt{k_1^2 - \lambda^2}$$

and $\mathrm{Im}(\tau_1) < 0$. The complementary solution is

$$\Pi^e_{z1,c} = -j\frac{(I\Delta L)}{4\pi\sigma_1}\int_0^\infty R(\lambda) e^{-j\tau_1 z} J_0(\lambda\rho)\lambda d\lambda$$

where $R(\lambda)$ represents the reflection coefficient.

The complete solution in Region 1 is

$$\Pi^e_{z1} = \Pi^e_{z1,p} + \Pi^e_{z1,c} = -j\frac{(I\Delta L)}{4\pi\sigma_1}\int_0^\infty \left[\frac{e^{-j\tau_1|z-h|}}{\tau_1} + R(\lambda)e^{-j\tau_1 z}\right] J_0(\lambda\rho)\lambda d\lambda \tag{8.7.3}$$

The transmitted field is expressed in the form

$$\Pi_{z2}^{e} = -j\frac{(I\Delta L)}{4\pi\sigma_1}\int_0^{\infty} T(\lambda)e^{j\tau_2 z}J_0(\lambda\rho)\lambda d\lambda \tag{8.7.4}$$

where

$$\tau_2 = \sqrt{k_2^2 - \lambda^2}$$

and $\mathrm{Im}(\tau_2) < 0$. $T(\lambda)$ is the transmission coefficient. The factors in front of the complementary and transmitted solutions were selected to be the same as the one in the particular solution for convenience.

The boundary conditions are the continuity of the tangential fields at the interface, which must be satisfied for all values of x and y. That is,

$$E_{x1} = E_{x2}|_{z=0} \Rightarrow \frac{\partial \Pi_{z1}^{e}}{\partial z} = \frac{\partial \Pi_{z2}^{e}}{\partial z}\bigg|_{z=0}$$

$$E_{y1} = E_{y2}|_{z=0} \Rightarrow \frac{\partial \Pi_{z1}^{e}}{\partial z} = \frac{\partial \Pi_{z2}^{e}}{\partial z}\bigg|_{z=0}$$

$$H_{x1} = H_{x2}|_{z=0} \Rightarrow \sigma_1 \Pi_{z1}^{e} = j\omega\varepsilon_o \Pi_{z2}^{e}\Big|_{z=0}$$

$$H_{y1} = H_{y2}|_{z=0} \Rightarrow \sigma_1 \Pi_{z1}^{e} = j\omega\varepsilon_o \Pi_{z2}^{e}\Big|_{z=0}$$

The first two boundary conditions are identical, as well as the last two. Also, observe that $n_1^2 \approx \sigma_1/j\omega\varepsilon_o$. Therefore, we obtain

$$\frac{\partial \Pi_{z1}^{e}}{\partial z} = \frac{\partial \Pi_{z2}^{e}}{\partial z}\bigg|_{z=0} \Rightarrow \frac{e^{-j\tau_1 h}}{\tau_1} - R(\lambda) = \frac{\tau_2}{\tau_1}T(\lambda)$$

and

$$\sigma_1 \Pi_{z1}^{e} = j\omega\varepsilon_o \Pi_{z2}^{e}|_{z=0} \Rightarrow n_1^2\left[\frac{e^{-j\tau_1 h}}{\tau_1} + R(\lambda)\right] = T(\lambda)$$

which are solved to give

$$R(\lambda) = \frac{\tau_1 - n_1^2\tau_2}{\tau_1 + n_1^2\tau_2}\frac{e^{-j\tau_1 h}}{\tau_1}$$

and

$$T(\lambda) = \frac{2n_1^2}{\tau_1 + n_1^2\tau_2} e^{-j\tau_1 h}$$

Expressing $R(\lambda)$ in the form

$$R(\lambda) = \frac{e^{-j\tau_1 h}}{\tau_1}\left[-1 + \frac{2\tau_1}{\tau_1 + n_1^2\tau_2}\right]$$

and substituting into (8.7.3) gives

$$\Pi_{z1}^{e} = \frac{(I\Delta L)}{\sigma_1}\frac{e^{-jk_1R_1}}{4\pi R_1} - \frac{(I\Delta L)}{\sigma_1}\frac{e^{-jk_1R_2}}{4\pi R_2} + Q$$

where

$$Q = -j\frac{(I\Delta L)}{2\pi\sigma_1}\int_0^\infty \left[\frac{1}{\tau_1 + n_1^2\tau_2} e^{-j\tau_1(z+h)}\right] J_0(\lambda\rho)\lambda d\lambda$$

The evaluation of this type of integral is discussed in Chapter 9

8.7.2 Magnetic Dipole

A summary of the relations for a magnetic dipole source are given. The Hertz vector $\mathbf{\Pi}^{m}$ in Region 1, from (1.7.9), satisfies

$$\begin{aligned}[\nabla^2 + k_1^2]\Pi_{z1}^{m}(x, y, z) &= -\frac{J_{m,z}}{j\omega\mu_o} = -\frac{(I_m\Delta L)}{j\omega\mu_o}\delta(x)\delta(y)\delta(z-h)\\ &= -(I\Delta S)\delta(x)\delta(y)\delta(z-h)\end{aligned}$$

and in the air region (Region 2):

$$[\nabla^2 + k_2^2]\Pi_{z2}^{m}(x, y, z) = 0$$

The complete solution in Region 1 is

$$\Pi_{z1}^{m} = -j\frac{(I\Delta S)}{4\pi}\int_0^\infty \left[\frac{e^{-j\tau_1|z-h|}}{\tau_1} + R^m(\lambda)e^{-j\tau_1 z}\right] J_0(\lambda\rho)\lambda d\lambda \qquad (8.7.5)$$

and for the transmitted field:

$$\Pi_{z2}^{m} = -j\frac{(I\Delta S)}{4\pi}\int_0^{\infty} T^m(\lambda)e^{j\tau_2 z}J_0(\lambda\rho)\lambda d\lambda$$

where $R^m(\lambda)$ and $T^m(\lambda)$ are the reflection and transmission coefficients, respectively.

The boundary conditions are:

$$E_{x1} = E_{x2}|_{z=0} \Rightarrow \Pi_{z1}^{m} = \Pi_{z2}^{m}\Big|_{z=0} \Rightarrow \frac{e^{-j\tau_1 h}}{\tau_1} + R^m(\lambda) = T^m(\lambda)$$

and

$$H_{x1} = H_{x2}|_{z=0} \Rightarrow \frac{\partial \Pi_{z1}^{m}}{\partial z} = \frac{\partial \Pi_{z2}^{m}}{\partial z}\Big|_{z=0} \Rightarrow \frac{e^{-j\tau_1 h}}{\tau_1} - R^m(\lambda) = \frac{\tau_2}{\tau_1}T^m(\lambda)$$

which gives

$$R^m(\lambda) = \frac{\tau_1 - \tau_2}{\tau_1 + \tau_2}\frac{e^{-j\tau_1 h}}{\tau_1} = \left(-1 + \frac{2\tau_2}{\tau_1 + \tau_2}\right)\frac{e^{-j\tau_1 h}}{\tau_1} \tag{8.7.6}$$

and

$$T^m(\lambda) = \frac{2}{\tau_1 + \tau_2}e^{-j\tau_1 h}$$

Substituting (8.7.6) into (8.7.5) shows that Π_{z1}^{m} can be expressed in the form

$$\Pi_{z1}^{m} = \frac{(I\Delta S)}{4\pi}\frac{e^{-jk_1R_1}}{R_1} - \frac{(I\Delta S)}{4\pi}\frac{e^{-jk_1R_2}}{R_2} + Q^m$$

where

$$Q^m = -j\frac{(I\Delta S)}{2\pi}\int_0^{\infty}\frac{1}{\tau_1 + \tau_2}e^{-j\tau_1(z+h)}J_0(\lambda\rho)\lambda d\lambda \tag{8.7.7}$$

If the source and observation point are close to the surface, then $R_1 \approx R_2$, $h = 0$, $z = 0$ and (8.7.7) can be evaluated using the method in Section 8.5.2.

8.8 RADIATION FROM AN APERTURE IN A PLANE

The geometry of an aperture in an infinite conducting plane is shown in Fig. 8.15. An aperture is commonly field-fed by a waveguide, either a rectangular or

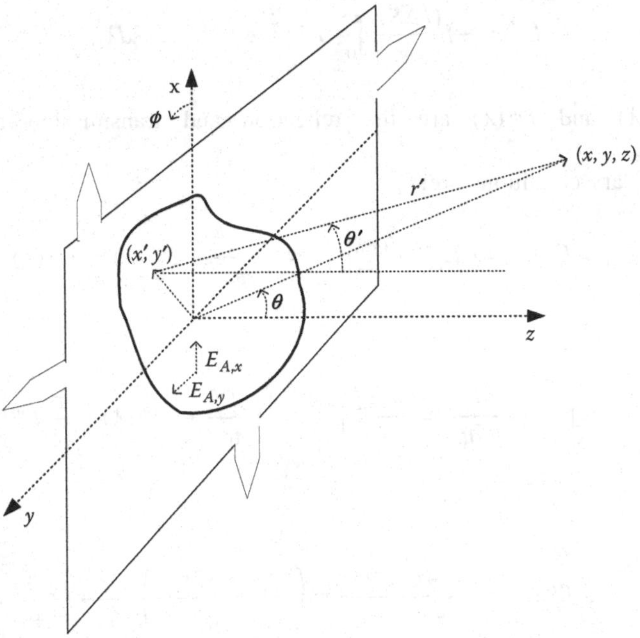

FIGURE 8.15 An aperture in a conducting plane.

cylindrical waveguide. Hence, we assume that the fields at the aperture are known and denoted by $E_{A,x}$ and $E_{A,y}$.

In the region $z > 0$ there are no sources and the fields can be calculated in terms of the **A, F, Π**^{e} or **Π**^{m} vectors. We will use the Hertz's vectors which satisfy

$$[\nabla^2 + k^2]\Pi_z^e = 0 \tag{8.8.1}$$

an

$$[\nabla^2 + k^2]\Pi_z^m = 0 \tag{8.8.2}$$

where $k = \omega\sqrt{\mu\varepsilon}$. The field components in terms of the Hertz's vectors are given by (4.2.7) to (4.2.12) and (4.2.25) to (4.2.30).

At $z = 0$ the electric fields must satisfy the boundary conditions at the aperture, namely

$$E_x(x, y, 0) = E_{A,x}(x', y') \tag{8.8.3}$$

and

$$E_y(x, y, 0) = E_{A,y}(x', y') \tag{8.8.4}$$

where the prime notation is used for aperture coordinates.

Since the x and y variables for $z > 0$ extend from minus infinity to plus infinity the Hertz's vectors in (8.8.1) and (8.8.2) can be Fourier transformed using

$$\Pi_z^{e}(\alpha_1, \alpha_2, z) = \int_{-\infty}^{\infty}\int_{-\infty}^{\infty} \Pi_z^{e}(x, y, z)\, e^{-j(\alpha_1 x + \alpha_2 y)}\, dx dy \tag{8.8.5}$$

and

$$\Pi_z^{m}(\alpha_1, \alpha_2, z) = \int_{-\infty}^{\infty}\int_{-\infty}^{\infty} \Pi_z^{m}(x, y, z)\, e^{-j(\alpha_1 x + \alpha_2 y)}\, dx dy \tag{8.8.6}$$

The inverse transforms are

$$\Pi_z^{e}(x, y, z) = \frac{1}{4\pi^2}\int_{-\infty}^{\infty}\int_{-\infty}^{\infty} \Pi_z^{e}(\alpha_1, \alpha_2, z)\, e^{j(\alpha_1 x + \alpha_2 y)}\, d\alpha_1 d\alpha_2 \tag{8.8.7}$$

and

$$\Pi_z^{m}(x, y, z) = \frac{1}{4\pi^2}\int_{-\infty}^{\infty}\int_{-\infty}^{\infty} \Pi_z^{m}(\alpha_1, \alpha_2, z)\, e^{j(\alpha_1 x + \alpha_2 y)}\, d\alpha_1 d\alpha_2 \tag{8.8.8}$$

Applying the Fourier transforms in (8.8.5) and (8.8.6) to (8.8.1) and (8.8.2) gives

$$\left[\frac{d^2}{dz^2} + \tau^2\right] \Pi_z^{e}(\alpha_1, \alpha_2, z) = 0 \tag{8.8.9}$$

and

$$\left[\frac{d^2}{dz^2} + \tau^2\right] \Pi_z^{m}(\alpha_1, \alpha_2, z) = 0 \tag{8.8.10}$$

where

$$\tau = \sqrt{k^2 - \alpha_1^2 - \alpha_2^2}$$

The solutions to (8.8.9) and (8.8.10) are

$$\Pi_z^{e}(\alpha_1, \alpha_2, z) = B(\alpha_1, \alpha_2) e^{-j\tau z}$$

and

$$\Pi_z^{\mathrm{m}}(\alpha_1, \alpha_2, z) = C(\alpha_1, \alpha_2) e^{-j\tau z}$$

where $\mathrm{Im}(\tau) < 0$. Substituting the above solutions into the inverse transforms in (8.8.7) and (8.8.8) gives

$$\Pi_z^{\mathrm{e}}(x, y, z) = \frac{1}{4\pi^2} \int_{-\infty}^{\infty} \int_{-\infty}^{\infty} B(\alpha_1, \alpha_2)\, e^{j(\alpha_1 x + \alpha_2 y - \tau z)} d\alpha_1 d\alpha_2 \tag{8.8.11}$$

and

$$\Pi_z^{\mathrm{m}}(x, y, z) = \frac{1}{4\pi^2} \int_{-\infty}^{\infty} \int_{-\infty}^{\infty} C(\alpha_1, \alpha_2)\, e^{j(\alpha_1 x + \alpha_2 y - \tau z)} d\alpha_1 d\alpha_2 \tag{8.8.12}$$

The tangential electric fields at $z = 0$ must be equal to the aperture values of $E_{A,x}(x', y')$ and $E_{A,y}(x', y')$. Since,

$$E_x = \frac{\partial^2 \Pi_z^{\mathrm{e}}}{\partial x \partial z} - j\omega\mu \frac{\partial \Pi_z^{\mathrm{m}}}{\partial y} = \frac{1}{4\pi^2} \int_{-\infty}^{\infty} \int_{-\infty}^{\infty} (\alpha_1 \tau B + \omega\mu\alpha_2 C) e^{j(\alpha_1 x + \alpha_2 y - \tau z)} d\alpha_1 d\alpha_2 \tag{8.8.13}$$

and

$$E_y = \frac{\partial^2 \Pi_z^{\mathrm{e}}}{\partial y \partial z} + j\omega\mu \frac{\partial \Pi_z^{\mathrm{m}}}{\partial x} = \frac{1}{4\pi^2} \int_{-\infty}^{\infty} \int_{-\infty}^{\infty} (\alpha_2 \tau B - \omega\mu\alpha_1 C) e^{j(\alpha_1 x + \alpha_2 y - \tau z)} d\alpha_1 d\alpha_2 \tag{8.8.14}$$

it follows from (8.8.13) and (8.8.14) that at $z = 0$:

$$E_{A,x} = \frac{1}{4\pi^2} \int_{-\infty}^{\infty} \int_{-\infty}^{\infty} (\alpha_1 \tau B + \omega\mu\alpha_2 C) e^{j(\alpha_1 x + \alpha_2 y)} d\alpha_1 d\alpha_2 \tag{8.8.15}$$

and

$$E_{A,y} = \frac{1}{4\pi^2} \int_{-\infty}^{\infty} \int_{-\infty}^{\infty} (\alpha_2 \tau B - \omega\mu\alpha_1 C) e^{j(\alpha_1 x + \alpha_2 y)} d\alpha_1 d\alpha_2 \tag{8.8.16}$$

which show that the integrands are the Fourier transforms of $E_{A,x}$ and $E_{A,y}$. That is,

$$\begin{aligned} \alpha_1 \tau B + \omega\mu\alpha_2 C &= \iint_{Ap} E_{A,x}(x', y') e^{-j(\alpha_1 x' + \alpha_2 y')} dx' dy' \\ &= \widehat{E_{A,x}}(\alpha_1, \alpha_2) \end{aligned} \tag{8.8.17}$$

and

$$\alpha_2 \tau B - \omega\mu\alpha_1 C = \iint_{Ap} E_{A,y}(x', y') e^{-j(\alpha_1 x' + \alpha_2 y')} dx' dy' \\ = \widehat{E_{A,y}}(\alpha_1, \alpha_2) \tag{8.8.18}$$

where the transforms are denoted by $\widehat{E_{A,x}}(\alpha_1, \alpha_2)$ and $\widehat{E_{A,y}}(\alpha_1, \alpha_2)$. The integration in (8.8.17) and (8.8.18) is over the aperture (denoted by Ap) since $E_{A,x}$ and $E_{A,y}$ are defined only over the aperture. Solving (8.8.17) and (8.8.18) gives

$$B = \frac{\alpha_1 \widehat{E_{A,x}}}{\tau(\alpha_1^2 + \alpha_2^2)} + \frac{\alpha_2 \widehat{E_{A,y}}}{\tau(\alpha_1^2 + \alpha_2^2)} \tag{8.8.19}$$

and

$$C = -\frac{\alpha_1 \widehat{E_{A,y}}}{\omega\mu(\alpha_1^2 + \alpha_2^2)} + \frac{\alpha_2 \widehat{E_{A,x}}}{\omega\mu(\alpha_1^2 + \alpha_2^2)} \tag{8.8.20}$$

Substituting (8.8.19) and (8.8.20) into (8.8.11) and (8.8.12) gives

$$\begin{aligned} \Pi_z^e(x, y, z) &= \frac{1}{4\pi^2} \int_{-\infty}^{\infty} \int_{-\infty}^{\infty} \frac{\alpha_1 \widehat{E_{A,x}}}{\tau(\alpha_1^2 + \alpha_2^2)} e^{j(\alpha_1 x + \alpha_2 y - \tau z)} d\alpha_1 d\alpha_2 \\ &+ \frac{1}{4\pi^2} \int_{-\infty}^{\infty} \int_{-\infty}^{\infty} \frac{\alpha_2 \widehat{E_{A,y}}}{\tau(\alpha_1^2 + \alpha_2^2)} e^{j(\alpha_1 x + \alpha_2 y - \tau z)} d\alpha_1 d\alpha_2 \\ &= -j\frac{1}{4\pi^2} \frac{\partial}{\partial x} \int_{-\infty}^{\infty} \int_{-\infty}^{\infty} \frac{\widehat{E_{A,x}}}{\tau(\alpha_1^2 + \alpha_2^2)} e^{j(\alpha_1 x + \alpha_2 y - \tau z)} d\alpha_1 d\alpha_2 \\ &- j\frac{1}{4\pi^2} \frac{\partial}{\partial y} \int_{-\infty}^{\infty} \int_{-\infty}^{\infty} \frac{\widehat{E_{A,y}}}{\tau(\alpha_1^2 + \alpha_2^2)} e^{j(\alpha_1 x + \alpha_2 y - \tau z)} d\alpha_1 d\alpha_2 \end{aligned} \tag{8.8.21}$$

and

$$\begin{aligned} \Pi_z^m(x, y, z) &= -\frac{1}{4\pi^2\omega\mu} \int_{-\infty}^{\infty} \int_{-\infty}^{\infty} \frac{\alpha_1 \widehat{E_{A,y}}}{(\alpha_1^2 + \alpha_2^2)} e^{j(\alpha_1 x + \alpha_2 y - \tau z)} d\alpha_1 d\alpha_2 \\ &+ \frac{1}{4\pi^2\omega\mu} \int_{-\infty}^{\infty} \int_{-\infty}^{\infty} \frac{\alpha_2 \widehat{E_{A,x}}}{(\alpha_1^2 + \alpha_2^2)} e^{j(\alpha_1 x + \alpha_2 y - \tau z)} d\alpha_1 d\alpha_2 \\ &= -\frac{1}{4\pi^2\omega\mu} \frac{\partial^2}{\partial x \partial z} \int_{-\infty}^{\infty} \int_{-\infty}^{\infty} \frac{\widehat{E_{A,y}}}{\tau(\alpha_1^2 + \alpha_2^2)} e^{j(\alpha_1 x + \alpha_2 y - \tau z)} d\alpha_1 d\alpha_2 \\ &+ \frac{1}{4\pi^2\omega\mu} \frac{\partial^2}{\partial y \partial z} \int_{-\infty}^{\infty} \int_{-\infty}^{\infty} \frac{\widehat{E_{A,x}}}{\tau(\alpha_1^2 + \alpha_2^2)} e^{j(\alpha_1 x + \alpha_2 y - \tau z)} d\alpha_1 d\alpha \end{aligned} \tag{8.8.22}$$

The forms in (8.8.21) and (8.8.22) show that the Hertz's vectors follow from the evaluation of two integrals, denoted by I_x and I_y, where

$$I_x = \int_{-\infty}^{\infty}\int_{-\infty}^{\infty} \frac{\widehat{E_{A,x}}}{\tau(\alpha_1^2+\alpha_2^2)} e^{j(\alpha_1 x+\alpha_2 y-\tau z)} d\alpha_1 d\alpha_2 \tag{8.8.23}$$

and

$$I_y = \int_{-\infty}^{\infty}\int_{-\infty}^{\infty} \frac{\widehat{E_{A,y}}}{\tau(\alpha_1^2+\alpha_2^2)} e^{j(\alpha_1 x+\alpha_2 y-\tau z)} d\alpha_1 d\alpha_2 \tag{8.8.24}$$

Substituting the transform form for $\widehat{E_{A,x}}$ and $\widehat{E_{A,y}}$ into (8.8.23) and (8.8.24) gives

$$I_x = \iint_{Ap} E_{A,x}(x', y') f(x, x', y, y', z) dx'dy' \tag{8.8.25}$$

and

$$I_y = \iint_{Ap} E_{A,y}(x', y') f(x, x', y, y', z) dx'dy' \tag{8.8.26}$$

where

$$f(x, x', y, y', z) = \int_{-\infty}^{\infty}\int_{-\infty}^{\infty} \frac{e^{j[\alpha_1(x-x')+\alpha_2(y-y')-\tau z]}}{\tau(\alpha_1^2+\alpha_2^2)} d\alpha_1 d\alpha_2 \tag{8.8.27}$$

When the point of observation is far from the aperture (which is usually the case), (8.8.27) can be evaluated using the method of steepest descent. Referring to Fig. 6.1, let

$$\begin{aligned}
\alpha_1 &= \lambda \cos\gamma \\
\alpha_2 &= \lambda \sin\gamma \\
x - x' &= |\rho - \rho'| \cos\varphi \\
y - y' &= |\rho - \rho'| \sin\varphi \\
\lambda^2 &= \alpha_1^2 + \alpha_2^2 \\
d\alpha_1 d\alpha_2 &= \lambda d\lambda d\gamma
\end{aligned}$$

Then, (8.8.27) reads

$$
\begin{aligned}
f(\rho, \phi, z) &= \int_0^{\infty} \frac{e^{-j\sqrt{k^2-\lambda^2}z}}{\lambda\sqrt{k^2-\lambda^2}} d\lambda \int_0^{2\pi} e^{j\lambda|\rho-\rho'|\cos(\varphi-\gamma)} d\gamma \\
&= 2\pi \int_0^{\infty} \frac{J_0(\lambda|\rho-\rho'|)e^{-j\sqrt{k^2-\lambda^2}z}}{\lambda\sqrt{k^2-\lambda^2}} d\lambda \\
&= \pi \int_{-\infty}^{\infty} \frac{H_0^{(2)}(\lambda|\rho-\rho'|)e^{-j\sqrt{k^2-\lambda^2}z}}{\lambda\sqrt{k^2-\lambda^2}} d\lambda \\
&\approx \pi e^{j\pi 4} \int_{-\infty}^{\infty} \sqrt{\frac{2}{\pi\lambda|\rho-\rho'|}} \frac{e^{-j\lambda|\rho-\rho'|}e^{-j\sqrt{k^2-\lambda^2}z}}{\lambda\sqrt{k^2-\lambda^2}} d\lambda
\end{aligned}
\tag{8.8.28}
$$

Next, let

$$
\begin{aligned}
\lambda &= k\sin\beta \\
|\rho-\rho'| &= r'\sin\theta' \\
z &= r'\cos\theta'
\end{aligned}
$$

and (8.8.28) becomes

$$
f(r', \theta') = \int_{\Gamma} G(\beta)e^{-jkr'\cos(\beta-\theta')} d\beta \tag{8.8.29}
$$

where

$$
G(\beta) = \frac{e^{j\pi/4}}{k\sin\beta}\sqrt{\frac{2\pi}{kr'\sin\beta\sin\theta'}}
$$

The saddle point in (8.8.29) occurs at $\beta = \theta'$, and we obtain

$$
f(r', \theta') \approx G(\theta')\sqrt{\frac{2\pi}{kr'}}e^{-j(kr'-\pi/4)} \approx j2\pi\frac{e^{-jkr'}}{k^2r'\sin^2\theta'} \tag{8.8.30}
$$

Thus, the evaluation of (8.8.27) is

$$
\int_{-\infty}^{\infty}\int_{-\infty}^{\infty} \frac{e^{j[\alpha_1(x-x')+\alpha_2(y-y')-\tau z]}}{\tau(\alpha_1^2+\alpha_2^2)} d\alpha_1 d\alpha_2 \approx j2\pi\frac{e^{-jkr'}}{k^2r'\sin^2\theta'} \tag{8.8.31}
$$

where

$$\begin{aligned} r' &= \sqrt{(x-x')^2+(y-y')^2+z^2} \\ &= \sqrt{(r\sin\theta\cos\phi - x')^2 + (r\sin\theta\sin\phi - y')^2 + r^2\cos^2\theta} \\ &\approx r\sqrt{1-\frac{2x'\sin\theta\cos\phi}{r}-\frac{2y'\sin\theta\cos\phi}{r}} \approx r - x'\sin\theta\cos\phi - y'\sin\theta\cos\phi \end{aligned} \tag{8.8.32}$$

The approximation $(x'^2 + y'^2) << r$ was used since the aperture dimensions are small and the point of observation is far from the aperture.

Substituting (8.8.31) into (8.8.25) and (8.8.26) gives

$$I_x \approx \frac{2\pi j}{k^2} \iint_{Ap} E_{A,x}(x', y') \frac{e^{-jkr'}}{r'\sin^2\theta'} dx'dy' \tag{8.8.33}$$

and

$$I_y \approx \frac{2\pi j}{k^2} \iint_{Ap} E_{A,y}(x', y') \frac{e^{-jkr'}}{r'\sin^2\theta'} dx'dy' \tag{8.8.34}$$

Using (8.8.32) in the exponent of (8.8.33) and (8.8.34), and in the denominator $r' \approx r$ and $\theta' \approx \theta$, we obtain

$$I_x \approx \frac{2\pi j}{k^2}\frac{e^{-jkr}}{r\sin^2\theta} \iint_{Ap} E_{A,x}(x', y')\, e^{jk\sin\theta(x'\cos\phi + y'\cos\phi)} dx'dy' \tag{8.8.35}$$

and

$$I_y \approx \frac{2\pi j}{k^2}\frac{e^{-jkr}}{r\sin^2\theta} \iint_{Ap} E_{A,y}(x', y')\, e^{jk\sin\theta(x'\cos\phi + y'\cos\phi)} dx'dy' \tag{8.8.36}$$

The substitution of (8.8.35) and (8.8.36) into (8.8.21) and (8.8.22) requires the following operations in the phase terms. That is,

$$\begin{cases} \dfrac{\partial}{\partial x}e^{-jkr} = \dfrac{\partial r}{\partial x}\dfrac{\partial}{\partial r}e^{-jkr} = -jke^{-jkr}\sin\theta\cos\phi \\ \dfrac{\partial}{\partial y}e^{-jkr} = \dfrac{\partial r}{\partial y}\dfrac{\partial}{\partial r}e^{-jkr} = -jke^{-jkr}\sin\theta\sin\phi \\ \dfrac{\partial}{\partial z}e^{-jkr} = \dfrac{\partial r}{\partial z}\dfrac{\partial}{\partial r}e^{-jkr} = -jke^{-jkr}\cos\theta \end{cases} \tag{8.8.37}$$

Hence, in the far field (8.8.21) and (8.8.22) read

$$\Pi_z^{\mathrm{e}}(x, y, z) \approx -j\frac{e^{-jkr}}{2\pi kr \sin\theta} \iint_{Ap} [E_{A,x}(x', y')\cos\phi + E_{A,y}(x', y')\sin\phi]\, e^{jk\sin\theta(x'\cos\phi + y'\sin\phi)}\, dx'dy' \qquad (8.8.38)$$

and

$$\Pi_z^{\mathrm{m}}(x, y, z) \approx -j\frac{\cos\theta\, e^{-jkr}}{2\pi\omega\mu r \sin\theta} \iint_{Ap} [E_{A,x}(x', y')\sin\phi - E_{A,y}(x', y')\cos\phi]\, e^{jk\sin\theta(x'\cos\phi + y'\sin\phi)}\, dx'dy' \qquad (8.8.39)$$

The field expressions follow from (4.2.7) to (4.2.12) and (4.2.25) to (4.2.30). Using (8.8.37), the far fields are given by

$$\begin{cases} E_x \approx -k^2 \cos\theta \sin\theta \cos\phi\, \Pi_z^{\mathrm{e}} - k\omega\mu \sin\theta \sin\phi\, \Pi_z^{\mathrm{m}} \\ E_y \approx -k^2 \cos\theta \sin\theta \sin\phi\, \Pi_z^{\mathrm{e}} + k\omega\mu \sin\theta \cos\phi\, \Pi_z^{\mathrm{m}} \\ E_z \approx k^2 \sin^2\theta\, \Pi_z^{\mathrm{e}} \\ H_x \approx k\omega\varepsilon \sin\theta \sin\phi\, \Pi_z^{\mathrm{e}} - k^2 \cos\theta \sin\theta \cos\phi\, \Pi_z^{\mathrm{m}} \\ H_y \approx -k\omega\varepsilon \sin\theta \cos\phi\, \Pi_z^{\mathrm{e}} - k^2 \cos\theta \sin\theta \sin\phi\, \Pi_z^{\mathrm{m}} \\ H_z \approx k^2 \sin^2\theta\, \Pi_z^{\mathrm{m}} \end{cases} \qquad (8.8.40)$$

The substitution of (8.8.38) and (8.8.39) into (8.8.40) completes the formal solution, except that the fields are expressed in rectangular coordinates. It is convenient to express the fields in spherical coordinates. To this end, the rectangular components of a vector (say T_x, T_y and T_z) are transformed into spherical components (say T_r, T_θ and T_ϕ) using

$$\begin{bmatrix} T_r \\ T_\theta \\ T_\phi \end{bmatrix} = \begin{bmatrix} \sin\theta\cos\phi & \sin\theta\sin\phi & \cos\theta \\ \cos\theta\cos\phi & \cos\theta\sin\phi & -\sin\theta \\ -\sin\phi & \cos\phi & 0 \end{bmatrix} \begin{bmatrix} T_x \\ T_y \\ T_z \end{bmatrix} \qquad (8.8.41)$$

From (8.8.40) into (8.8.41) one obtains

$$\begin{cases} E_\theta \approx -k^2 \sin\theta\, \Pi_z^{\mathrm{e}} \\ E_\phi \approx k\omega\mu \sin\theta\, \Pi_z^{\mathrm{m}} \\ E_r \approx 0 \\ H_\theta \approx -k^2 \sin\theta\, \Pi_z^{\mathrm{m}} \\ H_\phi \approx -k\omega\varepsilon \sin\theta\, \Pi_z^{\mathrm{e}} \\ H_r \approx 0 \end{cases} \qquad (8.8.42)$$

and substituting (8.8.38) and (8.8.39) into (8.8.42), the far fields are given by

$$\begin{cases} E_\theta \approx j\frac{k}{2\pi}\frac{e^{-jkr}}{r}\iint_{Ap}[E_{A,x}(x',y')\cos\phi + E_{A,y}(x',y')\sin\phi]e^{jk\sin\theta(x'\cos\phi+y'\sin\phi)}dx'dy' \\ E_\phi \approx -\frac{jk\cos\theta}{2\pi}\frac{e^{-jkr}}{r}\iint_{Ap}[E_{A,x}(x',y')\sin\phi \\ \qquad - E_{A,y}(x',y')\cos\phi]\,e^{jk\sin\theta(x'\cos\phi+y'\sin\phi)}dx'dy' \\ E_r \approx 0 \\ H_\theta = -\frac{E_\phi}{\eta} \\ H_\phi = \frac{E_\theta}{\eta} \\ H_r \approx 0 \end{cases} \tag{8.8.43}$$

where $\eta = \sqrt{\mu/\varepsilon}$.

8.8.1 Rectangular Aperture

The results in (8.8.43) are now applied to calculate the fields for the rectangular aperture shown in Fig. 8.16. Assume that the field over the aperture is constant and given by

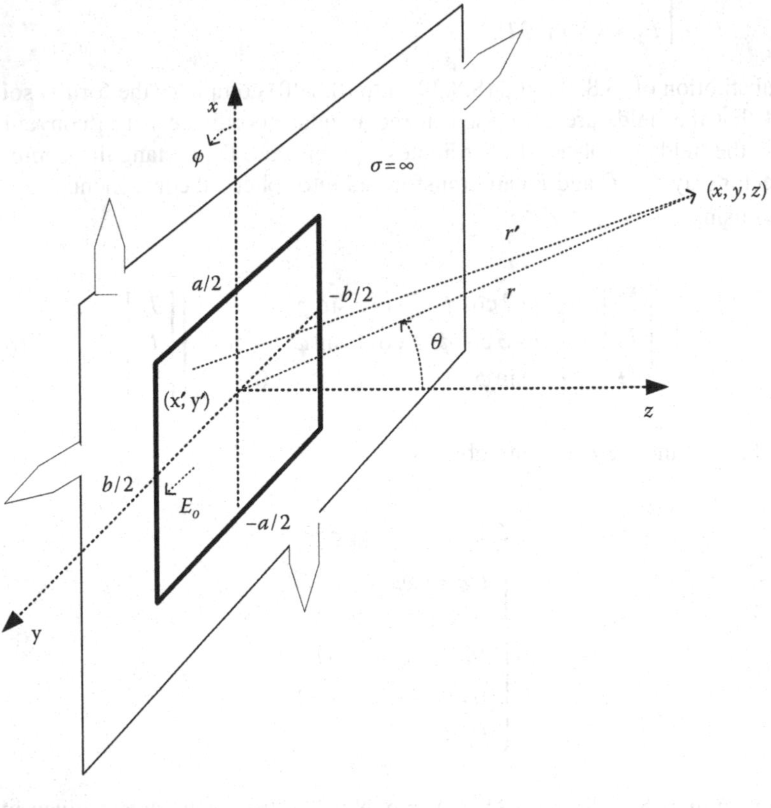

FIGURE 8.16 A rectangular aperture.

$$E_{A,y}(x', y') = E_o$$

Then, (8.8.43) shows that

$$\begin{aligned} E_\theta &\approx jE_o \frac{k}{2\pi}\frac{e^{-jkr}}{r} \sin\phi \left[\int_{-a/2}^{a/2} e^{jk(x' \sin\theta\cos\phi)} dx'\right]\left[\int_{-b/2}^{b/2} e^{jk(y' \sin\theta\sin\phi)} dy'\right] \\ &\approx jE_o \frac{kab}{2\pi}\frac{e^{-jkr}}{r} \sin\phi \left[\frac{\sin\left(\frac{ka}{2}\sin\theta\cos\phi\right)}{\frac{ka}{2}\sin\theta\cos\phi}\right]\left[\frac{\sin\left(\frac{kb}{2}\sin\theta\sin\phi\right)}{\frac{kb}{2}\sin\theta\sin\phi}\right] \end{aligned} \tag{8.8.44}$$

$$\begin{aligned} E_\phi &\approx jE_o \frac{k}{2\pi}\frac{e^{-jkr}}{r} \cos\theta\cos\phi \left[\int_{-a/2}^{a/2} e^{jk(x' \sin\theta\cos\phi)} dx'\right]\left[\int_{-b/2}^{b/2} e^{jk(y' \sin\theta\sin\phi)} dy'\right] \\ &\approx jE_o \frac{kab}{2\pi}\frac{e^{-jkr}}{r} \cos\theta\cos\phi \left[\frac{\sin\left(\frac{ka}{2}\sin\theta\cos\phi\right)}{\frac{ka}{2}\sin\theta\cos\phi}\right]\left[\frac{\sin\left(\frac{kb}{2}\sin\theta\sin\phi\right)}{\frac{kb}{2}\sin\theta\sin\phi}\right] \end{aligned} \tag{8.8.45}$$

$$H_\theta = -\frac{E_\phi}{\eta} \tag{8.8.46}$$

$$H_\phi = \frac{E_\theta}{\eta} \tag{8.8.47}$$

Equations (8.8.44) and (8.8.45) show that in the $\phi = 0$ plane of observation:

$$E_\theta = 0$$

and

$$E_\phi \approx jE_o \frac{kab}{2\pi}\frac{e^{-jkr}}{r}\left[\cos\theta \frac{\sin\left(\frac{ka}{2}\sin\theta\right)}{\frac{ka}{2}\sin\theta}\right]$$

Also, in the $\phi = \pi/2$ plane of observation:

$$E_\phi = 0$$

and

$$E_\theta \approx jE_o \frac{kab}{2\pi}\frac{e^{-jkr}}{r}\left[\frac{\sin\left(\frac{kb}{2}\sin\theta\right)}{\frac{kb}{2}\sin\theta}\right]$$

Usually a rectangular aperture is fed from a rectangular waveguide where the principal mode is the TE_{10} mode. In this case the field at the aperture is approximated by

$$E_{A,y} = E_o \cos\left(\frac{\pi}{a}x'\right)$$

and from (8.8.43) the far fields are:

$$E_\theta \approx jE_o \frac{k}{2\pi}\frac{e^{-jkr}}{r} \sin\phi \left[\int_{-a/2}^{a/2} \cos\left(\frac{\pi}{a}x'\right) e^{jkx' \sin\theta\cos\phi} dx'\right]\left[\int_{-b/2}^{b/2} e^{jky' \sin\theta \sin\phi} dy'\right]$$

$$\approx -jE_o \frac{kab}{4}\frac{e^{-jkr}}{r} \sin\phi \left[\frac{\cos\left(\frac{ka}{2}\sin\theta\cos\phi\right)}{\left(\frac{ka}{2}\sin\theta\cos\phi\right)^2 - \frac{\pi}{2}}\right]\left[\frac{\sin\left(\frac{kb}{2}\sin\theta\sin\phi\right)}{\frac{kb}{2}\sin\theta\sin\phi}\right]$$

$$E_\phi \approx -jE_o \frac{kab}{4}\frac{e^{-jkr}}{r} \cos\theta\cos\phi \left[\frac{\cos\left(\frac{ka}{2}\sin\theta\cos\phi\right)}{\left(\frac{ka}{2}\sin\theta\cos\phi\right)^2 - \frac{\pi}{2}}\right]\left[\frac{\sin\left(\frac{kb}{2}\sin\theta\sin\phi\right)}{\frac{kb}{2}\sin\theta\sin\phi}\right]$$

$$H_\theta = -\frac{E_\phi}{\eta}$$

$$H_\phi = \frac{E_\theta}{\eta}$$

The evaluation of the previous integral uses the relation:

$$\int_{x_1}^{x_2} \cos(\alpha x) e^{\beta x} dx = \frac{1}{\alpha^2 + \beta^2} e^{\beta x}\left[\beta\cos(\alpha x) + \alpha\sin(\alpha x)\right]\Big|_{x_1}^{x_2}$$

8.8.2 Circular Aperture

A circular aperture is shown in Fig. 8.17. Assume that the field over the aperture is constant and given by

$$E_{A,y}(x', y') = E_o$$

Then, in (8.8.43) with

$$x' = \rho' \cos\phi'$$

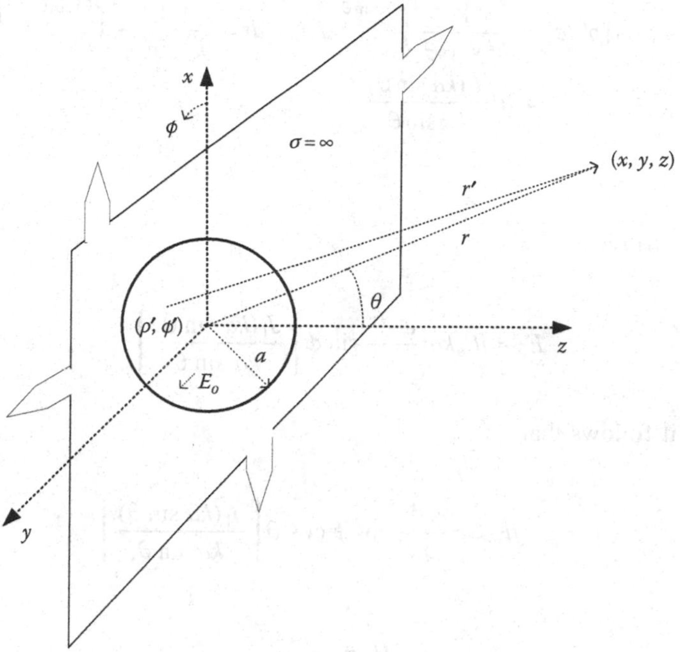

FIGURE 8.17 A circular aperture.

and

$$y' = \rho' \sin\phi'$$

it follows that

$$\begin{aligned} E_\theta &\approx j\frac{E_o k}{2\pi}\frac{e^{-jkr}}{r}\sin\phi \int_0^{2\pi}\int_0^a e^{jk\sin\theta(\rho'\cos\phi'\cos\phi+\rho'\sin\phi'\sin\phi)}\rho' d\rho' d\phi' \\ &\approx j\frac{E_o k}{2\pi}\frac{e^{-jkr}}{r}\sin\phi \int_0^a\left[\int_0^{2\pi} e^{jk\rho'\sin\theta\cos(\phi-\phi')}d\phi'\right]\rho' d\rho' \\ &\approx jE_o k\frac{e^{-jkr}}{r}\sin\phi \int_0^a [J_0(k\rho'\sin\theta)]\rho' d\rho' \end{aligned} \tag{8.8.48}$$

To evaluate the integral, let

$$t = k\rho'\sin\theta$$
$$dt = k\sin\theta d\rho'$$

to obtain

$$\int_0^a [J_0(k\rho' \sin\theta)]\rho' d\rho' = \frac{1}{k^2 \sin^2\theta}\int_0^{ka \sin\theta} J_0(t) t dt = \frac{1}{k^2 \sin^2\theta}\int_0^{ka \sin\theta} \frac{d[tJ_1(t)]}{dt} dt$$
$$= a^2 \frac{J_1(ka \sin\theta)}{ka \sin\theta}$$

Then, (8.8.48) reads

$$E_\theta \approx jE_o ka^2 \frac{e^{-jkr}}{r} \sin\phi \left[\frac{J_1(ka \sin\theta)}{ka \sin\theta}\right]$$

Similarly, it follows that

$$E_\phi \approx jE_o ka^2 \frac{e^{-jkr}}{r} \cos\phi \cos\theta \left[\frac{J_1(ka \sin\theta)}{ka \sin\theta}\right]$$

$$H_\theta = -\frac{E_\phi}{\eta}$$
$$H_\phi = \frac{E_\theta}{\eta}$$

8.9 RADIATION FROM APERTURES USING THE EQUIVALENCE PRINCIPLE

In antenna books the fields from apertures are usually analyzed using the equivalence principle. The electromagnetic field in a bounded, lossy region is uniquely specified by the sources in the region and the specification of either: (a) the tangential components of the electric field on the boundary, or (b) the tangential components of the magnetic field on the boundary, or (c) the tangential component of the electric field on part of the boundary and the tangential component of the magnetic field on the other part of the boundary. The previous statements follow from the Uniqueness Theorem.

The Equivalence Principle permits the source of the field to be represented by equivalent sources that produce the same radiated field. The Equivalence Principle that is discussed next is known as the Love's Equivalence Principle.

Consider the sources $\mathbf{J}_e$ and $\mathbf{J}_m$ producing the fields $\mathbf{E}_1$ and $\mathbf{H}_1$, as shown in Fig. 8.18a. The closed surface, shown in dashed lines, encloses the sources and divides the region into two parts, denoted by volumes V_1 and V_2. The closed surface in Fig. 8.18 is shown as an ellipse but it can any closed surface enclosing the sources. Next, remove the sources and introduce equivalent surface currents $\mathbf{J}_{e,s}$ and $\mathbf{J}_{m,s}$ that produce the fields $\mathbf{E}_1$ and $\mathbf{H}_1$ in V_1, and $\mathbf{E}_2$ and $\mathbf{H}_2$ in V_2, as shown in Fig. 8.18b. At the boundary S, the equivalent surface current sources must satisfy

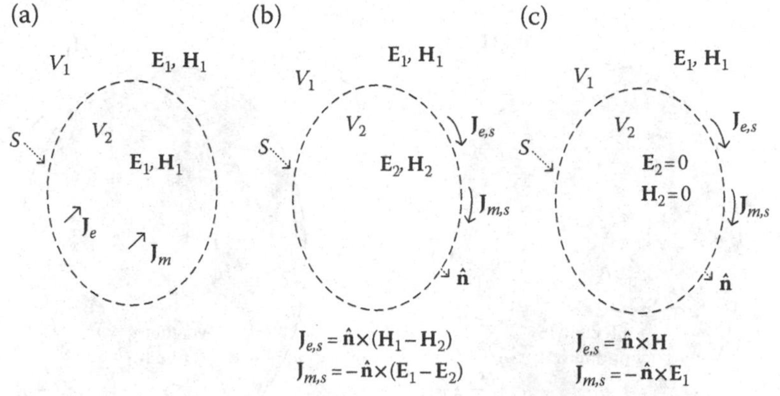

FIGURE 8.18 (a) Sources $\mathbf{J}_e$ and $\mathbf{J}_m$ producing the fields $\mathbf{E}_1$ and $\mathbf{H}_1$; (b) equivalent surface current; (c) equivalent surface current with $\mathbf{E}_2$ and $\mathbf{H}_2$ set to zero.

$$\mathbf{J}_{e,s} = \hat{\mathbf{n}} \times (\mathbf{H}_1 - \mathbf{H}_2)$$

and

$$\mathbf{J}_{m,s} = -\hat{\mathbf{n}} \times (\mathbf{E}_1 - \mathbf{E}_2)$$

The fields in V_2 are of no interest and can be set equal to zero. Then, as shown in Fig. 8.18c, with $\mathbf{E}_2 = \mathbf{H}_2 = 0$, the required surface currents are

$$\mathbf{J}_{e,s} = \hat{\mathbf{n}} \times \mathbf{H}_1$$

and

$$\mathbf{J}_{m,s} = -\hat{\mathbf{n}} \times \mathbf{E}_1$$

These surface currents produce the desired fields $\mathbf{E}_1$ and $\mathbf{H}_1$ in V_1. Thus, as far as region V_1 is concerned, this is an equivalent representation of the original sources.

If region V_2 in Fig. 8.18c is taken to be a perfect electric conductor, then $\mathbf{J}_{e,s} = 0$ and the fields $\mathbf{E}_1$ and $\mathbf{H}_1$ in V_1 are due only to $\mathbf{J}_{m,s}$, as shown in Fig. 8.19a. Similarly, if region V_2 in Fig. 8.18c is taken to be a perfect magnetic conductor, then $\mathbf{J}_{m,s} = 0$ and the fields $\mathbf{E}_1$ and $\mathbf{H}_1$ in V_1 are due to $\mathbf{J}_{e,s}$, as shown in Fig. 8.19b.

The application of the Equivalence Principle to radiation from an aperture is as follows. Consider the two-dimensional drawing of an aperture on a conducting plane as shown in Fig. 8.20a, where $\mathbf{E}_A$ is a known field at the aperture. In Fig. 8.20b the aperture region is divided into regions V_1 and V_2, where in V_2 the fields are set equal to zero. In Fig. 8.20b, the currents $\mathbf{J}_{e.s}$ and $\mathbf{J}_{m.s}$ are shown, where $\mathbf{J}_{m.s}$ only flows over the aperture. The dashed surface can now be replaced by a perfect electric conductor which makes $\mathbf{J}_{e,s} = 0$ and the fields in V_2 are due to the value of $\mathbf{J}_{m,s} = -\hat{\mathbf{n}} \times \mathbf{E}_A$ over the aperture, as shown in Fig. 8.20c. Since the equivalent source is located above a

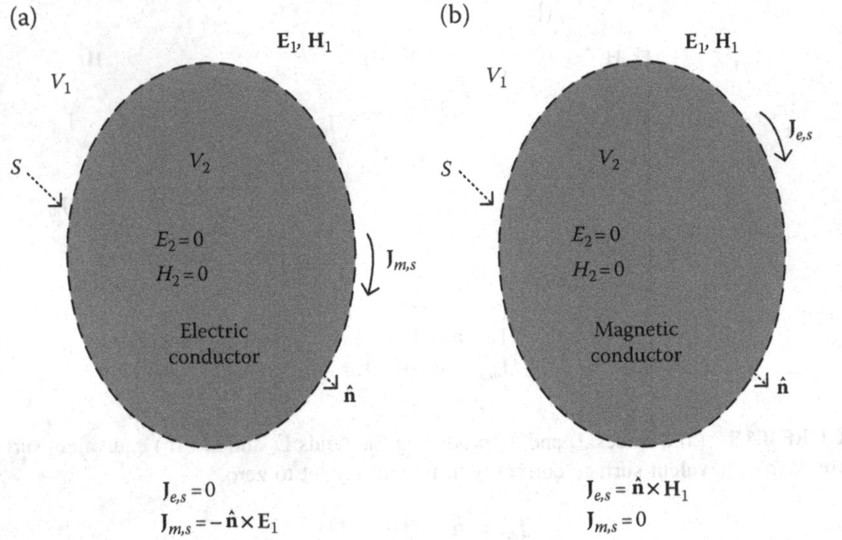

FIGURE 8.19 (a) Surface current $\mathbf{J}_{m,s}$ when V_2 is a perfect electric conductor; (b) surface current $\mathbf{J}_{e,s}$ when V_2 is a perfect magnetic conductor.

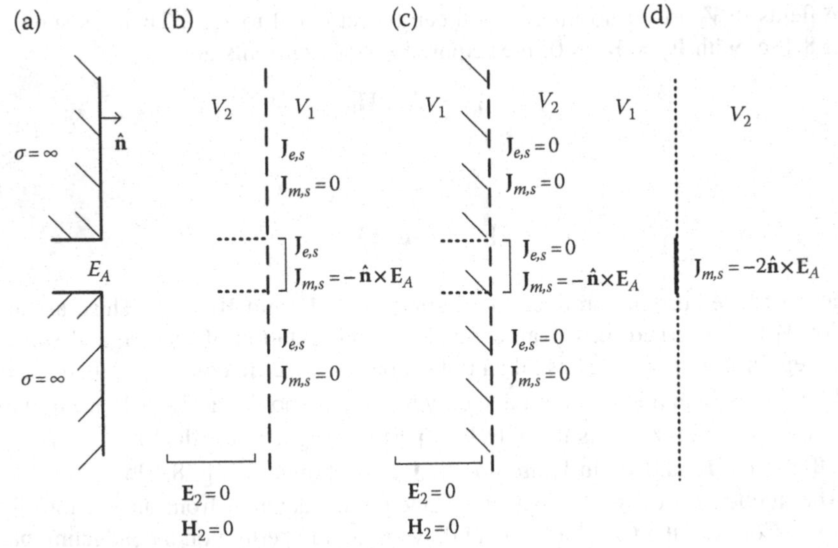

FIGURE 8.20 (a) Two-dimensional view of an aperture; (b) the aperture region divided into regions V_1 and V_2; (c) aperture surface replaced by a perfect electric conductor; (d) the image representation.

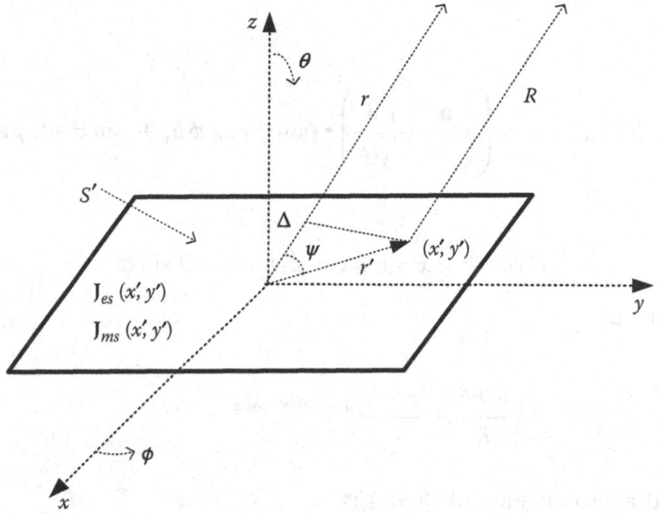

FIGURE 8.21 Geometry for the far-field approximations.

perfect electric conductor, the method of images shows that the fields in V_2 are due to a source given by $\mathbf{J}_{m,s} = -2\hat{\mathbf{n}} \times \mathbf{E}_A$, as shown in Fig. 8.20d.

In the case that the magnetic field over the aperture is given, say $\mathbf{H}_A$, replacing the dashed surface by a perfect magnetic conductor, which makes $\mathbf{J}_{m,s} = 0$, the fields in V_2 are due to the value of $\mathbf{J}_{e,s} = 2\hat{\mathbf{n}} \times \mathbf{H}_A$ over the aperture.

The equivalent source $\mathbf{J}_{m,s}$ in Fig. 8.20d radiates into unbounded media. Therefore, the fields can be calculated in terms of the $\mathbf{F}$ vector. That is,

$$\mathbf{F} = \frac{\varepsilon}{4\pi} \iint_{S'} \mathbf{J}_{m,s} \frac{e^{-jk|\mathbf{r}-\mathbf{r}'|}}{|\mathbf{r} - \mathbf{r}'|} dS' \tag{8.9.1}$$

The evaluation of (8.9.1) in the far field uses the approximation in (8.8.32) for the phase term and in the denominator: $|\mathbf{r} - \mathbf{r}'| \approx r$. The far-field approximations are illustrated in Fig. 8.21, where in the far field

$$R \approx r - \Delta \approx r - r' \cos \Psi$$

The angle between r and r' is Ψ. The far-field approximation is

$$\frac{e^{-jk|\mathbf{r}-\mathbf{r}'|}}{|\mathbf{r} - \mathbf{r}'|} = \frac{e^{-jkR}}{R} \approx \frac{e^{-jk(r-\Delta)}}{r - \Delta} \approx \frac{e^{-jkr}}{r} e^{jk\Delta} \approx \frac{e^{-jkr}}{r} e^{jkr' \cos \Psi} \tag{8.9.2}$$

Since

$$r' = \sqrt{x'^2 + y'^2}$$

and

$$\cos \Psi = \hat{\mathbf{a}}_{r'} \bullet \hat{\mathbf{a}}_r = \left(\frac{x'\hat{\mathbf{a}}_x + y'\hat{\mathbf{a}}_y}{\sqrt{x'^2 + y'^2}} \right) \bullet (\sin\theta \cos\phi\, \hat{\mathbf{a}}_x + \sin\theta \sin\phi\, \hat{\mathbf{a}}_y)$$

then

$$r' \cos \Psi = x' \sin\theta \cos\phi + y' \sin\theta \sin\phi$$

and (8.9.2) reads

$$\frac{e^{-jkR}}{R} \approx \frac{e^{-jkr}}{r} e^{jk \sin\theta (x' \cos\phi + y' \sin\phi)} \tag{8.9.3}$$

The far-field approximation of (8.9.1) is

$$\mathbf{F} \approx \frac{\varepsilon}{4\pi} \frac{e^{-jkr}}{r} \iint_{S'} \mathbf{J}_{m,s}(x', y')\, e^{jk \sin\theta (x' \cos\phi + y' \sin\phi)} dS' \tag{8.9.4}$$

Similarly, if the equivalent source $\mathbf{J}_{e,s}$ is used, then, the radiated field is expressed in terms of the $\mathbf{A}$ vector, as

$$\mathbf{A} = \frac{\mu}{4\pi} \iint_{S'} \mathbf{J}_{e,s} \frac{e^{-jk|\mathbf{r}-\mathbf{r}'|}}{|\mathbf{r} - \mathbf{r}'|} dS'$$

and the far-field approximation is

$$\mathbf{A} \approx \frac{\mu}{4\pi} \frac{e^{-jkr}}{r} \iint_{S'} \mathbf{J}_{e,s}(x', y')\, e^{jk \sin\theta (x' \cos\phi + y' \sin\phi)} dS'$$

Consider the rectangular aperture in Fig. 8.16, where the field over the aperture is given by

$$E_{A,y}(x', y') = E_o$$

Using the Equivalence Principle (see Fig. 8.20d), the equivalent surface current over the aperture is

$$\mathbf{J}_{m,s} = -2\hat{\mathbf{a}}_z \times E_{A,y}\hat{\mathbf{a}}_y = 2E_o\hat{\mathbf{a}}_x \tag{8.9.5}$$

Substituting (8.9.5) into (8.9.4) gives

$$F_x \approx 2E_o \frac{\varepsilon}{4\pi} \frac{e^{-jkr}}{r} \int_{y'=-b/2}^{b/2} \int_{x'=-a/2}^{a/2} e^{jk(x' \sin\theta \cos\phi + y' \sin\theta \cos\phi)} dx' dy'$$

$$F_x \approx E_o \frac{\varepsilon}{2\pi} \frac{e^{-jkr}}{r} \left[\int_{-a/2}^{a/2} e^{jk(x' \sin\theta \cos\phi)} dx' \right] \left[\int_{-b/2}^{b/2} e^{jk(y' \sin\theta \sin\phi)} dy' \right] \tag{8.9.6}$$

$$F_x \approx E_o \frac{\varepsilon ab}{2\pi} \frac{e^{-jkr}}{r} \left[\frac{\sin\left(\frac{ka}{2} \sin\theta \cos\phi\right)}{\frac{ka}{2} \sin\theta \cos\phi} \right] \left[\frac{\sin\left(\frac{kb}{2} \sin\theta \sin\phi\right)}{\frac{kb}{2} \sin\theta \sin\phi} \right]$$

In the far field, from (2.3.4):

$$\begin{cases} H_\theta \approx -j\omega F_\theta \\ H_\phi \approx -j\omega F_\phi \\ E_\theta = \eta H_\phi \\ E_\phi = -\eta H_\theta \end{cases} \tag{8.9.7}$$

where

$$\begin{cases} F_\theta = F_x \cos\theta \cos\phi \\ F_\phi = -F_x \sin\phi \end{cases} \tag{8.9.8}$$

Hence, from (8.9.6), (8.9.7), and (8.9.8), we obtain

$$E_\theta \approx \eta H_\phi \approx -j\omega\eta F_\phi = jE_o \frac{kab}{2\pi} \frac{e^{-jkr}}{r} \sin\phi \left[\frac{\sin\left(\frac{ka}{2} \sin\theta \cos\phi\right)}{\frac{ka}{2} \sin\theta \cos\phi} \right] \times \left[\frac{\sin\left(\frac{kb}{2} \sin\theta \sin\phi\right)}{\frac{kb}{2} \sin\theta \sin\phi} \right]$$

and

$$E_\phi \approx -\eta H_\theta \approx j\omega\mu\eta F_\theta = jE_o \frac{kab}{2\pi} \frac{e^{-jkr}}{r} \cos\theta \cos\phi \left[\frac{\sin\left(\frac{ka}{2} \sin\theta \cos\phi\right)}{\frac{ka}{2} \sin\theta \cos\phi} \right] \times \left[\frac{\sin\left(\frac{kb}{2} \sin\theta \sin\phi\right)}{\frac{kb}{2} \sin\theta \sin\phi} \right]$$

which are identical to (8.8.44) to (8.8.47).

Problems

P8.1. The TE modes in the parallel-plate waveguide in Fig. 8.1 were derived using the $\mathbf{\Pi}^{\mathrm{m}}$ vector (see (8.1.8)). Perform the analysis using the **F** vector and compare the results.

P8.2. The analysis of a line source above a lossy surface in Fig. 8.4 was done in terms of k_1 and k_2 (see (8.3.9) and (8.3.10)).

a. Perform the analysis in terms of the complex propagation constants γ_1 and γ_2, and show that for $\mathrm{Re}(\gamma_2 x) >> 1$:

$$E_{y1} \approx \frac{j\omega\mu_o I\gamma_1}{\pi\left(\gamma_2^2 - \gamma_1^2\right)x} K_1(\gamma_1 x)$$

b. Approximate E_{y1} when $|\gamma_1 x| >> 1$, and compare the result with (8.3.29).

c. Determine H_{z1} if $\mathrm{Re}(\gamma_2 x) >> 1$, and then if $|\gamma_1 x| >> 1$.

d. From the results in (b) and (c), show that $H_{z1} \approx E_{y1}/\eta_o$.

P8.3. From (8.5.23), determine the approximate location of the poles when $n_2' >> n_2''$.

P8.4. Perform the analysis of the magnetic dipole in Fig. 8.12 in terms of the **F** vector.

P8.5.

a. Verify (8.5.44).

b. Verify (8.5.45) and (8.5.46).

P8.6. Evaluate Q^m in (8.7.7).

P8.7. Verify (8.8.42).

P8.8. Use the Equivalent Principle to calculate the far fields when the electric field over the aperture in Fig. 8.17 is given by

$$E_{A,y} = E_o$$

9 Further Studies of Electromagnetic Waves in Cylindricals Geometries

9.1 DIFFRACTION BY A CONDUCTING CYLINDER

A plane-wave incident on a cylindrical structure is shown in Fig. 9.1. It is a two-dimensional problem since the cylinder is of infinite extent. The cylinder is assumed to be a perfect conductor, with radius a. For the incident wave shown in Fig. 9.1, the electric field is given by

$$E_z^i = Ce^{-j\beta x} = Ce^{-j\beta\rho \cos\phi}$$

where $\beta = \omega\sqrt{\mu\varepsilon}$. Using (4.5.3), the incident TM^z field is written in the form

$$E_z^i = Ce^{-j\beta\rho \cos\phi} = C \sum_{n=-\infty}^{\infty} j^{-n} J_n(\beta\rho)\, e^{jn\phi}$$

The associated magnetic field (i.e., $H_y^i = -E_z^i/\eta_o$) is shown in Fig. 9.1.

The incident field is diffracted by the cylinder. The diffracted field (E_z^d) satisfies

$$\left[\frac{\partial^2}{\partial\rho^2} + \frac{1}{\rho}\frac{\partial}{\partial\rho} + \frac{1}{\rho^2}\frac{\partial^2}{\partial\phi^2} + \beta^2\right] E_z^d = 0(\rho > a)$$

and the solution to E_z^d that represents an outgoing wave as $\rho \to \infty$ is

$$E_z^d = C \sum_{n=-\infty}^{\infty} a_n H_n^{(2)}(\beta\rho) e^{jn\phi} \tag{9.1.1}$$

The total field outside the cylinder is then

$$E_z = E_z^i + E_z^d = C \sum_{n=-\infty}^{\infty} [j^{-n} J_n(\beta\rho) + a_n H_n^{(2)}(\beta\rho)] e^{jn\phi} \tag{9.1.2}$$

DOI: 10.1201/9781003219729-9

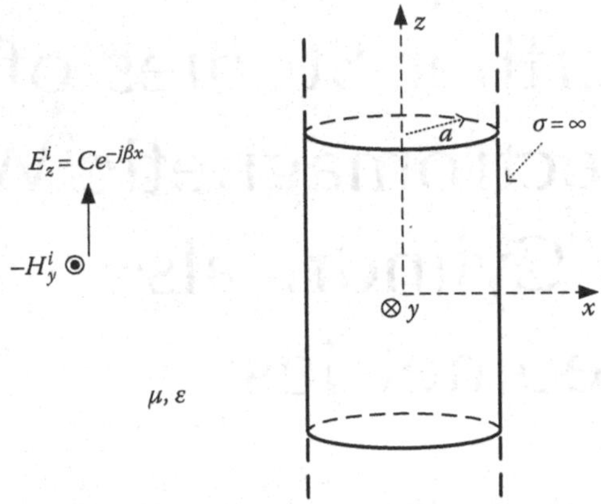

FIGURE 9.1 A plane wave incident on a conducting cylinder.

The boundary condition is

$$E_z = 0|_{\rho=a} \Rightarrow a_n = -j^{-n}\frac{J_n(\beta a)}{H_n^{(2)}(\beta a}$$

Therefore,

$$E_z = C\sum_{n=-\infty}^{\infty} j^{-n}\left[J_n(\beta\rho) - \frac{J_n(\beta a)}{H_n^{(2)}(\beta a)}H_n^{(2)}(\beta\rho)\right]e^{jn\phi} \tag{9.1.3}$$

From Maxwell's equations, the associated magnetic fields are

$$H_\rho = -\frac{1}{j\omega\mu}\frac{1}{\rho}\frac{\partial E_z}{\partial\phi} = -C\frac{1}{\omega\mu\rho}\sum_{n=-\infty}^{\infty} nj^{-n}\left[J_n(\beta\rho) - \frac{J_n(\beta a)}{H_n^{(2)}(\beta a)}H_n^{(2)}(\beta\rho)\right]e^{jn\phi}$$

and

$$H_\phi = \frac{1}{j\omega\mu}\frac{\partial E_z}{\partial\rho} = C\frac{\beta}{\omega\mu}\sum_{n=-\infty}^{\infty} j^{-n-1}\left[J_n'(\beta\rho) - \frac{J_n(\beta a)}{H_n^{(2)}(\beta a)}H_n^{(2)\prime}(\beta\rho)\right]e^{jn\phi}$$

The surface current density induced on the cylinder is given by

$$\mathbf{J}_{e,s} = \hat{\mathbf{n}} \times \mathbf{H}|_{\rho=a} = \hat{\mathbf{a}}_\rho \times (H_\rho\hat{\mathbf{a}}_\rho + H_\phi\hat{\mathbf{a}}_\phi)|_{\rho a} = H_\phi\hat{\mathbf{a}}_z|_{\rho=a}$$

Then, $\mathbf{J}_{e,s} = J_{e,s}\,\hat{\mathbf{a}}_z$ where

$$J_{e,s} = H_\phi|_{\rho=a} = C\frac{\beta}{\omega\mu}\sum_{n=-\infty}^{\infty}\frac{j^{-n-1}}{H_n^{(2)}(\beta a)}[H_n^{(2)}(\beta a)J_n{}'(\beta a) - J_n(\beta a)H_n^{(2)}{}'(\beta\rho)]e^{jn\phi}$$
$$= C\frac{2}{\omega\mu\pi a}\sum_{n=-\infty}^{\infty}\frac{j^{-n}}{H_n^{(2)}(\beta a)}e^{jn\phi}$$

and the Wronskian was used, namely

$$W[J_n(\beta a), H_n^{(2)}(\beta a)] = -2j/\pi\beta a \tag{9.1.4}$$

In the far field, the diffracted electric field in (9.1.3) is given by

$$E_z^d \approx -C\sqrt{\frac{2}{\pi\beta\rho}}\,e^{-(j\beta\rho-\pi/4)}\sum_{n=-\infty}^{\infty}\frac{J_n(\beta a)}{H_n^{(2)}(\beta a)}e^{jn\phi}$$

where the large argument approximation for $H_n^{(2)}(\beta\rho)$ was used. That is,

$$H_n^{(2)}(\beta\rho) \approx \sqrt{\frac{2}{\pi\beta\rho}}\,j^n e^{-j(\beta\rho-\pi/4)}$$

The magnitude of the incident field (i.e., $Ce^{-j\beta x}$) is C. Therefore, the magnitude ratio of the diffracted to incident field is:

$$\left|\frac{E_z^d}{E_z^i}\right| = \sqrt{\frac{2}{\pi\beta\rho}}\left|\sum_{n=-\infty}^{\infty}\frac{J_n(\beta a)}{H_n^{(2)}(\beta a)}e^{jn\phi}\right| = \sqrt{\frac{2}{\pi\beta\rho}}\left|\sum_{n=0}^{\infty}\varepsilon_n\frac{J_n(\beta a)}{H_n^{(2)}(\beta a)}\cos(n\phi)\right| \tag{9.1.5}$$

where ε_n is the Neumann's number.

From (9.1.5), the diffraction pattern due to a thin cylinder or at very low frequencies can be calculated. In general, for the cases that $\beta a \to 0$, using only the $n = 0$ term in (9.1.5):

$$\lim_{\beta a\to 0}\frac{J_0(\beta a)}{H_0^{(2)}(\beta a)} \approx \frac{1}{-jY_0(\beta a)} \approx \frac{j\pi}{2}\frac{1}{\ln(0.8905\beta a)}$$

gives

$$\lim_{\beta a\to 0}\left|\frac{E_z^d}{E_z^i}\right| \approx \sqrt{\frac{\pi}{2\beta\rho}}\frac{1}{\ln(0.8905\beta a)}$$

If the $n = 0$ and $n = 1$ terms in (9.1.5) are used, then with

$$\lim_{\beta a \to 0} \frac{J_1(\beta a)}{H_1^{(2)}(\beta a)} \approx \frac{J_1(\beta a)}{-Y_1(\beta a)} \approx \frac{\frac{\beta a}{2}}{\frac{j}{\pi}\left(\frac{2}{\beta a}\right)} = -j\pi\left(\frac{\beta a}{2}\right)^2$$

gives

$$\left|\frac{E_z^d}{E_z^i}\right| \approx \sqrt{\frac{\pi}{2\beta\rho}}\left|\frac{1}{\ln(0.8905\beta a)} - (\beta a)^2 \cos\phi\right| \approx \sqrt{\frac{\pi}{2\beta\rho}}\frac{1}{\ln(0.8905\beta a)}$$

which shows that if $\beta a \to 0$ the power is diffracted fairly uniformly around the cylinder.

In the case that the polarization is such that the incident field is TE^z, the magnetic field is given by:

$$H_z^i = Be^{-j\beta\rho\cos\phi} = B\sum_{n=-\infty}^{\infty} j^{-n} J_n(\beta\rho)e^{jn\phi}$$

The diffracted field is of the form

$$H_z^d = B\sum_{n=-\infty}^{\infty} b_n H_n^{(2)}(\beta\rho)e^{jn\phi} \tag{9.1.6}$$

Therefore,

$$H_z = H_z^i + H_z^d = B\sum_{n=-\infty}^{\infty} [j^{-n}J_n(\beta\rho) + b_n H_n^{(2)}(\beta\rho)]e^{jn\phi} \tag{9.1.7}$$

For this polarization, the boundary condition at the cylinder surface is

$$E_\phi = \frac{-1}{j\omega\varepsilon}\frac{\partial H_z}{\partial\rho} = 0\Bigg|_{\rho=a} \Rightarrow b_n = -j^{-n}\frac{J_n'(\beta a)}{H_n^{(2)\prime}(\beta a)}$$

and (9.1.7) reads

$$H_z = B\sum_{n=-\infty}^{\infty} j^{-n}\left[J_n(\beta\rho) - \frac{J_n'(\beta a)}{H_n^{(2)\prime}(\beta a)}H_n^{(2)}(\beta\rho)\right]e^{jn\phi} \tag{9.1.8}$$

The other field components are

$$E_\phi = -\frac{B\beta}{\omega\varepsilon}\sum_{n=-\infty}^{\infty} j^{-n-1}\left[J_n'(\beta\rho) - \frac{J_n'(\beta a)}{H_n^{(2)\prime}(\beta a)}H_n^{(2)\prime}(\beta\rho)\right]e^{jn\phi}$$

and

$$E_\rho = \frac{1}{j\omega\varepsilon}\frac{1}{\rho}\frac{\partial H_z(\beta\rho)}{\partial\phi} = \frac{B}{\omega\varepsilon}\frac{1}{\rho}\sum_{n=-\infty}^{\infty} nj^{-n-1}\left[J_n(\beta\rho) - \frac{J_n'(\beta a)}{H_n^{(2)\prime}(\beta a)}H_n^{(2)}(\beta\rho)\right]e^{jn\phi}$$

The surface current density induced on the cylinder is

$$\mathbf{J}_{e,s} = \hat{\mathbf{n}} \times \mathbf{H}|_{\rho=a} = \hat{\mathbf{a}}_\rho \times H_z\hat{\mathbf{a}}_z|_{\rho=a} = -H_z\hat{\mathbf{a}}_\phi|_{\rho=a}$$

Then, $\mathbf{J}_{e,s} = J_{e,s}\hat{\mathbf{a}}_\phi$ where

$$\begin{aligned} J_{e,s} = H_z|_{\rho=a} &= -B\sum_{n=-\infty}^{\infty}\frac{j^{-n}}{H_n^{(2)\prime}(\beta a)}[J_n(\beta a)H_n^{(2)\prime}(\beta a) - J_n'(\beta a)H_n^{(2)}(\beta\rho)]e^{jn\phi} \\ &= B\frac{2}{\pi\beta a}\sum_{n=-\infty}^{\infty}\frac{j^{-n+1}}{H_n^{(2)\prime}(\beta a)}e^{jn\phi} \end{aligned} \tag{9.1.9}$$

and the Wronskian in (9.1.4) was used.

In the far field, using the large argument approximation for $H_n^{(2)}(\beta\rho)$, the diffracted magnetic field in (9.1.8) is

$$H_z^d \approx -B\sqrt{\frac{2}{\pi\beta\rho}}e^{-(j\beta\rho-\pi/4)}\sum_{n=-\infty}^{\infty}\frac{J_n'(\beta a)}{H_n^{(2)\prime}(\beta a)}e^{jn\phi}$$

The magnitude of the diffracted to incident field is:

$$\left|\frac{H_z^d}{H_z^i}\right| = \sqrt{\frac{2}{\pi\beta\rho}}\left|\sum_{n=-\infty}^{\infty}\frac{J_n'(\beta a)}{H_n^{(2)\prime}(\beta a)}e^{jn\phi}\right| = \sqrt{\frac{2}{\pi\beta\rho}}\left|\sum_{n=0}^{\infty}\varepsilon_n\frac{J_n'(\beta a)}{H_n^{(2)\prime}(\beta a)}\cos(n\phi)\right| \tag{9.1.10}$$

From (9.1.10) the diffraction pattern can be calculated when the cylinder is thin and for frequencies such that $\beta a \to 0$. For $\beta a \to 0$, the pattern can be calculated using the $n = 0$ and $n = 1$ terms in (9.1.10) where:

$$\lim_{\beta a\to 0}\frac{J'_0(\beta a)}{H_0^{(2)\prime}(\beta a)} = \lim_{\beta a\to 0}\frac{-J_1(\beta a)}{-H_1^{(2)}(\beta a)} \approx \frac{-j\pi}{4}(\beta a)^2$$

$$\lim_{\beta a\to 0}\frac{J'_1(\beta a)}{H_1^{(2)\prime}(\beta a)} = \lim_{\beta a\to 0}\frac{\frac{1}{\beta a}J_1(\beta a) - J_2(\beta a)}{\frac{1}{\beta a}H_1^{(2)}(\beta a) - H_2^{(2)}(\beta a)} = \frac{\frac{1}{\beta a}\left(\frac{\beta a}{2}\right) - 0}{\frac{1}{\beta a}\left(\frac{j2}{\pi\beta a}\right) - j\frac{4}{\pi(\beta a)^2}} \approx \frac{j\pi}{4}(\beta a)^2$$

to obtain

$$\lim_{\beta a \to 0} \left| \frac{H_z^d}{H_z^i} \right| \approx \sqrt{\frac{2}{\pi \beta \rho}} \left| -j\frac{\pi}{4}(\beta a)^2 + j\frac{\pi}{2}(\beta a)^2 \cos\phi \right| = \sqrt{\frac{\pi}{8\beta\rho}}\,(\beta a)^2 |1 - 2\cos\phi|$$

In this case, the diffraction pattern is a function of ϕ.

Using the $n = 0$ and $n = 1$ terms in (9.1.10) the surface current when $\beta a \to 0$ is approximately given by

$$J_{e,s} \approx B(-1 + j2\beta a \cos\phi)$$

Another way of deriving the expression for the field in (9.1.3) is to use the Green function for outside a conducting cylinder that satisfies the same boundary condition as the E_z field. The procedure is as follows. For an electric current source of infinite length, amplitude I and located at (ρ_o, ϕ_o), (1.3.13) or (1.3.21) show that E_z satisfies

$$[\nabla^2 + \beta^2]E_z = j\omega\mu J_{e,z} = j\omega\mu I \frac{\delta(\rho - \rho_o)\delta(\phi - \phi_o)}{\rho}$$

where $\beta = \omega\sqrt{\mu\varepsilon}$. The associated Green function that satisfy $G_< = 0$ at $\rho = a$ (see (5.6.51)) is

$$G_<(\rho, \phi, \rho', \phi') = -\frac{j}{4} \sum_{n=-\infty}^{\infty} H_n^{(2)}(\beta\rho') \left[J_n(\beta\rho) - \frac{J_n(\beta a)}{H_n^{(2)}(\beta a)} H_n^{(2)}(\beta\rho) \right] e^{jn(\phi - \phi')}$$

Then, E_z is given by

$$\begin{aligned} E_z &= -j\omega\mu I G_<(\rho, \phi, \rho_o, \phi_o) \\ &= -\frac{\omega\mu I}{4} \sum_{n=-\infty}^{\infty} H_n^{(2)}(\beta\rho_o) \left[J_n(\beta\rho) - \frac{J_n(\beta a)}{H_n^{(2)}(\beta a)} H_n^{(2)}(\beta\rho) \right] e^{jn(\phi - \phi_o)} \qquad (\rho \le \rho_o) \end{aligned}$$

If the source is moved to a large distance, such that $\beta\rho_o$ is large, using:

$$H_n^{(2)}(\beta\rho_o) \approx \sqrt{\frac{2}{\pi\beta\rho_o}}\, j^n e^{-j(\beta\rho_o - \pi/4)}$$

the field E_z is

$$E_z = -\omega\mu I \sqrt{\frac{1}{8\pi\beta\rho_o}} e^{-j(\beta\rho_o - \pi/4)} \sum_{n=-\infty}^{\infty} j^n \left[J_n(\beta\rho) - \frac{J_n(\beta a)}{H_n^{(2)}(\beta a)} H_n^{(2)}(\beta\rho) \right] e^{jn(\phi - \phi_o)} \tag{9.1.11}$$

Let the source, which is at a large distance, be located at $\phi_o = \pi$ (i.e., $e^{-jn\pi} = (-1)^n$), and define the constant C as

$$C = -\omega\mu I \sqrt{\frac{1}{8\pi\beta\rho_o}} e^{-j(\beta\rho_o - \pi/4)}$$

Hence, (9.1.11) is expressed in the form

$$E_z = C \sum_{n=-\infty}^{\infty} j^{-n} \left[J_n(\beta\rho) - \frac{J_n(\beta a)}{H_n^{(2)}(\beta a)} H_n^{(2)}(\beta\rho) \right] e^{jn\phi}$$

which is identical to (9.1.3). The first term is recognized as the incident wave and the second as the diffracted field. Therefore, E_z can be expressed as

$$E_z = C \left[e^{-j\beta\rho\cos\phi} - \sum_{n=-\infty}^{\infty} j^{-n} \frac{J_n(\beta a)}{H_n^{(2)}(\beta a)} H_n^{(2)}(\beta\rho) e^{jn\phi} \right]$$

Similarly, using the Green function for the H_z polarization (see (5.6.52)) the derivation follows similar steps, resulting in:

$$H_z = B \left[e^{-j\beta\rho\cos\phi} - \sum_{n=-\infty}^{\infty} j^{-n} \frac{J_n'(\beta a)}{H_n^{(2)\prime}(\beta a)} H_n^{(2)}(\beta\rho) e^{jn\phi} \right] \tag{9.1.12}$$

For the case that $\beta a \gg 1$, which occur over a range of frequencies when the radius of the cylinder is large, the series for the field expressions in (9.1.3) and (9.1.8) converge very slowly. Convergence of the series in (9.1.3) and (9.1.8) require at least $2\beta a$ terms. The sum of a large number of terms involving Bessel functions of large argument and orders is a difficult computational task. For such cases, it will be convenient to convert the series to a fast-convergent series. This is done using the Watson transformation which converts a slowly convergent series into a fast-convergent series. The procedure is illustrated using (9.1.3). That is,

$$E_z = C \sum_{n=-\infty}^{\infty} j^{-n} \left[J_n(\beta\rho) - \frac{J_n(\beta a)}{H_n^{(2)}(\beta a)} H_n^{(2)}(\beta\rho) \right] e^{jn\phi} \tag{9.1.13}$$

From (7.1.12), (9.1.13) can be expressed as

$$E_z = \frac{1}{2j}\oint_C F_\nu(\beta\rho)\frac{e^{j\nu(\phi-\pi)}}{\sin(\nu\pi)}d\nu = \sum_{n=-\infty}^{\infty} F_n(\beta\rho)e^{jn\phi} \tag{9.1.14}$$

where C is the contour that encloses the poles of $\sin(\pi\nu)$, as shown in Fig. 7.3a, and

$$\begin{aligned} F_\nu(\beta\rho) &= Cj^{-\nu}\left[J_\nu(\beta\rho) - \frac{J_\nu(\beta a)}{H_\nu^{(2)}(\beta a)}H_\nu^{(2)}(\beta\rho)\right] \\ &= \frac{Ce^{-j\nu\pi/2}}{H_\nu^{(2)}(\beta a)}\left[H_\nu^{(2)}(\beta a)J_\nu(\beta\rho) - J_\nu(\beta a)H_\nu^{(2)}(\beta\rho)\right] \\ &= \frac{Ce^{-j\nu\pi/2}}{2H_\nu^{(2)}(\beta a)}\left[H_\nu^{(2)}(\beta a)H_\nu^{(1)}(\beta\rho) - H_\nu^{(1)}(\beta a)H_\nu^{(2)}(\beta\rho)\right] \end{aligned} \tag{9.1.15}$$

where $J_\nu(x) = [H_\nu^{(1)}(x) + H_\nu^{(2)}(x)]/2$ was used.

Since

$$H_{-\nu}^{(1)}(x) = e^{j\nu\pi}H_\nu^{(1)}(x)$$

and

$$H_{-\nu}^{(2)}(x) = e^{-j\nu\pi}H_\nu^{(2)}(x)$$

it follows that $F_\nu(\beta\rho)$ is an even function in ν. Equation (9.1.14) can be expressed as

$$E_z = \frac{1}{2j}\left[\int_{\infty+j\varepsilon}^{-\infty+j\varepsilon} F_\nu(\beta\rho)\frac{e^{j\nu(\phi-\pi)}}{\sin(\nu\pi)}d\nu + \int_{-\infty-j\varepsilon}^{\infty-j\varepsilon} F_\nu(\beta\rho)\frac{e^{j\nu(\phi-\pi)}}{\sin(\nu\pi)}d\nu\right]$$

Letting $\nu = -\nu$ in the first integral, and since $F_{-\nu}(\beta\rho) = F_\nu(\beta\rho)$, it follows that:

$$\begin{aligned} E_z &= \frac{1}{2j}\left[\int_{-\infty-j\varepsilon}^{\infty-j\varepsilon} F_\nu(\beta\rho)\frac{1}{\sin(\nu\pi)}\left[e^{-j\nu(\phi-\pi)} + e^{j\nu(\phi-\pi)}\right]d\nu\right] \\ &= \frac{1}{j}\left[\int_{-\infty-j\varepsilon}^{\infty-j\varepsilon} F_\nu(\beta\rho)\frac{\cos[\nu(\phi-\pi)]}{\sin(\nu\pi)}d\nu\right] \end{aligned} \tag{9.1.16}$$

The new contour runs slightly below the real axis and can be closed in the lower-half-plane (see Fig. 7.3b) provided that the contribution from the semicircle path vanishes as $|\nu| \to \infty$. Then, the integral (9.1.16) is evaluated in terms of the residues at the poles of $F_\nu(\beta\rho)$ in (9.1.15) in the lower-half plane, which are the zeroes of $H_\nu^{(2)}(\beta a)$. These poles are determined in terms of the zeroes of an Airy function according to (7.4.43), which we will denote them in this section as χ_p where

$p = 1, 2, 3, \ldots$ (i.e., $Ai(\chi_p) = 0$). That is, the zeroes (see (7.4.46)) occur at the values of ν given by

$$\nu_p = \beta a - \chi_p \left(\frac{\beta a}{2}\right)^{1/3} e^{-j\pi/3} \qquad (p = 1, 2, 3, \ldots)$$

Therefore, the evaluation of (9.1.16) in terms of its residues (i.e., where $H_\nu^{(2)}(\beta a) = 0$) is

$$E_z = C\pi \sum_{p=1}^{\infty} \frac{H_{\nu_p}^{(1)}(\beta a)}{\left.\frac{\partial H_\nu^{(2)}(\beta a)}{\partial \nu}\right|_{\nu=\nu_p}} \frac{\cos[\nu_p(\phi - \pi)]e^{-j\nu_p\pi/2}}{\sin(\nu_p\pi)} H_{\nu_p}^{(2)}(\beta\rho) \qquad (9.1.17)$$

The term $H_{\nu_p}^{(1)}(\beta a)$ in (9.1.17) can be expressed in terms of an Airy function. Using the Wronskian:

$$W\left[H_{\nu_p}^{(1)}(\beta a), H_{\nu_p}^{(2)}(\beta a)\right] = H_{\nu_p}^{(1)}(\beta a)H_{\nu_p}^{(2)\prime}(\beta a) - H_{\nu_p}^{(1)\prime}(\beta a)H_{\nu_p}^{(2)}(\beta a) = -\frac{4j}{\pi\beta a} \qquad (9.1.18)$$

it follows that with $H_\nu^{(2)}(\beta a) = 0$:

$$H_{\nu_p}^{(1)}(\beta a)H_{\nu_p}^{(2)\prime}(\beta a) = -\frac{4j}{\pi\beta a} \Rightarrow H_{\nu_p}^{(1)}(\beta a) = -\frac{4j}{\pi\beta a}\frac{1}{H_{\nu_p}^{(2)\prime}(\beta a)}$$

This expression, using (7.4.45), can be written in terms of the Airy function $Ai'(\chi_p)$. That is,

$$H_{\nu_p}^{(1)}(\beta a) \approx -\left(\frac{2}{\beta a}\right)^{1/3} \frac{e^{-j\pi/6}}{\pi Ai'(\chi_p)} \qquad (9.1.19)$$

Also, $\partial H_\nu^{(2)}(\beta a)/\partial\nu|_{\nu=\nu_p}$ in (9.1.17) can be expressed in terms of $Ai'(\chi_p)$. That is, from (7.4.43):

$$\left.\frac{\partial H_\nu^{(2)}(\beta a)}{\partial \nu}\right|_{\nu=\nu_p} \approx 2\left(\frac{2}{\beta a}\right)^{2/3} e^{-j\pi/3} Ai'(\chi_p) \qquad (9.1.20)$$

Substituting (9.1.19) and (9.1.20) into (9.1.17) results in:

$$E_z = -\frac{C}{2}\left(\frac{\beta a}{2}\right)^{1/3} e^{j\pi/6} \sum_{p=1}^{\infty} \frac{\cos[\nu_p(\phi - \pi)]e^{-j\nu_p\pi/2}}{[Ai'(\chi_p)]^2 \sin(\nu_p\pi)} H_{\nu_p}^{(2)}(\beta\rho) \qquad (9.1.21)$$

Since

$$\frac{1}{\sin(\nu_p \pi)} = \frac{2j}{e^{j\nu_p \pi}(1 - e^{-j\nu_p 2\pi})} = \frac{2j}{e^{j\nu_p \pi}} \sum_{n=0}^{\infty} e^{-j\nu_p 2\pi n} \approx 2je^{-j\nu_p \pi} \text{ for } n = 0$$

Then,

$$\frac{\cos[\nu_p(\phi - \pi)]}{\sin(\nu_p \pi)} e^{-j\pi\nu_p/2} \approx j\left[e^{-j\nu_p(\pi/2-\phi)} + e^{-j\nu_p(\phi+\pi/2)}\right] \tag{9.1.22}$$

The relation in (9.1.22) shows that the series in (9.1.21) converges with only a few terms in the region $-\pi/2 < \phi < \pi/2$ (i.e., when $\pi/2 - \phi > 0$ and $\phi + \pi/2 > 0$) due to the fact that imaginary part of ν_p as p increases has a significant negative imaginary part. That is, (9.1.21) converges in the shadow region of the cylinder. The fields in the shadow region (i.e., the back of the cylinder region) are the one of interest in many propagation calculations. Unfortunately, the series in (9.1.21) cannot be used to calculate the fields in the illuminated region (i.e., the fields in front of the cylinder).

It remains to be shown that the contribution from the semicircular path in the lower-half plane vanishes as $|\nu| \to \infty$. In this chapter several relations of the type shown in (9.1.16) occur, where the integrands involve combinations of Hankel functions. The analysis of the vanishing of these integrands along the semicircular path is discussed in Section 9.10.

The analysis of the H_z polarized incident wave on a conducting cylinder is similar. We will summarize the results. In (9.1.8) we obtained:

$$H_z = B \sum_{n=-\infty}^{\infty} j^{-n} \left[J_n(\beta\rho) - \frac{J_n'(\beta a)}{H_n^{(2)\prime}(\beta a)} H_n^{(2)}(\beta\rho) \right] e^{jn\phi} \tag{9.1.23}$$

which can be expressed as

$$H_z = \frac{1}{2j} \oint_C G_\nu(\beta\rho) \frac{e^{-j\nu(\phi-\pi)}}{\sin(\nu\pi)} d\nu = \sum_{n=-\infty}^{\infty} G_n(\beta\rho) e^{jn\phi} \tag{9.1.24}$$

where C is the closed contour shown in Fig. 7.3a and

$$\begin{aligned} G_\nu(\beta\rho) &= Bj^{-\nu} \left[J_\nu(\beta\rho) - \frac{J_\nu'(\beta a)}{H_\nu^{(2)\prime}(\beta a)} H_\nu^{(2)}(\beta\rho) \right] \\ &= \frac{Be^{-j\nu\pi/2}}{2H_\nu^{(2)\prime}(\beta a)} [H_\nu^{(2)\prime}(\beta a) H_\nu^{(1)}(\beta\rho) - H_\nu^{(1)\prime}(\beta a) H_\nu^{(2)}(\beta\rho)] \end{aligned} \tag{9.1.25}$$

Since $G_\nu(\beta\rho)$ is an even function in ν, (9.1.24) can be expressed as

$$H_z = \frac{1}{j}\left[\int_{-\infty-j\varepsilon}^{\infty-j\varepsilon} G_\nu(\beta\rho)\frac{\cos[\nu(\phi-\pi)]}{\sin(\nu\pi)}\,d\nu\right] \tag{9.1.26}$$

The new contour runs slightly below the real axis and can be closed in the lower half-plane. Then, the integral is evaluated by evaluating the residue of the poles associated with $G_\nu(\beta\rho)$ in (9.1.25), which are the zeros of $H_\nu^{(2)'}(\beta a)$. From (7.4.45) the zeros of $H_\nu^{(2)'}(\beta a)$ are determined from the zeros of the Airy function $Ai'(\chi_p{}') = 0$. Thus, the zeros of $H_\nu^{(2)'}(\beta a)$ occur at

$$\nu_p' = x - \chi_p'\left(\frac{x}{2}\right)^{1/3} e^{-j\pi/3}$$

Therefore, the evaluation of (9.1.26) in terms of its residues (i.e., where $H_\nu^{(2)'}(\beta a) = 0$) is:

$$H_z = B\pi\sum_{p=1}^{\infty}\frac{H_{\nu_p'}^{(1)'}(\beta a)}{\left.\frac{\partial H_\nu^{(2)'}(\beta a)}{\partial\nu}\right|_{\nu=\nu_p'}}\frac{\cos[\nu_p'(\phi-\pi)]e^{-j\nu_p'\pi/2}}{\sin(\nu_p'\pi)}H_{\nu_p'}^{(2)}(\beta\rho) \tag{9.1.27}$$

The term $H_{\nu_p'}^{(1)'}(\beta a)$ in (9.1.27) is evaluated using the Wronskian in (9.1.18). That is, with $H_{\nu_p'}^{(2)'}(\beta a) = 0$:

$$H_{\nu_p'}^{(1)'}(\beta a)H_{\nu_p'}^{(2)}(\beta a) = \frac{4j}{\pi\beta a} \Rightarrow H_{\nu_p'}^{(1)'}(\beta a) = \frac{4j}{\pi\beta a}\frac{1}{H_{\nu_p'}^{(2)}(\beta a}$$

This expression, using (7.4.43), can be written in terms of the Airy function $Ai(\chi_p')$, namely

$$H_{\nu_p'}^{(1)'}(\beta a) \approx \left(\frac{2}{\beta a}\right)^{2/3}\frac{e^{j\pi/6}}{\pi Ai(\chi_p')}$$

Also, in the denominator of (9.1.27):

$$\left.\frac{\partial H_\nu^{(2)'}(\beta a)}{\partial\nu}\right|_{\nu=\nu_p'} \approx \left(\frac{4}{\beta a}\right)Ai''(\chi_p'') \approx \left(\frac{4}{\beta a}\right)\chi_p' Ai(\chi_p')$$

where Airy differential equation (i.e., $Ai''(\chi_p') = \chi_p' Ai(\chi_p')$) was used. Therefore, (9.1.27) reads:

$$H_z = \frac{B}{2}\left(\frac{\beta a}{2}\right)^{1/3} e^{j\pi/6} \sum_{p=1}^{\infty} \frac{\cos[\nu_p'(\phi - \pi)]e^{-j\nu_p'\pi/2}}{\chi_p'[Ai(\chi_p')]^2 \sin(\nu_p'\pi)} H_{\nu_p'}^{(2)}(\beta\rho) \tag{9.1.28}$$

The series in (9.1.28), like that in (9.1.21), converges with only a few terms in the shadow region.

It is interesting to understand how the diffracted wave reaches the observation point in the shadow region. The physical interpretation is as follows. Consider the expression for E_z in (9.1.21). Using the Debye approximation (see Appendix C), the asymptotic form of $H_{\nu_p}^{(2)}$, valid for large values of $\beta\rho$ with $\beta\rho > |\nu_p|$, is

$$H_{\nu_p}^{(2)}(\beta\rho) \approx \sqrt{\frac{2}{\pi\sqrt{(\beta\rho)^2 - \nu_p^2}}}\, e^{-j\left[\sqrt{(\beta\rho)^2-\nu_p^2} - \nu_p \cos^{-1}(\nu_p/\beta\rho) - \pi/4\right]} \tag{9.1.29}$$

For the dominant terms in the series in (9.1.21), the approximation $\nu_p \approx \beta a$ is used. Then, in (9.1.29):

$$\sqrt{(\beta\rho)^2 - \nu_p^2} \approx \beta\sqrt{\rho^2 - a^2}$$

$$\cos^{-1}\left(\frac{\nu_p}{\beta\rho}\right) \approx \cos^{-1}\left(\frac{a}{\rho}\right)$$

and it follows that

$$H_{\nu_p}^{(2)}(\beta\rho) \approx \sqrt{\frac{2}{\pi\beta\sqrt{\rho^2 - a^2}}}\, e^{-j\left[\beta\sqrt{\rho^2-a^2} - \beta a \cos^{-1}(a/\rho) - \pi/4\right]} \tag{9.1.30}$$

The phase in (9.1.21) describes the propagation characteristics. Using (9.1.22) and (9.1.30), the dominant terms in (9.1.21) are approximated by

$$E_z \approx \frac{Ce^{-j\pi/12}}{\sqrt{2\pi\beta\sqrt{\rho^2 - a^2}}}\left(\frac{\beta a}{2}\right)^{1/3}$$
$$\times \sum_{p=1}^{\infty} \frac{e^{-j\beta\sqrt{\rho^2-a^2}} e^{-j\beta a[(\pi/2-\phi)-\cos^{-1}(a/\rho)]} + e^{-j\beta\sqrt{\rho^2-a^2}} e^{-j\beta a[(\phi+\pi/2)-\cos(a/\rho)]}}{[Ai'(\chi_p)]^2} \tag{9.1.31}$$

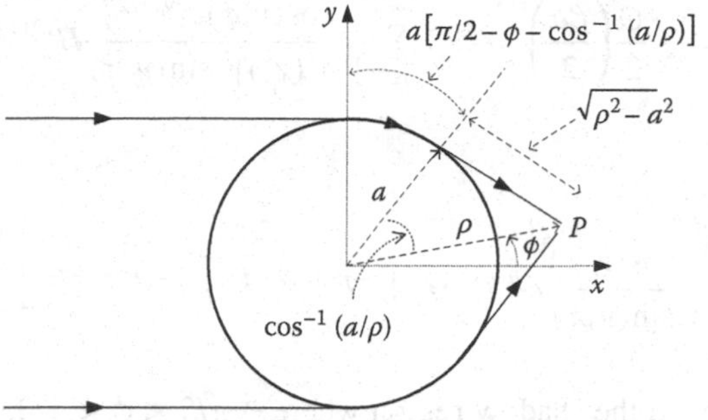

FIGURE 9.2 Creeping waves along the cylinder surface.

The physical interpretation of (9.1.31) is shown in Fig. 9.2. The figure shows two rays incident on the surface at $\pi/2$ and $-\pi/2$, respectively. Details of the upper ray as it travels to the observation point are shown. The ray travels a distance $a[\pi/2 - \phi - \cos^{-1}(a/\rho)]$ along the surface of the cylinder, and then a distance $\sqrt{\rho^2 - a^2}$, after leaving the surface, to the observation point P. A similar analysis applies to the lower ray. These waves that travel along the cylinder surface and produce the field in the shadow region are known as "creeping waves." The imaginary part of the roots ν_p account for the attenuation of these waves.

In order to determine the fields in the illuminated region (i.e., $\pi/2 < \phi < 3\pi/2$) we use the relation:

$$\cos\nu(\phi - \pi) = \sin\nu\phi\sin\nu\pi + \cos\nu\pi\cos\nu\phi = -je^{j\nu\phi}\sin\nu\pi + e^{j\nu\pi}\cos\nu\phi$$

Substituting into (9.1.16) shows that E_z is expressed in terms of two integrals, namely:

$$E_z = I_1 + I_2$$

where

$$I_1 = -\int_{-\infty-j\varepsilon}^{\infty-j\varepsilon} F_\nu(\beta\rho)\,e^{j\nu\phi}d\nu \tag{9.1.32}$$

and

$$I_2 = \frac{1}{j}\int_{-\infty-j\varepsilon}^{\infty-j\varepsilon} F_\nu(\beta\rho)\frac{\cos(\nu\phi)}{\sin(\nu\pi)}e^{j\nu\pi}d\nu \tag{9.1.33}$$

The evaluation of I_2 is similar to that in (9.1.16). That is, close the contour in the lower-half plane and evaluate (9.1.33) at the poles of $H_\nu^{(2)}(ka) = 0$. Hence:

$$I_2 = -\frac{C}{2}\left(\frac{\beta a}{2}\right)^{1/3} e^{j\pi/6} \sum_{p=1}^{\infty} \frac{\cos(\nu_p \phi) e^{j\nu_p \pi/2}}{[Ai'(\chi_p)]^2 \sin(\nu_p \pi)} H_{\nu_p}^{(2)}(\beta\rho) \qquad (9.1.34)$$

Since

$$\frac{\cos(\nu_p\phi)}{\sin(\nu_p\pi)} e^{j\pi\nu_p/2} \approx j\left[e^{-j\nu_p(\pi/2-\phi)} + e^{-j\nu_p(\phi+\pi/2)}\right]$$

(9.1.34) converges in the shadow region where $-\pi/2 < \phi < \pi/2$, since $\text{Im}(\nu_p) < 0$. The contribution of (9.1.34) to the field in the illuminated region is small.

Next, consider I_1 in (9.1.32), which can be evaluated using the method of stationary phase. The term with $H_\nu^{(1)}(\beta\rho)$ in the function $F_\nu(\beta\rho)$ (see (9.1.15)) has no zeros in the lower-half plane. Hence, the stationary points are due to the term in $F_\nu(\beta\rho)$ with $H_\nu^{(2)}(\beta\rho)$. That is,

$$I_1 = \frac{C}{2}\int_{-\infty-j\varepsilon}^{\infty-j\varepsilon} \frac{H_\nu^{(1)}(\beta a)}{H_\nu^{(2)}(\beta a)} H_\nu^{(2)}(\beta\rho) e^{j\nu(\phi-\pi/2)} d\nu \qquad (9.1.35)$$

To determine the stationary points, the Debye approximations for the Hankel functions are used (see Appendix C). Hence, for $\beta a \gg 1$:

$$\frac{H_\nu^{(1)}(\beta a)}{H_\nu^{(2)}(\beta a)} = \begin{cases} -je^{j2\left[\sqrt{(\beta a)^2-\nu^2}-\nu\cos^{-1}(\nu/\beta a)\right]} & (\nu < \beta a) \\ -1 & (\nu > \beta a) \end{cases} \qquad (9.1.36)$$

and for large ρ with $\beta\rho > \nu$:

$$H_\nu^{(2)}(\beta\rho) = \sqrt{\frac{2}{\pi\sqrt{(\beta\rho)^2-\nu^2}}}\, e^{-j\left[\sqrt{(\beta\rho)^2-\nu^2}-\nu\cos^{-1}(\nu/\beta\rho)-\pi/4\right]} \qquad (9.1.37)$$

Using (9.1.36) and (9.1.37), (9.1.35) is expressed in the form:

$$I_1 = F_< + F_>$$

where

$$F_< = \frac{Ce^{-j\pi/4}}{\sqrt{2\pi}}\int_{-\infty}^{\infty} \frac{1}{[(\beta\rho)^2-\nu^2]^{1/4}} e^{j\Psi_1} d\nu \qquad (\nu < \beta a < \beta\rho) \qquad (9.1.38)$$

$$F_> = -\frac{Ce^{j\pi/4}}{\sqrt{2\pi}}\int_{-\infty}^{\infty} \frac{1}{[(\beta\rho)^2-\nu^2]^{1/4}} e^{j\Psi_2} d\nu \qquad (\beta\rho > \nu > \beta a) \qquad (9.1.39)$$

$$\Psi_1 = 2\sqrt{(\beta a)^2 - \nu^2} - \sqrt{(\beta\rho)^2 - \nu^2} - 2\nu \cos^{-1}\left(\frac{\nu}{\beta a}\right) + \nu \cos^{-1}\left(\frac{\nu}{\beta\rho}\right) + \nu\phi - \nu\frac{\pi}{2} \tag{9.1.40}$$

and

$$\Psi_2 = -\sqrt{(\beta\rho)^2 - \nu^2} + \nu \cos^{-1}\left(\frac{\nu}{\beta\rho}\right) + \nu\phi - \nu\frac{\pi}{2} \tag{9.1.41}$$

The stationary point of (9.1.38), from (9.1.40), occurs at

$$\frac{\partial \Psi_1}{\partial \nu} \approx -2\cos^{-1}\left(\frac{\nu}{\beta a}\right) + \frac{\pi}{2} + \phi - \frac{\pi}{2} = 0 \Rightarrow \nu = \nu_1 \approx \beta a \cos\left(\frac{\phi}{2}\right)$$

where we used the relation

$$\frac{d}{dx}\cos^{-1}(ax) = \frac{-a}{\sqrt{1 - a^2x^2}}$$

The stationary point of (9.1.39), from (9.1.41), occurs at

$$\frac{\partial \Psi_2}{\partial \nu} \approx \cos^{-1}\left(\frac{\nu}{\beta\rho}\right) + \phi - \frac{\pi}{2} = 0 \Rightarrow \nu = \nu_2 \approx \beta\rho \cos\left(\frac{\pi}{2} - \phi\right) = \beta\rho \sin\phi$$

The stationary points along the path of integration are shown in Fig. 9.3. The poles of the integrand, which are the zeros of $H_\nu^{(2)}(\beta a)$ are also shown. The evaluation of (9.1.38) and (9.1.39) uses the result in (7.2.3) which requires (9.1.40) evaluated at ν_1 and (9.1.41) at ν_2, as well as their second derivatives. That is,

$$\Psi_1(\nu_1) \approx -\beta\left(\rho - 2a \sin\frac{\phi}{2}\right)$$

$$\Psi_2(\nu_2) \approx -\beta\rho \cos\phi$$

$$\left.\frac{\partial^2 \Psi_1}{\partial \nu^2}\right|_{\nu=\nu_1} \approx \frac{\partial}{\partial \nu}\left[-2\cos^{1}\left(\frac{\nu}{\beta a}\right) + \phi\right]\Bigg|_{\nu=\nu_1} \approx \frac{2}{\beta a \sin\frac{\phi}{2}}$$

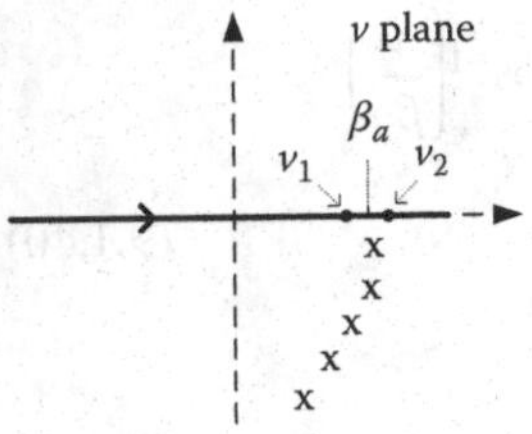

FIGURE 9.3 The stationary points along the path of integration.

$$\left.\frac{\partial^2 \Psi_2}{\partial \nu^2}\right|_{\nu=\nu_2} \approx \frac{\partial}{\partial \nu}\left[\cos^{-1}\left(\frac{\nu}{\beta\rho}\right)+\phi-\frac{\pi}{2}\right]\Bigg|_{\nu=\nu_2} \approx \frac{-1}{\beta\rho\cos\phi}$$

The path of integration is modified to pass through the stationary points at ν_1 and ν_2, as shown in Fig. 9.3. Hence, the evaluation of (9.1.38) is

$$F_< \approx \frac{Ce^{-j\pi/4}}{\sqrt{2\pi}}\frac{1}{[(\beta\rho)^2-\nu_1^2]^{1/4}}\sqrt{\frac{2\pi}{|\Psi_1''|^2}}\,e^{j\pi/4}e^{j\Psi_1(\nu_1)} \approx C\sqrt{\frac{a\sin\frac{\phi}{2}}{2\rho}}\,e^{-j\beta\left(\rho-2a\sin\frac{\phi}{2}\right)}$$

and the evaluation of (9.1.39) produces a field of the form $e^{-j\beta\rho\cos\phi}$. Adding $F_<$ and $F_>$, the behavior of the incident and reflected field is of the form

$$e^{-j\beta\rho\cos\phi}-\sqrt{\frac{a\sin\frac{\phi}{2}}{2\rho}}\,e^{-j\beta\left(\rho-2a\sin\frac{\phi}{2}\right)}$$

It is convenient in the illuminated region to express ϕ in terms of the angle ϕ_c, where $\phi_c = \pi - \phi$. The angle ϕ_c is shown in Fig. 9.4 and its range in the illuminated region is $-\pi/2 < \phi_c < \pi/2$. Hence, the behavior of the field in terms of ϕ_c reads:

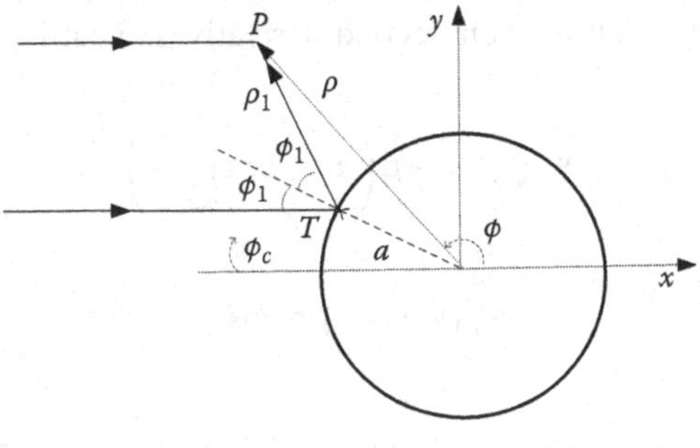

FIGURE 9.4 The incident and reflected ray in the illuminated region.

$$e^{j\beta\rho\cos\phi_c} - \sqrt{\frac{a\cos\frac{\phi_c}{2}}{2\rho}}\, e^{-j\beta\left(\rho - 2a\cos\frac{\phi_c}{2}\right)} \tag{9.1.42}$$

Equation (9.1.42) has the following physical significance. The first term represents the incident ray on the cylinder and the second term represents the reflected ray in the illuminated region, as shown in Fig. 9.4. An incident ray at point T has a phase of $e^{j\beta a\cos\phi_1}$. The ray reflects at the same angle ϕ_1, and travels a distance ρ_1 to the observation point P, adding to the phase a factor of $e^{-j\beta\rho_1}$. The phase of the reflected ray at P is $e^{-j\beta(\rho_1 - a\cos\phi_1)}$. At large distances from the cylinder:

$$\rho_1 \approx \rho - a\cos(\phi_c - \phi_1)$$

and $\phi_1 \approx \phi_c/2$ Thus, the phase at P is

$$e^{-j\beta(\rho_1 - a\cos\phi_1)} = e^{-j\beta(\rho - 2a\cos\phi_1)}$$

in agreement with (9.1.42).

The amplitude of the reflected wave is determined by the amplitude factor in the second term in (9.1.42). This factor is in agreement with the amplitude of the reflected wave in geometrical optics, where conservation of energy is used to determine the amplitude of the reflected wave.

9.2 DIFFRACTION BY A LOSSY DIELECTRIC CYLINDER

The case of a lossy dielectric cylinder is illustrated in Fig. 9.5. In this case, the TM^z fields inside the cylinder (denoted by Region 2) must be considered, and the appropriate boundary conditions applied at $\rho = a$.

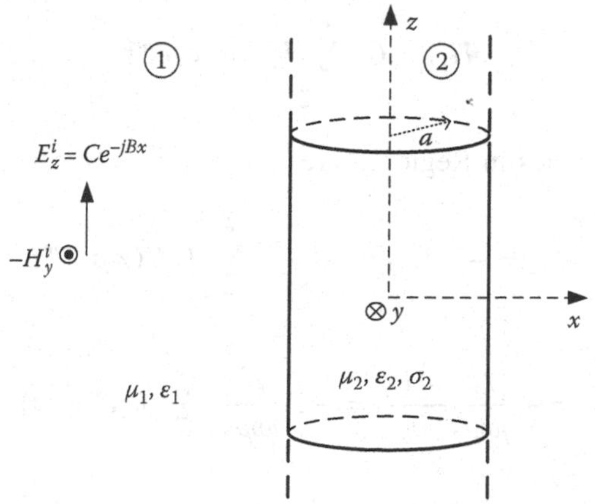

FIGURE 9.5 Diffraction of a plane wave by a lossy dielectric cylinder.

The electric field outside the cylinder (Region 1) is composed of the incident and diffracted field. That is,

$$E_{z1} = C \sum_{n=-\infty}^{\infty} \left[j^{-n} J_n(\beta_1 \rho) + a_n H_n^{(2)}(\beta_1 \rho)\right] e^{jn\phi} \tag{9.2.1}$$

where $\beta_1 = \omega\sqrt{\mu_1 \varepsilon_1}$. The H_ϕ component of the field is also needed to match the boundary condition, namely

$$H_{\phi 1} = \frac{1}{j\omega\mu_1}\frac{\partial E_{z1}}{\partial \rho} = C\frac{\beta_1}{j\omega\mu_1}\sum_{n=-\infty}^{\infty} \left[j^{-n} J_n'(\beta_1 \rho) + a_n H_n^{(2)\prime}(\beta_1 \rho)\right] e^{jn\phi} \tag{9.2.2}$$

In addition to H_ϕ there is an H_ρ component, given by

$$H_{\rho 1} = -\frac{1}{j\omega\mu_1}\frac{1}{\rho}\frac{\partial E_{z1}}{\partial \phi} = -C\frac{1}{\omega\mu_1 \rho}\sum_{n=-\infty}^{\infty} n\left[j^{-n} J_n(\beta_1 \rho) + a_n H_n^{(2)}(\beta_1 \rho)\right] e^{jn\phi} \tag{9.2.3}$$

Inside the cylinder the z component of the electric Hertz vector satisfies

$$\left[\frac{1}{\rho}\frac{\partial}{\partial \rho}\left(\rho\frac{\partial}{\partial \rho}\right) + \frac{1}{\rho^2}\frac{\partial^2}{\partial \phi^2} - \gamma_2^2\right]\Pi_{z2}^{\mathrm{e}} = 0 \qquad (\rho \le a) \tag{9.2.4}$$

where

$$\gamma_2^2 = j\omega\mu_2(\sigma_2 + j\omega\varepsilon_2)$$

The solution to (9.2.4) that is finite at $\rho = 0$ is the modified Bessel function $I_n(\gamma_2 \rho)$. Then,

$$\Pi_{z2}^{\mathrm{e}} = C\sum_{n=-\infty}^{\infty} b_n I_n(\gamma_2 \rho) e^{jn\phi}$$

and the associated fields in Region 2 are

$$E_{z2} = -\gamma_2^2 \Pi_{z2}^{\mathrm{e}} = -C\gamma_2^2 \sum_{n=-\infty}^{\infty} b_n I_n(\gamma_2 \rho) e^{jn\phi} \tag{9.2.5}$$

$$H_{\phi 2} = -\frac{\gamma_2^2}{j\omega\mu_2}\frac{\partial \Pi_{z2}^{\mathrm{e}}}{\partial \rho} = -C\frac{\gamma_2^3}{j\omega\mu_2}\sum_{n=-\infty}^{\infty} b_n I_n'(\gamma_2 \rho) e^{jn\phi} \tag{9.2.6}$$

and

$$H_{\rho 2} = \frac{\gamma_2^2}{j\omega\mu_2}\frac{1}{\rho}\frac{\partial \Pi_{z2}^{e}}{\partial \phi} = C\frac{\gamma_2^2}{\omega\mu_2\rho}\sum_{n=-\infty}^{\infty} nb_n I_n(\gamma_2\rho)e^{jn\phi} \tag{9.2.7}$$

The boundary conditions are

$$E_{z1} = E_{z2}|_{\rho=a} \Rightarrow \left[j^{-n}J_n(\beta_1 a) + a_n H_n^{(2)}(\beta_1 a)\right] = -\gamma_2^2 b_n I_n(\gamma_2 a)$$

and

$$H_{\phi 1} = H_{\phi 2}|_{\rho=a} \Rightarrow \frac{\beta_1}{j\omega\mu_1}\left[j^{-n}J_n'(\beta_1 a) + a_n H_n^{(2)\prime}(\beta_1 a)\right] = -\frac{\gamma_2^3}{j\omega\mu_2} b_n I_n'(\gamma_2 a)$$

whose solution for a_n and b_n are

$$a_n = j^{-n}\frac{J_n'(\beta_1 a)I_n(\gamma_2 a) - \frac{\mu_1\gamma_2}{\mu_2\beta_1}J_n(\beta_1 a)I_n'(\gamma_2 a)}{\frac{\mu_1\gamma_2}{\mu_2\beta_1}H_n^{(2)}(\beta_1 a)I_n'(\gamma_2 a) - H_n^{(2)\prime}(\beta_1 a)I_n(\gamma_2 a)} \tag{9.2.8}$$

and

$$b_n = -\frac{2j^{-n+1}}{\pi a\beta_1\gamma_2^2}\frac{1}{\frac{\mu_1\gamma_2}{\mu_2\beta_1}H_n^{(2)}(\beta_1 a)I_n'(\gamma_2 a) - H_n^{(2)\prime}(\beta_1 a)I_n(\gamma_2 a)} \tag{9.2.9}$$

where the Wronskian in (9.1.4) was used in the calculation of b_n. The substitution of (9.2.8) into (9.2.1) to (9.2.7) concludes the formal solution of this problem.

If Region 2 is lossless, the propagation constant in the dielectric is $\beta_2 = \omega\sqrt{\mu_2\varepsilon_2}$. Then, with $\gamma_2 = j\beta_2$ in (9.2.8) and (9.2.9), and since

$$I_n(jz) = j^{-n}J_n(z)$$
$$I_n{}'(jz) = j^{-n-1}J_n{}'(z)$$

it follows that for the lossless case, the constants are

$$a_n = j^{-n}\frac{J_n'(\beta_1 a)J_n(\beta_2 a) - \frac{\mu_1\beta_2}{\mu_2\beta_1}J_n(\beta_1 a)J_n'(\beta_2 a)}{\frac{\mu_1\beta_2}{\mu_2\beta_1}H_n^{(2)}(\beta_1 a)J_n'(\beta_2 a) - H_n^{(2)\prime}(\beta_1 a)J_n(\beta_2 a)}$$

and

$$b_n = \frac{2j}{\pi a \beta_1 \beta_2^2} \frac{1}{\frac{\mu_1 \beta_2}{\mu_2 \beta_1} H_n^{(2)}(\beta_1 a) J_n'(\beta_2 a) - H_n^{(2)\prime}(\beta_1 a) J_n(\beta_2 a)}$$

If Region 1 is air, the term $\mu_1\beta_2/\mu_2\beta_1 = \mu_o\beta_2/\mu_2\beta_o$ can be written as $\sqrt{\varepsilon_{r2}/\mu_{r2}}$. For example, from (9.2.5), the E_{z2} field is

$$E_{z2} = C\frac{2}{\pi a \beta_o} \sum_{n=-\infty}^{\infty} \frac{j^{-n+1}}{\sqrt{\frac{\varepsilon_{r2}}{\mu_{r2}}} H_n^{(2)}(\beta_0 a) J_n'(\beta_2 a) - H_n^{(2)\prime}(\beta_0 a) J_n(\beta_2 a)} J_n(\beta_2 \rho) e^{jn\phi}$$

9.3 CONDUCTING CYLINDER AND AN INFINITE-LENGTH CURRENT SOURCE

The geometry of the problem is shown in Fig. 9.6. The sinusoidal current is assumed to be of constant amplitude and located at (ρ_o, ϕ_o). Since the length of the line current is infinite, the fields do not vary with z.

The z component of the electric Hertz vector can be used to formulate the problem and obtain the fields for $\rho \geq a$. From (1.7.5):

$$\left[\frac{1}{\rho}\frac{\partial}{\partial \rho}\left(\rho \frac{\partial}{\partial \rho}\right) + \frac{1}{\rho^2}\frac{\partial^2}{\partial \phi^2} + \beta^2\right] \Pi_z^e(\rho, \phi) = -\frac{J_{e,z}(\rho, \phi)}{j\omega\varepsilon} = \frac{-I}{j\omega\varepsilon}\frac{\delta(\rho - \rho_o)\delta(\phi - \phi_o)}{\rho} \tag{9.3.1}$$

where $\beta = \omega\sqrt{\mu\varepsilon}$ and the fields follow from (4.4.11) to (4.4.16).

Since $E_z = \beta^2 \Pi_z^e$, the boundary condition is $\Pi_z^e = 0$ at $\rho = a$. Hence, the Green function associated with (9.3.1) that satisfies the boundary condition $G_< = 0$ at $\rho = a$ is given by (5.6.51), which for $\rho \geq \rho'$is

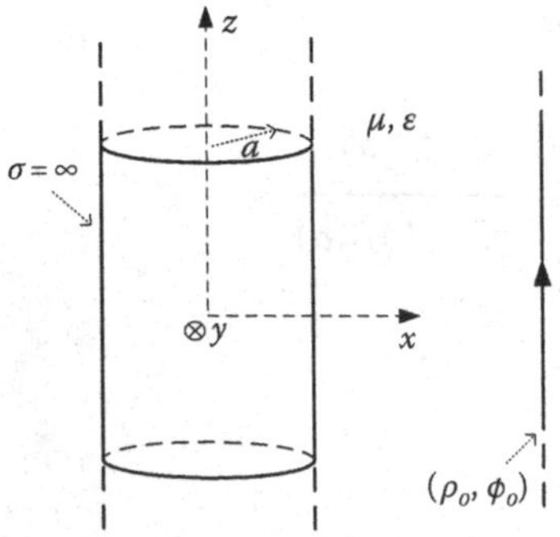

FIGURE 9.6 Conducting cylinder with a current source excitation.

$$G_>(\rho, \phi, \rho', \phi') = -\frac{j}{4} \sum_{n=-\infty}^{\infty} H_n^{(2)}(\beta\rho)\left[J_n(\beta\rho') - \frac{J_n(\beta a)}{H_n^{(2)}(\beta a)} H_n^{(2)}(\beta\rho')\right] e^{jn(\phi-\phi_o)} \tag{9.3.2}$$

Therefore, the solution to (9.3.1) for $\rho \geq \rho'$ is

$$\begin{aligned} \Pi_z^e &= \frac{1}{j\omega\varepsilon} \int_{\phi'=0}^{2\pi} \int_{\rho'=0}^{\infty} J_{e,z}(\rho', \phi')\, G_>(\rho, \phi, \rho', \phi')\rho' d\rho' d\phi' \\ &= \frac{I}{j\omega\varepsilon} \int_{\phi'=0}^{2\pi} \int_{\rho'=0}^{\infty} \frac{\delta(\rho' - \rho_o)\delta(\phi' - \phi_o)}{\rho'} G_>(\rho, \phi, \rho', \phi')\rho' d\rho' d\phi' \end{aligned} \tag{9.3.3}$$

Substituting (9.3.2) into (9.3.3):

$$\Pi_z^e = -\frac{I}{4\omega\varepsilon} \sum_{n=-\infty}^{\infty} H_n^{(2)}(\beta\rho)\left[J_n(\beta\rho_o) - \frac{J_n(\beta a)}{H_n^{(2)}(\beta a)} H_n^{(2)}(\beta\rho_o)\right] e^{jn(\phi-\phi_o)} \quad (\rho \geq \rho_o) \tag{9.3.4}$$

and for $a \leq \rho \leq \rho_o$ interchange ρ and ρ_o in (9.3.4).

From (4.4.16), the electric field is

$$E_z = \beta^2 \Pi_z^e = -\frac{\beta^2 I}{4\omega\varepsilon} \sum_{n=-\infty}^{\infty} H_n^{(2)}(\beta\rho)\left[J_n(\beta\rho_o) - \frac{J_n(\beta a)}{H_n^{(2)}(\beta a)} H_n^{(2)}(\beta\rho_o)\right] e^{jn(\phi-\phi_o)} \quad (\rho \geq \rho_o) \tag{9.3.5}$$

and from (4.4.11) and (4.4.12) the other field components are

$$\begin{aligned} H_\rho = \frac{j\omega\varepsilon}{\rho} \frac{\partial \Pi_z^e}{\partial \phi} = -\frac{I}{4\rho} \sum_{n=-\infty}^{\infty} nH_n^{(2)}(\beta\rho)\Bigg[J_n(\beta\rho_o) \\ - \frac{J_n(\beta a)}{H_n^{(2)}(\beta a)} H_n^{(2)}(\beta\rho_o)\Bigg] e^{jn(\phi-\phi_o)} \qquad (\rho \geq \rho_o) \end{aligned}$$

and

$$\begin{aligned} H_\phi = -j\omega\varepsilon \frac{\partial \Pi_z^e}{\partial \rho} = \frac{j\beta I}{4} \sum_{n=-\infty}^{\infty} H_n^{(2)\prime}(\beta\rho)\Bigg[J_n(\beta\rho_o) \\ - \frac{J_n(\beta a)}{H_n^{(2)}(\beta a)} H_n^{(2)}(\beta\rho_o)\Bigg] e^{jn(\phi-\phi_o)} \qquad (\rho \geq \rho_o) \end{aligned} \tag{9.3.6}$$

For $a \leq \rho \leq \rho_o$ interchange ρ and ρ_o in the expressions above.

In the far field, let

$$H_n^{(2)}(\beta\rho) \approx \sqrt{\frac{2}{\pi\beta\rho}}\, j^n e^{-j(\beta\rho-\pi/4)}$$

to obtain

$$E_z \approx -\frac{\beta^2 I}{4\omega\varepsilon}\sqrt{\frac{2}{\pi\beta\rho}}\, e^{-j(\beta\rho-\pi/4)} \sum_{n=-\infty}^{\infty} j^n \left[J_n(\beta\rho_o) - \frac{J_n(\beta a)}{H_n^{(2)}(\beta a)} H_n^{(2)}(\beta\rho_o)\right] e^{jn(\phi-\phi_o)}$$

and $H_\phi = -E_z/\eta$.

The current density on the surface of the cylinder is

$$\mathbf{J}_{e,s} = \hat{\mathbf{n}} \times \mathbf{H}|_{\rho=a} = \hat{\mathbf{a}}_\rho \times [H_\rho \hat{\mathbf{a}}_\rho + H_\phi \hat{\mathbf{a}}_\phi]|_{\rho=a} = H_\phi \hat{\mathbf{a}}_z|_{\rho=a}$$

From (9.3.6), using the expression for $a \le \rho \le \rho_o$ to evaluate H_ϕ, we obtain

$$\begin{aligned} J_{e,s} &= \frac{j\beta I}{4} \sum_{n=-\infty}^{\infty} H_n^{(2)}(\beta\rho_o)\left[J_n'(\beta a) - \frac{J_n(\beta a)}{H_n^{(2)}(\beta a)} H_n^{(2)\prime}(\beta a)\right] e^{jn(\phi-\phi_o)} \\ &= -\frac{I}{2\pi a} \sum_{n=-\infty}^{\infty} \frac{H_n^{(2)}(\beta\rho_o)}{H_n^{(2)}(\beta a)} e^{jn(\phi-\phi_o)} \end{aligned}$$

where the Wronskian in (9.1.4) was used.

This problem can also be analyzed using (2.2.11). That is, the potential A_z produced by the electric line source, denoted by the incident potential A_z^i is

$$A_z^i = -\frac{j\mu I}{4} H_0^{(2)}(\beta|\boldsymbol{\rho} - \boldsymbol{\rho}'|)$$

and

$$E_z^i = -j\omega A_z^i = -\frac{\omega\mu I}{4} H_0^{(2)}(\beta|\boldsymbol{\rho} - \boldsymbol{\rho}'|)$$

This is the incident electric field in the absence of the conducting dielectric. Using the addition theorem for Hankel functions in (5.6.44), E_z^i can be expressed as

$$E_z^i = -\frac{\omega\mu I}{4} \sum_{n=-\infty}^{\infty} J_n(\beta\rho) H_n^{(2)}(\beta\rho_o) e^{jn(\phi-\phi_o)} \qquad (\rho \le \rho_o)$$

and the diffracted field is given by

$$E_z^d = -\frac{\omega\mu I}{4} \sum_{n=-\infty}^{\infty} c_n H_n^{(2)}(\beta\rho) e^{jn(\phi-\phi_o)}$$

The boundary condition at $\rho = a$ requires that

$$E_z = E_z^i + E_z^d|_{\rho=a} = 0 \Rightarrow J_n(\beta a)H_n^{(2)}(\beta\rho_o) + c_n H_n^{(2)}(\beta a) = 0$$

or

$$c_n = -\frac{J_n(\beta a)}{H_n^{(2)}(\beta a)} H_n^{(2)}(\beta\rho_o)$$

Therefore, the total field for $a \le \rho \le \rho_o$ is given by

$$E_z = -\frac{\omega\mu I}{4} \sum_{n=-\infty}^{\infty} H_n^{(2)}(\beta\rho_o)\left[J_n(\beta\rho) - \frac{J_n(\beta a)}{H_n^{(2)}(\beta a)} H_n^{(2)}(\beta\rho)\right] e^{jn(\phi-\phi_o)} \qquad (a \le \rho \le \rho_o)$$

and for $\rho \ge \rho_o$ interchange ρ and ρ_o. This solution is identical to the one found using Green functions in (9.3.5), since $\omega\mu I/4 = \beta^2 I/4\omega\varepsilon$.

In the case of a magnetic current source the problem can be solved in terms of the magnetic Hertz vector Π_z^m or the F_z vector. The F_z vector satisfies:

$$\left[\frac{1}{\rho}\frac{\partial}{\partial\rho}\left(\rho\frac{\partial}{\partial\rho}\right) + \frac{1}{\rho^2}\frac{\partial^2}{\partial\phi^2} + \beta^2\right]F_z = -\varepsilon J_{m,z}(\rho,\phi) = -\varepsilon I_m \frac{\delta(\rho-\rho_o)\delta(\phi-\phi_o)}{\rho}$$

and the boundary condition, using (4.4.18), is

$$E_\phi = \frac{1}{\varepsilon}\frac{\partial F_z}{\partial\rho}\bigg|_{\rho=a} = 0 \Rightarrow \frac{\partial F_z}{\partial\rho}\bigg|_{\rho=a} = 0$$

The Green function that satisfy $dG_</d\rho = 0$ at $\rho = a$ (i.e., same boundary condition as F_z) is given by (5.6.52), which for $\rho \ge \rho'$:is

$$G_>^m(\rho,\phi,\rho',\phi') = \frac{-j}{4} \sum_{n=-\infty}^{\infty} H_n^{(2)}(\beta\rho)\left[J_n(\beta\rho') - \frac{J_n'(\beta a)}{H_n^{(2)\prime}(\beta a)} H_n^{(2)}(\beta\rho')\right] e^{jn(\phi-\phi_o)}$$

Therefore,

$$\begin{aligned} F_z &= \varepsilon I_m G_>^m(\rho,\phi,\rho_o,\phi_o) \\ &= -\frac{j\varepsilon I_m}{4} \sum_{n=-\infty}^{\infty} H_n^{(2)}(\beta\rho)\left[J_n(\beta\rho_o) - \frac{J'_n(\beta a)}{H_n^{(2)\prime}(\beta a)} H_n^{(2)}(\beta\rho_o)\right] e^{jn(\phi-\phi_o)} \qquad (\rho \ge \rho_o) \end{aligned}$$

From (4.4.22):

$$H_z = \frac{1}{j\omega\mu\varepsilon}\beta^2 F_z = -\frac{\beta^2 I_m}{4\omega\mu}\sum_{n=-\infty}^{\infty} H_n^{(2)}(\beta\rho)\left[J_n(\beta\rho_o) - \frac{J_n'(\beta a)}{H_n^{(2)\prime}(\beta a)}H_n^{(2)}(\beta\rho_o)\right]e^{jn(\phi-\phi_o)} \quad (\rho \geq \rho_o) \tag{9.3.7}$$

and the other field components are

$$E_\rho = -\frac{1}{\varepsilon}\frac{1}{\rho}\frac{\partial F_z}{\partial \phi} = -\frac{I_m}{4\rho}\sum_{n=-\infty}^{\infty} nH_n^{(2)}(\beta\rho)\left[J_n(\beta\rho_o) - \frac{J_n'(\beta a)}{H_n^{(2)\prime}(\beta a)}H_n^{(2)}(\beta\rho_o)\right]e^{jn(\phi-\phi_o)} \quad (\rho \geq \rho_o)$$

and

$$E_\phi = \frac{1}{\varepsilon}\frac{\partial F_z}{\partial \rho} = -j\frac{\beta I_m}{4}\sum_{n=-\infty}^{\infty} H_n^{(2)\prime}(\beta\rho)\left[J_n(\beta\rho_o) - \frac{J_n'(\beta a)}{H_n^{(2)\prime}(\beta a)}H_n^{(2)}(\beta\rho_o)\right]e^{jn(\phi-\phi_o)} \quad (\rho \geq \rho_o)$$

For $a \leq \rho \leq \rho_o$ interchange ρ and ρ_o in the expressions above.

In the far field, using the asymptotic form of $H_n^{(2)}(\beta\rho)$, (9.3.7) reads

$$H_z \approx -\frac{\beta^2 I_m}{4\omega\mu}\sqrt{\frac{2}{\pi\beta\rho}}\,e^{-j(\beta\rho-\pi/4)}\sum_{n=-\infty}^{\infty} j^n\left[J_n(\beta\rho_o) - \frac{J_n'(\beta a)}{H_n^{(2)\prime}(\beta a)}H_n^{(2)}(\beta\rho_o)\right]e^{jn(\phi-\phi_o)}$$

The current density on the surface of the cylinder is:

$$\mathbf{J}_s = \hat{\mathbf{n}} \times \mathbf{H}|_{\rho=a} = \hat{\mathbf{a}}_\rho \times H_z\hat{\mathbf{a}}_z|_{\rho=a} = -H_z\hat{\mathbf{a}}_\phi|_{\rho=a}$$

From (9.3.7), using the expression for $a \leq \rho \leq \rho_o$ gives:

$$J_\phi = -\frac{\beta^2 I_m}{4\omega\mu}\sum_{n=-\infty}^{\infty} H_n^{(2)}(\beta\rho_o)\left[J_n(\beta a) - \frac{J_n'(\beta a)}{H_n^{(2)\prime}(\beta a)}H_n^{(2)}(\beta\rho_o)\right]e^{jn(\phi-\phi_o)}$$

$$= -j\frac{I_m}{2\pi a\eta}\sum_{n=-\infty}^{\infty}\frac{H_n^{(2)}(\beta\rho_o)}{H_n^{(2)\prime}(\beta a)}e^{jn(\phi-\phi_o)}$$

where the Wronskian in (9.1.4) and $\omega\mu/\beta = \eta$ were used.

9.3.1 Conducting Cylinder and a Dipole Source

The geometry of the problem is shown in Fig. 9.7. The electric dipole is assumed to be in the z direction. The problem can be formulated in term of the electric Hertz vector. That is,

$$\left[\frac{1}{\rho}\frac{\partial}{\partial\rho}\left(\rho\frac{\partial}{\partial\rho}\right)+\frac{1}{\rho^2}\frac{\partial^2}{\partial\phi^2}+\frac{\partial^2}{\partial z^2}+k^2\right]\Pi_z^{\mathrm{e}}=-\frac{J_{e,z}(\rho,\phi)}{j\omega\varepsilon}$$
$$=-\frac{(I\Delta L)}{j\omega\varepsilon}\frac{\delta(\rho-\rho_o)\delta(\phi-\phi_o)\delta(z-z_o)}{\rho} \tag{9.3.8}$$

where $k=\omega\sqrt{\mu\varepsilon}$. The boundary condition follows from

$$E_z=0|_{\rho=a}\Rightarrow\Pi_z^{\mathrm{e}}=0|_{\rho=a}$$

Fourier transforming the z variable in (9.3.8), we obtain:

$$\left[\frac{1}{\rho}\frac{\partial}{\partial\rho}\left(\rho\frac{\partial}{\partial\rho}\right)+\frac{1}{\rho^2}\frac{\partial^2}{\partial\phi^2}+\tau^2\right]\Pi_z^{e}(\rho,\phi,\alpha)=-\frac{(I\Delta L)}{j\omega\varepsilon}e^{-j\alpha z_o}\frac{\delta(\rho-\rho_o)\delta(\phi-\phi_o)}{\rho} \tag{9.3.9}$$

where $\tau^2=k^2-\alpha^2$, and the inverse transform is

$$\Pi_z^{e}(\rho,\phi,z)=\frac{1}{2\pi}\int_{-\infty}^{\infty}\Pi_z^{e}(\rho,\phi,\alpha)e^{j\alpha z}d\alpha$$

The Green function associated with (9.3.9) satisfy

$$\left[\frac{1}{\rho}\frac{\partial}{\partial\rho}\left(\rho\frac{\partial}{\partial\rho}\right)+\frac{1}{\rho^2}\frac{\partial^2}{\partial\phi^2}+\tau^2\right]G(\rho,\phi,\rho',\phi')=-\frac{\delta(\rho-\rho')\delta(\phi-\phi')}{\rho} \tag{9.3.10}$$

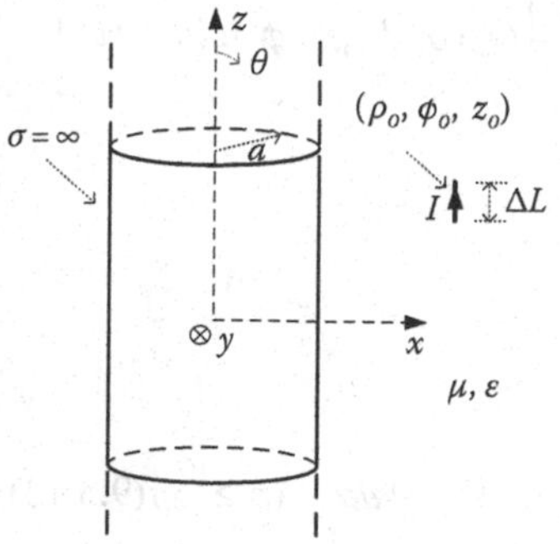

FIGURE 9.7 Conducting cylinder and an electric dipole source.

subject to the same boundary conditions that Π_z^e satisfy.

Since

$$\delta(\phi - \phi_o) = \frac{1}{2\pi} \sum_{n=-\infty}^{\infty} e^{jn(\phi - \phi_o)}$$

the Green function solution is of the form

$$G(\rho, \phi, \rho', \phi') = \frac{1}{2\pi} \sum_{n=-\infty}^{\infty} G(\rho, \rho') e^{jn(\phi - \phi')}$$

where, upon substitution into (9.3.10), $G(\rho, \rho')$ satisfies

$$\left[\frac{1}{\rho}\frac{\partial}{\partial \rho}\left(\rho \frac{\partial}{\partial \rho}\right) + \tau^2 - \frac{n^2}{\rho^2}\right] G(\rho, \rho') = -\frac{\delta(\rho - \rho')}{\rho} \tag{9.3.11}$$

The solution to (9.3.11) that satisfies $G_< = 0$ at $\rho = a$ was analyzed in Section 5.6 (see (5.6.51)), which for $\rho \geq \rho'$ and in terms of τ is

$$G_>(\rho, \rho') = -\frac{j\pi}{2} \sum_{n=-\infty}^{\infty} H_n^{(2)}(\tau\rho)\left[J_n(\tau\rho') - \frac{J_n(\tau a)}{H_n^{(2)}(\tau a)} H_n^{(2)}(\tau\rho')\right]$$

Therefore,

$$G_>(\rho, \phi, \rho', \phi') = -\frac{j}{4} \sum_{n=-\infty}^{\infty} H_n^{(2)}(\tau\rho)\left[J_n(\tau\rho') - \frac{J_n(\tau a)}{H_n^{(2)}(\tau a)} H_n^{(2)}(\tau\rho')\right] e^{jn(\phi - \phi')} \tag{9.3.12}$$

Using (9.3.12), the solution to Π_z^e in (9.3.9) for $\rho \geq \rho_o$ is

$$\begin{aligned}
&\Pi_z^e(\rho, \phi, \alpha) \\
&= \frac{(I\Delta L)}{j\omega\varepsilon} e^{-j\alpha z_o}\left[\int_{\rho'=0}^{\infty}\int_{\phi'=0}^{2\pi} \frac{\delta(\rho' - \rho_o)\delta(\phi' - \phi_o)}{\rho'} G_>(\rho, \phi, \rho', \phi')\rho' d\rho' d\phi'\right] \\
&= \frac{(I\Delta L)}{j\omega\varepsilon} e^{-j\alpha z_o} G_>(\rho, \phi, \rho_o, \phi_o) \qquad (\rho \geq \rho_o)
\end{aligned}$$

and forming the inverse Fourier transform gives

$$\Pi_z^e(\rho, \phi, z) = \frac{(I\Delta L)}{j\omega\varepsilon}\frac{1}{2\pi}\int_{-\infty}^{\infty} G_>(\rho, \phi, \rho_o, \phi_o) e^{j\alpha(z - z_o)} d\alpha \quad (\rho \geq \rho_o) \tag{9.3.13}$$

Substituting (9.3.12) into (9.2.13) gives for $\rho \geq \rho_o$:

$$\Pi_z^e = -\frac{(I\Delta L)}{8\pi\omega\varepsilon} \sum_{n=-\infty}^{\infty} e^{jn(\phi-\phi_o)} \int_{-\infty}^{\infty} H_n^{(2)}(\tau\rho)\left[J_n(\tau\rho_o) - \frac{J_n(\tau a)}{H_n^{(2)}(\tau a)} H_n^{(2)}(\tau\rho_o)\right] e^{j\alpha(z-z_o)} d\alpha \tag{9.3.14}$$

and the fields follow from (4.4.11) to (4.4.16).

The integral in (9.3.14) is evaluated in the far field (i.e., $k\rho \gg 1$) using the method of steepest descent. Letting $\alpha = k\cos\beta$ it follows from (7.4.66) that:

$$\int_{-\infty}^{\infty} F(\tau) H_n^{(2)}(\tau\rho) e^{j\alpha z} d\alpha \approx 2j^{n+1} F(k\sin\theta)\frac{e^{-jkr}}{r}$$

Hence, the far field evaluation of (9.3.14) is

$$\Pi_z^e \approx \frac{-j(I\Delta L)}{4\pi\omega\varepsilon}\frac{e^{-jkr}}{r} e^{-jkz_o\cos\theta} \sum_{n=-\infty}^{\infty} j^n \left[J_n(k\rho_o\sin\theta) - \frac{J_n(ka\sin\theta)}{H_n^{(2)}(ka\sin\theta)} H_n^{(2)}(k\rho_o\sin\theta)\right] e^{jn(\phi-\phi_o)}$$

In the case of a Hertzian dipole radiator with $r \gg z_o$ the term $e^{-jkz_o\cos\theta}$ is neglected. However, this term must be retained if the fields from a wire antenna is evaluated.

The electric and magnetic fields are

$$\begin{aligned} E_z &= k^2 \Pi_z^e \\ &\approx -j\frac{\omega\mu(I\Delta L)}{4\pi}\frac{e^{-jkr}}{r} e^{-jkz_o\cos\theta} \sum_{n=-\infty}^{\infty} j^n \\ &\times \left[J_n(k\rho_o\sin\theta) - \frac{J_n(ka\sin\theta)}{H_n^{(2)}(ka\sin\theta)} H_n^{(2)}(k\rho_o\sin\theta)\right] e^{jn(\phi-\phi_o)} \end{aligned}$$

and

$$\begin{aligned} H_\phi &= -j\omega\varepsilon\frac{\partial \Pi_z^e}{\partial\rho} \\ &\approx -\frac{(I\Delta L)k}{4\pi}\frac{e^{-jkr}}{r} e^{-jkz_o\cos\theta} \sin\theta \sum_{n=-\infty}^{\infty} j^n \\ &\times \left[J_n(k\rho_o\sin\theta) - \frac{J_n(ka\sin\theta)}{H_n^{(2)}(ka\sin\theta)} H_n^{(2)}(k\rho_o\sin\theta)\right] e^{jn(\phi-\phi_o)} \end{aligned}$$

where we used:

$$\frac{\partial}{\partial \rho}\frac{e^{-jkr}}{r} \approx \frac{1}{r}\frac{\partial r}{\partial \rho}\frac{\partial}{\partial r}e^{-jkr} \approx -jk \sin\theta \frac{e^{-jkr}}{r}$$

9.4 LOSSY DIELECTRIC CYLINDER AND AN INFINITE-LINE CURRENT SOURCE

The geometry of the problem is shown in Fig. 9.8. The current source is located in the air region (i.e., Region 1) and the lossy dielectric occupies Region 2. The **A** vector will be used to formulate the problem. The z component of the **A** vector in each region satisfy

$$\left[\frac{1}{\rho}\frac{\partial}{\partial \rho}\left(\rho \frac{\partial}{\partial \rho}\right) + \frac{1}{\rho^2}\frac{\partial^2}{\partial \phi^2} + k_1^2\right] A_{z1}(\rho, \phi) = -\mu_o J_{e,z}(\rho, \phi)$$

$$= -\mu_o I \frac{\delta(\rho - \rho_o)\delta(\phi - \phi_o)}{\rho} \quad (9.4.1)$$

and

$$\left[\frac{1}{\rho}\frac{\partial}{\partial \rho}\left(\rho \frac{\partial}{\partial \rho}\right) + \frac{1}{\rho^2}\frac{\partial^2}{\partial \phi^2} + k_2^2\right] A_{z2}(\rho, \phi) = 0 \quad (9.4.2)$$

where $k_1 = k_o = \omega\sqrt{\mu_o \varepsilon_o}$, and $k_2 = k_o n_2$. With $\mu_2 = \mu_o$:

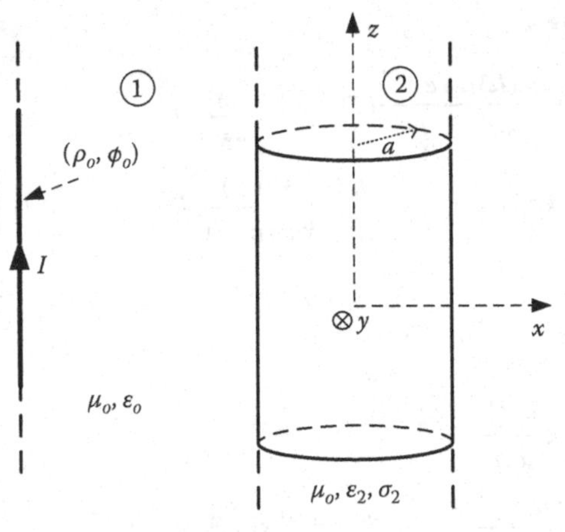

FIGURE 9.8 Lossy cylinder and an infinite-line current source.

$$n_2 = \sqrt{\varepsilon_{r2} - \frac{j\sigma_2}{\omega\varepsilon_o}}$$

The fields follow from (4.4.11) to (4.4.16). That is,

$$E_z = -j\omega A_z$$
$$H_\rho = \frac{1}{\mu_o}\frac{1}{\rho}\frac{\partial A_z}{\partial \phi}$$
$$H_\phi = -\frac{1}{\mu_o}\frac{\partial A_z}{\partial \rho}$$

The boundary conditions are the continuity of E_z and H_ϕ at $\rho = a$. Then, from the above relations

$$E_{z1} = E_{z2}\big|_{\rho=a} \Rightarrow A_{z1} = A_{z2}\big|_{\rho=a} \tag{9.4.3}$$

and

$$H_{\phi 1} = H_{\phi 2}\Big|_{\rho=a} \Rightarrow \frac{\partial A_{z1}}{\partial \rho} = \frac{\partial A_{z2}}{\partial \rho}\Big|_{\rho=a} \tag{9.4.4}$$

Since

$$\delta(\phi - \phi_0) = \frac{1}{2\pi}\sum_{n=-\infty}^{\infty} e^{jn(\phi-\phi_o)}$$

we let the solution to (9.4.1) be of the form

$$A_{z1} = \frac{1}{2\pi}\sum_{n=-\infty}^{\infty} F_1(\rho, \rho_o)\, e^{jn(\phi-\phi_o)} \tag{9.4.5}$$

Substituting these relations into (9.4.1) shows that $F_1(\rho, \rho_o)$ satisfy

$$\left[\frac{1}{\rho}\frac{\partial}{\partial \rho}\left(\rho\frac{\partial}{\partial \rho}\right) + k_1^2 - \frac{n^2}{\rho^2}\right]F_1(\rho, \rho_o) = -\mu_o I\frac{\delta(\rho - \rho_o)}{\rho} \tag{9.4.6}$$

The Green function associated with (9.4.6) satisfies

$$\left[\frac{1}{\rho}\frac{\partial}{\partial \rho}\left(\rho\frac{\partial}{\partial \rho}\right) + k_1^2 - \frac{n^2}{\rho^2}\right]G_1(\rho, \rho') = -\frac{\delta(\rho - \rho')}{\rho} \tag{9.4.7}$$

whose solutions are

$$G_1(\rho, \rho') = \begin{cases} G_{1,<}(\rho, \rho') = b_n J_n(k_1\rho) + c_n H_n^{(2)}(k_1\rho) & (a \le \rho \le \rho') \\ G_{1,>}(\rho, \rho') = d_n H_n^{(2)}(k_1\rho) & (\rho \ge \rho') \end{cases} \tag{9.4.8}$$

Therefore, the solution to (9.4.6) is

$$F_1(\rho, \rho_o) = \mu_o I \int_0^\infty \frac{\delta(\rho' - \rho_o)}{\rho'} G_1(\rho, \rho')\rho' d\rho' = \mu_o I G_1(\rho, \rho_o)$$

and from (9.4.5)

$$A_{z1} = \frac{\mu_o I}{2\pi} \sum_{n=-\infty}^{\infty} G_1(\rho, \rho_o) e^{jn(\phi - \phi_o)} \tag{9.4.9}$$

where $G_1(\rho, \rho_o)$ is the solution $G_1(\rho, \rho')$ evaluated at $\rho' = \rho_o$.

In the lossy dielectric region, the solution to (9.4.2) is of the form:

$$A_{z2} = \frac{\mu_o I}{2\pi} \sum_{n=-\infty}^{\infty} F_2(\rho)\, e^{jn(\phi - \phi_o)} \tag{9.4.10}$$

The factor in (9.4.10) was selected to be the same as that in (9.4.9). Substituting (9.4.10) into (9.4.2) gives:

$$\left[\frac{1}{\rho}\frac{\partial}{\partial \rho}\left(\rho \frac{\partial}{\partial \rho}\right) + \left(k_2^2 - \frac{n^2}{\rho^2}\right)\right] F_2 = 0 \tag{9.4.11}$$

The solution to (9.4.11) that is finite at $\rho = 0$ is

$$F_2 = e_n J_n(k_2\rho)$$

Then,

$$A_{z2} = \frac{\mu_o I}{2\pi} \sum_{n=-\infty}^{\infty} e_n J_n(k_2\rho)\, e^{jn(\phi - \phi_o)} \tag{9.4.12}$$

The boundary conditions in (9.4.3) and (9.4.4) are now applied to (9.4.9) and (9.4.12). Observe that for $G_1(\rho, \rho_o)$ in (9.4.9) one must use $G_{1,<}(\rho, \rho_o)$. Therefore,

$$A_{z1} = A_{z1}|_{\rho=a} \Rightarrow b_n J_n(k_1 a) + c_n H_n^{(2)}(k_1 a) = e_n J_n(k_2 a) \tag{9.4.13}$$

and

$$\frac{\partial A_{z1}}{\partial \rho} = \frac{\partial A_{z1}}{\partial \rho}\bigg|_{\rho=a} \Rightarrow b_n k_1 J_n{}' + c_n k_1 H_n^{(2)}{}'(k_1 a) = e_n k_2 J_n{}'(k_2 a) \quad (9.4.14)$$

Also, the Green functions in (9.4.8) must satisfy the conditions:

$$G_{1,<} = G_{1,>}\bigg|_{\rho=\rho'} \Rightarrow b_n J_n(k_1\rho') + c_n H_n^{(2)}(k_1\rho') = d_n H_n^{(2)}(k_1\rho') \quad (9.4.15)$$

and

$$\frac{\partial G_{1,>}}{\partial \rho} - \frac{\partial G_{1,<}}{\partial \rho}\bigg|_{\rho=\rho'} = -\frac{1}{\rho'} \Rightarrow d_n H_n^{(2)}{}'(k_1\rho') - b_n J_n{}'(k_1\rho') - c_n H_n^{(2)}{}'(k_1\rho') = -\frac{1}{k_1\rho'} \quad (9.4.16)$$

where is ρ' is evaluated at $\rho' = \rho_o$. The constants b_n, c_n, d_n and e_n follow from (9.4.13) to (9.4.16). From (9.4.15) and (9.4.16):

$$b_n = -j\frac{\pi}{2}H_n^{(2)}(k_1\rho_o) \quad (9.4.17)$$

and

$$d_n - c_n = -j\frac{\pi}{2}J_n(k_1\rho_o) \quad (9.4.18)$$

Then, from (9.4.13), (9.4.14) and (9.4.17), we obtain

$$c_n = j\frac{\pi}{2}H_n^{(2)}(k_1\rho_o)\left[\frac{J_n'(k_1 a)J_n(k_2 a) - n_2 J_n(k_1 a)J_n'(k_2 a)}{J_n(k_2 a)H_n^{(2)}{}'(k_1 a) - n_2 H_n^{(2)}(k_1 a)J_n'(k_2 a)}\right]$$

and from (9.4.18):

$$d_n = -j\frac{\pi}{2}\left\{J_n(k_1\rho_o) - H_n^{(2)}(k_1\rho_o)\left[\frac{J_n'(k_1 a)J_n(k_2 a) - n_2 J_n(k_1 a)J_n{}'(k_2 a)}{J_n(k_2 a)H_n^{(2)}{}'(k_1 a) - n_2 H_n^{(2)}(k_1 a)J_n'(k_2 a)}\right]\right\} \quad (9.4.19)$$

where $k_2/k_1 = k_o n_2/k_o = n_2$.

From (9.4.19) and (9.4.8), we obtain $G_{1,>}(\rho, \rho_o)$, and from (9.4.9) for $\rho \geq \rho_o$:

$$A_{z1} = \frac{\mu_o I}{2\pi} \sum_{n=-\infty}^{\infty} G_{1,>}(\rho, \rho_o) e^{jn(\phi-\phi_o)} = \frac{\mu_o I}{2\pi} \sum_{n=-\infty}^{\infty} d_n H_n^{(2)}(k_1\rho) e^{jn(\phi-\phi_o)}$$

$$= -\frac{j\mu_o I}{4} \sum_{n=-\infty}^{\infty} \left\{ J_n(k_1\rho_o) - H_n^{(2)}(k_1\rho_o) \left[\frac{J_n'(k_1 a) J_n(k_2 a) - n_2 J_n(k_1 a) J_n'(k_2 a)}{J_n(k_2 a) H_n^{(2)\prime}(k_1 a) - n_2 H_n^{(2)}(k_1 a) J_n'(k_2 a)} \right] \right\} H_n^{(2)}(k_1\rho) e^{jn(\phi-\phi_o)}$$

$$= -\frac{j\mu_o I}{4} \sum_{n=-\infty}^{\infty} \left\{ J_n(k_1\rho_o) - H_n^{(2)}\left(k_1\rho_o\right) \left[\frac{J_n(k_1 a) - \frac{J_n(k_2 a)}{n_2 J_n'(k_2 a)} J_n'(k_1 a)}{H_n^{(2)}(k_1 a) - \frac{J_n(k_2 a)}{n_2 J_n'(k_2 a)} H_n^{(2)\prime}(k_1 a)} \right] \right\} H_n^{(2)}(k_1\rho) e^{jn(\phi-\phi_o)} \tag{9.4.20}$$

A check on (9.4.20) is to let $\sigma_2 \to \infty$ (i.e., a conducting cylinder). Then, $|n_2| \to \infty$ and (9.4.20) reads:

$$A_{z1} = -\frac{j\mu_o I}{4} \sum_{n=-\infty}^{\infty} \left[J_n(k_1\rho_o) - \frac{J_n(k_1 a)}{H_n^{(2)}(k_1 a)} H_n^{(2)}(k_1\rho_o) \right] H_n^{(2)}(k_1\rho) e^{jn(\phi-\phi_o)}$$

which produces the same fields as those produced by Π_z^e in (9.3.4), since $A_z = j\omega\mu_o \varepsilon \Pi_z^e$.

The solution to this problem is considerably simplified if an impedance boundary condition can be used to determine the fields outside the cylinder. An impedance boundary conditions relates the tangential external field components at $\rho = a$. The impedance boundary condition is the surface impedance seen by a plane wave, or for this problem:

$$Z_s = \frac{E_z}{H_\phi} = -\eta_2$$

which in terms of $\mathbf{A} = A_z \hat{\mathbf{a}}_z$ is

$$H_\phi + \left. \frac{E_z}{\eta_2} \right|_{\rho=a} = 0 \Rightarrow \left. \frac{\partial A_{z1}}{\partial \rho} + \frac{j\omega\mu_o}{\eta_2} A_{z1} \right|_{\rho=a} = 0 \tag{9.4.21}$$

In Region 1, A_{z1} is given by (9.4.9) where the Green function satisfy (9.4.15) and (9.4.16). Hence, the b_n and $d_n - c_n$ relations in (9.4.17) and (9.4.18) apply. Next, using (9.4.9) and (9.4.8) the impedance boundary condition in (9.4.21) shows that

$$b_n k_1 J_n'(k_1 a) + c_n k_1 H_n^{(2)\prime}(k_1 a) + \frac{j\omega\mu_o}{\eta_2}\left[b_n J_n(k_1 a) + c_n H_n^{(2)}(k_1 a)\right] = 0$$

or

$$c_n = -b_n \frac{k_1 J_n'(k_1 a) + \frac{j\omega\mu_o}{\eta_2} J_n(k_1 a)}{k_1 H_n^{(2)\prime}(k_1 a) + \frac{j\omega\mu_o}{\eta_2} H_n^{(2)}(k_1 a)} = j\frac{\pi}{2} H_n^{(2)}(k_1\rho') \frac{k_1 J_n'(k_1 a) + \frac{j\omega\mu_o}{\eta_2} J_n(k_1 a)}{k_1 H_n^{(2)\prime}(k_1 a) + \frac{j\omega\mu_o}{\eta_2} H_n^{(2)}(k_1 a)}$$

where (9.4.17) was used for b_n. Then, d_n follows from (9.4.18). That is,

$$d_n = -\frac{j\pi}{2}\left\{J_n(k_1\rho') - H_n^{(2)}(k_1\rho')\left[\frac{J_n(k_1 a) + \frac{k_1\eta_2}{j\omega\mu_o} J_n'(k_1 a)}{H_n^{(2)}(k_1 a) + \frac{k_1\eta_2}{j\omega\mu_o} H_n^{(2)\prime}(k_1 a)}\right]\right\} \tag{9.4.22}$$

where

$$\frac{k_1\eta_2}{j\omega\mu_o} = \frac{\omega\sqrt{\mu_o\varepsilon_o}}{j\omega\mu_o}\sqrt{\frac{j\omega\mu_o}{\sigma_2 + j\omega\varepsilon_2}} = -j\frac{1}{\sqrt{\varepsilon_{r2} - \frac{j\sigma_2}{\varepsilon_o}}} = -\frac{j}{n_2}$$

Substituting (9.4.22) into (9.4.9) gives for $\rho \geq \rho_o$:

$$A_{z1} = -\frac{j\mu_o I}{4}\sum_{n=-\infty}^{\infty}\left\{J_n(k_1\rho_o) - H_n^{(2)}(k_1\rho_o)\left[\frac{J_n(k_1 a) - \frac{j}{n_2} J_n'(k_1 a)}{H_n^{(2)}(k_1 a) - \frac{j}{n_2} H_n^{(2)\prime}(k_1 a)}\right]\right\} H_n^{(2)}(k_1\rho)\, e^{jn(\phi-\phi_o)} \tag{9.4.23}$$

Comparing (9.4.23) with (9.4.20) shows that they are identical if

$$\frac{j}{n_2} = \frac{1}{n_2}\frac{J_n(k_2 a)}{J_n'(k_2 a)} \tag{9.4.24}$$

If $|k_2 a|$ is large, the cylinder surface resembles that of a plane surface. Let us examine the condition $|k_2 a| = k_o|n_2|a \gg 1$. When $k_o|n_2|a \gg 1$, and observing that $\text{Im}(n_2) < 0$, the asymptotic approximations of the Bessel functions in (9.4.24) show that

$$\frac{J_n(k_2 a)}{J_n'(k_2 a)} \approx \frac{H_n^{(2)}(k_2 a)}{H_n^{(2)\prime}(k_2 a)} \approx \frac{e^{j\pi/4}}{e^{-j\pi/4}} = j$$

and (9.4.24) is satisfied. Hence, if $k_o|n_2|a \gg 1$ the use of a surface impedance to represent the cylindrical surface is appropriate.

9.5 A WEDGE AND AN INFINITE-LENGTH CURRENT SOURCE

A wedge in the presence of an infinite-length current source is shown in Fig. 9.9. The current source extends from $-\infty < z < \infty$. The wedge consists of two metal surfaces having infinite conductivity at an angle Ψ. Since the electric current is in the z direction, the electric Hertz vector component Π_z^e can be used to formulate the problem. From (1.7.5):

$$\left[\frac{1}{\rho}\frac{\partial}{\partial\rho}\left(\rho\frac{\partial}{\partial\rho}\right)+\frac{1}{\rho^2}\frac{\partial^2}{\partial\phi^2}+\beta^2\right]\Pi_z^e(\rho,\phi)=-\frac{J_{e,z}(\rho,\phi)}{j\omega\varepsilon}=\frac{-I}{j\omega\varepsilon}\frac{\delta(\rho-\rho_o)\delta(\phi-\phi_o)}{\rho} \tag{9.5.1}$$

where $\beta=\omega\sqrt{\mu\varepsilon}$ and the fields follow from (4.4.11) to (4.4.16).

The Green function associated with (9.5.1) was obtained in the analysis of the infinite angular region in Section 5.6.4, where $G_<$ is givenby

$$G_<(\rho,\phi,\rho',\phi')=-\frac{j\pi}{\Psi}\sum_{n=1}^{\infty}J_{\nu_n}(\beta\rho)H_{\nu_n}^{(2)}(\beta\rho')\sin(\nu_n\phi)\sin(\nu_n\phi') \qquad (\rho\le\rho') \tag{9.5.2}$$

where

$$\nu_n=\frac{n\pi}{\Psi} \qquad (n=1,2,3,\ldots)$$

Therefore, the solution to (9.5.1) for $\rho\le\rho'$ is

$$\begin{aligned}\Pi_z^e&=\frac{1}{j\omega\varepsilon}\int_{\phi'=0}^{2\pi}\int_{\rho'=0}^{\infty}J_{e,z}(\rho',\phi')\,G_<(\rho,\phi,\rho',\phi')\rho'd\rho'd\phi'\\&=\frac{I}{j\omega\varepsilon}\int_{\phi'=0}^{2\pi}\int_{\rho'=0}^{\infty}\frac{\delta(\rho'-\rho_o)\delta(\phi'-\phi_o)}{\rho'}G_<(\rho,\phi,\rho',\phi')\rho'd\rho'd\phi'\end{aligned} \tag{9.5.3}$$

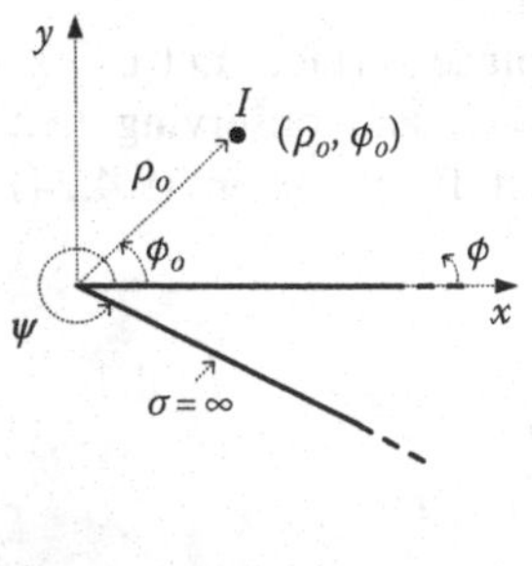

FIGURE 9.9 Two-dimensional view of a wedge with an infinite-length current excitation.

Substituting (9.5.2) into (9.5.3) gives:

$$\Pi_z^e = -\frac{I\pi}{\omega\varepsilon\Psi}\sum_{n=1}^{\infty} J_{\nu_n}(\beta\rho)H_{\nu_n}^{(2)}(\beta\rho_o)\sin(\nu_n\phi)\sin(\nu_n\phi_o) \qquad (\rho \le \rho_o) \quad (9.5.4)$$

and for $\rho \ge \rho_o$ interchange ρ and ρ_o in (9.5.4).

If the source is moved to infinity (i.e., $\beta\rho_o \to \infty$) the asymptotic relation

$$H_{\nu_n}^{(2)}(\beta\rho_o) \approx \sqrt{\frac{2}{\pi\beta\rho_o}}\, e^{-j(\beta\rho_o - \nu_n\pi/2 - \pi/4)}$$

shows that

$$\Pi_z^e = -\frac{I}{\omega\varepsilon\Psi}\sqrt{\frac{2\pi}{\beta}}\,\frac{e^{-j(\beta\rho_o - \pi/4)}}{\sqrt{\rho_o}}\sum_{n=1}^{\infty} e^{j\nu_n\pi/2}\sin(\nu_n\phi)\sin(\nu_n\phi_o)J_{\nu_n}(\beta\rho) \quad (9.5.5)$$

Since,

$$2\sin(\nu_n\phi)\sin(\nu_n\phi_o) = \cos[\nu_n(\phi - \phi_o)] - \cos[\nu_n(\phi + \phi_o)]$$

(9.5.5) can be expressed in the form

$$\Pi_z^e = -\frac{I}{\omega\varepsilon\Psi}\sqrt{\frac{\pi}{2\beta}}\,\frac{e^{-j(\beta\rho_o - \pi/4)}}{\sqrt{\rho_o}}\sum_{n=1}^{\infty} e^{j\nu_n\pi/2}\{\cos[\nu_n(\phi - \phi_o)] - \cos[\nu_n(\phi + \phi_o)]\}J_{\nu_n}(\beta\rho) \quad (9.5.6)$$

From (4.4.16) and (4.4.12) the far fields associated with Π_z^e in (9.5.6) are

$$E_z = \beta^2\Pi_z^e = -\frac{I\omega\mu}{\Psi}\sqrt{\frac{\pi}{2\beta}}\,\frac{e^{-j(\beta\rho_o - \pi/4)}}{\sqrt{\rho_o}}\sum_{n=1}^{\infty} e^{j\nu_n\pi/2}\{\cos[\nu_n(\phi - \phi_o)] - \cos[\nu_n(\phi + \phi_o)]\}J_{\nu_n}(\beta\rho) \quad (9.5.7)$$

and

$$H_\phi = -j\omega\varepsilon\frac{\partial\Pi_z^e}{\partial\rho} = j\frac{I}{\Psi}\sqrt{\frac{\pi\beta}{2}}\,\frac{e^{-j(\beta\rho_o - \pi/4)}}{\sqrt{\rho}}\sum_{n=1}^{\infty} e^{j\nu_n\pi/2}\{\cos[\nu_n(\phi - \phi_o)] - \cos[\nu_n(\phi + \phi_o)]\}J_{\nu_n}{}'(\beta\rho)$$

The ρ component of the magnetic field goes to zero in the far field (i.e., $H_\rho \to 0$).

As a check on the above relations, when $\Psi = \pi$ the wedge represents an infinite flat plane. Then, $\nu_n = n$ and since (see (4.5.3)):

$$e^{-j\beta\rho\cos(\phi\mp\phi_o)} = \sum_{n=-\infty}^{\infty} j^{-n} J_n(\beta\rho) e^{jn(\phi\mp\phi_o)} = \sum_{n=0}^{\infty} \varepsilon_n j^{-n} \cos[n(\phi \mp \phi_o)] J_n(\beta\rho)$$

it follows that (9.5.7) for this case reads

$$E_z = -I\omega\mu\sqrt{\frac{1}{8\pi\beta}}\,\frac{e^{-j(\beta\rho_o-\pi/4)}}{\sqrt{\rho_o}}\left[e^{j\beta\rho\cos(\phi-\phi_o)} - e^{j\beta\rho\cos(\phi+\phi_o)}\right]$$

which represents an incident and a reflected wave from a perfect conductor.

A wedge represents a half plane when $\Psi = 2\pi$. In this case: $\nu_n = n/2$ and (9.5.6) reads

$$\Pi_z^{\mathrm{e}} = -\frac{I}{\omega\varepsilon}\sqrt{\frac{1}{2\pi\beta}}\,\frac{e^{-j(\beta\rho_o-\pi/4)}}{\sqrt{\rho_o}}\sum_{n=1}^{\infty} e^{jn\pi/4}\left\{\cos\left[\frac{n}{2}(\phi-\phi_o)\right] - \cos\left[\frac{n}{2}(\phi+\phi_o)\right]\right\} J_{n/2}(\beta\rho)$$

In the case of a magnetic current source I_m, the magnetic Hertz vector component Π_z^{m}, for $\rho \le \rho_o$, is

$$\Pi_z^{\mathrm{m}} = -\frac{I_m\pi}{2\omega\mu\Psi}\sum_{n=0}^{\infty} \varepsilon_n J_{\nu_n}(\beta\rho) H_{\nu_n}^{(2)}(\beta\rho_o)\cos(\nu_n\phi)\cos(\nu_n\phi_o) \tag{9.5.8}$$

where ε_n is the Neumann's number.

9.5.1 A Wedge and a Dipole Source

A common reflector for an antenna consists of two flat metal plates, as shown in Fig. 9.10. This configuration is also known as a wedge or corner reflector. The dipole source is placed at the center and the reflectors are assumed of infinite lengths. In the case of an electric dipole current excitation in the z direction, the electric Hertz vector component Π_z^{e} satisfies

$$\left[\frac{1}{\rho}\frac{\partial}{\partial\rho}\left(\rho\frac{\partial}{\partial\rho}\right) + \frac{1}{\rho^2}\frac{\partial^2}{\partial\phi^2} + \frac{\partial^2}{\partial z^2} + k^2\right]\Pi_z^e = -\frac{J_{e,z}(\rho,\phi)}{j\omega\varepsilon}$$

$$= -\frac{(I\Delta L)}{j\omega\varepsilon}\frac{\delta(\rho-\rho_o)\delta(\phi-\phi_o)\delta(z-z_o)}{\rho}$$

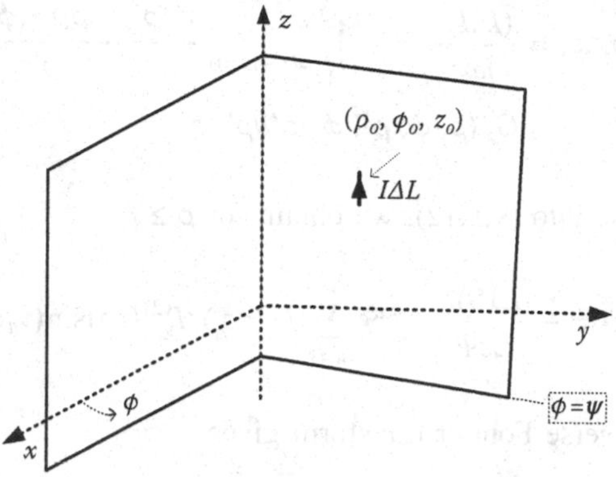

FIGURE 9.10 A corner reflector of infinite length excited by a dipole source.

where $k = \omega\sqrt{\mu\varepsilon}$. Fourier transforming the z variable, one obtains

$$\left[\frac{1}{\rho}\frac{\partial}{\partial\rho}\left(\rho\frac{\partial}{\partial\rho}\right) + \frac{1}{\rho^2}\frac{\partial^2}{\partial\phi^2} + \tau^2\right]\Pi_z^e(\rho, \phi, \alpha) = -\frac{(I\Delta L)}{j\omega\varepsilon}e^{-j\alpha z_o}\frac{\delta(\rho - \rho_o)\delta(\phi - \phi_o)}{\rho} \tag{9.5.9}$$

where $\tau^2 = k^2 - \alpha^2$. The inverse transform is:

$$\Pi_z^e(\rho, \phi, z) = \frac{1}{2\pi}\int_{-\infty}^{\infty}\Pi_z^e(\rho, \phi, \alpha)e^{j\alpha z}d\alpha$$

The Green function associated with (9.5.9) is

$$\left[\frac{1}{\rho}\frac{\partial}{\partial\rho}\left(\rho\frac{\partial}{\partial\rho}\right) + \frac{1}{\rho^2}\frac{\partial^2}{\partial\phi^2} + \tau^2\right]G(\rho, \phi, \rho', \phi') = -\frac{\delta(\rho - \rho')\delta(\phi - \phi')}{\rho} \tag{9.5.10}$$

whose solution for $\rho \geq \rho'$ (see Section 5.6.4) is

$$G_>(\rho, \phi, \rho', \phi') = -\frac{j\pi}{\Psi}\sum_{n=1}^{\infty}J_{\nu_n}(\tau\rho')H_{\nu_n}^{(2)}(\tau\rho)\sin(\nu_n\phi)\sin(\nu_n\phi') \tag{9.5.11}$$

where

$$\nu_n = \frac{n\pi}{\Psi} \qquad (n = 1, 2, 3, ...)$$

Therefore, the solution to (9.5.9) for $\rho \geq \rho'$ is

$$\Pi_z^{\mathrm{e}}(\rho, \phi, \alpha) = \frac{(I\Delta L)}{j\omega\varepsilon} e^{-j\alpha z_o} \int_{\phi'=0}^{2\pi} \int_{\rho'=0}^{\infty} \frac{\delta(\rho' - \rho_o)\delta(\phi' - \phi_o)}{\rho'} G_>(\rho, \phi, \rho', \phi')\rho' d\rho' d\phi' \tag{9.5.12}$$

Substituting (9.5.11) into (9.5.12), we obtain for $\rho \geq \rho_o$:

$$\Pi_z^{\mathrm{e}}(\rho, \phi, \alpha) = -\frac{(I\Delta L)\pi}{\omega\varepsilon\Psi} e^{-j\alpha z_o} \sum_{n=1}^{\infty} J_{\nu_n}(\tau\rho_o) H_{\nu_n}^{(2)}(\tau\rho) \sin(\nu_n\phi) \sin(\nu_n\phi_o)$$

and forming the inverse Fourier transform gives

$$\Pi_z^{e}(\rho, \phi, z) = -\frac{(I\Delta L)}{2\omega\varepsilon\Psi} \sum_{n=1}^{\infty} \sin(\nu_n\phi)\sin(\nu_n\phi_o) \int_{-\infty}^{\infty} J_{\nu_n}(\tau\rho_o) H_{\nu_n}^{(2)}(\tau\rho) e^{j\alpha(z-z_o)} d\alpha \tag{9.5.13}$$

The integral in (9.5.13) is evaluated in the far field (i.e., $k\rho \gg 1$) using the method of steepest descent. Letting $\alpha = k \cos \beta$ it follows from (7.4.66) that

$$\int_{-\infty}^{\infty} F(\tau) H_n^{(2)}(\tau\rho) e^{j\alpha z} d\alpha \approx 2j^{n+1} F(k \sin \theta) \frac{e^{-jkr}}{r}$$

Hence, (9.5.13) reads

$$\Pi_z^{\mathrm{e}}(\rho, \phi, z) \approx -\frac{(I\Delta L)}{\omega\varepsilon\Psi} \frac{e^{-jkr}}{r} e^{-jkz_o \cos\theta} \sum_{n=1}^{\infty} j^{n+1} \sin(\nu_n\phi)\sin(\nu_n\phi_o) J_{\nu_n}(k\rho_o \sin\theta)$$

The fields follow from (4.4.11) to (4.4.16). The field components that vary a $1/r$ are

$$H_\phi = -j\omega\varepsilon \frac{\partial \Pi_z^{\mathrm{e}}}{\partial \rho} = -j\omega\varepsilon \sin\theta \frac{\partial \Pi_z^{\mathrm{e}}}{\partial r}$$
$$\approx jk \frac{(I\Delta L)}{\Psi} \sin\theta \frac{e^{-jkr}}{r} e^{-jkz_o \cos\theta} \sum_{n=1}^{\infty} j^{n+1} \sin(\nu_n\phi)\sin(\nu_n\phi_o) J_{\nu_n}(k\rho_o \sin\theta)$$

and $E_\theta = H_\phi/\eta$.

In the case of a magnetic dipole source, the magnetic Hertz vector component Π_z^{m} is

$$\Pi_z^{\mathrm{m}}(\rho, \phi, z) \approx -j \frac{(I_m\Delta L)}{2\omega\mu\Psi} \frac{e^{-jkr}}{r} e^{-jkz_o \cos\theta} \sum_{n=0}^{\infty} \varepsilon_n j^n \cos(\nu_n\phi)\cos(\nu_n\phi_o) J_{\nu_n}(k\rho_o \sin\theta) \tag{9.5.14}$$

where ε_n is the Neumann's number.

9.6 VERTICAL ELECTRIC HERTZIAN DIPOLE ABOVE A LOSSY SURFACE

This problem was considered in Section 8.5 using rectangular coordinates. An electric Hertzian dipole above a lossy surface is shown Fig. 9.11, where Region 1 is the air region and Region 2 is the lossy region. In this section, the problem is solved in cylindrical coordinates and the resulting expressions are evaluated. The problem has angular symmetry since the field is symmetrical in ϕ (i.e., $\partial/\partial\phi = 0$), and the radiated field is a function of ρ and z. The expression for the $\mathbf{\Pi}^e$ vector in Region 1 is

$$\left[\frac{1}{\rho}\frac{\partial}{\partial\rho}\left(\rho\frac{\partial}{\partial\rho}\right) + \frac{\partial^2}{\partial z^2} + k_1^2\right]\Pi_{z1}^{e} = -\frac{(I\Delta L)}{j\omega\varepsilon_o}\frac{\delta(\rho)\delta(z-h)}{\rho} \qquad (9.6.1)$$

where

$$k_1 = k_o = \omega\sqrt{\mu_o\varepsilon_o}$$

In the lossy Region 2:

$$\left[\frac{1}{\rho}\frac{\partial}{\partial\rho}\left(\rho\frac{\partial}{\partial\rho}\right) + \frac{\partial^2}{\partial z^2} + k_2^2\right]\Pi_{z2}^{e} = 0 \qquad (9.6.2)$$

where, since $\mu_2 = \mu_o$, it follows that

$$k_2 = k_o n_2 = k_o\sqrt{\varepsilon_{r2} - j\frac{\sigma_2}{\omega\varepsilon_o}}$$

The particular solution of (9.6.1) can be expressed in terms of the Sommerfeld-Ott relation in (6.2.30), namely

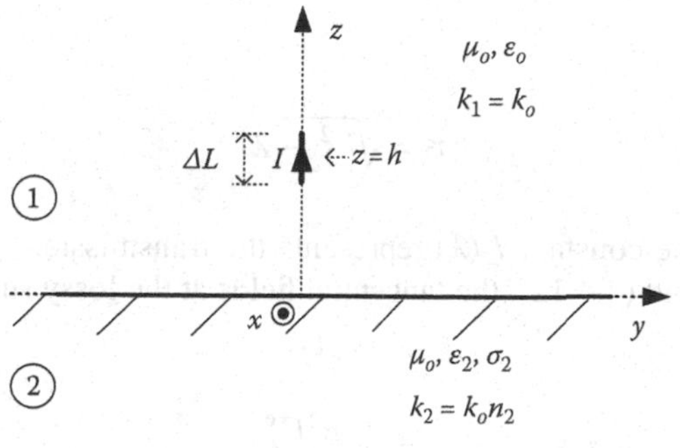

FIGURE 9.11 An electric dipole above a lossy surface.

$$\Pi^{e}_{z1,p} = \frac{(I\Delta L)}{j\omega\varepsilon_o}\frac{e^{-jk_1R_1}}{4\pi R_1} = -\frac{(I\Delta L)}{4\pi\omega\varepsilon_o}\int_0^\infty \frac{J_0(\lambda\rho)}{\tau_1}e^{-j\tau_1|z-h|}\lambda d\lambda \tag{9.6.3}$$

where

$$\tau_1 = \sqrt{k_1^2 - \lambda^2}$$

and

$$R_1 = |\mathbf{r} - \mathbf{r}'| = \sqrt{\rho^2 + (z-h)^2}$$

At the lossy surface, since $z < h$, the exponent in (9.6.3) is $|z - h| = (h - z)$.

The complementary solution can be expressed in terms of a reflection coefficient $R(\lambda)$ as

$$\Pi^{e}_{z1,c} = -\frac{(I\Delta L)}{4\pi\omega\varepsilon_o}\int_0^\infty R(\lambda)J_0(\lambda\rho)e^{-j\tau_1 z}\lambda d\lambda \tag{9.6.4}$$

The double value function τ_1 in (9.6.3) and (9.6.4) is selected such that the radiation condition is satisfied as $z \to \infty$, or $\mathrm{Im}(\tau_1) < 0$.

The total field in Region 1 is the sum of the particular and complementary solution, or:

$$\Pi^{e}_{z1} = \Pi^{e}_{z1,p} + \Pi^{e}_{z1c} = -\frac{(I\Delta L)}{4\pi\omega\varepsilon_o}\int_0^\infty \left[\frac{J_0(\lambda\rho)}{\tau_1}e^{-j\tau_1|z-h|} + R(\lambda)e^{-j\tau_1 z}\right]\lambda d\lambda \tag{9.6.5}$$

The transmitted fields in the lossy region are calculated in terms of Π^{e}_{z2}. That is,

$$\Pi^{e}_{z,2} = -\frac{(I\Delta L)}{4\pi\omega\varepsilon_o}\int_0^\infty T(\lambda)J_0(\lambda\rho)e^{j\tau_2 z}\lambda\, d\lambda \tag{9.6.6}$$

where

$$\tau_2 = \sqrt{k_2^2 - \lambda^2}$$

and $\mathrm{Im}(\tau_2) < 0$. The constant $T(\lambda)$ represents the transmission coefficient.

From (4.4.14) and (4.4.12), the tangential fields at the lossy surface are related to the Π^{e}_{z} by

$$E_\rho = \frac{\partial^2 \Pi^{e}_{z}}{\partial\rho\partial z}$$

and

$$H_\phi = \frac{k^2}{j\omega\mu_o}\frac{\partial \Pi_z^e}{\partial \rho}$$

The other field component is the E_z field, given by

$$E_z = \left(\frac{\partial^2}{\partial z^2} + k^2\right)\Pi_z^e$$

At $z = 0$, the tangential electric and magnetic fields are continuous for all values of ρ. Therefore,

$$E_{\rho 1} = E_{\rho 2}\Big|_{z=0} \Rightarrow \frac{\partial^2 \Pi_{z1}^e}{\partial\rho\partial z} = \frac{\partial^2 \Pi_{z2}^e}{\partial\rho\partial z}\Big|_{z=0} \Rightarrow \frac{\partial \Pi_{z1}^e}{\partial z} = \frac{\partial \Pi_{z2}^e}{\partial z}\Big|_{z=0} \tag{9.6.7}$$

and

$$H_{\phi 1} = H_{\phi 2}\Big|_{z=0} \Rightarrow k_1^2\frac{\partial \Pi_{z1}^e}{\partial \rho} = k_2^2\frac{\partial \Pi_{z2}^e}{\partial \rho}\Big|_{z=0} \Rightarrow k_1^2\Pi_{z1}^e = k_2^2\Pi_{z2}^e\Big|_{z=0} \tag{9.6.8}$$

Applying the boundary conditions (9.6.7) and (9.6.8) to (9.6.5) and (9.6.6):

$$e^{-j\tau_1 h} - \tau_1 R(\lambda) = \tau_2 T(\lambda) \Rightarrow \frac{e^{-j\tau_1 h}}{\tau_1} - R(\lambda) = \frac{\tau_2}{\tau_1}T(\lambda)$$

and

$$k_1^2\frac{e^{-j\tau_1 h}}{\tau_1} + k_1^2 R(\lambda) = k_2^2 T(\lambda) \Rightarrow \frac{e^{-j\tau_1 h}}{\tau_1} + R(\lambda) = \frac{k_2^2}{k_1^2}T(\lambda)$$

Solving the above equations gives

$$R(\lambda) = \frac{k_2^2\tau_1 - k_1^2\tau_2}{k_2^2\tau_1 + k_1^2\tau_2}\frac{e^{-j\tau_1 h}}{\tau_1} = \frac{n_2^2\tau_1 - \tau_2}{n_2^2\tau_1 + \tau_2}\frac{e^{-j\tau_1 h}}{\tau_1} \tag{9.6.9}$$

and

$$T(\lambda) = \frac{2k_1^2}{k_2^2\tau_1 + k_1^2\tau_2}e^{-j\tau_1 h} = \frac{2}{n_2^2\tau_1 + \tau_2}e^{-j\tau_1 h} \tag{9.6.10}$$

where $k_2/k_1 = k_o n_2/k_o = n_2$. Then, (9.6.5) and (9.6.6) read

$$\Pi_{z1}^{e} = \frac{(I\Delta L)}{j\omega\varepsilon_o}\frac{e^{-jk_1R_1}}{4\pi R_1} - \frac{(I\Delta L)}{4\pi\omega\varepsilon_o}\int_0^\infty \frac{1}{\tau_1}\left[\frac{n_2^2\tau_1 - \tau_2}{n_2^2\tau_1 + \tau_2}\right] J_0(\lambda\rho)e^{-j\tau_1(z+h)}\lambda d\lambda \quad (9.6.11)$$

and

$$\Pi_{z2}^{e} = -\frac{(I\Delta L)}{4\pi\omega\varepsilon_o}\int_0^\infty \left[\frac{2}{n_2^2\tau_1 + \tau_2}\right] J_0(\lambda\rho)e^{-j\tau_1 h}e^{j\tau_2 z}\lambda d\lambda \quad (9.6.12)$$

The evaluation of the integrals in (9.6.11) and (9.6.12) has been the subject of extensive research. These integrals can be evaluated using the stationary phase method and the method of steepest descent, taking into account the poles and branch points of the integrals. We will illustrate the evaluation of these integrals for some specific cases.

9.6.1 Fields in the Air Region

To evaluate the integral in (9.6.11), we first extend the range of integration to $-\infty < \lambda < \infty$. That is, let

$$J_0(\lambda\rho) = \frac{1}{2}[H_0^{(1)}(\lambda\rho) + H_0^{(2)}(\lambda\rho)]$$

and write the integral in in (9.6.11) in the form

$$\int_0^\infty F(\lambda)J_0(\lambda\rho)\lambda d\lambda = \frac{1}{2}\int_0^\infty F(\lambda)H_0^{(1)}(\lambda\rho)\lambda d\lambda + \frac{1}{2}\int_0^\infty F(\lambda)H_0^{(2)}(\lambda\rho)\lambda d\lambda \quad (9.6.13)$$

where $F(\lambda)$ represents the other terms in the integral. Observe that $F(\lambda)$ is an even function of λ. In (9.6.13), let $\lambda = -\lambda$ in the integral with $H_0^{(1)}(\lambda\rho)$. That is,

$$\begin{aligned}\frac{1}{2}\int_0^\infty F(\lambda)H_0^{(1)}(\lambda\rho)\lambda d\lambda &= \frac{1}{2}\int_0^{-\infty} F(-\lambda)H_0^{(1)}(-\lambda\rho)\lambda d\lambda \\ &= -\frac{1}{2}\int_0^{-\infty} F(\lambda)H_0^{(2)}(\lambda\rho)\lambda d\lambda \\ &= \frac{1}{2}\int_{-\infty}^{0} F(\lambda)H_0^{(2)}(\lambda\rho)\lambda d\lambda,\end{aligned} \quad (9.6.14)$$

where we used the identity

$$H_0^{(1)}(-z) = -H_0^{(2)}(z)$$

Therefore, using (9.6.14), (9.6.13) can be written as

$$\int_0^{\infty} F(\lambda) J_0(\lambda\rho)\lambda d\lambda = \frac{1}{2}\int_{-\infty}^{\infty} F(\lambda) H_0^{(2)}(\lambda\rho)\lambda d\lambda$$

The complementary solution in (9.6.11) is then:

$$\Pi_{z1,c}^{e} = -\frac{(I\Delta L)}{8\pi\omega\varepsilon_o}\int_{-\infty}^{\infty}\frac{1}{\tau_1}\left[\frac{n_2^2\tau_1 - \tau_2}{n_2^2\tau_1 + \tau_2}\right] H_0^{(2)}(\lambda\rho) e^{-j\tau_1(z+h)}\lambda d\lambda \tag{9.6.15}$$

A solution to (9.6.15) when the observation point is in the far field uses the large argument approximation for the Hankel function. Observing that

$$\lambda H_0^{(2)}(\lambda\rho) \approx \sqrt{\frac{2\lambda}{\pi\rho}}\, e^{-j(\lambda\rho - \pi/4)}$$

(9.6.15) reads

$$\Pi_{z1,c}^{e} = -\frac{(I\Delta L)}{8\pi\omega\varepsilon_o}\sqrt{\frac{2}{\pi\rho}}\, e^{j\pi/4}\int_{-\infty}^{\infty}\frac{1}{\tau_1}\left[\frac{n_2^2\tau_1 - \tau_2}{n_2^2\tau_1 + \tau_2}\right] e^{-j\lambda\rho} e^{-j\tau_1(z+h)}\sqrt{\lambda}\, d\lambda \tag{9.6.16}$$

This integral can be evaluated using the method of stationary phase, or the steepest descent method.

The integral in (9.6.16) is of the form

$$\Pi_{z1,c}^{e} = \int_{-\infty}^{\infty} G(\lambda)\, e^{-j\lambda\rho} e^{-j\tau_1(z+h)} d\lambda \tag{9.6.17}$$

where $G(\lambda)$ represents the other terms. The integral has branch points and poles singularities. The branch points occur at $\lambda = \pm k_1$, $\lambda = \pm k_2$, and $\lambda = 0$. Those associated with k_1 lie on the real axis. Letting k_1 have a small imaginary part (i.e., $k_1 = k'_1 - j\varepsilon$), the branch points are moved slightly away from the path of integration. Later, we will find the location of the poles and analyze their effect on the value of the integral. The path of integration and branch points are shown in Fig. 9.12. The plot is not to scale since k_1 should be on or slightly away from the $\mathrm{Re}(\lambda)$ axis. The branch point at $\lambda = 0$ is due to the Hankel function, and the associated cut (shown as a wiggle line) is drawn a little above the axis. The analysis of the branch cuts associated with k_1 and k_2 were analyzed in Section 8.5.1, and they follow the paths $\mathrm{Im}(\tau_1) = 0$ and $\mathrm{Im}(\tau_2) = 0$.

In the case of a perfectly conducting surface, (9.6.16) should represent the Hertz's vector component due to an image source at $z = -h$. The geometry of the electric dipole and its image is shown in Fig. 9.13.

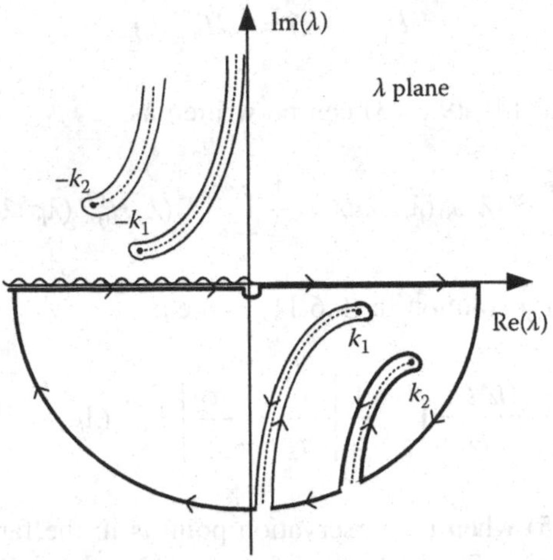

FIGURE 9.12 Path of integration in the λ plane.

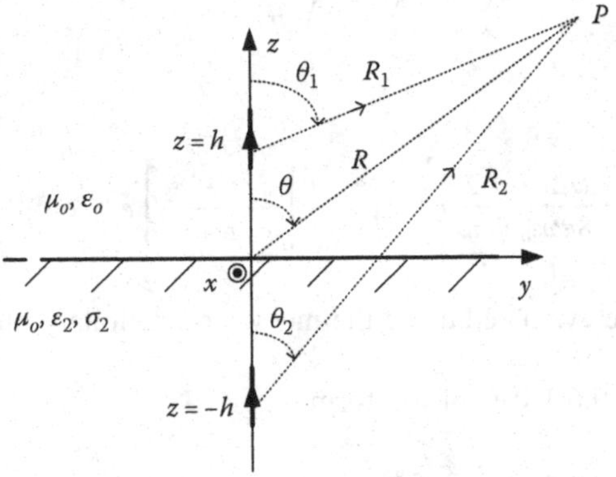

FIGURE 9.13 A dipole source and its image. When P is far away: $R \approx R_1 \approx R_2, \theta \approx \theta_1 \approx \theta_2$ and $R \gg h$.

Letting

$$z + h = R_2 \cos \theta_2$$
$$\rho = R_2 \sin \theta_2$$

where

$$R_2 = \sqrt{\rho^2 + (z + h)^2}$$

the exponent in (9.6.17) becomes

$$e^{-j\left(\lambda R_2 \sin\theta_2+\sqrt{k_1^2-\lambda^2}\,R_2\cos\theta_2\right)} = e^{jR_2 f(\lambda)}$$

where

$$f(\lambda) = -\lambda \sin\theta_2 - \sqrt{k_1^2-\lambda^2}\cos\theta_2$$

The integral in (9.6.17) can be evaluated using either the stationary phase or the steepest descent methods. The stationary phase method is considered first.

The stationary point follows from

$$\frac{df(\lambda)}{d\lambda} = \frac{d}{d\lambda}\left(\lambda\sin\theta_2 + \sqrt{k_1^2-\lambda^2}\cos\theta_2\right) = 0 \rightarrow \sin\theta_2 + \frac{\lambda}{\sqrt{k_1^2-\lambda^2}}\cos\theta_2 = 0$$

or

$$\tan\theta_2 = \frac{\lambda}{\sqrt{k_1^2-\lambda^2}} \Rightarrow \lambda = \lambda_s = k_1\sin\theta_2$$

Then, at the stationary point:

$$f(\lambda_s) = -k_1$$

and

$$\left.\frac{d^2f(\lambda)}{d\lambda^2}\right|_{\lambda=\lambda_s} = \frac{1}{k_1\cos^2\theta_2}$$

The stationary phase evaluation of (9.6.17) follows from:

$$\begin{aligned}\Pi^{e}_{z1,c} &= \int_{-\infty}^{\infty} G(\lambda)e^{jR_2 f(\lambda)}d\lambda \approx G(\lambda_s)\sqrt{\frac{2\pi}{R_2|f''(\lambda_s)|}}\,e^{j\pi/4}e^{-jR_2 f(\lambda_s)} \\ &\approx G(\lambda_s)\sqrt{\frac{2\pi k_1}{R_2}}\cos\theta_2\,e^{j\pi/4}e^{-jk_1R_2}\end{aligned} \tag{9.6.18}$$

where, from (9.6.16), with $\tau_1 = k_1\cos\theta_2$ and $\tau_2 = k_1\sqrt{n_2^2-\sin^2\theta_2}$, $G(\lambda_s)$ is given by

$$G(\lambda_s) = -\frac{(I\Delta L)}{8\pi\omega\varepsilon_o}\sqrt{\frac{2}{\pi k_1 R_2}}\frac{e^{j\pi/4}}{\cos\theta_2}\left[\frac{n_2^2\cos\theta_2 - \sqrt{n_2^2 - \sin^2\theta_2}}{n_2^2\cos\theta_2 + \sqrt{n_2^2 - \sin^2\theta_2}}\right] \tag{9.6.19}$$

In the far field, it is seen that $\theta \approx \theta_1 \approx \theta_2$ and (9.6.19) reads:

$$G(\lambda_s) = -\frac{(I\Delta L)}{8\pi\omega\varepsilon_o}\sqrt{\frac{2}{\pi k_1 R_2}}\frac{e^{j\pi/4}}{\cos\theta}R(\theta) \tag{9.6.20}$$

where

$$R(\theta) = \frac{n_2^2\cos\theta - \sqrt{n_2^2 - \sin^2\theta}}{n_2^2\cos\theta + \sqrt{n_2^2 - \sin^2\theta}}$$

is recognized as the plane-wave reflection coefficient for a TM plane wave.

Substituting (9.6.20) into (9.6.18) gives

$$\Pi_{z1,c}^e = \frac{(I\Delta L)}{j\omega\varepsilon_o}R(\theta)\frac{e^{-jk_1R_2}}{4\pi R_2}$$

and the total field is given by

$$\Pi_{z1}^e = \frac{(I\Delta L)}{j\omega\varepsilon_o}\left[\frac{e^{-jk_1R_1}}{4\pi R_1} + R(\theta)\frac{e^{-jk_1R_2}}{4\pi R_2}\right] \tag{9.6.21}$$

The geometrical meaning of (9.6.21) is illustrated in Fig. 9.13 where the reflected field appears to originate from the image source and is affected by the reflection coefficient of the surface. In the case of a perfectly conducting surface: $R(\theta) = 1$ and (9.6.21) represent the field due to the source and its image component at $z = -h$. At large distances, $R \approx R_1 \approx R_2$, $\theta \approx \theta_1 \approx \theta_2$ and $R \gg h$. Hence, with $R(\theta) = 1$, (9.6.21) reads

$$\Pi_{z1}^e = 2\frac{(I\Delta L)}{j\omega\varepsilon_o}\frac{e^{-jk_1R}}{4\pi R}$$

The expression in (9.6.21) also shows that when the observation point is along the surface; then, with $R_1 \approx R_2$ and $\theta \to \pi/2$ it follows that $R(\theta) \to -1$ and (9.6.21) goes to zero. Hence, (9.6.21) cannot be used in this situation.

Next, the method of steepest descent is used to evaluate (9.6.16). To this end, let

$$\lambda = k_1 \sin\beta \tag{9.6.22}$$

and making the far-field approximation $\theta \approx \theta_1 \approx \theta_2$:

$$z + h = R_2 \cos\theta$$

$$\rho = R_2 \sin\theta$$

The radial distance was left as R_2 to identify the image term. Then,

$$\tau_1 = \sqrt{k_1^2 - \lambda^2} = k_1 \cos\beta$$
$$\tau_2 = \sqrt{k_2^2 - \lambda^2} = k_1 \sqrt{n_2^2 - \sin^2\beta}$$

and

$$e^{-j\lambda\rho} e^{-j\sqrt{k_1^2-\lambda^2}\,(z+h)} = e^{-jk_1R_2 \sin\beta \sin\theta} e^{-jk_1R_2\cos\beta\cos\theta} = e^{-jk_1R_2\cos(\beta-\theta)}$$

So (9.6.16) reads

$$\Pi^{e}_{z1,c} = -\frac{(I\Delta L)}{8\pi\omega\varepsilon_o}\sqrt{\frac{2k_1}{\pi R_2 \sin\theta}}\, e^{j\pi/4} \int_\Gamma \left[\frac{n_2^2\cos\beta - \sqrt{n_2^2 - \sin^2\beta}}{n_2^2\cos\beta + \sqrt{n_2^2 - \sin^2\beta}}\right] \times \sqrt{\sin\beta}\; e^{-jk_1R_2\cos(\beta-\theta)} d\beta \qquad (9.6.23)$$

The transformation (9.6.22) and the resulting path Γ in the β plane (discussed in Section 7.4) are shown in Fig. 9.14. The relation (9.6.23) is of the form

$$\Pi^{e}_{z1} = \int_\Gamma G(\beta) e^{-jk_1R_2\cos(\beta-\theta)} d\beta = \int_\Gamma G(\beta) e^{k_1R_2 f(\beta)} d\beta \qquad (9.6.24)$$

where

$$G(\beta) = -\frac{(I\Delta L)}{8\pi\omega\varepsilon_o}\sqrt{\frac{2k_1}{\pi R_2 \sin\theta}}\, e^{j\pi/4} \left[\frac{n_2^2\cos\beta - \sqrt{n_2^2 - \sin\beta}}{n_2^2\cos\beta + \sqrt{n_2^2 - \sin\beta}}\right] \sqrt{\sin\beta} \qquad (9.6.25)$$

and

$$f(\beta) = -j\cos(\beta - \theta)$$

The poles of the integral in the β plane are located at

$$n_2^2\cos\beta + \sqrt{n_2^2 - \sin^2\beta} = 0$$

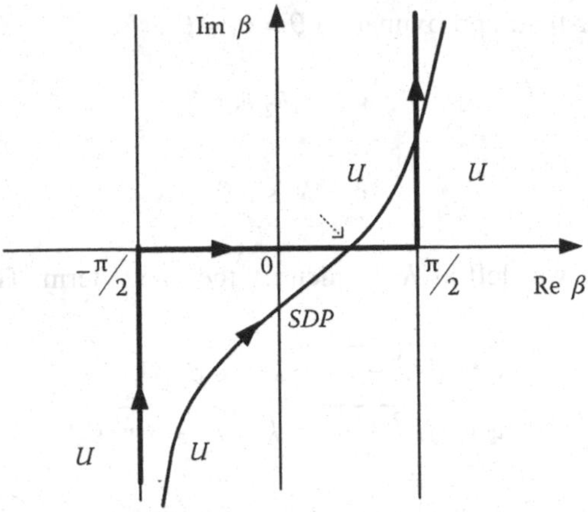

FIGURE 9.14 Path of integration Γ in the β plane and the SDP.

and the branch point due to $\lambda = \pm k_2$ are located at

$$\sin\beta = \pm n_2$$

The branch points due to $\lambda = \pm k_1$ do not appear in the β plane.

The evaluation of (9.6.24) when the poles and branch point do not affect the change from the path Γ to the SDP, and are not close to the saddle point is presented. In Section 7.4 it was shown that the saddle point of (9.6.24) occurs at $\beta = \beta_s = \theta$. At the saddle point: $f(\beta_s) = -j$ and $f''(\beta_s) = j = e^{j\pi/2}$ which shows that the inclination of the steepest descent path (SDP) as it crosses the real β axis is $\phi = \pi/4$.

The evaluation of (9.6.24) using the method of steepest descent is

$$\Pi^e_{z1,c} = \int_{SDP} G(\beta)e^{-jk_1R_2\cos(\beta-\theta)}d\beta \approx G(\theta)e^{-jk_1R_2}e^{j\pi/4}\sqrt{\frac{2\pi}{k_1R_2}}$$

Substituting (9.6.25) for $G(\theta)$ gives

$$\Pi^e_{z1,c} \approx \frac{(I\Delta L)}{j\omega\varepsilon_o}R(\theta)\frac{e^{-jk_1R_2}}{4\pi R_2}$$

where $R(\theta)$ is the plane-wave reflection coefficient in (9.6.20). The analysis has shown that the term in brackets in (9.6.16) is the equivalent to the plane wave reflection coefficient. That is,

$$R(\theta) = \frac{n_2^2\tau_1 - \tau_2}{n_2^2\tau_1 + \tau_2} = \frac{n_2^2\cos\theta - \sqrt{n_2^2 - \sin^2\theta}}{n_2^2\cos\theta + \sqrt{n_2^2 - \sin^2\theta}} \qquad (9.6.26)$$

The total field is given by

$$\Pi_z^e = \frac{(I\Delta L)}{j\omega\varepsilon_o}\left[\frac{e^{-jk_1R_1}}{4\pi R_1} + R(\theta)\frac{e^{-jk_1R_2}}{4\pi R_2}\right]$$

which is identical to (9.6.21).

For the fields close to the surface: $R \approx R_1 \approx R_2$, $\theta \to \pi/2$, $R(\theta) \to -1$ and, as previously discussed, Π_z^e goes to zero. Hence, for this case a more detailed analysis of the integral in (9.6.16) is needed. Such analysis requires the location of the poles and branch points of (9.6.16) in the β plane, and how they affect the integration.

To evaluate the field close to the surface (i.e., as $\theta \to \pi/2$), there are two common methods to write the total field in (9.6.16). One method is to write $R(\lambda)$ as

$$R(\lambda) = \frac{e^{-j\tau_1 h}}{\tau_1}\left[1 - \frac{2\tau_2}{n_2^2\tau_1 + \tau_2}\right] \tag{9.6.27}$$

Then, substituting (9.6.27) into (9.6.16), Π_{z1}^e is expressed as

$$\Pi_{z1}^e = \frac{(I\Delta L)}{j\omega\varepsilon_o}\frac{e^{-jk_1R_1}}{4\pi R_1} + \frac{(I\Delta L)}{j\omega\varepsilon_o}\frac{e^{-jk_1R_2}}{4\pi R_2} + P \tag{9.6.28}$$

where in the far field:

$$P = \frac{(I\Delta L)}{4\pi\omega\varepsilon_o}\sqrt{\frac{2}{\pi\rho}}\, e^{j\pi/4}\int_{-\infty}^{\infty}\frac{1}{\tau_1}\left[\frac{\tau_2}{n_2^2\tau_1 + \tau_2}\right]e^{-j\lambda\rho}e^{-j\tau_1(z+h)}\sqrt{\lambda}\,d\lambda \tag{9.6.29}$$

The other method is to write $R(\lambda)$as

$$R(\lambda) = \frac{e^{-j\tau_1 h}}{\tau_1}\left[-1 + \frac{2n_2^2\tau_1}{n_2^2\tau_1 + \tau_2}\right] \tag{9.6.30}$$

Then, substituting (9.6.30) into (9.6.16), Π_{z1}^e is expressed as

$$\Pi_{z1}^e = \frac{(I\Delta L)}{j\omega\varepsilon_o}\frac{e^{-jk_1R_1}}{4\pi R_1} - \frac{(I\Delta L)}{j\omega\varepsilon_o}\frac{e^{-jk_1R_2}}{4\pi R_2} + Q \tag{9.6.31}$$

where in the far field

$$Q = -\frac{(I\Delta L)}{4\pi\omega\varepsilon_o}n_2^2\sqrt{\frac{2}{\pi\rho}}\, e^{j\pi/4}\int_{-\infty}^{\infty}\left[\frac{1}{n_2^2\tau_1 + \tau_2}\right]e^{-j\lambda\rho}e^{-j\tau_1(z+h)}\sqrt{\lambda}\,d\lambda \tag{9.6.32}$$

In the representation in (9.6.31) at the surface of the earth the first two terms cancel out, and one is left with the evaluation of Q.

The evaluation of (9.6.29) or (9.6.32) lead to the same final results for Π^e_{z1} when $\theta \approx \pi/2$. We will evaluate (9.6.32). The evaluation of this integral involves the modified method of steepest descent since when the observation point is close the surface (i.e., when $\theta \approx \theta_1 \approx \theta_2 \approx \pi/2$) a pole is close to the saddle point. In some cases, the evaluation of this integral also requires the analysis of a branch point singularity and its contribution to the results.

Using the transformation

$$\lambda = k_1 \sin\beta$$

and with

$$z + h = R_2 \cos\theta$$
$$\rho = R_2 \sin\theta$$

it follows that

$$\tau_1 = k_1 \cos\beta$$
$$\tau_2 = k_1\sqrt{n_2^2 - \sin^2\beta}$$

and (9.6.32) reads

$$Q = K\int_\Gamma F(\beta)e^{-jk_1R_2\cos(\beta-\theta)}d\beta = K\int_\Gamma F(\beta)e^{k_1R_2 f(\beta)}d\beta \tag{9.6.33}$$

where

$$K = -\frac{(I\Delta L)}{4\pi\omega\varepsilon_o}n_2^2\sqrt{\frac{2k_1}{\pi R_2\sin\theta}}\,e^{j\pi/4}$$

$$F(\beta) = \frac{\sqrt{\sin\beta}\cos\beta}{n_2^2\cos\beta + \sqrt{n_2^2 - \sin^2\beta}} \tag{9.6.34}$$

and

$$f(\beta) = -j\cos(\beta - \theta)$$

The saddle point of (9.6.33) occurs at $\beta = \beta_s = \theta$ (i.e., at $f'(\beta) = 0$). At the saddle point: $f(\theta) = -j$. The path of steepest descent is along the constant imaginary part of $f(\beta)$, where

$$f(\beta) = -j\cos(\beta - \theta) = -j[\cos(\beta' - \theta)\cosh\beta'' + j\sin(\beta' - \theta)\sinh\beta'']$$

Since $\text{Im}f(\beta) = \text{Im}f(\beta_s) = -1$, it follows that the path is described by

$$\cos(\beta' - \theta)\cosh\beta'' = 1$$

Along this path the real part of the exponential decreases very rapidly, which is $e^{k_1R_2\sin(\beta'-\theta)\sinh\beta''}$.

To evaluate (9.6.33), let

$$f(\beta) = f(\theta) - s^2 \Rightarrow -j\cos(\beta - \theta) = -j - s^2 \tag{9.6.35}$$

and (9.6.33) reads

$$Q = Ke^{-jk_1R_2}\int_{-\infty}^{\infty} F(s)\frac{d\beta}{ds}e^{-k_1R_2s^2}\,ds \tag{9.6.36}$$

where s is a real variable and, as discussed in Section 7.4, the limits of integration can be extended from $-\infty$ to ∞.

From (9.6.35),

$$\cos(\beta - \theta) = 1 - js^2$$

and in the vicinity of the saddle point (i.e., $\beta \approx \theta$ or $s \approx 0$):

$$\cos(\beta - \theta) \approx 1 - (\beta - \theta)^2/2 \Rightarrow (\beta - \theta)^2/2 = js^2$$

With $\beta = \beta' + j\beta''$ we obtain

$$(\beta' - \theta)^2 - \beta''^2 = 0 \Rightarrow \beta' - \theta = \pm\beta'' \Rightarrow \beta'' = (\beta' - \theta) \tag{9.6.37}$$

and

$$\beta''(\beta' - \theta) = s^2 \tag{9.6.38}$$

The sign in (9.6.37) was chosen as positive, since (9.6.38) shows that if $\beta' < \theta$ then $\beta'' < 0$, and if $\beta' > \theta$ then $\beta'' > 0$. Hence, (9.6.37) shows that the path of integration intersects the real β axis at an angle of $\pi/4$. A fact that is illustrated in Fig. 9.14 where the angle of inclination is $\phi = \pi/4$.

The integral in (9.6.33) has branch points at

$$\tau_2 = \sqrt{n_2^2 - \sin^2\beta_B} = 0 \Rightarrow \sin\beta_B = \pm n_2 \tag{9.6.39}$$

The branch points due to $\lambda = \pm k_1$ in the λ plane do not appear in the β plane.

Since τ_2 appears in (9.6.12) as $e^{j\tau_2 z}$ where $z < 0$, then as $z \to -\infty$ the imaginary part of τ_2 must be negative to satisfy the radiation condition, or

$$\mathrm{Im}(\tau_2) = \mathrm{Im}\left[\sqrt{n_2^2 - \sin^2\beta}\right] < 0$$

This condition defines the upper Riemann surface.

The branch cut associated with a given branch point goes from the branch point to ∞ along the curve given by

$$\mathrm{Im}\left[\sqrt{n_2^2 - \sin^2\beta}\right] = 0$$

The location of the branch points in (9.6.39) depend on the value of n_2.

The poles of the integral in (9.6.33) are located at

$$n_2^2\cos\beta_p + \sqrt{n_2^2 - \sin^2\beta_p} = 0 \Rightarrow \sin\beta_p = \frac{\pm n_2}{\sqrt{n_2^2 + 1}} \tag{9.6.40}$$

The branch points in (9.6.39) and poles in (9.6.40) must be determined for a problem once n_2 is given. We will consider the case when Region 2 represents specific Earth surfaces. Typical values of the constitutive parameters for dry ground are: σ_2 around 0.01 mS/m, ε_{r2} around 15, and $\mu_{r2} = 1$. At a frequency of 100 kHz, it follows that $n_2 \approx 30(1 - j)$. In comparison, for sea water, typical values are: σ_2 around 0.004 mS/m, ε_{r2} around 80, and $\mu_{r2} = 1$, and typical value of at 100 kHz is $n_2 \approx 600(1 - j)$. For both cases at lower frequencies (say 10 kHz) the magnitude of n_2 increases, while at higher frequencies (say 1 MHz) the magnitude of n_2 decreases, but stays at a high value.

Using $n_2 \approx 30(1 - j)$ in (9.6.39), in the range $-\pi \le \beta' \le \pi$ the branch points are

$$\beta_B = \pm\sin^{-1}[30(1 - j)] = \begin{cases} \beta_{B1} = \dfrac{\pi}{4} - j4.441 \\ \beta_{B2} = -\dfrac{\pi}{4} + j4.441 \\ \beta_{B3} = \dfrac{3\pi}{4} + j4.441 \\ \beta_{B4} = -\dfrac{3\pi}{4} - j4.441 \end{cases}$$

For other typical values of n_2 in dry ground the real part of β_B remains about the same and the imaginary parts varies but the locations of the branch points stay far from any pole, as the following analysis for the poles locations show.

Since the magnitude of n_2 is large, (9.6.40) is approximated by

$$\pm \sin\beta_p = \frac{n_2}{\sqrt{n_2^2+1}} = \frac{1}{\sqrt{1+\frac{1}{n_2^2}}} \approx 1 - \frac{1}{2n_2^2}$$

and width

$$\pm \sin\beta_p = \cos\left(\frac{\pi}{2} \mp \beta_p\right) \approx 1 - \frac{\left(\frac{\pi}{2} \mp \beta_p\right)^2}{2}$$

it follows that the poles are located at

$$\frac{\left(\frac{\pi}{2} \mp \beta_p\right)^2}{2} \approx \frac{1}{2n_2^2} \Rightarrow \beta_p \approx \pm\frac{\pi}{2} \pm \frac{1}{n_2} \tag{9.6.41}$$

Typical values of n_2 for dry ground are of the form $n_2 = n_2'(1-j) = |n_2|e^{-j\pi/4}$ where $|n_2| \gg 1$, and (9.6.41) shows that the location of the poles are at

$$\beta_p \approx \pm\frac{\pi}{2} \pm \frac{e^{j\pi/4}}{|n_2|} = \begin{cases} \beta_{p1} = \frac{\pi}{2} + \frac{e^{j\pi/4}}{|n_2|} \\ \beta_{p2} = \frac{\pi}{2} - \frac{e^{j\pi/4}}{|n_2|} \\ \beta_{p3} = -\frac{\pi}{2} + \frac{e^{j\pi/4}}{|n_2|} \\ \beta_{p4} = -\frac{\pi}{2} - \frac{e^{j\pi/4}}{|n_2|} \end{cases}$$

For dry ground, since $|n_2| \approx 30$, the poles are close to $\pm\pi/2$. Figure 9.15 shows the path of integration, the branch points, and the poles in the β plane. Only the branch cuts associated with β_{B1} and β_{B2} are shown. It is seen that for $\beta = \theta \approx \pi/2$ (i.e., at the Earth's surface), the saddle point is close to the upper Riemann sheet pole β_{p1}, while the pole β_{p2} is in the lower Riemann sheet. The branch point β_{B2} is in the upper Riemann sheet, but far from the saddle point. The modified method of steepest descent must be used since the pole β_{p1} is very close to the saddle point at $\theta \approx \pi/2$.

Using the modified method of steepest descent, (9.6.36) is expressed in the form

$$\begin{aligned} Q &= Ke^{-jk_1R_2}\int_{-\infty}^{\infty} F(s)\frac{d\beta}{ds}e^{-k_1R_2s^2}ds \\ &= Ke^{-jk_1R_2}\int_{-\infty}^{\infty}\left[F(s)\frac{d\beta}{ds} - \frac{a_{-1}(s_p)}{s-s_p}\right]e^{-k_1R_2s^2}ds + Ke^{-jk_1R_2}\int_{-\infty}^{\infty}\left[\frac{a_{-1}(s_p)}{s-s_p}\right]e^{-k_1R_2s^2}ds \end{aligned} \tag{9.6.42}$$

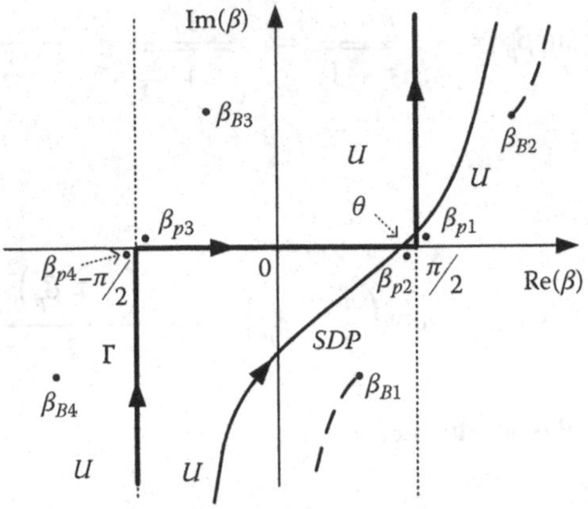

FIGURE 9.15 Path of integration Γ and SDP when θ approaches $\pi/2$. The pole β_{p1} is close to the saddle point when $\theta \approx \pi/2$.

where $a_{-1}(s_p)$ is the residue at the pole s_p. The location of s_p in the s plane follows from the transformation of the β to the s plane. That is,

$$f(\beta) = f(\beta_s) - s^2 \Rightarrow \cos(\beta - \theta) = 1 - js^2$$

and it follows (see (7.4.72)) that the transformation reads

$$s = \sqrt{2}\, e^{-j\pi/4} \sin\left(\frac{\beta - \theta}{2}\right) \tag{9.6.43}$$

Referring to (7.4.75) and (7.4.76), it follows that from (9.6.43):

$$\frac{d\beta}{ds} = \frac{2j}{\sqrt{s^2 + j2}} \tag{9.6.44}$$

The integrand in the first integral in (9.6.42) is an analytic function in the s plane up to the singularities of $d\beta/ds$, which occurs at $s = \pm\sqrt{2}\, e^{-j\pi/4}$ and are far from the saddle point in the s plane. The saddle point corresponds to the origin in the s plane (i.e., $s = 0$). This integral can be evaluated by expanding the integrand in a Taylor expansion in even power of s, since for odd power of s the integral is zero. That is,

$$F(s)\frac{d\beta}{ds} - \frac{a_{-1}(s_p)}{s - s_p} = \sum_{n=0}^{\infty} B_{2n} s^{2n}$$

where

$$B_{2n} = \frac{1}{n!}\frac{d^n}{d\beta^n}\left[F(s)\frac{d\beta}{ds} - \frac{a_{-1}(s_p)}{s - s_p}\right]\Bigg|_{s=0}$$

The first term of the Taylor expansion is

$$B_0 = F(0)\frac{d\beta}{ds}\bigg|_{s=0} + \frac{a_{-1}(s_p)}{s_p} \tag{9.6.45}$$

where s_p is given by (9.6.43) with $\beta = \beta_p$.

The term $F(0)$ in (9.6.45) in terms of s corresponds to $F(\beta)$ evaluated at $\beta = \theta$. From (9.6.44):

$$\frac{d\beta}{ds}\bigg|_{s=0} = \sqrt{2}\, e^{j\pi/4}$$

and in (9.6.45):

$$F(0)\frac{d\beta}{ds}\bigg|_{s=0} = \frac{\sqrt{\sin\theta}\cos\theta}{n_2^2\cos\theta + \sqrt{n_2^2 - \sin^2\theta}}\sqrt{2}\, e^{j\pi/4} \tag{9.6.46}$$

The residue in (9.6.45) is

$$\begin{aligned} a_{-1}(s_p) &= \frac{\sqrt{\sin\beta}\cos\beta}{\frac{d}{d\beta}\left(n_2^2\cos\beta + \sqrt{n_2^2 - \sin^2\beta}\right)\frac{d\beta}{ds}}\frac{d\beta}{ds}\Bigg|_{\beta=\beta_p} \\ &= -\frac{\sqrt{\sin\beta_p}\cos\beta_p}{n_2^2\sin\beta_p + \sin\beta_p\cos\beta_p(n_2^2 - \sin^2\beta_p)^{-1/2}} \end{aligned} \tag{9.6.47}$$

Since at the pole location:

$$\sqrt{n_2^2 - \sin^2\beta_p} = -n_2^2\cos\beta_p$$

and $\beta_p \approx \pi/2$ it follows that (9.6.47) is approximated by

$$a_{-1}(s_p) = -\sqrt{\sin\beta_p}\frac{n_2^2}{n_2^4 - 1}\frac{\cos\beta_p}{\sin\beta_p} \approx \frac{n_2}{n_2^4 - 1} \approx \frac{1}{n_2^3} \tag{9.6.48}$$

Substituting (9.6.46) and (9.6.48) into (9.6.45), and then into the first integral in (9.6.42) shows that the analytic part can be approximated by

$$
\begin{aligned}
&Ke^{-jk_1R_2}\int_{-\infty}^{\infty}\left[F(s)\frac{d\beta}{ds}-\frac{a_{-1}(s_p)}{s-s_p}\right]e^{-k_1R_2s^2}ds \\
&\approx Ke^{-jk_1R_2}\left[F(0)\frac{d\beta}{ds}\bigg|_{s=0}+\frac{a_{-1}(s_p)}{s_p}\right]\int_{-\infty}^{\infty}e^{-k_1R_2s^2}ds \\
&\approx Ke^{-jk_1R_2}\sqrt{\frac{2\pi}{k_1R_2}}\left[e^{j\pi/4}\frac{\sqrt{\sin\theta}\cos\theta}{n_2^2\cos\theta+\sqrt{n_2^2-\sin^2\theta}}+\frac{1}{\sqrt{2}\,s_pn_2^3}\right]
\end{aligned}
\tag{9.6.49}
$$

The second integral in (9.6.42) is expressed in terms of the complementary error function (see Section 7.5). That is,

$$
\begin{aligned}
&Ke^{-jk_1R_2}\int_{-\infty}^{\infty}\frac{a_{-1}}{s-s_p}e^{-k_1R_2s^2}ds = Ke^{-jk_1R_2}a_{-1}(s_p)\int_{-\infty}^{\infty}\frac{e^{-k_1R_2s^2}}{s-s_p}ds \\
&\approx Ke^{-jk_1R_2}\frac{1}{n_2^3}\left[-j\pi e^{-k_1R_2s_p^2}\operatorname{erfc}\left(j\sqrt{k_1R_2s_p^2}\right)\right] \qquad (\mathrm{Im}(s_p)<0)
\end{aligned}
\tag{9.6.50}
$$

The fact that $\mathrm{Im}(s_p) < 0$ in dry ground follows from (9.6.43), where

$$
s_p = \sqrt{2}\,e^{-j\pi/4}\sin\left(\frac{\beta_{p1}-\theta}{2}\right)
$$

With $\theta \approx \pi/2$ and

$$
\beta_{p1} \approx \frac{\pi}{2} + \frac{1}{n_2}
$$

it follows that

$$
s_p \approx \sqrt{2}\,e^{-j\pi 4}\sin\left(\frac{1}{2n_2}\right) \approx \frac{e^{-j\pi 4}}{\sqrt{2}\,n_2}
\tag{9.6.51}
$$

Since

$$
n_2 = \sqrt{\varepsilon_{r2} - \frac{j\sigma_2}{\omega\varepsilon_o}}
$$

the phase of n_2 for $\sigma_2/\omega\varepsilon_o \to \infty$ (i.e., a large value) approaches $-\pi/4$, and for

$\sigma_2/\omega\varepsilon_o \to 0$ it approaches 0. Then, (9.6.51) shows that for $\sigma_2/\omega\varepsilon_o \to \infty$ the phase of s_p approach 0, and for $\sigma_2/\omega\varepsilon_o \to 0$ the phase approach $-\pi/4$. Thus, the phase of s_p are between $-\pi/4$ and 0, or $\text{Im}(s_p) < 0$. For the case considered, where $n_2 \approx |n_2|e^{-j\pi/4}$, (9.6.51) is approximated by

$$s_p \approx \frac{1}{\sqrt{2}\,|n_2|}$$

with the understanding that s_p has a small imaginary part such that $\text{Im}(s_p) < 0$.

Substituting (9.6.49) and (9.6.50) into (9.6.42), the value of Q is

$$\begin{aligned} Q &= Ke^{-jk_1R_2}\sqrt{\frac{2\pi}{k_1R_2}}\left(e^{j\pi/4}\frac{\sqrt{\sin\theta}\cos\theta}{n_2^2\cos\theta+\sqrt{n_2^2-\sin^2\theta}}+\frac{1}{\sqrt{2}\,s_pn_2^3}\right) \\ &\qquad - Ke^{-jk_1R_2}\frac{1}{n_2^3}\left[j\pi e^{-k_1R_2s_p^2}\operatorname{erfc}\left(j\sqrt{k_1R_2s_p^2}\right)\right] \\ &= Ke^{-jk_1R_2}\sqrt{\frac{2\pi}{k_1R_2}}\left(e^{j\pi/4}\frac{\sqrt{\sin\theta}\cos\theta}{n_2^2\cos\theta+\sqrt{n_2^2-\sin^2\theta}}\right) \\ &\qquad + Ke^{-jk_1R_2}\frac{1}{\sqrt{2}\,s_pn_2^3}\sqrt{\frac{2\pi}{k_1R_2}}\left[1-j\sqrt{\pi k_1R_2s_p^2}\,e^{-k_1R_2s_p^2}\operatorname{erfc}\left(j\sqrt{k_1R_2s_p^2}\right)\right] \end{aligned} \tag{9.6.52}$$

Define the parameter p, known as the numerical distanceas

$$p = k_1R_2s_p^2 \tag{9.6.53}$$

Then, (9.6.52) reads

$$\begin{aligned} Q &= Ke^{-jk_1R_2}\sqrt{\frac{2\pi}{k_1R_2}}\left(e^{j\pi/4}\frac{\sqrt{\sin\theta}\cos\theta}{n_2^2\cos\theta+\sqrt{n_2^2-\sin^2\theta}}\right) \\ &\quad + Ke^{-jk_1R_2}\frac{1}{\sqrt{2}\,s_pn_2^3}\sqrt{\frac{2\pi}{k_1R_2}}\left[1-j\sqrt{\pi p}\,e^{-p}\operatorname{erfc}\left(j\sqrt{p}\right)\right] \end{aligned} \tag{9.6.54}$$

Define the term $F(p)$, known as the Sommerfeld attenuation function, as

$$F(p) = 1 - j\sqrt{\pi p}\,e^{-p}\operatorname{erfc}\left(j\sqrt{p}\right) \tag{9.6.55}$$

where $\text{Im}(p) < 0$ (see (9.6.53)). Since,

$$\operatorname{erfc}\left(j\sqrt{p}\right) = \frac{2}{\sqrt{\pi}}\int_{j\sqrt{p}}^{\infty} e^{-x^2}dx == 1 - \operatorname{erf}\left(j\sqrt{p}\right)$$

(9.6.55) can also be expressed in the form

$$F(p) = 1 - j\sqrt{\pi p}\,e^{-p} + j2\sqrt{p}\,e^{-p}\int_0^{j\sqrt{p}} e^{-x^2}dx = 1 - j\sqrt{\pi p}\,e^{-p} - 2\sqrt{p}\,e^{-p}\int_0^{\sqrt{p}} e^{x^2}dx$$

Using (9.6.54) and (9.6.55), Π_{z1} in (9.6.31) reads

$$\begin{aligned}\Pi_{z1} = &\frac{(I\Delta L)}{j\omega\varepsilon_o}\frac{e^{-jk_1R_1}}{4\pi R_1} - \frac{(I\Delta L)}{j\omega\varepsilon_o}\frac{e^{-jk_1R_2}}{4\pi R_2}\\ &+ Ke^{-jk_1R_2}\sqrt{\frac{2\pi}{k_1R_2}}\left(e^{j\pi/4}\frac{\sqrt{\sin\theta}\cos\theta}{n_2^2\cos\theta + \sqrt{n_2^2 - \sin^2\theta}}\right)\\ &+ Ke^{-jk_1R_2}\frac{1}{\sqrt{2}\,s_p n_2^3}\sqrt{\frac{2\pi}{k_1R_2}}F(p)\end{aligned} \tag{9.6.56}$$

Substituting for K in (9.6.56), the second and third term can be combined and (9.6.56) is expressed in the form

$$\Pi_{z1} = \frac{(I\Delta L)}{j\omega\varepsilon_o}\frac{e^{-jk_1R_1}}{4\pi R_1} + R(\theta)\frac{(I\Delta L)}{j\omega\varepsilon_o}\frac{e^{-jk_1R_2}}{4\pi R_2} - \frac{(I\Delta L)}{2\pi\omega\varepsilon_o\sqrt{\sin\theta}}\frac{e^{j\pi/4}}{\sqrt{2}\,s_p n_2}\frac{e^{-jk_1R_2}}{R_2}F(p) \tag{9.6.57}$$

where

$$R(\theta) = \frac{n_2^2\cos\theta - \sqrt{n_2^2 - \sin^2\theta}}{n_2^2\cos\theta + \sqrt{n_2^2 - \sin^2\theta}}$$

is a Fresnel reflection coefficient.

At the ground surface: $\theta = \pi/2$. Then, $R(\theta) = -1$ and the radial distances are $R \approx R_1 \approx R_2$. Hence, the first two terms in (9.6.57) cancel out. Since $\sqrt{2}\,s_p n_2 \approx e^{-j\pi/4}$, Π_{z1}^{e} at the Earth's surface is

$$\Pi_{z1}^{e} \approx 2\frac{(I\Delta L)}{j\omega\varepsilon_o}\frac{e^{-jk_1R}}{4\pi R}F(p) \tag{9.6.58}$$

which is valid for $k_1R \gg 1$, large value of $|n_2|$, and $\mathrm{Im}(p) < 0$.

The vertical electric field at the Earth's surface, since $k_1 = k_o = \omega\sqrt{\mu_o\varepsilon_o}$, is

$$E_{z1} = k_1^2 \Pi_z^e \approx -2j\omega\mu_o (I\Delta L) \frac{e^{-jk_1R}}{4\pi R} F(p)$$

For $|p| < <1$, the uniform convergent expansion of erfc(z) (see Appendix G) shows that $F(p)$ is given by

$$F(p) \approx 1 - j\sqrt{\pi p}\, e^{-p} - 2pe^{-p} + j\sqrt{\pi}\, p^{3/2} e^{-p} + \ldots \approx 1 - j\sqrt{\pi p}$$

This relation approaches $F(p) \to 1$ when $p \to 0$, which corresponds to a perfect conducting surface. In this case, (9.6.58) show that the fields are due to the source plus its image. That is,

$$\Pi_{z1} = 2\frac{(I\Delta L)}{j\omega\varepsilon_o} \frac{e^{-jk_1R}}{4\pi R} \tag{9.6.59}$$

Equations (9.6.56) and (9.6.58) show that $F(p)$ represents an attenuation factor that accounts for the surface effect. It reduces the field in (9.6.58) by the factor $F(p)$. The field strength depends on the numerical distance p and the corresponding value of the attenuation function $F(p)$. Since the surface impedance is related to the index of refraction, the numerical distance can also be expressed in terms of the surface impedance (left to the problems).

From the asymptotic expansion of the complementary error function:

$$\text{erfc}(z) \approx \frac{e^{-z^2}}{\sqrt{\pi}\, z} \left[1 - \frac{1}{2z^2} + \frac{1 \times 3}{(2z^2)^2} + \ldots \right] \qquad (\arg(z) < 3\pi/4)$$

it follows that for large distances, say $|p| > 10$:

$$\text{erfc}\left(j\sqrt{p}\right) \approx -j\frac{e^p}{\sqrt{\pi p}} \left(1 + \frac{1}{2p}\right)$$

and the attenuation function in (9.6.55) is approximated by

$$F(p) \approx -\frac{1}{2p} \approx -\frac{1}{2k_o R\, s_p^2}$$

In this case, the vertical electric field at the Earth surface is

$$E_{z1} = k_1^2 \Pi_z^e = j\frac{\eta_o (I\Delta L)\, |n_2|^2}{2\pi} \frac{e^{-jk_1R}}{R^2}$$

which shows that the vertical electric field along the surface varies as e^{-jkR}/R^2.

9.6.2 Fields in the Lossy Region

The transmitted fields, also known as the refracted field, follow from the analysis of (9.6.12), namely:

$$\Pi^{e}_{z2} = -\frac{(I\Delta L)}{4\pi\omega\varepsilon_o}\int_0^\infty \left[\frac{2}{n_2^2\tau_1+\tau_2}\right] J_0(\lambda\rho)e^{-j\tau_1 h}e^{j\tau_2 z}\lambda d\lambda$$
$$= -\frac{(I\Delta L)}{8\pi\omega\varepsilon_o}\int_{-\infty}^\infty \left[\frac{2}{n_2^2\tau_1+\tau_2}\right] H_0^{(2)}(\lambda\rho)e^{-j\tau_1 h}e^{j\tau_2 z}\lambda d\lambda$$

The evaluation of this integral when the point of observation is far from the surface can be performed using the method of steepest descent. Using the asymptotic approximation for the Hankel function gives

$$\Pi^{e}_{z2} = -\frac{(I\Delta L)}{8\pi\omega\varepsilon_o}\sqrt{\frac{2}{\pi\rho}}e^{-j\pi/4}\int_{-\infty}^\infty \left[\frac{2}{n_2^2\tau_1+\tau_2}\right] e^{-j\lambda\rho}e^{-j\tau_1 h}e^{j\tau_2 z}\sqrt{\lambda}\, d\lambda \tag{9.6.60}$$
$$= \int_{-\infty}^{\infty} G(\lambda)e^{f(\lambda)}d\lambda$$

where

$$G(\lambda) = -\frac{(I\Delta L)}{4\pi\omega\varepsilon_o}\sqrt{\frac{2}{\pi\rho}}e^{-j\pi/4}\left[\frac{\sqrt{\lambda}}{n_2^2\tau_1+\tau_2}\right]$$

and

$$f(\lambda) = -j(\rho\lambda + \tau_1 h - \tau_2 z)$$

Using the method of steepest descent, let

$$\lambda = k_1 \sin\beta$$

Then,

$$\tau_1 = \sqrt{k_1^2 - (k_1\sin\beta)^2} = k_1\cos\beta$$
$$\tau_2 = \sqrt{k_2^2 - (k_1\sin\beta)^2} = k_1\sqrt{n_2^2 - \sin^2\beta}$$

and

$$f(\beta) = -j\left(\rho k_1\sin\beta + h k_1\cos\beta - z k_1\sqrt{n_2^2-\sin^2\beta}\right)$$

The saddle point occurs at

$$f'(\beta) = -j\left(\rho k_1 \cos\beta - hk_1 \sin\beta + zk_1 \frac{\sin\beta\cos\beta}{\sqrt{n_2^2 - \sin^2\beta}}\right) = 0$$

$$= -jk_1 \cos\beta\left(\rho - h\tan\beta + z\frac{\sin\beta}{\sqrt{n_2^2 - \sin^2\beta}}\right) = 0$$

It is observed that if $\beta = \beta_s = \theta_1'$, where $\theta_1' = \pi - \theta_1$, then

$$\rho - h\tan\theta_1' + z\frac{\sin\theta_1'}{\sqrt{n_2^2 - \sin^2\theta_1'}} = 0 \Rightarrow \rho - h\tan\theta_1' + z\tan\theta_2' = 0 \quad (9.6.61)$$

where Snell's law was used (i.e., $\sin\theta_1' = n_2 \sin\theta_2'$). Referring to Fig. 9.16:

$$\rho = \rho_1 + \rho_2 = h\tan\theta_1' - z\tan\theta_2'$$

which shows that (9.6.61) is satisfied.

The saddle point evaluation of (9.6.60) is

$$\Pi_z^e = \int_{-\infty}^{\infty} G(\lambda)\, e^{f(\lambda)} d\lambda \approx G(\lambda_s)\, e^{f(\lambda_s)} \sqrt{\frac{2\pi}{|f''(\lambda_s)|}}\, e^{j\pi/4} \quad (9.6.62)$$

where $\lambda_s = k_1 \sin\theta_1'$.

The phase of the wave as it travels from the dipole to the observation point P is

$$f(\lambda_s) = -j(\lambda_s\rho + \tau_1 h - \tau_2 z) = -j[\lambda_s(\rho_1 + \rho_2) + \tau_1 h + \tau_2 z_o],$$

Since,

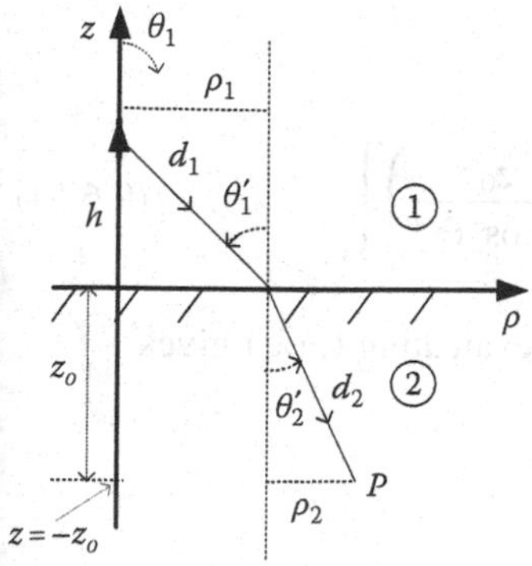

FIGURE 9.16 Incident and refracted wave.

$$\rho_1 = d_1 \sin\theta_1'$$
$$\rho_2 = d_2 \sin\theta_2'$$
$$h = d_1 \cos\theta_1'$$

and

$$z_0 = d_2 \cos\theta_2'$$

it follows that

$$\begin{aligned} f(\lambda_s) &= -j\,[k_1 d_1 \sin^2\theta_1' + k_1 d_2 \sin\theta_1' \sin\theta_2' + k_1 d_1 \cos^2\theta_1' + k_2 d_2 \cos^2\theta_2'] \\ &= -jk_1 d_1 - jk_2 d_2 \end{aligned}$$

where in the second term we used Snell's law. The phase in (9.6.62) shows that

$$e^{f(\lambda_s)} = e^{-jk_1 d_1} e^{-jk_2 d_2} \tag{9.6.63}$$

In other words, the wave from the dipole is incident on the surface at θ_1', traveling a distance d_1 and refracts at an angle θ_2', traveling a distance d_2 to the observation point.

In (9.6.62), the evaluation of $f''(\lambda_s)$ is as follows. Since $d\tau_1/d\lambda = -\lambda/\tau_1$ and $d\tau_2/d\lambda = -\lambda/\tau_2$; then:

$$f'(\lambda) = -j\left(\rho - \frac{\lambda}{\tau_1}h - \frac{\lambda}{\tau_2}z_o\right)$$

$$f''(\lambda) = jh\left[\lambda\frac{d}{d\lambda}\left(\frac{1}{\tau_1}\right) + \frac{1}{\tau_1}\right] + jz_o\left[\lambda\frac{d}{d\lambda}\left(\frac{1}{\tau_2}\right) + \frac{1}{\tau_2}\right] = jh\left[\frac{\lambda^2}{\tau_1^3} + \frac{1}{\tau_1}\right] + jz_o\left[\frac{\lambda^2}{\tau_2^3} + \frac{1}{\tau_2}\right]$$

Hence,

$$f''(\lambda_s) = j\left[\left(\frac{h}{k_1\cos^3\theta_1'}\right) + \left(\frac{z_0}{k_2\cos^3\theta_2'}\right)\right] \tag{9.6.64}$$

Substituting (9.6.63) and (9.6.64) into (9.6.62), and evaluating $G(\lambda_s)$ gives

$$\Pi_{z2}^{e} = -\frac{I\Delta L}{2\pi\omega\varepsilon_o\sqrt{\rho}} \frac{\sqrt{\sin\theta_1{}'}}{(n_2^2\cos\theta_1{}' + n_2\cos\theta_2{}')\sqrt{\frac{h}{\cos^3\theta_1{}'} + \frac{z_o}{n\cos^3\theta_2{}'}}} e^{-j(k_1d_1+k_2d_2)}$$

which can be expressed in the form

$$\Pi_{z2}^{e} = -\frac{I\Delta L}{4\pi\omega\varepsilon_o} T(\theta_1{}') \frac{\sqrt{\sin\theta_1{}'}}{\sqrt{\rho}\cos\theta_1{}'\sqrt{\frac{h}{\cos^3\theta_1{}'} + \frac{z_o}{n_2\cos^3\theta_2{}'}}} e^{-j(k_1d_1+k_2d_2)} \quad (9.6.65)$$

where

$$T(\theta_1{}') = \frac{2\cos\theta_1{}'}{n_2^2\cos\theta_1{}' + \sqrt{n_2^2 - \sin^2\theta_1{}'}}$$

is the Fresnel transmission coefficient for an incident angle $\theta_1{}'$. When the observation point approaches the surface (i.e., $\theta_1{}' \approx \pi/2$), $\Pi_{z,2}^{e} \to 0$ and (9.6.65) fail to predict the fields. This occurs when the saddle point is close to the singularities of (9.6.60). Equation (9.6.65) is valid at large distances, and for observation points far from the surface.

9.7 VERTICAL MAGNETIC HERTZIAN DIPOLE ABOVE A LOSSY SURFACE

The geometry for a magnetic Hertzian dipole above a lossy surface is the same as that in Fig. 9.11 except that I is replaced by I_m. This problem was considered in Section 8.5.2 using rectangular coordinates. In this section, the problem is analyzed using cylindrical coordinates and in terms of the propagation constants γ_1 and γ_2. The fields are derived in terms of the $\mathbf{\Pi}^{m}$vector, which in Region 1 satisfies

$$\left[\frac{1}{\rho}\frac{\partial}{\partial\rho}\left(\rho\frac{\partial}{\partial\rho}\right) + \frac{\partial^2}{\partial z^2} - \gamma_1^2\right]\Pi_{z1}^{m} = -\frac{(I_m\Delta L)}{j\omega\mu_o}\frac{\delta(\rho)\delta(z-h)}{\rho} \quad (9.7.1)$$

where

$$\gamma_1 = j\omega\sqrt{\mu_o\varepsilon_o}$$

In the lossy Region 2, with $\mu_1 = \mu_o$,

$$\left[\frac{1}{\rho}\frac{\partial}{\partial\rho}\left(\rho\frac{\partial}{\partial\rho}\right) + \frac{\partial^2}{\partial z^2} - \gamma_2^2\right]\Pi_{z2}^{m} = 0 \quad (9.7.2)$$

where

$$\gamma_2 = \sqrt{j\omega\mu_o(\sigma_2 + j\omega\varepsilon_2)}$$

The particular solution of (9.7.1) can be expressed in terms of the Sommerfeld-Ott relation in (6.2.30), namely

$$\Pi^{m}_{z1,p} = \frac{(I_m\Delta L)}{j\omega\mu_o}\frac{e^{-\gamma_1 R_1}}{4\pi R_1} = -\frac{(I_m\Delta L)}{4\pi\omega\mu_o}\int_0^\infty \frac{J_0(\lambda\rho)}{\tau_1}e^{-\tau_1|z-h|}\lambda d\lambda \tag{9.7.3}$$

where

$$\tau_1 = \sqrt{\gamma_1^2 + \lambda^2}$$

and

$$R_1 = |\mathbf{r} - \mathbf{r}'| = \sqrt{\rho^2 + (z-h)^2}$$

The radiation condition requires that $\mathrm{Re}(\tau_1) > 0$ as $z \to \infty$.

The complementary solution can be expressed in terms of a reflection coefficient $R^m(\lambda)$ as

$$\Pi^{m}_{z1,c} = -\frac{(I_m\Delta L)}{4\pi\omega\mu_o}\int_0^\infty R^m(\lambda)J_0(\lambda\rho)\,e^{-\tau_1 z}\lambda d\lambda \tag{9.7.4}$$

The total field in Region 1 is the sum of the particular and complementary solution, or

$$\Pi^{m}_{z1} = \Pi^{m}_{z1,p} + \Pi^{m}_{z1,c} \tag{9.7.5}$$

The transmitted fields in the Region 2 is written in terms of a constant $T^m(\lambda)$, which represents a transmission coefficient. That is,

$$\Pi^{m}_{z2} = -\frac{(I_m\Delta L)}{4\pi\omega\mu_o}\int_0^\infty T^m(\lambda)J_0(\lambda\rho)e^{\tau_2 z}\lambda d\lambda \tag{9.7.6}$$

where

$$\tau_2 = \sqrt{\gamma_2^2 + \lambda^2}$$

and $\mathrm{Re}(\tau_2) > 0$.

At $z = 0$, the tangential components of the fields must be continuous, which from (4.4.17) and (4.4.21) in terms of Π_z^{m} are

$$E_{\rho 1} = E_{\rho 2}\Big|_{z=0} \Rightarrow \Pi_{z1}^{\mathrm{m}} = \Pi_{z2}^{\mathrm{m}}\Big|_{z=0}$$

and

$$H_{\phi 1} = H_{\phi 2}\Big|_{z=0} \Rightarrow \frac{\partial \Pi_{z1}^{\mathrm{m}}}{\partial z} = \frac{\partial \Pi_{z2}^{\mathrm{m}}}{\partial z}\Bigg|_{z=0}$$

Appling the boundary conditions to (9.7.5) and (9.7.6) gives

$$\frac{e^{-\tau_1 h}}{\tau_1} + R^m(\lambda) = T^m(\lambda)$$

and

$$\frac{e^{-\tau_1 h}}{\tau_1} - R^m(\lambda) e^{-\tau_1 h} = \frac{\tau_2}{\tau_1} T^m(\lambda)$$

Solving for R^m and T^m gives

$$R^m = \frac{e^{-\tau_1 h}}{\tau_1}\left(\frac{\tau_1 - \tau_2}{\tau_1 + \tau_2}\right) = \left(-1 + \frac{2\tau_1}{\tau_1 + \tau_2}\right)\frac{e^{-\tau_1 h}}{\tau_1} \tag{9.7.7}$$

and

$$T^m = \left(\frac{2}{\tau_1 + \tau_2}\right) e^{-\tau_1 h}$$

Substituting (9.7.7) into (9.7.5) gives

$$\Pi_{z1}^{\mathrm{m}} = \frac{(I_m \Delta L)}{j\omega\mu_o}\left[\frac{e^{-\gamma_1 R_1}}{4\pi R_1} - \frac{e^{-\gamma_1 R_2}}{4\pi R_2} + Q^m\right] \tag{9.7.8}$$

where

$$Q^m = \frac{1}{2\pi}\int_0^{\infty} \frac{1}{\tau_1 + \tau_2} J_0(\lambda\rho)\, e^{-\tau_1 (z+h)} \lambda d\lambda \tag{9.7.9}$$

Consider the case where the radiator and observation points are close to the surface. For this case, $z \approx h \approx 0$ and $R_1 \approx R_2$. Then, in (9.7.8) the first two cancel and the fields follow from

$$\Pi_{z1}^{m} = \frac{(I_m\Delta L)}{j\omega\mu_o}Q^m = \frac{(I\Delta S)}{2\pi}\int_0^\infty \frac{1}{\tau_1+\tau_2}J_0(\lambda\rho)\lambda d\lambda \qquad (9.7.10)$$

where the relation between a loop radiator and a magnetic dipole was used. That is,

$$(I_m\Delta L) = j\omega\mu_o(I\Delta S)$$

where I is the current in the loop and ΔS the loop area.

The integral in (9.7.10) can be evaluated in closed form. Observing that

$$\frac{1}{\tau_1+\tau_2} = \frac{\tau_1-\tau_2}{(\tau_1+\tau_2)(\tau_1-\tau_2)} = \frac{\tau_1-\tau_2}{\gamma_1^2-\gamma_2^2}$$

Then, (9.7.10) reads

$$\Pi_{z1}^{m} = \frac{(I\Delta S)}{2\pi}\frac{1}{\left(\gamma_1^2-\gamma_2^2\right)}\left[\int_0^\infty \tau_1 J_0(\lambda\rho)\lambda d\lambda - \int_0^\infty \tau_2 J_0(\lambda\rho)\lambda d\lambda\right] \qquad (9.7.11)$$

Since

$$\gamma_1^2 - \gamma_2^2 = k_o^2(n_2^2-1)$$

the result in (9.7.11) is similar to that in (8.5.41), and $E_{\phi 1}$ and H_{z1} are given in (8.5.45) and (8.5.46).

9.8 VERTICAL ELECTRIC DIPOLE IN A THREE-LAYER REGION

A vertical electric dipole in a three-layer region is shown in Fig. 9.17. In Region 1, the z component of the Hertz vector satisfies (9.6.1). The particular solution in Region 1 is

$$\Pi_{z1,p}^{e} = \frac{(I\Delta L)}{j\omega\varepsilon_o}\frac{e^{-jk_1R_1}}{4\pi R_1} = -\frac{(I\Delta L)}{4\pi\omega\varepsilon_o}\int_0^\infty \frac{J_0(\lambda\rho)}{\tau_1}e^{-j\tau_1|z-h|}\lambda d\lambda$$

where

$$\tau_1 = \sqrt{k_1^2-\lambda^2}$$

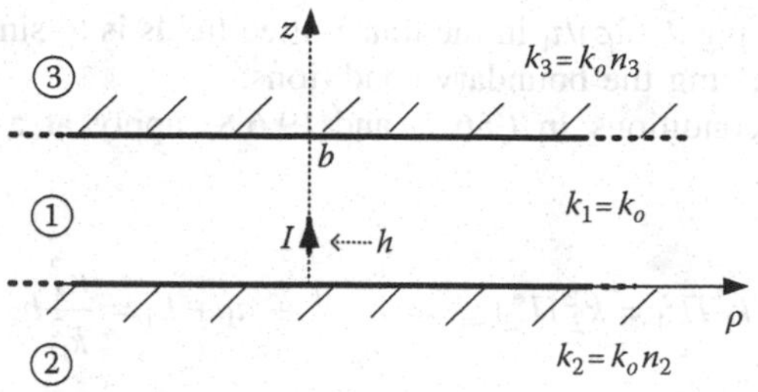

FIGURE 9.17 A three-layer region.

and $k_1 = k_o = \omega\sqrt{\mu_o \varepsilon_o}$. The complementary solution is written in the form

$$\Pi^{e}_{z1,c} = -\frac{(I\Delta L)}{4\pi\omega\varepsilon_o}\int_0^{\infty}\left[a_1(\lambda)e^{-j\tau_1 z} + b_1(\lambda)e^{j\tau_1 z}\right]\frac{J_0(\lambda\rho)}{\tau_1}\lambda d\lambda$$

Hence,

$$\Pi^{e}_{z1} = \Pi^{e}_{z1,p} + \Pi^{e}_{z1,c} = -\frac{(I\Delta L)}{4\pi\omega\varepsilon_o}\int_0^{\infty}\left[e^{-j\tau_1|z-h|} + a_1(\lambda)e^{-j\tau_1 z} + b_1(\lambda)e^{j\tau_1 z}\right]\frac{J_0(\lambda\rho)}{\tau_1}\lambda d\lambda \tag{9.8.1}$$

In Regions 2 and 3, the solutions are

$$\Pi^{e}_{z2} = -\frac{(I\Delta L)}{4\pi\omega\varepsilon_o}\int_0^{\infty}\left[b_2(\lambda)e^{j\tau_2 z}\right]\frac{J_0(\lambda\rho)}{\tau_1}\lambda d\lambda$$

and

$$\Pi^{e}_{z3} = -\frac{(I\Delta L)}{4\pi\omega\varepsilon_o}\int_0^{\infty}\left[a_3(\lambda)e^{-j\tau_3(z-b)}\right]\frac{J_0(\lambda\rho)}{\tau_1}\lambda d\lambda$$

where

$$\tau_2 = \sqrt{k_2^2 - \lambda^2}$$

$$\tau_3 = \sqrt{k_3^2 - \lambda^2}$$

$$k_2 = k_o n_2$$

and

$$k_3 = k_o n_3$$

The reason for writing $J_0(\lambda\rho)/\tau_1$ in the transmitted fields is to simplify the resulting relations when applying the boundary conditions.

The boundary conditions in (9.6.7) and (9.6.8) apply at $z = 0$ and at $z = b$. That is,

$$k_1^2 \Pi_{z1}^e = k_2^2 \Pi_{z2}^e|_{z=0} \Rightarrow e^{-j\tau_1 h} + a_1 + b_1 = \frac{k_2^2}{k_1^2} b_2$$

$$\frac{\partial \Pi_{z1}^e}{\partial z} = \frac{\partial \Pi_{z2}^e}{\partial z}\bigg|_{z=0} \Rightarrow e^{-j\tau_1 h} - a_1 + b_1 = \frac{\tau_2}{\tau_1} b_2$$

$$k_1^2 \Pi_{z1}^e = k_3^2 \Pi_{z3}^e|_{z=b} \Rightarrow e^{-j\tau_1 (b-h)} + a_1 e^{-j\tau_1 b} + b_1 e^{j\tau_1 b} = \frac{k_3^2}{k_1^2} a_3$$

$$\frac{\partial \Pi_{z1}^e}{\partial z} = \frac{\partial \Pi_{z3}^e}{\partial z}\bigg|_{z=b} \Rightarrow e^{-j\tau_1 (b-h)} + a_1 e^{-j\tau_1 b} - b_1 e^{j\tau_1 b} = \frac{\tau_3}{\tau_1} a_3$$

which can be expressed as

$$\begin{bmatrix} 1 & 1 & -n_2^2 & 0 \\ 1 & -1 & \tau_2/\tau_1 & 0 \\ e^{-j\tau_1 b} & e^{j\tau_1 b} & 0 & -n_3^2 \\ e^{-j\tau_1 b} & -e^{j\tau_1 b} & 0 & -\tau_3/\tau_1 \end{bmatrix} \begin{bmatrix} a_1 \\ b_1 \\ b_2 \\ a_3 \end{bmatrix} = \begin{bmatrix} -e^{-j\tau_1 h} \\ e^{-j\tau_1 h} \\ -e^{-j\tau_1 (b-h)} \\ -e^{-j\tau_1 (b-h)} \end{bmatrix} \quad (9.8.2)$$

and solved for the constants.

An alternate way to solve for the constants in (9.8.2) is to observe that the term in bracket in (9.8.1) represent a plane wave term. From (9.8.1), the incident wave into the lower region is

$$e^{-j\tau_1 (h-z)} + b_1(\lambda) e^{j\tau_1 z}$$

and the reflected wave is

$$a_1(\lambda) e^{-j\tau_1 z}$$

The ratio of these waves must be equal to the to the plane-wave reflection coefficient at the lower region (i.e., at $z = 0$), denoted by R_{12}. That is,

$$R_{12} = \frac{a_1}{e^{-j\tau_1 h} + b_1} \quad (9.8.3)$$

which from (9.6.26) must be equal to

$$R_{12} = \frac{n_2^2\tau_1 - \tau_2}{n_2^2\tau_1 + \tau_2} \tag{9.8.4}$$

Similarly, from (9.8.1) the incident wave into the upper region is

$$e^{-j\tau_1(z-h)} + a_1(\lambda)e^{-j\tau_1 z}$$

and the reflected wave is

$$b_1(\lambda)e^{j\tau_1 z}$$

The ratio of these waves must be equal to the to the plane-wave reflection coefficient at the upper region (i.e., at $z = b$), denoted by R_{13}. That is,

$$R_{13} = \frac{b_1 e^{j\tau_1 h}}{e^{-j\tau_1(b-h)} + a_1 e^{-j\tau_1 b}} \tag{9.8.5}$$

where

$$R_{13} = \frac{n_3^2\tau_1 - \tau_3}{n_3^2\tau_1 + \tau_3} \tag{9.8.6}$$

Solving (9.8.3) and (9.8.5) for a_1 and b_1 in terms of R_{12} and R_{13} gives

$$a_1 = R_{12}\frac{e^{-j\tau_1 h} + R_{13}e^{-j\tau_1(2b-h)}}{1 - R_{12}R_{13}e^{-j2\tau_1 b}} \tag{9.8.7}$$

and

$$b_1 = R_{13}\frac{1 + R_{12}e^{-j2\tau_1 h}}{1 - R_{12}R_{13}e^{-j2\tau_1 b}}e^{-j\tau_1(2b-h)} \tag{9.8.8}$$

where R_{12} and R_{13} are given by (9.8.4) and (9.8.6).

Similarly, for the transmission coefficient at $z = 0$, denoted by T_{12}, is

$$T_{12} = \frac{b_2}{e^{-j\tau_1 h} + b_1} = \frac{2\tau_1}{n_2^2\tau_1 + \tau_2}$$

and the transmission coefficient at $z = b$, denoted by T_{13}, is

$$T_{13} = \frac{a_3}{e^{-j\tau_1(b-h)} + a_1 e^{-j\tau_1 b}} = \frac{2\tau_1}{n_3^2\tau_1 + \tau_3}$$

Solving for b_2 and a_3 gives:

$$b_2 = T_{12}e^{-j\tau_1 h}\left[\frac{1 + R_{13}e^{-j2\tau_1(b-h)}}{1 - R_{12}R_{13}e^{-j2\tau_1 b}}\right]$$

and

$$a_3 = T_{13}e^{-j\tau_1(b-h)}\left[\frac{1 + R_{12}e^{-j2\tau_1 h}}{1 - R_{12}R_{13}e^{-j2\tau_1 b}}\right]$$

Substituting (9.8.7) and (9.8.8) into (9.8.1) gives

$$\Pi_{z1}^{e} = -\frac{(I\Delta L)}{4\pi\omega\varepsilon_o}\int_0^\infty [e^{-j\tau_1|z-h|} + R_{12}\frac{e^{-j\tau_1 h} + R_{13}e^{-j\tau_1(2b-h)}}{1 - R_{12}R_{13}e^{-j2\tau_1 b}}e^{-j\tau_1 z} + R_{13}\frac{1 + R_{12}e^{-j2\tau_1 h}}{1 - R_{12}R_{13}e^{-j2\tau_1 b}}e^{-j\tau_1(2b-h)}e^{j\tau_1 z}]\frac{J_0(\lambda\rho)}{\tau_1}\lambda d\lambda$$

which can be expressed in the form

$$\Pi_{z1}^{e} = -\frac{(I\Delta L)}{4\pi\omega\varepsilon_o}\int_0^\infty F(\lambda)\frac{J_0(\lambda\rho)}{\tau_1}\lambda d\lambda$$

where

$$F(\lambda) = \begin{cases} \dfrac{(e^{j\tau_1 h} + R_{12}e^{-j\tau_1 h})(e^{j\tau_1(b-z)} + R_{13}e^{-j\tau_1(b-z)})}{e^{j\tau_1 b}(1 - R_{12}R_{13}e^{-j2\tau_1 b})} & (z > h) \\ \dfrac{(e^{j\tau_1 z} + R_{12}e^{-j\tau_1 z})(e^{j\tau_1(b-h)} + R_{13}e^{-j\tau_1(b-h)})}{e^{j\tau_1 b}(1 - R_{12}R_{13}e^{-j2\tau_1 b})} & (z < h) \end{cases} :$$

If the boundaries at $z = 0$ and $z = b$ are perfect conductors (i.e., $n_2 = n_3 = \infty$), then $R_{12} = R_{13} = 1$ and it follows that

$$F(\lambda) = \begin{cases} -2j\dfrac{\cos(\tau_1 h)\cos[\tau_1(b-z)]}{\sin(\tau_1 b)} & (z > h) \\ -2j\dfrac{\cos(\tau_1 z)\cos[\tau_1(b-h)]}{\sin(\tau_1 b)} & (z < h) \end{cases}$$

For this case, Π_{z1}^{e} reads

$$\Pi_{z1}^{e} = \begin{cases} j\dfrac{(I\Delta L)}{2\pi\omega\varepsilon_o}\displaystyle\int_0^{\infty} \dfrac{\cos(\tau_1 h)\cos[\tau_1(b-z)]}{\tau_1 \sin(\tau_1 b)} J_0(\lambda\rho)\lambda d\lambda & (z > h) \\ j\dfrac{(I\Delta L)}{2\pi\omega\varepsilon_o}\displaystyle\int_0^{\infty} \dfrac{\cos(\tau_1 z)\cos[\tau_1(b-h)]}{\tau_1 \sin(\tau_1 b)} J_0(\lambda\rho)\lambda d\lambda & (z < h) \end{cases}$$

or

$$\Pi_{z1}^{e} = \begin{cases} j\dfrac{(I\Delta L)}{4\pi\omega\varepsilon_o}\displaystyle\int_{-\infty}^{\infty} \dfrac{\cos(\tau_1 h)\cos[\tau_1(b-z)]}{\tau_1 \sin(\tau_1 b)} H_0^{(2)}(\lambda\rho)\lambda d\lambda & (z > h) \\ j\dfrac{(I\Delta L)}{4\pi\omega\varepsilon_o}\displaystyle\int_{-\infty}^{\infty} \dfrac{\cos(\tau_1 z)\cos[\tau_1(b-h)]}{\tau_1 \sin(\tau_1 b)} H_0^{(2)}(\lambda\rho)\lambda d\lambda & (z < h) \end{cases} \tag{9.8.9}$$

Equation (9.8.9) is evaluated using residue theory by closing the contour in the lower-half plane with a large-radius semicircle. The contribution from the semicircular path must vanish as its radius goes to infinity. We have seen that $H_0^{(2)}(\lambda\rho)$ decays exponentially along the semicircular path in the lower-half plane. The vanishing of the integrand in (9.8.9) along this semicircular path is left to the problems.

The poles in (9.8.9) are located at

$$\sin(\tau_1 b) = 0 \Rightarrow \tau_1 = \frac{n\pi}{b} \qquad (n = 0, \pm 1, \pm 2, \ldots)$$

or at

$$\lambda_n = \pm\sqrt{k_1^2 - \tau_1^2} = \pm\sqrt{k_1^2 - \left(\frac{n\pi}{b}\right)^2} \tag{9.8.10}$$

An analysis of (9.8.10) illustrates the distribution of the poles. To this end, let k_1 have a small imaginary part (i.e., $k_1 = k_1{}' - k_1{}''$ with $k_1{}' \gg k_1{}''$). Then, (9.8.10) shows that

$$\lambda_n = \begin{cases} \pm\sqrt{k_1'^2 - \left(\dfrac{n\pi}{b}\right)^2} \mp j\dfrac{k_1{}'k_1{}''}{\sqrt{k_1'^2 - \left(\frac{n\pi}{b}\right)^2}} & \left(k_1{}' > \dfrac{n\pi}{b}\right) \\ \pm\dfrac{k_1{}'k_1{}''}{\left(\frac{n\pi}{b}\right)} \mp j\left(\dfrac{n\pi}{b}\right) & \left(k_1{}' < \dfrac{n\pi}{b}\right) \end{cases}$$

The above relations show that if Region 1 is lossless (i.e., $k_1{}'' = 0$), the location of the poles when $k_1 > n\pi/b$ (i.e., $n < k_1 b/\pi$) are on the real axis, and if $k_1 < n\pi/b$ (i.e.,

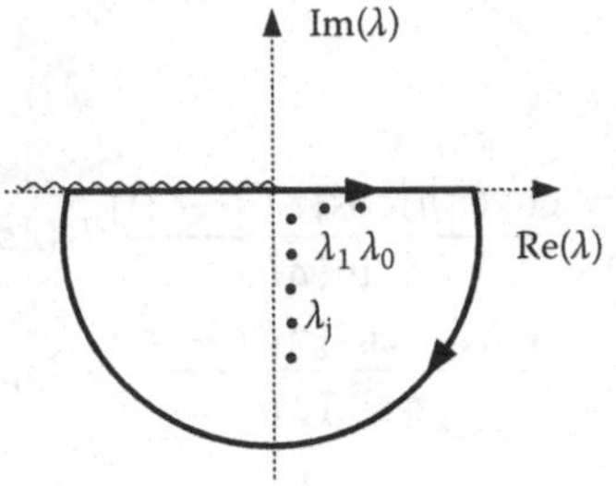

FIGURE 9.18 Path of integration.

$n > k_1 b/\pi$), the poles are on the imaginary axis. If Region 1 has a small loss, the location of the poles in the lower-half plane is shown in Fig. 9.18.

There is the branch point due to the Hankel function and also branch points at $\lambda = \pm k_1$ (i.e., $\tau_1 = 0$). The integrands in (9.8.9) are an even function of τ_1 so the branch cut contribution is zero, since one side cancels the other. Thus, the branch cut is not shown in Fig. 9.18. The path of integration enclosing the poles can be closed in the lower-half plane since the contribution from the semicircle vanish as its radius goes to infinity.

The integral in (9.8.9) for $z < h$ is of the form

$$\Pi_z^e = \int_{-\infty}^{\infty} \frac{G(\lambda)}{\sin(\tau_1 b)} d\lambda \tag{9.8.11}$$

where $G(\lambda)$ represents the other terms in (9.8.9). Using residue theory:

$$\int_{-\infty}^{\infty} \frac{G(\lambda)}{\sin(\tau_1 b)} d\lambda = -2\pi j \frac{G(\lambda_n)}{\left.\frac{d[\sin(\tau_1 b)]}{d\lambda}\right|_{\lambda=\lambda_n}} = 2\pi j \frac{G(\lambda_n)}{(-1)^n \frac{\lambda_n b}{\tau_1(\lambda_n)}}$$

where $\tau_1(\lambda_n) = n\pi/b$ and

$$G(\lambda_n) = j\frac{(I\Delta L)}{4\pi\omega\varepsilon_o}\frac{\lambda_n}{\tau_1(\lambda_n)}(-1)^n \cos\left(\frac{n\pi z}{b}\right)\cos\left(\frac{n\pi h}{b}\right)H_0^{(2)}(\lambda_n \rho) \qquad (z < h)$$

Hence, (9.8.9) for $z < h$ reads:

$$\begin{aligned} \Pi_z^e &= -\frac{(I\Delta L)}{2\omega\varepsilon_o b}\sum_{n=-\infty}^{\infty} \cos\left(\frac{n\pi z}{b}\right)\cos\left(\frac{n\pi h}{b}\right)H_0^{(2)}(\lambda_n\rho) \\ &= -\frac{(I\Delta L)}{2\omega\varepsilon_o b}\sum_{n=0}^{\infty} \varepsilon_n \cos\left(\frac{n\pi z}{b}\right)\cos\left(\frac{n\pi h}{b}\right)H_0^{(2)}(\lambda_n\rho) \end{aligned} \tag{9.8.12}$$

where ε_n is Neumann's number. The result for $z > h$ is the same. Hence, (9.8.12) represents the fields for any value of z. At distances far from the source the asymptotic form of the Hankel function can be used. The asymptotic form of the Hankel function show that the roots with imaginary parts do not contribute much to the sum.

9.9 LATERAL WAVES

Lateral waves occur when a dipole radiates from a region having a large magnitude value of the index of refraction to a region with a smaller magnitude value of the index of refraction. The geometry of the problem is shown in Fig. 9.19. For example, when Region 1 represents the Earth's ground or a dense forest environment and Region 2 is air. For this case: $|n_1| > 1$, $n_2 = 1$ and $|n_{21}| = |n_2/n_1| < 1$. Another example is when Region 1 is air and Region 2 represents the ionosphere. The index of refraction of the Ionosphere for frequencies much higher than the collision frequency ν, and neglecting any anisotropy, is given by

$$n_2 = \sqrt{\varepsilon_{r2}} \approx \sqrt{1 - \left(\frac{\omega_p}{\omega}\right)^2} + j\frac{\nu\omega_p^2}{2\omega^3[1 - (\omega_p/\omega)^2]^{1/2}}$$

where ω_p is the plasma frequency. Thus, $n_2 = n_2' + jn_2''$ where $n_2' \gg n_2''$ and in the ideal case $n_2'' \to 0$. With $n_2'' \to 0$:

$$n_2 \approx \sqrt{1 - \left(\frac{\omega_p}{\omega}\right)^2}$$

In this case: $n_1 = 1$, $n_2 < 1$ and $n_{21} = n_2/n_1 < 1$.

The analysis will show that for certain angles of incidence the contribution from the branch-point integration in the reflected field is the important one and represents a lateral wave.

From the result in (9.6.11) the total field in Region 1 is given by

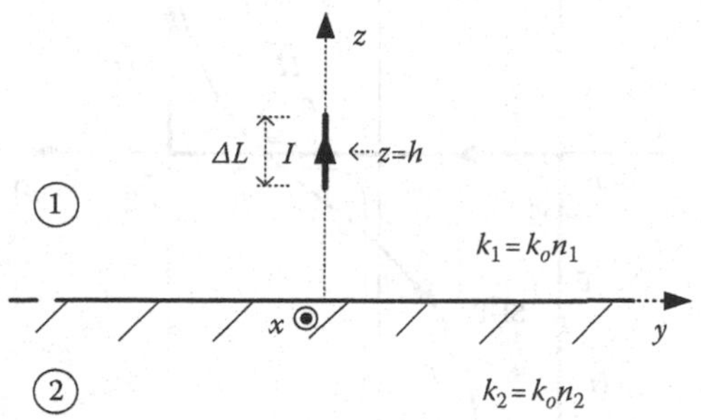

FIGURE 9.19 The dipole is located in the dense Region 1 and Region 2 is such that $|n_{21}| = |n_2/n_1| < 1$.

$$\Pi^{e}_{z1} = \frac{(I\Delta L)}{j\omega\varepsilon_o}\frac{e^{-jk_1R_1}}{4\pi R_1} - \frac{(I\Delta L)}{4\pi\omega\varepsilon_o}\int_0^{\infty}\frac{1}{\tau_1}\left[\frac{n_{21}^2\tau_1 - \tau_2}{n_{21}^2\tau_1 + \tau_2}\right]J_0(\lambda\rho)e^{-j\tau_1(z+h)}\lambda d\lambda$$

where $|n_{21}| = |n_2/n_1| < 1$. The complementary solution represents the reflected part of the field and as shown in (9.6.23) it is expressed as

$$\Pi^{e}_{z1,c} = -\frac{(I\Delta L)}{8\pi\omega\varepsilon_o}\sqrt{\frac{2k_1}{\pi R_2 \sin\theta}}\, e^{j\pi/4}\int_{\Gamma}\left[\frac{n_{21}^2\cos\beta - \sqrt{n_{21}^2 - \sin^2\beta}}{n_{21}^2\cos\beta + \sqrt{n_{21}^2 - \sin^2\beta}}\right] \times \sqrt{\sin\beta}\, e^{-jk_1R_2\cos(\beta-\theta)}d\beta \qquad (9.9.1)$$

Figure 9.20 shows the path of integration Γ in the β plane, the path of steepest descent (SDP) and the branch cut when $\theta < \theta_B$. The angle θ_B is the value of θ when the SDP passes thru β_B. The branch points are located at

$$\sin^2\beta_B = n_{21}^2 \Rightarrow \beta_B = \pm\sin^{-1} n_{21}$$

and the cut is taken along $\mathrm{Im}(\tau_2) = 0$. The branch point is on the real axis if n_2 is real, and slightly away if n_2 is complex with $n_2' \gg n_2''$. The branch point and cut are drawn in Fig. 9.20 for the case where $n_2' \gg n_2''$. On the branch cut:

$$\mathrm{Im}(\tau) = \mathrm{Im}\sqrt{n_{21}^2 - \sin^2\beta} = 0$$

and following a similar procedure to that in (8.5.20) and (8.5.21) shows that the equation for the cut is a hyperbola with asymptotes $\beta' \approx n_{21}{}' n_{21}{}'' e^{-|\beta''|}$ as $\beta'' \to \pm\infty$.

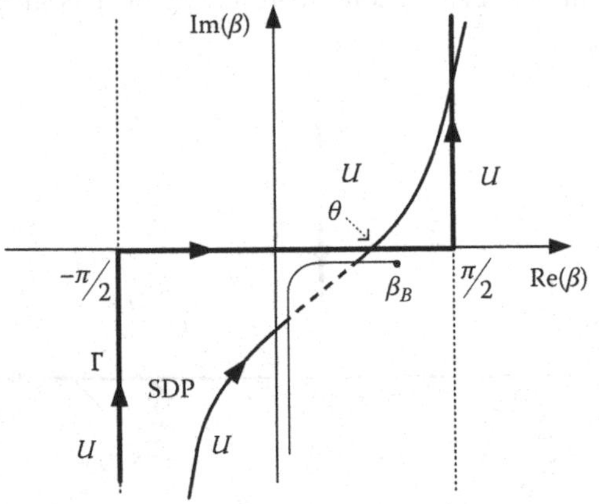

FIGURE 9.20 Path of integration and steepest descent path for $\theta < \theta_B$.

Figure 9.20 shows that for an observation point such that $\theta < \theta_B$ the steepest descent path crosses the branch cut twice and ends in the upper Riemann sheet. For this case, the branch point does not affect the result and the saddle point evaluation of (9.9.1) was done in Section 9.6, leading to

$$\Pi_{z1,c}^{e} = \frac{(I\Delta L)}{j\omega\varepsilon_o} R(\theta) \frac{e^{-jk_1R_2}}{4\pi R_2} \tag{9.9.2}$$

The total field in Region 1 is

$$\Pi_{z1}^{e} = \frac{(I\Delta L)}{j\omega\varepsilon_o} \frac{e^{-jk_1R_1}}{4\pi R_1} + \frac{(I\Delta L)}{j\omega\varepsilon_o} R(\theta) \frac{e^{-jk_1R_2}}{4\pi R_2}$$

If the observation point is such that $\theta > \theta_B$, the path of integration can be changed to the path Γ_1 which encloses the branch cut and passes through the saddle point, as shown in Fig. 9.21a . The path Γ_1 can then be changed to the path shown in Fig. 9.21b, which is equal to the path of steepest descent (SDP) plus the path around the branch cut (BC). That is,

$$\int_{\Gamma_1} \ldots = \int_{SDP} \ldots + \int_{BC} \ldots \tag{9.9.3}$$

In the case that the index of refraction in both regions are real and $n_{21} = n_2/n_1 < 1$, the angle $\theta = \theta_B$ is related to the critical angle where total reflection occurs. From Snell's law the critical angle occurs when the transmitted angle is 90° and is given by

$$\sin\theta_c = n_{21}$$

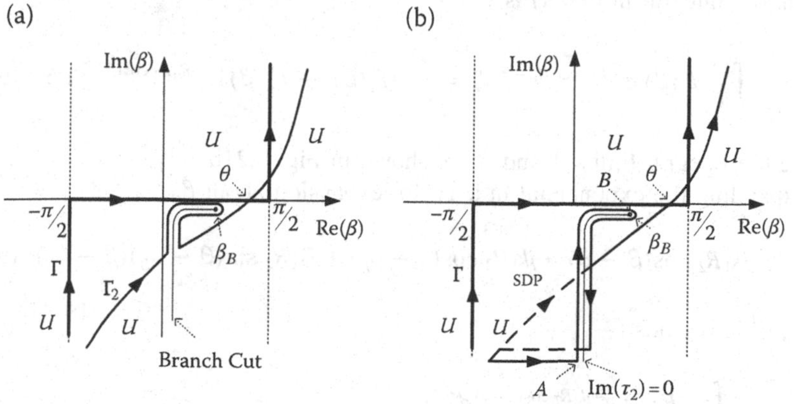

FIGURE 9.21 (a) Path of integration Γ_1 for $\theta > \theta_B$; (b) steepest descent path and path around the branch cut for $\theta > \theta_B$.

which is the same relation for the branch point (i.e., $\beta_B = \theta_c$). Thus, for an angle of observation such that $\theta < \theta_B$, the angle of incidence is less than the critical angle, and for $\theta > \theta_B$ the angle of incidence is greater than the critical angle.

The SDP integral in (9.9.3) is given by (9.9.2). The branch point integral in (9.9.3) is now considered. Let I_{BC} represent the branch cut integral. That is,

$$I_{BC} = -\frac{(I\Delta L)}{8\pi\omega\varepsilon_o}\sqrt{\frac{2k_1}{\pi R_2 \sin\theta}}\, e^{j\pi/4}\int_{BC} F(\beta)\, e^{-jk_1 R_2 \cos(\beta-\theta)}\, d\beta \qquad (9.9.4)$$

where

$$F(\beta) = \frac{n_{21}^2 \cos\beta - \sqrt{n_{21}^2 - \sin^2\beta}}{n_{21}^2 \cos\beta + \sqrt{n_{21}^2 - \sin^2\beta}}\sqrt{\sin\beta} \qquad (9.9.5)$$

Referring to the discussion in Section 8.5, to the left of the branch cut $\mathrm{Re}(\tau_2) > 0$ and to the right $\mathrm{Re}(\tau_2) < 0$. Denote the value of $F(\beta)$ to the left of the branch cut by $F_1(\beta)$:

$$F_1(\beta) = \frac{n_{21}^2 \cos\beta - \sqrt{n_{21}^2 - \sin^2\beta}}{n_{21}^2 \cos\beta + \sqrt{n_{21}^2 - \sin^2\beta}}\sqrt{\sin\beta}$$

and to the right of the branch cut by $F_2(\beta)$:

$$F_2(\beta) = \frac{n_{21}^2 \cos\beta + \sqrt{n_{21}^2 - \sin^2\beta}}{n_{21}^2 \cos\beta - \sqrt{n_{21}^2 - \sin^2\beta}}\sqrt{\sin\beta}$$

Then, the integral in (9.9.4) is

$$\int_{BC} F(\beta)\, e^{-jk_1 R_2 \cos(\beta-\theta)}\, d\beta = \int_A^B [F_1(\beta) - F_2(\beta)]\, e^{-jk_1 R_2 \cos(\beta-\theta)}\, d\beta \qquad (9.9.6)$$

where the integral limits A and B are shown in Fig. 9.21b.

Expanding the exponential in a Taylor expansion about β_B:

$$jk_1 R_2 \cos(\beta - \theta) \approx jk_1 R_2 \cos(\theta - \beta_B) + jk_1 R_2 \sin(\theta - \beta_B)(\beta - \beta_B) \qquad (9.9.7)$$

Then, (9.9.6) reads

$$\int_{BC} F(\beta)\, e^{-jk_1 R_2 \cos(\beta-\theta)}\, d\beta$$
$$= e^{-jk_1 R_2 \cos(\theta-\beta_B)} \int_A^B [F_1(\beta) - F_2(\beta)]\, e^{-jk_1 R_2 \sin(\theta-\beta_B)(\beta-\beta_B)}\, d\beta \qquad (9.9.8)$$

The expansion in (9.9.7) and the integral in (9.9.8) suggest the use of the transformation:

$$\beta - \beta_B = -js$$

This transformation changes the path of integration to the s plane in a way that most of the contribution comes from the value of s close to zero (i.e., steepest descent path), which corresponds to values of β close to the branch point. The resulting path of integration in the s plane, by analytic continuation, can be moved to be along the real axis extending from ∞ to 0. Then, changing the limits of the integral to 0 to ∞, (9.9.8) reads

$$\int_{BC} F(\beta)\, e^{-jk_1R_2\cos(\beta-\theta)} d\beta = -je^{-jk_1R_2\cos(\theta-\beta_B)} \int_0^\infty [F_2(s) - F_1(s)] e^{-k_1R_2\sin(\theta-\beta_B)\, s}\, ds \tag{9.9.9}$$

Next, the term $F_2(s) - F_1(s)$ needs to be expanded around $s = 0$ (or $\beta = \beta_B$), which reads

$$F_2(\beta) - F_1(\beta) = \frac{4n_{21}^2 \cos\beta\sqrt{n_{21}^2 - \sin^2\beta}}{n_{21}^4\cos^2\beta - (n_{21}^2 - \sin^2\beta)}\sqrt{\sin\beta} \tag{9.9.10}$$

Around $\beta = \beta_B$, the first term of the Taylor expansion shows that

$$n_{21}^2 - \sin^2\beta \approx -2\sin\beta_B\cos\beta_B(\beta - \beta_P)$$

which in terms of s is

$$n_{21}^2 - \sin^2\beta \approx (2j\, n_{21}\cos\beta_B)\, s \;\Rightarrow\; \sqrt{n_{21}^2 - \sin^2\beta} \approx \sqrt{2n_{21}\cos\beta_B}\; e^{j\pi/4}\sqrt{s}$$

Hence, (9.9.10) near $s = 0$ can be approximated by

$$F_2(s) - F_1(s) \approx \frac{4n_{21}^2\cos\beta_B\sqrt{2n_{21}\cos\beta_B}\, e^{j\pi/4}}{n_{21}^4\cos^2\beta_B}\sqrt{n_{21}}\sqrt{s} \approx \frac{4\sqrt{2}\, e^{j\pi/4}}{n_{21}\sqrt{\cos\beta_B}}\sqrt{s}$$

and (9.9.9) reads

$$\int_{BC} F(\beta)\, e^{-jk_1R_2\cos(\beta-\theta)} d\beta$$

$$= -j\, e^{-jk_1R_2\cos(\theta-\beta_B)} \frac{4\sqrt{2}}{n_{21}\sqrt{\cos\beta_B}}\, e^{j\pi/4}\int_0^\infty \sqrt{s}\, e^{-k_1R_2\sin(\theta-\beta_B)\, s}\, ds$$

Using the integral

$$\int_0^{\infty} \sqrt{s}\, e^{-\alpha s} ds = \frac{\sqrt{2\pi}}{(2\alpha)^{3/2}} \qquad (\alpha > 0)$$

it follows that

$$\int_{BC} F(\beta)\, e^{-jk_1R_2\cos(\beta-\theta)} d\beta = -j\, e^{j\pi/4} e^{-jk_1R_2\cos(\theta-\beta_B)} \frac{2\sqrt{2\pi}}{n_{21}\sqrt{\cos\beta_B}} \frac{1}{[k_1R_2\sin(\theta-\beta_B)]^{3/2}}$$

and (9.9.4) reads

$$I_{BC} = -\frac{(I\Delta L)}{2\pi\omega\varepsilon_o k_1 n_{21}} \sqrt{\frac{1}{R_2\sin\theta\cos\beta_B}} \frac{e^{-jk_1R_2\cos(\theta-\beta_B)}}{[R_2\sin(\theta-\beta_B)]^{3/2}} \qquad (9.9.11)$$

The exponent in (9.9.11) is interpreted in terms of a lateral wave. Referring to Fig. 9.22:

$$z = d_2\cos\beta_B$$
$$h = d_0\cos\beta_B$$
$$\rho = d_0\sin\beta_B + d_1 + d_2\sin\beta_B$$

and since: $\rho = R_2\sin\theta$ and $z + h = R_2\cos\theta$, the exponent can be expressed in the form

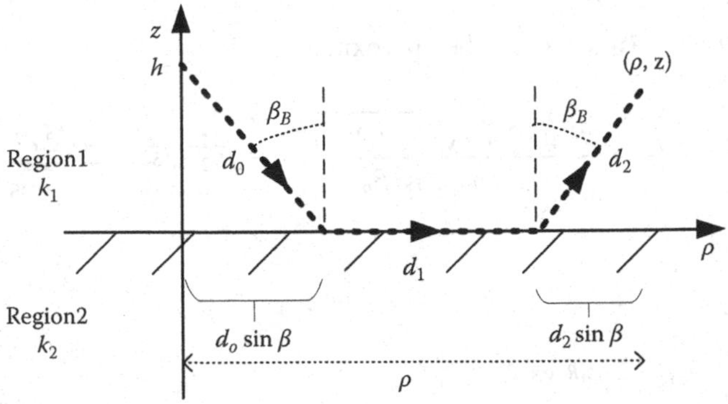

FIGURE 9.22 The lateral wave.

$$\begin{aligned} k_1 R_2 \cos(\theta - \beta_B) &= k_1 R_2 \cos\theta \cos\beta_B + k_1 R_2 \sin\theta \sin\beta_B \\ &= k_1 (z+h)\cos\beta_B + k_1 \rho \sin\beta_B \\ &= k_1 d_2 + k_1 d_0 + k_1 d_1 \sin\beta_B \\ &= k_1 d_0 + k_1 n_2 d_1 + k_1 d_2 \end{aligned}$$

Also, in the denominator:

$$\begin{aligned} R_2 \sin(\theta - \beta_B) &= R_2 \sin\theta \cos\beta_B - R_2 \cos\theta \sin\beta_B \\ &= \rho \cos\beta_B - (z+h)\sin\beta_B \\ &= d_1 \cos\beta_B \end{aligned}$$

and

$$\cos^2\beta_B = 1 - \sin^2\beta = 1 - n_{21}^2$$

Then, (9.9.11) reads

$$\Pi_{z1,c}^{e} = I_{BC} = -\frac{(I\Delta L)}{2\pi\omega\varepsilon_o k_1 n_2 (1 - n_2^2)} \frac{1}{d_1^{3/2}\sqrt{\rho}} e^{-j(k_1 d_0 + k_1 n_2 d_1 + k_1 d_2)} \qquad (9.9.12)$$

Equation (9.9.12) shows that the wave propagates a distance d_0 in free space, a distance d_1 along the surface with the propagation constant $k_2 = k_1 n$, and then a distance d_2 to the observation point. As it travels through the surface it radiates energy into the upper region. The wave attenuates according to $d_1^{3/2}\sqrt{\rho}$, which shows that (9.9.12) is infinite as $d_1 \to 0$. This happens because as $d_1 \to 0$ the saddle point gets closer to the branch point and the separation of the saddle point and branch point integration cannot be performed separately.

9.10 AN INFINITE-LENGTH SLOT IN A CIRCULAR CYLINDER

Consider the case of a thin-slot aperture of infinite-length in a conducting cylinder, as shown in Fig. 9.23. Such a radiator approximates a long-slot cut in the wall of a circular waveguide. The waveguide is excited so the field along the slot has a constant value of its ϕ component (i.e., say a TE_{11} mode). The field across the aperture is denoted by

$$E_\phi(a, \phi) = E_A \qquad (-\phi_o \le \phi \le \phi_o) \qquad (9.10.1)$$

where E_A represents the constant value of the field over the aperture.

Equations (4.4.15) and (4.4.18) show that an E_ϕ field with no z variation requires the use of the F_z or Π_z^m potentials. In the region $\rho > a$, there are no sources and because of the infinite-length slot the fields are independent z. From (4.4.17) to (4.4.22), the fields in terms of Π_z^m are

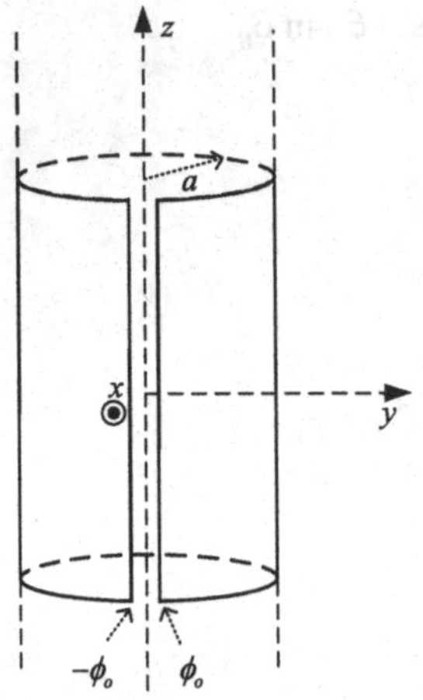

FIGURE 9.23 A thin slot of infinite length in a conducting cylinder.

$$E_\rho = -\frac{j\omega\mu}{\rho}\frac{\partial \Pi_z^{\mathrm{m}}}{\partial \phi}$$

$$E_\phi = j\omega\mu\frac{\partial \Pi_z^{\mathrm{m}}}{\partial \rho}$$

$$H_z = k^2 \Pi_z^{\mathrm{m}}$$

where $k = \omega\sqrt{\mu\varepsilon}$ and Π_z^{m} satisfies

$$\left[\frac{1}{\rho}\frac{\partial}{\partial \rho}\left(\rho\frac{\partial}{\partial \rho}\right) + \frac{1}{\rho^2}\frac{\partial^2}{\partial \phi^2} + k^2\right]\Pi_z^{\mathrm{m}} = 0 \tag{9.10.2}$$

The solution to (9.10.2), observing that the fields must be periodic in ϕ with period of 2π, is of the form

$$\Pi_z^{\mathrm{m}} = \sum_{n=-\infty}^{\infty} F_n(\rho)\, e^{jn\phi} \tag{9.10.3}$$

Substituting (9.10.3) into (9.10.2) shows that

$$\left[\frac{1}{\rho}\frac{\partial}{\partial \rho}\left(\rho\frac{\partial}{\partial \rho}\right) - \frac{n^2}{\rho^2} + k^2\right]F_n(\rho) = 0$$

whose solution that decays as $\rho \to \infty$ is $H_n^{(2)}(k\rho)$. Hence, the solution in (9.10.3) reads

$$\Pi_z^{\mathrm{m}} = \sum_{n=-\infty}^{\infty} B_n H_n^{(2)}(k\rho)\, e^{jn\phi}$$

The electric field E_ϕ must satisfy the boundary condition at the aperture in (9.10.1). Since,

$$E_\phi = j\omega\mu \frac{\partial \Pi_z^m}{\partial \rho} = j\omega\mu k \sum_{n=-\infty}^{\infty} B_n H_n^{(2)\prime}(k\rho)\, e^{jn\phi}$$

Then, at the aperture:

$$E_\phi(a, \phi) = E_A = j\omega\mu k \sum_{n=-\infty}^{\infty} B_n H_n^{(2)\prime}(ka)\, e^{jn\phi} \qquad -\phi_o \le \phi \le \phi_o$$

which is a Fourier expansion, whose coefficients follow from

$$j\omega\mu k B_n H_n^{(2)\prime}(ka) = \frac{1}{2\pi}\int_{-\phi_o}^{\phi_o} E_A\, e^{-jn\phi} d\phi = \frac{E_A}{\pi n} \sin n\phi_o \tag{9.10.4}$$

Associated with E_A, a voltage V can be defined across the aperture, whose length is $a(2\phi_o)$. That is,

$$E_A = \frac{V}{a(2\phi_o)}$$

and from (9.10.4):

$$B_n = -j\frac{V}{2\pi a\omega\mu k} \frac{\sin n\phi_o}{n\phi_o} \frac{1}{H_n^{(2)\prime}(ka)}$$

Hence, the Π_z^m component is

$$\Pi_z^m = -j\frac{V}{2\pi a\omega\mu k} \sum_{n=-\infty}^{\infty} \frac{\sin n\phi_o}{n\phi_o} \frac{H_n^{(2)}(k\rho)}{H_n^{(2)\prime}(ka)}\, e^{jn\phi} \tag{9.10.5}$$

If the slot is an infinitesimal small slot (i.e., a delta slot):

$$\lim_{\phi_o \to 0} \frac{\sin n\phi_o}{n\phi_o} = 1$$

and (9.10.5) reads

$$\Pi_z^m = -j\frac{V}{2\pi a\omega\mu k} \sum_{n=-\infty}^{\infty} \frac{H_n^{(2)}(k\rho)}{H_n^{(2)\prime}(ka)}\, e^{jn\phi}$$

The resulting expressions for the fields are

$$H_z = -j\frac{V}{2\pi a\eta}\sum_{n=-\infty}^{\infty}\frac{H_n^{(2)}(k\rho)}{H_n^{(2)\prime}(ka)}e^{jn\phi} \tag{9.10.6}$$

$$E_\rho = -j\frac{V}{2\pi ak\rho}\sum_{n=-\infty}^{\infty}n\frac{H_n^{(2)}(k\rho)}{H_n^{(2)\prime}(ka)}e^{jn\phi}$$

$$E_\phi = \frac{V}{2\pi a}\sum_{n=-\infty}^{\infty}\frac{H_n^{(2)\prime}(k\rho)}{H_n^{(2)\prime}(ka)}e^{jn\phi}$$

For cylinders with a small radius, the previous series converge with only a few terms. However, for large cylinders, such that $ka \gg 1$, the series converges very slowly. The Watson transformation is used to obtain an expression for the fields in terms of a fast-convergent series.

Consider the H_z field in (9.10.6). Using (7.1.12), it can be expressed as

$$\begin{aligned}H_z &= -j\frac{V}{2\pi a\eta}\frac{1}{2j}\oint_C\frac{H_\nu^{(2)}(k\rho)}{H_\nu^{(2)\prime}(ka)}\frac{e^{j\nu(\phi-\pi)}}{\sin(\nu\pi)}d\nu\\ &= -\frac{V}{4\pi a\eta}\left[\int_{\infty+j\varepsilon}^{-\infty+j\varepsilon}\frac{H_\nu^{(2)}(k\rho)}{H_\nu^{(2)\prime}(ka)}\frac{e^{j\nu(\phi-\pi)}}{\sin(\nu\pi)}d\nu + \int_{-\infty-j\varepsilon}^{\infty-j\varepsilon}\frac{H_\nu^{(2)}(k\rho)}{H_\nu^{(2)\prime}(ka)}\frac{e^{j\nu(\phi-\pi)}}{\sin(\nu\pi)}d\nu\right]\end{aligned} \tag{9.10.7}$$

where C is the closed contour that encloses the zeros of $\sin(\nu\pi)$, as shown in Fig. 7.3a. Letting $\nu = -\nu$ in the first integral and using

$$H_{-\nu}^{(2)}(z) = e^{-j\nu\pi}H_\nu^{(2)}(z)$$

and

$$H_{-\nu}^{(2)\prime}(z) = e^{-j\nu\pi}H_\nu^{(2)\prime}(z)$$

the first integral reads

$$\begin{aligned}\int_{\infty+j\varepsilon}^{-\infty+j\varepsilon}\frac{H_\nu^{(2)}(k\rho)}{H_\nu^{(2)\prime}(ka)}\frac{e^{j\nu(\phi-\pi)}}{\sin(\nu\pi)}d\nu &= \int_{-\infty-j\varepsilon}^{\infty-j\varepsilon}\frac{H_{-\nu}^{(2)}(k\rho)}{H_{-\nu}^{(2)\prime}(ka)}\frac{e^{-j\nu(\phi-\pi)}}{\sin(\nu\pi)}d\nu\\ &= \int_{-\infty-j\varepsilon}^{\infty-j\varepsilon}\frac{H_\nu^{(2)}(k\rho)}{H_\nu^{(2)\prime}(ka)}\frac{e^{-j\nu(\phi-\pi)}}{\sin(\nu\pi)}d\nu\end{aligned}$$

Hence, (9.10.7) reads

$$H_z = -\frac{V}{4\pi a\eta_o}\left[\int_{-\infty-j\varepsilon}^{\infty-j\varepsilon}\frac{H_\nu^{(2)}(k\rho)}{H_\nu^{(2)\prime}(ka)}\frac{1}{\sin(\nu\pi)}\left[e^{j\nu(\phi-\pi)}+e^{-j\nu(\phi-\pi)}\right]d\nu\right]$$
$$= -\frac{V}{2\pi a\eta_o}\int_{-\infty-j\varepsilon}^{\infty-j\varepsilon}\frac{H_\nu^{(2)}(k\rho)}{H_\nu^{(2)\prime}(ka)}\frac{\cos[\nu(\phi-\pi)]}{\sin(\nu\pi)}d\nu \qquad (9.10.8)$$

Closing the contour as shown in Fig. 7.3b, the evaluation of (9.10.8) is $-2\pi j$ times the residues of an integrand at the zeros of $H_\nu^{(2)\prime}(ka)$ in the lower-half plane. That is,

$$H_z = j\frac{V}{a\eta_o}\sum_{p=1}^{\infty}\frac{H_{\nu_p'}^{(2)}(k\rho)}{\left.\frac{\partial H_\nu^{(2)\prime}(ka)}{\partial\nu}\right|_{\nu=\nu_p'}}\frac{\cos[\nu_p'(\phi-\pi)]}{\sin(\nu_p'\pi)} \qquad (9.10.9)$$

In (9.10.8) we need to analyze the behavior of $H_\nu^{(2)}(k\rho)/H_\nu^{(2)\prime}(ka)$ along the large semicircle to show that its value goes to zero as the radius of the circle goes to infinity. Then, the roots of $H_\nu^{(2)\prime}(ka)=0$ need to be determined, and finally evaluate (9.10.9).

The evaluation of the roots of $H_\nu^{(2)\prime}(ka)=0$ uses the expansion of the Hankel functions in terms of Airy functions (see (7.4.42) to (7.4.45)). The zeros of $H_\nu^{(2)\prime}(x)$ are given by

$$\nu_p{}' = x - a_p{}'\left(\frac{x}{2}\right)^{1/3}e^{-j\pi/3} \qquad (p = 1, 2, 3, \ldots) \qquad (9.10.10)$$

where a_p are the zeros of $Ai'(a_p')=0$.

The contour of integration and the poles are shown in Fig. 9.24. For the first few poles, the real parts of the poles are close to $\nu_p' \approx x \approx ka$. As $|\nu|$ gets larger, the approximation for large values of $|\nu|$ of $H_\nu^{(2)\prime}(ka)$ must be used, showing that the poles approach the imaginary axis.

In (9.10.9):

$$\left.\frac{\partial H_\nu^{(2)\prime}(ka)}{\partial\nu}\right|_{\nu=\nu_p'} = \left(\frac{4}{ka}\right)Ai''(a_p') = \left(\frac{4a_p'}{ka}\right)Ai(a_p')$$

where the relation $Ai''(a_p') = a_p'Ai(a_p')$ was used. Hence, (9.10.19) reads

$$H_z = j\frac{Vk}{4\eta_o}\sum_{p=1}^{\infty}\frac{H_{\nu_p'}^{(2)}(k\rho)}{a_p'Ai\left(a_p'\right)}\frac{\cos[\nu_p'(\phi-\pi)]}{\sin(\nu_p'\pi)} \qquad (9.10.11)$$

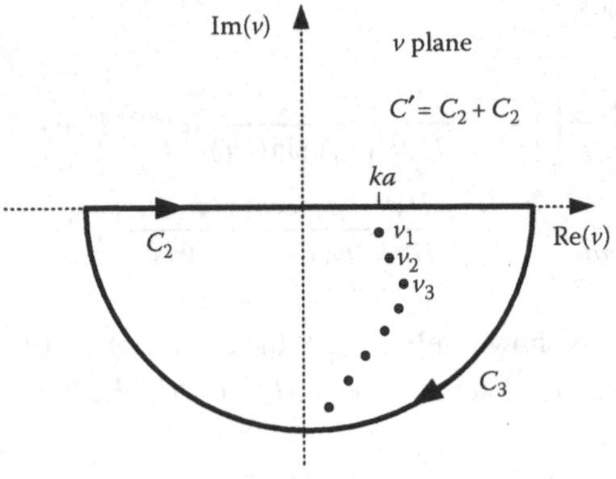

FIGURE 9.24 Path of integration and typical distribution of the poles.

Far from the cylinder, using the asymptotic form of the Hankel function in (9.10.11):

$$H_z = j\frac{Vk}{4\eta_o}\sqrt{\frac{2}{\pi k\rho}}\,e^{-j(\beta\rho-\pi/4)}\sum_{p=1}^{\infty}\frac{\cos[\nu_p'(\phi-\pi)]e^{j\nu_p'\pi/2}}{a_p' Ai(a_p')\sin(\nu_p'\pi)} \tag{9.10.12}$$

The relation in (9.1.22) applies to the angular terms in (9.10.12) which show that the series converges in the shadow region, where $\pi/2 < \phi < 3\pi/2$. It only requires a few terms for the sum in (9.10.12) to converge in this region since the imaginary part of ν_p' has a significant negative value.

In this chapter, there are integrands involving Hankel functions whose behavior along the large semicircular path in the lower-half plane must vanish. For example, in (9.10.8) the term $H_\nu^{(2)}(k\rho)/H_\nu^{(2)\prime}(ka)$ must go to zero along the semicircular path in the lower-half plane. In (9.1.16), it is the term $H_\nu^{(2)}(k\rho)H_\nu^{(1)}(ka)/H_\nu^{(2)\prime}(ka)$ that must go to zero along the large semicircle.

To analyze the integrand in (9.10.8), let $\nu = Re^{j\theta}$ and use the approximation for large values of $|\nu|$ (see Appendix C), namely

$$H_\nu^{(2)}(x) \approx \begin{cases} j\sqrt{\dfrac{2}{\pi\nu}}\left(\dfrac{2\nu}{ex}\right)^{\nu} & \left(-\dfrac{\pi}{2} < \theta < \dfrac{\pi}{2}\right) \\ \sqrt{\dfrac{2}{\pi\nu}}\left(\dfrac{ex}{2\nu}\right)^{\nu} & \left(\dfrac{\pi}{2} < \theta < \dfrac{3\pi}{2}\right) \end{cases}$$

and

$$H_\nu^{(2)\prime}(x) \approx \begin{cases} -\dfrac{\nu}{x} H_\nu^{(2)}(x) & \left(-\dfrac{\pi}{2} < \theta < \dfrac{\pi}{2}\right) \\ \dfrac{\nu}{x} H_\nu^{(2)}(x) & \left(\dfrac{\pi}{2} < \theta < \dfrac{3\pi}{2}\right) \end{cases}$$

Therefore,

$$\left|\frac{H_\nu^{(2)}(k\rho)}{H_\nu^{(2)\prime}(ka)}\right| \approx \begin{cases} \left|\dfrac{ka}{\nu}\left(\dfrac{a}{\rho}\right)^\nu\right| = \dfrac{ka}{R} e^{R\cos\theta \ln(a/\rho)} & \left(-\dfrac{\pi}{2} < \theta < \dfrac{\pi}{2}\right) \\ \left|\dfrac{ka}{\nu}\left(\dfrac{\rho}{a}\right)^\nu\right| = \dfrac{ka}{R} e^{R\cos\theta \ln(\rho/a)} & \left(\dfrac{\pi}{2} < \theta < \dfrac{3\pi}{2}\right) \end{cases} \tag{9.10.13}$$

In the region $-\pi/2 < \theta < \pi/2$: $\cos\theta > 0$ and since $\ln(a/\rho) < 0$ then $R\cos\theta\ln(a/\rho) < 0$. In the region $\pi/2 < \theta < 3\pi/2$: $\cos\theta < 0$ and since $\ln(\rho/a) > 0$ then $R\cos\theta\ln(a/\rho) < 0$. Hence, in the lower-half plane, (9.10.13) goes to 0 as $R \to \infty$. Also, the term

$$\left|\frac{\cos\nu(\pi-\theta)}{\sin\nu\pi}\right| \approx \left|\frac{e^{j\nu(\pi-\theta)}}{e^{j\nu\pi}}\right| \approx e^{R\theta\sin\theta}$$

goes to 0 as $R \to \infty$ in the lower-half plane where $\pi < \theta < 2\pi$. Thus, the integrand in (9.10.8) vanishes as $R \to \infty$ in the semicircular part of the path.

The analysis of the fields in the illuminated region is similar to that in Section 9.1. In the illuminated region (i.e., $-\pi/2 < \phi < \pi/2$) use the relation:

$$\cos\nu(\phi-\pi) = \sin\nu\phi\sin\nu\pi + \cos\nu\pi\cos\nu\phi = -je^{j\nu\phi}\sin\nu\pi + e^{j\nu\pi}\cos\nu\phi$$

Substituting into (9.10.8) shows that H_z is expressed in terms of two integrals, namely

$$\begin{aligned} H_z &= -\frac{V}{2\pi a\eta_o}\int_{-\infty-j\varepsilon}^{\infty-j\varepsilon} \frac{H_\nu^{(2)}(k\rho)}{H_\nu^{(2)\prime}(ka)} \frac{\cos[\nu(\phi-\pi)]}{\sin(\nu\pi)} d\nu \\ &= \underbrace{-j\frac{V}{2\pi a\eta_o}\int_{-\infty-j\varepsilon}^{\infty-j\varepsilon} \frac{H_\nu^{(2)}(k\rho)}{H_\nu^{(2)\prime}(ka)} e^{j\nu\phi} d\nu}_{I_1} + \underbrace{\frac{V}{2\pi a\eta_o}\int_{-\infty-j\varepsilon}^{\infty-j\varepsilon} \frac{H_\nu^{(2)}(k\rho)}{H_\nu^{(2)\prime}(ka)} \frac{\cos(\nu\phi)}{\sin(\nu\pi)} e^{j\nu\pi} d\nu}_{I_2} \end{aligned} \tag{9.10.14}$$

The evaluation of I_2 is similar to that in (9.1.33) and its contribution to the field in the illuminated region is small (left to the problems).

Next, consider I_1 in (9.10.14):

$$H_z \approx I_1 = -j\frac{V}{2\pi a\eta_o}\int_{-\infty-j\varepsilon}^{\infty-j\varepsilon}\frac{H_\nu^{(2)}(k\rho)}{H_\nu^{(2)\prime}(ka)}e^{j\nu\phi}d\nu \tag{9.10.15}$$

Equation (9.10.15) is now evaluated using the method of stationary phase. To determine the stationary points, the Debye approximations for the Hankel functions are used. Hence,

$$\frac{H_\nu^{(2)}(k\rho)}{H_\nu^{(2)\prime}(ka)} = \begin{cases} j\dfrac{ka}{\sqrt{k\rho\sqrt{(ka)^2-\nu^2}}}e^{j\Psi_1} & (\nu < ka < k\rho) \\ j\dfrac{kae^{j\pi/4}}{\sqrt{k\rho\sqrt{\nu^2-(ka)^2}}}e^{j\Psi_2} & (k\rho > \nu > ka) \end{cases} \tag{9.10.16}$$

where

$$\Psi_1 = \sqrt{(ka)^2-\nu^2} - \sqrt{(k\rho)^2-\nu^2} - \nu\cos^{-1}\left(\frac{\nu}{ka}\right) + \nu\cos^{-1}\left(\frac{\nu}{k\rho}\right) + \nu\phi \tag{9.10.17}$$

and

$$\Psi_2 = -\sqrt{(k\rho)^2-\nu^2} - j\sqrt{\nu^2-(ka)^2} + j\nu\cosh^{-1}\left(\frac{\nu}{ka}\right) + \nu\cos^{-1}\left(\frac{\nu}{k\rho}\right) + \nu\phi \tag{9.10.18}$$

The procedure for evaluating (9.10.15) is similar to that used in the evaluation of (9.1.35). In (9.10.15), one stationary point occurs at

$$\frac{\partial\Psi_1}{\partial\nu} \approx -\cos^{-1}\left(\frac{\nu}{ka}\right) + \cos^{-1}\left(\frac{\nu}{k\rho}\right) + \phi = 0 \Rightarrow \nu = \nu_1 \approx ka\sin\phi \tag{9.10.19}$$

The other stationary point follows from (9.10.18), which occurs at a complex value of ν (left to the problems). Since a complex value ν cannot be associated with a real angle ϕ, the only stationary point is that of ν_1.

The stationary point evaluation of (9.10.15) requires (9.10.17) evaluated at ν_1 as well as its second derivative. That is,

$$\Psi_1(\nu_1) \approx -k(\rho - a\sin\phi)$$

and

$$\left.\frac{\partial^2 \Psi_1}{\partial \nu^2}\right|_{\nu=\nu_1} \approx \frac{1}{ka\cos\phi}$$

The path of integration is modified to pass through the stationary points at ν_1 and the evaluation of (9.10.15) is

$$H_z \approx -j\frac{V}{2\pi a\eta}\frac{jka}{\sqrt{k\rho}\sqrt{ka\cos\phi}}\sqrt{\frac{2\pi}{|\Psi_1''(\nu_1)|}}e^{j\Psi_1}e^{j\pi/4} \approx \frac{V}{\eta}\sqrt{\frac{k}{2\pi\rho}}e^{-jk(\rho-a\cos\phi)}e^{j\pi/4}$$

(9.10.20)

Equation (9.10.20) is recognized as the far field from a conducting slot in a conducting plane (see (8.4.10)).

9.11 RADIATION FROM AN APERTURE IN A CIRCULAR CYLINDER

An aperture in a circular cylinder is shown in Fig. 9.25. The aperture has finite dimensions and the fields in the aperture are known and given by

$$E_\phi(a, \phi, z) = E_{A,\phi}(a, \phi', z')$$
$$E_z(a, \phi, z) = E_{A,z}(a, \phi', z')$$

where the aperture coordinates are denoted using primed notation.

The problem can be analyzed in terms of the Hertz's vectors components Π_z^{e} and Π_z^{m}. First, consider the wave equation satisfied by Π_z^{e}, namely

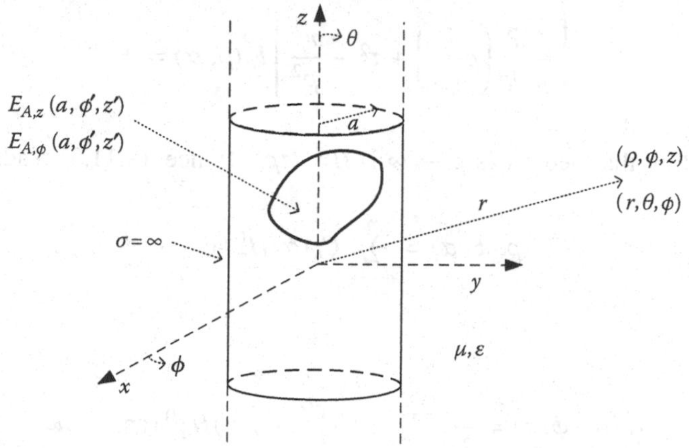

FIGURE 9.25 An aperture in a circular conducting cylinder.

$$\left[\frac{1}{\rho}\frac{\partial}{\partial\rho}\left(\rho\frac{\partial}{\partial\rho}\right)+\frac{1}{\rho^2}\frac{\partial^2}{\partial\phi^2}+\frac{\partial^2}{\partial z^2}+k^2\right]\Pi_z^{\mathrm{e}}=0 \tag{9.11.1}$$

where $k = \omega\sqrt{\mu\varepsilon}$.

Since the cylinder is infinite in length (i.e., $-\infty \leq z \leq \infty$), a Fourier transform on the z variable is used. That is,

$$\Pi_z^{\mathrm{e}}(\rho,\phi,\alpha)=\int_{-\infty}^{\infty}\Pi_z^{\mathrm{e}}(\rho,\phi,z)e^{-j\alpha z}dz$$

$$\Pi_z^{\mathrm{e}}(\rho,\phi,z)=\frac{1}{2\pi}\int_{-\infty}^{\infty}\Pi_z^{\mathrm{e}}(\rho,\phi,\alpha)e^{j\alpha z}d\alpha$$

and (9.11.1) reads

$$\left[\frac{1}{\rho}\frac{\partial}{\partial\rho}\left(\rho\frac{\partial}{\partial\rho}\right)+\frac{1}{\rho^2}\frac{\partial^2}{\partial\phi^2}+\tau^2\right]\Pi_z^{\mathrm{e}}(\rho,\phi,\alpha)=0 \tag{9.11.2}$$

where

$$\tau=\sqrt{k^2-\alpha^2}$$

and $\mathrm{Im}(\tau) < 0$.

The solution to (9.11.2) is of the form

$$\Pi_z^{\mathrm{e}}=\sum_{n=-\infty}^{\infty}F_n(\rho,\alpha)\,e^{jn\phi} \tag{9.11.3}$$

Substituting (9.11.3) into (9.11.2) shows that

$$\left[\frac{1}{\rho}\frac{\partial}{\partial\rho}\left(\rho\frac{\partial}{\partial\rho}\right)+\tau^2-\frac{n^2}{\rho^2}\right]F_n(\rho,\alpha)=0 \tag{9.11.4}$$

whose solution that decays as $\rho\to\infty$ is $H_n^{(2)}(\tau\rho)$. Hence, (9.11.3) reads

$$\Pi_z^{\mathrm{e}}(\rho,\phi,\alpha)=\sum_{n=-\infty}^{\infty}C_n(\alpha)H_n^{(2)}(\tau\rho)\,e^{jn\phi} \tag{9.11.5}$$

and

$$\Pi_z^{\mathrm{e}}(\rho,\phi,z)=\frac{1}{2\pi}\sum_{n=-\infty}^{\infty}e^{jn\phi}\int_{-\infty}^{\infty}C_n(\alpha)H_n^{(2)}(\tau\rho)e^{j\alpha z}d\alpha \tag{9.11.6}$$

To apply the boundary condition at $\rho = a$, the expression for E_z is needed. From (4.4.16) and (9.11.5):

$$E_z = \left(\frac{\partial^2}{\partial z^2} + k^2\right)\Pi_z^e = \frac{1}{2\pi}\sum_{n=-\infty}^{\infty} e^{jn\phi}\int_{-\infty}^{\infty} C_n(\alpha)(k^2 - \alpha^2)H_n^{(2)}(\tau\rho)e^{j\alpha z}d\alpha$$
$$= \frac{1}{2\pi}\sum_{n=-\infty}^{\infty} e^{jn\phi}\int_{-\infty}^{\infty} C_n(\alpha)\tau^2 H_n^{(2)}(\tau\rho)e^{j\alpha z}d\alpha$$

At $\rho = a$:

$$E_z(a, \phi, z) = E_{A,z}(a, \phi', z') = \sum_{n=-\infty}^{\infty}\left[\frac{1}{2\pi}\int_{-\infty}^{\infty} C_n(\alpha)\tau^2 H_n^{(2)}(\tau a)e^{j\alpha z}d\alpha\right]e^{jn\phi}$$

which represents a Fourier expansion of $E_{A,z}(a, \phi', z')$. Hence, the terms in brackets are the Fourier coefficients given by

$$\int_{-\infty}^{\infty} C_n(\alpha)\tau^2 H_n^{(2)}(\tau a)e^{j\alpha z}d\alpha = \int_0^{2\pi} E_{A,z}(a, \phi', z')e^{-jn\phi'}d\phi'$$

Note that $E_{A,z}$ is only non-zero over the aperture range of ϕ', so the above limits from 0 to 2π will later change to the aperture limits for ϕ'. Furthermore, the integral on the left is an inverse Fourier transform which shows that the integral on the right must be its Fourier transform. That is,

$$C_n(\alpha)\tau^2 H_n^{(2)}(\tau a) = \frac{1}{2\pi}\int_{-\infty}^{\infty}\left[\int_0^{2\pi} E_{A,z}(a, \phi', z')e^{-jn\phi'}d\phi'\right]e^{-j\alpha z}dz$$
$$= \frac{1}{2\pi}\int_0^{2\pi} \widehat{E}_{A,z}(a, \phi', \alpha)e^{-jn\phi'}d\phi'$$

where

$$\widehat{E}_{A,z}(a, \phi', \alpha) = \int_{-\infty}^{\infty} E_{A,z}(a, \phi', z')e^{-j\alpha z}dz' \tag{9.11.7}$$

Again, the limits from $-\infty$ to ∞ will be change to the aperture limits for z'. Hence,

$$C_n(\alpha) = \frac{1}{2\pi\tau^2 H_n^{(2)}(\tau a)}\int_0^{2\pi} \widehat{E}_{A,z}(a, \phi', \alpha)e^{-jn\phi'}d\phi' \tag{9.11.8}$$

and (9.11.6) reads

$$\Pi_z^e(\rho,\phi,z)=\frac{1}{4\pi^2}\sum_{n=-\infty}^{\infty}e^{jn\phi}\int_{-\infty}^{\infty}\frac{H_n^{(2)}(\tau\rho)}{\tau^2H_n^{(2)}(\tau a)}e^{j\alpha z}\left[\int_0^{2\pi}\widehat{E_{A,z}}(a,\phi',\alpha)e^{-jn\phi'}d\phi'\right]d\alpha \tag{9.11.9}$$

Substituting (9.11.7) into (9.11.9):

$$\Pi_z^e(\rho,\phi,z)=\frac{1}{4\pi^2}\sum_{n=-\infty}^{\infty}e^{jn\phi}\iint_{Ap}E_{A,z}(a,\phi',z')e^{-jn\phi'}I_1\,d\phi'dz' \tag{9.11.10}$$

where the integration is over the aperture (denoted by Ap) where $E_{A,z}(a,\phi',z')$ is defined. The term I_1 is a function of $\rho,\ z,\ z'$ and given by

$$I_1=\int_{-\infty}^{\infty}\frac{H_n^{(2)}(\tau\rho)}{\tau^2H_n^{(2)}(\tau a)}e^{j\alpha(z-z')}d\alpha \tag{9.11.11}$$

The integration limits over ϕ' and z' in (9.11.10) denote the integration over the aperture dimensions where the source $E_{A,z}(a,\phi',z')$ is defined. On the surface of the conducting part of the cylinder the field E_z is zero.

The integral I_1 can be evaluated in the far field using the method of steepest descent. The integral has branch points at $\tau=0$ and $\tau=\pm k$, and poles where $H_n^{(2)}(\tau a)=0$. Also, the integration must be performed in the upper Riemann surface where $\text{Im}(\tau)<0$. The zeroes of $H_n^{(2)}(\tau a)$ as a function of τ occur only when $\text{Im}(\tau)>0$. Therefore, they are not on the upper Riemann surface.

To evaluate I_1 in (9.1.11) in the far field, the method of steepest descent is used. Equation (9.11.11) is of the form

$$I_1=\int_{-\infty}^{\infty}F(\tau)H_n^{(2)}(\tau\rho)e^{j\alpha z}d\alpha \tag{9.11.12}$$

where

$$F(\tau)=\frac{e^{-j\alpha z'}}{\tau^2H_n^{(2)}(\tau a)}$$

To evaluate (9.11.12), let

$$\alpha=k\cos\beta$$
$$\rho=R\sin\theta$$
$$z=R\cos\theta$$
$$\tau=k\sin\beta$$

and use the asymptotic approximation of the Hankel function. Referring to (7.4.66), the evaluation of (9.11.12) is

$$I_1 \approx 2j^{n+1} \frac{e^{jkz' \cos \theta}}{(k \sin \theta)^2 H_n^{(2)}(ka \sin \theta)} \frac{e^{-jkr}}{r} \tag{9.11.13}$$

and (9.11.10) reads

$$\Pi_z^{e}(\rho, \phi, z) = \frac{1}{2\pi^2 (k \sin \theta)^2} \frac{e^{-jkr}}{r} \sum_{n=-\infty}^{\infty} j^{n+1} \frac{e^{jn\phi}}{H_n^{(2)}(ka \sin \theta)} \times \int_{z'} \int_{\phi'} E_{A,z}(a, \phi', z') e^{jkz' \cos \theta} e^{-jn\phi'} d\phi' dz' \tag{9.11.14}$$

Next, the evaluation of Π_z^{m} is performed. The Hertz vector Π_z^{m} satisfies (9.11.1), and following similar steps leading to (9.11.6), we write

$$\Pi_z^{m}(\rho, \phi, z) = \frac{1}{2\pi} \sum_{n=-\infty}^{\infty} e^{jn\phi} \int_{-\infty}^{\infty} B_n(\alpha) H_n^{(2)}(\tau\rho) e^{j\alpha z} d\alpha \tag{9.11.15}$$

where the constant $B_n(\alpha)$ is determined by the boundary condition on E_ϕ at $\rho = a$, which requires the expression for $E_\phi(\rho, \phi, z)$. From (4.4.15) and (4.4.18):

$$E_\phi = \frac{1}{\rho} \frac{\partial^2 \Pi_z^{e}}{\partial \phi \, \partial z} + j\omega\mu \frac{\partial \Pi_z^{m}}{\partial \rho}$$

and using (9.11.6) and (9.11.15):

$$E_\phi = \frac{1}{2\pi} \sum_{n=-\infty}^{\infty} e^{jn\phi} \int_{-\infty}^{\infty} \left[-\frac{n\alpha}{\rho} C_n(\alpha) H_n^{(2)}(\tau\rho) + j\omega\mu B_n(\alpha) \tau H_n^{(2)\prime}(\tau\rho) \right] e^{j\alpha z} d\alpha$$

Applying the boundary condition at $\rho = a$:

$$E_\phi(a, \phi, z) = E_{A,\phi}(a, \phi', z') = \sum_{n=-\infty}^{\infty} \left\{ \frac{1}{2\pi} \int_{-\infty}^{\infty} \left[-\frac{n\alpha}{a} C_n(\alpha) H_n^{(2)}(\tau a) + j\omega\mu B_n(\alpha) \tau H_n^{(2)\prime}(\tau a) \right] e^{j\alpha z} d\alpha \right\} e^{jn\phi}$$

which represents a Fourier expansion of $E_{A,\phi}(a, \phi', z')$. Hence, the Fourier coefficients are given by

$$\int_{-\infty}^{\infty} \left[-\frac{n\alpha}{a} C_n(\alpha) H_n^{(2)}(\tau a) + j\omega\mu B_n(\alpha) \tau H_n^{(2)\prime}(\tau a) \right] e^{j\alpha z} d\alpha = \int_0^{2\pi} E_{A,\phi}(a, \phi', z') e^{-jn\phi'} d\phi'$$

The integral on the left is a Fourier transform which shows that the integral on the right must be its inverse. That is,

$$
\begin{aligned}
&-\frac{n\alpha}{a}C_n(\alpha)H_n^{(2)}(\tau a)+j\omega\mu B_n(\alpha)\tau H_n^{(2)\prime}(\tau a)\\
&=\frac{1}{2\pi}\int_{-\infty}^{\infty}\left[\int_0^{2\pi}E_{A,\phi}(a,\phi',z')e^{-jn\phi'}d\phi'\right]e^{-j\alpha z'}dz'\\
&=\frac{1}{2\pi}\int_0^{2\pi}\widehat{E}_{A,\phi}(a,\phi',\alpha)e^{-jn\phi'}d\phi'
\end{aligned}
\tag{9.11.16}
$$

where

$$
\widehat{E}_{A,\phi}(a,\phi',\alpha)=\int_{-\infty}^{\infty}E_{A,\phi}(a,\phi',z')e^{-j\alpha z'}dz'
$$

Substituting (9.11.8) into (9.11.16) gives

$$
B_n(\alpha)=\frac{1}{2\pi j\omega\mu\tau H_n^{(2)\prime}(\tau a)}\int_0^{2\pi}\left[\frac{n\alpha}{a\tau^2}E_{A,z}(a,\phi',\alpha)+E_{A,\phi}(a,\phi',z')\right]e^{-jn\phi'}d\phi'
$$

and (9.11.15) reads

$$
\begin{aligned}
\Pi_z^{\mathrm{m}}(\rho,\phi,z)&=\frac{1}{4\pi^2 j\omega\mu}\sum_{n=-\infty}^{\infty}e^{jn\phi}\int_{-\infty}^{\infty}\left\{\int_0^{2\pi}\left[\frac{n\alpha}{a\tau^2}\widehat{E}_{A,z}(a,\phi',\alpha)\right.\right.\\
&\qquad\left.\left.+\widehat{E}_{A,\phi}(a,\phi',\alpha)\right]e^{-jn\phi'}d\phi'\right\}\frac{H_n^{(2)}(\tau\rho)}{\tau H_n^{(2)\prime}(\tau a)}e^{j\alpha z}d\alpha\\
&=\frac{1}{4\pi^2 j\omega\mu}\sum_{n=-\infty}^{\infty}e^{jn\phi}\iint_{Ap}E_{A,z}(a,\phi',z')e^{-jn\phi'}\\
&\quad\times\left[\int_{-\infty}^{\infty}\frac{n\alpha}{a}\frac{H_n^{(2)}(\tau\rho)}{\tau^3 H_n^{(2)\prime}(\tau a)}e^{j\alpha(z-z')}d\alpha\right]d\phi' dz'\\
&\quad+\frac{1}{4\pi^2 j\omega\mu}\sum_{n=-\infty}^{\infty}e^{jn\phi}\iint_{Ap}E_{A,\phi}(a,\phi',z')e^{-jn\phi'}\\
&\quad\times\left[\int_{-\infty}^{\infty}\frac{H_n^{(2)}(\tau\rho)}{\tau H_n^{(2)\prime}(\tau a)}e^{j\alpha(z-z')}d\alpha\right]d\phi' dz'
\end{aligned}
$$

which can be expressed in the form

$$\Pi_z^{\mathrm{m}}(\rho, \phi, z) = \frac{-1}{4\pi^2 \omega\mu} \sum_{n=-\infty}^{\infty} e^{jn\phi} \iint_{Ap} e^{-jn\phi'} \left[\frac{n}{a} E_{A,z}(a, \phi', z') \frac{\partial I_2}{\partial z} + jE_{A,\phi}(a, \phi', z') \left(\frac{\partial^2}{\partial z^2} + k^2 \right) I_2 \right] d\phi' dz' \tag{9.11.17}$$

where

$$I_2 = \int_{-\infty}^{\infty} \frac{H_n^{(2)}(\tau\rho)}{\tau^3 H_n^{(2)\prime}(\tau a)} e^{j\alpha(z-z')} d\alpha \tag{9.11.18}$$

The integral I_2 in (9.11.18) is similar in form to that in (9.11.12). Hence, using (7.4.66), its evaluation in the far field using the method of steepest descent is

$$\int_{-\infty}^{\infty} \frac{H_n^{(2)}(\tau\rho)}{\tau^3 H_n^{(2)\prime}(\tau a)} e^{j\alpha(z-z')} d\alpha \approx 2j^{n+1} \frac{e^{jkz' \cos\theta}}{(k \sin\theta)^3 H_n^{(2)\prime}(ka \sin\theta)} \frac{e^{-jkr}}{r} \tag{9.11.19}$$

and (9.11.17) reads

$$\Pi_z^{\mathrm{m}}(\rho, \phi, z) = \frac{-1}{2\pi^2 \omega\mu k \sin\theta} \frac{e^{-jkr}}{r} \sum_{n=-\infty}^{\infty} j^n \frac{e^{jn\phi}}{H_n^{(2)\prime}(ka \sin\theta)} \times \left[\frac{n \cos\theta}{ka \sin^2\theta} \iint_{Ap} E_{A,z}(a, \phi', z') e^{jkz' \cos\theta} e^{-jn\phi'} d\phi' dz' + \iint_{Ap} E_{A,\phi}(a, \phi', z') e^{jkz' \cos\theta} e^{-jn\phi'} d\phi' dz' \right] \tag{9.11.20}$$

where (9.11.21) was used.

Equations (9.11.14) and (9.11.20) are the solutions for the calculations of the fields from an aperture on the cylindrical surface in Fig. 9.25. It is convenient to express the far fields in spherical coordinates. Using (4.4.11) to (4.4.22), the fields in cylindrical coordinates are obtained. The expressions require derivatives with respect to ρ and z in (9.11.14) and (9.11.20) of e^{-jkr}/r. In the far field:

$$\frac{\partial}{\partial\rho}\left(\frac{e^{-jkr}}{r}\right) \approx \frac{1}{r}\frac{\partial r}{\partial\rho}\frac{\partial e^{-jkr}}{\partial r} = -jk \sin\theta \frac{e^{-jkr}}{r}$$

and

$$\frac{\partial}{\partial z}\left(\frac{e^{-jkr}}{r}\right) \approx \frac{1}{r}\frac{\partial r}{\partial z}\frac{\partial e^{-jkr}}{\partial r} = -jk \cos\theta \frac{e^{-jkr}}{r} \tag{9.11.21}$$

To convert the electric fields to spherical coordinates, use the transformations:

$$\begin{cases} E_r = E_\rho \sin\theta + E_z \cos\theta \\ E_\theta = E_\rho \cos\theta - E_z \sin\theta \\ E_\phi = E_\phi \end{cases}$$

Hence, the far fields follow from:

$$\begin{cases} E_\theta = \dfrac{H_\phi}{\eta} \approx -k^2 \sin\theta\, \Pi_z^{\mathrm{e}} \\ E_\phi = -\dfrac{H_\theta}{\eta} \approx k\omega\mu \sin\theta\, \Pi_z^{\mathrm{m}} \end{cases} \quad (9.11.22)$$

and, of course, $E_r \approx 0$ and $H_r \approx 0$.

As an example of the field calculations using (9.11.20), consider the aperture shown in Fig. 9.26 which consists of an axial slot of length L and angular width ϕ_o. The electric field on the aperture is given by

$$E_{A,\phi} = \frac{V}{a\phi_o} \cos\frac{\pi z'}{L} \qquad \begin{pmatrix} -\phi_o/2 \le \phi \le \phi_o/2 \\ -L/2 \le z \le L/2 \end{pmatrix}$$

which, of course, satisfy $E_{A,\phi} = 0$ at $z' = \pm L/2$.

From (9.11.20), associate it with the aperture field excitation $E_{A,\phi}$, $\Pi_z^{\mathrm{m}}(\rho, \phi, z)$ is given by

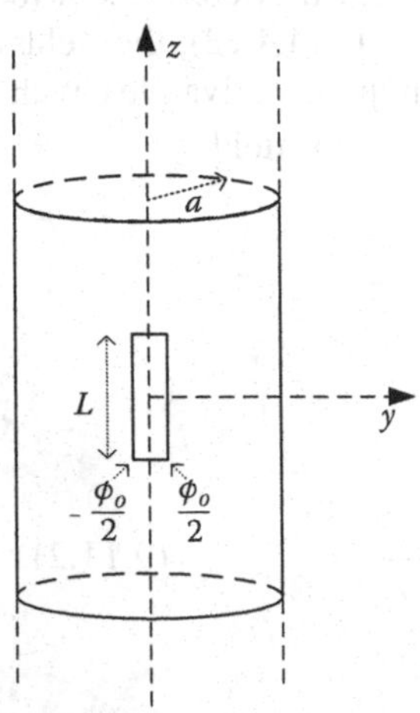

FIGURE 9.26 An axial slot aperture in a conducting cylinder.

$$\Pi_z^{\mathrm{m}}(\rho, \phi, z) = \frac{1}{2\pi^2 \omega\mu k \sin\theta} \frac{e^{-jkr}}{r} \sum_{n=-\infty}^{\infty} j^n \frac{e^{jn\phi}}{H_n^{(2)\prime}(ka \sin\theta)} \left(\frac{V}{a\phi_o}\right)$$
$$\int_{-L/2}^{L/2} \int_{-\phi_o/2}^{\phi_o/2} \cos\left(\frac{\pi z'}{L}\right) e^{jkz'\cos\theta} e^{-jn\phi'} d\phi' dz'$$

Using

$$\int_{-L/2}^{L/2} \cos\left(\frac{\pi z'}{L}\right) e^{jkz'\cos\theta} dz' = \left(\frac{\pi L}{2}\right) \frac{\cos\left(\frac{kL}{2}\cos\theta\right)}{\left(\frac{\pi}{2}\right)^2 - \left(\frac{kL}{2}\cos\theta\right)^2}$$

it follows from (9.11.22) that

$$E_\phi = -\frac{H_\theta}{\eta} \approx k\omega\mu \sin\theta\, \Pi_z^{\mathrm{m}}$$
$$= \frac{VL}{4\pi a} \frac{e^{-jkr}}{r} \left\{ \frac{\cos[(kL/2)\cos\theta]}{(\pi/2)^2 - [(kL/2)\cos\theta]^2} \right\} \sum_{n=-\infty}^{\infty} j^n \frac{e^{jn\phi}}{H_n^{(2)\prime}(ka \sin\theta)} \frac{\sin(n\phi_o/2)}{(n\phi_o/2)}$$

As a second example (left to the problems), the Hertz vector Π_z^{m} in (9.10.5) and the field H_z in (9.10.6) for the infinite-length aperture in Fig. 9.19 can be calculated using (9.11.17).

Problems

P9.1. Derive (9.1.12) using the appropriate Green function for the TE case.

P9.2. Write the expressions for the diffracted fields in the geometry described in Fig. 9.5.

P9.3. Derive (9.5.8).

P9.4. Derive (9.5.14).

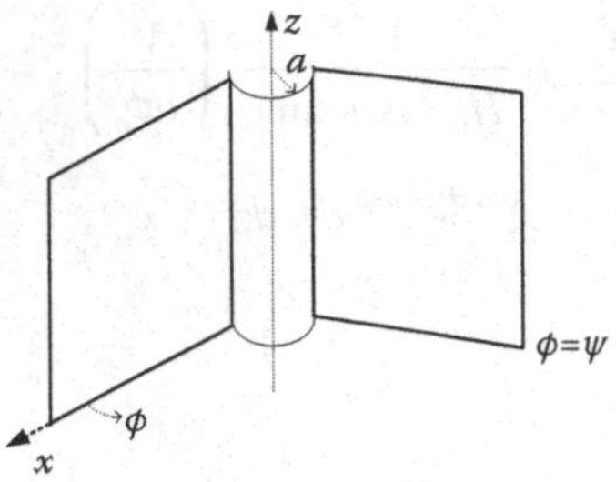

FIGURE P9.5

P9.5.

a. The wedge shown in Fig. P9.5 is excited by an infinite-length current source located in front of the wedge at $(\rho_o,\ \phi_o)$. Determine the radiated fields.

b. Determine the radiated fields if the source in front of the wedge is an electric dipole situated at $(\rho_o,\ \phi_o,\ z_o)$.

P9.6.

a. Show that the numerical distance in (9.6.53) in terms of the surface impedance is given by

$$p = -j\frac{k_1 R_2}{2} z_g^2$$

where $z_g = \eta_2/\eta_o$ is the normalized ground impedance.

b. Determine the range of the phase of z_g.

P9.7. Verify (9.6.64).

P9.8. Show that the exponential behavior of the integrand in (9.8.9) along the semicircular path in Fig. 9.18 goes to zero.

P9.9. Verify that (9.10.11) converges when $\pi/2 < \phi < 3\pi/2$.

P9.10. Show that for $|\nu| \to \infty$:

$$\begin{cases} |H_\nu^{(1)}(\beta a)| \approx \dfrac{1}{\sqrt{R}} e^{-\theta R \sin\theta} e^{R\cos\theta \ln(2R/e\beta a)} & (-\pi/2 < \theta < \pi/2) \\ |H_\nu^{(1)}(\beta a)| \approx \dfrac{1}{\sqrt{R}} e^{\theta R \sin\theta} e^{-R\cos\theta \ln(2R/e\beta a)} & (\pi/2 < \theta < 3\pi/2) \end{cases}$$

and verify that the integrand in (9.1.16) vanishes as $R \to \infty$ in the semicircular part of the path.

P9.11. Evaluate I_2 in (9.10.14) and show that its contribution in the illuminated region is small.

P9.12. Show that the stationary point associated with Ψ_2 in (9.10.18) occurs at a complex value of ν.

P9.13. Verify (9.11.19).

P9.14. Determine the radiated field from the annular slot shown in Fig. P9.14. The field over the aperture at $\rho = a$ has a value given by $E_{A,z}$.

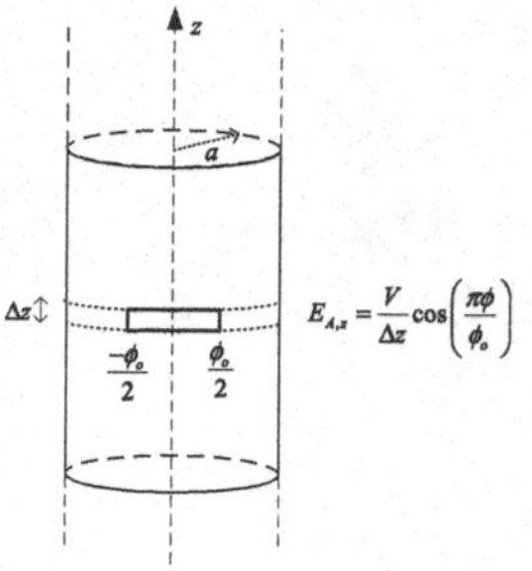

FIGURE P9.14

P9.15. Derive the Hertz vector Π_z^{m} in (9.10.5) and the field H_z in (9.10.6) using (9.11.20).

10 Further Studies of Electromagnetic Waves in Spherical Geometries

10.1 DIFFRACTION BY A CONDUCTING SPHERE

A plane wave incident on a conducting sphere is shown in Fig. 10.1. The plane wave is traveling in the z direction and is given by

$$\mathbf{E} = E_x \hat{\mathbf{a}}_x = Ce^{-j\beta z}\hat{\mathbf{a}}_x = Ce^{-j\beta r \cos\theta}\hat{\mathbf{a}}_x$$

and

$$\mathbf{H} = H_y \hat{\mathbf{a}}_y = \frac{C}{\eta}e^{-j\beta z}\hat{\mathbf{a}}_y = \frac{C}{\eta}e^{-j\beta r \cos\theta}\hat{\mathbf{a}}_y$$

where $\beta = \omega\sqrt{\mu\varepsilon}$ and $\eta = (\mu/\varepsilon)^{1/2}$.

To express the incident field in spherical coordinates, one lets:

$$\hat{\mathbf{a}}_x = \sin\theta\cos\phi\,\hat{\mathbf{a}}_r + \cos\theta\cos\phi\,\hat{\mathbf{a}}_\theta - \sin\phi\,\hat{\mathbf{a}}_\phi$$

and

$$e^{-j\beta r\cos\theta} = \sum_{n=0}^{\infty} j^{-n}(2n+1)j_n(\beta r)P_n(\cos\theta)$$

where in terms of the Schelkunoff-Bessel function $\hat{j}_n(\beta r)$:

$$e^{-j\beta r\cos\theta} = \frac{1}{\beta r}\sum_{n=0}^{\infty} j^{-n}(2n+1)\hat{j}_n(\beta r)P_n(\cos\theta)$$

Then, the radial component the incident electric field is

$$\begin{aligned}
E_r^i &= C\sin\theta\cos\phi e^{-j\beta r\cos\theta} \\
&= C\frac{\cos\phi}{j\beta r}\frac{\partial e^{-j\beta r\cos\theta}}{\partial\theta} \\
&= C\frac{\cos\phi}{j(\beta r)^2}\sum_{n=0}^{\infty} j^{-n}(2n+1)\hat{j}_n(\beta r)\frac{dP_n(\cos\theta)}{d\theta} \\
&= C\frac{\cos\phi}{j(\beta r)^2}\sum_{n=1}^{\infty} j^{-n}(2n+1)\hat{j}_n(\beta r)P_n^1(\cos\theta) \qquad (10.1.1)
\end{aligned}$$

DOI: 10.1201/9781003219729-10

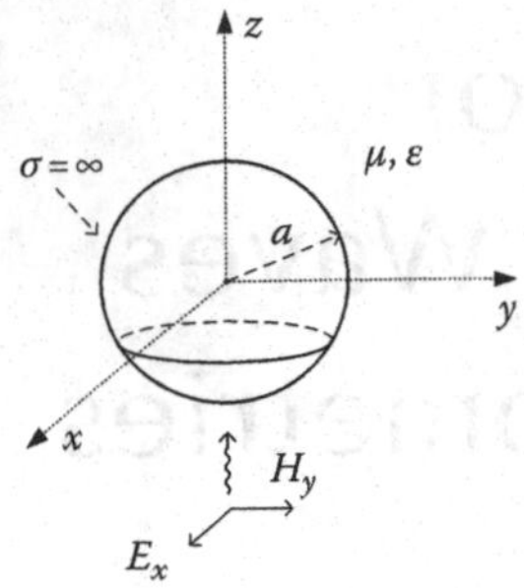

FIGURE 10.1   A plane wave incident on a conducting sphere.

where we used the relation:

$$\frac{dP_n(\cos\theta)}{d\theta} = P_n^1(\cos\theta)$$

and the sum begins with $n = 1$ since $P_0^1(\cos\theta) = 0$. The other components of the incident electric field are

$$E_\theta^i = C\cos\theta\cos\phi\, e^{-j\beta r\cos\theta}$$

$$= C\frac{\cos\theta\cos\phi}{\beta r}\sum_{n=0}^{\infty} j^{-n}(2n+1)\hat{j}_n(\beta r)P_n(\cos\theta)$$

and

$$E_\phi^i = -C\sin\phi\, e^{-j\beta r\cos\theta}$$

$$= -C\frac{\sin\phi}{\beta r}\sum_{n=0}^{\infty} j^{-n}(2n+1)\hat{j}_n(\beta r)P_n(\cos\theta)$$

Next, the form of the Debye potential associated with the incident electric field is determined. The Debye potential associated with E_r^i, denoted by U^i, follows from (4.6.14), namely

$$E_r^i = \left(\frac{\partial^2}{\partial r^2} + \beta^2\right)(rU^i) \tag{10.1.2}$$

The form of U^i is such that when substituted into (10.1.2), produces (10.1.1), letting

$$rU^i = C\frac{\cos\phi}{j\beta^2}\sum_{n=1}^{\infty} a_n\hat{j}_n(\beta r)P_n^1(\cos\theta) \tag{10.1.3}$$

and since (see Appendix C):

$$\left[\frac{d^2}{dr^2} + \beta^2\right]\hat{j}_n(\beta r) = \frac{n(n+1)}{r^2}\hat{j}_n(\beta r) \tag{10.1.4}$$

it follows that substituting (10.1.3) into (10.1.2) and using (10.1.4) that

$$E_r^i = C\frac{\cos\phi}{j(\beta r)^2}\sum_{n=1}^{\infty} a_n n(n+1)\hat{j}_n(\beta r)P_n^1(\cos\theta) \tag{10.1.5}$$

Comparing (10.1.5) with (10.1.1) shows that

$$a_n = j^{-n}\frac{(2n+1)}{n(n+1)}$$

Therefore, the Debye potential U^i is given by

$$U^i = C\frac{\cos\phi}{j\beta^2 r}\sum_{n=1}^{\infty} j^{-n}\frac{(2n+1)}{n(n+1)}\hat{j}(\beta r)P_n^1(\cos\theta) \tag{10.1.6}$$

As a check, substituting (10.1.6) into (10.1.2) and using (10.1.4) results in the desired E_r^i in (10.1.1).

The Debye potential associated with H_r^i, denoted by V^i, follows from (4.6.17), namely:

$$H_r^i = \left(\frac{\partial^2}{\partial r^2} + \beta^2\right)(rV^i) \tag{10.1.7}$$

and the expression for the incident magnetic field H_r^i in spherical coordinates. Since,

$$\hat{\mathbf{a}}_y = \sin\theta\sin\phi\,\hat{\mathbf{a}}_r + \cos\theta\sin\phi\hat{\mathbf{a}}_\theta + \cos\phi\hat{\mathbf{a}}_\phi$$

it follows that

$$H_y^i = \frac{E_x^i}{\eta} \Rightarrow H_r^i = \frac{E_x^i}{\eta}\sin\theta\sin\phi$$

or

$$\begin{aligned}
H_r^i &= \frac{C}{\eta}\sin\theta\sin\phi e^{-j\beta r\cos\theta}\\
&= C\frac{1}{j\beta r\eta}\sin\phi\frac{\partial e^{-j\beta r\cos\theta}}{\partial\theta}\\
&= \frac{C}{j(\beta r)^2\eta}\sin\phi\sum_{n=1}^{\infty} j^{-n}(2n+1)\hat{j}_n(\beta r)P_n^1(\cos\theta)
\end{aligned}$$

Therefore, the Debye potential V^i is given by

$$V^i = C\frac{\sin\phi}{j\beta^2 r\eta}\sum_{n=1}^{\infty} j^{-n}\frac{(2n+1)}{n(n+1)}\hat{j}\,(\beta r)P_n^1(\cos\theta) \tag{10.1.8}$$

which can be verified by substituting (10.1.8) into (10.1.7) and using (10.1.4).

The conducting sphere produces a diffracted field. The diffracted Debye potentials are given by

$$U^d = C\frac{\cos\phi}{j\beta^2 r}\sum_{n=1}^{\infty} f_n\hat{h}_n^{(2)}(\beta r)P_n^1(\cos\theta) \tag{10.1.9}$$

and

$$V^d = C\frac{\sin\phi}{j\beta^2 r\eta}\sum_{n=1}^{\infty} g_n\hat{h}_n^{(2)}(\beta r)P_n^1(\cos\theta) \tag{10.1.10}$$

For convenience in applying the boundary conditions, the factors in front of (10.1.9) and (10.1.10) are the same as those in the incident waves.

The constants f_n and g_n are determined by the boundary condition on the conducting sphere. That is,

$$E_\theta = E_\phi = 0|_{r=a}$$

where from (4.6.15) and (4.6.16):

$$E_\theta = \frac{1}{r}\frac{\partial^2}{\partial r\partial\theta}(rU^i + rU^d) - \frac{j\omega\mu}{r\sin\theta}\frac{\partial}{\partial\phi}(rV^i + rV^d) \tag{10.1.11}$$

and

$$E_\phi = \frac{1}{r\sin\theta}\frac{\partial^2}{\partial r\partial\phi}(rU^i + rU^d) + \frac{j\omega\mu}{r}\frac{\partial}{\partial\theta}(rV^i + rV^d) \tag{10.1.12}$$

Since the boundary conditions must be satisfied for any θ and ϕ values, it follows from either (10.1.11) or (10.1.12) that the boundary conditions can be stated as

$$\frac{\partial}{\partial r}(rU^i + rU^d) = 0|_{r=a} \tag{10.1.13}$$

and

$$rV^i + rV^d = 0|_{r=a} \tag{10.1.14}$$

Applying the boundary condition (10.1.13), it follows from (10.1.6) and (10.1.9) that:

$$j^{-n}\frac{(2n+1)}{n(n+1)}\hat{j}_n{}'(\beta a) + f_n\hat{h}_n^{(2)}{}'(\beta a) = 0 \Rightarrow f_n = -j^{-n}\frac{(2n+1)}{n(n+1)}\frac{\hat{j}_n{}'(\beta a)}{\hat{h}_n^{(2)}{}'(\beta a)}$$

and from (10.1.14), (10.1.8) and (10.1.10):

$$j^{-n}\frac{(2n+1)}{n(n+1)}\hat{j}_n(\beta a) + g_n\hat{h}_n^{(2)}(\beta a) = 0 \Rightarrow g_n = -j^{-n}\frac{(2n+1)}{n(n+1)}\frac{\hat{j}_n(\beta a)}{\hat{h}_n^{(2)}(\beta a)}$$

The expressions for the Debye potentials are

$$U = U^i + U^d = \frac{C\cos\phi}{j\beta^2 r}\sum_{n=1}^{\infty} j^{-n}\frac{(2n+1)}{n(n+1)}\left[\hat{j}_n(\beta r) - \frac{\hat{j}_n{}'(\beta a)}{\hat{h}_n^{(2)}{}'(\beta a)}\hat{h}_n^{(2)}(\beta r)\right]P_n^1(\cos\theta) \tag{10.1.15}$$

and

$$V = V^i + V^d = \frac{C\sin\phi}{j\beta^2 r\eta}\sum_{n=1}^{\infty} j^{-n}\frac{(2n+1)}{n(n+1)}\left[\hat{j}_n(\beta r) - \frac{\hat{j}_n(\beta a)}{\hat{h}_n^{(2)}(\beta a)}\hat{h}_n^{(2)}(\beta r)\right]P_n^1(\cos\theta) \tag{10.1.16}$$

The diffracted fields far from the sphere are of particular importance in radar applications and target recognition. The diffracted electric fields are

$$\begin{aligned} E_r^d &= \left(\frac{\partial^2}{\partial r^2} + \beta^2\right)(rU^d) \\ &= -\frac{C\cos\phi}{j\beta^2 r^2}\sum_{n=1}^{\infty} j^{-n}(2n+1)\frac{\hat{j}_n{}'(\beta a)}{\hat{h}_n^{(2)}{}'(\beta a)}\hat{h}_n^{(2)}(\beta r)P_n^1(\cos\theta) \end{aligned} \tag{10.1.17}$$

$$\begin{aligned} E_\theta^d &= \frac{1}{r}\frac{\partial^2}{\partial r\partial\theta}(rU^d) - \frac{j\omega\mu}{r\sin\theta}\frac{\partial}{\partial\phi}(rV^d) \\ &= -\frac{C\cos\phi}{j\beta r}\sum_{n=1}^{\infty} j^{-n}\frac{(2n+1)}{n(n+1)}\frac{\hat{j}_n{}'(\beta a)}{\hat{h}_n^{(2)}{}'(\beta a)}\hat{h}_n^{(2)}{}'(\beta r)\frac{dP_n^1(\cos\theta)}{d\theta} \\ &\quad + \frac{C\cos\phi}{\beta r}\sum_{n=1}^{\infty} j^{-n}\frac{(2n+1)}{n(n+1)}\frac{\hat{j}_n(\beta a)}{\hat{h}_n^{(2)}(\beta a)}\hat{h}_n^{(2)}(\beta r)\frac{P_n^1(\cos\theta)}{\sin\theta} \end{aligned} \tag{10.1.18}$$

$$E_\phi^d = \frac{1}{r\sin\theta}\frac{\partial^2}{\partial r \partial\phi}(rU^d) + \frac{j\omega\mu}{r}\frac{\partial}{\partial\theta}(rV^d)$$
$$= \frac{C\sin\phi}{j\beta r}\sum_{n=1}^{\infty} j^{-n}\frac{(2n+1)}{n(n+1)}\frac{\hat{J}_n{}'(\beta a)}{\hat{h}_n^{(2)\prime}(\beta a)}\hat{h}_n^{(2)\prime}(\beta r)\frac{P_n^1(\cos\theta)}{\sin\theta}$$
$$- \frac{C\sin\phi}{\beta r}\sum_{n=1}^{\infty} j^{-n}\frac{(2n+1)}{n(n+1)}\frac{\hat{J}_n(\beta a)}{\hat{h}_n^{(2)}(\beta a)}\hat{h}_n^{(2)}(\beta r)\frac{dP_n^1(\cos\theta)}{d\theta} \qquad (10.1.19)$$

The result in (10.1.17) used (10.1.4). Also, in (10.1.18) and (10.1.19) the Legendre's function derivative can be expressed as

$$\frac{dP_n^1(\cos\theta)}{d\theta} = -\sin\theta P_n^{1\prime}(\cos\theta)$$

where the prime denotes the derivative with respect the argument.

In the far field, using the large argument approximations:

$$\hat{h}_n^{(2)}(\beta r) \approx j^{n+1}e^{-j\beta r} \qquad (10.1.20)$$

and

$$\hat{h}_n^{(2)\prime}(\beta r) = \frac{d\hat{h}_n^{(2)}(\beta r)}{d(\beta r)} \approx j^{n+1}(-j)e^{-j\beta r} = j^n e^{-j\beta r} \qquad (10.1.21)$$

it follows from (10.1.17) that E_r^d varies as $1/r^2$ and can be neglected in the far field. The fields that vary as $1/r$ are the angular fields. Looking at (10.1.18) and (10.1.19):

$$E_\theta^d \approx -\frac{C\cos\phi}{j\beta}\frac{e^{-j\beta r}}{r}\sum_{n=1}^{\infty}\frac{(2n+1)}{n(n+1)}\left[\frac{\hat{J}_n{}'(\beta a)}{\hat{h}_n^{(2)\prime}(\beta a)}\frac{dP_n^1(\cos\theta)}{d\theta} + \frac{\hat{J}_n(\beta a)}{\hat{h}_n^{(2)}(\beta a)}\frac{P_n^1(\cos\theta)}{\sin\theta}\right] \qquad (10.1.22)$$

and

$$E_\phi^d \approx \frac{C\sin\phi}{j\beta}\frac{e^{-j\beta r}}{r}\sum_{n=1}^{\infty}\frac{(2n+1)}{n(n+1)}\left[\frac{\hat{J}_n{}'(\beta a)}{\hat{h}_n^{(2)\prime}(\beta a)}\frac{P_n^1(\cos\theta)}{\sin\theta} + \frac{\hat{J}_n(\beta a)}{\hat{h}_n^{(2)}(\beta a)}\frac{dP_n^1(\cos\theta)}{d\theta}\right] \qquad (10.1.23)$$

The far-field magnetic components are

$$H_\phi^d = \frac{E_\theta^d}{\eta} \qquad (10.1.24)$$

and

$$H_\theta^d = -\frac{E_\phi^d}{\eta} \tag{10.1.25}$$

The diffracted far fields can also be determined if the diffracted Debye potential in (10.1.15) and (10.1.16) are first calculated in the far-field, namely:

$$U^d \approx -\frac{C\cos\phi}{\beta^2}\frac{e^{-j\beta r}}{r}\sum_{n=1}^{\infty}\frac{(2n+1)}{n(n+1)}\frac{\hat{j}_n'(\beta a)}{\hat{h}_n^{(2)'}(\beta a)}P_n^1(\cos\theta) \tag{10.1.26}$$

and

$$V^d \approx -\frac{C\sin\phi}{\beta^2\eta}\frac{e^{-j\beta r}}{r}\sum_{n=1}^{\infty}\frac{(2n+1)}{n(n+1)}\frac{\hat{j}_n(\beta a)}{\hat{h}_n^{(2)}(\beta a)}P_n^1(\cos\theta) \tag{10.1.27}$$

Substituting (10.1.26) and (10.1.27) into (10.1.18) and (10.1.19) results in the far-field relations in (10.1.22) to (10.1.25).

The field diffracted into the direction of the incoming plane wave is obtained by setting $\theta = \pi$ and $\phi = \pi$ in (10.1.22). That is, $\hat{a}_x = \hat{a}_\theta$ at $\theta = \pi$ and $\phi = \pi$, which shows that

$$E_x^d = E_\theta^d|_{\substack{\theta=\pi \\ \phi=\pi}}$$

At $\theta = \pi$ the Legendre functions in (10.1.22) are

$$\left.\frac{P_n^1(\cos\theta)}{\sin\theta}\right|_{\theta=\pi} = (-1)^{n+1}\frac{n(n+1)}{2}$$

and

$$\left.\frac{dP_n^1(\cos\theta)}{d\theta}\right|_{\theta=\pi} = (-1)^n\frac{n(n+1)}{2}$$

Hence,

$$E_x^d = E_\theta^d\bigg|_{\substack{\theta=\pi \\ \phi=\pi}} = \frac{C}{j\beta}\frac{e^{-j\beta r}}{r}\sum_{n=1}^{\infty}(-1)^n\frac{(2n+1)}{2}\left[\frac{\hat{j}_n'(\beta a)}{\hat{h}_n^{(2)'}(\beta a)} - \frac{\hat{j}_n(\beta a)}{\hat{h}_n^{(2)}(\beta a)}\right]$$

$$= \frac{C}{2\beta}\frac{e^{-j\beta r}}{r}\sum_{n=1}^{\infty}\frac{(-1)^n(2n+1)}{\hat{h}_n^{(2)'}(\beta a)\hat{h}_n^{(2)}(\beta a)} \tag{10.1.28}$$

where the Wronskian

$$W\,[\hat{j}_n(\beta a),\, \hat{h}_n^{(2)}(\beta a)] = -j \tag{10.1.29}$$

was used.

In radar work the cross section of a target is an important parameter. The cross section of a target is a measure of how detectable is the target, or the target ability to reflect the incoming signal towards the radar. Several cross-section definitions are used, such as the backscattering cross section, the total cross section, and the scattering cross section. The backscattering cross section (σ) (also known as the echo area) is defined as

$$\sigma = \lim_{r\to\infty}\left[4\pi r^2 \frac{|\mathbf{E}^d|^2}{|\mathbf{E}_o|^2}\right]$$

where $\mathbf{E}_o$ represents the power in the incident plane wave (i.e., $|C|^2$) and $\mathbf{E}^d = E_x^d\,\hat{\mathbf{a}}_x$ where E_x^d is given by (10.1.28). Therefore,

$$\sigma = \lim_{r\to\infty}\left[4\pi r^2 \frac{|E_x^d|^2}{|C|^2}\right] = \frac{\lambda^2}{4\pi}\left|\sum_{n=1}^{\infty} \frac{(-1)^n(2n+1)}{\hat{h}_n^{(2)\prime}(\beta a)\hat{h}_n^{(2)}(\beta a)}\right|^2 \tag{10.1.30}$$

For small radius, such that $\beta a << 1$, using the $n = 1$ term in (10.1.30) gives

$$\sigma = \frac{\lambda^2}{4\pi}\frac{9}{|\hat{h}_1^{(2)\prime}(\beta a)\hat{h}_1^{(2)}(\beta a)|^2} \tag{10.1.31}$$

Using the small argument approximations (see Appendix C):

$$\hat{h}_1^{(2)}(\beta a) \approx j\frac{1}{\beta a} \tag{10.1.32}$$

and

$$\hat{h}_1^{\prime(2)}(\beta a) \approx -j\frac{1}{(\beta a)^2} \tag{10.1.33}$$

Hence, (10.1.31) reads

$$\sigma = \frac{9\lambda^2}{4\pi}(\beta a)^6 = \frac{9}{4\pi\lambda^4}(2\pi a)^6$$

which is referred as the "Rayleigh scattering law." This relation shows that the scattering cross section of small conducting spheres varies as λ^{-4}. The Rayleigh scattering law was used to explain the blue color of the sky.

10.2 DIFFRACTION BY A DIELECTRIC SPHERE

A plane wave incident on a dielectric sphere is shown in Fig. 10.2. This problem is referred to as "Mie scattering." The region $r \leq a$ is a good dielectric characterized by ε_2 and μ_2. In this case, there is a field in the dielectric region and the Debye potentials in the two regions are written in the forms:

$$U_1 = U_1^i + U_1^d = C\frac{\cos\phi}{j\beta_1^2 r}\sum_{n=1}^{\infty} j^{-n}\frac{(2n+1)}{n(n+1)}[\hat{J}_n(\beta_1 r) + a_n \hat{h}_n^{(2)}(\beta_1 r)]P_n^1(\cos\theta)$$

$$V_1 = V_1^i + V_1^d = C\frac{\sin\phi}{j\beta_1^2 r\eta_1}\sum_{n=1}^{\infty} j^{-n}\frac{(2n+1)}{n(n+1)}[\hat{J}_n(\beta_1 r) + b_n \hat{h}_n^{(2)}(\beta_1 r)]P_n^1(\cos\theta)$$

$$U_2 = C\frac{\cos\phi}{j\beta_1^2 r}\sum_{n=1}^{\infty} j^{-n}\frac{(2n+1)}{n(n+1)}[c_n\hat{J}_n(\beta_2 r)]P_n^1(\cos\theta)$$

$$V_2 = C\frac{\sin\phi}{j\beta_1^2 r\eta_1}\sum_{n=1}^{\infty} j^{-n}\frac{(2n+1)}{n(n+1)}[d_n\hat{J}_n(\beta_2 r)]P_n^1(\cos\theta)$$

where $\beta_1 = \omega\sqrt{\mu_1\varepsilon_1}$ and $\beta_2 = \omega\sqrt{\mu_2\varepsilon_2}$. Again, the factors in front of the potentials are selected for convenience when applying the boundary conditions.

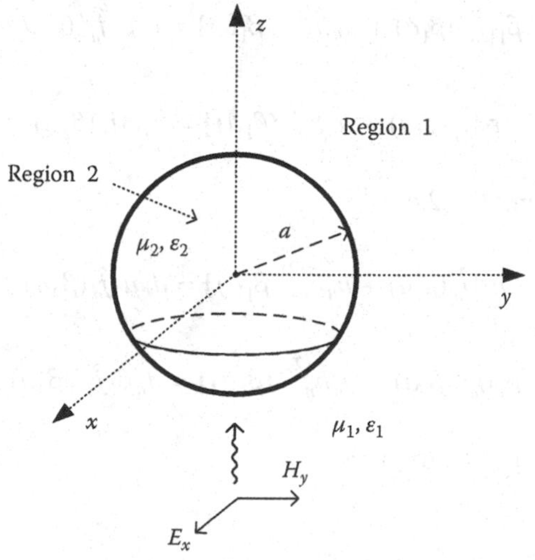

FIGURE 10.2 A plane wave incident on a dielectric sphere.

At $r = a$, the tangential electric and magnetic fields must be continuous. Then,

$$E_{\theta,1} = E_{\theta,2}|_{r=a}$$

and

$$H_{\theta,1} = H_{\theta,2}|_{r=a}$$

The continuity of the tangential electric field E_θ at $r = a$ (see (4.6.15)) shows that

$$\frac{\partial}{\partial r}(rU_1) = \frac{\partial}{\partial r}(rU_2)|_{r=a} \tag{10.2.1}$$

and

$$\mu_1(rV_1) = \mu_2(rV_2)|_{r=a} \tag{10.2.2}$$

The continuity of the tangential magnetic field H_θ at $r = a$ (see (4.6.18)) requires that

$$\varepsilon_1(rU_1) = \varepsilon_2(rU_2)|_{r=a} \tag{10.2.3}$$

and

$$\frac{\partial}{\partial r}(rV_1) = \frac{\partial}{\partial r}(rV_2)|_{r=a} \tag{10.2.4}$$

From (10.2.1) and (10.2.3) we obtain:

$$\beta_1[\hat{j}_n{}'(\beta_1 a) + a_n \hat{h}_n^{(2)\prime}(\beta_1 a)] = c_n \beta_2 \hat{j}_n{}'(\beta_2 a) \tag{10.2.5}$$

$$\varepsilon_1[\hat{j}_n(\beta_1 a) + a_n \hat{h}_n^{(2)}(\beta_1 a)] = c_n \varepsilon_2 \hat{j}_n(\beta_2 a) \tag{10.2.6}$$

and from (10.2.2) and (10.2.4):

$$\mu_1[\hat{j}_n(\beta_1 a) + b_n \hat{h}_n^{(2)}(\beta_1 a)] = d_n \mu_2 \hat{j}_n(\beta_2 a) \tag{10.2.7}$$

$$\beta_1[\hat{j}_n{}'(\beta_1 a) + b_n \hat{h}_n^{(2)\prime}(\beta_1 a)] = d_n \beta_2 \hat{j}_n{}'(\beta_2 a) \tag{10.2.8}$$

Solving (10.2.5) and (10.2.6) gives

$$a_n = \frac{\hat{J}_n(\beta_2 a)\hat{J}_n{}'(\beta_1 a) - \eta_{21}\hat{J}_n{}'(\beta_2 a)\hat{J}_n(\beta_1 a)}{\eta_{21}\hat{J}_n{}'(\beta_2 a)\hat{H}_n^{(2)}(\beta_1 a) - \hat{J}_n(\beta_2 a)\hat{H}_n^{(2)\prime}(\beta_1 a)} \tag{10.2.9}$$

$$c_n = \frac{j\varepsilon_{12}}{\eta_{21}\hat{J}_n{}'(\beta_2 a)\hat{H}_n^{(2)}(\beta_1 a) - \hat{J}_n(\beta_2 a)\hat{H}_n^{(2)\prime}(\beta_1 a)}$$

where the Wronskian in (10.1.29) was used and

$$\frac{\beta_2 \varepsilon_1}{\beta_1 \varepsilon_2} = \sqrt{\frac{\mu_2 \varepsilon_1}{\mu_1 \varepsilon_2}} = \frac{\eta_2}{\eta_1} = \eta_{21}$$

From (10.2.7) and (10.2.8) we obtain:

$$b_n = \frac{\hat{J}_n(\beta_2 a)\hat{J}_n{}'(\beta_1 a) - \eta_{12}\hat{J}_n{}'(\beta_2 a)\hat{J}_n(\beta_1 a)}{\eta_{12}\hat{J}_n{}'(\beta_2 a)\hat{H}_n^{(2)}(\beta_1 a) - \hat{J}_n(\beta_2 a)\hat{H}_n^{(2)\prime}(\beta_1 a)} \tag{10.2.10}$$

$$d_n = \frac{j\mu_{12}}{\eta_{12}\hat{J}_n{}'(\beta_2 a)\hat{H}_n^{(2)}(\beta_1 a) - \hat{J}_n(\beta_2 a)\hat{H}_n^{(2)\prime}(\beta_1 a)}$$

Using (10.1.20), the diffracted Debye potentials in the far field are

$$U_1^d = C\frac{\cos\phi}{\beta_1^2}\frac{e^{-j\beta_1 r}}{r}\sum_{n=1}^{\infty}\frac{(2n+1)}{n(n+1)}a_n P_n^1(\cos\theta)$$

and

$$V_1^d = C\frac{\sin\phi}{\beta_1^2\eta_1}\frac{e^{-j\beta_1 r}}{r}\sum_{n=1}^{\infty}\frac{(2n+1)}{n(n+1)}b_n P_n^1(\cos\theta)$$

Hence, the diffracted electric fields in the far field are

$$\begin{aligned} E_\theta^d &= \frac{1}{r}\frac{\partial^2}{\partial r\partial\theta}(rU_1^d) - \frac{j\omega\mu}{r\sin\theta}\frac{\partial}{\partial\phi}(rV_1^d) \\ &= C\frac{\cos\phi}{j\beta_1}\frac{e^{-j\beta_1 r}}{r}\sum_{n=1}^{\infty}\frac{(2n+1)}{n(n+1)}\left[a_n\frac{dP_n^1(\cos\theta)}{d\theta} + b_n\frac{P_n^1(\cos\theta)}{\sin\theta}\right] \end{aligned}$$

and

$$E_\phi^d = \frac{1}{r\sin\theta}\frac{\partial^2}{\partial r\partial\phi}(rU^d) + \frac{j\omega\mu}{r}\frac{\partial}{\partial\theta}(rV^d)$$
$$= -C\frac{\sin\phi}{j\beta_1}\frac{e^{-j\beta_1 r}}{r}\sum_{n=1}^{\infty}\frac{(2n+1)}{n(n+1)}\left[a_n\frac{P_n^1(\cos\theta)}{\sin\theta} + b_n\frac{dP_n^1(\cos\theta)}{d\theta}\right]$$

where a_n and b_n are given by (10.2.9) and (10.2.10). The far-field magnetic fields are

$$H_\theta^d = -\frac{E_\phi^d}{\eta_1}$$

and

$$H_\phi^d = \frac{E_\theta^d}{\eta_1}$$

10.3 A VERTICAL ELECTRIC DIPOLE ABOVE A CONDUCTING SPHERE

A VED above a conducting sphere is shown in Fig. 10.3, where the region outside the sphere is air. The Debye potential U satisfies (4.6.13), namely

$$[\nabla^2 + k_o^2]U = -\frac{J_{e,r}}{j\omega\varepsilon_o r} = -\frac{1}{j\omega\varepsilon_o r}\frac{(I\Delta L)\delta(r-b)\delta(\theta)}{2\pi r^2\sin\theta} \tag{10.3.1}$$

where $k_o = \omega\sqrt{\mu_o\varepsilon_o}$. The symmetry shows that the fields have no ϕ dependence (i.e., $\partial/\partial\phi = 0$), and from (4.6.14) to (4.6.19):

$$E_r = \left(\frac{\partial^2}{\partial r^2} + k_o^2\right)(rU)$$

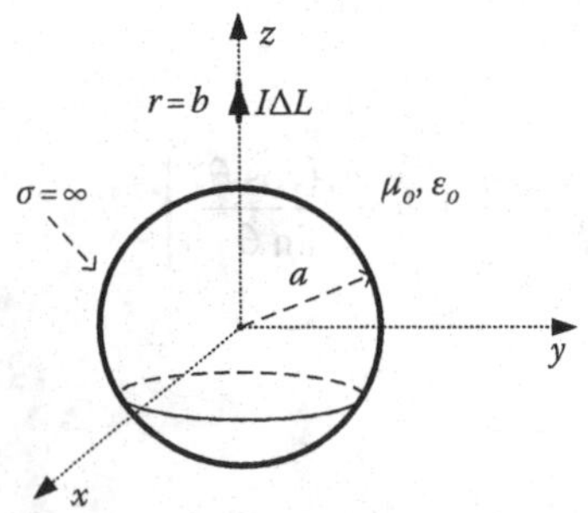

FIGURE 10.3 A VED above a spherical conductor.

$$E_\theta = \frac{1}{r}\frac{\partial^2}{\partial r\partial\theta}(rU) \tag{10.3.2}$$

and

$$H_\phi = -\frac{j\omega\varepsilon_o}{r}\frac{\partial(rU)}{\partial\theta}$$

which represents a TM^r field configuration. The components E_ϕ, H_r and H_θ are zero.

The particular solution to (10.3.1) is found using the Green function of free space, namely:

$$U^p = \frac{1}{j\omega\varepsilon_o}\int_V \frac{J_{e,r}(r')}{r'}G(\mathbf{r},\mathbf{r}')dV' \tag{10.3.3}$$

Therefore,

$$\begin{aligned}
U^p &= \frac{(I\Delta L)}{j\omega\varepsilon_o}\int_{V'}\frac{e^{-jk_o|\mathbf{r}-\mathbf{r}'|}}{4\pi|\mathbf{r}-\mathbf{r}'|}\frac{1}{r'}\frac{\delta(r'-b)\delta(\theta')}{2\pi r'^2\sin\theta'}r'^2\sin\theta'dr'd\theta'd\phi' \\
&= \frac{(I\Delta L)}{j\omega\varepsilon_o b}\frac{e^{-jk_oR}}{4\pi R}
\end{aligned} \tag{10.3.4}$$

where

$$R = |\mathbf{r}-\mathbf{r}'| = \sqrt{r^2 + r'^2 - 2rr'\cos(\theta-\theta')}\,\Big|_{\substack{r'=b \\ \theta'=0}}$$

$$= \sqrt{r^2 + b^2 - 2rb\cos\theta}$$

Since,

$$\frac{e^{-jk_oR}}{4\pi R} = \frac{-j}{4\pi k_o rb}\sum_{n=0}^{\infty}(2n+1)P_n(\cos\theta)\begin{cases}\hat{j}_n(k_or)\hat{h}_n^{(2)}(k_ob) & (r\le b) \\ \hat{j}_n(k_ob)\hat{h}_n^{(2)}(k_or) & (r\ge b)\end{cases}$$

(10.3.4) can be expressed in the form

$$U^p = \frac{-(I\Delta L)\eta_o}{4\pi k_o^2 b^2 r}\sum_{n=0}^{\infty}(2n+1)P_n(\cos\theta)\begin{cases}\hat{j}_n(k_or)\hat{h}_n^{(2)}(k_ob) & (r\le b) \\ \hat{j}_n(k_ob)\hat{h}_n^{(2)}(k_or) & (r\ge b)\end{cases}$$

The complementary solution is of the form

$$U^c = \frac{-(I\Delta L)\eta_o}{4\pi k_o^2 b^2 r} \sum_{n=0}^{\infty} a_n (2n+1)\hat{h}_n^{(2)}(k_o r) P_n(\cos\theta)$$

and the total solution $U = U^p + U^c$ is given by

$$U = \begin{cases} -\frac{(I\Delta L)\eta_o}{4\pi k_o^2 b^2 r} \sum_{n=0}^{\infty} (2n+1) P_n(\cos\theta) \left[a_n \hat{h}_n^{(2)}(k_o r) + \hat{j}_n(k_o r)\hat{h}_n^{(2)}(k_o b) \right] & (a \le r \le b) \\ -\frac{(I\Delta L)\eta_o}{4\pi k_o^2 b^2 r} \sum_{n=0}^{\infty} (2n+1) P_n(\cos\theta) \left[a_n \hat{h}_n^{(2)}(k_o r) + \hat{j}_n(k_o b)\hat{h}_n^{(2)}(k_o r) \right] & (r \ge b) \end{cases} \quad (10.3.5)$$

The boundary condition on the surface of the conducting sphere must be satisfied for all θ. Therefore, from (10.3.2):

$$E_\theta = 0|_{r=a} \Rightarrow \frac{\partial(rU)}{\partial r} = 0|_{r=a} = 0 \quad (10.3.6)$$

and from (10.3.5) we obtain

$$a_n \hat{h}_n^{(2)\prime}(k_o a) + \hat{j}_n{}'(k_o a)\hat{h}_n^{(2)}(k_o b) = 0 \Rightarrow a_n = -\frac{\hat{j}_n{}'(k_o a)}{\hat{h}_n^{(2)\prime}(k_o a)} \hat{h}_n^{(2)}(k_o b) \quad (10.3.7)$$

The total solution is then::

$$U = \begin{cases} \frac{-(I\Delta L)\eta_o}{4\pi k_o^2 b^2 r} \sum_{n=0}^{\infty} (2n+1) P_n(\cos\theta) \left[\hat{j}_n(k_o r) - \frac{\hat{j}_n{}'(k_o a)}{\hat{h}_n^{(2)\prime}(k_o a)} \hat{h}_n^{(2)}(k_o r) \right] \hat{h}_n^{(2)}(k_o b) & (a \le r \le b) \\ \frac{-(I\Delta L)\eta_o}{4\pi k_o^2 b^2 r} \sum_{n=0}^{\infty} (2n+1) P_n(\cos\theta) \left[\hat{j}_n(k_o b) - \frac{\hat{j}_n{}'(k_o a)}{\hat{h}_n^{(2)\prime}(k_o a)} \hat{h}_n^{(2)}(k_o b) \right] \hat{h}_n^{(2)}(k_o r) & (r \ge b) \end{cases} \quad (10.3.8)$$

If the dipole is slightly above the surface (i.e., $b \approx a$), (10.3.8) reads

$$U = j\frac{(I\Delta L)\eta_o}{4\pi k_o^2 a^2 r} \sum_{n=0}^{\infty} (2n+1) \frac{\hat{h}_n^{(2)}(k_o r)}{\hat{h}_n^{(2)\prime}(k_o a)} P_n(\cos\theta) \qquad (r \ge b) \quad (10.3.9)$$

where the Wronskian in (10.1.29) was used.

In cases where $k_o a$ is small and $k_o r >> 1$ (i.e., in the far field), using the large argument approximation in (10.1.20) shows that (10.3.9) reads

$$U = -\frac{(I\Delta L)\eta_o}{4\pi k_o^2 a^2}\frac{e^{-jk_o r}}{r}\sum_{n=0}^{\infty} j^n \frac{(2n+1)}{\hat{h}_n^{(2)\prime}(k_o a)} P_n(\cos\theta)$$

From (10.3.2), the far-field form of E_θ, is

$$E_\theta \approx j\frac{(I\Delta L)\eta_o}{4\pi k_o a^2}\frac{e^{-jk_o r}}{r}\sum_{n=0}^{\infty} j^n \frac{(2n+1)}{\hat{h}_n^{(2)\prime}(k_o a)} P_n^1(\cos\theta)$$

and $H_\phi = E_\theta/\eta_o$.

If the VED in Fig. 10.3 is located at (b, θ', ϕ') (i.e., inclined with respect to the axis shown), (5.6.66) shows that the Green function is

$$\frac{e^{-jk_o|\mathbf{r}-\mathbf{r}'|}}{4\pi|\mathbf{r}-\mathbf{r}'|} = \frac{-j}{k_o r b}\sum_{n=0}^{\infty} Y_{nm}(\theta,\phi)Y_{nm}^*(\theta',\phi')\begin{cases}\hat{j}_n(k_o r)\hat{h}_n^{(2)}(k_o b) & (r \le b)\\ \hat{j}_n(k_o b)\hat{h}_n^{(2)}(k_o r) & (r \ge b)\end{cases}$$

For this case, the particular solution is

$$U^p = \frac{(I\Delta L)}{j\omega\varepsilon_o b}\frac{e^{-jk_o|\mathbf{r}-\mathbf{r}'|}}{4\pi|\mathbf{r}-\mathbf{r}'|}$$

$$= \frac{-(I\Delta L)\eta_o}{k_o^2 b^2 r}\sum_{n=0}^{\infty} Y_{nm}(\theta,\phi)Y_{nm}^*(\theta',\phi')\begin{cases}\hat{j}_n(k_o r)\hat{h}_n^{(2)}(k_o b) & (r \le b)\\ \hat{j}_n(k_o b)\hat{h}_n^{(2)}(k_o r) & (r \ge b)\end{cases}$$

and the complementary solution is

$$U^c = \frac{-(I\Delta L)\eta_o}{k_o^2 b^2 r}\sum_{n=0}^{\infty} a_n h_n^{(2)}(k_o r)Y_{nm}(\theta,\phi)Y_{nm}^*(\theta',\phi').$$

Applying the boundary condition in (10.3.6) shows that a_n is given by (10.3.7). Therefore, the total solution is

$$U = \begin{cases}\frac{-(I\Delta L)\eta_o}{k_o^2 b^2 r}\sum_{n=0}^{\infty}\left[\hat{j}_n(k_o r) - \frac{\hat{j}_n{}'(k_o a)}{\hat{h}_n^{(2)\prime}(k_o a)}\hat{h}_n^{(2)}(k_o r)\right]\hat{h}_n^{(2)}(k_o b)Y_{nm}(\theta,\phi)Y_{nm}^*(\theta',\phi') & (a \le r \le b)\\ \frac{-(I\Delta L)\eta_o}{k_o^2 b^2 r}\sum_{n=0}^{\infty}\left[\hat{j}_n(k_o b) - \frac{\hat{j}_n{}'(k_o a)}{\hat{h}_n^{(2)\prime}(k_o a)}\hat{h}_n^{(2)}(k_o b)\right]\hat{h}_n^{(2)}(k_o r)Y_{nm}(\theta,\phi)Y_{nm}^*(\theta',\phi') & (r \ge b)\end{cases} \tag{10.3.10}$$

It is seen that if $\theta' = 0$ and $\phi = \phi' = 0$ (i.e., no ϕ dependence); then,

$$Y_{nm}(\theta, \phi) Y^*_{nm}(\theta', \phi')\Bigg|_{\substack{\theta'=0 \\ \phi=\phi'=0}} = \frac{2n+1}{4\pi} P_n(\cos\theta)$$

and (10.3.10) is equivalent to (10.3.8).

The sums in (10.3.8) to (10.3.10) can be used to calculate the fields for small values of $k_o a$ and $k_o r$. For large values of $k_o a$ and $k_o r$ these sums converge very slowly. For example, if the conducting sphere represents the Earth; then, $r \approx a \approx 6{,}300$ km and if $f = 30$ kHz (i.e., $\lambda = 10$ km) results in $k_o r \approx k_o a = 3{,}958$. The factor $\hat{h}_n^{(2)}(k_o r)/\hat{h}_n^{(2)'}(k_o a)$ in (10.3.9) is fairly independent of n until $n \approx k_o a$. At these values of n, the Debye approximation and third-order approximation to the Hankel functions are used, and the sum begins to converge. It has been estimated that up to $2k_o a$ are needed, with terms above $2k_o a$ being negligible. These sums can be converted to a fast-convergent series using the Watson transformation.

Consider the Debye potential U in (10.3.9). This sum can be expressed in the form:

$$U = \sum_{n=0}^{\infty} (2n+1) f(n) P_n(\cos\theta) \tag{10.3.11}$$

where $f(n)$ is

$$f(n) = j\frac{(I\Delta L)\eta_o}{4\pi k_o^2 a^2 r} \frac{\hat{h}_n^{(2)}(k_o r)}{\hat{h}_n^{(2)'}(k_o a)}$$

Using a Watson transformation, (10.3.11) can be represented by the following contour integral:

$$U = j\oint_C \frac{\nu}{\cos \nu\pi} f\left(\nu - \frac{1}{2}\right) P_{\nu-1/2}(-\cos\theta)\, d\nu \tag{10.3.12}$$

where the contour $C = C_1 + C_2$, shown in Fig. 10.4a, encloses the positive real-axis poles of $\cos \nu\pi$, located at $\nu = n + 1/2$ for $n = 0, 1, 2, \ldots$. Using the residue theorem to evaluate (10.3.12) along C shows that

$$\begin{aligned} U &= (2\pi j) j \sum_{n=0}^{\infty} \frac{\nu f\left(\nu - \frac{1}{2}\right)}{-\pi \sin \nu\pi} P_{\nu-1/2}(-\cos\theta)\Bigg|_{\nu=n+1/2} \\ &= 2\sum_{n=0}^{\infty} \frac{\left(n+\frac{1}{2}\right) f(n)}{(-1)^n} P_n(-\cos\theta) \end{aligned}$$

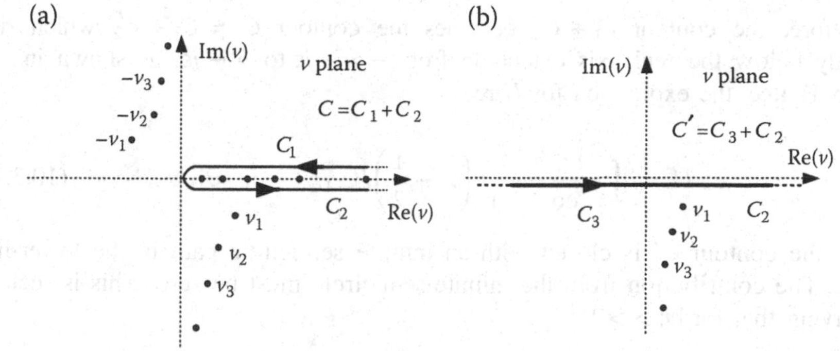

FIGURE 10.4 (a) Contour C encloses the poles of $\cos(\nu\pi)$ along the real axis; (b) the equivalent contour $C' = C_3 + C_2$.

and since $P_n(-x) = (-1)^n P_n(x)$, U is given by

$$U = \sum_{n=0}^{\infty} (2n+1) f(n) P_n(\cos\theta)$$

which is (10.3.11).

In (10.3.12), since:

$$P_{-\nu-1/2}(\cos\theta) = P_{\nu-1/2}(\cos\theta)$$
$$\hat{h}^{(2)}_{-\nu-1/2}(x) = e^{-j\nu\pi}\hat{h}^{(2)}_{\nu-1/2}(x)$$

and

$$\hat{h}^{(2)\prime}_{-\nu-1/2}(x) = e^{-j\nu\pi}\hat{h}^{(2)\prime}_{\nu-1/2}(x)$$

it follows that $f(\nu - 1/2)$ and the Legendre functions are even function of ν. The term $\nu/\cos\nu\pi$ is odd and, therefore, the integrand in (10.3.12) is odd. Letting $\nu = -\nu$ in the integral along C_1, we obtain

$$j\int_{C_1} \frac{\nu}{\cos\nu\pi} f\left(\nu - \frac{1}{2}\right) P_{\nu-1/2}(-\cos\theta)\,d\nu$$
$$= j\int_{-\infty-j\varepsilon}^{0-j\varepsilon} \frac{-\nu}{\cos(-\nu\pi)} f\left(-\nu - \frac{1}{2}\right) P_{-\nu-1/2}(-\cos\theta)(-d\nu)$$
$$= j\int_{-\infty-j\varepsilon}^{0-j\varepsilon} \frac{\nu}{\cos(\nu\pi)} f\left(\nu - \frac{1}{2}\right) P_{\nu-1/2}(-\cos\theta)\,d\nu$$
$$= j\int_{C_3} \frac{\nu}{\cos\nu\pi} f\left(\nu - \frac{1}{2}\right) P_{\nu-1/2}(-\cos\theta)(d\nu)$$

Therefore, the contour $C_1 + C_2$ becomes the contour $C' = C_3 + C_2$ which runs slightly below the real axis extending from $-\infty - j\varepsilon$ to $\infty - j\varepsilon$, as shown in Fig. 10.4b. Hence, the expression for U reads

$$U = j\int_{C'} \frac{\nu}{\cos(\nu\pi)} f\left(\nu - \frac{1}{2}\right) P_{\nu-1/2}(-\cos\theta)\,d\nu \tag{10.3.13}$$

Next, the contour C' is closed with an infinite semicircle path in the lower-half plane. The contribution from the infinite semicircle must be zero. This is seen by observing that for $|\nu| >> 1$:

$$\begin{aligned} P_{\nu-1/2}(-\cos\theta) &\approx \sqrt{\frac{2}{\pi\nu\sin\theta}}\cos[\nu(\pi-\theta)-\pi/4] \\ &\approx \sqrt{\frac{1}{2\pi\nu\sin\theta}}[e^{j\nu(\pi-\theta)}e^{-j\pi/4} + e^{-j\nu(\pi-\theta)}e^{j\pi/4}] \end{aligned}$$

where in the lower-half plane: $\mathrm{Im}(\nu) < 0$. For $\theta < \pi$, and θ not near 0 or π, the main contribution is from the first exponential term. Hence,

$$P_{\nu-1/2}(-\cos\theta) \approx \sqrt{\frac{1}{2\pi\nu\sin\theta}}\,e^{j\nu(\pi-\theta)}e^{-j\pi/4}$$

Also,

$$\cos(\nu\pi) \approx \frac{e^{j\nu\pi}}{2}$$

showing that

$$\frac{P_{\nu-1/2}(-\cos\theta)}{\cos(\nu\pi)} \approx \sqrt{\frac{2}{\pi\nu\sin\theta}}\,e^{-j\nu\theta}e^{-j\pi/4} \tag{10.3.14}$$

Then, with $\nu = Re^{j\theta}$ and $R \to \infty$:

$$\begin{aligned} \left|\frac{P_{\nu-1/2}(-\cos\theta)}{\cos(\nu\pi)}\right| &\approx \left|\sqrt{\frac{2}{\pi\nu\sin\theta}}\,e^{-j\nu\theta}e^{-j\pi/4}\right| \\ &\approx \sqrt{\frac{1}{R}}\,e^{R\sin\theta} \end{aligned} \tag{10.3.15}$$

Since $\sin\theta < 0$ in the lower-half plane, this term goes to zero along the infinite semicircle.

The other term to consider in $f(\nu - 1/2)$ is the behavior of $\hat{h}^{(2)}_{\nu-1/2}(k_o r)/\hat{h}^{(2)\prime}_{\nu-1/2}(k_o a)$ as $|\nu| >> 1$ in the lower-half plane. From the results in Section 9.10 it follows that this term vanishes as $R \to \infty$ in the semicircular part of the path. Hence, the contribution from the semicircular path is zero.

The value of the integral in (10.3.13) is equal to the sum of the residues at the poles of $f(\nu - 1/2)$ (i.e., those of $\hat{h}^{(2)'}_{\nu-1/2}(k_o a)$ in the lower-half plane). That is,

$$U = \frac{(I\Delta L)\eta_o}{2k_o^2 a^2 r} \sum_{p=1}^{\infty} \frac{\nu_p}{\cos(\nu_p \pi)} P_{\nu_p - 1/2}(-\cos\theta) \frac{\hat{h}^{(2)}_{\nu_p-1/2}(k_o r)}{\left.\frac{\partial \hat{h}^{(2)'}_{\nu-1/2}(k_o a)}{\partial \nu}\right|_{\nu=\nu_p}} \tag{10.3.16}$$

where the poles of $\hat{h}^{(2)'}_{\nu-1/2}(k_o a)$ from (7.4.47) occur at

$$\nu = \nu_p{}' = k_o a - \chi_p{}' \left(\frac{k_o a}{2}\right)^{1/3} e^{-j\pi/3}$$

The term $\chi_p{}'$ denotes the zeroes of $Ai'(\chi_p{}') = 0$.

The evaluation of (10.3.16) is done using the asymptotic approximations of the Hankel functions in terms of the Airy functions, or in terms of the Fok-Airy functions. In Section 10.4, a more general problem is considered and the evaluation of the electric field associated with (10.3.16) is obtained.

10.4 A VERTICAL ELECTRIC DIPOLE ABOVE A SPHERICAL SURFACE

The geometry of the problem for a VED above spherical surface is shown in Fig. 10.5. The spherical surface will be described by a surface impedance. This model represents a spherical object that satisfies a surface-impedance boundary condition, such as the surface of the Earth. The problem can be formulated in terms of the Debye potential U. That is,

$$[\nabla^2 + k_o^2]U = -\frac{J_{e,r}}{j\omega\varepsilon_o r} = -\frac{1}{j\omega\varepsilon_o r} \frac{(I\Delta L)\delta(r-b)\delta(\theta)}{2\pi r^2 \sin\theta} \tag{10.4.1}$$

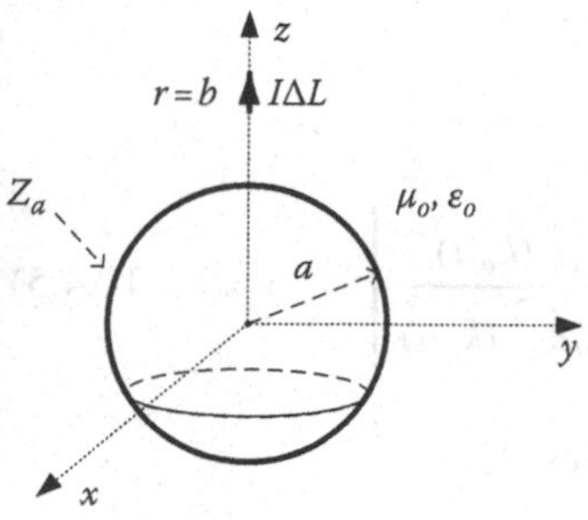

FIGURE 10.5 A VED above a spherical surface described by its surface impedance.

where $k_o = \omega\sqrt{\mu_o \varepsilon_o}$. The non-zero fields, from (4.6.14) to (4.6.19) with $\partial/\partial\phi = 0$, are

$$\begin{cases} E_r = \left(\frac{\partial^2}{\partial r^2} + k_o^2\right)(rU) \\ E_\theta = \frac{1}{r}\frac{\partial^2}{\partial r \partial\theta}(rU) \\ H_\phi = -\frac{j\omega\varepsilon_o}{r}\frac{\partial(rU)}{\partial\theta} \end{cases} \tag{10.4.2}$$

The particular and complementary solutions are given by (10.3.5). That is,

$$U = \begin{cases} -\frac{(I\Delta L)\eta_o}{4\pi k_o^2 b^2 r}\sum_{n=0}^{\infty}(2n+1)P_n(\cos\theta)\left[a_n\hat{h}_n^{(2)}(k_o r) + \hat{j}_n(k_o r)\hat{h}_n^{(2)}(k_o b)\right] & (a \le r \le b) \\ -\frac{(I\Delta L)\eta_o}{4\pi k_o^2 b^2 r}\sum_{n=0}^{\infty}(2n+1)P_n(\cos\theta)\left[a_n\hat{h}_n^{(2)}(k_o r) + \hat{j}_n(k_o b)\hat{h}_n^{(2)}(k_o r)\right] & (r \ge b) \end{cases} \tag{10.4.3}$$

The coefficient a_n is determined by the surface impedance boundary condition, denoted by Z_a. That is,

$$Z_a = -\frac{E_\theta}{H_\phi}\bigg|_{r=a} \Rightarrow \frac{\partial(rU)}{\partial r} = Z_a j\omega\varepsilon_o(rU)\bigg|_{r=a} \tag{10.4.4}$$

Applying (10.4.4) to (10.4.3) gives

$$k_o\left[a_n\hat{h}_n^{(2)\prime}(k_o a) + \hat{j}_n{}'(k_o a)\hat{h}_n^{(2)}(k_o b)\right] = Z_a j\omega\varepsilon_o\left[a_n\hat{h}_n^{(2)}(k_o a) + \hat{j}_n(k_o a)\hat{h}_n^{(2)}(k_o b)\right]$$

Since $k_o/\omega\varepsilon_o = \eta_o$ and defining a normalized surface impedance as

$$z_a = Z_a/\eta_o$$

we obtain

$$a_n = -\hat{h}_n^{(2)}(k_o b)\left[\frac{\hat{j}_n{}'(k_o a) - jz_a\hat{j}_n(k_o a)}{\hat{h}_n^{(2)\prime}(k_o a) - jz_a\hat{h}_n^{(2)}(k_o a)}\right] \tag{10.4.5}$$

Substituting (10.4.5) into (10.4.3), for $a \le r \le b$, gives

$$U = -\frac{(I\Delta L)\eta_o}{4\pi k_o^2 b^2 r} \sum_{n=0}^{\infty} (2n+1)P_n(\cos\theta)\hat{j}_n(k_o r)\hat{h}_n^{(2)}(k_o b)$$
$$\times\left\{1 - \frac{\hat{h}_n^{(2)}(k_o r)}{\hat{j}_n(k_o r)}\left[\frac{\hat{j}_n{}'(k_o a) - jz_a\hat{j}_n(k_o a)}{\hat{h}_n^{(2)\prime}(k_o a) - jz_a\hat{h}_n^{(2)}(k_o a)}\right]\right\} \tag{10.4.6}$$

For $r \geq b$ interchange r and b in (10.4.6).

For cases where $k_o a >> 1$ a spherical wave striking the surface at a grazing angle resembles a plane wave striking the surface at a grazing angle. Hence, from (3.2.36) the surface impedance in some cases can be approximated by

$$Z_a \approx \eta_2\sqrt{1 - \left(\frac{k_o}{k_2}\right)^2} \tag{10.4.7}$$

where η_2 is the intrinsic impedance of the region $r \leq a$ and k_2 its propagation constant.

The approximation in (10.4.7) can be analyzed if the exact solution is obtained and compared with (10.4.6). That is, write the Debye potential in the region $r \leq a$ as

$$U_2 = -\frac{(I\Delta L)\eta_o}{4\pi k_o^2 b^2 r} \sum_{n=0}^{\infty} (2n+1)P_n(\cos\theta)\,[b_n\hat{j}_n(k_2 r)] \qquad (r \leq a)$$

where $k_2 = k_o n_2$ and let $\mu_2 = \mu_o$. The Debye potential in the region $r \geq a$ is given by (10.4.3) and we will refer to this potential as U_1.

The boundary conditions at $r = a$ are

$$E_{\theta 1} = E_{\theta 2}|_{r=a} \Rightarrow \frac{\partial}{\partial r}(rU_1) = \frac{\partial}{\partial r}(rU_2)|_{r=a}$$

$$H_{\phi 1} = H_{\phi 2}|_{r=a} \Rightarrow \gamma_1^2(rU_1) = \gamma_2^2(rU_2)|_{r=a} \Rightarrow (rU_1) = n_2^2(rU_2)|_{r=a}$$

where $\gamma_1 = -jk_o$ and $\gamma_2 = -jk_o n_2$. Applying the boundary conditions to U_1 and U_2:

$$a_n\hat{h}_n^{(2)\prime}(k_o a) + \hat{j}_n{}'(k_o a)\hat{h}_n^{(2)}(k_o b) = b_n n_2\hat{j}_n{}'(k_2 a)$$

$$a_n\hat{h}_n^{(2)}(k_o a) + \hat{j}_n(k_o a)\hat{h}_n^{(2)}(k_o b) = b_n n_2^2\hat{j}_n(k_2 a)$$

Solving for the constant a_n:

$$a_n = -\hat{h}_n^{(2)}(k_o b)\frac{\hat{j}_n(k_2 a)\hat{j}_n{}'(k_o a) - \frac{1}{n_2}\hat{j}_n{}'(k_2 a)\hat{j}_n(k_o a)}{\hat{j}_n(k_2 a)\hat{h}_n^{(2)\prime}(k_o a) - \frac{1}{n_2}\hat{j}_n{}'(k_2 a)\hat{h}_n^{(2)}(k_o a)} \tag{10.4.8}$$

Substituting (10.4.8) into (10.4.3) gives:

$$U_1 = -\frac{(I\Delta L)\eta_o}{4\pi k_o^2 b^2 r}\sum_{n=0}^{\infty}(2n+1)P_n(\cos\theta)\hat{j}_n(k_o r)\hat{h}_n^{(2)}(k_o b)$$
$$\times\left\{1-\frac{\hat{h}_n^{(2)}(k_o r)}{\hat{j}_n(k_o r)}\left[\frac{\hat{j}_n{}'(k_o a)-\frac{1}{n_2}\frac{\hat{j}_n{}'(k_2 a)}{\hat{j}_n(k_2 a)}\hat{j}_n(k_o a)}{\hat{h}_n^{(2)\prime}(k_o a)-\frac{1}{n_2}\frac{\hat{j}_n{}'(k_2 a)}{\hat{j}_n(k_2 a)}\hat{h}_n^{(2)}(k_o a)}\right]\right\}(r\geq b)\quad(10.4.9)$$

A comparison of (10.4.6) with (10.4.9) shows that

$$jz_a = \frac{1}{n_2}\frac{\hat{j}_n{}'(k_2 a)}{\hat{j}_n(k_2 a)} \qquad (10.4.10)$$

In (10.4.9), for $k_o a >> 1$ and $r \approx a \approx b$ the sum begins to converge when $n \approx k_o a$. Using the Debye approximations in (10.4.10) it follows that

$$j\frac{Z_a}{\eta_o} \approx \frac{1}{n_2}\frac{\hat{h}_n^{(1)\prime}(k_2 a)}{\hat{h}_n^{(1)}(k_2 a)} \approx \frac{1}{n_2}j\left[1-\left(\frac{n}{k_2 a}\right)^2\right]^{1/2} \Rightarrow Z_a \approx \eta_2\sqrt{1-\left(\frac{k_o}{k_2}\right)^2}$$
$$(10.4.11)$$

where $\eta_2 = \eta_o/n_2$. This analysis shows when the use of (10.4.11) is appropriate.

Next, the model in Fig. 10.5 is used to determine the fields from a VED above the Earth's ground surface. In this case, Z_a represents the ground impedance. To analyze this problem, it is convenient to express (10.4.3) in terms of spherical Hankel functions. Substituting:

$$\hat{j}_n(k_o r) = \frac{\hat{h}_n^{(1)}(k_o r) + \hat{h}_n^{(2)}(k_o r)}{2}$$

in (10.4.3) we obtain the form

$$U = \begin{cases} -\frac{(I\Delta L)\eta_o}{8\pi k_o^2 b^2 r}\sum_{n=0}^{\infty}(2n+1)P_n(\cos\theta)\left[b_n\hat{h}_n^{(2)}(k_o r)+\hat{h}_n^{(1)}(k_o r)\hat{h}_n^{(2)}(k_o b)\right] & (a\leq r\leq b)\\ -\frac{(I\Delta L)\eta_o}{8\pi k_o^2 b^2 r}\sum_{n=0}^{\infty}(2n+1)P_n(\cos\theta)\left[b_n\hat{h}_n^{(2)}(k_o r)+\hat{h}_n^{(1)}(k_o b)\hat{h}_n^{(2)}(k_o r)\right] & (r\geq b)\end{cases}$$
$$(10.4.12)$$

where b_n is a new constant, which in terms of the constant in (10.4.3) is $b_n = 2a_n + h_n^{(2)}(k_o b)$.

Applying (10.4.4) to (10.4.12) gives

$$b_n = -\hat{h}_n^{(2)}(k_o b)\left[\frac{\hat{h}_n^{(1)\prime}(k_o a) - jz_a \hat{h}_n^{(1)}(k_o a)}{\hat{h}_n^{(2)\prime}(k_o a) - jz_a \hat{h}_n^{(2)}(k_o a)}\right] = \hat{h}_n^{(2)}(k_o b) R_a^s \quad (10.4.13)$$

where R_a^s is known as the spherical reflection coefficient and given by

$$R_a^s = -\left[\frac{\hat{h}_n^{(1)\prime}(k_o a) - jz_a \hat{h}_n^{(1)}(k_o a)}{\hat{h}_n^{(2)\prime}(k_o a) - jz_a \hat{h}_n^{(2)}(k_o a)}\right] = \frac{\hat{h}_n^{(1)}(k_o a)}{\hat{h}_n^{(2)}(k_o a)} R_a \quad (10.4.14)$$

where

$$R_a = -\left[\frac{ln'\hat{h}_n^{(1)}(k_o a) - jz_a}{ln'\hat{h}_n^{(2)}(k_o a) - jz_a}\right] \quad (10.4.15)$$

and the logarithmic derivative is

$$ln'\hat{h}_n^{(t)}(k_o a) = \frac{1}{\hat{h}_n^{(t)}(x)} \frac{d\hat{h}_n^{(t)}(x)}{dx}\bigg|_{x=k_o a} \qquad (t = 1 \text{ or } 2)$$

The term in (10.4.15) can be related to a planar Fresnel reflection coefficient when the Debye approximations are used (left to the problems).

Substituting (10.4.13) and (10.4.14) into (10.4.12), the Debye potential for $a \leq r \leq b$ is

$$U = -\frac{(I\Delta L)\eta_o}{8\pi k_o^2 b^2 r} \sum_{n=0}^{\infty} (2n+1) P_n(\cos\theta)\, \hat{h}_n^{(1)}(k_o r)\hat{h}_n^{(2)}(k_o b)\left[1 + \frac{\hat{h}_n^{(2)}(k_o r)}{\hat{h}_n^{(1)}(k_o r)} R_g^s\right] \quad (10.4.16)$$

For $r \geq b$ interchange r and b.

For the case that the surface is a perfect conductor (i.e., $z_a = 0$), from (10.4.14):

$$R_a^s = -\frac{\hat{h}_n^{(1)\prime}(k_o a)}{\hat{h}_n^{(2)\prime}(k_o a)}$$

and in this case, for $a \leq r \leq b$, (10.4.16) reads

$$U = -\frac{(I\Delta L)\eta_o}{8\pi k_o^2 b^2 r} \sum_{n=0}^{\infty} (2n+1)\, P_n(\cos\theta)\hat{h}_n^{(1)}(k_o r)\hat{h}_n^{(2)}(k_o b)\left[1 - \frac{\hat{h}_n^{(1)\prime}(k_o a)\, \hat{h}_n^{(2)}(k_o r)}{\hat{h}_n^{(2)\prime}(k_o a)\, \hat{h}_n^{(1)}(k_o r)}\right]$$

$$= -\frac{(I\Delta L)\eta_o}{8\pi k_o^2 b^2 r} \sum_{n=0}^{\infty} (2n+1) P_n(\cos\theta)\, \frac{\hat{h}_n^{(2)}(k_o b)}{\hat{h}_n^{(2)\prime}(k_o a)}\left[\hat{h}_n^{(2)\prime}(k_o a)\hat{h}_n^{(1)}(k_o r) - \hat{h}_n^{(1)\prime}(k_o a)\hat{h}_n^{(2)}(k_o r)\right]$$

and for $r \geq b$:

$$U = -\frac{(I\Delta L)\eta_o}{8\pi k_o^2 b^2 r} \sum_{n=0}^{\infty} (2n+1) P_n(\cos\theta) \frac{\hat{h}_n^{(2)}(k_o r)}{\hat{h}_n^{(2)\prime}(k_o a)} \times \left[\hat{h}_n^{(2)\prime}(k_o a) \hat{h}_n^{(1)}(k_o b) - \hat{h}_n^{(1)\prime}(k_o a) \hat{h}_n^{(2)}(k_o b) \right] \tag{10.4.17}$$

In (10.4.17), if the dipole is slightly above the surface (i.e., $b \approx a$), the Wronskian shows that

$$W = \left[\hat{h}_n^{(1)}(k_o a), \hat{h}_n^{(2)}(k_o a) \right] = -2j$$

and the expression for U that is valid for $r \geq b$ is

$$U = j \frac{(I\Delta L)\eta_o}{4\pi k_o^2 a^2 r} \sum_{n=0}^{\infty} (2n+1) \frac{\hat{h}_n^{(2)}(k_o r)}{\hat{h}_n^{(2)\prime}(k_o a)} P_n(\cos\theta) \tag{10.4.18}$$

which agrees with (10.3.9).

When $z_a \neq 0$, the radial component of the electric field along the Earth's surface follows from (10.4.16) and (10.4.2). That is,

$$E_r = -\frac{(I\Delta L)\eta_o}{8\pi k_o^2 b^2 r^2} \sum_{n=0}^{\infty} n(n+1)(2n+1) P_n(\cos\theta)\, \hat{h}_n^{(1)}(k_o r) \hat{h}_n^{(2)}(k_o b) \times \left[1 + \frac{\hat{h}_n^{(2)}(k_o r)}{\hat{h}_n^{(1)}(k_o r)} R_a^s \right] \tag{10.4.19}$$

The sum in (10.4.19) converges very slowly for large values of the arguments of the Schelkunoff-Bessel functions. The Watson transformation is used to obtain a fast-convergent series. The sum in (10.4.19) can be expressed in the form

$$E_r = K \sum_{n=0}^{\infty} n(n+1)(2n+1) f(n) P_n(\cos\theta) \tag{10.4.20}$$

where

$$K = -\frac{(I\Delta L)\eta_o}{8\pi k_o^2 b^2 r^2}$$

and $f(n)$ represents the other terms in (10.4.19). Equation (10.4.20) can be represented by the following contour integral:

$$E_r = jK \oint_C \frac{\nu(\nu^2 - 1/4)}{\cos \nu\pi} f\left(\nu - \frac{1}{2}\right) P_{\nu-1/2}(-\cos\theta)\,d\nu \qquad (10.4.21)$$

where the contour $C = C_1 + C_2$, shown in Fig. 10.4a, encloses the poles of $\cos \nu\pi$ (located at $\nu = n + 1/2$ for $n = 0, 1, 2, \ldots$) along the positive real axis. Hence, forming $2\pi j$ times the sum of the residues at these poles it follows that (10.4.21) is identical to (10.4.20).

In (10.4.21):

$$\begin{aligned}
P_{-\nu-1/2}(\cos\theta) &= P_{\nu-1/2}(\cos\theta) \\
\hat{h}^{(1)}_{-\nu-1/2}(x) &= e^{j\nu\pi}\hat{h}^{(1)}_{\nu-1/2}(x) \\
\hat{h}^{(1)'}_{-\nu-1/2}(x) &= e^{j\nu\pi}\hat{h}^{(1)'}_{\nu-1/2}(x) \\
\hat{h}^{(2)}_{-\nu-1/2}(x) &= e^{-j\nu\pi}\hat{h}^{(2)}_{\nu-1/2}(x) \\
\hat{h}^{(2)'}_{-\nu-1/2}(x) &= e^{-j\nu\pi}\hat{h}^{(2)'}_{\nu-1/2}(x)
\end{aligned}$$

and it follows that $f(\nu - 1/2)$ is an even function of ν. The term $\nu(\nu^2 - 1/4)/\cos\nu\pi$ is odd and, therefore, the integrand in (10.4.21) is odd. Letting $\nu = -\nu$ in the integral along C_1 it follows that

$$E_r = jK \int_{C'} \frac{\nu(\nu^2 - 1/4)}{\cos \nu\pi} f\left(\nu - \frac{1}{2}\right) P_{\nu-1/2}(-\cos\theta)\,d\nu \qquad (10.4.22)$$

where C' is the path of integration just below the axis shown in Fig. 10.4b and

$$\begin{aligned}
f(\nu - 1/2) &= \hat{h}^{(1)}_{\nu-1/2}(k_o r)\hat{h}^{(2)}_{\nu-1/2}(k_o b)\left[1 + \frac{\hat{h}^{(2)}_{\nu-1/2}(k_o r)}{\hat{h}^{(1)}_{\nu-1/2}(k_o r)} R_a^s\right] \\
&= \hat{h}^{(1)}_{\nu-1/2}(k_o r)\hat{h}^{(2)}_{\nu-1/2}(k_o b) - \hat{h}^{(2)}_{\nu-1/2}(k_o r)\hat{h}^{(2)}_{\nu-1/2}(k_o b) \\
&\quad \times \frac{\hat{h}^{(1)'}_{\nu-1/2}(k_o a) - jz_a \hat{h}^{(1)}_{\nu-1/2}(k_o a)}{\hat{h}^{(2)'}_{\nu-1/2}(k_o a) - jz_a \hat{h}^{(2)}_{\nu-1/2}(k_o a)}
\end{aligned} \qquad (10.4.23)$$

The integral in (10.4.22) is evaluated by closing the path of integration with an infinite-radius semicircle in the lower-half-plane, where the contribution from this portion of the contour vanishes as $|\nu| \to \infty$ in the lower-half plane. Hence, the integral is equal to the sum of the residues at the poles of $f(\nu - 1/2)$ in the lower-half plane. Only the term R_a^s in (10.4.23) has poles. The location of the poles are determined by the solution of

$$\hat{h}^{(2)'}_{\nu-1/2}(k_o a) - jz_a \hat{h}^{(2)}_{\nu-1/2}(k_o a) = 0 \qquad (10.4.24)$$

For $z_a = 0$ the poles are given by $\hat{h}^{(2)\prime}_{\nu-1/2}(k_o a) = 0$ and as $|z_a|$ increases the poles approaches those of $\hat{h}^{(2)}_{\nu-1/2}(k_o a) = 0$. For these two cases the pole locations follow from (7.1.21) and (7.1.22). Since $k_o a$ is large, the first few poles are close to $k_o a$. Pole locations for other values of z_a are discussed later.

The Bessel and Legendre functions in (10.4.22) can be replaced by appropriate asymptotic expansions. For the Schelkunoff-Bessel functions we will use the Fok-Airy functions $w_1(t)$ and $w_2(t)$ (see Appendix D), and for the Legendre function we use (10.3.14).

The functions $\hat{h}^{(1)}_{\nu-1/2}(k_o a)$ and $\hat{h}^{(2)}_{\nu-1/2}(k_o a)$ and their derivatives are related to $w_1(t)$ and $w_2(t)$ by:

$$\hat{h}^{(1)}_{\nu-1/2}(k_o a) \approx -j\left(\frac{k_o a}{2}\right)^{1/6} w_1(t)$$

$$\hat{h}^{(2)}_{\nu-1/2}(k_o a) \approx j\left(\frac{k_o a}{2}\right)^{1/6} w_2(t)$$

$$\hat{h}^{(1)\prime}_{\nu-1/2}(k_o a) \approx j\left(\frac{k_o a}{2}\right)^{-1/6} w_1{}'(t)$$

$$\hat{h}^{(2)\prime}_{\nu-1/2}(k_o a) \approx -j\left(\frac{k_o a}{2}\right)^{-1/6} w_2{}'(t)$$

where

$$t = \left(\frac{k_o a}{2}\right)^{-1/3} (\nu - k_o a)$$

Also,

$$\begin{aligned}\hat{h}^{(1)}_{\nu-1/2}(k_o r) &\approx -j\left(\frac{k_o r}{2}\right)^{1/6} w_1\left[\left(\frac{k_o r}{2}\right)^{-1/3} (\nu - k_o r)\right] \\ &\approx -j\left(\frac{k_o r}{2}\right)^{1/6} w_1\left[\left(\frac{k_o r}{2}\right)^{-1/3}\left(k_o a - \left(\frac{k_o a}{2}\right)^{1/3} t - k_o r\right)\right] \\ &\approx -j\left(\frac{k_o r}{2}\right)^{1/6} w_1(m_r t - y_r)\end{aligned}$$

where

$$m_r = \left(\frac{a}{r}\right)^{1/3}$$

$$y_r = \left(\frac{k_o r}{2}\right)^{-1/3} k_o(r-a)$$

Similarly,

$$\hat{h}_n^{(2)}(k_o b) \approx j\left(\frac{k_o b}{2}\right)^{1/6} w_2(m_b t - y_b)$$

where

$$m_b = \left(\frac{a}{b}\right)^{1/3}$$

$$y_b = \left(\frac{k_o b}{2}\right)^{-1/3} k_o(b-a)$$

For the case that $r \approx a \approx b$ with $r > a$ and $b > a$, the factors m_r and m_b are set to one (i.e., $m_r \approx m_b \approx 1$).

In (10.4.22), letting

$$\nu = k_o a + \left(\frac{k_o a}{2}\right)^{1/3} t \Rightarrow d\nu = \left(\frac{k_o a}{2}\right)^{1/3} dt \qquad (10.4.25)$$

the equation can be written in terms of the variable t. In terms of t:

$$\begin{aligned} \frac{P_{\nu-1/2}(-\cos\theta)}{\cos(\nu\pi)} &\approx \sqrt{\frac{2}{\pi\nu\sin\theta}}\, e^{-j\nu\theta} e^{-j\pi/4} \\ &\approx \sqrt{\frac{2}{\pi\nu\sin\theta}}\, e^{-jka\theta} e^{-jxt} e^{-j\pi/4} \end{aligned} \qquad (10.4.26)$$

where

$$x = \left(\frac{k_a a}{2}\right)^{1/3} \theta$$

In (10.4.21) for the ν terms:

$$\frac{\nu(\nu^2 - 1/4)}{\sqrt{\nu}} \approx \nu^{5/2} \approx (k_o a)^{5/2}\left[1 + \frac{5}{4}\left(\frac{2}{k_o a}\right)^{2/3} t\right] \approx (k_o a)^{5/2}$$

Substituting (10.4.23), (10.4.25) and (10.4.26) into (10.4.22) shows that

$$E_r = jK\sqrt{\frac{2}{\pi \sin\theta}}\, e^{-jk_o a\theta} e^{-j\pi/4} (k_o a)^{5/2} \left(\frac{k_o a}{2}\right)^{1/3} \int_{C'} e^{-jxt} f(t)\,dt \quad (10.4.27)$$

where (10.4.23) in terms if t, with $r \approx a \approx b$, is

$$f(t) \approx \left(\frac{k_o a}{2}\right)^{1/3} \left\{ w_1(t - y_r) w_2(t - y_b) - w_2(t - y_r) w_2(t - y_b) \left[\frac{w_1'(t) - q w_1(t)}{w_2'(t) - q w_2(t)}\right] \right\} \quad (10.4.28)$$

and

$$q = -j\left(\frac{k_o a}{2}\right)^{1/3} z_a$$

As previously mentioned, the contour C' is closed in the lower-half plane, which is now in the t plane. The poles in the lower-half plane are those of $f(t)$. From (10.4.27) the poles occur at $t = t_p$ where

$$w_2'(t_p) - q w_2(t_p) = 0 \qquad (p = 1, 2, 3, \ldots) \quad (10.4.29)$$

and the evaluation of (10.4.27) is

$$E_r \approx jK\sqrt{\frac{2}{\pi \sin\theta}}\, e^{-jk_o a\theta} e^{-j\pi/4} (k_o a)^{5/2} \left(\frac{k_o a}{2}\right)^{2/3} \times (2\pi j)(-1) \sum_{p=1}^{\infty} e^{-jxt_p} w_2(t_p - y_b) w_2(t_p - y_r) \frac{w_1'(t_p) - q w_1(t_p)}{\frac{\partial}{\partial t}[w_2'(t) - q w_2(t)]|_{t=t_p}} \quad (10.4.30)$$

In (10.4.30):

$$\begin{aligned}\frac{\partial}{\partial t}[w_2'(t) - q w_2(t)]|_{t=t_p} &= w_2''(t_p) - q w_2'(t_p) \\ &= t_p w_2(t_p) - q[q w_2(t_p)] \\ &= (t_p - q^2) w_2(t_p)\end{aligned}$$

where we used the Airy equation: $w_2''(t_p) = t_p w_2(t_p)$. Also, in (10.4.30) using (10.4.29):

$$w_1'(t_p) - qw_1(t_p) = w_1'(t_p) - \frac{w_2'(t_p)}{w_2(t_p)} w_1(t_p)$$

$$= \frac{w_1'(t_p)w_2(t_p) - w_2'(t_p)w_1(t_p)}{w_2(t_p)} = \frac{-2j}{w_2(t_p)}$$

Substituting these relations and K in (10.4.30):

$$E_r \approx -j\frac{(I\Delta L)\eta_o k_o}{2a}\sqrt{\frac{1}{\pi \sin\theta}}\, e^{-jk_o a\theta} e^{-j\pi/4}\left(\frac{k_o a}{2}\right)^{1/6}$$
$$\times \sum_{p=1}^{\infty} \frac{e^{-jxt_p}}{t_p - q^2} \frac{w_2(t_p - y_b)}{w_2(t_p)} \frac{w_2(t_p - y_r)}{w_2(t_p)} \quad (10.4.31)$$

The expression (10.4.31) commonly appears in the form:

$$E_r \approx -j\frac{(I\Delta L)\eta_o k_o}{4\pi a\sqrt{\theta \sin\theta}} e^{-jk_o a\theta}\, V_o$$

where

$$V_o = 2(\pi x)^{1/2} e^{-j\pi/4} \sum_{p=1}^{\infty} \frac{e^{-jxt_p}}{t_p - q^2} \frac{w_2(t_p - y_b)}{w_2(t_p)} \frac{w_2(t_p - y_r)}{w_2(t_p)} \quad (10.4.32)$$

The term V_o is known as the attenuation function. If $r = a = b$, then $y_b = y_r = 0$ and (10.4.32) reads

$$V_o = 2(\pi x)^{1/2} e^{-j\pi/4} \sum_{p=1}^{\infty} \frac{e^{-jxt_p}}{t_p - q^2}$$

In the case that $\sin\theta \approx \theta$ and with $d = a\theta$, E_r is given by

$$E_r \approx -j\frac{(I\Delta L)\eta_o k_o}{4\pi}\frac{e^{-jk_o d}}{d} V_o \approx -j30(I\Delta L)k_o \frac{e^{-jk_o d}}{d} V_o \quad (10.4.33)$$

The reason for writing E_r in the form shown in (10.4.33) is that for θ values such that $\sin\theta \approx \theta$ the factor multiplying V_o is the E_θ field produced by a dipole at $\theta = \pi/2$ (see (2.1.9)), which is the negative of E_r.

The characteristic of the propagating modes in (10.4.31) are determined by the solution of (10.4.29). For $q = 0$, the roots of $w_2'(t_p) = 0$, denoted by t_p^0, determine the modes. For $q = \infty$, the roots of $w_2(t_p) = 0$, denoted by t_p^∞, determine the modes. These roots are listed in Appendix D. For other values of q differentiating (10.4.29) with respect to q gives

$$w_2{}'(t_p)\frac{dt_p}{dq} - qw_2{}'\frac{dt_p}{dq} - w_2(t_p) = 0$$

$$t_p w_2(t_p)\frac{dt_p}{dq} - q^2 w_2(t_p)\frac{dt_p}{dq} - w_2(t_p) = 0 \Rightarrow \frac{dt_p}{dq} = \frac{1}{t_p - q^2} \qquad (10.4.34)$$

The differential equation in (10.4.34) can be solved for any value of q using numerical integration where t_p^0 or t_p^∞ are use as initial condition. It is also observed that a series solution of (10.4.34) can be obtained using the initial condition t_p^0 for $q = 0$. That is,

$$t_p(q) = t_p^0 + \frac{1}{t_p^0}q - \frac{1}{2\left(t_p^0\right)^3}q^2 + \left[\frac{1}{3\left(t_p^0\right)^2} + \frac{1}{2\left(t_p^0\right)^5}\right]q^3 - \ldots$$

which is valid for $|q/\sqrt{t_p}| < 1$. If the initial condition is taken as t_p^∞ for $q = \infty$, then

$$t_p(q) = t_p^\infty + \frac{1}{q} + \frac{t_p^\infty}{3}\frac{1}{q^3} + \frac{1}{4}\frac{1}{q^4} + \frac{\left(t_p^\infty\right)^2}{5}\frac{1}{q^5} + \frac{7t_p^\infty}{18}\frac{1}{q^6} + \ldots.$$

which is valid for $|q/\sqrt{t_p}| > 1$.

10.5 A VERTICAL ELECTRIC DIPOLE IN A SPHERICAL WAVEGUIDE—ZONAL HARMONICS SOLUTION

The geometry of the problem for a VED in a spherical waveguide is shown in Fig. 10.6. An example of a spherical waveguide is the Earth-Ionosphere waveguide, where a dipole source excites the fields that propagate between the Earth's ground and the Ionosphere.

The problem can be formulated in terms of the Debye potential A_r in (4.6.9) or the Debye potential U in (4.6.13). Using (4.6.9), we write

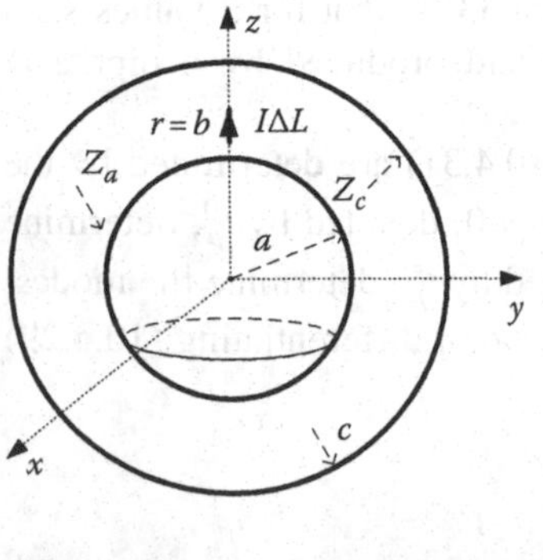

FIGURE 10.6 A VED in the spherical waveguide.

$$[\nabla^2 + k_o^2]\frac{A_r}{r} = \frac{-\mu_o J_{e,r}}{r} \tag{10.5.1}$$

where

$$J_{e,r} = \frac{(I\Delta L)\delta(r-b)\delta(\theta)}{2\pi r^2 \sin\theta}$$

Observing the symmetry in ϕ (i.e., $\partial/\partial\phi = 0$), from (4.6.14) to (4.6.19) the field components are given by

$$E_r = \frac{1}{j\omega\mu_o\varepsilon_o}\left(\frac{\partial^2}{\partial r^2} + k_o^2\right)A_r \tag{10.5.2}$$

$$E_\theta = \frac{1}{j\omega\mu_o\varepsilon_o\, r}\frac{\partial^2 A_r}{\partial r\partial\theta} \tag{10.5.3}$$

$$H_\phi = \frac{-1}{\mu_o r}\frac{\partial A_r}{\partial\theta} \tag{10.5.4}$$

The particular solution to (10.5.1) is

$$\frac{A_r^p}{r} = \frac{\mu_o(I\Delta L)}{4\pi}\int_{V'}\frac{e^{-jk_o|\mathbf{r}-\mathbf{r}'|}}{|\mathbf{r}-\mathbf{r}'|}\frac{1}{r'}\frac{\delta(r'-b)\delta(\theta')}{2\pi r'^2\sin\theta}r'^2\sin\theta dr'd\theta'd\phi'$$

which gives

$$A_r^p = \mu_o(I\Delta L)\frac{r}{b}\frac{e^{-jk_oR}}{4\pi R} \tag{10.5.5}$$

where $R = |\mathbf{r}-\mathbf{r}'| = (r^2 + b^2 - 2rb\cos\theta)^{1/2}$.

Since,

$$\frac{e^{-jk_oR}}{4\pi R} = \frac{-j}{4\pi k_o rb}\sum_{n=0}^{\infty}(2n+1)P_n(\cos\theta)\begin{cases}\hat{J}_n(k_or)\hat{H}_n^{(2)}(k_ob) & (r\le b)\\ \hat{J}_n(k_ob)\hat{H}_n^{(2)}(k_or) & (r\ge b)\end{cases} \tag{10.5.6}$$

(10.5.5) can be expressed in the form

$$A_r^p = \frac{\mu_o(I\Delta L)}{j4\pi k_o b^2}\sum_{n=0}^{\infty}(2n+1)P_n(\cos\theta)\begin{cases}\hat{J}_n(k_or)\hat{H}_n^{(2)}(k_ob) & (r\le b)\\ \hat{J}_n(k_ob)\hat{H}_n^{(2)}(k_or) & (r\ge b)\end{cases} \tag{10.5.7}$$

Letting $u_r = k_or$ and $u_b = k_ob$, (10.5.7) is written as

$$A_r^p = \frac{\mu_o k_o (I\Delta L)}{j4\pi u_b^2} \sum_{n=0}^{\infty} (2n+1) P_n(\cos\theta) \begin{cases} \hat{j}_n(u_r)\hat{h}_n^{(2)}(u_b) & (r \le b) \\ \hat{j}_n(u_b)\hat{h}_n^{(2)}(u_r) & (r \ge b) \end{cases} \tag{10.5.8}$$

The complementary solution is written as

$$A_r^c = \frac{\mu_o k_o (I\Delta L)}{j4\pi u_b^2} \sum_{n=0}^{\infty} (2n+1) P_n(\cos\theta) \left[b_n \hat{j}_n(u_r) + c_n \hat{h}_n^{(2)}(u_r) \right]$$

where b_n and c_n are constants that are determined by the boundary conditions at $r = a$ and $r = c$. The total solution is then:

$$A_r = A_r^p + A_r^c$$

or

$$A_r = \begin{cases} \frac{k_o \mu_o (I\Delta L)}{j4\pi u_b^2} \sum_{n=0}^{\infty} (2n+1) P_n(\cos\theta) \left\{ \hat{j}_n(u_r)\hat{h}_n^{(2)}(u_b) + \left[b_n \hat{j}_n(u_r) + c_n \hat{h}_n^{(2)}(u_r) \right] \right\} (a \le r \le b) \\ \frac{k_o \mu_o (I\Delta L)}{j4\pi u_b^2} \sum_{n=0}^{\infty} (2n+1) P_n(\cos\theta) \left\{ \hat{j}_n(u_b)\hat{h}_n^{(2)}(u_r) + \left[b_n \hat{j}_n(u_r) + c_n \hat{h}_n^{(2)}(u_r) \right] \right\} (b \le r \le c) \end{cases} \tag{10.5.9}$$

The E_r field component follows from (10.5.2). That is,

$$E_r = \begin{cases} \frac{-\eta_o k_o^2 (I\Delta L)}{4\pi u_b^2 u_r^2} \sum_{n=0}^{\infty} n(n+1)(2n+1) P_n(\cos\theta) \left\{ \hat{j}_n(u_r)\hat{h}_n^{(2)}(u_b) + \left[b_n \hat{j}_n(u_r) + c_n \hat{h}_n^{(2)}(u_r) \right] \right\} & (a \le r \le b) \\ \frac{-\eta_o k_o^2 (I\Delta L)}{4\pi u_b^2 u_r^2} \sum_{n=0}^{\infty} n(n+1)(2n+1) P_n(\cos\theta) \left\{ \hat{j}_n(u_b)\hat{h}_n^{(2)}(u_r) + \left[b_n \hat{j}_n(u_r) + c_n \hat{h}_n^{(2)}(u_r) \right] \right\} & (b \le r \le c) \end{cases} \tag{10.5.10}$$

where we used (see Appendix C):

$$\left[\frac{d^2}{dr^2} + k_o^2 \right] \begin{Bmatrix} \hat{j}_n(k_o r) \\ \hat{h}_n^{(2)}(k_o r) \end{Bmatrix} = \frac{n(n+1)}{r^2} \begin{Bmatrix} \hat{j}_n(k_o r) \\ \hat{h}_n^{(2)}(k_o r) \end{Bmatrix} \tag{10.5.11}$$

The E_θ field component follows from (10.5.3). That is,

$$E_\theta = \begin{cases} \frac{-\eta_o k_o^2 (I\Delta L)}{4\pi u_b^2 u_r} \sum_{n=0}^{\infty} (2n+1) P_n^1(\cos\theta) \left\{ \hat{j}_n{}'(u_r)\hat{h}_n^{(2)}(u_b) + \left[b_n \hat{j}_n{}'(u_r) + c_n \hat{h}_n^{(2)\prime}(u_r) \right] \right\} (a \le r \le b) \\ \frac{-\eta_o k_o^2 (I\Delta L)}{4\pi u_b^2 u_r} \sum_{n=0}^{\infty} (2n+1) P_n^1(\cos\theta) \left\{ \hat{j}_n(u_b)\hat{h}_n^{(2)\prime}(u_r) + \left[b_n \hat{j}_n{}'(u_r) + c_n \hat{h}_n^{(2)\prime}(u_r) \right] \right\} (b \le r \le c) \end{cases} \tag{10.5.12}$$

and the H_ϕ component follows from (10.5.4):

$$H_\phi = \begin{cases} \frac{jk_o^2 (I\Delta L)}{4\pi u_b^2 u_r} \sum_{n=0}^{\infty} (2n+1) P_n^1(\cos\theta) \left\{ \hat{j}_n(u_r)\hat{h}_n^{(2)}(u_b) + \left[b_n \hat{j}_n(u_r) + c_n \hat{h}_n^{(2)}(u_r) \right] \right\} (a \le r \le b) \\ \frac{jk_o^2 (I\Delta L)}{4\pi u_b^2 u_r} \sum_{n=0}^{\infty} (2n+1) P_n^1(\cos\theta) \left\{ \hat{j}_n(u_b)\hat{h}_n^{(2)}(u_r) + \left[b_n \hat{j}_n(u_r) + c_n \hat{h}_n^{(2)}(u_r) \right] \right\} (b \le r \le c) \end{cases} \tag{10.5.13}$$

Next, consider the case where the boundary conditions at $r = a$ and at $r = c$ are considered to be good conductors. For this case, (10.5.3) show that:

$$E_\theta|_{r=a} = 0 \Rightarrow \left. \frac{\partial A_r}{\partial r} \right|_{r=a} = 0$$

and from (10.5.9) or (10.5.12) it follows that

$$\hat{j}_n{}'(u_a)\hat{h}_n^{(2)}(u_b) + \left[b_n \hat{j}_n{}'(u_a) + c_n \hat{h}_n^{(2)\prime}(u_a) \right] = 0 \tag{10.5.14}$$

where $u_a = k_o a$.

At $r = c$, the boundary condition is

$$E_\theta|_{r=c} = 0 \Rightarrow \left. \frac{\partial A_r}{\partial r} \right|_{r=c} = 0$$

From (10.5.9) or (10.5.12) it follows :that

$$\hat{j}_n(u_b)\hat{h}_n^{(2)\prime}(u_c) + \left[b_n \hat{j}_n{}'(u_c) + c_n \hat{h}_n^{(2)\prime}(u_c) \right] = 0 \tag{10.5.15}$$

where $u_c = k_o c$.

Solving (10.5.14) and (10.5.15) gives:

$$b_n = \frac{\hat{h}_n^{(2)\prime}(u_a)\hat{j}_n(u_b)\hat{h}_n^{(2)\prime}(u_c) - \hat{j}_n{}'(u_a)\hat{h}_n^{(2)}(u_b)\hat{h}_n^{(2)\prime}(u_c)}{\hat{j}_n{}'(u_a)\hat{h}_n^{(2)\prime}(u_c) - \hat{h}_n^{(2)\prime}(u_a)\hat{j}_n{}'(u_c)} \tag{10.5.16}$$

and

$$c_n = \frac{\hat{j}_n{}'(u_a)\hat{h}_n^{(2)}(u_b)\hat{j}_n{}'(u_c) - \hat{j}_n{}'(u_a)\hat{j}_n(u_b)\hat{h}_n^{(2)}{}'(u_c)}{\hat{j}_n{}'(u_a)\hat{h}_n^{(2)}{}'(u_c) - \hat{h}_n^{(2)}{}'(u_a)\hat{j}_n{}'(u_c)} \tag{10.5.17}$$

Substituting (10.5.16) and (10.5.17) into (10.5.9) to (10.5.13) completes the field formulation of the problem.

A more realistic situation is to model the waveguide surfaces by impedance boundary conditions. Denoting the surface impedance at $r = a$ by Z_a and that at $r = c$ by Z_c, the boundary conditions are

$$Z_a = -\frac{E_\theta}{H_\phi}\bigg|_{r=a} \Rightarrow \frac{1}{k_o}\frac{\partial A_r}{\partial r} = j\frac{Z_a}{\eta_o}A_r\bigg|_{r=a} \tag{10.5.18}$$

and

$$Z_c = \frac{E_\theta}{H_\phi}\bigg|_{r=c} \Rightarrow \frac{1}{k_o}\frac{\partial A_r}{\partial r} = -j\frac{Z_c}{\eta_o}A_r\bigg|_{r=c} \tag{10.5.19}$$

Different surfaces require appropriate forms Z_a and Z_c. For example, if the surface at $r = a$ represents the Earth's ground, the surface impedance can be approximated by that of a parallel polarized plane wave incident at grazing incidence. That is,

$$Z_a \approx \eta_a\sqrt{1 - \left(\frac{k_o}{k_a}\right)^2} \tag{10.5.20}$$

where k_a and η_a are the propagation constant and intrinsic impedance of the surface.

Similarly, in certain cases the surface impedance at $r = c$ can be represented by

$$Z_c \approx \eta_c\sqrt{1 - \left(\frac{k_o}{k_c}\right)^2} \tag{10.5.21}$$

and k_c and η_c are the propagation constant and intrinsic impedance of the surface. In the case that the surface at $r = c$ represents the Ionosphere the form of Z_c depends on the model used for the ionospheric plasma and the frequency of operation.

Using (10.5.18) and (10.5.19), it follows from (10.5.9) that

$$\hat{J}_n{}'(u_a)\hat{H}_n^{(2)}(u_b) + \left[b_n\hat{J}_n{}'(u_a) + c_n\hat{H}_n^{(2)\prime}(u_a)\right] = j\frac{Z_a}{\eta_o}\left\{\hat{J}_n(u_a)\hat{H}_n^{(2)}(u_b) + \left[b_n\hat{J}_n(u_a) + c_n\hat{H}_n^{(2)}(u_a)\right]\right\}$$

and

$$\hat{J}_n(u_b)\hat{H}_n^{(2)\prime}(u_c) + \left[b_n\hat{J}_n{}'(u_c) + c_n\hat{H}_n^{(2)\prime}(u_c)\right] = -j\frac{Z_c}{\eta_o}\left\{\hat{J}_n(u_b)\hat{H}_n^{(2)}(u_c) + \left[b_n\hat{J}_n(u_c) + c_n\hat{H}_n^{(2)}(u_c)\right]\right\}$$

which can be solved for b_n and c_n. Defining the normalized impedances:

$$z_a = \frac{Z_a}{\eta_o}$$

and

$$z_c = \frac{Z_c}{\eta_o}$$

the equations are written in the form

$$b_n\left[\hat{J}_n{}'(u_a) - jz_a\hat{J}_n(u_a)\right] + c_n\left[\hat{H}_n^{(2)\prime}(u_a) - jz_a\hat{H}_n^{(2)}(u_a)\right] = -\hat{H}_n^{(2)}(u_b)\left[\hat{J}_n{}'(u_a) - jz_a\hat{J}_n(u_a)\right]$$

$$b_n\left[\hat{J}_n{}'(u_c) + jz_c\hat{J}_n(u_c)\right] + c_n\left[\hat{H}_n^{(2)\prime}(u_c) + jz_c\hat{H}_n^{(2)}(u_c)\right] = -\hat{J}_n(u_b)\left[\hat{H}_n^{(2)\prime}(u_c) + jz_c\hat{H}_n^{(2)}(u_c)\right]$$

whose solutions are conveniently expressed using determinants as

$$b_n = \frac{\begin{vmatrix} -\hat{H}_n^{(2)}(u_b)\left[\hat{J}_n{}'(u_a) - jz_a\hat{J}_n(u_a)\right] & \hat{H}_n^{(2)\prime}(u_a) - jz_a\hat{H}_n^{(2)}(u_a) \\ -\hat{J}_n(u_b)\left[\hat{H}_n^{(2)\prime}(u_c) + jz_c\hat{H}_n^{(2)}(u_c)\right] & \hat{H}_n^{(2)\prime}(u_c) + jz_c\hat{H}_n^{(2)}(u_c) \end{vmatrix}}{\begin{vmatrix} \hat{J}_n{}'(u_a) - jz_a\hat{J}_n(u_a) & \hat{H}_n^{(2)\prime}(u_a) - jz_a\hat{H}_n^{(2)}(u_a) \\ \hat{J}_n{}'(u_c) + jz_c\hat{J}_n(u_c) & \hat{H}_n^{(2)\prime}(u_c) + jz_c\hat{H}_n^{(2)}(u_c) \end{vmatrix}} \tag{10.5.22}$$

$$c_n = \frac{\begin{vmatrix} \hat{J}_n{}'(u_a) - jz_a\hat{J}_n(u_a) & -\hat{h}_n^{(2)}(u_b)\left[\hat{J}_n{}'(u_a) - jz_a\hat{J}_n(u_a)\right] \\ \hat{J}_n{}'(u_c) + jz_c\hat{J}_n(u_c) & -\hat{J}_n(u_b)\left[\hat{h}_n^{(2)\prime}(u_c) + jz_c\hat{h}_n^{(2)}(u_c)\right] \end{vmatrix}}{\begin{vmatrix} \hat{J}_n{}'(u_a) - jz_a\hat{J}_n(u_a) & \hat{h}_n^{(2)\prime}(u_a) - jz_a\hat{h}_n^{(2)}(u_a) \\ \hat{J}_n{}'(u_c) + jz_c\hat{J}_n(u_c) & \hat{h}_n^{(2)\prime}(u_c) + jz_c\hat{h}_n^{(2)}(u_c) \end{vmatrix}} \tag{10.5.23}$$

Equations (10.5.9) to (10.5.13) describe the field formulation of the problem, where the constants b_n and c_n are given by (10.5.22) and (10.5.23). In the limit that $z_a = z_c = 0$, (10.5.22) and (10.5.23) reduce to (10.5.16) and (10.5.17).

The series in (10.5.9) to (10.5.13) converge slowly for large values of u_r, u_a and u_c, since it takes about $2k_o a$ terms to obtain convergence. However, the use of the Watson transformation produces practical results in many cases. For example, in the calculation of the fields in the Earth-Ionosphere waveguide the Watson transformation produces good results in the LF/VLF band. However, it does not work well for the calculation of the fields in the ELF band. Researchers have been able to manipulate the infinite sum in (10.5.9) to (10.5.13) to produce practical values for the fields in many situations, especially in the ELF band. The type of algorithms used to perform the sum are referred as speed-up numerical convergence algorithms.

In order to use the Watson transformation, it is convenient to first write A_r in (10.5.9) only in terms of spherical Hankel functions, as was done in (10.4.12). Hence, in terms of Schelkunoff-Bessel functions (10.5.9) reads

$$A_r = \begin{cases} \frac{\mu_o k_o (I\Delta L)}{j8\pi u_b^2} \sum_{n=0}^{\infty} (2n+1)P_n(\cos\theta)\left\{\hat{h}_n^{(1)}(u_r)\hat{h}_n^{(2)}(u_b) + \left[d_n\hat{h}_n^{(1)}(u_r) + e_n\hat{h}_n^{(2)}(u_r)\right]\right\} (a \le r \le b) \\ \frac{\mu_o k_o (I\Delta L)}{j8\pi u_b^2} \sum_{n=0}^{\infty} (2n+1)P_n(\cos\theta)\left\{\hat{h}_n^{(1)}(u_b)\hat{h}_n^{(2)}(u_r) + \left[d_n\hat{h}_n^{(1)}(u_r) + e_n\hat{h}_n^{(2)}(u_r)\right]\right\} (b \le r \le c) \end{cases} \tag{10.5.24}$$

Applying the boundary conditions in (10.5.18) and (10.5.19) to (10.5.24) gives

$$\hat{h}_n^{(1)\prime}(u_a)\hat{h}_n^{(2)}(u_b) + \left[d_n\hat{h}_n^{(1)\prime}(u_a) + e_n\hat{h}_n^{(2)\prime}(u_a)\right] = j\frac{Z_a}{\eta_o}\left\{\hat{h}_n^{(1)}(u_a)\hat{h}_n^{(2)}(u_b) + \left[d_n\hat{h}_n^{(1)}(u_a) + e_n\hat{h}_n^{(2)}(u_a)\right]\right\}$$

and

$$\hat{h}_n^{(1)}(u_b)\hat{h}_n^{(2)\prime}(u_c) + \left[d_n\hat{h}_n^{(1)\prime}(u_c) + e_n\hat{h}_n^{(2)\prime}(u_c)\right] = -j\frac{Z_c}{\eta_o}\left\{\hat{h}_n^{(1)}(u_b)\hat{h}_n^{(2)}(u_c) + \left[d_n\hat{h}_n^{(1)}(u_c) + e_n\hat{h}_n^{(2)}(u_c)\right]\right\}$$

which can be written in the form

$$d_n X_1 + e_n X_2 = -h_n^{(2)}(u_b) X_1 \tag{10.5.25}$$

and

$$d_n Y_1 + e_n Y_2 = -h_n^{(1)}(u_b) Y_2 \tag{10.5.26}$$

where

$$\begin{aligned} X_1 &= h_n^{(1)\prime}(u_a) - jz_a h_n^{(1)}(u_a) \\ X_2 &= h_n^{(2)\prime}(u_a) - jz_a h_n^{(2)}(u_a) \\ Y_1 &= h_n^{(1)\prime}(u_c) + jz_c h_n^{(1)}(u_c) \\ Y_2 &= h_n^{(2)\prime}(u_c) + jz_c h_n^{(2)}(u_c) \end{aligned}$$

The solutions to (10.5.25) and (10.5.26) are

$$\begin{aligned} d_n &= \frac{-h_n^{(2)}(u_b) X_1 Y_2 + h_n^{(1)}(u_b) X_2 Y_2}{X_1 Y_2 - X_2 Y_1} \\ &= \frac{h_n^{(2)}(u_b)\frac{X_1 Y_2}{X_2 Y_1} - h_n^{(1)}(u_b)\frac{Y_2}{Y_1}}{\left(1 - \frac{X_1 Y_2}{X_2 Y_1}\right)} \end{aligned} \tag{10.5.27}$$

and

$$\begin{aligned} e_n &= \frac{-h_n^{(1)}(u_b) X_1 Y_2 + h_n^{(2)}(u_b) X_1 Y_1}{X_1 Y_2 - X_2 Y_1} \\ &= \frac{h_n^{(1)}(u_b)\frac{X_1 Y_2}{X_2 Y_1} - h_n^{(2)}(u_b)\frac{X_1}{X_2}}{\left(1 - \frac{X_1 Y_2}{X_2 Y_1}\right)} \end{aligned} \tag{10.5.28}$$

Substituting (10.5.27) and (10.5.28) into (10.5.24) shows that the potential for $a \le r \le b$ is expressed in the form:

$$\begin{aligned} A_r = {} & \frac{\mu_o k_o (I\Delta L)}{j8\pi u_b^2} \sum_{n=0}^{\infty} (2n+1) P_n(\cos\theta) \\ & \times \left[\frac{\hat{h}_n^{(1)}(u_r)\hat{h}_n^{(2)}(u_b) - \hat{h}_n^{(1)}(u_r)\hat{h}_n^{(1)}(u_b)\frac{Y_2}{Y_1} + \hat{h}_n^{(2)}(u_r)\hat{h}_n^{(1)}(u_b)\frac{X_1 Y_2}{X_2 Y_1} - \hat{h}_n^{(2)}(u_r)\hat{h}_n^{(2)}(u_b)\frac{X_1}{X_2}}{1 - \frac{X_1 Y_2}{X_2 Y_1}} \right] \end{aligned}$$

or

$$A_r = \frac{\mu_o k_o (I\Delta L)}{j8\pi u_b^2} \sum_{n=0}^{\infty} (2n+1) P_n(\cos\theta) \hat{h}_n^{(1)}(u_r) \hat{h}_n^{(2)}(u_b) \left\{ \frac{\left[1 - \frac{\hat{h}_n^{(2)}(u_r)}{\hat{h}_n^{(1)}(u_r)} \frac{X_1}{X_2}\right]\left[1 - \frac{\hat{h}_n^{(1)}(u_b)}{\hat{h}_n^{(2)}(u_b)} \frac{Y_2}{Y_1}\right]}{1 - \frac{X_1 Y_2}{X_2 Y_1}} \right\} \tag{10.5.29}$$

Using the definition of the spherical reflection coefficients:

$$R_a^s = -\frac{X_1}{X_2} = -\left[\frac{\hat{h}_n^{(1)\prime}(u_a) - jz_a \hat{h}_n^{(1)}(u_a)}{\hat{h}_n^{(2)\prime}(u_a) - jz_a \hat{h}_n^{(2)}(u_a)}\right] = \frac{\hat{h}_n^{(1)}(u_a)}{\hat{h}_n^{(2)}(u_a)} R_a \tag{10.5.30}$$

$$R_c^s = -\frac{Y_2}{Y_1} = -\left[\frac{\hat{h}_n^{(2)\prime}(u_c) + jz_c \hat{h}_n^{(2)}(u_c)}{\hat{h}_n^{(1)\prime}(u_c) + jz_c \hat{h}_n^{(1)}(u_c)}\right] = \frac{\hat{h}_n^{(2)}(u_c)}{\hat{h}_n^{(1)}(u_c)} R_c \tag{10.5.31}$$

where

$$R_a = -\frac{ln'\hat{h}_n^{(1)}(u_a) - jz_a}{ln'\hat{h}_n^{(2)}(u_a) - jz_a}$$

$$R_c = -\frac{ln'\hat{h}_n^{(2)}(u_c) + jz_c}{ln'\hat{h}_n^{(1)}(u_c) + jz_c}$$

we finally obtain for $a \le r \le b$:

$$A_r = \frac{\mu_o k_o (I\Delta L)}{j8\pi u_b^2} \sum_{n=0}^{\infty} (2n+1) P_n(\cos\theta) \hat{h}_n^{(1)}(u_r) \hat{h}_n^{(2)}(u_b) \left\{ \frac{\left[1 + \frac{\hat{h}_n^{(2)}(u_r)}{\hat{h}_n^{(1)}(u_r)} R_a^s\right]\left[1 + \frac{\hat{h}_n^{(1)}(u_b)}{\hat{h}_n^{(2)}(u_b)} R_c^s\right]}{1 - R_a^s R_c^s} \right\} \tag{10.5.32}$$

For $b \le r \le c$ interchange u_r and u_b in (10.5.32).

The fields follow from (10.5.2) to (10.5.4). For example, using (10.5.32) in (10.5.2), the radial electric field for $a \le r \le b$ is given by

$$E_r = -\frac{\eta_o k_o^2 (I\Delta L)}{8\pi u_b^2 u_r^2} \sum_{n=0}^{\infty} n(n+1)(2n+1) P_n(\cos\theta) \times \hat{h}_n^{(1)}(u_r) \hat{h}_n^{(2)}(u_b) \left\{ \frac{\left[1 + \frac{\hat{h}_n^{(2)}(u_r)}{\hat{h}_n^{(1)}(u_r)} R_a^s\right]\left[1 + \frac{\hat{h}_n^{(1)}(u_b)}{\hat{h}_n^{(2)}(u_b)} R_c^s\right]}{1 - R_a^s R_c^s} \right\} \tag{10.5.33}$$

Applying the Watson transformation to (10.5.33) produces a rapidly convergent series and the field can be determined. Let

$$f(n) = \hat{h}_n^{(1)}(u_r)\hat{h}_n^{(2)}(u_b)\frac{\left[1 + \dfrac{\hat{h}_n^{(2)}(u_r)}{\hat{h}_n^{(1)}(u_r)}R_a^s\right]\left[1 + \dfrac{\hat{h}_n^{(1)}(u_b)}{\hat{h}_n^{(2)}(u_b)}R_c^s\right]}{1 - R_a^s R_c^s}$$

In terms of the Watson transformation (10.5.33) is expressed as

$$E_r = jK\oint_C \frac{\nu(\nu^2 - 1/4)}{\cos \nu\pi} f\left(\nu - \frac{1}{2}\right) P_{\nu-1/2}(-\cos\theta)d\nu \tag{10.5.34}$$

where the path of integration C is shown in Fig. 10.3a, and

$$K = -\frac{\eta_o k_o^2 (I\Delta L)}{8\pi u_b^2 u_r^2}$$

If $f(\nu - 1/2)$ is an even function, the integrand is odd and the contour C can be change to the contour C' shown in Fig. 10.4b. Closing the contour with an infinite semicircle in the lower-half plane shows that the integral can be evaluated in terms of the poles of $f(\nu - 1/2)$ (i.e., the poles of $1 - R_a^s R_c^s$). Hence, the evaluation of (10.5.34) can be performed following similar steps as those used in the evaluation of (10.4.22), although it is a more involved evaluation.

In the case of a VED in the Earth-Ionosphere waveguide a procedure that has been used to evaluate (10.5.34) is to express (10.5.34), with $r \approx a \approx b$, as

$$E_r \approx jK\int_{C'} \frac{\nu^3}{\cos \nu\pi} P_{\nu-1/2}(-\cos\theta)\hat{h}_{\nu-1/2}^{(2)}(u_a)\hat{h}_{\nu-1/2}^{(1)}(u_a)\,(1 + R_a)\frac{1 + pR_c}{1 - pR_aR_c}d\nu \tag{10.5.35}$$

where

$$p = \frac{h_{\nu-1/2}^{(1)}(u_a)h_{\nu-1/2}^{(2)}(u_c)}{h_{\nu-1/2}^{(2)}(u_a)h_{\nu-1/2}^{(1)}(u_c)}$$

Then, if $|pR_aR_c| < < 1$ along the path of integration, $1/(1 - pR_aR_c)$ is expanded in an ascending series:

$$\begin{aligned}\frac{1 + pR_c}{1 - pR_aR_c} &= (1 + pR_c)\left[1 + \sum_{j=1}^{\infty} (pR_aR_c)^j\right] \\ &= 1 + (1 + pR_a)\sum_{j=1}^{\infty} (R_a)^{j-1}(pR_c)^j \end{aligned} \tag{10.5.36}$$

Substituting (10.5.36) into (10.5.35):

$$E_r \approx jK \int_{C'} \frac{\nu^3}{\cos \nu\pi} P_{\nu-1/2}(-\cos\theta)\hat{h}^{(2)}_{\nu-1/2}(u_a)\hat{h}^{(1)}_{\nu-1/2}(u_a)(1+R_a)d\nu$$
$$+ jK \sum_{j=1}^{\infty} \int_{C'} \frac{\nu^3}{\cos \nu\pi} P_{\nu-1/2}(-\cos\theta)\hat{h}^{(2)}_{\nu-1/2}(u_a)\hat{h}^{(1)}_{\nu-1/2}(u_a)(1+R_a)^2(R_a)^{j-1}(pR_c)^j d\nu \tag{10.5.37}$$

The first integral represents the ground wave whose evaluation was done in Section 10.4. The second terms are known as the wave-hops integrals. On the contour of integration, R_c varies slowly and can be replaced by a constant Fresnel reflection coefficient. Hence, the wave-hops are written in term of an ionospheric reflection coefficient times a path integral I_j, and (10.5.37) can be expressed as

$$E_r = E_{gw} + \sum_{j=1}^{\infty} (R_c)^j I_j$$

where E_{gw} is the ground wave and the path integral is

$$I_j = jK \int_{C'} \frac{\nu^3}{\cos \nu\pi} P_{\nu-1/2}(-\cos\theta)\hat{h}^{(2)}_{\nu-1/2}(u_a)\hat{h}^{(1)}_{\nu-1/2}(u_a)\,(1+R_a)^2(R_a)^{j-1}p^j d\nu$$

The meaning of the wave-hops is shown in Fig. 10.7. The first hop (i.e., $j = 1$) is the wave that reflects from the ionosphere once. The second hop reflects twice from the ionosphere and once from the ground, and so on. The path integrals have been evaluated using three different methods depending on the location of the receiver. Associated with the j-hop there is an angle of incidence on the Earth's surface, say τ_j. In the illuminated region (i.e., for $\tau_j < \pi/2$), the path integral is evaluated using the method of steepest descent. For $\tau_j > \pi/2$, either a residue series method or a numerical integration method is used. When the path integrals are evaluated using the saddle-point method, the wave-hops are recognized to be identical to the corresponding geometrical optics ray.

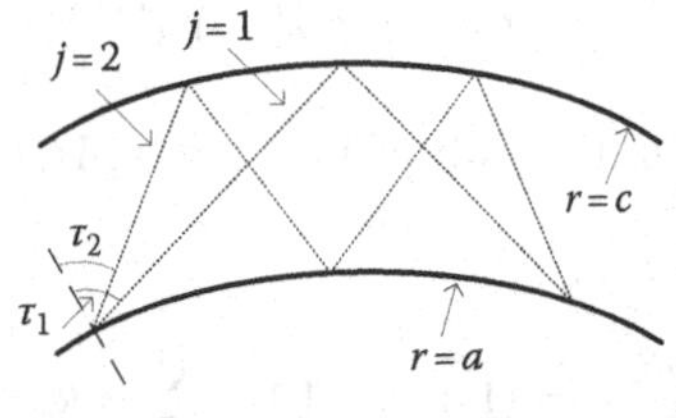

FIGURE 10.7 Geometrical representation for the first two wave-hops.

10.6 A VERTICAL MAGNETIC DIPOLE IN A SPHERICAL WAVEGUIDE—ZONAL HARMONICS SOLUTION

The geometry of the problem for a VMD in a spherical waveguide is shown in Fig. 10.8. The problem can be formulated in terms of the electric Debye potential F_r in (4.6.7), namely

$$[\nabla^2 + k_o^2]\frac{F_r}{r} = \frac{-\varepsilon_o J_{m,r}}{r} \tag{10.6.1}$$

where the field components, from (4.6.14) to (4.6.19), are given by

$$H_r = \frac{1}{j\omega\mu_o\varepsilon_o}\left(\frac{\partial^2}{\partial r^2} + k_o^2\right)F_r \tag{10.6.2}$$

$$H_\theta = \frac{1}{j\omega\mu_o\varepsilon_o r}\frac{\partial^2 F_r}{\partial r\partial\theta} \tag{10.6.3}$$

$$E_\phi = \frac{1}{\varepsilon_o r}\frac{\partial F_r}{\partial\theta} \tag{10.6.4}$$

The particular solution of (10.6.1) is

$$F_r^p = \varepsilon_o(I_m\Delta L)\frac{r}{b}\frac{e^{-jk_oR}}{4\pi R}$$

where $R = (r^2 + b^2 - 2rb\cos\theta)^{1/2}$. Using (10.5.6), F_r^p is expressed in the form

$$F_r^p = \frac{\varepsilon_o k_o(I_m\Delta L)}{j4\pi u_b^2}\sum_{n=0}^{\infty}(2n+1)P_n(\cos\theta)\begin{cases}\hat{j}_n(u_r)\hat{h}_n^{(2)}(u_b) & (r \le b)\\ \hat{j}_n(u_b)\hat{h}_n^{(2)}(u_r) & (r \ge b)\end{cases}$$

where $u_r = kr$ and $u_b = kb$.

The complementary solution is written as:

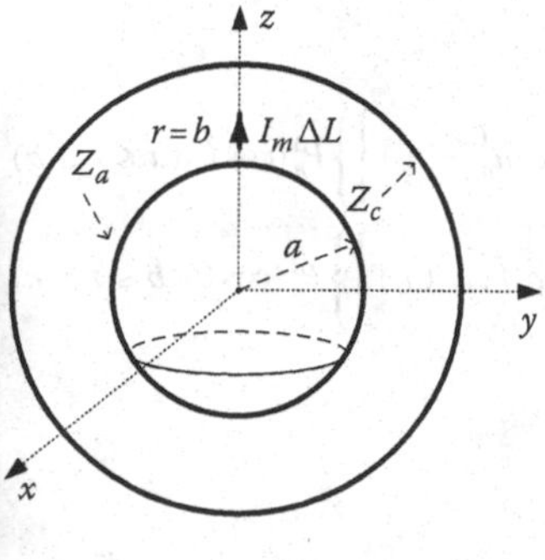

FIGURE 10.8 A VMD in a spherical waveguide.

$$F_r^c = \frac{\varepsilon_o k_o (I_m \Delta L)}{j4\pi u_b^2} \sum_{n=0}^{\infty} (2n+1) P_n(\cos\theta) \left[b_n \hat{j}_n(u_r) + c_n \hat{h}_n^{(2)}(u_r) \right]$$

where b_n and c_n are constants determined by the boundary conditions. The total solution is then

$$F_r = \begin{cases} \frac{\varepsilon_o k_o (I_m \Delta L)}{j4\pi u_b^2} \sum_{n=0}^{\infty} (2n+1) P_n(\cos\theta) \left\{ \hat{j}_n(u_r) \hat{h}_n^{(2)}(u_b) + \left[b_n \hat{j}_n(u_r) + c_n \hat{h}_n^{(2)}(u_r) \right] \right\} (a \le r \le b) \\ \frac{\varepsilon_o k_o (I_m \Delta L)}{j4\pi u_b^2} \sum_{n=0}^{\infty} (2n+1) P_n(\cos\theta) \left\{ \hat{j}_n(u_b) \hat{h}_n^{(2)}(u_r) + \left[b_n \hat{j}_n(u_r) + c_n \hat{h}_n^{(2)}(u_r) \right] \right\} (b \le r \le c) \end{cases} \tag{10.6.5}$$

The H_r field component follows from (10.6.2) and (10.6.5). That is:

$$H_r = \begin{cases} \frac{-k_o^2 (I_m \Delta L)}{4\pi\eta_o u_b^2 u_r^2} \sum_{n=0}^{\infty} n(n+1)(2n+1) P_n(\cos\theta) \left\{ \hat{j}_n(u_r) \hat{h}_n^{(2)}(u_b) \right. \\ \qquad \left. + \left[b_n \hat{j}_n(u_r) + c_n \hat{h}_n^{(2)}(u_r) \right] \right\} \quad (a \le r \le b) \\ \frac{-k_o^2 (I \Delta L)}{4\pi\eta_o u_b^2 u_r^2} \sum_{n=0}^{\infty} n(n+1)(2n+1) P_n(\cos\theta) \left\{ \hat{j}_n(u_b) \hat{h}_n^{(2)}(u_r) \right. \\ \qquad \left. + \left[b_n \hat{j}_n(u_r) + c_n \hat{h}_n^{(2)}(u_r) \right] \right\} \quad (b \le r \le c) \end{cases} \tag{10.6.6}$$

The other field components follow from (10.6.3) and (10.6.4). That is

$$E_\phi = \begin{cases} \frac{k_o^2 (I_m \Delta L)}{j4\pi u_b^2 u_r} \sum_{n=0}^{\infty} (2n+1) P_n^1(\cos\theta) \left\{ \hat{j}_n(u_r) \hat{h}_n^{(2)}(u_b) + \left[b_n \hat{j}_n(u_r) + c_n \hat{h}_n^{(2)}(u_r) \right] \right\} (a \le r \le b) \\ \frac{k_o^2 (I_m \Delta L)}{j4\pi u_b^2 u_r} \sum_{n=0}^{\infty} (2n+1) P_n^1(\cos\theta) \left\{ \hat{j}_n(u_b) \hat{h}_n^{(2)}(u_r) + \left[b_n \hat{j}_n(u_r) + c_n \hat{h}_n^{(2)}(u_r) \right] \right\} (b \le r \le c) \end{cases}$$

and

$$H_\theta = \begin{cases} \frac{-k_o^2 (I_m \Delta L)}{4\pi\eta_o u_b^2 u_r} \sum_{n=0}^{\infty} (2n+1) \left\{ \hat{j}_n{}'(u_r) \hat{h}_n^{(2)}(u_b) + \left[b_n \hat{j}_n{}'(u_r) + c_n \hat{h}_n^{(2)\prime}(u_r) \right] \right\} P_n^1(\cos\theta) (a \le r \le b) \\ \frac{-k_o^2 (I_m \Delta L)}{4\pi\eta_o u_b^2 u_r} \sum_{n=0}^{\infty} (2n+1) \left\{ \hat{j}_n(u_b) \hat{h}_n^{(2)\prime}(u_r) + \left[b_n \hat{j}_n{}'(u_r) + c_n \hat{h}_n^{(2)\prime}(u_r) \right] \right\} P_n^1(\cos\theta) (b \le r \le c) \end{cases}$$

If the boundaries at $r = a$ and $r = c$ are considered to be good conductors, then

$$E_\phi|_{r=a} = 0 \Rightarrow F_r|_{r=a} = 0$$

and

$$E_\phi|_{r=c} = 0 \Rightarrow F_r|_{r=c} = 0$$

which from (10.6.5) results in the following relations:

$$\hat{j}_n(u_a)\hat{h}_n^{(2)}(u_b) + \left[b_n\hat{j}_n(u_a) + c_n\hat{h}_n^{(2)}(u_a) \right] = 0 \tag{10.6.7}$$

and

$$\hat{j}_n(u_b)\hat{h}_n^{(2)}(u_c) + \left[b_n\hat{j}_n(u_c) + c_n\hat{h}_n^{(2)}(u_c) \right] = 0 \tag{10.6.8}$$

where $u_a = k_o a$ and $u_c = k_o c$. Solving (10.6.7) and (10.6.8) gives

$$b_n = \frac{\hat{h}_n^{(2)}(u_a)\hat{j}_n(u_b)\hat{h}_n^{(2)}(u_c) - \hat{j}_n(u_a)\hat{h}_n^{(2)}(u_b)\hat{h}_n^{(2)}(u_c)}{\hat{j}_n(u_a)\hat{h}_n^{(2)}(u_c) - \hat{h}_n^{(2)}(u_a)\hat{j}_n(u_c)}$$

and

$$c_n = \frac{\hat{j}_n(u_a)\hat{h}_n^{(2)}(u_b)\hat{j}_n(u_c) - \hat{j}_n(u_a)\hat{j}_n(u_b)\hat{h}_n^{(2)}(u_c)}{\hat{j}_n(u_a)\hat{h}_n^{(2)}(u_c) - \hat{h}_n^{(2)}(u_a)\hat{j}_n(u_c)}$$

If impedance boundary conditions are used, denoted by $\tilde{Z}_a$ and $\tilde{Z}_c$, to model the ground-air and ionosphere-air interfaces. Then,

$$\tilde{Z}_a = \left.\frac{E_\phi}{H_\theta}\right|_{r=a} \Rightarrow k_o F_r = -j\left.\frac{\tilde{Z}_a}{\eta_o}\frac{\partial F_r}{\partial r}\right|_{r=a} \tag{10.6.9}$$

and

$$\tilde{Z}_c = -\left.\frac{E_\phi}{H_\theta}\right|_{z=c} \Rightarrow k_o F_r = j\left.\frac{\tilde{Z}_c}{\eta_o}\frac{\partial F_r}{\partial r}\right|_{r=c} \tag{10.6.10}$$

Different surfaces require appropriate forms $\tilde{Z}_a$ and $\tilde{Z}_c$. For example, if the surface at $r = a$ represents the Earth's ground the surface impedance can be approximated by that of a perpendicular polarized plane wave incident at grazing incidence. That is,

$$\tilde{Z}_a = \frac{\eta_a}{\sqrt{1 - \left(\frac{k_o}{k_a}\right)^2}}$$

Similarly, in certain cases the surface impedance at $r = c$ can be represented by

$$\tilde{Z}_c \approx \frac{\eta_c}{\sqrt{1 - \left(\frac{k_o}{k_c}\right)^2}}$$

In the case that the surface at $r = c$ represents the Ionosphere the form of $\tilde{Z}_c$ depends on the model used for the ionospheric plasma and the frequency of operation.

Using (10.6.5), it follows from (10.6.9) and (10.6.10) that

$$\hat{j}_n(u_a)\hat{h}_n^{(2)}(u_b) + \left[b_n\hat{j}_n(u_a) + c_n\hat{h}_n^{(2)}(u_a)\right]$$
$$= -j\frac{\tilde{Z}_a}{\eta_o}\left\{\hat{j}_n{}'(u_a)\hat{h}_n^{(2)}(u_b) + \left[b_n\hat{j}_n{}'(u_a) + c_n\hat{h}_n^{(2)\prime}(u_a)\right]\right\}$$

and

$$\hat{j}_n(u_b)\hat{h}_n^{(2)}(u_c) + \left[b_n\hat{j}_n(u_c) + c_n\hat{h}_n^{(2)}(u_c)\right]$$
$$= j\frac{\tilde{Z}_c}{\eta_o}\left\{\hat{j}_n(u_b)\hat{h}_n^{(2)\prime}(u_c) + \left[b_n\hat{j}_n{}'(u_c) + c_n\hat{h}_n^{(2)\prime}(u_c)\right]\right\}$$

which can be solved for b_n and c_n. Defining the normalized impedances $\tilde{z}_a = \tilde{Z}_a/\eta_o$ and $\tilde{z}_c = \tilde{Z}_c/\eta_o$, we obtain:

$$b_n = \frac{\begin{vmatrix} -\hat{h}_n^{(2)}(u_b)\left[\hat{j}_n(u_a) + \tilde{z}_a\hat{j}_n{}'(u_a)\right] & \hat{h}_n^{(2)}(u_a) + j\tilde{z}_a\hat{h}_n^{(2)\prime}(u_a) \\ -\hat{j}_n(u_b)\left[\hat{h}_n^{(2)}(u_c) - j\tilde{z}_c\hat{h}_n^{(2)\prime}(u_c)\right] & \hat{h}_n^{(2)}(u_c) - j\tilde{z}_c\hat{h}_n^{(2)\prime}(u_c) \end{vmatrix}}{\begin{vmatrix} \hat{j}_n(u_a) + j\tilde{z}_a\hat{j}_n{}'(u_a) & \hat{h}_n^{(2)}(u_a) + j\tilde{z}_a\hat{h}_n^{(2)\prime}(u_a) \\ \hat{j}_n(u_c) - j\tilde{z}_c\hat{j}_n{}'(u_c) & \hat{h}_n^{(2)}(u_c) - j\tilde{z}_c\hat{h}_n^{(2)\prime}(u_c) \end{vmatrix}} \tag{10.6.11}$$

$$c_n = \frac{\begin{vmatrix} \hat{j}_n(u_a) + j\tilde{z}_a\hat{j}_n{}'(u_a) & -\hat{h}_n^{(2)}(u_b)\left[\hat{j}_n(u_a) + j\tilde{z}_a\hat{j}_n{}'(u_a)\right] \\ \hat{j}_n(u_c) - j\tilde{z}_c\hat{j}_n{}'(u_c) & -\hat{j}_n(u_b)\left[\hat{h}_n^{(2)}(u_c) - j\tilde{z}_c\hat{h}_n^{(2)}{}'(u_c)\right] \end{vmatrix}}{\begin{vmatrix} \hat{j}_n(u_a) + j\tilde{z}_a\hat{j}_n{}'(u_a) & \hat{h}_n^{(2)}(u_a) + j\tilde{z}_a\hat{h}_n^{(2)}{}'(u_a) \\ \hat{j}_n(u_c) - j\tilde{z}_c\hat{j}_n{}'(u_c) & \hat{h}_n^{(2)}(u_c) - j\tilde{z}_c\hat{h}_n^{(2)}{}'(u_c) \end{vmatrix}} \qquad (10.6.12)$$

Equation (10.6.5) describes the potential F_r formulation of the problem, where the constants b_n and c_n are given by (10.6.11) and (10.6.12). The fields follow from (10.6.2) to (10.6.4).

The potential F_r can also be written only in terms of Hankel functions as was done in (10.4.12). That is,

$$F_r = \begin{cases} \dfrac{\varepsilon_0 k_0 (I_m \Delta L)}{j8\pi u_b^2} \sum_{n=0}^{\infty} (2n+1) P_n(\cos\theta)\left\{\hat{h}_n^{(1)}(u_r)\hat{h}_n^{(2)}(u_b)\right. \\ \qquad\qquad \left. + \left[d_n\hat{h}_n^{(1)}(u_r) + e_n\hat{h}_n^{(2)}(u_r)\right]\right\} \quad (a \le r \le b) \\[2em] \dfrac{\varepsilon_0 k_0 (I_m \Delta L)}{j8\pi u_b^2} \sum_{n=0}^{\infty} (2n+1) P_n(\cos\theta)\left\{\hat{h}_n^{(1)}(u_b)\hat{h}_n^{(2)}(u_r)\right. \\ \qquad\qquad \left. + \left[d_n\hat{h}_n^{(1)}(u_r) + e_n\hat{h}_n^{(2)}(u_r)\right]\right\} \quad (b \le r \le c) \end{cases}$$

$$(10.6.13)$$

Applying the boundary conditions in (10.6.9) and (10.6.10) results in two equations similar in form to that in (10.5.25) and (10.5.26). That is,

$$d_n X_1 + e_n X_2 = -h_n^{(2)}(u_b) X_1$$

and

$$d_n Y_1 + e_n Y_2 = -h_n^{(1)}(u_b) Y_2$$

where for the present analysis:

$$X_1 = h_n^{(1)}(u_a) + j\tilde{z}_a h_n^{(1)}{}'(u_a)$$
$$X_2 = h_n^{(2)}(u_a) + j\tilde{z}_a h_n^{(2)}{}'(u_a)$$
$$Y_1 = h_n^{(1)}(u_c) - j\tilde{z}_c h_n^{(1)}{}'(u_c)$$
$$Y_2 = h_n^{(2)}(u_c) - j\tilde{z}_c h_n^{(2)}{}'(u_c)$$

Following a similar procedure as that in (10.5.25) to (10.5.32) shows that F_r for $a \le r \le b$ can be expressed in the form:

$$F_r = \frac{\varepsilon_o k_o (I_m \Delta L)}{j8\pi u_b^2} \sum_{n=0}^{\infty} (2n+1) P_n(\cos\theta) \hat{h}_n^{(1)}(u_r) \hat{h}_n^{(2)}(u_b) \times \left\{ \frac{\left[1 + \frac{\hat{h}_n^{(2)}(u_r)}{\hat{h}_n^{(1)}(u_r)} \tilde{R}_a^s\right]\left[1 + \frac{\hat{h}_n^{(1)}(u_b)}{\hat{h}_n^{(2)}(u_b)} \tilde{R}_c^s\right]}{1 - \tilde{R}_a^s \tilde{R}_c^s} \right\} \quad (10.6.14)$$

where the spherical reflection coefficients are

$$\tilde{R}_a^s = -\left[\frac{\hat{h}_n^{(1)\prime}(k_o a) - \frac{j}{\tilde{z}_a}\hat{h}_n^{(1)}(k_o a)}{\hat{h}_n^{(2)\prime}(k_o a) - \frac{j}{\tilde{z}_a}\hat{h}_n^{(2)}(k_o a)}\right] \quad (10.6.15)$$

and

$$\tilde{R}_c^s = -\left[\frac{\hat{h}_n^{(2)\prime}(u_c) + \frac{j}{\tilde{z}_c}\hat{h}_n^{(2)}(u_c)}{\hat{h}_n^{(1)\prime}(u_c) + \frac{j}{\tilde{z}_c}\hat{h}_n^{(1)}(u_c)}\right] \quad (10.6.16)$$

For $b \le r \le c$ interchange u_r and u_b in (10.6.14).

The fields follow from (10.6.2) to (10.6.4). For example, the radial magnetic field for $a \le r \le b$ is given by:

$$H_r = -\frac{k_o^2 (I_m \Delta L)}{8\pi\eta_o u_b^2 u_r^2} \sum_{n=0}^{\infty} n(n+1)(2n+1) \times P_n(\cos\theta) \hat{h}_n^{(2)}(u_b) \hat{h}_n^{(1)}(u_r) \left\{ \frac{\left[1 + \frac{\hat{h}_n^{(2)}(u_r)}{\hat{h}_n^{(1)}(u_r)} \tilde{R}_a^s\right]\left[1 + \frac{\hat{h}_n^{(1)}(u_b)}{\hat{h}_n^{(2)}(u_b)} \tilde{R}_c^s\right]}{1 - \tilde{R}_a^s \tilde{R}_c^s} \right\} \quad (10.6.17)$$

The application of the Watson transformation to (10.6.17) is similar to the procedure described in Sections 10.4 and 10.5.

10.7 A HORIZONTAL ELECTRIC DIPOLE IN A SPHERICAL WAVEGUIDE—ZONAL HARMONICS SOLUTION

The geometry of the problem for a HED in a spherical waveguide is shown in Fig. 10.9. The electric dipole points in the x direction; therefore, the problem can be formulated in terms of the electric Hertz vector $\mathbf{\Pi}^e = \Pi_x^e \hat{\mathbf{a}}_x$, which satisfy the relation:

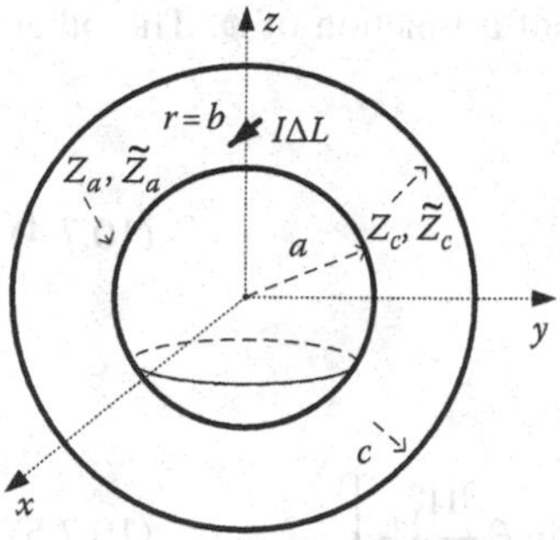

FIGURE 10.9 A HED in a spherical waveguide.

$$[\nabla^2 + k_o^2]\Pi_x^e = \frac{-(I\Delta L)}{j\omega\varepsilon_o}\delta(x)\delta(y)\delta(z-b) \tag{10.7.1}$$

The particular solution of (10.7.1) is

$$\Pi_{x,p}^e = \frac{(I\Delta L)}{j\omega\varepsilon_o}\frac{e^{-jk_oR}}{4\pi R} \tag{10.7.2}$$

where $R = (r^2 + b^2 - 2rb\cos\theta)^{1/2}$.

Equation (10.7.2) can be written in spherical coordinates as:

$$\mathbf{\Pi}_p^e = \Pi_{x,p}^e\hat{\mathbf{a}}_x = \Pi_{x,p}^e[\sin\theta\cos\phi\hat{\mathbf{a}}_r + \cos\theta\cos\phi\hat{\mathbf{a}}_\theta - \sin\phi\hat{\mathbf{a}}_\phi]$$

which shows that

$$\begin{aligned}\Pi_{r,p}^e &= \Pi_{x,p}^e \sin\theta\cos\phi\\ \Pi_{\theta,p}^e &= \Pi_{x,p}^e \cos\theta\cos\phi\\ \Pi_{\phi,p}^e &= -\Pi_{x,p}^e \sin\phi\end{aligned}$$

The magnetic field components follow from

$$\mathbf{H} = j\omega\varepsilon_o\nabla\times\mathbf{\Pi}^e$$

Expanding the curl in spherical coordinates shows that the particular solution for the radial component of **H** is

$$\begin{aligned}H_r^p &= \frac{j\omega\varepsilon_o}{r\sin\theta}\left[\frac{\partial}{\partial\theta}\left(\Pi_{\phi,p}^e\sin\theta\right) - \frac{\partial\Pi_{\theta,p}^e}{\partial\phi}\right]\\ &= \frac{j\omega\varepsilon_o}{r\sin\theta}\left[-\sin\phi\frac{\partial}{\partial\theta}\left(\Pi_{x,p}^e\sin\theta\right) - \cos\theta\frac{\partial}{\partial\phi}\left(\Pi_{x,p}^e\cos\phi\right)\right]\\ &= -\frac{j\omega\varepsilon_o}{r}\sin\phi\frac{\partial\Pi_{x,p}^e}{\partial\theta}\end{aligned} \tag{10.7.3}$$

where we use the fact that $\partial \Pi^e_{x,p}/\partial \phi = 0$ since R is not a function of ϕ. The other components are

$$H^p_\theta = j\omega\varepsilon_o \sin\phi \frac{\partial \Pi^e_{x,p}}{\partial r} \tag{10.7.4}$$

and

$$H^p_\phi = j\omega\varepsilon_o \cos\phi \left[\cos\theta \frac{\partial \Pi^e_{x,p}}{\partial r} - \sin\theta \frac{\partial \Pi^e_{x,p}}{\partial \theta} \right] \tag{10.7.5}$$

Substituting (10.7.2) into (10.7.3):

$$H^p_r = -\frac{(I\Delta L)}{r} \sin\phi \frac{\partial}{\partial \theta} \left(\frac{e^{-jk_o R}}{4\pi R} \right)$$

Using (10.5.6) the spherical harmonics expression for H^p_r is

$$H^p_r = j\frac{(I\Delta L)}{4\pi u_b r^2} \sin\phi \sum_{n=0}^{\infty} (2n+1) P^1_n(\cos\theta) \begin{cases} \hat{j}_n(u_r)\hat{h}^{(2)}_n(u_b) & (a \leq r \leq b) \\ \hat{j}_n(u_b)\hat{h}^{(2)}_n(u_r) & (b \leq r \leq c) \end{cases} \tag{10.7.6}$$

where $u_r = k_o r$ and $u_b = k_o b$.

The Debye potential associated with the particular solution, denoted by V^p, is related to H^p_r by

$$H^p_r = \left(\frac{\partial^2}{\partial r^2} + k^2_o \right)(rV^p) \tag{10.7.7}$$

and it follows from (10.7.6) and (10.7.7) that the form of V^p for the HED is

$$rV^p = j\frac{(I\Delta L)}{4\pi u_b} \sin\phi \sum_{n=0}^{\infty} b_n P^1_n(\cos\theta) \begin{cases} \hat{j}_n(u_r)\hat{h}^{(2)}_n(u_b) & (a \leq r \leq b) \\ \hat{j}_n(u_b)\hat{h}^{(2)}_n(u_r) & (b \leq r \leq c) \end{cases} \tag{10.7.8}$$

Substituting (10.7.8) into (10.7.7), and using (10.5.11), gives

$$H^p_r = j\frac{(I\Delta L)}{4\pi u_b r^2} \sin\phi \sum_{n=0}^{\infty} b_n n(n+1) P^1_n(\cos\theta) \begin{cases} \hat{j}_n(u_r)\hat{h}^{(2)}_n(u_b) & (a \leq r \leq b) \\ \hat{j}_n(u_b)\hat{h}^{(2)}_n(u_r) & (b \leq r \leq c) \end{cases} \tag{10.7.9}$$

Comparing (10.7.6) with (10.7.9) shows that:

$$b_n = \frac{(2n+1)}{n(n+1)}$$

Therefore, the Debye potential V^p is given by

$$V^p = j\frac{(I\Delta L)}{4\pi u_b r}\sin\phi\sum_{n=0}^{\infty}\frac{(2n+1)}{n(n+1)}P_n^1(\cos\theta)\begin{cases}\hat{j}_n(u_r)\hat{h}_n^{(2)}(u_b) & (a \le r \le b)\\ \hat{j}_n(u_b)\hat{h}_n^{(2)}(u_r) & (b \le r \le c)\end{cases} \qquad (10.7.10)$$

The complementary solution for the Debye potential V is

$$V^c = j\frac{(I\Delta L)}{4\pi u_b r}\sin\phi\sum_{n=0}^{\infty}\frac{(2n+1)}{n(n+1)}P_n^1(\cos\theta)\left[f_n\hat{j}_n(u_r) + g_n\hat{h}_n^{(2)}(u_r)\right]$$

The derivation for the Debye potential U^p for a HED follows the expression of E_r^p. From Maxwell equation:

$$\mathbf{E}^p = \frac{1}{j\omega\varepsilon_o}\nabla\times\mathbf{H}^p$$

performing the curl operation on the components of $\mathbf{H}^p$ in (10.7.3), (10.7.4) and (10.7.5), shows that E_r^p is

$$\begin{aligned}E_r^p &= \frac{\cos\phi}{r}\left[\frac{\partial}{\partial\theta}\left(\cos\theta\frac{\partial\Pi_{x,p}^e}{\partial r}\right) - \frac{\sin\theta}{r}\frac{\partial^2\Pi_{x,p}^e}{\partial\theta^2}\right] - \frac{\cos\phi}{r}\left(\sin\theta\frac{\Pi_{x,p}^e}{\partial r} + \frac{2\cos\theta}{r}\frac{\partial\Pi_{x,p}^e}{\partial\theta}\right)\\ &= \frac{\cos\phi}{r}\frac{\partial}{\partial\theta}\left(\cos\theta\frac{\partial\Pi_{x,p}^e}{\partial r} - \frac{\sin\theta}{r}\frac{\partial\Pi_{x,p}^e}{\partial\theta}\right) - \frac{\cos\phi}{r}\left(\sin\theta\frac{\Pi_{x,p}^e}{\partial r} + \frac{\cos\theta}{r}\frac{\partial\Pi_{x,p}^e}{\partial\theta}\right)\end{aligned} \qquad (10.7.11)$$

Since (with $(I\Delta L)=1$ in the next relations):

$$\begin{aligned}\frac{\partial\Pi_{x,p}^e}{\partial r} &= \frac{1}{j\omega\varepsilon_o 4\pi}\frac{\partial}{\partial r}\left(\frac{e^{-jkR}}{R}\right) = -\frac{(jk_oR+1)}{j\omega\varepsilon_o 4\pi R^3}e^{-jk_oR}(r - b\cos\theta)\\ \frac{\partial\Pi_{x,p}^e}{\partial\theta} &= \frac{1}{j\omega\varepsilon_o 4\pi}\frac{\partial}{\partial\theta}\left(\frac{e^{-jkR}}{R}\right) = -\frac{(jk_oR+1)}{4\pi j\omega\varepsilon_o R^3}e^{-jk_oR}rb\sin\theta\end{aligned}$$

the terms in parenthesis in (10.7.11) are written as

$$\cos\theta\frac{\partial\Pi_{x,p}^e}{\partial r} - \frac{\sin\theta}{r}\frac{\partial\Pi_{x,p}^e}{\partial\theta} = -\frac{(jk_oR+1)}{j\omega\varepsilon_o 4\pi R^3}e^{-jk_oR}(r\cos\theta - b) \qquad (10.7.12)$$

and

$$\sin\theta \frac{\partial \Pi^{e}_{x,p}}{\partial r} + \frac{\cos\theta}{r}\frac{\partial \Pi^{e}_{x,p}}{\partial \theta} = -\frac{(jk_oR+1)}{j\omega\varepsilon_o 4\pi R^3} e^{-jk_oR}\, r \tag{10.7.13}$$

Observing that

$$\frac{\partial \Pi^{e}_{x,p}}{\partial b} = \frac{1}{j\omega\varepsilon_o 4\pi}\frac{\partial}{\partial b}\left(\frac{e^{-jkR}}{R}\right) = -\frac{(jk_oR+1)}{4\pi R^3} e^{-jk_oR}(b - r\cos\theta)$$

it follows that (10.7.12) and (10.7.13) can be expressed in the forms

$$\cos\theta \frac{\partial \Pi^{e}_{x,p}}{\partial r} - \frac{\sin\theta}{r}\frac{\partial \Pi^{e}_{x,p}}{\partial \theta} = -\frac{\partial \Pi^{e}_{x,p}}{\partial b}$$

$$\sin\theta \frac{\partial \Pi^{e}_{x,p}}{\partial r} + \frac{\cos\theta}{r}\frac{\partial \Pi^{e}_{x,p}}{\partial \theta} = \frac{1}{b}\frac{\partial \Pi^{e}_{x,p}}{\partial \theta}$$

and (10.7.11) reads

$$E_r^p = -\frac{\cos\phi}{r}\frac{\partial}{\partial\theta}\left(\frac{\partial \Pi^{e}_{x,p}}{\partial b} + \frac{1}{b}\Pi^{e}_{x,p}\right)$$
$$= -\frac{\cos\phi}{rb}\frac{\partial^2}{\partial\theta\partial b}\left(b\Pi^{e}_{x,p}\right) \tag{10.7.14}$$

which provides a convenient form to find E_r^p. Substituting (10.7.2) into (10.7.14) and using the spherical harmonic expansion in (10.5.6) gives:

$$E_r^p = \frac{(I\Delta L)\eta_o}{4\pi u_b r^2}\cos\phi \sum_{n=0}^{\infty}(2n+1)P_n^1(\cos\theta)\begin{cases} \hat{J}_n(u_r)\hat{h}_n^{(2)\prime}(u_b) & (a \le r \le b) \\ \hat{J}_n{}'(u_b)\hat{h}_n^{(2)}(u_r) & (b \le r \le c)\end{cases} \tag{10.7.15}$$

The Debye potential U^p and E_r^p are related by

$$E_r^p = \left(\frac{\partial^2}{\partial r^2} + k_o^2\right)(rU^p) \tag{10.7.16}$$

Therefore, from (10.7.15) and (10.7.16), U^p is of the form

$$rU^p = \frac{(I\Delta L)\eta_o}{4\pi u_b}\cos\phi \sum_{n=0}^{\infty} c_n P_n^1(\cos\theta)\begin{cases} \hat{J}_n(u_r)\hat{h}_n^{(2)\prime}(u_b) & (a \le r \le b) \\ \hat{J}_n{}'(u_b)\hat{h}_n^{(2)}(u_r) & (b \le r \le c)\end{cases} \tag{10.7.17}$$

Substituting (10.7.17) into (10.7.16) and using (10.5.13) gives

$$E_r^p = \frac{\eta_o(I\Delta L)}{4\pi u_b r^2} \cos\phi \sum_{n=0}^{\infty} c_n n(n+1) P_n^1(\cos\theta) \begin{cases} \hat{j}_n(u_r)\hat{h}_n^{(2)\prime}(u_b) & (a \le r \le b) \\ \hat{j}_n{}'(u_b)\hat{h}_n^{(2)}(u_r) & (b \le r \le c) \end{cases} \quad (10.7.18)$$

Comparing (10.7.15) with (10.7.18) shows that

$$c_n = \frac{(2n+1)}{n(n+1)}$$

Therefore, the particular solution for the Debye potential is

$$U^p = \frac{(I\Delta L)\eta_o}{4\pi u_b r} \cos\phi \sum_{n=0}^{\infty} \frac{(2n+1)}{n(n+1)} P_n^1(\cos\theta) \begin{cases} \hat{j}_n(u_r)\hat{h}_n^{(2)\prime}(u_b) & (a \le r \le b) \\ \hat{j}_n{}'(u_b)\hat{h}_n^{(2)}(u_r) & (b \le r \le c) \end{cases}$$

The complementary solution for the potential U is:

$$U^c = \frac{(I\Delta L)\eta_o}{4\pi u_b r} \cos\phi \sum_{n=0}^{\infty} \frac{(2n+1)}{n(n+1)} P_n^1(\cos\theta) \left[d_n \hat{j}_n(u_r) + e_n \hat{h}_n^{(2)}(u_r) \right]$$

The total Debye potentials in the region $a \le r \le c$ are

$$\begin{aligned}
U &= U^p + U^c \\
&= \frac{(I\Delta L)\eta_o}{4\pi u_b r} \cos\phi \sum_{n=0}^{\infty} \frac{(2n+1)}{n(n+1)} P_n^1(\cos\theta) \\
&\times \begin{cases} \hat{j}_n(u_r)\hat{h}_n^{(2)\prime}(u_b) + \left[d_n \hat{j}_n(u_r) + e_n \hat{h}_n^{(2)}(u_r) \right] & (a \le r \le b) \\ \hat{j}_n{}'(u_b)\hat{h}_n^{(2)}(u_r) + \left[d_n \hat{j}_n(u_r) + e_n \hat{h}_n^{(2)}(u_r) \right] & (b \le r \le c) \end{cases}
\end{aligned} \quad (10.7.19)$$

and

$$\begin{aligned}
V &= V^p + V^c \\
&= j\frac{(I\Delta L)}{4\pi u_b r} \sin\phi \sum_{n=0}^{\infty} \frac{(2n+1)}{n(n+1)} P_n^1(\cos\theta) \\
&\times \begin{cases} \hat{j}_n(u_r)\hat{h}_n^{(2)}(u_b) + \left[f_n \hat{j}_n(u_r) + g_n \hat{h}_n^{(2)}(u_r) \right] & (a \le r \le b) \\ \hat{j}_n(u_b)\hat{h}_n^{(2)}(u_r) + \left[f_n \hat{j}_n(u_r) + g_n \hat{h}_n^{(2)}(u_r) \right] & (b \le r \le c) \end{cases}
\end{aligned} \quad (10.7.20)$$

The fields follow from (4.6.14) and (4.6.19), which in this case depend on U and V.

The boundary conditions at the spherical surfaces can be stated in terms of surface impedances. Since U and V produce different field polarizations we will denote the ground and ionosphere impedances due to U by Z_a and Z_c, and those due to V by $\tilde{Z}_a$ and $\tilde{Z}_c$. Therefore,

$$E_\theta = -Z_s H_\phi|_{r=a} \Rightarrow \begin{cases} \left.\dfrac{\partial}{\partial r}(rU) = j\omega\varepsilon_o Z_a (rU)\right|_{r=a} \\ \left.\dfrac{\partial}{\partial r}(rV) = \dfrac{j\omega\mu_o}{\tilde{Z}_a}(rV)\right|_{r=a} \end{cases} \tag{10.7.21}$$

and

$$E_\theta = Z_s H_\phi|_{r=c} \Rightarrow \begin{cases} \left.\dfrac{\partial}{\partial r}(rU) = -j\omega\varepsilon_o Z_c (rU)\right|_{r=c} \\ \left.\dfrac{\partial}{\partial r}(rV) = -\dfrac{j\omega\mu_o}{\tilde{Z}_c}(rV)\right|_{r=c} \end{cases} \tag{10.7.22}$$

Applying the boundary conditions in (10.7.21) and (10.7.22) to the Debye potential U in (10.7.19) gives

$$\begin{aligned} &\hat{J}_n{}'(u_a)\hat{h}_n^{(2)\prime}(u_b) + \left[d_n\hat{J}_n{}'(u_a) + e_n\hat{h}_n^{(2)\prime}(u_a)\right] \\ &= j\frac{Z_a}{\eta_o}\left\{\hat{J}_n(u_a)\hat{h}_n^{(2)\prime}(u_b) + \left[d_n\hat{J}_n(u_a) + e_n\hat{h}_n^{(2)\prime}(u_a)\right]\right\} \end{aligned} \tag{10.7.23}$$

and

$$\begin{aligned} &\hat{J}_n{}'(u_b)\hat{h}_n^{(2)\prime}(u_c) + \left[d_n\hat{J}_n{}'(u_c) + e_n\hat{h}_n^{(2)\prime}(u_c)\right] \\ &= -j\frac{Z_c}{\eta_o}\left\{\hat{J}_n{}'(u_b)\hat{h}_n^{(2)}(u_c) + \left[d_n\hat{J}_n(u_c) + e_n\hat{h}_n^{(2)}(u_c)\right]\right\} \end{aligned} \tag{10.7.24}$$

The solution to the constants in (10.7.23) and (10.7.24) can be expressed as

$$d_n = \frac{\begin{vmatrix} -\hat{h}_n^{(2)\prime}(u_b)\left[\hat{J}_n{}'(u_a) - jz_a\hat{J}_n(u_a)\right] & \hat{h}_n^{(2)\prime}(u_a) - jz_a\hat{h}_n^{(2)}(u_a) \\ -\hat{J}_n{}'(u_b)\left[\hat{h}_n^{(2)\prime}(u_c) + jz_c\hat{h}_n^{(2)}(u_c)\right] & \hat{h}_n^{(2)\prime}(u_c) + jz_c\hat{h}_n^{(2)}(u_c) \end{vmatrix}}{\begin{vmatrix} \hat{J}_n{}'(u_a) - jz_a\hat{J}_n(u_a) & \hat{h}_n^{(2)\prime}(u_a) - jz_a\hat{h}_n^{(2)}(u_a) \\ \hat{J}_n{}'(u_c) + jz_c\hat{J}_n(u_c) & \hat{h}_n^{(2)\prime}(u_c) + jz_c\hat{h}_n^{(2)}(u_c) \end{vmatrix}} \tag{10.7.25}$$

and

$$e_n = \frac{\begin{vmatrix} \hat{j}_n'(u_a) - jz_a\hat{j}_n(u_a) & -\hat{h}_n^{(2)\prime}(u_b)[\hat{j}_n'(u_a) - jz_a\hat{j}_n(u_a)] \\ \hat{j}_n'(u_c) + jz_c\hat{j}_n(u_c) & -\hat{j}_n'(u_b)[\hat{h}_n^{(2)\prime}(u_c) + jz_c\hat{h}_n^{(2)}(u_c)] \end{vmatrix}}{\begin{vmatrix} \hat{j}_n'(u_a) - jz_a\hat{j}_n(u_a) & \hat{h}_n^{(2)\prime}(u_a) - jz_a\hat{h}_n^{(2)}(u_a) \\ \hat{j}_n'(u_c) + jz_c\hat{j}_n(u_c) & \hat{h}_n^{(2)\prime}(u_c) + jz_c\hat{h}_n^{(2)}(u_c) \end{vmatrix}} \tag{10.7.26}$$

Applying the boundary conditions in (10.7.21) and (10.7.22) to the Debye potential V in (10.7.20) gives

$$\begin{aligned} &\hat{j}_n'(u_a)\hat{h}_n^{(2)}(u_b) + [f_n\hat{j}_n'(u_a) + g_n\hat{h}_n^{(2)\prime}(u_a)] \\ &\quad = j\frac{\eta_o}{\bar{Z}_a}\left\{\hat{j}_n(u_a)\hat{h}_n^{(2)}(u_b) + [f_n\hat{j}_n(u_a) + g_n\hat{h}_n^{(2)}(u_a)]\right\} \end{aligned} \tag{10.7.27}$$

and

$$\begin{aligned} &\hat{j}_n(u_b)\hat{h}_n^{(2)\prime}(u_c) + [f_n\hat{j}_n'(u_c) + g_n\hat{h}_n^{(2)\prime}(u_c)] \\ &\quad = -j\frac{\eta_o}{\bar{Z}_c}\left\{\hat{j}_n(u_b)\hat{h}_n^{(2)}(u_c) + [f_n\hat{j}_n(u_c) + g_n\hat{h}_n^{(2)}(u_c)]\right\} \end{aligned} \tag{10.7.28}$$

The solution of (10.7.27) and (10.7.28) can be expressed as:

$$f_n = \frac{\begin{vmatrix} -\hat{h}_n^{(2)\prime}(u_b)[j\bar{z}_a\hat{j}_n'(u_a) + \hat{j}_n(u_a)] & j\bar{z}_a\hat{h}_n^{(2)\prime}(u_a) + \hat{h}_n^{(2)}(u_a) \\ -\hat{j}_n'(u_b)[j\bar{z}_c\hat{h}_n^{(2)\prime}(u_c) - \hat{h}_n^{(2)}(u_c)] & j\bar{z}_c\hat{h}_n^{(2)\prime}(u_c) - \hat{h}_n^{(2)}(u_c) \end{vmatrix}}{\begin{vmatrix} j\bar{z}_a\hat{j}_n'(u_a) + \hat{j}_n(u_a) & j\bar{z}_a\hat{h}_n^{(2)\prime}(u_a) + \hat{h}_n^{(2)}(u_a) \\ j\bar{z}_c\hat{j}_n' - \hat{j}_n(u_c) & j\bar{z}_c\hat{h}_n^{(2)\prime}(u_c) - \hat{h}_n^{(2)}(u_c) \end{vmatrix}} \tag{10.7.29}$$

and

$$e_n = \frac{\begin{vmatrix} j\bar{z}_a\hat{j}_n'(u_a) + \hat{j}_n(u_a) & -\hat{h}_n^{(2)\prime}(u_b)[j\bar{z}_a\hat{j}_n'(u_a) + \hat{j}_n(u_a)] \\ j\bar{z}_c\hat{j}_n'(u_c) - \hat{j}_n(u_c) & -\hat{j}_n'(u_b)[j\bar{z}_c\hat{h}_n^{(2)\prime}(u_c) - \hat{h}_n^{(2)}(u_c)] \end{vmatrix}}{\begin{vmatrix} j\bar{z}_a\hat{j}_n'(u_a) + \hat{j}_n(u_a) & j\bar{z}_a\hat{h}_n^{(2)\prime}(u_a) + \hat{h}_n^{(2)}(u_a) \\ j\bar{z}_c\hat{j}_n'(u_c) - \hat{j}_n(u_c) & j\bar{z}_c\hat{h}_n^{(2)\prime}(u_c) - \hat{h}_n^{(2)}(u_c) \end{vmatrix}} \tag{10.7.30}$$

The relations (10.7.25), (10.7.26), (10.7.29), and (10.7.30) when substituted into (10.7.19) and (10.7.20) complete the solution of the problem. The fields follow from (4.6.14) to (4.6.19). The series representations of the fields converge readily for small values of u_a, u_b and u_r. However, the series is difficult to sum for large values of u_a, u_b and u_r. One such case, is when the spherical model represents the Earth-Ionosphere waveguide. For such cases, the series can be summed using the Watson transformation. In addition to the Watson transformation, summations techniques, known as speed-up numerical convergence algorithms, have been developed that work well for cases where u_a, u_b, and u_r have large values.

Next, the application of the Watson transformation to sum the series associated with the U and V potentials is discussed. The U and V potentials in (10.7.19) and (10.7.20) can be expressed in terms of Hankel functions as

$$U = \frac{(I\Delta L)\eta_o}{8\pi u_b r}\cos\phi \sum_{n=0}^{\infty} \frac{(2n+1)}{n(n+1)} P_n^1(\cos\theta) \times \begin{cases} \hat{h}_n^{(1)}(u_r)\hat{h}_n^{(2)\prime}(u_b) + [d_n\hat{h}_n^{(1)}(u_r) + e_n\hat{h}_n^{(2)}(u_r)] & (a \le r \le b) \\ \hat{h}_n^{(1)\prime}(u_b)\hat{h}_n^{(2)}(u_r) + [d_n\hat{h}_n^{(1)}(u_r) + e_n\hat{h}_n^{(2)}(u_r)] & (b \le r \le c) \end{cases} \tag{10.7.31}$$

and

$$V = j\frac{(I\Delta L)}{8\pi u_b r}\sin\phi \sum_{n=0}^{\infty} \frac{(2n+1)}{n(n+1)} P_n^1(\cos\theta) \times \begin{cases} \hat{h}_n^{(1)}(u_r)\hat{h}_n^{(2)}(u_b) + [f_n\hat{h}_n^{(1)}(u_r) + g_n\hat{h}_n^{(2)}(u_r)] & (a \le r \le b) \\ \hat{h}_n^{(1)}(u_b)\hat{h}_n^{(2)}(u_r) + [f_n\hat{h}_n^{(1)}(u_r) + g_n\hat{h}_n^{(2)}(u_r)] & (b \le r \le c) \end{cases} \tag{10.7.32}$$

Of course the d_n, e_n, f_n and g_n constants in (10.7.31) and (10.7.32) are different from those in (10.7.19) and (10.7.20). Applying the boundary conditions in (10.7.21) and (10.7.22) to the above Debye potential U in (10.7.31) gives

$$\hat{h}_n^{(1)\prime}(u_a)\hat{h}_n^{(2)\prime}(u_b) + [d_n\hat{h}_n^{(1)\prime}(u_a) + e_n\hat{h}_n^{(2)\prime}(u_a)] = j\frac{Z_a}{\eta_o}\left\{\hat{h}_n^{(1)}(u_a)\hat{h}_n^{(2)\prime}(u_b) + [d_n\hat{h}_n^{(1)}(u_a) + e_n\hat{h}_n^{(2)}(u_a)]\right\}$$

and

$$\hat{h}_n^{(1)\prime}(u_b)\hat{h}_n^{(2)\prime}(u_c) + [d_n\hat{h}_n^{(1)\prime}(u_c) + e_n\hat{h}_n^{(2)\prime}(u_c)] = -j\frac{Z_c}{\eta_o}\left\{\hat{h}_n^{(1)\prime}(u_b)\hat{h}_n^{(2)}(u_c) + [d_n\hat{h}_n^{(1)}(u_c) + e_n\hat{h}_n^{(2)}(u_c)]\right\}$$

which can be expressed in the form:

$$d_n X_1 + e_n X_2 = -h_n^{(2)'}(u_b) X_1$$

and

$$d_n Y_1 + e_n Y_2 = -h_n^{(1)'}(u_b) Y_2$$

where for the present analysis:

$$\begin{aligned} X_1 &= h_n^{(1)'}(u_a) - jz_a h_n^{(1)}(u_a) \\ X_2 &= h_n^{(2)'}(u_a) - jz_a h_n^{(2)}(u_a) \\ Y_1 &= h_n^{(1)'}(u_c) + jz_c h_n^{(1)}(u_c) \\ Y_2 &= h_n^{(2)'}(u_c) + jz_c h_n^{(2)}(u_c) \end{aligned}$$

Following a similar procedure as that in (10.5.25) to (10.5.32) shows that U for $a \le r \le b$ can be expressed in the form

$$U = \frac{(I\Delta L)\eta_o}{8\pi u_b r} \cos\phi \sum_{n=0}^{\infty} \frac{(2n+1)}{n(n+1)} P_n^1(\cos\theta) \hat{h}_n^{(1)}(u_r) \hat{h}_n^{(2)'}(u_b) \times \left\{ \frac{\left[1 + \frac{\hat{h}_n^{(2)}(u_r)}{\hat{h}_n^{(1)}(u_r)} R_a^s\right]\left[1 + \frac{\hat{h}_n^{(1)'}(u_b)}{\hat{h}_n^{(2)'}(u_b)} R_c^s\right]}{1 - R_a^s R_c^s} \right\}$$

where R_a^s and R_c^s are given by (10.5.30) and (10.5.31).

The radial electric field follows from (10.7.16). That is,

$$E_r = \frac{(I\Delta L)\eta_o}{8\pi u_b r^2} \cos\phi \sum_{n=0}^{\infty} (2n+1) P_n^1(\cos\theta) \hat{h}_n^{(1)}(u_r) \hat{h}_n^{(2)'}(u_b) \times \left\{ \frac{\left[1 + \frac{\hat{h}_n^{(2)}(u_r)}{\hat{h}_n^{(1)}(u_r)} R_a^s\right]\left[1 + \frac{\hat{h}_n^{(1)'}(u_b)}{\hat{h}_n^{(2)'}(u_b)} R_c^s\right]}{1 - R_a^s R_c^s} \right\} \quad (10.7.33)$$

Applying the boundary conditions in (10.7.21) and (10.7.22) to the Debye potential V in (10.7.32) gives

$$\hat{h}_n^{(1)'}(u_a)\hat{h}_n^{(2)}(u_b) + [f_n\hat{h}_n^{(1)'}(u_a) + g_n\hat{h}_n^{(2)'}(u_a)] = j\frac{\eta_o}{\tilde{Z}_a}\left\{\hat{h}_n^{(1)}(u_a)\hat{h}_n^{(2)}(u_b) + [f_n\hat{h}_n^{(1)}(u_a) + g_n\hat{h}_n^{(2)}(u_a)]\right\}$$

and

$$\hat{h}_n^{(1)}(u_b)\hat{h}_n^{(2)'}(u_c) + [f_n\hat{h}_n^{(1)'}(u_c) + g_n\hat{h}_n^{(2)'}(u_c)] = -j\frac{\eta_o}{\tilde{Z}_c}\left\{\hat{h}_n^{(1)}(u_b)\hat{h}_n^{(2)}(u_c) + [f_n\hat{h}_n^{(1)}(u_c) + g_n\hat{h}_n^{(2)}(u_c)]\right\}$$

which can be expressed in the forms

$$d_n X_1 + e_n X_2 = -h_n^{(2)}(u_b)X_1$$

and

$$d_n Y_1 + e_n Y_2 = -h_n^{(1)}(u_b)Y_2$$

where for the present analysis:

$$X_1 = h_n^{(1)'}(u_a) - \frac{j}{\tilde{z}_a}h_n^{(1)}(u_a)$$

$$X_2 = h_n^{(2)'}(u_a) - \frac{j}{\tilde{z}_a}h_n^{(2)}(u_a)$$

$$Y_1 = h_n^{(1)'}(u_c) + \frac{j}{\tilde{z}_c}h_n^{(1)}(u_c)$$

$$Y_2 = h_n^{(2)'}(u_c) + \frac{j}{\tilde{z}_c}h_n^{(2)}(u_c)$$

Following a similar procedure as that in (10.5.25) to (10.5.32) shows that V for $a \le r \le b$ can be expressed in the form

$$V = j\frac{(I\Delta L)}{8\pi u_b r}\sin\phi\sum_{n=0}^{\infty}\frac{(2n+1)}{n(n+1)}P_n^1(\cos\theta)\hat{h}_n^{(1)}(u_r)\hat{h}_n^{(2)}(u_b)\times\left\{\frac{\left[1+\dfrac{\hat{h}_n^{(1)}(u_r)}{\hat{h}_n^{(2)}(u_r)}\tilde{R}_a^s\right]\left[1+\dfrac{\hat{h}_n^{(2)}(u_b)}{\hat{h}_n^{(1)}(u_b)}\tilde{R}_c^s\right]}{1-\tilde{R}_a^s\tilde{R}_c^s}\right\}$$

where $\tilde{R}_a^s$ and $\tilde{R}_c^s$ are given by (10.6.15) and (10.6.16).

The radial magnetic field follows from (10.7.7). That is,

$$H_r = j\frac{(I\Delta L)}{8\pi u_b r^2}\sin\phi\sum_{n=0}^{\infty}(2n+1)P_n^1(\cos\theta)\hat{h}_n^{(1)}(u_r)\hat{h}_n^{(2)}(u_b)$$

$$\times\left\{\frac{\left[1+\frac{\hat{h}_n^{(1)}(u_r)}{\hat{h}_n^{(2)}(u_r)}\tilde{R}_a^s\right]\left[1+\frac{\hat{h}_n^{(2)}(u_b)}{\hat{h}_n^{(1)}(u_b)}\tilde{R}_c^s\right]}{1-\tilde{R}_a^s\tilde{R}_c^s}\right\} \qquad (10.7.34)$$

The Watson transformation applied to (10.7.33) and (10.7.34) is used to obtain a fast-convergent series. Consider the radial electric field in (10.7.33). Letting

$$f(n) = \frac{\hat{h}_n^{(1)}(u_r)\hat{h}_n^{(2)'}(u_b)}{1-R_a^sR_c^s}\left[1+\frac{\hat{h}_n^{(2)}(u_r)}{\hat{h}_n^{(1)}(u_r)}R_a^s\right]\left[1+\frac{\hat{h}_n^{(1)'}(u_b)}{\hat{h}_n^{(2)'}(u_b)}R_c^s\right] \qquad (10.7.35)$$

(10.7.33) in terms of the Watson transformation is expressed as

$$E_r = -jK\oint_C\frac{\nu}{\cos\nu\pi}f\left(\nu-\frac{1}{2}\right)P_{\nu-1/2}^1(-\cos\theta)d\nu \qquad (10.7.36)$$

where

$$K = \frac{(I\Delta L)\eta_o}{8\pi u_b r^2}\cos\phi$$

and C is the contour enclosing the zeroes of $\cos(\nu\pi)$ along the real axis, as shown in Fig. 10.4a. To show that (10.7.36) and (10.7.33) are equal, add the residues and use:

$$P_n^1(-\cos\theta) = (-1)^{n+1}P_n^1(\cos\theta)$$

In (10.7.36), $\nu/\cos(\nu\pi)$ is odd and the Legendre function is even in ν so the contour C can be changed to the contour C' in Fig. 10.4b provided that $f(\nu-1/2)$ is an even function. This condition is satisfied if the ground and ionosphere impedances are approximated by planar reflection coefficients.

As was done for the VED in Section 10.4, the Fok-Airy approximations for the spherical Hankel functions are used. Using a change of variable in (10.7.36), namely

$$\nu = k_o a + \left(\frac{k_o a}{2}\right)^{1/3} t$$

and using the approximation

$$\frac{P^1_{\nu-1/2}(-\cos\theta)}{\cos(\nu\pi)} \approx -\sqrt{\frac{2\nu}{\pi\sin\theta}}\, e^{-j\nu\theta}e^{j\pi/4} \approx -\sqrt{\frac{2k_o a}{\pi\sin\theta}}\, e^{-jk_o a\theta}e^{-jxt}e^{j\pi/4}$$

where

$$x = \left(\frac{k_a a}{2}\right)^{1/3}\theta$$

it follows that (10.7.36) reads

$$E_r = -jK\sqrt{\frac{2k_o a}{\pi\sin\theta}}\, e^{-jk_o a\theta}e^{j\pi/4}(k_o a)\left(\frac{k_o a}{2}\right)^{1/3}\int_{C'} e^{-jxt}f(t)dt \qquad (10.7.37)$$

where $f(t)$ follows from (10.7.35). The expression for $f(t)$ in terms of the Fok-Airy functions is obtained as follows. Let

$$f(t) = -\frac{N_1(t)N_2(t)}{1 - R_a^s(t)R_c^s(t)}$$

where

$$N_1(\nu - 1/2) = h^{(1)}_{\nu-1/2}(k_o r) + h^{(2)}_{\nu-1/2}(k_o r)R_a^s \Rightarrow$$

$$N_1(t) = -j\left(\frac{k_o a}{2}\right)^{1/6}\left\{w_1(t - y_r) - w_2(t - y_r)\frac{w_1{}'(t) - qw_1(t)}{w_2{}'(t) - qw_2(t)}\right\}$$

$$N_2(\nu - 1/2) = h^{(2)\prime}_{\nu-1/2}(k_o b) + h^{(1)\prime}_{\nu-1/2}(k_o b)R_c^s \Rightarrow$$

$$N_2(t) = -j\left(\frac{k_o b}{2}\right)^{-1/6}\left\{w_2{}'(t - y_b) - w_1{}'(t - y_b)\frac{w_2{}'(t - y_c) + q_i w_2(t - y_c)}{w_1{}'(t - y_c) + q_i w_1(t - y_c)}\right\}$$

$$R_a^s = -\frac{h^{(1)\prime}_{\nu-1/2}(k_o a) - jz_a h^{(1)}_{\nu-1/2}(k_o a)}{h^{(2)\prime}_{\nu-1/2}(k_o a) - jz_a h^{(2)}_{\nu-1/2}(k_o a)} \Rightarrow R_a^s(t) \approx \frac{w_1{}'(t) - q_a w_1(t)}{w_2{}'(t) - q_a w_2(t)}$$

$$R_c^s = -\frac{h^{(2)\prime}_{\nu-1/2}(k_o c) + jz_c h^{(2)}_{\nu-1/2}(k_o c)}{h^{(1)\prime}_{\nu-1/2}(k_o c) + jz_c h^{(1)}_{\nu-1/2}(k_o c)} \Rightarrow R_c^s(t) \approx \frac{w_2{}'(t - y_c) + q_c w_2(t - y_c)}{w_1{}'(t - y_c) + q_c w_1(t - y_c)}$$

$$q_a = -j\left(\frac{k_o a}{2}\right)^{1/3} z_a$$

$$q_c = -j\left(\frac{k_o a}{2}\right)^{1/3} z_c$$

The evaluation of (10.7.37) is done by closing the contour in the lower-half plane and forming $-2\pi j$ times the sum of the residues, which are the poles of $f(t)$. That is, the poles are determined from:

$$1 - R_a^s(t)R_c^s(t) = 1 - \left[\frac{w'_1(t) - q_a w_1(t)}{w'_2(t) - q_a w_2(t)}\right]\left[\frac{w'_2(t - y_c) + q_c w_2(t - y_c)}{w'_1(t - y_c) + q_c w_1(t - y_c)}\right] = 0$$

(10.7.38)

This rather cumbersome relation determines the values of t, denoted by t_p, that determine the propagating modes. Therefore, the evaluation of (10.7.37) is

$$E_r = -jK\sqrt{\frac{2k_o a}{\pi \sin\theta}}\, e^{-jk_o a\theta} e^{j\pi/4}(k_o a)\left(\frac{k_o a}{2}\right)^{1/3} (-2\pi j) \sum_{t_p=1}^{\infty} \frac{N_1(t_p)N_2(t_p)}{-\frac{\partial}{\partial t}(R_a^s R_c^s)|_{t=t_p}} e^{-jxt_p}$$

(10.7.39)

In the denominator of (10.7.39):

$$\frac{\partial R_a^s}{\partial t} = \frac{[w_2'(t) - q_a w_2(t)][w_1''(t) - q_a w_1'(t)] - [w_1'(t) - q_a w_1(t)][w_2''(t) - q_a w_2'(t)]}{[w_2'(t) - q_a w_2(t)]^2}$$

$$= 2j\frac{t - q_a^2}{[w_2'(t) - q_a w_2(t)]^2}$$

where Airy equation and the Wronskian were used, namely

$$w_k''(t) = t w_k(t) \qquad (k = 1 \text{ or } 2)$$

and

$$W[w_1(t), w_2(t)] = 2j$$

Similarly,

$$\frac{\partial R_c^s}{\partial t} = -2j\frac{t - y_c - q_c^2}{[w_1'(t - y_c) + q_c w_1(t - y_c)]^2}$$

Hence, the denominator reads

$$- R_a^s \frac{\partial R_c^s}{\partial t} - R_c^s \frac{\partial R_a^s}{\partial t} = 2j \frac{w_1'(t) - q_a w_1(t)}{w_2'(t) - q_a w_2(t)} \frac{t - y_c - q_c^2}{[w_1'(t - y_c) + q_c w_1(t - y_c)]^2}$$
$$- 2j \frac{w_2'(t - y_c) + q_c w_2(t - y_c)}{w_1'(t - y_c) + q_c w_1(t - y_c)} \frac{t - q_a^2}{[w_2'(t) - q_a w_2(t)]^2} \tag{10.7.40}$$

From the expressions for $N_1(t)$, $N_2(t)$ and (10.7.40), let

$$G(t_p) = \frac{N_1(t_p) N_2(t_p)}{-\frac{\partial}{\partial t}(R_a^s R_c^s)|_{t=t_p}}$$
$$= \frac{\left(\frac{1}{2j}\right)\left\{w_2(t_p) - w_1(t_p)\frac{w_2'(t_p) - q_a w_2(t_p)}{w_1'(t_p) - q_a w_1(t_p)}\right\}\left\{w_1'(t_p) - w_2'(t_p)\frac{w_1'(t_p - y_c) + q_c w_1(t_p - y_c)}{w_2'(t_p - y_c) + q_c w_2(t_p - y_c)}\right\}}{\frac{t_p - q_a^2}{[w_1'(t_p) - q_a w_1(t_p)][w_2'(t_p) - q_a w_2(t_p)]} + \frac{t_p - y_c - q_c^2}{[w_1'(t_p - y_c) + q_c w_1(t_p - y_c)][w_2'(t_p - y_c) + q_c w_2(t_p - y_c)]}}$$

and (10.7.39) can be expressed in the form

$$E_r = \frac{(I\Delta L)\eta_o}{2r^2} \sqrt{\frac{1}{\pi \sin\theta}}\, e^{-jk_o a\theta} e^{j\pi/4} \left(\frac{k_o a}{2}\right)^{5/6} \cos\phi \sum_{t_p=1}^{\infty} e^{-jxt_p} G(t_p) \tag{10.7.41}$$

The derivation for H_r in (10.7.28) in terms of the Watson transformation is similar (left to the problems).

10.8 A VERTICAL ELECTRIC DIPOLE IN THE EARTH-IONOSPHERE WAVEGUIDE—MODAL SOLUTION

In this section the analysis of a VED in a spherical waveguide is presented using a modal solution. Specifically, the analysis is applied to the classical problem of a VED in the Earth-Ionosphere waveguide. The fields from a VED in a spherical waveguide were analyzed using a series of zonal harmonics in Section 10.5, whose solution can be used to analyze the fields in the Earth-Ionosphere waveguide. In the zonal-harmonic solution the fields were written in terms of spherical Bessel functions and Legendre functions of integer order. In the modal solution, the fields produced by the VED in the Earth-Ionosphere waveguide requires the determination of the eigenvalues ν of Legendre functions, which represent the wave numbers in the θ direction, that satisfy the impedance boundary conditions at the ground and ionosphere surfaces. The resulting fields are written in terms of spherical Bessel functions and Legendre functions of complex order. This method produces good results in the ELF to LF bands.

The spherical model of the Earth-Ionosphere waveguide is shown in Fig. 10.5. The dipole is located at $r' = b$, $\theta' = 0$ and the height of the waveguide is $h = c - a$. The TM fields produced by the VED modes satisfy (4.6.13). That is,

$$[\nabla^2 + k_o^2]U = -\frac{1}{j\omega\varepsilon_o}\frac{J_{e,r}}{r} = -\frac{1}{j\omega\varepsilon_o}\frac{(I\Delta L)\delta(\mathbf{r} - \mathbf{r}')}{r} \tag{10.8.1}$$

where

$$\delta(\mathbf{r} - \mathbf{r}') = \frac{\delta(r - b)\delta(\theta)}{2\pi r^2 \sin\theta}$$

The fields are given by (4.6.14) to (4.6.19). Since the fields are symmetrical in ϕ (i.e., $\partial/\partial\phi = 0$):

$$E_r = \left(\frac{\partial^2}{\partial r^2} + k_o^2\right)(rU) \tag{10.8.2}$$

$$E_\theta = \frac{1}{r}\frac{\partial^2(rU)}{\partial r \partial\theta} \tag{10.8.3}$$

$$H_\phi = -\frac{j\omega\varepsilon_o}{r}\frac{\partial(rU)}{\partial\theta} \tag{10.8.4}$$

The modes solution to the homogeneous equation in (10.8.1) is written in the form

$$U(r, \theta) = \sum_\nu [B_\nu h_\nu^{(1)}(k_o r) + C_\nu h_\nu^{(2)}(k_o r)]P_\nu(-\cos\theta)$$

or

$$U(r, \theta) = \sum_\nu B_\nu [h_\nu^{(1)}(k_o r) + d_\nu h_\nu^{(2)}(k_o r)]P_\nu(-\cos\theta) \tag{10.8.5}$$

where in (10.8.5) the constant B_ν is determined by the source excitation and the constant d_ν by the boundary conditions. The summation is over all the modes ν, where ν will not be an integer. The reason for selecting the solution $P_\nu(-\cos\theta)$ (instead of $P_\nu(\cos\theta)$) is that it is singular at $\theta = 0$ where the dipole is located, and well behaved at $\theta = \pi$. If the Earth's ground and Ionosphere are considered to be homogeneous and isotropic, the following surface impedances boundary conditions apply:

$$Z_a = -\frac{E_\theta}{H_\phi}\bigg|_{r=a} = \frac{\frac{1}{r}\frac{\partial^2(rU)}{\partial r\partial\theta}}{\frac{j\omega\varepsilon_o}{r}\frac{\partial(rU)}{\partial\theta}}\Bigg|_{r=a} = \frac{1}{j\omega\varepsilon_o a}\frac{1}{U}\frac{\partial(rU)}{\partial r}\bigg|_{r=a} \tag{10.8.6}$$

and

$$Z_c = \frac{E_\theta}{H_\phi}\bigg|_{r=c} = -\frac{1}{j\omega\varepsilon_o c}\frac{1}{U}\frac{\partial(rU)}{\partial r}\bigg|_{r=c} \tag{10.8.7}$$

where Z_a and Z_c are the surface impedances at the air-ground and air-ionosphere interfaces, respectively. The impedances are assumed to have the same constant values for all the modes. Also, in (10.8.6) and (10.8.7) we used the fact that the boundary conditions hold (i.e., are continuous) for all values of θ.

Defining normalized surface impedances as

$$z_a = \frac{Z_a}{\eta_o}$$

and

$$z_c = \frac{Z_c}{\eta_o}$$

(10.8.6) and (10.8.7) are written in the form

$$jz_a U|_{r=a} = \frac{1}{k_o a}\frac{\partial(rU)}{\partial r}\bigg|_{r=a} \tag{10.8.8}$$

and

$$-jz_c U|_{r=c} = \frac{1}{k_o c}\frac{\partial(rU)}{\partial r}\bigg|_{r=c} \tag{10.8.9}$$

Substituting (10.8.5) into (10.8.8) gives

$$jz_a[h_\nu^{(1)}(k_o a) + d_\nu h_\nu^{(2)}(k_o a)] = \frac{1}{k_o a}\left\{\frac{\partial[rh_\nu^{(1)}(k_o r)]}{\partial r}\bigg|_{r=a} + d_\nu\frac{\partial[rh_\nu^{(2)}(k_o r)]}{\partial r}\bigg|_{r=a}\right\}$$

Solving for d_ν the resulting expression can be written in the form

$$d_\nu = -\frac{\left[\frac{1}{u_a}\frac{\partial[uh_\nu^{(1)}(u)]}{\partial u}\bigg|_{u=u_a} - jz_a h_\nu^{(1)}(u_a)\right]}{\left[\frac{1}{u_a}\frac{\partial[uh_\nu^{(2)}(u)]}{\partial u}\bigg|_{u=u_a} - jz_a h_\nu^{(2)}(u_a)\right]} \tag{10.8.10}$$

where $u = k_o r$ and $u_a = k_o a$.

A second expression for d_ν is obtained by substituting (10.8.5) into (10.8.9). That is,

$$d_\nu = -\frac{\left[\frac{1}{u_c}\frac{\partial[uh_\nu^{(1)}(u)]}{\partial u}\bigg|_{u=u_c} + jz_c h_\nu^{(1)}(u_c)\right]}{\left[\frac{1}{u_c}\frac{\partial[uh_\nu^{(2)}(u)]}{\partial u}\bigg|_{u=u_c} + jz_c h_\nu^{(2)}(u_c)\right]} \tag{10.8.11}$$

where $u_c = k_o c$.

Equating (10.8.10) and (10.8.11) shows that

$$\frac{\left[\frac{1}{u_a}\frac{\partial[uh_\nu^{(1)}(u)]}{\partial u}\bigg|_{u=u_a} - jz_a h_\nu^{(1)}(u_a)\right]\left[\frac{1}{u_c}\frac{\partial[uh_\nu^{(2)}(u)]}{\partial u}\bigg|_{u=u_c} + jz_c h_\nu^{(2)}(u_c)\right]}{\left[\frac{1}{u_a}\frac{\partial[uh_\nu^{(2)}(u)]}{\partial u}\bigg|_{u=u_a} - jz_a h_\nu^{(2)}(u_a)\right]\left[\frac{1}{u_c}\frac{\partial[uh_\nu^{(1)}(u)]}{\partial u}\bigg|_{u=u_c} + jz_c h_\nu^{(1)}(u_c)\right]} = 1 \tag{10.8.12}$$

which is the modal (or characteristic) equation. That is, the equation whose solution gives the eigenvalues of ν, denoted by ν_o, ν_1, ν_3, ν_4, ... Therefore, (10.8.5) is specifically written as

$$U(r, \Theta) = \sum_{n=0}^{\infty} B_{\nu_n} R_{\nu_n}(u) P_{\nu_n}(-\cos\theta) \tag{10.8.13}$$

where the radial part of the Debye potential is

$$R_{\nu_n}(u) = h_{\nu_n}^{(1)}(u) + d_{\nu_n} h_{\nu_n}^{(2)}(u) \tag{10.8.14}$$

The reason for denoting the first mode with $n = 0$ is that in the case of a parallel waveguide the first TM mode starts $n = 0$.

In (10.8.13), B_{ν_n} depends on the dipole-source strength. It is called the mode excitation coefficient for the VED. To evaluate the excitation coefficient, multiply both sides of (10.8.13) by $R_{\nu_m}(u)$ and integrate between u_a and u_c. The radial functions are orthogonal if they satisfy the boundary conditions in (10.8.8) and (10.8.9) (see Appendix H). That is,

$$\int_{u_a}^{u_c} R_{\nu_n}(u) R_{\nu_m}(u)\, du = N(\nu_n)\delta_{nm}$$

where

$$N(\nu_n) = \int_{u_a}^{u_i} [R_{\nu_n}(u)]^2 du$$

Then, from (10.8.13), B_{ν_n} is given by

$$B_{\nu_n} = \frac{1}{N(\nu_n) P_{\nu_n}(-\cos\Theta)} \int_{u_a}^{u_c} U(r, \Theta)\, R_{\nu_n}(u) du \tag{10.8.15}$$

This expression is evaluated by letting $\theta \to 0$ and $r \to b$. Close to the dipole source, $U(r, \theta)$ is given by the particular solution:

$$U^p(r, \theta) = \frac{(I\Delta L)}{j\omega\varepsilon_o b} \frac{e^{-jk_o R}}{4\pi R}$$

where

$$R = |\mathbf{r} - \mathbf{r}'| = \sqrt{r^2 + b^2 - 2rb\cos\theta}$$

Also, close to the dipole $R_{\nu_n}(u) \approx R_{\nu_n}(k_o b) = R_{\nu_n}(u_b)$, where $u_b = k_o b$. Therefore, (10.8.15) with $du = k_o dr$ can be approximated by

$$B_{\nu_n} \approx \lim_{\Theta\to 0} \frac{k_o R_{\nu_n}(u_b)}{N(\nu_n)} \left[\frac{\int_{b(1-\varepsilon)}^{b(1+\varepsilon)} U^p(r, \Theta)\, dr}{P_{\nu_n}(-\cos\Theta)} \right] = \lim_{\Theta\to 0} \frac{k_o R_{\nu_n}(u_b)}{N(\nu_n)} \frac{(I\Delta L)}{j4\pi\omega\varepsilon_o b} \left[\frac{\int_{b(1-\varepsilon)}^{b(1+\varepsilon)} \frac{e^{-jk_o R}}{R} dr}{P_{\nu_n}(-\cos\Theta)} \right] \tag{10.8.16}$$

which shows that the excitation coefficient is determined by the source term, according to (10.8.16).

To evaluate the numerator of (10.8.16) let $r = b(1 + \eta)$ where η is a small number $\eta << 1$, such that $-\varepsilon < \eta < \varepsilon$, and $\theta \to 0$. Then, the following approximations can be made:

$$\begin{aligned} R &= b\sqrt{(1+\eta)^2 - 2(1+\eta)\cos\theta + 1} \\ &\approx b\sqrt{(1+\eta)^2 - 2(1+\eta)(1 - \theta^2/2) + 1} \\ &\approx b\sqrt{\eta^2 + (1+\eta)\theta^2} \\ &\approx b\sqrt{\eta^2 + \theta^2} \end{aligned}$$

Hence, the numerator in the integrand of (10.8.16) can be approximated by $e^{-jk_oR} \approx 1$ and the denominator by $R \approx b\sqrt{\eta^2 + \theta^2}$. Next, taking the limit as $\theta \to 0$, with a finite value for ε (such that $\theta \ll \varepsilon$) and since $dr = b\,d\eta$ one obtains

$$\lim_{\theta\to 0}\int_{b(1-\varepsilon)}^{b(1+\varepsilon)} \frac{e^{-jk_oR}}{R}dr \approx \lim_{\theta\to 0}\int_{-\varepsilon}^{\varepsilon} \frac{1}{\sqrt{\eta^2+\theta^2}}d\eta = \ln\left[\eta + \sqrt{\eta^2+\theta^2}\right]\Bigg|_{-\varepsilon}^{\varepsilon}$$

$$= \ln\frac{\varepsilon + \sqrt{\varepsilon^2+\theta^2}}{-\varepsilon + \sqrt{\varepsilon^2+\theta^2}}$$

$$\approx \ln\frac{\varepsilon + \varepsilon + \theta^2/2\varepsilon}{-\varepsilon + \varepsilon + \theta^2/2\varepsilon}$$

$$= \ln\frac{2\varepsilon + \theta^2/2\varepsilon}{\theta^2/2\varepsilon}$$

$$\approx \ln\left(\frac{4\varepsilon^2}{\theta^2}\right)$$

$$\approx -\ln\theta^2 \tag{10.8.17}$$

In the denominator of (10.8.16), the following approximation is used (see Appendix E):

$$\lim_{\theta\to 0} P_{\nu_n}(-\cos\theta) \approx \frac{\sin\nu_n\pi}{\pi}\ln\theta^2 \tag{10.8.18}$$

Substituting (10.8.17) and (10.8.18) into (10.8.16) gives

$$B_{\nu_n} = j\frac{\eta_o(I\Delta L)\,R_{\nu_n}(u_b)}{4bN(\nu_n)\sin\nu_n\pi} \tag{10.8.19}$$

where $\eta_o = k_o/\omega\varepsilon_o$.
Substituting (10.8.19) into (10.8.13) the Debye potential is

$$U(r,\theta) = j\frac{\eta_o(I\Delta L)}{4b}\sum_{n=0}^{\infty}\frac{R_{\nu_n}(u_b)}{N(\nu_n)\sin\nu_n\pi}R_{\nu_n}(u)P_{\nu_n}(-\cos\theta)$$

and the fields, from (10.8.2) to (10.8.4), are given by

$$E_r = j\frac{\eta_o(I\Delta L)}{4rb}\sum_{n=0}^{\infty}\frac{\nu_n(\nu_n+1)R_{\nu_{\nu n}}(u_b)}{N(\nu_n)\sin\nu_n\pi}R_{\nu_n}(u)P_{\nu_n}(-\cos\theta)$$

$$E_\theta = j\frac{\eta_o(I\Delta L)}{4br}\sum_{n=0}^{\infty}\frac{R_{\nu_{\nu n}}(u_b)}{N(\nu_n)\sin\nu_n\pi}\frac{\partial}{\partial u}\left[uR_{\nu_n}(u)\right]\frac{\partial P_{\nu_n}(-\cos\theta)}{\partial\theta} \tag{10.8.20}$$

$$H_\phi = \frac{k_o(I\Delta L)}{4b}\sum_{n=0}^{\infty}\frac{R_{\nu_{\nu_n}}(u_b)}{N(\nu_n)\sin\nu_n\pi}R_{\nu_n}(u)\frac{\partial P_{\nu_n}(-\cos\theta)}{\partial\theta}$$

The field E_r was obtained using (see Appendix C):

$$\left(\frac{\partial^2}{\partial r^2}+k_o^2\right)\left[rh_{\nu_n}^{(j)}(k_o r)\right] = \frac{\nu_n(\nu_n+1)}{r^2}\left[rh_{\nu_n}^{(j)}(k_o r)\right] = \frac{\nu_n(\nu_n+1)}{r}h_{\nu_n}^{(j)}(k_o r)\ (j=1,2) \tag{10.8.21}$$

If the VED and the observation point are located on the Earth's surface, the above expressions are evaluated with $r = b = a$. In the ELF band the fundamental propagating mode is the $n = 0$ mode. This mode is called the "quasi-TEM mode."

The solution of (10.8.12) for the eigenvalues of ν is done using numerical methods. However, for some cases a closed form solution is possible. One such case is the so called "thin-shell approximation," where $(b - a)/r < < 1$. This approximation produces good results in the ELF and VLF bands. Another case is the so called "Earth-flattening approximation," where the angle θ is replaced by $\rho = a\theta$ (i.e., the great-circle distance). This approximation produces good results when the distance ρ is small compared to the radius of the Earth.

10.9 A VERTICAL MAGNETIC DIPOLE IN THE EARTH-IONOSPHERE WAVEGUIDE—MODAL SOLUTION

The TE fields produced by the VMD satisfy (4.6.12). That is,

$$[\nabla^2 + k_o^2]V = -\frac{1}{j\omega\mu_o}\frac{J_{m,r}}{r} = -\frac{1}{j\omega\mu_o}\frac{(I_m\Delta L)\delta(\mathbf{r}-\mathbf{r}')}{r} \tag{10.9.1}$$

where

$$\delta(\mathbf{r}-\mathbf{r}') = \frac{\delta(r-b)\delta(\theta)}{2\pi r^2\sin\theta}$$

The solution to the homogeneous equation in (10.10.1), with no ϕ variation, is

$$V(r,\theta) = \sum_\nu F_\nu\left[h_\nu^{(1)}(k_o r) + g_\nu h_\nu^{(2)}(k_o r)\right]P_\nu(-\cos\theta) \tag{10.9.2}$$

where F_ν is the excitation factor and g_ν is a constant determined by the boundary conditions.

From (4.6.14) to (4.6.19), the fields are

$$H_r = \left(\frac{\partial^2}{\partial r^2}+k_o^2\right)(rV) \tag{10.9.3}$$

$$H_\theta = \frac{1}{r}\frac{\partial^2(rV)}{\partial r \partial \theta} \tag{10.9.4}$$

$$E_\phi = \frac{j\omega\mu_o}{r}\frac{\partial(rV)}{\partial \theta} \tag{10.9.5}$$

The surface impedances boundary conditions for the VMD are

$$\tilde{Z}_a = \left.\frac{E_\phi}{H_\theta}\right|_{r=a} = \left.\frac{\frac{j\omega\mu_o}{r}\frac{\partial(rV)}{\partial\theta}}{\frac{1}{r}\frac{\partial^2(rV)}{\partial r\partial\theta}}\right|_{r=a} = j\omega\mu_o a \left.\frac{V}{\frac{\partial(rV)}{\partial r}}\right|_{r=a} \tag{10.9.6}$$

and

$$\tilde{Z}_c = -\left.\frac{E_\phi}{H_\theta}\right|_{r=c} = -j\omega\mu_o c \left.\frac{V}{\frac{\partial(rV)}{\partial r}}\right|_{r=c} \tag{10.9.7}$$

As previously seen, in some cases the surface impedance of the ground for the TE case in many cases is the same as that of TM case.

In terms of the normalized surface impedances (10.9.6) and (10.9.7) read:

$$\left.\frac{jV}{z_a}\right|_{r=a} = \left.\frac{1}{k_o a}\frac{\partial(rV)}{\partial r}\right|_{r=a} \tag{10.9.8}$$

and

$$\left.\frac{jV}{z_c}\right|_{r=c} = -\left.\frac{1}{k_o c}\frac{\partial(rV)}{\partial r}\right|_{r=c} \tag{10.9.9}$$

Substituting (10.9.2) into (10.9.8) gives:

$$\frac{j}{z_a}\left[h_\nu^{(1)}(k_o a) + g_\nu h_\nu^{(2)}(k_o a)\right] = \frac{1}{k_o a}\left\{\left.\frac{\partial[rh_\nu^{(1)}(k_o r)]}{\partial r}\right|_{r=a} + g_\nu \left.\frac{\partial[rh_\nu^{(2)}(k_o r)]}{\partial r}\right|_{r=a}\right\}$$

Solving for g_ν we obtain:

$$g_\nu = -\frac{\left.\frac{1}{u_a}\frac{\partial[uh_\nu^{(1)}(u)]}{\partial u}\right|_{u=u_a} - \frac{j}{z_a}h_\nu^{(1)}(u_a)}{\left.\frac{1}{u_a}\frac{\partial[uh_\nu^{(2)}(u)]}{\partial u}\right|_{u=u_a} - \frac{j}{z_a}h_\nu^{(2)}(u_a)} \tag{10.9.10}$$

where $u = k_o r$ and $u_a = k_o a$.

A second expression for g_ν is obtained by substituting (10.9.2) into (10.9.9). That is,

$$g_\nu = -\frac{\left.\frac{1}{u_c}\frac{\partial[uh_\nu^{(1)}(u)]}{\partial u}\right|_{u=u_c} + \frac{j}{z_c}h_\nu^{(1)}(u_c)}{\left.\frac{1}{u_c}\frac{\partial[uh_\nu^{(2)}(u)]}{\partial u}\right|_{u=u_c} + \frac{j}{z_c}h_\nu^{(2)}(u_c)} \tag{10.9.11}$$

where $u_c = k_o c$.

Equating (10.9.10) and (10.9.11) gives

$$\frac{\left[\left.\frac{1}{u_a}\frac{\partial[uh_\nu^{(1)}(u)]}{\partial u}\right|_{u=u_a} - \frac{j}{z_a}h_\nu^{(1)}(u_a)\right]\left[\left.\frac{1}{u_c}\frac{\partial[uh_\nu^{(2)}(u)]}{\partial u}\right|_{u=u_c} + \frac{j}{z_c}h_\nu^{(2)}(u_c)\right]}{\left[\left.\frac{1}{u_a}\frac{\partial[uh_\nu^{(2)}(u)]}{\partial u}\right|_{u=u_a} - \frac{j}{z_a}h_\nu^{(2)}(u_a)\right]\left[\left.\frac{1}{u_c}\frac{\partial[uh_\nu^{(1)}(u)]}{\partial u}\right|_{u=u_c} + \frac{j}{z_c}h_\nu^{(1)}(u_c)\right]} = 1 \tag{10.9.12}$$

which is the modal equation for the TE modes. That is, the equation whose solution gives the values of the eigenvalues of $\tilde{\nu}$, denoted by $\tilde{\nu}_1$, $\tilde{\nu}_2$, $\tilde{\nu}_3 \ldots$ for the TE modes. Therefore, (10.9.2) is written as

$$V(r, \theta) = \sum_{n=1}^{\infty} F_{\tilde{\nu}_n} R_{\tilde{\nu}_n}(u) P_{\tilde{\nu}_n}(-\cos\theta) \tag{10.9.13}$$

where

$$R_{\tilde{\nu}_n}(u) = h_{\tilde{\nu}_n}^{(1)}(u) + g_{\tilde{\nu}_n} h_{\tilde{\nu}_n}^{(2)}(u)$$

The reason for denoting the first mode with $n = 1$ is that in the case of a parallel waveguide the first TE mode starts $n = 1$.

In (10.9.13), $F_{\tilde{\nu}_n}$ depends on the dipole source strength. It is called the mode excitation coefficient for the VMD. The evaluation of the excitation coefficient in (10.9.13) is similar to that in (10.8.15). Hence, $F_{\tilde{\nu}_n}$ is given by

$$F_{\tilde{\nu}_n} = \lim_{\theta \to 0} \frac{1}{N(\tilde{\nu}_n) P_{\tilde{\nu}_n}(-\cos\theta)} \int_{u_a}^{u_c} V(r, \theta) R_{\tilde{\nu}_n}(u) du \tag{10.9.14}$$

where

$$N(\tilde{\nu}_n) = \int_{u_a}^{u_c} [R_{\tilde{\nu}_n}(u)]^2 du$$

The particular solution for the magnetic dipole in free space is given by

$$V^p(r,\theta) = \frac{(I_m \Delta L)}{j\omega\mu_o b}\frac{e^{-jk_oR}}{4\pi R} = \frac{(I\Delta S)}{b}\frac{e^{-jk_oR}}{4\pi R}$$

where we used: $(I_m \Delta L) = j\omega\mu_o (I\Delta S)$.

Referring to the derivation of (10.8.19), it follows from (10.9.14) that

$$F_{\tilde{\nu}_n} = -\frac{k_o (I\Delta S) R_{\tilde{\nu}_n}(u_b)}{4bN(\tilde{\nu}_n)\sin \tilde{\nu}_n \pi} \tag{10.9.15}$$

From (10.9.13) and (10.9.15) the Debye potential is:

$$V(r,\theta) = -\frac{k_o (I\Delta S)}{4b}\sum_{n=1}^{\infty}\frac{R_{\tilde{\nu}_n}(u_b)}{N(\tilde{\nu}_n)\sin \tilde{\nu}_n \pi} R_{\tilde{\nu}_n}(u) P_{\tilde{\nu}_n}(-\cos\theta)$$

and the field components, from (10.9.3) to (10.9.5), are

$$H_r = -\frac{k_o (I\Delta S)}{4br}\sum_{n=1}^{\infty}\frac{\tilde{\nu}_n(\tilde{\nu}_n+1) R_{\tilde{\nu}_n}(u_b)}{N(\tilde{\nu}_n)\sin \tilde{\nu}_n \pi} R_{\tilde{\nu}_n}(u) P_{\tilde{\nu}_n}(-\cos\theta)$$

$$H_\theta = -\frac{k_o (I\Delta S)}{4br}\sum_{n=1}^{\infty}\frac{R_{\tilde{\nu}_n}(u_b)}{N(\tilde{\nu}_n)\sin \tilde{\nu}_n \pi}\frac{\partial}{\partial u}\left[u R_{\tilde{\nu}_n}(u)\right]\frac{\partial P_{\tilde{\nu}_n}(-\cos\theta)}{\partial\theta}$$

$$E_\phi = -\frac{j\eta_o k_o^2 (I\Delta S)}{4b}\sum_{n=1}^{\infty}\frac{R_{\tilde{\nu}_n}(u_b)}{N(\tilde{\nu}_n)\sin \tilde{\nu}_n \pi} R_{\tilde{\nu}_n}(u)\frac{\partial P_{\tilde{\nu}_n}(-\cos\theta)}{\partial\theta} \tag{10.9.16}$$

If the VMD and the observation point are located on the Earth's surface, the above expressions are evaluated with $r \approx a \approx b$.

10.10 A HORIZONTAL ELECTRIC DIPOLE IN THE EARTH-IONOSPHERE WAVEGUIDE—MODAL SOLUTION

The field produced by a horizontal electric dipole can be calculated using the reciprocity theorem. The reciprocity theorem finds application in circuit theory and in antenna theory. In circuit theory the reciprocity theorem basically states that in a linear circuit if a voltage in a branch (say at the input of the circuit) produces a current in another branch (say I_o at the output of the circuit), then the same voltage applied at the output will produce the current I_o at the input. In antenna theory, the theorem states that if a current in antenna 1 induces a voltage V_o in antenna 2, the same current in antenna 2 induces the same voltage V_o in antenna 1.

In terms of fields, in a linear region if a current source 1 (I_1) produces an electric field ($\mathbf{E}_2$) at the location of source 2, and current source 2 (I_2) produces an electric field ($\mathbf{E}_1$) at the location of source 1, then:

$$\int \mathbf{E}_1 \bullet I_2 \, d\mathbf{l}_2 = \int \mathbf{E}_2 \bullet I_1 \, d\mathbf{l}_1 \tag{10.10.1}$$

where dl_1 and dl_2 are the differential element of length of the current elements.

Since in field theory we can also have magnetic currents, the general form of (10.10.1) is

$$\iiint_V (\mathbf{E}_1 \bullet \mathbf{J}_{e,2} - \mathbf{H}_1 \bullet \mathbf{J}_{m,2}) dV = \iiint_V (\mathbf{E}_2 \bullet \mathbf{J}_{e,1} - \mathbf{H}_2 \bullet \mathbf{J}_{m,2}) dV$$

which can be written as

$$\int \mathbf{E}_1 \bullet I_2 d\mathbf{l}_2 - \mathbf{H}_1 \bullet I_{m,2} \, d\mathbf{l}_2 = \int \mathbf{E}_2 \bullet I_1 d\mathbf{l}_1 - \mathbf{H}_2 \bullet I_{m,1} \, d\mathbf{l}_1 \tag{10.10.2}$$

where I_1 and I_2 are the electric currents that produce the fields $\mathbf{E}_1$ and $\mathbf{E}_2$, respectively; and $I_{m,1}$ and $I_{m,2}$ are magnetic currents that produces $\mathbf{H}_1$ and $\mathbf{H}_2$, respectively.

In Fig. 10.10a a horizontal electric dipole (HED) along the x axis with dipole moment $I_1 \Delta L_1$ is located at $(x, y, z) = (0, 0, z_r)$. The HED produces the field $\mathbf{E}_1^{he}$ (*he* for horizontal electric). A VED along the z axis with dipole moment $I_2 \Delta L_2$ is located at $(x', y', z') = (0, 0, z'_s)$. The VED produces the field $\mathbf{E}_2^{ve}$ (*ve* for vertical electric). The origin of the rectangular coordinate systems for the HED and VED lie on the Earth's surface, with the $x, \; y$ and $x', \; y'$ planes being tangent to the surface. The great-circle distance is equal to the radius of the Earth times the angle between the origins of the VED and HED.

The VED produces an electric field along the x axis at the HED location given by $E_{2,x}^{ve}(0, 0, z_r)$, which induces a voltage in the HED (i.e., given by $E_{2,x}^{ve} \Delta L_1$). The notation $E_{2,x}^{ve}(0, 0, z_r)$ means the field from the vertical electric dipole in the x direction evaluated at the coordinates associated with $x = 0, y = 0, z = z_r$. Similarly, the HED produces an electric field along z' axis at the location of the VED given by $E_{1,z'}^{he}(0, 0, z'_s)$, which induces the voltage $E_{1,z'}^{he} \Delta L_2$ in the VED. Therefore, from (10.10.1):

$$E_{1,z'}^{he}(0, 0, z'_s)(I_2 \Delta L_2) = E_{2,x}^{ve}(0, 0, z_r)(I_1 \Delta L_1)$$

If the currents and lengths are equal (i.e., with $(I_1 \Delta L_1) = (I_2 \Delta L_2)$), the relation reads:

$$E_{1,z'}^{he}(0, 0, z'_s) = E_{2,x}^{ve}(0, 0, z_r) \tag{10.10.3}$$

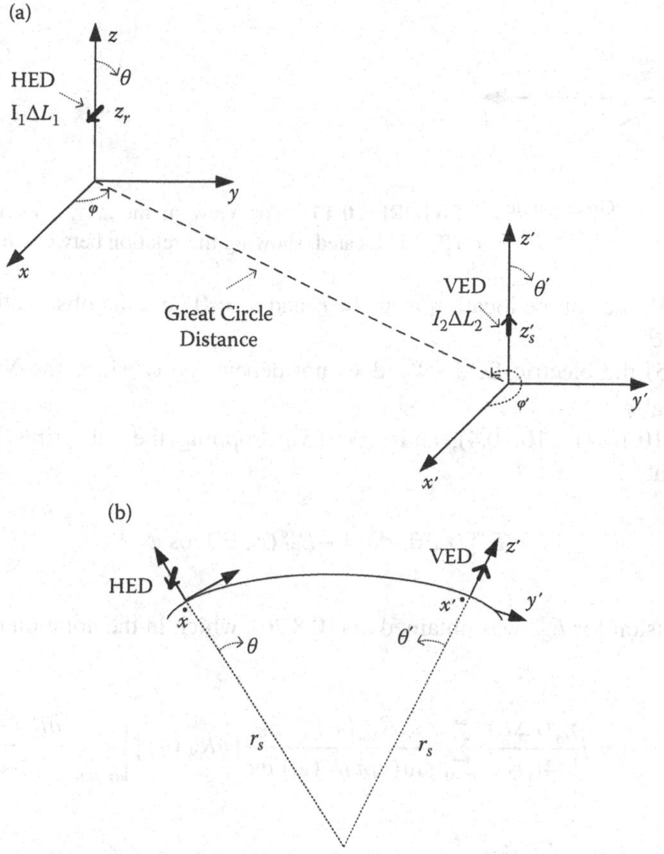

FIGURE 10.10 (a) Model for the determination of the HED fields using the reciprocity theorem; (b) spherical view.

Next, (10.10.3) is expressed in spherical coordinates. For the left-hand side, $E^{he}_{1,z'}$ can be expressed in terms of the radial component due to the HED. Hence, at the location of the VED (i.e., at (r_s, θ, ϕ)) it follows that

$$E^{he}_{1,z'}(0, 0, z'_s) = E^{he}_{1,r}(r_s, \theta, \phi) \tag{10.10.4}$$

In (10.10.4) the source location (i.e., the HED) is at $r = r_r$, $\theta = 0$ and $\phi = 0$, and the observation point at $r = r_s$, θ and ϕ.

For the right-hand side of (10.10.3), as illustrated in Fig. 10.11, the component along the x axis produced by the VED is related to its θ' component by:

$$E^{ve}_{2,x}(0, 0, z_r) = -E^{ve}_{2,\theta'}(r_r, \theta')\cos\phi \tag{10.10.5}$$

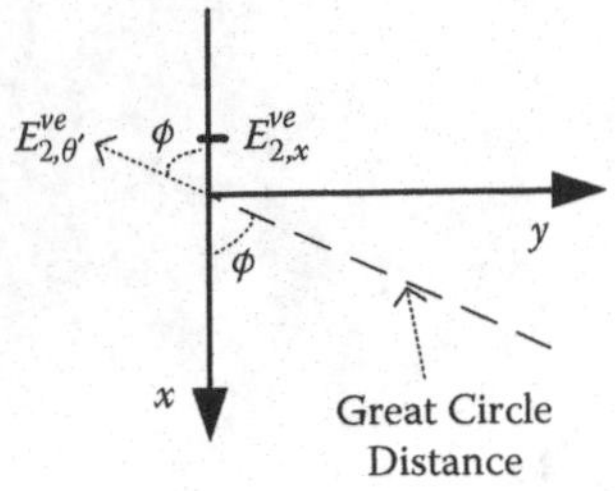

FIGURE 10.11 Top view of the x, y, z plane where the HED is located, showing the relation between $E_{2,\theta'}^{ve}$ and $E_{2,x}^{ve}$.

In (10.10.5), the source location is at $r = r_s$ and $\theta' = 0$, and the observation point at $r = r_r$ and θ'

In (10.10.5) the electric field $E_{2,\theta'}^{ve}$ does not depend on ϕ' since the VED field is symmetrical ϕ'.

From (10.10.3), (10.10.4), and (10.10.5), dropping the subscripts 1 and 2, it follows that

$$E_r^{he}(r_s, \theta, \phi) = -E_{\theta'}^{ve}(r_r, \theta')\cos\phi \tag{10.10.6}$$

The expression for $E_{\theta'}^{ve}$ was obtained in (10.8.20), which in the notation of (10.10.6) reads

$$E_{\theta'}^{ve}(r_r, \theta') = j\frac{\eta_o(I\Delta L)}{4r_s r_r}\sum_{n=0}^{\infty}\frac{R_{v_{vn}}(u_s)}{\sin(v_n\pi)N(v_n)}\frac{\partial}{\partial u}\left[uR_{v_n}(u)\right]\Big|_{u=k_o r_r}\frac{\partial P_{v_n}(-\cos\theta')}{\partial\theta'} \tag{10.10.7}$$

Substituting (10.10.7) into (10.10.6) gives:

$$E_r^{he}(r_s, \theta, \phi) = -j\frac{\eta_o(I\Delta L)}{4r_s r_r}\cos\phi\sum_{n=1}^{\infty}\frac{R_{v_v}(u_s)}{\sin(v_n\pi)N(v_n)}\frac{\partial}{\partial u}\left[uR_{v_n}(u)\right]\Big|_{u=k_o r_r}\frac{\partial P_{v_n}(-\cos\theta')}{\partial\theta'} \tag{10.10.8}$$

which applies to a HED at $(r_r, 0, 0)$ and the field point at (r_s, θ, ϕ). If the HED is located at $(r_s, 0, 0)$ and the observation point at an arbitrary location (r, θ, ϕ) then replacing r_s by r and r_r by r_s in (10.10.8), and using the fact that the θ derivatives change sign when the source and observation point are interchanged (i.e., $\partial/\partial\theta' = -\partial/\partial\theta$) gives

$$E_r^{he}(r, \theta, \phi) = j\frac{\eta_o(I\Delta L)}{4r\, r_s}\cos\phi\sum_{n=1}^{\infty}\frac{R_{v_n}(u)}{\sin(v_n\pi)N(v_n)}\frac{\partial}{\partial u}\left[uR_{v_n}(u)\right]\Big|_{u=k_o r_s} \times\frac{\partial P_{v_n}(-\cos\theta)}{\partial\theta} \tag{10.10.9}$$

The field expression in (10.10.9) can be used to derive the form of the Debye potential U for a HED dipole. Since,

$$E_r^{he} = \left(\frac{\partial^2}{\partial r^2} + k_o^2\right)(rU) \tag{10.10.10}$$

the form of the Debye potential should be such that when substituted in (10.10.10) results in (10.10.9). Hence, letting

$$U = \cos\phi \sum_{n=0}^{\infty} C_{\nu_n} R_{\nu_n}(u) \frac{\partial P_{\nu_n}(-\cos\theta)}{\partial\theta} \tag{10.10.11}$$

where C_{ν_n} is to be determined. Substituting (10.10.11) into (10.10.10), and using (10.8.21), gives

$$E_r^{he} = \frac{\cos\phi}{r} \sum_{n=0}^{\infty} C_{\nu_n} \nu_n(\nu_n + 1) R_{\nu_n}(u) \frac{\partial}{\partial\theta} P_{\nu_n}(-\cos\theta) \tag{10.10.12}$$

Comparing (10.10.9) with (10.10.12) shows that

$$C_{\nu_n} = \frac{j\eta_o(I\Delta L)}{4r_s} \frac{1}{\nu_n(\nu_n + 1)\sin(\nu_n\pi)N(\nu_n)} \frac{\partial}{\partial u}\left[uR_{\nu_n}(u)\right]\Big|_{u=k_o r_s}$$

and the Debye potential in (10.10.11) is then

$$U = \frac{j\eta_o(I\Delta L)}{4r_s} \cos\phi \sum_{n=0}^{\infty} \frac{R_{\nu_n}(u)}{\nu_n(\nu_n + 1)\sin(\nu_n\pi)N(\nu_n)} \frac{\partial}{\partial u}\left[uR_{\nu_n}(u)\right]\Big|_{u=k_o r_s} \frac{\partial}{\partial\theta} P_{\nu_n}(-\cos\theta) \tag{10.10.13}$$

To calculate the other field components, the form of the Debye potential V is needed. To calculate this potential the form of H_r^{he} is needed. The H_r^{he} component can be calculated using the reciprocity between a HED and a VMD. Consider Fig. 10.10 where a VMD, with magnetic dipole moment $I_{m2}\Delta L_2$, is placed at the location of the VED.

The magnetic field produced by the HED at the location of the VMD is $H_{1,z'}^{he}(0, 0, z'_s)$, and the field produced by the VMD at the HED location is $E_{2,x}^{vm}(0, 0, z_r)$ (vm refers to vertical magnetic). Therefore, from (10.10.2):

$$-H_{1,z'}^{he}(0, 0, z'_s)(I_{m2}\Delta L_2) = E_{2,x}^{vm}(0, 0, z_r)(I_1\Delta L_1) \tag{10.10.14}$$

Next, (10.10.14) is expressed in spherical coordinates. In the left-hand side:

$$H_{1,z'}^{he}(0, 0, z'_s) = H_{1,r}^{he}(r_s, \theta, \phi) \tag{10.10.15}$$

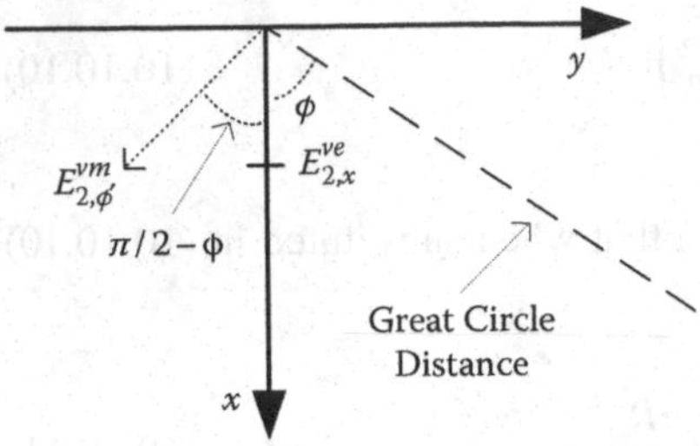

FIGURE 10.12 Top view of the x, y, z plane where the HED is located, showing the relation between $E_{2,\phi'}^{vm}$ and $E_{2,x}^{vm}$.

In (10.10.15) the source location is at $r = r_r$, $\theta = 0$ and $\phi = 0$, and the observation point at $r = r_s$, θ and ϕ.

For the right-hand side of (10.10.14), as shown in Fig. 10.12, $E_{2,x}^{vm}$ can be expressed in terms of the $E_{2,\phi'}^{vm}$ component of the VMD. That is,

$$E_{2,x}^{vm}(0, 0, z_r) = E_{2,\phi'}^{vm}(r_r, \theta')\cos\left(\frac{\pi}{2} - \phi\right) = E_{2,\phi'}^{vm}(r_r, \theta')\sin\phi \quad (10.10.16)$$

A magnetic dipole and a small electric current loop are related by $(I_{m2}\Delta L_2) = j\omega\mu_o(I_2\Delta S_2)$. Then, from (10.10.14), (10.10.15) and (10.10.16), dropping the subscripts 1 and 2 in the field terms, one obtains

$$H_r^{he}(r_s, \theta, \phi) = \frac{-1}{j\omega\mu_o}\left(\frac{I_1\Delta L_1}{I_2\Delta S_2}\right)E_{\phi'}^{vm}(r_r, \theta')\sin\phi \quad (10.10.17)$$

The field $E_{\phi'}^{vm}$ was found in (10.9.16), which in the notation of (10.10.7) reads:

$$E_{\phi'}^{vm}(r_r, \theta') = -\frac{j\eta_o k_o^2(I_2\Delta S_2)}{4r_s}\sum_{n=1}^{\infty}\frac{R_{\tilde{\nu}_n}(u_s)}{\sin(\tilde{\nu}_n\pi)N(\tilde{\nu}_n)}R_{\tilde{\nu}_n}(u_r)\frac{\partial P_{\tilde{\nu}_n}(-\cos\theta')}{\partial\theta'} \quad (10.10.18)$$

where $u_s = k_o r_s$ and $u_r = k_o r_r$.

Substituting (10.10.18) into (10.10.17) gives:

$$H_r^{he}(r_s, \theta, \phi) = \frac{(I_1\Delta L_1)k_o}{4r_s}\sin\phi\sum_{n=1}^{\infty}\frac{R_{\tilde{\nu}_n}(u_s)}{\sin(\tilde{\nu}_n\pi)N(\tilde{\nu}_n)}R_{\tilde{\nu}_n}(u_r)\frac{\partial P_{\tilde{\nu}_n}(-\cos\theta')}{\partial\theta'} \quad (10.10.19)$$

which applies to a HED at $(r_r, 0, 0)$ and the field point at (r_s, θ, ϕ). If the HED is located at $(r_s, 0, 0)$ and the field point at (r, θ, ϕ) then replacing r_s by r and r_r by r_s in (10.10.19), and using the fact that $\partial/\partial\theta' = -\partial/\partial\theta$ gives

$$H_r^{he}(r, \theta, \phi) = \frac{-(I\Delta L)k_o}{4r}\sin\phi\sum_{n=1}^{\infty}\frac{R_{\tilde{\nu}_n}(u_s)}{\sin(\tilde{\nu}_n\pi)N(\tilde{\nu}_n)}R_{\tilde{\nu}_n}(u)\frac{\partial P_{\tilde{\nu}_n}(-\cos\theta)}{\partial\theta} \quad (10.10.20)$$

where $u = k_o r$, and the dipole moment is written as $I\Delta L$.

Expression (10.10.20) is now used to derive the form of the Debye potential V for a HED dipole. Since,

$$H_r^{he} = \left(\frac{\partial^2}{\partial r^2} + k_o^2\right)(rV) \tag{10.10.21}$$

the form of the Debye potential, according to (10.10.20) and (10.10.21), is

$$V = \sin\phi \sum_{n=1}^{\infty} B_{\tilde{\nu}_n} R_{\tilde{\nu}_n}(u) \frac{\partial}{\partial\theta} P_{\tilde{\nu}_n}(-\cos\theta) \tag{10.10.22}$$

where $B_{\tilde{\nu}_n}$ is to be determined.

Substituting (10.10.22) into (10.10.21) gives

$$H_r^{he} = \frac{\sin\phi}{r} \sum_{n=1}^{\infty} B_{\tilde{\nu}_n} \tilde{\nu}_n(\tilde{\nu}_n + 1) R_{\tilde{\nu}_n}(u) \frac{\partial}{\partial\theta} P_{\tilde{\nu}_n}(-\cos\theta) \tag{10.10.23}$$

Comparing (10.10.20) with (10.10.23) shows that

$$B_{\tilde{\nu}_n} = -\frac{(I\Delta L)k_o}{4} \frac{R_{\tilde{\nu}_n}(u_s)}{\tilde{\nu}_n(\tilde{\nu}_n + 1)\sin(\tilde{\nu}_n\pi) N(\tilde{\nu}_n)}$$

The Debye potential in (10.10.22) is then

$$V = -\frac{k_o(I\Delta L)}{4} \sin\phi \sum_{n=1}^{\infty} \frac{R_{\tilde{\nu}_n}(u_s)}{\tilde{\nu}_n(\tilde{\nu}_n + 1)\sin(\tilde{\nu}_n\pi) N(\tilde{\nu}_n)} R_{\tilde{\nu}_n}(u) \frac{\partial}{\partial\theta} P_{\tilde{\nu}_n}(-\cos\theta) \tag{10.10.24}$$

With the Debye potentials in (10.10.13) and (10.10.24) the remaining field components are calculated from (4.6.14) to (4.6.19). For example,

$$E_\theta^{he} = \frac{1}{r}\frac{\partial^2(rU)}{\partial r \partial\theta} - \frac{j\omega\mu_o}{\sin\theta}\frac{\partial V}{\partial\phi}$$

and using (10.3.13) and (10.3.24) shows that

$$\begin{aligned} E_\theta^{he} = {} & \frac{j\eta_o(I\Delta L)}{4rr_s} \cos\phi \sum_{n=0}^{\infty} \frac{1}{\nu_n(\nu_n + 1)\sin(\nu_n\pi) N(\nu_n)} \frac{\partial}{\partial u}[uR_{\nu_n}(u)]\bigg|_{u=u_s} \times \frac{\partial}{\partial u} \\ & [uR_{\nu_n}(u)] \frac{\partial^2}{\partial\theta^2} P_{\nu_n}(-\cos\theta) \\ & + \frac{j\omega\mu_o k_o(I\Delta L)}{4\sin\theta} \cos\phi \sum_{n=1}^{\infty} \frac{R_{\tilde{\nu}_n}(u_s)}{\tilde{\nu}_n(\tilde{\nu}_n + 1)\sin(\tilde{\nu}_n\pi) N(\tilde{\nu}_n)} R_{\tilde{\nu}_n}(u) \frac{\partial}{\partial\theta} P_{\tilde{\nu}_n}(-\cos\theta) \end{aligned} \tag{10.10.25}$$

Similar manipulations produce the expressions of E_ϕ^{he}, H_θ^{he}, and H_ϕ^{he}.

Problems

P10.1 Verify (10.1.32) and (10.1.33).

P10.2 The conducting sphere in Fig. 10.1 is covered with a dielectric layer of radius $r = b$, as shown in Fig. P10.2. Determine the diffracted electric fields.

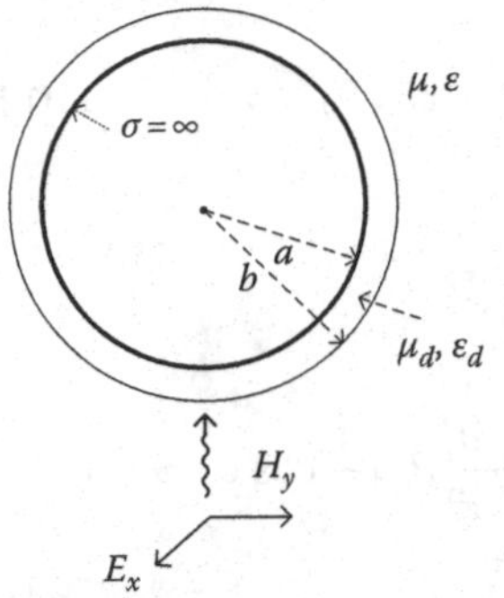

FIGURE P10.2

P10.3 Figure 10.5 shows that the region $r > a$ is lossless. If the region is lossy with parameters σ, μ, and ε, instead of (10.4.1) the equation written in terms of the complex propagation constant γ is

$$(\nabla^2 - \gamma^2)U = -\frac{(I\Delta L)}{(\sigma + j\omega\varepsilon)r}\,\frac{\delta(r-b)\delta(\theta)\delta(\phi)}{2\pi r^2 \sin\theta}$$

a. Determine the particular solution for the Debye potential U, using the Green function in the form

$$\frac{e^{-\gamma R}}{4\pi R} = \frac{1}{4\pi\gamma b r}\sum_{n=0}^{\infty}(2n+1)P_n(\cos\theta)\hat{k}_n(\gamma b)\hat{i}_n(\gamma r)\ (r \le b)$$

where $\hat{k}_n(z)$ and $\hat{i}_n(z)$ are the modified spherical Schelkunoff-Bessel functions.

b. Show that the solution in terms of the surface impedance is given by

$$U = \frac{(I\Delta L)}{4\pi(\sigma + j\omega\varepsilon)\gamma r b^2} \times \sum_{n=0}^{\infty}(2n+1)P_n(\cos\theta)[\hat{k}_n(\gamma b)\hat{i}_n(\gamma r) + a_n\hat{k}_n(\gamma r)]\ (r < b)$$

and determine a_n.

P10.4 Show that (10.4.32) can be expressed in the form

$$V_o = -2(\pi x)^{1/2} e^{-j\pi/4} \sum_{p=1}^{\infty} \frac{e^{-jxt}}{1 - \frac{t_p}{q^2}} \frac{w_2(t_p - y_b)}{w_2'(t_p)} \frac{w_2(t_p - y_r)}{w_2'(t_p)}$$

P10.5 Express (10.3.16) in terms of the Fok-Airy functions.

P10.6 The reflection coefficient R_a in (10.4.15) reads

$$R_a = -\left[\frac{ln'\hat{h}_\nu^{(1)}(k_o a) - jz_a}{ln'\hat{h}_\nu^{(2)}(k_o a) - jz_a} \right]$$

Using the Debye approximations for $\hat{h}_\nu^{(t)'}(k_o a)/\hat{h}_\nu^{(t)}(k_o a)$ ($t = 1$ or 2), show that R_a reduces to a planar reflection coefficient when $\nu/k_o a$ is identified as the sine of the angle of incidence.

P10.7 In (10.4.30), the expression for E_r was derived. Determine H_ϕ.

P10.8 Verify (10.6.14).

P10.9 In (10.7.41), the expression for E_r was obtained. Starting with (10.7.34), apply the Watson transformation and obtain the expression for II_r.

P10.10

a. Fig. P10.10 shows a small aperture on a conducting sphere at θ_o. Assume that the electric field in the circular slot is only due to E_θ and is independent of ϕ. Show that the resulting fields can be derived in term of A_r.
b. Determine the radiated E_θ field if the field at the aperture is given by

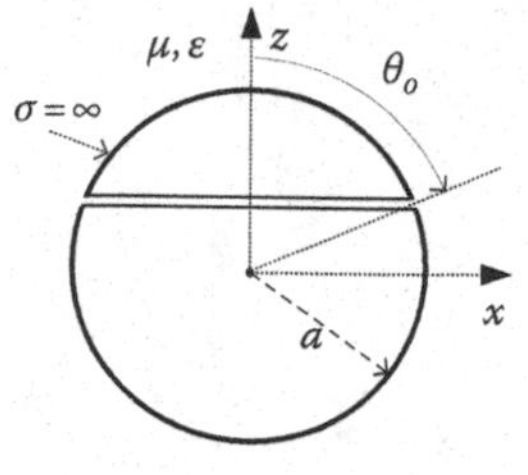

FIGURE P10.10

$$E_\theta|_{r=a} = \frac{V}{a}\delta(\theta - \theta_o)$$

c. Show the form of E_θ if $\theta_o = \pi/2$.

P10.11 In (10.10.25), the expression for E_θ^{he} was derived. Determine the expression for H_θ^{he}.

P10.12 If in Fig. 10.10 the HED is replaced by a horizontal magnetic dipole (HMD), explain how the reciprocity theorem is used to obtain the fields due to the HMD.

Appendix A: Nomenclature and Units

System of International Units (SI units)

Basic SI units: Meter (m)
Kilogram (kg)
Second (s)
Ampere (A)

SI derived units with special names and symbol:

SYMBOL	NAME	UNITS
C	Capacitance in Farad	F (F = C/V = s/Ω)
$\mathcal{E}$	Energy in Joule	J (J = N·m)
F	Force in Newton	N (N = m·kg/s^2)
f	Frequency in Hertz	Hz (Hz = 1/s)
G	Conductance in Siemen	S (S = A/V)
L	Inductance in Henry	H (H = Wb/A = Ω·s)
q_e	Electric charge in Coulomb	C (C = A·s)
R	Resistance in Ohm	Ω (Ω = V/A)
V	Electric potential in Volt	V (Wb/s = A·Ω)
W	Power in Watt	W (W = J/s)
ψ	Magnetic flux in Weber	Wb (Wb = V·s)

Other symbols and unit used in this book:

SYMBOL	NAME	UNITS
A	Magnetic vector potential	Wb/m
B	Magnetic field density vector	Wb/m^2
c	Speed of light	≈3 × 10^8 m/s
D	Displacement field density vector	C/m^2
E	Electric field intensity vector	V/m
F	Electric vector potential	C/m
H	Magnetic field intensity vector	A/m
I	Electric current	A

(*Continued*)

SYMBOL	NAME	UNITS
I_m	Magnetic current	V = Wb/s
$\mathbf{J}$	Electric current density	A/m^2
$\mathbf{J}_e$	Impressed electric current density	A/m^2
$\mathbf{J}_m$	Magnetic current density	V/m^2
$\mathbf{J}_{ms}$	Surface magnetic current density	V/m
$\mathbf{J}_M$	Magnetization current density	A/m^2
$\mathbf{J}_p$	Polarization current density	A/m^2
$\mathbf{J}_d$	Displacement current density	A/m^2
$\mathbf{J}_c$	Conduction current density	A/m^2
k	Complex propagation constant	1/m
k'	Propagation constant (in radians/m)	1/m
k''	Attenuation constant (in nepers/m)	1/m
k_o	Propagation constant in free space	1/m
$\mathbf{M}$	Magnetization vector	A/m
$\mathbf{M}_m$	Impressed magnetization vector	A/m
$\mathbf{m}$	Magnetic dipole moment	A·m^2
n	Index of refraction	no units
$\mathbf{P}$	Electric polarization vector	C/m^2
$\mathbf{P}$	Polarization vector	C/m^2 = A·s/m^2
$\mathbf{P}_e$	Impressed polarization vector	C/m^2 = A·s/m^2
$\mathbf{P}$	Poynting vector (time average)	W/m^2
$\mathbf{p}$	Electric dipole moment	C·m
q_e	Electric charge	C
q_m	Magnetic charge	Wb
T	Period	s
U	Electric Debye potential	V
V	Magnetic Debye potential	A
v_p	Phase velocity	m/s
Z_s	Surface impedance	Ω
z_s	Normalized surface impedance	no units
$\mathbf{\Pi}^e$	Electric Hertz's vector	V·m
$\mathbf{\Pi}^m$	Magnetic Hertz's vector	A·m
α	Attenuation constant (in nepers/m)	1/m
β_o	Propagation constant in free space	1/m
β	Propagation constant (in radians/m)	1/m
γ	Complex propagation constant	1/m
ε_o	Permittivity of free space	8.854×10^{-12} F/m
ε_c	Complex permittivity	F/m
ε_{cd}	Complex permittivity of dielectric	F/m
ε_r	Relative permittivity	no units
ε_{cr}	Complex relative permittivity	no unit
ε	Permittivity	F/m
η_o	Impedance of free space	377 Ω

SYMBOL	NAME	UNITS
η	Complex impedance	Ω
λ	Wavelength	m
μ_o	Permeability of free space	$4\pi \times 10^{-7}$ H/m
μ	Permeability	H/m
μ_r	Relative Permeability	no units
ρ_f	Electric free charge density	C/m^3
ρ_e	Electric impressed charge density	C/m^3
ρ	Electric charge density	C/m^3
ρ_m	Magnetic charge density	Wb/m^3
ρ_{ms}	Surface magnetic charge density	Wb/m^2
ρ_p	Polarization charge density	C/m^3
σ	Conductivity	S/m = 1/Ω m
σ_s	Static conductivity	S/m = 1/Ω m
σ_{eff}	Effective conductivity	S/m = 1/Ω m
φ	Scalar electric potential	V
φ_m	Scalar magnetic potential	A
χ_e	Electric susceptibility	no units
χ_m	Magnetic susceptibility	no units
ω	Frequency in radians/s	1/s

One should always check that the units of an equation match. The matching of the units does not mean that the equation is correct. However, it certainly tells you that if the units do not match, the equation is incorrect.

Appendix B: Vector and Other Identities

RECTANGULAR COORDINATES:

$$\nabla\psi = \frac{\partial\psi}{\partial x}\hat{\mathbf{a}}_x + \frac{\partial\psi}{\partial y}\hat{\mathbf{a}}_y + \frac{\partial\psi}{\partial z}\hat{\mathbf{a}}_z$$

$$\nabla^2\psi = \nabla\cdot\nabla\psi = \frac{\partial^2\psi}{\partial x^2} + \frac{\partial^2\psi}{\partial y^2} + \frac{\partial^2\psi}{\partial z^2}$$

$$\nabla\cdot\mathbf{A} = \frac{\partial A_x}{\partial x} + \frac{\partial A_y}{\partial y} + \frac{\partial A_z}{\partial z}$$

$$\nabla\times\mathbf{A} = \left(\frac{\partial A_z}{\partial y} - \frac{\partial A_y}{\partial z}\right)\hat{\mathbf{a}}_x + \left(\frac{\partial A_x}{\partial z} - \frac{\partial A_z}{\partial x}\right)\hat{\mathbf{a}}_y + \left(\frac{\partial A_y}{\partial x} - \frac{\partial A_x}{\partial y}\right)\hat{\mathbf{a}}_z$$

$$\nabla^2\mathbf{A} = \nabla^2 A_x\hat{\mathbf{a}}_x + \nabla^2 A_y\hat{\mathbf{a}}_y + \nabla^2 A_z\hat{\mathbf{a}}_z$$

CYLINDRICAL COORDINATES:

$$\nabla\psi = \frac{\partial\psi}{\partial\rho}\hat{\mathbf{a}}_\rho + \frac{1}{\rho}\frac{\partial\psi}{\partial\phi}\hat{\mathbf{a}}_\phi + \frac{\partial\psi}{\partial z}\hat{\mathbf{a}}_z$$

$$\nabla^2\psi = \nabla\cdot\nabla\psi = \frac{1}{\rho}\frac{\partial}{\partial\rho}\left(\rho\frac{\partial\psi}{\partial\rho}\right) + \frac{1}{\rho^2}\frac{\partial^2\psi}{\partial\phi^2} + \frac{\partial^2\psi}{\partial z^2}$$

$$\nabla\cdot\mathbf{A} = \frac{1}{\rho}\frac{\partial}{\partial\rho}(\rho A_\rho) + \frac{1}{\rho}\frac{\partial A_\phi}{\partial\phi} + \frac{\partial A_z}{\partial z}$$

$$\nabla\times\mathbf{A} = \left(\frac{1}{\rho}\frac{\partial A_z}{\partial\phi} - \frac{\partial A_\phi}{\partial z}\right)\hat{\mathbf{a}}_\rho + \left(\frac{\partial A_\rho}{\partial z} - \frac{\partial A_z}{\partial\rho}\right)\hat{\mathbf{a}}_\phi + \left[\frac{1}{\rho}\frac{\partial}{\partial\rho}(\rho A_\phi) - \frac{1}{\rho}\frac{\partial A_\rho}{\partial\phi}\right]\hat{\mathbf{a}}_z$$

$$\nabla^2\mathbf{A} = \left(\nabla^2 A_\rho - \frac{A_\rho}{\rho^2} - \frac{2}{\rho^2}\frac{\partial A_\phi}{\partial \phi}\right)\hat{\mathbf{a}}_\rho + \left(\nabla^2 A_\phi - \frac{A_\phi}{\rho^2} + \frac{2}{\rho^2}\frac{\partial A_\rho}{\partial \phi}\right)\hat{\mathbf{a}}_\phi + \nabla^2 A_z \hat{\mathbf{a}}_z$$

SPHERICAL COORDINATES:

$$\nabla\psi = \frac{\partial\psi}{\partial r}\hat{\mathbf{a}}_r + \frac{1}{r}\frac{\partial\psi}{\partial\theta}\hat{\mathbf{a}}_\theta + \frac{1}{r\sin\theta}\frac{\partial\psi}{\partial\phi}\hat{\mathbf{a}}_\phi$$

$$\nabla^2\psi = \nabla\cdot\nabla\psi = \frac{1}{r^2}\frac{\partial}{\partial r}\left(r^2\frac{\partial\psi}{\partial r}\right) + \frac{1}{r^2\sin\theta}\frac{\partial}{\partial\theta}\left(\sin\theta\frac{\partial\psi}{\partial\theta}\right) + \frac{1}{r^2\sin^2\theta}\frac{\partial^2\psi}{\partial\phi^2}$$

$$\nabla\cdot\mathbf{A} = \frac{1}{r^2}\frac{\partial}{\partial r}(r^2 A_r) + \frac{1}{r\sin\theta}\frac{\partial}{\partial\theta}(\sin\theta A_\theta) + \frac{1}{r\sin\theta}\frac{\partial A_\phi}{\partial\phi}$$

$$\nabla\times\mathbf{A} = \frac{1}{r\sin\theta}\left[\frac{\partial}{\partial\theta}(A_\phi\sin\theta) - \frac{\partial A_\theta}{\partial\phi}\right]\hat{\mathbf{a}}_r + \frac{1}{r}\left[\frac{1}{\sin\theta}\frac{\partial A_r}{\partial\phi} - \frac{\partial}{\partial r}(rA_\phi)\right]\hat{\mathbf{a}}_\theta + \frac{1}{r}\left[\frac{\partial}{\partial r}(rA_\theta) - \frac{\partial A_r}{\partial\theta}\right]\hat{\mathbf{a}}_\phi$$

$$\begin{aligned}\nabla^2\mathbf{A} = &\left(\nabla^2 A_r - \frac{2A_r}{r^2} - \frac{2\cot\theta}{r^2}A_\theta - \frac{2}{r^2}\frac{\partial A_\theta}{\partial\phi} - \frac{2}{r^2\sin\theta}\frac{\partial A_\phi}{\partial\phi}\right)\hat{\mathbf{a}}_r \\ &+ \left(\nabla^2 A_\theta + \frac{2}{r^2}\frac{\partial A_r}{\partial\theta} - \frac{1}{r^2\sin^2\theta}A_\theta - \frac{2\cos\theta}{r^2\sin^2\theta}\frac{\partial A_\phi}{\partial\phi}\right)\hat{\mathbf{a}}_\theta \\ &+ \left(\nabla^2 A_\phi + \frac{2}{r^2\sin\theta}\frac{\partial A_r}{\partial\phi} - \frac{1}{r^2\sin^2\theta}A_\phi + \frac{2\cos\theta}{r^2\sin^2\theta}\frac{\partial A_\theta}{\partial\phi}\right)\hat{\mathbf{a}}_\phi\end{aligned}$$

VECTOR IDENTITIES:

$$\nabla\cdot(\nabla\times\mathbf{A}) = 0$$

$$\nabla\times\nabla\psi = 0$$

$$\nabla\times\nabla\times\mathbf{A} = \nabla(\nabla\cdot\mathbf{A}) - \nabla^2\mathbf{A}$$

$$\nabla\cdot(\psi\mathbf{A}) = \psi\,\nabla\cdot\mathbf{A} + \mathbf{A}\cdot\nabla\psi$$
$$\nabla\times(\psi\mathbf{A}) = \psi\,\nabla\times\mathbf{A} + \nabla\psi\times\mathbf{A}$$
$$\nabla\cdot(\mathbf{A}\times\mathbf{B}) = \mathbf{B}\cdot\nabla\times\mathbf{A} - \mathbf{A}\cdot\nabla\times\mathbf{B}$$
$$\mathbf{A}\cdot\mathbf{B}\times\mathbf{C} = \mathbf{B}\cdot\mathbf{C}\times\mathbf{A} = \mathbf{C}\cdot\mathbf{A}\times\mathbf{B}$$
$$\mathbf{A}\times\mathbf{B}\times\mathbf{C} = (\mathbf{A}\cdot\mathbf{C})\mathbf{B} - (\mathbf{A}\cdot\mathbf{B})\mathbf{C}$$

COORDINATES TRANSFORMATIONS:

$$x = \rho\cos\phi = r\sin\theta\cos\phi$$
$$y = \rho\sin\phi = r\sin\theta\sin\phi$$
$$z = r\cos\theta$$
$$\rho = r\sin\theta$$
$$\rho = \sqrt{x^2 + y^2}$$
$$r = \sqrt{x^2 + y^2 + z^2} = \sqrt{\rho^2 + z^2}$$
$$\phi = \tan^{-1}\frac{y}{x}$$
$$\theta = \tan^{-1}\frac{\rho}{z}$$

VECTOR TRANSFORMATIONS:

$$\hat{\mathbf{a}}_x = \cos\phi\,\hat{\mathbf{a}}_\rho - \sin\phi\,\hat{\mathbf{a}}_\phi$$
$$\hat{\mathbf{a}}_y = \sin\phi\,\hat{\mathbf{a}}_\rho + \cos\phi\,\hat{\mathbf{a}}_\phi$$
$$\hat{\mathbf{a}}_\rho = \cos\phi\,\hat{\mathbf{a}}_x + \sin\phi\,\hat{\mathbf{a}}_y$$
$$\hat{\mathbf{a}}_\phi = -\sin\phi\,\hat{\mathbf{a}}_x + \cos\phi\,\hat{\mathbf{a}}_y$$
$$\hat{\mathbf{a}}_x = \sin\theta\cos\phi\,\hat{\mathbf{a}}_r + \cos\theta\cos\phi\,\hat{\mathbf{a}}_\theta - \sin\phi\,\hat{\mathbf{a}}_\phi$$
$$\hat{\mathbf{a}}_y = \sin\theta\sin\phi\,\hat{\mathbf{a}}_r + \cos\theta\sin\phi\,\hat{\mathbf{a}}_\theta + \cos\phi\,\hat{\mathbf{a}}_\phi$$
$$\hat{\mathbf{a}}_z = \cos\theta\,\hat{\mathbf{a}}_r - \sin\theta\,\hat{\mathbf{a}}_\theta$$
$$\hat{\mathbf{a}}_\rho = \sin\theta\,\hat{\mathbf{a}}_r + \cos\theta\,\hat{\mathbf{a}}_\theta$$
$$\hat{\mathbf{a}}_r = \sin\theta\cos\phi\,\hat{\mathbf{a}}_x + \sin\theta\sin\phi\,\hat{\mathbf{a}}_y + \cos\phi\,\hat{\mathbf{a}}_z$$
$$\hat{\mathbf{a}}_\theta = \cos\theta\cos\phi\,\hat{\mathbf{a}}_x + \cos\theta\sin\phi\,\hat{\mathbf{a}}_y - \sin\theta\,\hat{\mathbf{a}}_z$$
$$\hat{\mathbf{a}}_r = \sin\theta\,\hat{\mathbf{a}}_\rho + \cos\theta\,\hat{\mathbf{a}}_z$$
$$\hat{\mathbf{a}}_\theta = \cos\theta\,\hat{\mathbf{a}}_\rho - \sin\theta\,\hat{\mathbf{a}}_z$$

$$A_x = A_\rho \cos\phi - A_\phi \sin\phi$$
$$A_y = A_\rho \sin\phi + A_\phi \cos\phi$$
$$A_\rho = A_x \cos\phi + A_y \sin\phi$$
$$A_\phi = -A_x \sin\phi + A_y \cos\phi$$
$$A_x = A_r \sin\theta \cos\phi + A_\theta \cos\theta \cos\phi - A_\phi \sin\phi$$
$$A_y = A_r \sin\theta \sin\phi + A_\theta \cos\theta \sin\phi + A_\phi \cos\phi$$
$$A_z = A_r \cos\theta - A_\theta \sin\theta$$
$$A_\rho = A_r \sin\theta + A_\theta \cos\theta$$
$$A_r = A_x \sin\theta \cos\phi + A_y \sin\theta \sin\phi + A_z \cos\phi$$
$$A_\theta = A_x \cos\theta \cos\phi + A_y \cos\theta \sin\phi - A_z \sin\theta$$
$$A_r = A_\rho \sin\theta + A_z \cos\theta$$
$$A_\theta = A_\rho \cos\theta - A_z \sin\theta$$

Appendix C: Bessel Functions

C.1 BESSEL FUNCTIONS OF THE FIRST AND SECOND KIND

Bessel's differential equation is

$$x^2\frac{d^2y}{dx^2} + x\frac{dy}{dx} + (x^2 - \nu^2)y = 0 \tag{C.1.1}$$

whose general solution for non-integer values of ν is

$$y(x) = AJ_\nu(x) + BJ_{-\nu}(x)$$

where

$$J_\nu(x) = \sum_{k=0}^{\infty} \frac{(-1)^k (x/2)^{2k+\nu}}{k!(k+\nu)!} \tag{C.1.2}$$

The functions $J_\nu(x)$ and $J_{-\nu}(x)$ are known as Bessel functions of the first kind of argument x and order ν and $-\nu$, respectively. The functions $J_\nu(x)$ and $J_{-\nu}(x)$ with $\nu \neq n$ are linearly independent. The Wronskian is

$$W[J_\nu(x), J_{-\nu}(x)] = \frac{-2\sin(\nu\pi)}{\pi x}$$

For integer values of ν ($\nu = n$) the functions $J_{-n}(x)$ and $J_n(x)$ are linearly dependent since

$$J_{-n}(x) = (-1)^n J_n(x)$$

The quantities $k!$ and $(k+\nu)!$ in (C.1.2) can be written in terms of the gamma function. That is, $k! = \Gamma(k+1)$ and $(k+\nu)! = \Gamma(k+\nu+1)$. The gamma function $\Gamma(x)$ is defined for all real values of x, except at $x = 0, -1, -2, -3, ...$ where it has an infinite value. The factorial function $x!$, where

$$x! = \Gamma(x+1)$$

is defined for all real values of x, except at negative integers where it has an infinite value. For $x = n$, where n is an integer greater than -1, the factorial function reads

$$n! = n(n-1)(n-2)(n-3)........1 \qquad (n > 0)$$

and

$$0! = 1$$

For integer values of ν, a second solution of (C.1.1) is the functions $Y_n(x)$, known as the Bessel function of the second kind or Neumann function, which is also denoted by $N_n(x)$. The functions $Y_n(x)$ is derived by forming a linear combination of $J_\nu(x)$ and $J_{-\nu}(x)$ as

$$Y_\nu(x) = \frac{J_\nu(x)\cos\nu\pi - J_{-\nu}(x)}{\sin\nu\pi} \tag{C.1.3}$$

which is a solution for $\nu \neq n$. For $\nu = n$, taking the limit of (C.1.3) gives

$$Y_n(x) = \lim_{\nu\to n} \frac{J_\nu(x)\cos\nu\pi - J_{-\nu}(x)}{\sin\nu\pi}$$

The limit exists and the second solution for $\nu = n$ is $Y_n(x)$. Therefore, the general solution to (C.1.1) for any value of ν is then

$$y(x) = AJ_\nu(x) + BY_\nu(x)$$

The functions $J_\nu(x)$ and $Y_\nu(x)$ are linearly independent for any ν.

Wronskian relations are

$$W[J_\nu(x), Y_\nu(x)] = J_\nu(x)Y_\nu{}'(x) - Y_\nu(x)J_\nu{}'(x) = \frac{2}{\pi x}$$

$$J_{\nu+1}(x)Y_\nu(x) - Y_{\nu+1}(x)J_\nu(x) = \frac{2}{\pi x}$$

Graphical plots of the Bessel functions $J_n(x)$ and $Y_n(x)$ for $n = 0$ and 1 are shown in Fig. C.1. The function $Y_n(x)$ has a natural logarithm singularity at $x = 0$.

Small argument approximations for $x \to 0$ are

$$J_o(x) \approx 1$$

$$Y_0(x) \approx \frac{2}{\pi}\ln(0.8905x)$$

$$J_\nu(x) \approx \frac{1}{\Gamma(\nu+1)}\left(\frac{x}{2}\right)^\nu \qquad (\nu \neq -1, -2, -3, \ldots)$$

$$Y_\nu(x) \approx -\frac{\Gamma(\nu)}{\pi}\left(\frac{2}{x}\right)^\nu \qquad (\nu \neq 0, -1, -2, \ldots)$$

$$J_0{}'(x) = -J_1(x) \approx \frac{x}{2}$$

$$Y_0{}'(x) = -Y_1(x) \approx \frac{2}{\pi x}$$

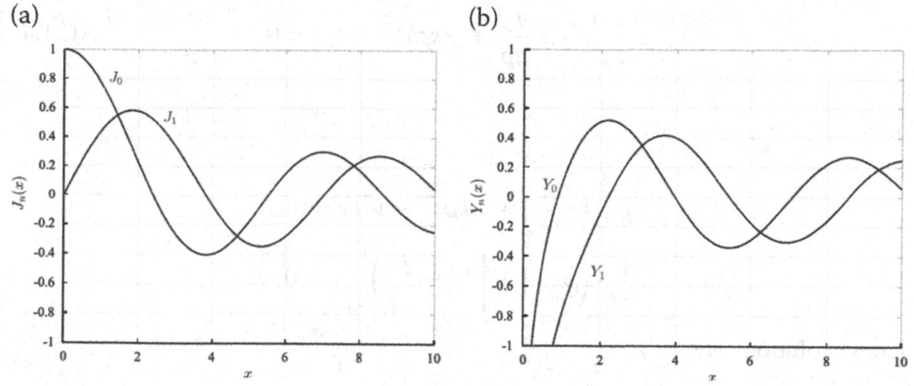

FIGURE C.1 (a) Plot of $J_0(x)$ and $J_1(x)$; (b) plot of $Y_0(x)$ and $Y_1(x)$.

The zeros of the Bessel function $J_n(x)$, denoted by χ_{nl} (i.e., $J_n(\chi_{nl}) = 0$), are listed in Fig. C.2a for several integer orders. The parameter l ($l = 1, 2, 3, ..$) denotes the zeros associated with the order n. The zeros of $J_n'(x)$ (i.e., $J_n'(\tilde{\chi}_{nl}) = 0$) are denoted by $\tilde{\chi}_{nl}$ and listed in Fig. C.2b.

Using the variable transformation $x = k\rho$, Bessel's differential equation appears in the form

(a)

χ_{nl}	$l = 1$	$l = 2$	$l = 3$
$n = 0$	2.405	5.520	8.654
$n = 1$	3.832	7.106	10.174
$n = 2$	5.136	8.417	11.620
$n = 3$	6.380	9.761	13.015

(b)

$\tilde{\chi}_{nl}$	$l = 1$	$l = 2$	$l = 3$
$n = 0$	3.832	7.016	10.174
$n = 1$	1.841	5.332	8.536
$n = 2$	3.054	6.706	9.970
$n = 3$	4.201	8.015	11.346

FIGURE C.2 (a) Zeros of $J_n(x)$ (i.e., $J_n(\chi_{nl}) = 0$); (b) zeros of $J_n'(x)$ (i.e., $J_n'(\tilde{\chi}_{nl}) = 0$).

$$\rho^2 \frac{d^2y}{d\rho^2} + \rho \frac{dy}{d\rho} + [(k\rho)^2 - \nu^2]y = 0 \tag{C.1.4}$$

or

$$\rho \frac{d}{d\rho}\left(\rho \frac{dy}{d\rho}\right) + [(k\rho)^2 - \nu^2]y = 0$$
$$\frac{1}{\rho}\frac{d}{d\rho}\left(\rho \frac{dy}{d\rho}\right) + \left[k^2 - \left(\frac{\nu}{\rho}\right)^2\right]y = 0$$

whose solutions are

$$y(\rho) = \begin{cases} AJ_\nu(k\rho) + BJ_{-\nu}(k\rho) & (\nu \neq n) \\ AJ_\nu(k\rho) + BY_\nu(k\rho) & (\text{any}\nu) \end{cases}$$

The sinusoidal behavior of the Bessel functions for large values of x with ν fixed $(x > > \nu)$ is

$$J_\nu(x) \approx \sqrt{\frac{2}{\pi x}} \cos\left(x - \frac{\nu\pi}{2} - \frac{\pi}{4}\right) \tag{C.1.5}$$

and

$$Y_\nu(x) \approx \sqrt{\frac{2}{\pi x}} \sin\left(x - \frac{\nu\pi}{2} - \frac{\pi}{4}\right) \tag{C.1.6}$$

For large values of ν with x fixed $(\nu > > x)$, the approximations are

$$J_\nu(x) \approx \frac{1}{\sqrt{2\pi\nu}}\left(\frac{ex}{2\nu}\right)^\nu \tag{C.1.7}$$

$$Y_\nu(x) \approx -\sqrt{\frac{2}{\pi\nu}}\left(\frac{2\nu}{ex}\right)^\nu \tag{C.1.8}$$

The Bessel functions $J_\nu(x)$ and $Y_\nu(x)$ satisfies a series of recurrence relations:

$$\begin{cases} \frac{d}{dx}[x^\nu B_\nu(x)] = x^\nu B_{\nu-1}(x) \\ \frac{d}{dx}[x^{-\nu} B_\nu(x)] = -x^{-\nu} B_{\nu+1}(x) \\ B_\nu{}'(x) = B_{\nu-1}(x) - \frac{\nu}{x}B_\nu(x) \\ B_\nu{}'(x) = \frac{\nu}{x}B_\nu(x) - B_{\nu+1}(x) \Rightarrow B_0{}'(x) = -B_1(x) \\ 2B_\nu{}'(x) = B_{\nu-1}(x) - B_{\nu+1}(x) \\ \frac{2\nu}{x}B_\nu(x) = B_{\nu-1}(x) + B_{\nu+1}(x) \end{cases} \tag{C.1.9}$$

where $B_\nu(x)$ represents either $J_\nu(x)$ or $Y_\nu(x)$.

Other identities are

$$\begin{aligned} J_n(-x) &= (-1)^n J_n(x) \\ Y_n(-x) &= (-1)^n [Y_n(x) + j2J_n(x)] \\ Y_{-n}(x) &= (-1)^n Y_n(x) \end{aligned}$$

Generating functions for Bessel junctions are

$$\begin{aligned} e^{\frac{x}{2}\left(t-\frac{1}{t}\right)} &= \Sigma_{n=-\infty}^{\infty} J_n(x)\, t^n \\ e^{-jx\cos\theta} &= \Sigma_{n=-\infty}^{\infty} j^{-n} J_n(x)\, e^{jn\theta} \\ e^{-jx\sin\theta} &= \Sigma_{n=-\infty}^{\infty} (-1)^n J_n(x)\, e^{jn\theta} \end{aligned}$$

Integral representations are

$$J_\nu(x) = \frac{1}{\pi}\int_0^\pi \cos(x\sin\theta - \nu\theta)\,d\theta - \frac{\sin\nu\pi}{\pi}\int_0^\infty e^{-x\sinh\theta-\nu\theta}\,d\theta \quad \text{(C.1.10)}$$

$$J_n(x) = \frac{1}{\pi}\int_0^\pi \cos(x\sin\theta - n\theta)\,d\theta = \frac{1}{2\pi}\int_{-\pi}^{\pi} \cos(x\sin\theta - n\theta)\,d\theta$$

$$\begin{aligned} J_n(x) &= \frac{e^{-jn(\alpha+\pi/2)}}{2\pi}\int_0^{2\pi} e^{j[x\cos(\theta-\alpha)]} e^{jn\theta}\,d\theta \\ J_n(x) &= \frac{1}{2\pi}\int_{-\pi}^{\pi} e^{j(x\sin\theta-n\theta)}\,d\theta = \frac{1}{2\pi}\int_0^{2\pi} e^{j(x\sin\theta-n\theta)}\,d\theta \\ J_n(x) &= \frac{1}{2\pi j^n}\int_0^{2\pi} e^{j(x\cos\theta+n\theta)}\,d\theta = \frac{j^n}{2\pi}\int_0^{2\pi} e^{-j(x\cos\theta+n\theta)}\,d\theta \\ J_n(x) &= \frac{1}{2\pi j^n}\int_0^{2\pi} \cos(n\theta)\, e^{jx\cos\theta}\,d\theta = \frac{1}{\pi j^n}\int_0^{\pi} \cos(n\theta)\, e^{jx\cos\theta}\,d\theta \end{aligned}$$

If $J_\nu(\lambda_i x)$ and $J_\nu(\lambda_j x)$ satisfy homogeneous boundary conditions at $x = a$, then

$$\int_0^a xJ_\nu(\lambda_i x)J_\nu(\lambda_j x)\,dx = \begin{cases} 0 \text{ for } \lambda_i \neq \lambda_j \\ \frac{a^2}{2}\left\{[J_\nu'(\lambda_i a)]^2 + \left(1 - \frac{\nu^2}{\lambda_i^2 a^2}\right)J_\nu^2(\lambda_i a)\right\} \text{ for } \lambda_i = \lambda_j \end{cases}$$

where $J_\nu'(\lambda_i a) = dJ_\nu(\lambda_i x)/d(\lambda_i x)$ evaluated at $x = a$.

Common homogeneous boundary conditions in electromagnetics are either $J_n(\lambda_i a) = 0$ or $J_n'(\lambda_i a) = 0$. In the case that $J_n(\lambda_i a) = 0$ (i.e., $\lambda_i a = \chi_{nl}$), the previous relation reads

$$\int_0^a x J_n^2(\lambda_i x)dx = \frac{a^2}{2}[J_n'(\chi_{nl})]^2 = \frac{a^2}{2}J_{n+1}^2(\chi_{nl})$$

C.2 HANKEL FUNCTIONS

The Hankel functions $H_\nu^{(1)}(x)$ and $H_\nu^{(2)}(x)$ are defined as

$$H_\nu^{(1)}(x) = J_\nu(x) + jY_\nu(x)$$

and

$$H_\nu^{(2)}(x) = J_\nu(x) - jY_\nu(x)$$

They are called Hankel functions of the first and second kind of order ν, respectively; sometimes called Bessel functions of the third kind.

Using (C.1.3), it follows that

$$H_\nu^{(1)}(x) = j\csc\nu\pi\,[J_\nu(x)e^{-j\nu\pi} - J_{-\nu}(x)]$$

and

$$H_\nu^{(2)}(x) = -j\csc\nu\pi\,[J_\nu(x)e^{j\nu\pi} - J_{-\nu}(x)]$$

Some properties are

$$\begin{aligned}
&H_{-n}^{(1)}(x) = (-1)^n J_n(x) + j(-1)^n Y_n(x) = (-1)^n H_n^{(1)}(x)\\
&H_{-n}^{(2)}(x) = (-1)^n J_n(x) - j(-1)^n Y_n(x) = (-1)^n H_n^{(2)}(x)\\
&H_n^{(1)}(-x) = (-1)^{n+1} H_n^{(2)}(x)\\
&H_\nu^{(1)}(x) = e^{-j\nu\pi} H_{-\nu}^{(1)}(x)\\
&H_\nu^{(2)}(x) = e^{j\nu\pi} H_{-\nu}^{(2)}(x)
\end{aligned}$$

The Hankel functions are linearly independent solutions of the Bessel's differential equation in (C.1.1). For the Bessel equation in (C.1.4) the solutions are $H_\nu^{(1)}(k\rho)$ and $H_\nu^{(2)}(k\rho)$. Common solutions to Bessel's equation in (C.1.1) in terms of Hankel functions are

$$\begin{aligned}
y(x) &= AH_\nu^{(1)}(x) + BH_\nu^{(2)}(x)\\
y(x) &= AJ_\nu(x) + BH_\nu^{(1)}(x)\\
y(x) &= AJ_\nu(x) + BH_\nu^{(2)}(x)
\end{aligned}$$

The Wronskians are

$$W[H_\nu^{(1)}(x), H_\nu^{(2)}(x)] = \frac{-j4}{\pi x}$$
$$W[J_\nu(x), H_\nu^{(1)}(x)] = \frac{j2}{\pi x}$$
$$W[J_\nu(x), H_\nu^{(2)}(x)] = \frac{-j2}{\pi x}$$

The Hankel functions satisfy the recurrence relations in (C.1.9) (i.e., replace $B_\nu(x)$ by either $H_\nu^{(1)}(x)$or $H_\nu^{(2)}(x)$).

Small argument approximations (i.e., as $x \to 0$) are

$$H_0^{(1)}(x) \approx jY_0(x) \approx \frac{j2}{\pi}\ln(0.8905x)$$
$$H_0^{(2)}(x) \approx -jY_0(x) \approx -\frac{j2}{\pi}\ln(0.8905x)$$
$$H_\nu^{(1)}(x) \approx jY_\nu(x) \approx -\frac{j}{\pi}\Gamma(\nu)\left(\frac{2}{x}\right)^\nu$$
$$H_\nu^{(2)}(x) \approx -jY_\nu(x) \approx \frac{j}{\pi}\Gamma(\nu)\left(\frac{2}{x}\right)^\nu$$
$$H_0^{(1)\prime}(x) = -H_1^{(1)}(x) \approx j\frac{2}{\pi x}$$
$$H_0^{(2)\prime}(x) = -H_1^{(2)}(x) \approx -j\frac{2}{\pi x}$$

There are several asymptotic expansions for the Hankel functions. For $x >> \nu$ and $x \to \infty$:

$$H_\nu^{(1)}(x) \approx \sqrt{\frac{2}{\pi x}}\, e^{j(x-\nu\pi/2-\pi/4)} \tag{C.2.1}$$

$$H_\nu^{(1)\prime}(x) \approx \sqrt{\frac{2}{\pi x}}\, e^{j(x-\nu\pi/2+\pi/4)} \tag{C.2.2}$$

and

$$H_\nu^{(2)}(x) \approx \sqrt{\frac{2}{\pi x}}\, e^{-j(x-\nu\pi/2-\pi/4)} \tag{C.2.3}$$

$$H_\nu^{(2)\prime}(x) \approx \sqrt{\frac{2}{\pi x}}\, e^{-j(x-\nu\pi/2+\pi/4)} \tag{C.2.4}$$

The Debye approximations (or second-order approximations) apply when the values of x and ν are large, but is not valid when $x \approx \nu$. Specifically, the Debye approximation applies when

$$|\nu^2 - x^2| >> x^{4/3}$$

For large values of x and ν with $x > \nu$:

$$H_\nu^{(1)}(x) \approx \sqrt{\frac{2}{\pi (x^2 - \nu^2)^{1/2}}}\, e^{j[(x^2-\nu^2)^{1/2} - \nu \cos^{-1}(\nu/x) - \pi/4]} \qquad (x > \nu)$$

$$H_\nu^{(2)}(x) \approx \sqrt{\frac{2}{\pi (x^2 - \nu^2)^{1/2}}}\, e^{-j[(x^2-\nu^2)^{1/2} - \nu \cos^{-1}(\nu/x) - \pi/4]} \qquad (x > \nu)$$

$$\begin{aligned} H_\nu^{(1)\prime}(x) &\approx j\sqrt{1 - \left(\frac{\nu}{x}\right)^2}\, H_\nu^{(1)}(x) = j\frac{\sqrt{x^2 - \nu^2}}{x} H_\nu^{(1)}(x) \\ &\approx j\sqrt{\frac{2}{\pi}}\frac{(x^2 - \nu^2)^{1/4}}{x} e^{j[(x^2-\nu^2)^{1/2} - \nu \cos^{-1}(\nu/x) - \pi/4]} \qquad (x > \nu) \end{aligned}$$

$$\begin{aligned} H_\nu^{(2)\prime}(x) &\approx -j\sqrt{1 - \left(\frac{\nu}{x}\right)^2}\, H_\nu^{(2)}(x) \approx -j\frac{\sqrt{x^2 - \nu^2}}{x} H_\nu^{(2)}(x) \\ &\approx -j\sqrt{\frac{2}{\pi}}\frac{(x^2 - \nu^2)^{1/4}}{x} e^{-j[(x^2-\nu^2)^{1/2} - \nu \cos^{-1}(\nu/x) - \pi/4]} \qquad (x > \nu) \end{aligned}$$

Observe that the above approximations for $x >> \nu$ reduce to (C.2.1) to (C.2.4).

For large values of x and ν with $\nu > x$:

$$H_\nu^{(1)}(x) \approx -j\sqrt{\frac{2}{\pi (\nu^2 - x^2)^{1/2}}}\, e^{\left[\nu \cosh^{-1}(\nu/x) - (\nu^2 - x^2)^{1/2}\right]} \qquad (\nu > x) \quad \text{(C.2.5)}$$

$$H_\nu^{(2)}(x) \approx j\sqrt{\frac{2}{\pi (\nu^2 - x^2)^{1/2}}}\, e^{\left[\nu \cosh^{-1}(\nu/x) - (\nu^2 - x^2)^{1/2}\right]} \qquad (\nu > x) \quad \text{(C.2.6)}$$

$$H_\nu^{(1)\prime}(x) \approx -\frac{(\nu^2 - x^2)^{1/2}}{x} H_\nu^{(1)}(x) \qquad (\nu > x)$$

$$H_\nu^{(2)\prime}(x) \approx -\frac{(\nu^2 - x^2)^{1/2}}{x} H_\nu^{(2)}(x) \qquad (\nu > x)$$

Third-order approximations for large x and ν with $\nu \approx x$ are expressed in terms of the Airy function $Ai(x)$:

$$\begin{cases} H_\nu^{(1)}(x) \approx 2\left(\frac{2}{x}\right)^{1/3} e^{-j\pi/3} Ai\left[-(\nu - x)\left(\frac{2}{x}\right)^{1/3} e^{-j\pi/3}\right] \\ H_\nu^{(1)\prime}(x) \approx -2\left(\frac{2}{x}\right)^{2/3} e^{j\pi/3} Ai'\left[-(\nu - x)\left(\frac{2}{x}\right)^{1/3} e^{-j\pi/3}\right] \end{cases} \quad \text{(C.2.7)}$$

$$\begin{cases} H_\nu^{(2)}(x) \approx 2\left(\frac{2}{x}\right)^{1/3} e^{j\pi/3} Ai\left[-(\nu - x)\left(\frac{2}{x}\right)^{1/3} e^{j\pi/3}\right] \\ H_\nu^{(2)\prime}(x) \approx -2\left(\frac{2}{x}\right)^{2/3} e^{-j\pi/3} Ai'\left[-(\nu - x)\left(\frac{2}{x}\right)^{1/3} e^{j\pi/3}\right] \end{cases} \quad \text{(C.2.8)}$$

$$J_\nu(x) \approx \left(\frac{2}{x}\right)^{1/3} Ai\left[(\nu - x)\left(\frac{2}{x}\right)^{1/3}\right]$$

$$J_\nu'(x) \approx -\left(\frac{2}{x}\right)^{2/3} Ai'\left[(\nu - x)\left(\frac{2}{x}\right)^{1/3}\right]$$

Figure C.3 shows a plot of $|H_\nu^{(2)}(10)|$ and the behavior of the Debye and third-order approximations around $\nu \approx x$. The Debye approximation fails at $\nu = x$ and for values of ν close to x. The third-order approximation is applicable in the region where $\nu \approx x$.

For large values of ν with x fixed ($\nu >> x$), the following approximations follow from (C.1.7) and (C.1.8):

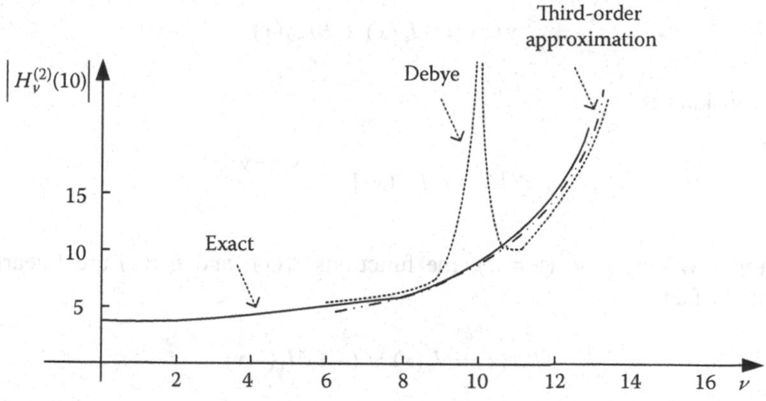

FIGURE C.3 Typical plot of a Hankel function showing the behavior of the Debye and third-order approximation.

$$H_\nu^{(2)}(x) \approx -H_\nu^{(1)}(x) \approx j\sqrt{\frac{2}{\pi\nu}}\left(\frac{2\nu}{ex}\right)^\nu \tag{C.2.9}$$

C.3 MODIFIED BESSEL FUNCTIONS

The modified Bessel functions are solutions to the modified Bessel equation:

$$x^2\frac{d^2y}{dx^2} + x\frac{dy}{dx} - (x^2 + \nu^2)y = 0 \tag{C.3.1}$$

This equation follows form (C.1.4) if $k = e^{j\pi/2}$. Therefore, the solution to (C.3.1) is

$$y(x) = AJ_\nu(xe^{j\pi/2}) + BY_\nu(xe^{j\pi/2})$$

The above solutions have the disadvantage of not being real for real x. However, if one of the independent solutions is defined as

$$I_\nu(x) = j^{-\nu}J_\nu(jx)$$

Then, $I_\nu(x)$ is real for real x. The function $I_\nu(x)$ is called the modified Bessel function of the first kind of order ν. The series representation of $I_\nu(x)$ is

$$I_\nu(x) = \sum_{k=0}^{\infty} \frac{(x/2)^{2k+\nu}}{k!(k+\nu)!}$$

The general solution of (C.3.1) for non-integer values of ν is

$$y(x) = AI_\nu(x) + BI_{-\nu}(x)$$

The Wronskian is

$$W[I_\nu(x), I_{-\nu}(x)] = \frac{2\sin\pi\nu}{\pi x}$$

For integer values of ν, ($\nu = n$), the functions $I_n(x)$ and $I_{-n}(x)$ are linearly dependent. In fact,

$$I_{-n}(x) = I_n(x) = (-1)^n I_n(-x)$$

For arbitrary ν, the second linearly independent solution, known as the modified Bessel function of the second kind of order $\nu$is $K_\nu(x)$, defined by

$$K_\nu(x) = \frac{\pi}{2}\frac{I_{-\nu}(x) - I_\nu(x)}{\sin(\nu\pi)}$$

which for $\nu = n$, the limit exists and is evaluated by

$$K_n(x) = \lim_{\nu\to n}\frac{\pi}{2}\frac{I_{-\nu}(x) - I_\nu(x)}{\sin(\nu\pi)}$$

Hence, the second solution for $\nu = n$ is $K_n(x)$. The function $K_n(x)$ satisfies

$$K_{-n}(x) = K_n(x)$$

The general solution to (C.3.1) for any value of ν is then

$$y(x) = A I_\nu(x) + BK_\nu(x)$$

The Wronskian is

$$W[I_\nu(x), K_\nu(x)] = -\frac{1}{x}$$

Letting $x=kp$ in (C.3.1), the modified Bessel equation reads

$$\rho^2\frac{d^2y}{d\rho^2} + \rho\frac{dy}{d\rho} - (k^2\rho^2 + \nu^2)y = 0$$

whose solutions are of the form $I_\nu(k\rho)$ and $K_\nu(k\rho)$.

Plots of the functions $I_n(x)$ and $K_n(x)$ are given in Fig. C.4. The function $K_n(x)$ has a natural logarithm singularity at $x = 0$.

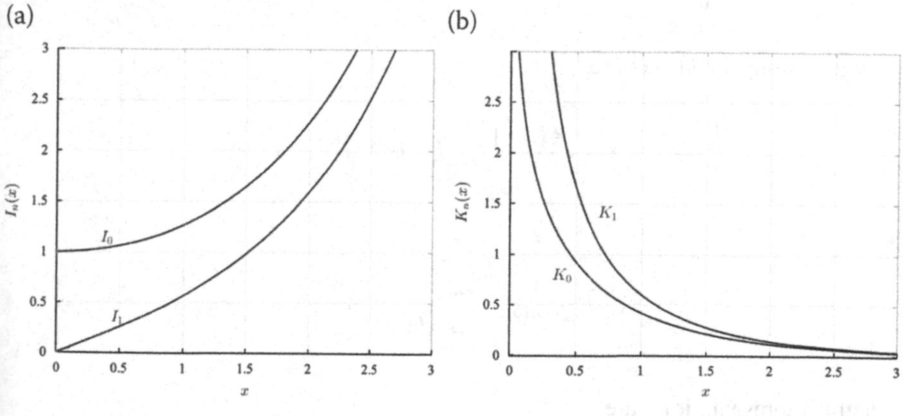

FIGURE C.4 (a) Plot of $I_0(x)$ and $I_1(x)$; (b) plot of $K_0(x)$ and $K_1(x)$.

The small argument approximations are

$$I_0(x) \approx 1$$
$$K_0(x) \approx -\ln(0.8905x)$$
$$I_\nu(x) = \frac{1}{\Gamma(\nu+1)}\left(\frac{x}{2}\right)^\nu \qquad (\nu \neq -1, -2, -3, ...)$$
$$K_\nu(x) = \frac{\Gamma(\nu)}{2}\left(\frac{2}{x}\right)^\nu \qquad (\nu > 0)$$

The modified Bessel functions satisfy a series of recurrence relations:

$$\frac{d}{dx}[x^\nu I_\nu(x)] = x^\nu I_{\nu-1}(x)$$
$$\frac{d}{dx}[x^{-\nu} I_\nu(x)] = x^{-\nu} I_{\nu+1}(x)$$
$$I_\nu{}'(x) = I_{\nu-1}(x) - \frac{\nu}{x} I_\nu(x)$$
$$I_\nu{}'(x) = \frac{\nu}{x} I_\nu(x) + I_{\nu+1}(x) \Rightarrow I_0{}'(x) = I_1(x)$$
$$2I_\nu{}'(x) = I_{\nu-1}(x) + I_{\nu+1}(x)$$
$$\frac{2\nu}{x} I_\nu(x) = I_{\nu-1}(x) - I_{\nu+1}(x)$$

and

$$\frac{d}{dx}[x^\nu K_\nu(x)] = -x^\nu K_{\nu-1}(x)$$
$$\frac{d}{dx}[x^{-\nu} K_\nu(x)] = -x^{-\nu} K_{\nu+1}(x)$$
$$K_\nu{}'(x) = -K_{\nu-1}(x) - \frac{\nu}{x} K_\nu(x)$$
$$K_\nu{}'(x) = \frac{\nu}{x} K_\nu(x) - K_{\nu+1}(x) \Rightarrow K_0{}'(x) = -K_1(x)$$
$$2K_\nu{}'(x) = -K_{\nu-1}(x) - K_{\nu+1}(x)$$
$$\frac{2\nu}{x} K_\nu(x) = -K_{\nu-1}(x) + K_{\nu+1}(x)$$

The generating functions are

$$e^{\frac{x}{2}\left(t+\frac{1}{t}\right)} = \sum_{n=-\infty}^{\infty} I_n(x)\, t^n$$

and

$$e^{x\cos\theta} = \sum_{n=-\infty}^{\infty} I_n(x)\, e^{jn\theta}$$

Integral representations are

$$I_\nu(x) = \frac{1}{\pi}\int_0^\pi e^{x\cos\theta}\cos(\nu\theta)\,d\theta - \frac{\sin\nu\pi}{\pi}\int_0^\infty e^{-x\cosh\theta - \nu\theta}\,d\theta \tag{C.3.2}$$

$$I_n(x) = \frac{1}{\pi}\int_0^\pi e^{x\cos\theta}\cos(n\theta)\,d\theta$$

$$K_\nu(x) = \int_0^\infty e^{-x\cosh\theta}\cosh(\nu\theta)\,d\theta \tag{C.3.3}$$

The behavior of the modified Bessel functions for large values of x with ν fixed are

$$I_\nu(x) \approx \frac{e^x}{\sqrt{2\pi x}} \tag{C.3.4}$$

$$K_\nu(x) \approx \sqrt{\frac{\pi}{2x}}\,e^{-x} \tag{C.3.5}$$

For large values of ν with x fixed ($\nu >> x$) the approximations are

$$I_\nu(x) \approx \sqrt{\frac{1}{2\pi x}}\left(\frac{ex}{2\nu}\right)^\nu$$

$$K_\nu(x) \approx \sqrt{\frac{\pi}{2\nu}}\left(\frac{2\nu}{ex}\right)^\nu$$

Other identities are

$$I_\nu(x) = j^{-\nu}J_\nu(jx) \tag{C.3.6}$$

$$I_\nu(x) = j^{\nu}J_\nu(-jx) \tag{C.3.7}$$

$$I_\nu(jx) = j^{\nu}J_\nu(x) \tag{C.3.8}$$

$$I_\nu(-jx) = j^{-\nu}J_\nu(x) \tag{C.3.9}$$

$$I_\nu(e^{\pm j\pi}x) = e^{\pm j\nu\pi}I_\nu(x)$$

$$K_\nu(x) = \frac{\pi}{2}j^{\nu+1}H_\nu^{(1)}(jx) \tag{C.3.10}$$

$$K_\nu(x) = \frac{\pi}{2}(-j)^{\nu+1}H_\nu^{(2)}(-jx) \tag{C.3.11}$$

$$K_\nu(jx) = \frac{\pi}{2}(-j)^{\nu+1}H_\nu^{(2)}(x) \tag{C.3.12}$$

C.4 SPHERICAL BESSEL FUNCTIONS

The spherical Bessel differential equation is

$$x^2\frac{d^2y}{dx^2} + 2x\frac{dy}{dx} + [x^2 - \nu(\nu+1)]y = 0 \tag{C.4.1}$$

Letting

$$y(x) = \frac{F(x)}{\sqrt{x}}$$

(C.4.1) becomes

$$x^2\frac{d^2F}{dx^2} + x\frac{dF}{dx} + \left[x^2 - \left(\nu + \frac{1}{2}\right)^2\right]F = 0$$

which is recognized as Bessel's differential equation of order $\nu + 1/2$. Therefore, the solutions of (C.4.1) can be written in various forms, such as

$$y(x) = A\frac{J_{\nu+1/2}(x)}{\sqrt{x}} + B\frac{Y_{\nu+1/2}(x)}{\sqrt{x}}$$

$$y(x) = A\frac{J_{\nu+1/2}(x)}{\sqrt{x}} + B\frac{H^{(2)}_{\nu+1/2}(x)}{\sqrt{x}}$$

The spherical Bessel functions $j_\nu(x)$, $y_\nu(x)$, $h_\nu^{(1)}(x)$, and $h_\nu^{(2)}(x)$ are defined as

$$j_\nu(x) = \sqrt{\frac{\pi}{2x}}J_{\nu+1/2}(x)$$

$$y_\nu(x) = \sqrt{\frac{\pi}{2x}}Y_{\nu+1/2}(x)$$

$$h_\nu^{(1)}(x) = \sqrt{\frac{\pi}{2x}}H^{(1)}_{\nu+1/2}(x)$$

$$h_\nu^{(2)}(x) = \sqrt{\frac{\pi}{2x}}H^{(2)}_{\nu+1/2}(x)$$

Therefore, typical solution to (C.4.1) are written in the forms:

$$y(x) = Aj_\nu(x) + By_\nu(x)$$

$$y(x) = Aj_\nu(x) + Bh_\nu^{(2)}(x)$$

$$y(x) = Ah_\nu^{(1)}(x) + Bh_\nu^{(2)}(x)$$

The Wronskians are

$$W[j_\nu(x), y_\nu(x)] = \frac{1}{x^2}$$
$$W[j_\nu(x), h_\nu^{(2)}(x)] = -\frac{j}{x^2}$$
$$W[h_\nu^{(1)}(x), h_\nu^{(2)}(x)] = -\frac{j2}{x^2}$$

Letting $x = kr$ and $y = R$, the spherical Bessel equation reads

$$r^2\frac{d^2R}{dr^2} + 2r\frac{dR}{dr} + [(kr)^2 - \nu(\nu+1)]R = 0$$

or

$$\frac{d}{dr}\left(r^2\frac{dR}{dr}\right) + [(kr)^2 - \nu(\nu+1)]R = 0 \qquad \text{(C.4.2)}$$

which are the common form for the radial wave equation in spherical coordinates. The solutions to (C.4.2) are of the form $j_\nu(kr)$, $y_\nu(kr)$, $h_\nu^{(1)}(kr)$, and $h_\nu^{(2)}(kr)$.

For $\nu = n$, the spherical Bessel functions satisfy

$$j_n(-x) = (-1)^n j_n(x)$$
$$y_n(-x) = (-1)^{n+1} y_n(x)$$
$$h_n^{(1)}(-x) = (-1)^n h_n^{(2)}(x)$$
$$h_n^{(2)}(-x) = (-1)^n h_n^{(1)}(x)$$

Some specific values are

$$J_{1/2}(x) = \sqrt{\frac{2}{\pi x}}\sin x \Rightarrow j_0(x) = \frac{\sin x}{x}$$
$$Y_{1/2}(x) = -\sqrt{\frac{2}{\pi x}}\cos x \Rightarrow y_0(x) = -\frac{\cos x}{x}$$
$$H_{1/2}^{(1)}(x) = -jH_{-1/2}^{(1)}(x) = -j\sqrt{\frac{2}{\pi x}}\,e^{jx} \Rightarrow h_0^{(1)}(x) = -j\frac{e^{jx}}{x}$$
$$H_{1/2}^{(2)}(x) = jH_{-1/2}^{(2)}(x) = j\sqrt{\frac{2}{\pi x}}\,e^{-jx} \Rightarrow h_0^{(2)}(x) = j\frac{e^{-jx}}{x}$$
$$j_1(x) = \frac{\sin x}{x^2} - \frac{\cos x}{x}$$
$$y_1(x) = -\frac{\sin x}{x} - \frac{\cos x}{x^2}$$
$$h_1^{(1)}(x) = -e^{jx}\left(\frac{1}{x} + j\frac{1}{x^2}\right)$$
$$h_1^{(2)}(x) = -e^{-jx}\left(\frac{1}{x} - j\frac{1}{x^2}\right)$$

(a)

K_{nl}	$l=1$	$l=2$	$l=3$
$n=0$	3.142	6.283	9.425
$n=1$	4.493	7.725	10.904
$n=2$	5.763	9.095	12.323
$n=3$	6.988	10.417	13.698

(b)

$\tilde{K}_{nl}$	$l=1$	$l=2$	$l=3$
$n=0$	0	4.494	7.725
$n=1$	2.744	6.117	9.317
$n=2$	3.870	7.443	10.713
$n=3$	4.973	8.722	12.064

FIGURE C.5 (a) Zeros of $j_n(x)$ (i.e., $j_n(\kappa_{nl}) = 0$); (b) zeros of $j_n'(x)$ (i.e., $j'_n(\tilde{\kappa}_{nl}) = 0$).

The zeros of the Bessel function $j_n(x)$, denoted by κ_{nl} (i.e., $j_n(\kappa_{nl}) = 0$), are listed in Fig. C.5a for several integer orders. The parameter l ($l = 1, 2, 3, ..$) denotes the zeros associated with the order n. The zeros of $j_n'(x)$ are denoted by $\tilde{\kappa}_{nl}$ (i.e., $j_n'(\tilde{\kappa}_{nl}) = 0$) and listed in Fig. C.5b.

Denoting any of the spherical Bessel functions by $b_\nu(x)$, the recursion relations for the spherical Bessel functions are:

$$\frac{d}{dx}[x^{\nu+1}b_\nu(x)] = x^{\nu+1}b_{\nu-1}(x)$$
$$\frac{d}{dx}[x^{-\nu}b_\nu(x)] = -x^{-\nu}b_{\nu+1}(x)$$
$$b_\nu'(x) = b_{\nu-1}(x) - \frac{\nu+1}{x}b_\nu(x)$$
$$b_\nu'(x) = \frac{\nu}{x}b_\nu(x) - b_{\nu+1}(x) \Rightarrow b_0'(x) = -b_1(x)$$
$$\frac{2\nu+1}{x}b_\nu(x) = b_{\nu-1}(x) + b_{\nu+1}(x)$$

The small argument approximations for $x \to 0$ are

$$j_n(x) \approx \frac{n!}{(2n+1)!}(2x)^n$$

$$y_n(x) \approx -2\frac{(2n)!}{n!}\frac{1}{(2x)^{n+1}}$$

$$h_n^{(1)}(x) \approx jy_n(x) \Rightarrow h_1^{(1)}(x) \approx -j\frac{1}{x^2}$$
$$h_n^{(2)}(x) \approx -jy_n(x) \Rightarrow h_1^{(2)}(x) \approx j\frac{1}{x^2}$$

The asymptotic approximation for large x are

$$j_\nu(x) \approx \frac{1}{x}\sin\left(x - \frac{\nu\pi}{2}\right)$$
$$y_\nu(x) \approx -\frac{1}{x}\cos\left(x - \frac{\nu\pi}{2}\right)$$
$$h_\nu^{(1)}(x) \approx (-j)^{\nu+1}\frac{e^{jx}}{x} = -\frac{j}{x}e^{j\left(x-\frac{\nu\pi}{2}\right)}$$
$$h_\nu^{(2)}(x) \approx j^{\nu+1}\frac{e^{-jx}}{x} = \frac{j}{x}e^{-j\left(x-\frac{\nu\pi}{2}\right)}$$

An orthogonal relation for the spherical Bessel functions when $j_\nu(\lambda_i a) = 0$ is

$$\int_0^a j_\nu(\lambda_i r) j_\nu(\lambda_j r) r^2 dr = \frac{a^3}{2}[j_{\nu+1}(\lambda_i a)]^2 \delta_{i,j}$$

The Riccati-Bessel functions, which in electromagnetic applications are known as the spherical Schelkunoff-Bessel functions, are denoted using the caret notation $\hat{j}_\nu(x)$, $\hat{y}_\nu(x)$, $\hat{h}_\nu^{(1)}(x)$, and $\hat{h}_\nu^{(2)}(x)$. They are defined by

$$\hat{j}_\nu(x) = xj_\nu(x) = \sqrt{\frac{\pi x}{2}}J_{\nu+1/2}(x)$$
$$\hat{y}_\nu(x) = xy_\nu(x) = \sqrt{\frac{\pi x}{2}}Y_{\nu+1/2}(x)$$
$$\hat{h}_\nu^{(1)}(x) = xh_\nu^{(1)}(x) = \sqrt{\frac{\pi x}{2}}H_{\nu+1/2}^{(1)}(x)$$
$$\hat{h}_\nu^{(2)}(x) = xh_\nu^{(2)}(x) = \sqrt{\frac{\pi x}{2}}H_{\nu+1/2}^{(2)}(x)$$

These functions satisfy the equation:

$$\frac{d^2\hat{y}}{dx^2} + \left[1 - \frac{\nu(\nu+1)}{x^2}\right]\hat{y} = 0 \tag{C.4.3}$$

where $\hat{y}(x)$ represents any of the spherical Schelkunoff-Bessel functions. Letting $x = kr$, the equation reads

$$\frac{d^2\hat{y}}{dr^2} + \left[k^2 - \frac{\nu(\nu+1)}{r^2}\right]\hat{y} = 0 \tag{C.4.4}$$

where $\hat{y}(kr)$ represents: $\hat{j}_\nu(kr)$, $\hat{y}_\nu(kr)$, $\hat{h}_\nu^{(1)}(kr)$ and $\hat{h}_\nu^{(2)}(kr)$. The zeros of the spherical Schelkunoff-Bessel functions are the same as those of the spherical Bessel functions.

The Wronskian are

$$W[\hat{j}_\nu(x), \hat{y}_\nu(x)] = 1$$
$$W[\hat{j}_\nu(x), \hat{h}_\nu^{(2)}(x)] = -j$$
$$W[\hat{h}_\nu^{(1)}(x), \hat{h}_\nu^{(2)}(x)] = -2j$$

The asymptotic forms of the spherical Schelkunoff-Bessel functions are

$$\hat{j}_\nu(x) \approx \sin\left(x - \frac{\nu\pi}{2}\right)$$
$$\hat{y}_\nu(x) \approx -\cos\left(x - \frac{\nu\pi}{2}\right)$$
$$\hat{h}_\nu^{(1)}(x) \approx (-j)^{\nu+1}e^{jx} = -j\,e^{j\left(x-\frac{\nu\pi}{2}\right)}$$
$$\hat{h}_\nu^{(2)}(x) \approx j^{\nu+1}e^{-jx} = j\,e^{-j\left(x-\frac{\nu\pi}{2}\right)}$$

Small argument approximations follow from those of $j_n(x)$, $y_n(x)$, $h_n^{(1)}(x)$ and $h_n^{(2)}(x)$, since $\hat{j}_n(x) = x j_n(x)$, $\hat{y}_n(x) = x y_n(x)$, $\hat{h}_n^{(1)}(x) = x h_n^{(1)}(x)$, and $\hat{h}_n^{(2)}(x) = x h_n^{(2)}(x)$.

The Debye approximations (or second order approximations) of the spherical Bessel functions or the spherical Schelkunoff-Bessel functions are conveniently expressed in term of order $\nu - 1/2$. That is, for large values of x:

$$\hat{h}_{\nu-1/2}^{(1)}(x) = x h_{\nu-1/2}^{(1)}(x) \approx \left[1 - \left(\frac{\nu}{x}\right)^2\right]^{-1/4} e^{j\left[(x^2-\nu^2)^{1/2} - \nu\cos^{-1}(\nu/x) - \pi/4\right]} \qquad (x > \nu)$$

$$\hat{h}_{\nu-1/2}^{(2)}(x) = x h_{\nu-1/2}^{(2)}(x) \approx \left[1 - \left(\frac{\nu}{x}\right)^2\right]^{-1/4} e^{-j\left[(x^2-\nu^2)^{1/2} - \nu\cos^{-1}(\nu/x) - \pi/4\right]} \qquad (x > \nu)$$

$$h_{\nu-1/2}^{(1)\prime}(x) \approx j\sqrt{1 - \left(\frac{\nu}{x}\right)^2}\, h_{\nu-1/2}^{(1)}(x) \qquad (x > \nu)$$

$$h_{\nu-1/2}^{(2)\prime}(x) \approx -j\sqrt{1 - \left(\frac{\nu}{x}\right)^2}\, h_{\nu-1/2}^{(2)}(x) \qquad (x > \nu)$$

Debye approximations for $x < \nu$ follow from (C.2.5) and (C.2.6).

Third-order approximations in terms of Airy functions follow from (C.2.7) and (C.2.8). Third-order approximations are also conveniently expressed in terms of the Fok-Airy functions and given in Appendix D.

The modified spherical Bessel functions $i_\nu(x)$ and $k_\nu(x)$ are defined by

$$i_\nu(x) = \sqrt{\frac{\pi}{2x}} I_{\nu+1/2}(x)$$
$$k_\nu(x) = \sqrt{\frac{\pi}{2x}} K_{\nu+1/2}(x)$$

which satisfy the equation

$$x^2\frac{d^2y}{dx^2} + 2x\frac{dy}{dx} - [x^2 + \nu(\nu+1)]y = 0$$

where $y(x)$ represents either $i_\nu(x)$ or $k_\nu(x)$. The Wronskian is

$$W[i_\nu(x), k_\nu(x)] = -\frac{\pi}{2x^2}$$

Some specific values are

$$I_{1/2}(x) = \sqrt{\frac{2}{\pi x}} \sinh x \Rightarrow i_0(x) = \frac{\sinh x}{x}$$
$$I_{-1/2}(x) = \sqrt{\frac{2}{\pi x}} \cosh x \Rightarrow i_{-1}(x) = \frac{\cosh x}{x}$$
$$K_{1/2}(x) = K_{-1/2}(x) = \sqrt{\frac{\pi}{2x}} e^{-x} \Rightarrow k_0(x) = \frac{\pi}{2x}e^{-x}$$

Asymptotic approximations for $x \to \infty$ are

$$i_\nu(x) \approx \frac{1}{2x}e^x$$
$$k_\nu(x) \approx \frac{\pi}{2x}e^{-x}$$

Some recursion relations are

$$i_{n-1}(x) - i_{n+1}(x) = \frac{2n+1}{x} i_n(x)$$
$$k_{n-1}(x) - k_{n+1}(x) = -\frac{2n+1}{x} k_n(x)$$

Some properties are

$$i_n(x) = j^{-n} j_n(jx) = j^n j_n(-jx)$$
$$k_n(x) = -\frac{\pi}{2} j^n h_n^{(1)}(jx) = -\frac{\pi}{2} j^{-n} h_n^{(2)}(-jx)$$

Modified spherical Schelkunoff-Bessel functions are defined as

$$\hat{i}_\nu(x) = x\, i_\nu(x) = \sqrt{\frac{\pi x}{2}}\, I_{\nu+1/2}(x)$$
$$\hat{k}_\nu(x) = \frac{2}{\pi} x\, k_\nu(x) = \sqrt{\frac{2x}{\pi}}\, K_{\nu+1/2}(x)$$

which satisfy the equation

$$\frac{d^2\hat{y}}{dx^2} - \left[1 + \frac{\nu(\nu+1)}{x^2}\right]\hat{y} = 0$$

where $\hat{y}$ represents either $\hat{i}_\nu(x)$ or $\hat{k}_\nu(x)$. Note the way that $\hat{k}_\nu(x)$ is defined.

C.5 COMPLEX ARGUMENTS

If the argument of the Bessel functions is complex, denoted by z, the Bessel function $J_\nu(z)$ in (C.1.2) has a branch point singularity at $z = 0$ due to the term $(z/2)^{\nu/2}$ when ν is not an integer. The singularity requires the use of a branch cut extending from $-\infty$ to 0 along the real negative axis. In the cut plane, the Bessel function $J_\nu(z)$ is an analytic function of $z = re^{j\theta}$. For example, for $\nu = 1/2$, the multiple value term $(z/2)^{1/2}$ in (C.1.2) is made single valued if $-\pi < \theta \leq \pi$. The upper Riemann surface (or principal value) corresponds to $-\pi < \theta \leq \pi$. Similar considerations apply to $J_{-\nu}(z)$.

When $\nu = n$, the functions $J_{\pm n}(z)$ are analytic everywhere. At a fixed value of z in the cut plane, $J_{\pm\nu}(z)$ are analytic functions of ν.

Equation (C.1.3) shows that the upper Riemann surface of $Y_\nu(x)$ is the upper surface associated with $J_{\pm\nu}(z)$. That is, a branch cut extending from $-\infty$ to 0 along the real negative axis, such that $-\pi < \theta \leq \pi$ defines the upper Riemann surface. At a fixed value of z in the cut plane, $Y_\nu(x)$ is an analytic function of ν.

The functions $H_\nu^{(1)}(z)$ and $H_\nu^{(2)}(z)$ also require a branch cut from $-\infty$ to 0 along the real negative axis, such that $-\pi < \theta \leq \pi$ defines the upper Riemann surface. At a fixed value of z in the cut plane, $H_\nu^{(1)}(z)$ and $H_\nu^{(2)}(z)$ are analytic functions of ν.

The expressions in Sections C.1 to C.4 usually extend to complex values by replacing x by $z = re^{\theta}$, where z is in the upper Riemann surface, or $-\pi < \theta \leq \pi$. However, some relations have a different range of validity in the z plane. For example:

(C.1.5) in terms of z is valid for $|z| >> \nu$ and $-\pi < \theta < \pi$
(C.1.6) in terms of z is valid for $|z| >> \nu$ and $-\pi < \theta < \pi$
(C.1.10) in terms of z is valid for $\text{Re}(z) > 0$
(C.2.1) in terms of z is valid for $|z| >> \nu$ and $-\pi < \theta < 2\pi$
(C.2.2) in terms of z is valid for $|z| >> \nu$ and $-\pi < \theta < 2\pi$
(C.2.3) in terms of z is valid for $|z| >> \nu$ and $-2\pi < \theta < \pi$
(C.2.4) in terms of z is valid for $|z| >> \nu$ and $-2\pi < \theta < \pi$

(C.3.2) in terms of z is valid for $-\pi/2 < \theta < \pi/2$
(C.3.3) in terms of z is valid for $-\pi/2 < \theta < \pi/2$
(C.3.4) in terms of z is valid for $-\pi/2 < \theta < \pi/2$
(C.3.5) in terms of z is valid for $-3\pi/2 < \theta < 3\pi/2$
(C.3.6) in terms of z is valid for $-\pi \le \theta \le \pi/2$
(C.3.7) in terms of z is valid for $-\pi/2 \le \theta \le \pi$
(C.3.8) in terms of z is valid for $-\pi \le \theta \le \pi/2$
(C.3.9) in terms of z is valid for $-\pi/2 \le \theta \le \pi$
(C.3.10) in terms of z is valid for $-\pi \le \theta \le \pi/2$
(C.3.11) in terms of z is valid for $-\pi/2 \le \theta \le \pi$
(C.3.12) in terms of z is valid for $-\pi \le \theta \le \pi/2$

An asymptotic approximation that applies for complex values ν as $|\nu| \to \infty$ for a given x for (C.1.7) is

$$J_\nu(x) \approx \frac{1}{\sqrt{2\pi\nu}}\left(\frac{ex}{2\nu}\right)^\nu \qquad (|\arg(\nu) < \pi/2|)$$

and for (C.2.9):

$$H_\nu^{(1)}(x) \approx \begin{cases} -j\sqrt{\frac{2}{\pi\nu}}\left(\frac{2\nu}{ex}\right)^\nu & (|\arg(\nu)| < \pi/2) \\ \sqrt{\frac{2}{\pi\nu}}\left(\frac{2\nu}{ex}\right)^\nu & (\pi/2 < \arg(\nu) < 3\pi/2) \end{cases}$$

Analytic continuation formulas are

$$\begin{aligned} J_\nu(ze^{\pm j\pi}) &= e^{\pm j\pi\nu}J_\nu(z) \\ Y_\nu(ze^{j\pi}) &= e^{-j\pi\nu}Y_\nu(z) + j2\cos(\nu\pi)J_\nu(z) \\ H_\nu^{(1)}(ze^{j\pi}) &= -e^{-j\pi\nu}H_\nu^{(2)}(z) \\ H_\nu^{(2)}(ze^{-j\pi}) &= -e^{j\pi\nu}H_\nu^{(1)}(z) \end{aligned} \qquad \text{(C.5.1)}$$

In the above relations, if z is in the upper Riemann surface, that is $z = re^{j\theta}$ where $-\pi < \theta \le \pi$, the argument of $ze^{jm\pi} = re^{j(\theta+m\pi)}$ where $m = \pm 1$ remains in the upper Riemann surface if $-\pi < \theta + m\pi \le \pi$. For example, if z is a point in the third or fourth quadrant (i.e., $-\pi < \theta \le 0$), then with $m = 1$ it follows that $ze^{j\pi} = re^{j(\theta+\pi)}$ is a point in the first or second quadrant of the upper Riemann surface. Similarly, if z is a point in the first or second quadrant (i.e., $(0 < \theta \le \pi)$, then with $m = -1$ it follows that $ze^{-j\pi} = re^{j(\theta-\pi)}$ is a point in the third or fourth quadrant of the upper Riemann surface. The analytic continuation in (C.5.1) remains in the upper Riemann surface if

$$J_\nu(ze^{j\pi}) = e^{j\nu\pi} J_\nu(z) \qquad (-\pi < \theta \le 0)$$
$$J_\nu(ze^{-j\pi}) = e^{-j\nu\pi} J_\nu(z) \qquad (0 < \theta \le \pi)$$

C.6 THE SOMMERFELD INTEGRALS

The integral solution

$$J_n(x) = \frac{e^{-jn\pi/2}}{2\pi} \int_{-\pi}^{\pi} e^{j(x\sin\theta + n\theta)} d\theta \tag{C.6.1}$$

suggest a solution of the form

$$J_\nu(x) = \frac{e^{-j\pi\nu/2}}{2\pi} \int_{C_0} e^{j(x\cos\beta + \nu\beta)} d\beta \tag{C.6.2}$$

where β is a complex variable and C_0 is a contour such that (C.6.2) satisfies Bessel's equation, and when $\nu = n$ it is equal to (C.6.1). Similarly, for the Hankel functions:

$$H_\nu^{(1)}(x) = \frac{e^{-j\pi\nu/2}}{\pi} \int_{C_1} e^{j(x\cos\beta + \nu\beta)} d\beta \tag{C.6.3}$$

and

$$H_\nu^{(2)}(x) = \frac{e^{-j\pi\nu/2}}{\pi} \int_{C_2} e^{j(x\cos\beta + \nu\beta)} d\beta \tag{C.6.4}$$

The relations (C.6.2), (C.6.3), and (C.6.4) satisfy Bessel's equation if

$$\mathrm{Re}[j(x\cos\beta + \nu\beta)] < 0 \tag{C.6.5}$$

as $\beta'' \to \pm\infty$ along the contours C_0, C_1 and C_2. To analyze the regions where (C.6.5) is satisfied, we write

$$\begin{aligned} j(x\cos\beta + \nu\beta) &= j[x\cos(\beta' + j\beta'') + \nu(\beta' + j\beta'')] \\ &= j[x\cos\beta'\cosh\beta'' - jx\sin\beta'\sinh\beta'' + \nu\beta' + j\nu\beta''] \end{aligned}$$

Hence,

$$\mathrm{Re}[j(x\cos\beta + \nu\beta)] = x\sin\beta'\sinh\beta'' - \nu\beta'' \tag{C.6.6}$$

Consider the range $-\pi < \beta' < 2\pi$. As $\beta'' \to \infty$, since

$$\sinh\beta'' \approx e^{\beta''}/2 >> \nu\beta''$$

it follows from (C.6.6) that

$$\begin{cases} \mathrm{Re}[j(x\cos\beta+\nu\beta)] < 0 \text{ if } \sin\beta' > 0 \Rightarrow 0 < \beta' < \pi \\ \mathrm{Re}[j(x\cos\beta+\nu\beta)] > 0 \text{ if } \sin\beta' < 0 \Rightarrow -\pi < \beta' < 0,\ \pi < \beta' < 2\pi \end{cases}$$

As $\beta'' \to -\infty$, since

$$\sinh\beta'' \approx -e^{-\beta''}/2$$

it follows from (C.6.6) that

$$\begin{cases} \mathrm{Re}[j(x\cos\beta+\nu\beta)] > 0 \text{ if } \sin\beta' < 0 \Rightarrow -\pi < \beta' < 0,\ \pi < \beta' < 2\pi \\ \mathrm{Re}[j(x\cos\beta+\nu\beta)] < 0 \text{ if } \sin\beta' > 0 \Rightarrow 0 < \beta' < \pi \end{cases}$$

The above considerations show that (C.6.2), (C.6.3), and (C.6.4) converge if the integration paths end in the regions denoted by $\mathrm{Re}(\zeta) < 0$, where $\zeta = j(x\cos\beta + \nu\beta)$, in Fig. C.6.

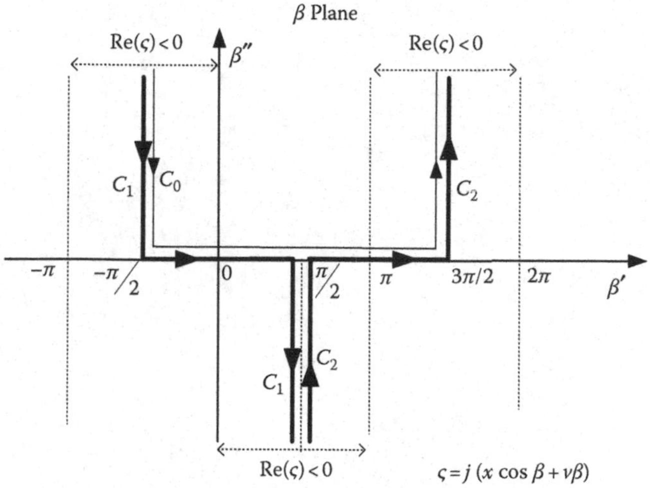

FIGURE C.6 Path of integrations for the integrals of $J_\nu(x)$, $H_\nu^{(1)}(x)$ and $H_\nu^{(2)}(x)$.

Appendix D: Airy Functions

The Airy differential equation is

$$\frac{d^2y}{dx^2} - xy = 0 \tag{D.1}$$

whose solutions are the Airy functions of the first and second kind, denoted by $Ai(x)$ and $Bi(x)$. That is,

$$y(x) = c_1 Ai(x) + c_2 Bi(x)$$

The series forms of the solutions are

$$Ai(x) = af(x) - bg(x)$$
$$Bi(x) = \sqrt{3}\,[af(x) + bg(x)]$$

where

$$f(x) = 1 + \frac{1}{3!}x^3 + \frac{1\cdot 4}{6!}x^6 + \frac{1\cdot 4\cdot 7}{9!}x^9 + \ldots$$
$$g(x) = x + \frac{2}{4!}x^4 + \frac{2\cdot 5}{7!}x^7 + \frac{2\cdot 5\cdot 8}{10!}x^{10} + \ldots$$

$$a = \frac{1}{3^{2/3}\Gamma(2/3)} = 0.35503$$

and

$$b = \frac{1}{3^{1/3}\Gamma(1/3)} = 0.25882$$

Plots of the Airy functions are shown in Fig. D.1.

Closed forms for the solutions are

$$Ai(x) = \sum_{k=0}^{\infty} \frac{x^{3k}}{3^{2k+2/3}k!\Gamma(k+2/3)} - \sum_{k=0}^{\infty} \frac{x^{3k+1}}{3^{2k+4/3}k!\Gamma(k+4/3)}$$

and

$$Bi(x) = \sqrt{3}\left[\sum_{k=0}^{\infty} \frac{x^{3k}}{3^{2k+2/3}k!\Gamma(k+2/3)} + \sum_{k=0}^{\infty} \frac{x^{3k+1}}{3^{2k+4/3}k!\Gamma(k+4/3)}\right]$$

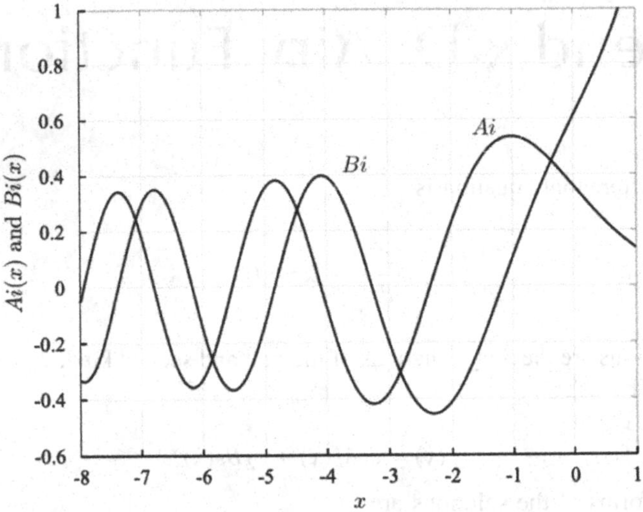

FIGURE D.1 Plots of the Airy functions.

The Wronskian is:

$$W[Ai(x), Bi(x)] = \frac{1}{\pi}$$

The Airy functions are related to the Bessel functions since the solution of (D.1) can be written in terms of Bessel functions of order 1/3. Some relations between the Airy functions and the Bessel functions are

$$\begin{aligned}
Ai(x) &= \frac{x^{1/2}}{3}[I_{-1/3}(\varsigma) - I_{1/3}(\varsigma)] = \frac{1}{\pi}\sqrt{\frac{x}{3}}K_{1/3}(\varsigma) \\
Bi(x) &= \frac{x^{1/2}}{3}[I_{-1/3}(\varsigma) + I_{1/3}(\varsigma)] \\
Ai(-x) &= \frac{x^{1/2}}{3}[J_{-1/3}(\varsigma) + J_{1/3}(\varsigma)] \\
Bi(-x) &= \frac{x^{1/2}}{3}[J_{-1/3}(\varsigma) - J_{1/3}(\varsigma)]
\end{aligned}$$

$$\begin{aligned}
Ai(x) &= \frac{1}{2}\sqrt{\frac{x}{3}}e^{j2\pi/3}H_{1/3}^{(1)}(\varsigma e^{j\pi/2}) \\
Ai(x) &= \frac{1}{2}\sqrt{\frac{x}{3}}e^{-j2\pi/3}H_{1/3}^{(2)}(\varsigma e^{-j\pi/2}) \\
Ai'(x) &= \frac{1}{2}\frac{x}{\sqrt{3}}e^{-j\pi/6}H_{2/3}^{(1)}(\varsigma e^{j\pi/2}) \\
Ai'(x) &= \frac{1}{2}\frac{x}{\sqrt{3}}e^{j\pi/6}H_{2/3}^{(2)}(\varsigma e^{-j\pi/2})
\end{aligned}$$

where

$$\varsigma = \frac{2}{3}x^{3/2}$$

Also:

$$Bi(x) \pm jAi(x) = 2e^{\pm j\pi/6}Ai(xe^{\pm j2\pi/3}) \tag{D.2}$$

and

$$Bi'(x) \pm jAi'(x) = -2e^{\mp j\pi/6}Ai'(xe^{\pm j2\pi/3}) \tag{D.3}$$

where the argument $xe^{\pm j2\pi/3}$ can also be expressed as $-\ xe^{\mp j\pi/3}$.

The asymptotic expressions as $x \to \infty$ are

$$Ai(x) \approx \frac{1}{2\sqrt{\pi}x^{1/4}}e^{-\frac{2}{3}x^{3/2}} \tag{D.4}$$

$$Ai'(x) \approx -\frac{x^{1/4}}{2\sqrt{\pi}}e^{-\frac{2}{3}x^{3/2}} \tag{D.5}$$

$$Bi(x) \approx \frac{1}{\sqrt{\pi}x^{1/4}}e^{\frac{2}{3}x^{3/2}} \tag{D.6}$$

$$Bi'(x) \approx \frac{x^{1/4}}{\sqrt{\pi}}e^{\frac{2}{3}x^{3/2}} \tag{D.7}$$

$$Ai(-x) \approx \frac{1}{\sqrt{\pi}x^{1/4}}\cos\left(\frac{2}{3}x^{3/2} - \frac{\pi}{4}\right) = \frac{1}{\sqrt{\pi}x^{1/4}}\sin\left(\frac{2}{3}x^{3/2} + \frac{\pi}{4}\right) \tag{D.8}$$

$$Bi(-x) \approx -\frac{1}{\sqrt{\pi}x^{1/4}}\sin\left(\frac{2}{3}x^{3/2} - \frac{\pi}{4}\right) = \frac{1}{\sqrt{\pi}x^{1/4}}\cos\left(\frac{2}{3}x^{3/2} + \frac{\pi}{4}\right) \tag{D.9}$$

It is to be noted that the expressions (D.8) and (D.9) do not match exactly the zeros the Airy functions as $x \to \infty$. However, they are still useful in many applications.

The zeros of the Airy functions $Ai(x)$ and the corresponding value of $Ai'(x)$, and the zeros of $Ai'(x)$ and the corresponding value of $Ai(x)$are listed in Fig. D.2.

The Airy functions are analytic functions of z. Hence, except for (D.4) to (D.9), the previous relations are valid with x replaced by $z = re^{j\theta}$. The asymptotic relations in (D.4) to (D.9) have a restricted range of validity:

(a)

$Ai(a_n) = 0$	$Ai'(a_n)$
$a_1 = -2.338$	0.701
$a_2 = -4.088$	-0.803
$a_3 = -5.521$	0.865
$a_4 = -6.787$	-0.911
$a_5 = -7.944$	0.947
$a_6 = -9.023$	-0.978

(b)

$Ai'(a'_n) = 0$	$Ai(a'_n)$
$a'_1 = -1.019$	0.536
$a'_2 = -3.248$	-0.419
$a'_3 = -4.820$	0.380
$a'_4 = -6.163$	-0.358
$a'_5 = -7.372$	0.342
$a'_6 = -8.488$	-0.330

FIGURE D.2 (a) The zeros of $Ai(x)$ denoted by a_n for $n = 1, 2, 3, \ldots$, and the corresponding values of $Ai'(a_n)$; (b) The zeros of $Ai'(x)$ denoted by a_n' for $n = 1, 2, 3, \ldots$, and the corresponding values of $Ai(a_n')$.

(D.4) is valid for $|\theta| < \pi$.
(D.5) is valid for $|\theta| < \pi$.
(D.6) is valid for $|\theta| < \pi/3$.
(D.7) is valid for $|\theta| < \pi/3$.
(D.8) is valid for $|\theta| < 2\pi/3$.
(D.9) is valid for $|\theta| < 2\pi/3$.

The solutions to Airy's differential equation can also be written in terms of contour integrals. Write (D.1) as

$$\frac{d^2y}{dt^2} - ty = 0 \tag{D.10}$$

For $Ai(t)$ the solution to (D.10) is

$$Ai(t) = \frac{1}{2\pi j} \int_{C_1} e^{tz - z^3/3} dz \tag{D.11}$$

To show that (D.11) is a solution, substitute (D.11) into (D.10). That is

$$\begin{aligned} \left(\frac{d^2}{dt^2} - t \right) Ai(t) &= \frac{1}{2\pi j} \int_{C_1} (z^2 - t) e^{tz - z^3/3} dz \\ &= -\frac{1}{2\pi j} \int_{C_1} \frac{d}{dz} \left(e^{tz - z^3/3} \right) dz \\ &= -\frac{1}{2\pi j} e^{tz - z^3/3} \Big|_a^b \end{aligned}$$

which require that

$$e^{tz - z^3/3} \Big|_a^b = 0 \tag{D.12}$$

where a and b are the end points of the contour C_1, taken at $|z| \rightarrow \infty$. The condition (D.12) is satisfied at the end points if

$$e^{-z^3/3} = e^{-1/3[\mathrm{Re}(z^3) + j\,\mathrm{Im}(z^3)]} \rightarrow 0 \text{ as } |z| \rightarrow \infty$$

Letting $z = re^{j\theta}$, convergence at the end points require that:

$$\mathrm{Re}(z^3) = r^3 \cos(3\theta) > 0 \tag{D.13}$$

which is satisfied if the end points are in the regions $-\pi/6 < \theta < \pi/6$, $\pi/2 < \theta < 5\pi/6$ and $-5\pi/6 < \theta < -\pi/2$. The three regions where (D.13) is satisfied are shown in Fig. D.3. The contour C_1 goes from $\infty e^{-j2\pi/3}$ to $\infty e^{j2\pi/3}$. Two other contours are shown, C_2 goes from $\infty e^{j2\pi/3}$ to ∞, and C_3 from ∞ to $\infty e^{-j2\pi/3}$. Since the integral in (D.11) is analytic, it follows that the integral added over the contours C_1+C_2+C_3 is zero.

The integral representation for $Bi(t)$ is

$$Bi(t) = \frac{1}{2\pi} \int_{C_2} e^{tz - z^3/3} dz - \frac{1}{2\pi} \int_{C_3} e^{tz - z^3/3} dz \tag{D.14}$$

The integrands in (D.11) and (D.14) are analytic so the contours can be changed as long as they end in their respective regions of convergence. For example the

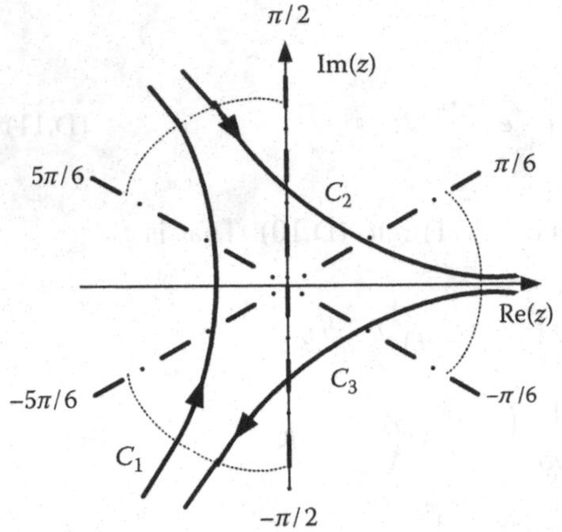

FIGURE D.3 Region where $\mathrm{Re}(z^3) > 0$, and the contours C_1, C_2, and C_3.

contour C_1 can be moved to the imaginary axis provided that the end points begin at $\theta < -\pi/2$ and end at $\theta > \pi/2$.

Changing z to jx (x real) in (D.11) and (D.14), it follows that:

$$Ai(t) = \frac{1}{2\pi}\int_{-\infty}^{\infty} e^{j(tx+x^3/3)}dx = \frac{1}{\pi}\int_0^{\infty}\cos(tx + x^3/3)dx$$

and

$$Bi(t) = \frac{1}{\pi}\int_0^{\infty}\left[e^{tx-x^3/3} + \sin(tx + x^3/3)\right]dx$$

In diffraction problems the functions $w_1(t)$ and $w_2(t)$, known as the Fok-Airy (or Fock-Airy) functions, are also found. They were originally defined by Fok as

$$w_1(t) = \frac{1}{\sqrt{\pi}}\int_{\Gamma_1} e^{tz-z^3/3}dz \tag{D.15}$$

and

$$w_2(t) = \frac{1}{\sqrt{\pi}}\int_{\Gamma_2} e^{tz-z^3/3}dz \tag{D.16}$$

where Γ_1 and Γ_2 are shown in Fig. D.4. However, in the literature one also finds the definition interchanged (i.e., $w_1(t)$ defined as $w_2(t)$, and $w_2(t)$ defined as $w_1(t)$). We will adhere to the original definition by Fok in (D.15) and (D.16).

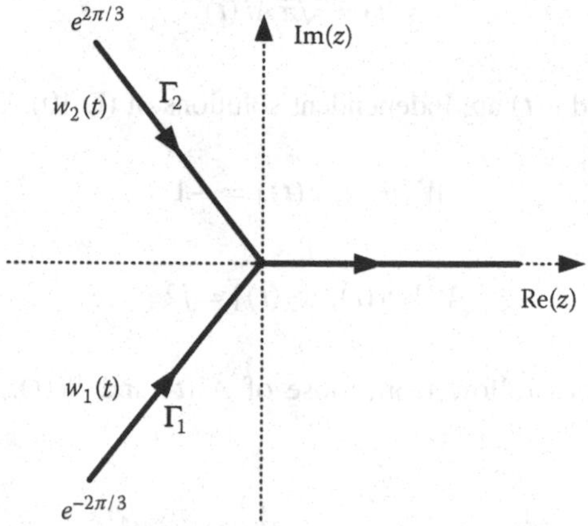

FIGURE D.4 Paths of integration Γ_1 and Γ_2.

From (D.11), (D.14), (D.15), and (D.16), observing that the contour C_2 is the same as Γ_2, C_3 is the opposite of Γ_1, and C_1 is $\Gamma_1 - \Gamma_2$ it follows that:

$$j2\sqrt{\pi}Ai(t) = w_1(t) - w_2(t) \quad \Rightarrow \quad Ai(t) = \frac{1}{j2\sqrt{\pi}}[w_1(t) - w_2(t)]$$

$$2\sqrt{\pi}Bi(t) = w_2(t) + w_1(t) \quad \Rightarrow \quad Bi(t) = \frac{1}{2\sqrt{\pi}}[w_1(t) + w_2(t)]$$

or

$$w_1(t) = \sqrt{\pi}\,[Bi(t) + jAi(t)] \tag{D.17}$$

$$w_2(t) = \sqrt{\pi}\,[Bi(t) - jAi(t)] \tag{D.18}$$

The real and imaginary parts of $w_1(z)$ and $w_2(z)$ are usually denoted by $u(z)$ and $v(z)$.

That is,

$$w_1(t) = u(t) + jv(t)$$
$$w_2(t) = u(t) - jv(t)$$

where

$$u(t) = \sqrt{\pi}Bi(t)$$

and

$$v(t) = \sqrt{\pi} Ai(t)$$

The functions $u(t)$ and $v(t)$ are independent solutions of (D.10). The Wronskians are

$$W[u(t), v(t)] = -1$$

$$W[w_1(t), w_2(t)] = j2$$

Asymptotic expressions follow from those of $Ai(t)$ and $Bi(t)$. For large negative values of x:

$$\begin{aligned}
w_1(x) &= (-x)^{-1/4} e^{j\left[\frac{2}{3}(-x)^{3/2} + \pi/4\right]} \\
w_1{}'(x) &= (-x)^{1/4} e^{j\left[\frac{2}{3}(-x)^{3/2} - \pi/4\right]} \\
w_2(x) &= (-x)^{-1/4} e^{-j\left[\frac{2}{3}(-x)^{3/2} + \pi/4\right]} \\
w_2{}'(x) &= (-x)^{1/4} e^{-j\left[\frac{2}{3}(-x)^{3/2} - \pi/4\right]}
\end{aligned}$$

For large positive values of x:

$$\begin{aligned}
u(x) &\approx x^{-1/4} e^{\frac{2}{3}x^{3/2}} \\
u'(x) &\approx x^{1/4} e^{\frac{2}{3}x^{3/2}} \\
v(x) &\approx \frac{1}{2} x^{-1/4} e^{-\frac{2}{3}x^{3/2}} \\
v'(x) &\approx -\frac{1}{2} x^{-1/4} e^{-\frac{2}{3}x^{3/2}} \\
w_1(x) &\approx w_2(x) \approx u(x) \approx x^{-1/4} e^{\frac{2}{3}x^{3/2}} \\
w_1{}'(x) &\approx w_2{}'(x) \approx u'(x) \approx x^{1/4} e^{\frac{2}{3}x^{3/2}}
\end{aligned}$$

The above asymptotic expansions in terms of z (i.e, $z = re^{j\theta}$) are valid for $|\theta| < \pi/3$.
The functions $w_1(t)$ and $w_2(t)$ are related to the Hankel functions by

$$\begin{aligned}
w_1(t) &= \sqrt{\frac{\pi}{3}} (-t)^{1/2} e^{j2\pi/3} H_{1/3}^{(1)} \left[\frac{2}{3}(-t)^{3/2}\right] \\
w_2(t) &= \sqrt{\frac{\pi}{3}} (-t)^{1/2} e^{-j2\pi/3} H_{1/3}^{(2)} \left[\frac{2}{3}(-t)^{3/2}\right]
\end{aligned}$$

Asymptotic expressions for large values of x and ν of the Bessel functions in terms of the Fok-Airy functions are

$$t = \left(\frac{2}{x}\right)^{1/3} (\nu - x)$$

$$J_\nu(x) \approx \frac{1}{\sqrt{\pi}}\left(\frac{2}{x}\right)^{1/3} v(t)$$
$$H_\nu^{(1)}(x) \approx -j\frac{1}{\sqrt{\pi}}\left(\frac{2}{x}\right)^{1/3} w_1(t)$$
$$H_\nu^{(2)}(x) \approx j\frac{1}{\sqrt{\pi}}\left(\frac{2}{x}\right)^{1/3} w_2(t)$$

and for the spherical Schelkunoff-Bessel functions:

$$\hat{h}_{\nu-1/2}^{(1)}(x) \approx -j\left(\frac{x}{2}\right)^{1/6} w_1(t)$$
$$\hat{h}_{\nu-1/2}^{(2)}(x) \approx j\left(\frac{x}{2}\right)^{1/6} w_2(t)$$
$$\hat{h}_{\nu-1/2}^{(1)\prime}(x) \approx j\left(\frac{x}{2}\right)^{-1/6} w_1{}'(t)$$
$$\hat{h}_{\nu-1/2}^{(2)\prime}(x) \approx -j\left(\frac{x}{2}\right)^{-1/6} w_2{}'(t)$$

$$\hat{j}_{\nu-1/2}(x) \approx \left(\frac{x}{2}\right)^{1/6} v(t)$$
$$\hat{j}_{\nu-1/2}'(x) \approx -\left(\frac{x}{2}\right)^{-1/6} v'(t)$$

The above asymptotic approximations are valid for $t < (x/2)^{2/3}$.

The zeros of Fok-Airy functions and those of the Airy functions are related. From (D.2), (D.3), (D.17), and (D.18):

$$w_1(t) = \sqrt{\pi}\,[Bi(t) + jAi(t)] = 2\sqrt{\pi}\,e^{j\pi/6}Ai(-te^{-j\pi/3})$$
$$w_2(t) = \sqrt{\pi}\,[Bi(t) - jAi(t)] = 2\sqrt{\pi}\,e^{-j\pi/6}Ai(-te^{j\pi/3})$$
$$w_1{}'(t) = \sqrt{\pi}\,[Bi'(t) + jAi'(t)] = -2\sqrt{\pi}\,e^{-j\pi/6}Ai(-te^{-j\pi/3})$$
$$w_2{}'(t) = \sqrt{\pi}\,[Bi'(t) - jAi'(t)] = -2\sqrt{\pi}\,e^{j\pi/6}Ai(-te^{j\pi/3})$$

which show that the zeros of $w_1(t)$, denoted by t_n (i.e., $w_1(t_n) = 0$), are those of $Ai(-t_n e^{-j\pi/3})$. The zeros of $Ai(x)$ are listed in Fig. D.2. Hence,

$$Ai(-t_n e^{-j\pi/3}) = Ai(a_n) = 0 \Rightarrow t_n = -a_n e^{j\pi/3} \qquad (n = 1, 2, 3, ...)$$

Similarly, the zeros of $w_1'(t)$, denoted by t_n', are those of $Ai(-t_n' e^{-j\pi/3})$, or:

$$Ai(-t_n{}' e^{-j\pi/3}) = Ai(a_n') = 0 \Rightarrow t_n' = -a_n' e^{j\pi/3} \qquad (n = 1, 2, 3, ...)$$

The zeros of the Fok-Airy functions are listed in Fig. D.5. The zeros of $w_2(t)$ and $w_2{}'(t)$ are the complex conjugate of the zeros of $w_1(t)$ and $w_1{}'(t)$, respectively.

$w_1(t_n=0)$	$w_1'(t_n'=0)$
$t_1=2.338\ e^{j\pi/3}$	$t_1'=1.019\ e^{j\pi/3}$
$t_2=4.088\ e^{j\pi/3}$	$t_2'=3.248\ e^{j\pi/3}$
$t_3=5.521\ e^{j\pi/3}$	$t_3'=4.820\ e^{j\pi/3}$
$t_4=6.787\ e^{j\pi/3}$	$t_4'=6.163\ e^{j\pi/3}$
$t_5=7.944\ e^{j\pi/3}$	$t_5'=7.372\ e^{j\pi/3}$
$t_6=9.023\ e^{j\pi/3}$	$t_6'=8.488\ e^{j\pi/3}$

$w_2(t_n=0)$	$w_2'(t_n'=0)$
$t_1=2.338\ e^{-j\pi/3}$	$t_1'=1.019\ e^{-j\pi/3}$
$t_2=4.088\ e^{-j\pi/3}$	$t_2'=3.248\ e^{-j\pi/3}$
$t_3=5.521\ e^{-j\pi/3}$	$t_3'=4.820\ e^{-j\pi/3}$
$t_4=6.787\ e^{-j\pi/3}$	$t_4'=6.163\ e^{-j\pi/3}$
$t_5=7.944\ e^{-j\pi/3}$	$t_5'=7.372\ e^{-j\pi/3}$
$t_6=9.023\ e^{-j\pi/3}$	$t_6'=8.488\ e^{-j\pi/3}$

FIGURE D.5 The zeros of $w_1(t)$, $w_1'(x)$, $w_2(t)$ and $w_2'(t)$. The zeros are denoted t_n and t_n'.

Appendix E: Legendre Functions

E.1 LEGENDRE'S EQUATION

Legendre's differential equation is

$$(1 - x^2)\frac{d^2y}{dx^2} - 2x\frac{dy}{dx} + \nu(\nu + 1)y = 0 \tag{E.1.1}$$

or with $x = \cos\theta$:

$$\frac{1}{\sin\theta}\frac{d}{d\theta}\left(\sin\theta\frac{dy}{d\theta}\right) + \nu(\nu + 1)y = 0$$

which can also be expressed as

$$\frac{d^2y}{d\theta^2} + \cot\theta\frac{dy}{d\theta} + \nu(\nu + 1)y = 0$$

Legendre's differential equation appears in the solution to the wave equation in spherical coordinates, where the range of θ is $0 \le \theta \le \pi$, which in terms of x the range is $-1 \le x \le 1$.

The series solution to (E.1.1) is

$$y(x) = c_o y_1(x) + c_1 y_2(x)$$

where

$$y_1(x) = 1 - \frac{\nu(\nu + 1)}{2!}x^2 + \frac{(\nu - 2)\nu(\nu + 1)(\nu + 3)}{4!}x^4 + ...$$

and

$$y_2(x) = x - \frac{(\nu - 1)(\nu + 2)}{3!}x^3 + \frac{(\nu - 1)(\nu - 3)(\nu + 2)(\nu + 4)}{5!}x^5 + ...$$

For positive integer values of ν (i.e., $\nu = n, n = 0, 1, 2, 3, ...$), one of the series $y_1(x)$ or $y_2(x)$ is finite, and yields a polynomial solution, known as the Legendre function of the first kind $P_n(x)$. These polynomials, of course, converge for all values of x. Only positive integers need to be considered because (E.1.1) does not change if n is replaced by $-n-1$, so solutions for negative integers are readily obtained. The other series remains an infinite series, and is known as the Legendre function of the second kind $Q_n(x)$. The function $Q_n(x)$ converges for $|x|<1$, and produces an infinite value at $x=\pm 1$. If ν is not an integer, the solutions $y_1(x)$ or $y_2(x)$ are infinite polynomials.

If n is even, the series $y_1(x)$ produces the polynomials $P_0(x),\ P_2(x), \ldots$; and if n is odd, the series $y_2(x)$ produces the polynomials $P_1(x),\ P_3(x), \ldots$. Each polynomial is normalized to one at $x=1$, showing that the coefficients c_o and c_1 for the nth Legendre polynomial are

$$c_o = \frac{(-1)^{n/2} n!}{2^n\left[\left(\frac{n}{2}\right)!\right]^2} \qquad \text{(for even } n\text{)}$$

and

$$c_1 = \frac{(-1)^{(n-1)/2} n!}{2^{n-1}\left[\left(\frac{n-1}{2}\right)!\right]^2} \qquad \text{(for odd } n\text{)}$$

Some of the Legendre polynomials are

$$\begin{aligned} P_0(x) &= 1 & P_0(\cos\theta) &= 1 \\ P_1(x) &= x & P_1(\cos\theta) &= \cos\theta \\ P_2(x) &= \tfrac{1}{2}(3x^2-1) & P_2(\cos\theta) &= \tfrac{1}{2}(3\cos^2\theta - 1) \\ P_3(x) &= \tfrac{1}{2}(5x^2-2x) & P_3(\cos\theta) &= \tfrac{1}{2}(5\cos^2\theta - 2\cos\theta) \end{aligned}$$

The Legendre polynomials are given by the Rodrigues' formula:

$$P_n(x) = \frac{1}{2^n n!}\frac{d^n}{dx^n}(x^2-1)^n$$

The Legendre functions of the second kind follow from

$$Q_n(x) = \frac{(-1)^{n/2} 2^n\left[\left(\frac{n}{2}\right)!\right]^2}{n!} y_2(x) \qquad \text{(for even } n\text{)}$$

and

$$Q_n(x) = \frac{(-1)^{(n+1)/2} 2^{n-1}\left[\left(\frac{n-1}{2}\right)!\right]^2}{n!} y_1(x) \qquad \text{(for odd } n\text{)}$$

which can also be expressed as

$$Q_n(x) = \frac{1}{2}P_n(x)\ln\frac{1+x}{1-x} - \sum_{j=1}^{n}\frac{1}{j}P_{j-1}(x)P_{n-j}(x)$$

Some solutions for $Q_n(x)$, where $|x| < 1$, are

$$Q_0(x) = \frac{1}{2}\ln\frac{1+x}{1-x} \qquad Q_0(\cos\theta) = \ln\left[\cot\left(\frac{\theta}{2}\right)\right]$$

$$Q_1(x) = \frac{x}{2}\ln\frac{1+x}{1-x} - 1 \qquad Q_1(\cos\theta) = \cos\theta\ln\left[\cot\left(\frac{\theta}{2}\right)\right] - 1$$

$$Q_2(x) = \frac{3x^2-1}{4}\ln\frac{1+x}{1-x} - \frac{3x}{2} \qquad Q_2(\cos\theta) = P_2(\cos\theta)\ln\left[\cot\left(\frac{\theta}{2}\right)\right] - \frac{3}{2}P_1(\cos\theta)$$

The solutions to (E.1.1) when $\nu \neq n$ can be expressed as

$$y(x) = AP_\nu(x) + BP_\nu(-x) \tag{E.1.2}$$

or

$$y(x) = AP_\nu(x) + BQ_\nu(x)$$

For $\nu = n$, we have that:

$$P_n(-x) = (-1)^n P_n(x)$$

which shows that for $\nu = n$ the solutions in (E.1.2) are linearly dependent. For $\nu = n$, the linearly independent solutions are $P_n(x)$ and $Q_n(x)$. Hence,

$$y(x) = AP_n(x) + BQ_n(x)$$

Since (E.1.1) is invariant if $x \to -x$, the solution can also be given by:

$$y(x) = AP_n(-x) + BQ_n(-x)$$

Plots of the Legendre's polynomials $P_n(x)$ for $n = 1, 2, 3, 4$ are shown in Fig. E.1a, and the Legendre's functions $Q_n(x)$ for $n = 0, 1, 2, 3, 4, 5$ in Fig. E.1b. Since $Q_n(\pm 1) = \infty$, the only solution that is valid for integer orders in $-1 \leq x \leq 1$ (or $0 \leq \theta \leq \pi$) is $P_n(x)$.

In terms of $x = \cos\theta$ and $\nu \neq n$, the solution $P_\nu(-\cos\theta)$ is singular at $\theta = 0$ and finite at $\theta = \pi$. The solution $Q_\nu(-\cos\theta)$ is singular at $\theta = 0$ and at $\theta = \pi$.

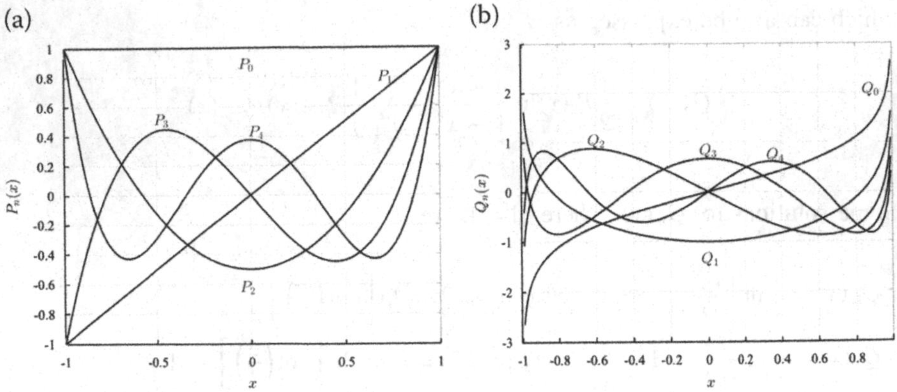

FIGURE E.1 (a) Plot of $P_n(x)$ for $n = 0, 1, 2, 3, 4$; (b) plot of $Q_n(x)$ for $n = 0, 1, 2, 3, 4$.

The Wronskians are

$$W[P_\nu(x), Q_\nu(x)] = \frac{1}{1-x^2}$$
$$W[P_\nu(x), P_\nu(-x)] = \frac{-2\sin\nu\pi}{\pi(1-x^2)}$$

Some properties of the Legendre polynomials are

$$P_n(1) = 1$$
$$P_n(-1) = (-1)^n$$
$$P_n(-x) = (-1)^n P_n(x)$$
$$P_n'(x)|_{x=1} = \frac{1}{2}n(n+1)$$
$$P_n'(x)|_{x=-1} = \frac{1}{2}(-1)^{n-1}n(n+1)$$

and some properties of the Legendre function of the second kind are

$$Q_n(-x) = (-1)^{n+1}Q_n(x)$$
$$Q_n(1) = \infty$$
$$Q_n(-1) = -\infty \qquad \text{(for even } n\text{)}$$
$$Q_n(-1) = \infty \qquad \text{(for odd } n\text{)}$$

The generating function is

$$\frac{1}{\sqrt{1-2tx+t^2}} = \sum_{n=0}^{\infty} P_n(x)\, t^n$$

There are many recursion relations, such as

$$(2n + 1)P_n(x) = P_{n+1}'(x) - P_{n-1}'(x)$$
$$P_{n+1}'(x) = (n + 1)P_n(x) + xP_n'(x)$$

The function $Q_n(x)$ satisfies the same recursion relations.

The approximation of $P_\nu(-\cos\theta)$ for $\theta \to 0$ is

$$P_\nu(-\cos\theta) \underset{\lim \theta\to 0}{\approx} \frac{\sin \nu\pi}{\pi} \ln \theta^2 \tag{E.1.3}$$

and for large values of ν

$$P_\nu(-\cos\theta) \approx \sqrt{\frac{2}{\pi(\nu + 1/2)\sin\theta}} \cos\left[(\nu + 1/2)(\pi - \theta) - \frac{\pi}{4}\right]$$

The orthogonality relation is

$$\int_{-1}^{1} P_n(x)P_p(x)dx = \int_0^\pi P_n(\cos\theta)P_p(\cos\theta)\sin\theta d\theta = \frac{2}{2n + 1}\delta_{np}$$

This property is used to expand a function $f(\theta)$ in terms of Legendre polynomials in the range 0 to π as

$$f(\theta) = \sum_{n=0}^{\infty} a_n P_n(\cos\theta)$$

where

$$a_n = \frac{2n + 1}{2} \int_0^\pi f(\theta)P_n(\cos\theta)\sin\theta d\theta$$

E.2 ASSOCIATED LEGENDRE EQUATION

The associated Legendre equation is

$$(1 - x^2)\frac{d^2y}{dx^2} - 2x\frac{dy}{dx} + \left[\nu(\nu + 1) - \frac{m^2}{1 - x^2}\right]y = 0$$
$$\frac{d}{dx}\left[(1 - x^2)\frac{dy}{dx}\right] + \left[\nu(\nu + 1) - \frac{m^2}{1 - x^2}\right]y = 0, \tag{E.2.1}$$

or with $x = \cos\theta$

$$\frac{1}{\sin\theta}\frac{d}{d\theta}\left(\sin\theta\frac{dy}{d\theta}\right)+\left[\nu(\nu+1)-\frac{m^2}{\sin^2\theta}\right]y=0$$

which can also be expressed as

$$\frac{d^2y}{d\theta^2}+\cot\theta\frac{dy}{d\theta}+\left[\nu(\nu+1)-\frac{m^2}{\sin^2\theta}\right]y=0$$

In this book, only the cases where m is an integer are considered, which in electromagnetics problems is most common.

The solutions to (E.2.1) are the associated Legendre functions of the first and second kind, namely $P_\nu^m(x)$ and $Q_\nu^m(x)$. That is,

$$y(x)=AP_\nu^m(x)+B\,Q_\nu^m(x)$$

where ν is the degree and m is the order. For $\nu = n$, the only solution that is valid for $-1 \le x \le 1$ is $P_n^m(x)$ (or $P_n^m(\cos\theta)$ for $0 \le \theta \le \pi$). At $x = \pm 1$ the $Q_n^m(x)$ function has an infinite value (or $Q_n^m(\cos\theta)$ is infinite at $\theta = 0$ and $\theta = \pi$). The functions $P_n^m(-x)$ (or $P_n^m(-\cos\theta)$) and $Q_n^m(-x)$ (or $Q_n^m(-\cos\theta)$) are also solutions, as well as $P_{-\nu-1}^m(\pm x)$ and $Q_{-\nu-1}^m(\pm x)$, since (E.2.1) is invariant under the transformations $\nu \to -\nu - 1$ and $x \to -x$. The equation is also invariant to the transformation $m \to -m$, leading to solutions with order $-m$.

If $\nu \ne n$, the solution can be expressed as

$$y(x)=AP_\nu^m(x)+B\,P_\nu^m(-x)$$

The solution $P_\nu^m(x)$ is singular at $x = -1$ (or $P_\nu^m(\cos\theta)$ is singular at $\theta = \pi$). Hence, it cannot be used if $x = -1$, or $\theta = \pi$ is in the angular region of interest. Similarly, $P_\nu^m(-x)$ is singular at $x = 1$ (or $P_\nu^m(-\cos\theta)$ is singular at $\theta = 0$). Hence, it cannot be used if $x = 1$ or $\theta = 0$ is in the angular region of interest.

The Wronskian is

$$W\left[P_\nu^m(x),\,Q_\nu^m(x)\right]=\frac{\Gamma(\nu+m-1)}{\Gamma(\nu-m+1)}\frac{1}{1-x^2}$$

For $\nu = n$ and with m an integer such that $0 \le m \le n$, the solutions $P_n^m(x)$ are polynomials, known as the associated Legendre polynomials. These polynomials are given by the following form of the Rodrigues' formula:

$$P_n^m(x)=(-1)^m(1-x^2)^{m/2}\frac{d^mP_n(x)}{dx^m}=\frac{(-1)^m}{2^nn!}(1-x^2)^{m/2}\frac{d^{n+m}}{dx^{n+m}}(x^2-1)^n$$

The $Q_n^m(x)$ functions follow from

$$Q_n^m(x) = (-1)^m (1 - x^2)^{m/2} \frac{d^m Q_n(x)}{dx^m}$$

Some authors omit the factor $(-1)^m$ in the above formulas.

In terms of $x = \cos\theta$, Rodrigues' formula reads

$$P_n^m(\cos\theta) = (-1)^m \sin^m\theta \frac{d^m P_n(\cos\theta)}{d(\cos\theta)^m} = (-\sin\theta)^{m-1} \frac{d^m P_n(\cos\theta)}{d\theta^m}$$

and it follows that

$$P_n^1(\cos\theta) = \frac{dP_n(\cos\theta)}{d\theta}$$

Some properties are

$$P_n^m(x) = 0 \qquad \text{(for } m > n\text{)}$$
$$Q_n^m(x) = 0 \qquad \text{(for } m > n\text{)}$$
$$P_0^0(x) = P_0(x) = 1$$
$$P_n^m(\pm 1) = \begin{cases} 1 & \text{(for } m = 0\text{)} \\ 0 & \text{(for } m > 0\text{)} \end{cases}$$
$$P_n^m(-x) = (-1)^{n+m} P_n^m(x)$$
$$Q_n^m(-x) = (-1)^{n+m+1} Q_n^m(x)$$
$$P_{-\nu-1}^m(x) = P_\nu^m(x)$$
$$P_\nu^{-m}(x) = (-1)^m \frac{\Gamma(\nu-m+1)}{\Gamma(\nu+m+1)} P_\nu^m(x) \Rightarrow P_n^{-m}(x) = (-1)^m \frac{(n-m)!}{(n+m)!} P_n^m(x)$$
$$Q_\nu^{-m}(x) = (-1)^m \frac{\Gamma(\nu-m+1)}{\Gamma(\nu+m+1)} Q_\nu^m(x) \Rightarrow Q_n^{-m}(x) = (-1)^m \frac{(n-m)!}{(n+m)!} Q_n^m(x)$$

Some of the associated Legendre polynomials $P_n^m(x)$ and $P_n^m(\cos\theta)$ are:

$$P_n^0(x) = P_n(x) \qquad P_n^0(\cos\theta) = P_n(\cos\theta)$$
$$P_1^1(x) = -(1 - x^2)^{1/2} \qquad P_1^1(\cos\theta) = -\sin\theta$$
$$P_2^1(x) = -3x(1 - x^2)^{1/2} \qquad P_2^1(\cos\theta) = -3\cos\theta\sin\theta$$
$$P_2^2(x) = 3(1 - x^2) \qquad P_2^2(\cos\theta) = 3\sin^2\theta$$

and closed forms for $Q_n^m(x)$:

$$Q_1^1(x) = -(1 - x^2)\left[\frac{1}{2}\ln\frac{1+x}{1-x} + \frac{x}{1-x^2}\right]$$
$$Q_2^1(x) = -(1 - x^2)\left[\frac{3x}{2}\ln\frac{1+x}{1-x} + \frac{3x^2-2}{1-x^2}\right]$$

There are many recursion relations, such as

$$(n - m + 1)P_{n+1}^m(x) = (2n + 1)x P_n^m(x) - (n + m)P_{n-1}^m(x)$$
$$(1 - x^2)P_n^{m'}(x) = -n x P_\nu^m(x) + (n + m)P_{n-1}^m(x)$$

The function $Q_n^m(x)$ satisfies the same recursion relations.

The orthogonality relation is

$$\int_{-1}^{1} P_n^m(x)P_p^m(x)dx = \int_0^\pi P_n^m(\cos\theta)P_p^m(\cos\theta)\sin\theta d\theta = \frac{2}{2n+1}\frac{(n+m)!}{(n-m)!}\delta_{np} \tag{E.2.2}$$

This property is used to expand a function $f(\theta, \phi)$ in terms of associated Legendre polynomials in the range 0 to π.

E.3 SPHERICAL HARMONICS

The angular parts of the wave equation in spherical coordinates when n and m are integers satisfy:

$$\frac{d^2\Phi}{d\phi^2} + m^2\Phi = 0$$

and

$$\frac{1}{\sin\theta}\frac{d}{d\theta}\left(\sin\theta\frac{d\Theta}{d\theta}\right) + \left[n(n+1) - \frac{m^2}{\sin^2\theta}\right]\Theta = 0$$

whose angular solutions for $0 \le \theta \le \pi$ are of the form $P_n^m(\cos\theta)\cos m\phi$ and $P_n^m(\cos\theta)\sin m\phi$. These terms are referred as spherical harmonics (or surface spherical harmonics). They are also called tesseral harmonics when $m < n$, and sectoral harmonics when $n = m$. The orthogonality condition for the associated Legendre polynomials in (E.2.2) and those of $\sin m\phi$ and $\cos m\phi$ are used to expand a function $f(\theta, \phi)$ in terms of spherical harmonics in the range 0 to π as

$$f(\theta, \phi) = \sum_{n=0}^{\infty} a_{n0}P_n(\cos\theta) + \sum_{m=1}^{\infty} [a_{nm}\cos(m\phi) + b_{nm}\sin(m\phi)]P_n^m(\cos\theta)$$

where

$$a_{n0} = \frac{2n+1}{4\pi}\int_{\theta=0}^{\pi}\int_{\phi=0}^{2\pi} f(\theta, \phi)P_n(\cos\theta)\sin\theta d\theta d\phi$$

$$a_{nm} = \frac{2n+1}{2\pi}\frac{(n-m)!}{(n+m)!}\int_{\theta=0}^{\pi}\int_{\phi=0}^{2\pi} f(\theta,\phi)P_n^m(\cos\theta)\cos(m\phi)\sin\theta d\theta d\phi$$

and

$$b_{nm} = \frac{2n+1}{2\pi}\frac{(n-m)!}{(n+m)!}\int_{\theta=0}^{\pi}\int_{\phi=0}^{2\pi} f(\theta,\phi)P_n^m(\cos\theta)\sin(m\phi)\sin\theta d\theta d\phi$$

The spherical harmonics also appear in terms of the spherical harmonic function $Y_{nm}(\theta, \phi)$. The normalized definition of the $Y_{nm}(\theta, \phi)$ functions are

$$Y_{n0}^e(\theta,\phi) = \sqrt{\frac{2n+1}{4\pi}}P_n(\cos\theta)$$

$$Y_{nm}^e(\theta,\phi) = \sqrt{\frac{2n+1}{2\pi}\frac{(n-m)!}{(n+m)!}}P_n^m(\cos\theta)\cos m\phi \qquad (0 < m \le n)$$

and

$$Y_{nm}^o(\theta,\phi) = \sqrt{\frac{2n+1}{2\pi}\frac{(n-m)!}{(n+m)!}}P_n^m(\cos\theta)\sin m\phi \qquad (1 \le m \le n)$$

where the superscript e and o refer to even in ϕ and odd in ϕ, respectively.

The functions $Y_{nm}^e(\theta, \phi)$ and $Y_{nm}^o(\theta, \phi)$ satisfy the orthogonal relations:

$$\int_{\theta=0}^{\pi}\int_{\phi=0}^{2\pi} Y_{nm}^e(\theta,\phi)Y_{pq}^e(\theta,\phi)\sin\theta d\phi d\theta = \delta_{np}\delta_{mq}$$

$$\int_{\theta=0}^{\pi}\int_{\phi=0}^{2\pi} Y_{nm}^o(\theta,\phi)Y_{pq}^o(\theta,\phi)\sin\theta d\phi d\theta = \delta_{np}\delta_{mq}$$

The functions $Y_{nm}(\theta, \phi)$ is also found in the form

$$Y_{nm}(\theta,\phi) = \sqrt{\frac{2n+1}{4\pi}\frac{(n-m)!}{(n+m)!}}P_n^m(\cos\theta)e^{jm\phi} \qquad (n = 0, 1, 2, \dots; -n \le m \le n)$$

which satisfies the associated Legendre differential equation:

$$\left[\frac{1}{\sin\theta}\frac{\partial}{\partial\theta}\left(\sin\theta\frac{\partial}{\partial\theta}\right) + \frac{1}{\sin^2\theta}\frac{\partial^2}{\partial\phi^2} + n(n+1)\right]Y_{nm}(\theta,\phi) = 0$$

The orthogonal relation is

$$\int_{\theta=0}^{\pi} \int_{\phi=0}^{2\pi} Y_{nm}(\theta, \phi) Y_{pq}^{*}(\theta, \phi) \sin\theta d\theta d\phi = \delta_{np}\delta_{mq}$$

Specific expressions are

$$Y_{00} = \sqrt{\frac{1}{4\pi}}$$

$$Y_{10} = \sqrt{\frac{3}{4\pi}} \cos\theta$$

$$Y_{11} = -\sqrt{\frac{3}{8\pi}} \sin\theta e^{j\phi}$$

$$Y_{21} = -\sqrt{\frac{15}{8\pi}} \sin\theta \cos\theta e^{j\phi}$$

and they satisfy the property

$$Y_{n,-m}(\theta, \phi) = (-1)^m Y_{nm}^{*}(\theta, \phi)$$

The expansion of a function $f(\theta, \phi)$ in terms the spherical harmonics $Y_{nm}(\theta, \phi)$ is

$$f(\theta, \phi) = \sum_{n=0}^{\infty} \sum_{m=-n}^{n} a_{nm} Y_{nm}(\theta, \phi)$$

where

$$a_{nm} = \int_{\theta=0}^{\pi} \int_{\phi=0}^{2\pi} f(\theta, \phi) Y_{nm}^{*}(\theta, \phi) \sin\theta d\theta d\phi$$

The addition theorem for the spherical harmonics is:

$$\begin{aligned} P_n(\cos\gamma) &= \frac{4\pi}{2n+1} \textstyle\sum_{m=-n}^{n} Y_{nm}(\theta, \phi)\, Y_{nm}^{*}(\theta', \phi') \\ &= \textstyle\sum_{m=-n}^{n} \frac{(n-m)!}{(n+m)!} P_n^m(\cos\theta) P_n^m(\cos\theta') e^{jm(\phi-\phi')} \\ &= \textstyle\sum_{m=0}^{n} \varepsilon_m \frac{(n-m)!}{(n+m)!} P_n^m(\cos\theta) P_n^m(\cos\theta') \cos\left[m(\phi - \phi')\right] \end{aligned}$$

where ε_m is the Neumann's number and

$$\cos\gamma = \cos\theta\cos\theta' + \sin\theta\sin\theta'\cos(\phi - \phi')$$

The angle γ is the angle between the vectors $\mathbf{r}$ and $\mathbf{r'}$ where

$$\hat{\mathbf{a}}_r = \sin\theta\cos\phi\, \hat{\mathbf{a}}_x + \sin\theta\sin\phi\, \hat{\mathbf{a}}_y + \cos\theta\, \hat{\mathbf{a}}_z$$

$$\hat{\mathbf{a}}_{r'} = \sin\theta'\cos\phi'\, \hat{\mathbf{a}}_x + \sin\theta'\sin\phi'\, \hat{\mathbf{a}}_y + \cos\theta'\, \hat{\mathbf{a}}_z$$

and

$$\hat{\mathbf{a}}_r \cdot \hat{\mathbf{a}}_{r'} = \cos\gamma$$

The completeness relation is

$$\frac{\delta(\theta - \theta')\delta(\phi - \phi')}{\sin\theta} = \sum_{n=0}^{\infty} \sum_{m=-n}^{n} Y_{nm}(\theta, \phi) Y_{nm}^*(\theta', \phi')$$

E.4 COMPLEX ORDER

In some problems, the evaluation of $P_\nu(x)$ for complex values of ν is required. For complex values of ν the solution $P_\nu(x)$ in (E.1.2) can be expressed in terms of the hypergeometric function. That is:

$$P_\nu(x) = F\left(-\nu, \nu + 1; 1; \frac{1 - x}{2}\right) \tag{E.4.1}$$

where the hypergeometric function is given by

$$F(a, d; c; x) = \frac{\Gamma(c)}{\Gamma(a)\Gamma(b)} \sum_{n=0}^{\infty} \frac{\Gamma(a + n)\Gamma(b + n)}{\Gamma(c + n)} \frac{x^n}{n!} \tag{E.4.2}$$

The region of convergence of (E.4.1) is $|x - 1| < 2$, and that of (E.4.2) is $|x| < 1$. The hypergeometric function in (E.4.2) is conveniently written as

$$F(a, d; c; x) = \sum_{n=0}^{\infty} \frac{(a)_n (b)_n}{(c)_n} \frac{x^n}{n!} \tag{E.4.3}$$

where the notation

$$(\alpha)_n = \frac{\Gamma(\alpha + n)}{\Gamma(n)} \qquad (n = 0, 1, 2, \ldots)$$

is the Pochhammer's notation.

The function $P_\nu(-\cos\theta)$ in terms of (E.4.3) is

$$P_\nu(-\cos\theta) = F\left(-\nu, \nu + 1; 1; \frac{1 + \cos\theta}{2}\right) = \sum_{n=0}^{\infty} (-\nu)_n (\nu + 1)_n \frac{(1 + \cos\theta)^n}{2^n (n!)^2} \tag{E.4.4}$$

which converges for $0 < \theta \le \pi$. The derivative of (E.4.4) is:

$$\begin{aligned} \frac{d}{d\theta} P_\nu(-\cos\theta) &= \frac{\nu(\nu+1)}{2} F\left(-\nu + 1, \nu + 2; 2; \frac{1 + \cos\theta}{2}\right) \\ &= \sin\theta \frac{\nu(\nu+1)}{2} \textstyle\sum_{n=0}^{\infty} (-\nu + 1)_n (\nu + 2)_n \frac{(1 + \cos\theta)^n}{2^n n!\,(n+1)!} \end{aligned}$$

Appendix F: The Transformations $\lambda = k \sin \beta$ and $\lambda = k \cos \beta$

The transformation $\lambda = k \sin \beta$ from the λ plane to the β plane is first analyzed for real values of k. The λ and β planes are shown in Fig. F.1a, where in the first quadrant of the λ plane: $\lambda' > 0$ and $\lambda'' > 0$. To analyze the mapping of the first quadrant, we write

$$\lambda' + j\lambda'' = k \sin(\beta' + j\beta'')$$

or

$$\lambda' = k \sin\beta' \cosh\beta'' \tag{F.1}$$

and

$$\lambda'' = k \cos\beta' \sinh\beta'' \tag{F.2}$$

If $\lambda' > 0$, (F.1) requires that $\sin\beta' > 0$ since $\cosh\beta'' > 0$ for β'' positive or negative. The condition $\sin\beta' > 0$ is satisfied by a variety of β' values, such as $0 < \beta' < \pi$, $2\pi < \beta' < 3\pi$, $-2\pi < \beta' < -\pi$, etc. If $\lambda'' > 0$, (F.2) shows that $\sinh\beta''$ is positive when $\beta'' > 0$ and negative when $\beta'' < 0$. Then, if $\sinh\beta'' > 0$ (i.e., $\beta'' > 0$), we must require that $\cos\beta' > 0$, which is satisfied by a variety of β' values, such as $-\pi/2 < \beta' < \pi/2$, $3\pi/2 < \beta' < 5\pi/2$, etc. If $\sinh\beta'' < 0$ (i.e., $\beta'' < 0$), then we require that $\cos\beta' < 0$, which is satisfied by a variety of β' values, such as $-3\pi/2 < \beta' < -\pi/2$, $\pi/2 < \beta' < 3\pi/2$, etc.

From the previous information, the mapping of the first quadrant in the λ plane to the β plane in the region $0 \le \beta' \le \pi$ is illustrated in Fig. F.1b. The figure also shows the mapping of several points (i.e., **a**, **b**, and **c**). The origin in the λ plane (i.e., $\lambda' = \lambda'' = 0$) maps into $\beta' = \beta'' = 0$ (see point **a**). The point $\lambda' = k$, $\lambda'' = 0$ maps into $\beta' = \pi/2$, $\beta'' = 0$ (see point **b**). For values of $\lambda' > k$ along the real axis (see point **c**), (F.1) and (F.2) show that $\beta' = \pi/2$ and since $\cosh\beta''$ is an even function, there are two values of β'' that will satisfy (F.1). As the point **c** moves to infinity (i.e., $\lambda' \to \infty$, $\lambda'' = 0$), it maps into $\beta' = \pi/2$, $\beta'' \to \pm\infty$.

The analysis of the mapping of the second, third, and fourth quadrant is similar. The complete mapping is shown in Fig. F.1b in the region $-\pi < \beta' < \pi$.

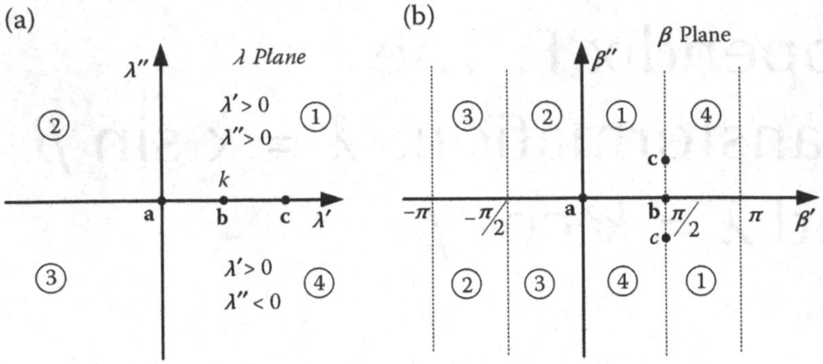

FIGURE F.1 (a) The λ plane; (b) mapping of the first and fourth quadrants in the λ plane to the β plane in the region $0 < \beta' < \pi$, and the mapping of the second and fourth quadrants.

A path of integration in the λ plane is usually along the real axis extending from $-\infty$ to ∞. The mapping of this path into the β plane comes with a requirement, such as $\mathrm{Im}[\sqrt{k^2 - \lambda^2}] < 0$, which in terms of the transformation $\lambda = k \sin\beta$ reads

$$\mathrm{Im}[k\cos\beta] < 0 \tag{F.3}$$

Since

$$\cos(\beta' + j\beta'') = \cos\beta' \cosh\beta'' - j \sin\beta' \sinh\beta''$$

the condition (F.3) requires that

$$\sin\beta' \sinh\beta'' > 0 \tag{F.4}$$

To satisfy (F.4), if $\beta'' > 0$ then $0 < \beta' < \pi$, and if $\beta'' < 0$, then $-\pi < \beta' < 0$. The regions where (F.4) is satisfied are called the upper Riemann surfaces (U), and the region where (F.4) is not satisfied is the lower Riemann surfaces (L). These surfaces are shown in Fig. F.2 in the range $-\pi < \beta' < \pi$, as well as the resulting path of integration in the β plane, which is in the upper Riemann surface. Observe that the point $\lambda' = -k$, $\lambda'' = 0$ maps into $\beta' = -\pi/2$, $\beta'' = 0$, and for $\lambda' < -k$, the path begins at $\beta' = -\pi/2$, $\beta'' \to -\infty$ (i.e., in the upper Riemann surface of the second and third quadrant mapping).

The mapping in Fig. F.2, of course, repeats for values of β' greater than π, and values of β' smaller than $-\pi$.

The other common transformation is $\lambda = k\cos\beta$. Since $\sin(\beta + \pi/2) = \cos\beta$, it follows that the mapping is similar to that of $\lambda = k\sin\beta$, except that is shifted by $\pi/2$, as shown in Fig. F.3. For this transformation

$$\lambda' = k\cos\beta' \cosh\beta'' \tag{F.5}$$

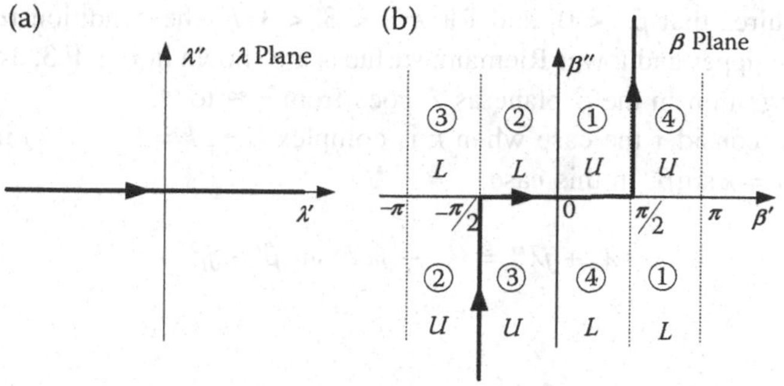

FIGURE F.2 The transformation $\lambda = k\sin\beta$ showing the upper and lower Riemann surfaces and the mapping of path of integration from the λ to the β plane.

and

$$\lambda'' = -k\sin\beta'\sinh\beta'' \tag{F.6}$$

The origin in the λ plane transforms to $\beta' = \pi/2$, $\beta'' = 0$, the point at $\lambda' = k$, $\lambda'' = 0$ transforms to the origin the β plane, and the point $\lambda' = -k$, $\lambda'' = 0$ transforms to

$$\beta' = \pi,\ \ \beta'' = 0.$$

The upper Riemann surface for the transformation $\lambda = k\cos\beta$ is defined by

$$\mathrm{Im}(k\sin\beta) < 0 \ \Rightarrow\ k\cos\beta'\sinh\beta'' < 0$$

Therefore, for $-\pi/2 < \beta' < \pi/2$, the above condition for the upper Riemann

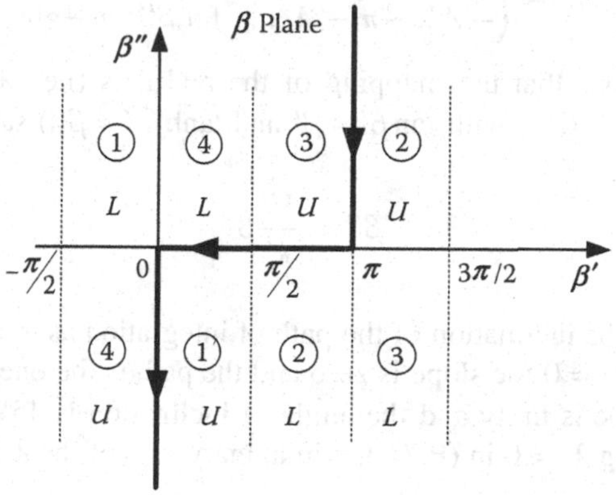

FIGURE F.3 The transformation $\lambda = k\cos\beta$ and the path of integration.

surface requires that $\beta'' < 0$, and for $\pi/2 < \beta' < 3\pi/2$ the condition requires that $\beta'' > 0$. The upper and lower Riemann surfaces are shown in Fig. F.3, as well as the path of integration in the β plane as λ goes from $-\infty$ to ∞.

Next, we consider the case when k is complex (i.e., $k = k' - jk''$) in the transformation $\lambda = k\sin\beta$. In this case,

$$\lambda' + j\lambda'' = (k' - jk'')\sin(\beta' + j\beta'')$$

Therefore,

$$\lambda' = k'\sin\beta'\cosh\beta'' + k''\cos\beta'\sinh\beta'' \tag{F.7}$$

and

$$\lambda'' = k'\cos\beta'\sinh\beta'' - k''\sin\beta'\cosh\beta'' \tag{F.8}$$

Setting $\lambda'' = 0$ in (F.8), the real axis of the λ plane (which represents the path of integration) transforms to

$$\tan\beta' = \frac{k'}{k''}\tanh\beta'' \tag{F.9}$$

or

$$\tan\beta'\coth\beta'' = \tan A_o \Rightarrow A_o = \tan^{-1}\left(\frac{k'}{k''}\right) \tag{F.10}$$

Since $\coth\beta'' \to \pm 1$ as $\beta'' \to \pm\infty$, the relation (F.10) has asymptotes given by

$$\beta' = \begin{cases} A_o,\ \pi + A_o,\ \ldots \text{ for } \beta'' \to \infty \\ -A_o,\ -\pi - A_o,\ \ldots \text{ for } \beta'' \to -\infty \end{cases}$$

Equation (F.9) shows that the mapping of the real axis (i.e., $\lambda'' = 0$) around the origin in the β plane (i.e., with $\tan\beta' \approx \beta'$ and $\tanh\beta'' \approx \beta''$) satisfies

$$\beta'' = \frac{k''}{k'}\beta' \tag{F.11}$$

which determines the inclination of the path of integration as it crosses the real axis in the β plane. If $k'' = 0$ the slope is zero and the path is the one shown in Fig. F.2. If $k' = k''$, the slope is unity and the angle of inclination is 45°.

Similarly, setting $\lambda' = 0$ in (F.7), the imaginary axis of the λ plane transforms to

$$\tan\beta' = -\frac{k''}{k'}\tanh\beta'' \tag{F.12}$$

or

$$\tan\beta' \coth\beta'' = -\tan B_o \Rightarrow B_o = \tan^{-1}\left(\frac{k'}{k''}\right) \tag{F.13}$$

This relation has asymptotes given by

$$\beta' = \begin{cases} -B_o, -\pi - B_o, \ldots \text{ for } \beta'' \to \infty \\ B_o, \pi + B_o, \ldots \text{ for } \beta'' \to -\infty \end{cases}$$

From (F.12), the shape of the path $\lambda' = 0$ around the origin is given by

$$\beta'' = -\frac{k'}{k''}\beta' \tag{F.14}$$

The relations (F.11) and (F.14) show that the mappings of the real and imaginary axis are at 90° from each other at the origin, as expected. A typical mapping based on (F.10) to (F.14) is shown in Fig. F.4.

The upper Riemann surface in the mapping shown in Fig. F.4 is determined by the condition:

$$\text{Im}[k\cos\beta] < 0$$

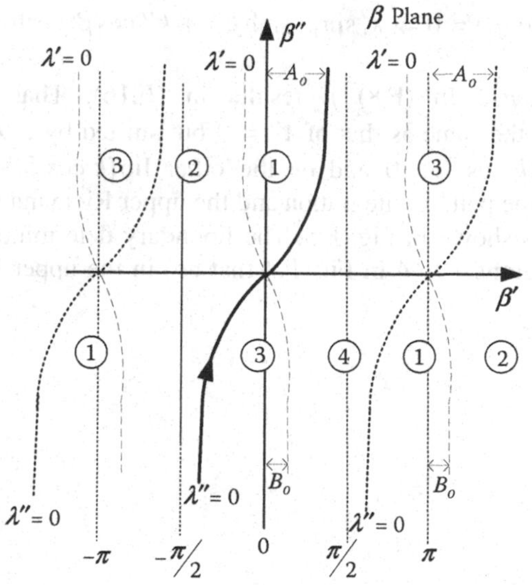

FIGURE F.4 A typical mapping of the real and imaginary axis of the λ plane for $k = k' - jk''$.

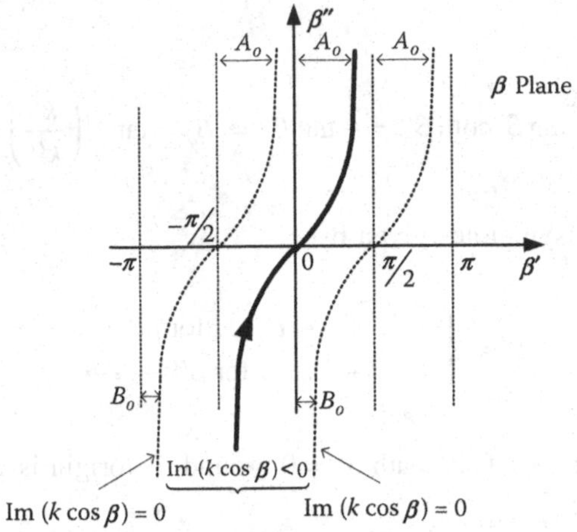

FIGURE F.5 The transformation $\lambda = k\sin\beta$ showing the upper and lower Riemann surfaces and the path of integration in the β plane for complex values of k.

which reads

$$\text{Im}[(k' - jk'')\cos(\beta' + j\beta'')] = -k'\sin\beta'\sinh\beta'' - k''\cos\beta'\cosh\beta'' < 0 \tag{F.15}$$

From (F.15) the boundary that separates the region $\text{Im}[k\cos\beta] < 0$ from $\text{Im}[k\cos\beta] > 0$ is determined by

$$\text{Im}(k\cos\beta) = 0 \Rightarrow k'\sin\beta'\sinh\beta'' + k''\cos\beta'\cosh\beta'' = 0 \tag{F.16}$$

Letting $\beta' \to \beta' \pm \pi/2$ in (F.8) it results in (F.16). That is, the boundary $\text{Im}[k\cos\beta] = 0$ is the same as that of $\lambda'' = 0$ but shifted by $\pm$ 90°. On one side of this boundary $\text{Im}[k\cos\beta] > 0$ and on the other $\text{Im}[k\cos\beta] < 0$. The resulting mapping showing the path of integration and the upper Riemann surface (i.e., where $\text{Im}[k\cos\beta] < 0$) is shown in Fig. F.5. The boundary determined by (F.16) shows the parts of the regions 1 to 4 in Fig. F.4 that are in the upper Riemann surface.

Appendix G: Error Function

The error function is defined by

$$\mathrm{erf}(x) = \frac{2}{\sqrt{\pi}} \int_0^x e^{-t^2} dt$$

A plot of the error function is shown in Fig. G.1. It is seen that

$$\mathrm{erf}(\infty) = \frac{2}{\sqrt{\pi}} \int_0^\infty e^{-t^2} dt = 1 \qquad \text{(G.1)}$$

$$\begin{aligned} \mathrm{erf}(-\infty) &= -1 \\ \mathrm{erf}(0) &= 0 \\ \mathrm{erf}(-x) &= -\mathrm{erf}(x) \end{aligned}$$

which shows that the error function is an odd function.

From (G.1) it follows that

$$\frac{2}{\sqrt{\pi}} \int_{-\infty}^\infty e^{-t^2} dt = 2 \Rightarrow \int_{-\infty}^\infty e^{-t^2} dt = \sqrt{\pi}$$

and

$$\frac{2}{\sqrt{\pi}} \int_0^\infty e^{-\alpha t^2} dt = \frac{1}{\sqrt{\alpha}} \Rightarrow \int_{-\infty}^\infty e^{-\alpha t^2} dt = \sqrt{\frac{\pi}{\alpha}}$$

The complementary error function is defined by:

$$\mathrm{erfc}(x) = \frac{2}{\sqrt{\pi}} \int_x^\infty e^{-t^2} dt$$

Since

$$\frac{2}{\sqrt{\pi}} \int_0^x e^{-t^2 dt} + \frac{2}{\sqrt{\pi}} \int_x^\infty e^{-t^2 dt} = \frac{2}{\sqrt{\pi}} \int_0^\infty e^{-t^2 dt} = 1$$

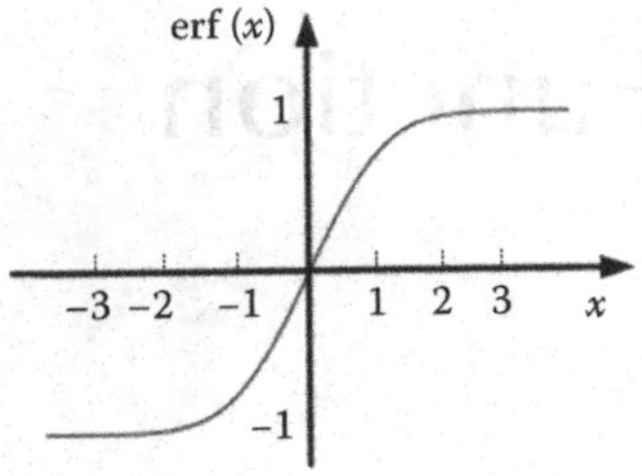

FIGURE G.1 Error function plot.

it follows that

$$\mathrm{erf}(x) + \mathrm{erfc}(x) = 1$$

Also,

$$\mathrm{erfc}(0) = 1$$
$$\mathrm{erfc}(\infty) = 0$$
$$\mathrm{erfc}(-x) = \frac{2}{\sqrt{\pi}} \int_{-x}^{\infty} e^{-t^2} dt = 1 - \mathrm{erf}(-x) = 1 + \mathrm{erf}(x)$$
$$\mathrm{erfc}(-x) = 2 - \mathrm{erfc}(x)$$

The error and complementary functions as a function of a complex variable z are analytic functions. The error function $\mathrm{erf}(z)$ is given by

$$\mathrm{erf}(z) = \frac{2}{\sqrt{\pi}} \int_{0}^{z} e^{-t^2} dt$$

where the integral is along a path from 0 to z. Letting $t \rightarrow -t$, shows that

$$\frac{2}{\sqrt{\pi}} \int_{0}^{z} e^{-t^2} dt = -\frac{2}{\sqrt{\pi}} \int_{0}^{-z} e^{-t^2} dt = \frac{2}{\sqrt{\pi}} \int_{-z}^{0} e^{-t^2} dt$$

or

$$\mathrm{erf}(z) = -\mathrm{erf}(-z)$$

The complementary error function for a complex argument is

$$\mathrm{erfc}(z) = \frac{2}{\sqrt{\pi}} \int_{z}^{\infty} e^{-t^2} dt$$

and it follows that

$$\text{erf}(z) + \text{erfc}(z) = 1 \tag{G.2}$$

Also,

$$\text{erfc}(-z) = \frac{2}{\sqrt{\pi}} \int_{-z}^{\infty} e^{-t^2} dt = 1 - \text{erf}(-z) = 1 + \text{erf}(z)$$

Some relations satisfied by the error and the complementary error function are:

$$\text{erf}(jz) = \frac{2j}{\sqrt{\pi}} \int_{0}^{z} e^{t^2} dt$$

and

$$\text{erfc}(-jz) = \frac{2}{\sqrt{\pi}} \int_{-jz}^{\infty} e^{-t^2} dt = 1 - \text{erf}(-jz) = 1 + \text{erf}(jz) \tag{G.3}$$

In (G.3), the real part of $-jz$ is positive if $\text{Im}(z) > 0$. Hence, when the real part of $-jz$ goes to ∞ the value of the integral is zero. It also follows from the analytical property of (G.3) that

$$\begin{aligned}\text{erfc}(-jz) &= \frac{2}{\sqrt{\pi}} \int_{-jz}^{\infty} e^{-t^2} dt = \frac{2}{\sqrt{\pi}} \int_{-\infty}^{\infty} e^{-t^2} dt - \frac{2}{\sqrt{\pi}} \int_{-\infty}^{-jz} e^{-t^2} dt \\ &= 2 + \frac{2}{\sqrt{\pi}} \int_{\infty}^{jz} e^{-t^2} dt = 2 - \frac{2}{\sqrt{\pi}} \int_{jz}^{\infty} e^{-t^2} dt \\ &= 2 - \text{erfc}(jz)\end{aligned}$$

For $|z| < 1$, the expansions is uniform and given by

$$\text{erf}(z) = \frac{2}{\sqrt{\pi}}\left(z - \frac{z^3}{3(1!)} + \frac{z^5}{5(2!)} - \frac{z^7}{7(3!)}..\right) = \frac{2}{\sqrt{\pi}} \sum_{n=0}^{\infty} \frac{(-1)^n z^{2n+1}}{n!(2n+1)}$$

For $|z| >> 1$,with $z = re^{j\theta}$, the asymptotic expansion can be expressed in various forms, namely

$$\begin{aligned}\operatorname{erf}(z) &= 1 - \frac{e^{-z^2}}{\sqrt{\pi} z}\left(1 - \frac{1}{2z^2} + \frac{1\cdot 3}{(2z^2)^2} - \frac{1\cdot 3\cdot 5}{(2z^2)^3}\cdots\right) \\ &= 1 - \frac{e^{-z^2}}{\sqrt{\pi} z}\sum_{n=0}^{\infty}\frac{(-1)^n(2n-1)\,!!}{2^n}\frac{1}{z^{2n}} \\ &= 1 - \frac{e^{-z^2}}{\sqrt{\pi} z}\sum_{n=0}^{\infty}\frac{(-1)^n(2n)\,!}{2^{2n}n\,!}\frac{1}{z^{2n}} \\ &= 1 - \frac{e^{-z^2}}{\pi z}\sum_{n=0}^{\infty}(-1)^n\Gamma\left(n+\frac{1}{2}\right)\frac{1}{z^{2n}} \qquad (|\theta| < 3\pi/4)\end{aligned} \tag{G.4}$$

In (G.4), the relations:

$$\Gamma\left(n + \frac{1}{2}\right) = \frac{(2n-1)!!}{2^n}\sqrt{\pi}$$

and

$$(2n-1)!! = \frac{(2n)!}{2^n n!}(n = 0, 1, 2, 3, ...)$$

were used. The term $(2n - 1)!!$ is the double factorial product of all odd numbers up to $(2n - 1)$, where $(-1)!!=1$.

Also,

$$\operatorname{erf}(-jz) = 1 - j\frac{e^{z^2}}{\pi z}\sum_{n=0}^{\infty}\Gamma\left(n+\frac{1}{2}\right)\frac{1}{z^{2n}} \qquad (|\theta| < \pi/4) \tag{G.5}$$

The asymptotic expansion of erfc(z) follows from (G.2) and (G.4), and that of erfc ($-jz$) follows from (G.3) and (G.5).

Appendix H: Orthogonality of Radial Solutions in Mode Theory

The radial solution $R(u)$ where $u = k_o r$ associated with the Debye potential for the TM modes in (10.8.13) consists of spherical Hankel functions, which for a given mode $\nu = \nu_n\,(n = 1, 2, 3, \ldots)$ satisfy

$$\frac{d}{du}\left(u^2\frac{dR_\nu}{du}\right) + [u^2 - \nu(\nu + 1)]R_\nu = 0$$

and can be expressed as

$$u\frac{d^2}{du^2}(uR_\nu) + [u^2 - \nu(\nu + 1)]R_\nu = 0 \qquad \text{(H.1)}$$

To derive the orthogonality property of the TM modes, consider another radial solution R_μ whose eigenvalues are denoted by μ. This radial function for a given mode $\mu = \mu_n$ $(n = 1, 2, 3, \ldots)$ satisfies

$$u\frac{d^2}{du^2}(uR_\mu) + [u^2 - \mu(\mu + 1)]R_\mu = 0 \qquad \text{(H.2)}$$

The boundary conditions satisfied by these radial functions are (see (10.8.8) and (10.8.9)):

$$\left.\frac{1}{u}\frac{d(uR)}{du}\right|_{r=a} = jz_a R|_{r=a} \qquad \text{(H.3)}$$

and

$$\left.\frac{1}{u}\frac{d(uR)}{du}\right|_{r=c} = -jz_c R|_{r=c} \qquad \text{(H.4)}$$

where R represents R_ν and R_μ. At $r = a$: $u = u_a = k_o a$ and at $r = c$: $u = u_c = k_o c$. The surface impedances $z_a = Z_a/\eta_o$ and $z_c = Z_c/\eta_o$ are assumed to be independent of the mode.

Multiplying (H.1) by R_μ and (H.2) by R_ν and subtracting the equations gives

$$uR_\mu \frac{d^2}{du^2}(uR_\nu) - uR_\nu \frac{d^2}{du^2}(uR_\mu) = [\nu(\nu + 1) - \mu(\mu + 1)]R_\mu R_\nu \qquad \text{(H.5)}$$

Integrating (H.5) between u_a and u_c gives

$$\int_{u_a}^{u_c} \left[uR_\mu \frac{d^2}{du^2}(uR_\nu) - uR_\nu \frac{d^2}{du^2}(uR_\mu) \right] du = [\nu(\nu + 1) - \mu(\mu + 1)] \int_{u_a}^{u_c} R_\mu(u) R_\nu(u)\, du$$

The integral on the left contains two terms which can be integrated by parts to obtain

$$\int_{u_a}^{u_c} R_\mu(u) R_\nu(u)\, du = \frac{\left[uR_\mu \frac{d}{du}(uR_\nu) - uR_\nu \frac{d}{du}(uR_\mu) \right] \Bigg|_{u_a}^{u_c}}{[\nu(\nu + 1) - \mu(\mu + 1)]} \qquad \text{(H.6)}$$

The numerator of (H.6) vanishes when the boundary conditions in (H.3) and (H.4) are used. For example, at the upper limit:

$$u_c R_\mu \frac{d}{du}(u_c R_\nu) - u_c R_\nu \frac{d}{du}(u_c R_\mu) = -u_c R_\mu [jz_c u_c R_\nu] + u_c R_\nu [jz_c u_c R_\mu] = 0$$

The integral in (H.6) is equal to zero except for $\nu = \mu$, where the denominator is also zero. Therefore, (H.6) reads

$$\int_{u_a}^{u_c} R_\nu(u) R_\mu(u)\, du = N(\nu)\, \delta_{\nu\mu}$$

where $N(\nu)$ is the value of the integral for $\nu = \mu$. That is,

$$N(\nu) = \int_{u_a}^{u_c} [R_\nu(u)]^2 du$$

The evaluation of $N(\nu)$ where $\nu = \nu_n$ (i.e., $N(\nu_n)$) is usually done numerically, because it requires the integration of spherical Bessel functions of order ν_n. This expression has been evaluated in closed form for specific conditions, such as the so-called thin-shell approximation (i.e., when $(c - a)/r \ll 1$) of the Earth-Ionosphere waveguide.

Index

Printed in the United States
by Baker & Taylor Publisher Services